URBAN TRANSIT

URBAN TRANSIT:
OPERATIONS, PLANNING, AND ECONOMICS

Vukan R. Vuchic

WILEY
JOHN WILEY & SONS, INC.

This book is printed on acid-free paper. ♾

Published by John Wiley & Sons, Inc., Hoboken, New Jersey
Published simultaneously in Canada

Library of Congress Cataloging-in-Publication Data:

Vuchic, Vukan R.
Urban Transit: operations, planning, and economics/Vukan R. Vuchic.
p. cm.
Includes bibliographical references and index.
ISBN 0-471-63265-1 (cloth)
1. Urban transportation. 2. Local transit. 3. Transportation—Planning.
I. Title.
HE305.V83 2004
388.4′068—dc22

2004013844

Printed in the United States of America

10 9 8 7 6 5 4 3

This book is dedicated to my students in transportation programs at the University of Pennsylvania and other universities where I have taught. As educators, researchers, engineers, planners, and civic leaders, they have applied their knowledge and enthusiasm to develop efficient urban transportation systems and livable cities. Their contributions are recognized in many cities around the world.

CONTENTS

PREFACE

Transportation, being one of the basic functions in urbanized areas, influences the form of cities and their livability—their economic, social and environmental characteristics. The usage of transit systems, or urban public transportation, decreased considerably when auto ownership began to grow. However, in recent decades many cities have made large investments in transit to reverse this trend. Experience has shown that transit has great significance for reducing traffic congestion, offering alternative means of travel, and contributing greatly to the quality of urban life. In developing countries, the role of transit is even greater than in the industrialized countries because it serves a greater number of people and offers capacities that highway systems cannot provide in rapidly growing cities.

In recent years, transit systems have benefitted from many physical and operational innovations. Yet literature on transit systems physical characteristics, planning, mode selection, and urban transportation policies is rather limited. Particularly missing are publications that present theoretical concepts as well as practical methods for their implementation in real world.

It has been the intention of this author to write a "transit trilogy"—a set of books on three major aspects of transit and urban transportation: transit systems, operations, and transit role in cities. The first book in this series was *Urban Public Transportation Systems and Technology* (Prentice-Hall, 1981), which presents a systematic classification and definitions of urban transportation modes, with detailed descriptions and their characteristics. Many concepts and terms from those definitions have become standard in the field. That book is now out of print, but the demand for it suggests there is a need for an updated edition. That will be the author's next task.

A broad review of urban transportation policies, systems, and impacts of the selection of private versus public transportation modes on cities is presented in *Transportation for Livable Cities* (CUPR Press, Rutgers, 1999). The relationships among the private automobile, transit, and other modes are systematically reviewed. The importance of transit for livability of cities is given particular attention. This book also defines and discusses such important contemporary concepts as balanced transportation and livability of cities.

The present book covers transit system operations and planning and represents a logical continuation of the first book. Thus, it falls between the preceding two books and completes the transit trilogy.

This book is organized in three parts. Part I, Chapters 1 to 5, covers transit systems operations and networks. Chapter 1 defines concepts of transit systems operations and presents the methodology for scheduling of transit units (vehicles) and personnel (runcutting) for transit lines. Chapter 2 covers a number of aspects of transit operations, including line capacity, transit speeds, types of stops, accelerated services, and scheduling for special geometric forms of lines. Chapter 3 presents an introduction to applications of modeling and optimization in the analysis of transit systems operations and design. Chapter 4 gives an extensive review and analysis of transit lines and networks: their forms, characteristics, and methods of evaluation. Chapter 5 describes determination of the number and locations of stations on rail transit lines.

Part II, Chapters 6 through 9, discusses transit agency operations, economics, and organization. Chapter 6 presents transit system management and operation, including internal organization of agencies, relations with personnel (labor) and passengers, accessibility requirements, information, and marketing. Chapter 7 covers fares: their types and classification, characteristics, and methods of collection. Chapter 8 reviews methods and sources of transit financing. Role of different government levels in financing and efforts to increase efficiency of operations are also discussed. Chapter 9 presents an overview of types of transit organizations and government regulation, including a review of numerous recent innovative concepts in public-private cooperation and separation of policy bodies from transit operating public agencies or private companies.

Part III of the book, chapters 10 through 12, focuses on transit system planning. Chapter 10 discusses the process and techniques of transit planning. Unlike many academic books on planning that focus on travel demand estimation only, this chapter covers the entire planning process, from the establishment of goals and policies to demand estimation, transit mode selection, and network design. Chapters 11 and 12 focus on the important phase of transit planning—selection of transit mode. A theoretical basis and general methodology of mode selection in Chapter 11 is followed in Chapter 12 by a systematic comparison of various aspects of transit modes. This part of planning—comparison and selection of transit modes—is of great relevance and importance for many cities around the world.

Each chapter has a set of exercises, including numerical as well as conceptual and other essay-type questions. References and relevant further reading are also given for each chapter. A bibliography of selected books relevant to the overall contents of the book is given at the end of the book.

Each equation in this book is numbered and accompanied by definitions of dimensions of all symbols. In Chapters 1 through 4, which have many equations, lists of symbols are included with their definitions.

This book, like its predecessors, is written with SI measures, the system that has been adopted in all countries except the United States. In this country, Congress mandated the transition to SI, but then, when this effort began to be implemented by many agencies, particularly the U.S. Department of Transportation, Congress reversed itself by canceling the previous requirements. This important effort to replace the obsolete English system of measures by the standard contemporary one was thus discontinued. For readers not yet familiar with the SI, conversion factors are presented in a very convenient form in Appendix I.

To facilitate understanding of the numerous abbreviations used in the book, these are listed and defined in Appendix II. Appendix III presents definitions of selected terms related to transit systems. Answers to selected exercises are given in Appendix IV. The answers are brief or include only selected ones from many required computed numbers. This allows the users to see whether their computational procedure is correct, and yet requires computation of all results.

Like the previous two books in the transit trilogy, this book is intended to introduce theoretical concepts and relationships in transit systems followed by practical methodologies for operations, planning, and analysis. The combination of theoretical basis and direct applications makes the book useful both in academic circles as well as to persons planning and operating transit systems. The intended readers thus include graduate and upper undergraduate students of transportation engineering and planning courses at universities and in continuing education courses, as well as transit planners and operators in transit agencies, metropolitan planning organizations, city governments, consulting firms, public interest organizations, and others.

The international orientation and coverage of the first two books found a very good response from the readers in many countries, and it is continued in this book. Although focused mostly on the United States and other industrialized countries, many examples and presented methodologies are applicable to any city, in developed as well as in developing countries, always, naturally, with required consideration of local factors.

ACKNOWLEDGMENTS

The author gratefully acknowledges the cooperation of many individuals and organizations in preparing materials, writing, and reviewing the text for this book, while retaining the responsibility for the accuracy of its contents.

I want to thank many colleagues who have reviewed chapters of this manuscript and provided valuable comments, corrections, and additional materials. All chapters were reviewed by Wolfgang S. Homburger, now retired from the Institute of Transportation Studies at the University of California in Berkeley. He applied his diligent, detailed, and constructive pen, well known throughout the field of transit and traffic engineering, with which he edited numerous books. Hans Leopold, the author's colleague from Hamburg, applied his deep understanding of the field through precise reviews and constructive suggestions. Jeffrey M. Casello, who first studied these materials in my courses, then read all the text with particular attention to students' points of view.

Most of the chapters were also reviewed by Thor K. Haatveit, a leading public transport official in Norway, who had also assisted in developing technical materials when he was a graduate student at the University of Pennsylvania. Similar contributions were made by Professor Shinya Kikuchi, Richard M. Stanger, and Dr. Eric C. Bruun. Angel Molinero contributed his inventive skills to develop analyses and then used them also in his excellent book on this topic published in Mexico. Many other author's students contributed to this text during their studies through their course papers and research on topics selected for the materials prepared for this book over a number of years. Professors Yong Eun Shin, Nikola Krstanoski, and Young-Jae Lee deserve particular credit and my gratitude in this respect.

Louis J. Gambaccini, a leading transit official, gave valuable comments on chapters on transit management, financing, and organization. Professor George M. Smerk reviewed the chapter on transit organization and policy. Professor David E. Boyce made valuable contributions through many discussions about the transportation planning process (Chapter 10). John Schumann provided on many occasions, extensive and reliable materials used in this book. Takis Salpeas gave excellent insights into various phases of planning and operations of major investment projects by several leading transit agencies (BART and WMATA).

I benefitted greatly from numerous professional discussions, comments on the topics in this book, and information on transit in different countries during the years of writing this book by Professors Günter Girnau and Adolf Müller-Hellmann (Germany), Hideo Nakamura (Japan), Franco de Falco and Antonio Musso (Italy), Heinrich Brändli (Switzerland), Tony Ridley and Tony Young (UK), Nenad Jovanović, Snežana Filipović, and the late Radovan Banković (Yugoslavia-Serbia), Jean-Claude Ziv (France), and many others.

On many occasions, I obtained valuable information, materials, exhibits, and data from several transit associations: American Public Transportation Association (APTA) (U.S.), Associação Nacional de Transporte Pùblicos (ANTP) (Brazil), Verband Deutscher Verkehrsunternehmen (VDV) (Germany), and transit agencies: BART and Muni (San Francisco), BVG (Berlin), NYCT (New York), OS (Oslo), SEPTA (Philadelphia), TRTA (Tokyo), WMATA (Washington), and many others.

I want to mention especially the support and inspiration I enjoyed through several decades of cooperation with the International Association of Public Transport (UITP, Brussels) and its present President Dr. Wolfgang Meyer and General Secretary Johannes Rat. Similarly influential have been my contacts with Professor Günter Girnau in Cologne. Our discussions about many topics, from the classical materials on transit operations produced by the late Dr. Friedrich Lehner through their continuous enhancements to current issues have been very helpful in my keeping up with the latest trends and developments in transit and urban transportation around the world.

During the writing of this book, I have enjoyed enthusiastic support from my former students, individually as well as through their organization, TSEAC, which has also supported the entire Program of Transportation Systems Engineering at the University of Pennsylvania. I feel that there is mutual support between my writing of these books and the efforts of dozens of my former students on improving transportation in many cities around the world. That feeling and my gratitude to them resulted in the dedication of this book.

Finally, during the most concentrated effort to complete this manuscript, I enjoyed high-quality, precise, and dedicated work by my team of assistants/doctoral students: Dr. Jeffrey M. Casello, Mario Semmler, Huafang Cui, and Christopher Puchalsky. They deserve my sincere thanks.

On the personal side, the large multiyear effort of writing this book was continuously supported and intellectually enhanced by my wife Dr. Radmila Vuchic, as well as by our children—Monika, Boris, Lili, and Victor—and their families, who readily accepted sacrifices when "Dad had to write the book." The husband/father deeply appreciates this family's understanding and support.

Vukan R. Vuchic
University of Pennsylvania

URBAN TRANSIT

PART I

TRANSIT SYSTEMS OPERATIONS AND NETWORKS

1

TRANSIT OPERATIONS AND SERVICE SCHEDULING

List of Symbols

Symbol	Unit	Definition	Symbol	Unit	Definition
a, $a(s)$	prs/km/h	Number of alighting passengers per kilometer of line per hour	P_{15}	prs/15 min	Maximum 15-min passenger volume
A, B	prs/h	Cumulative number of alighting and boarding passengers per hour, respectively	P_{av}	prs/h	Average passenger volume
b, $b(s)$	prs/km/h	Number of boarding passengers per kilometer of line per hour	P_c	sps-km/h^2	Productive capacity
c	veh/h	Vehicle line capacity	P_d	prs/h	Design passenger volume
C	sps/h	Offered transit line capacity	P_{max}	prs/h	Maximum passenger volume (usually on MLS)
C_o	sps/h	Scheduled line capacity	P_t	prs/h	Total number of passengers on line
C_s	sps/h	Capacity of a line with stations	PHC	—	Peak hour coefficient
C_v, C_{TU}	sps/veh, sps/TU	Vehicle and TU capacity, respectively	S, S_i	km	Station spacing
C_w	sps/h	Capacity of a line section without stations	S_{od}	km	Origin-destination distance
f	TU/h	Frequency of service	t	s	Time variable
h, h'	min/TU	Headway	t_a, t_b	min	Acceleration and braking time, respectively
h_p	min/TU	Policy headway	t_d	min	Deadhead time
h_s, h_w	min/TU	Station and way headway, respectively	t_f	min	Transfer time
ℓ	km	Passenger travel distance on line	t_o	min	Passenger on-line travel time
L	km	Line length	t_p	min	Crew paid time
L_p	km	Total distance from depot to a line	t_r	min	Running time
MLS	—	Station spacing with maximum passenger volume	t_s	min	Station standing (dwell) time
n	veh/TU	Number of vehicles per TU	t_t	min	Terminal time
N, N'	veh	Number of vehicles operating on line	t_w	min	Waiting time
N_f	veh	Total number of vehicles—fleet size	T, T'	min	Cycle time
N_{TU}	TU	Number of TUs	T_ℓ	s	Time lost per stopping at one station
p, $p(s)$	prs/h	Passenger volume variable	T_o	min	Operating or travel time
P	prs/h	Utilized capacity	T_{od}	min	Origin-destination travel time
			T_p	min	Platform time
			T_s	min	Station-to-station travel time
			v, V	m/s, km/h	Speed variable
			V_a	km/h	Access speed
			V_c	km/h	Cycle speed
			V_d, V_g, V_ℓ	km/h	Line design, programmed and legal speeds, respectively
			V_{max}	km/h	Maximum technical speed

List of Symbols (*continued*)

Symbol	Unit	Definition	Symbol	Unit	Definition
V_o	km/h	Operating or travel speed	η	—	Operating personnel efficiency coefficient
V_{od}	km/h	Origin-destination speed	η_a	—	Personnel attendence coefficient
V_p	km/h	Platform speed	η_s, η_t	—	Coefficient of run-cutting and schedule efficiency, respectively
V_r, V_s	km/h	Running and station-to-station speeds, respectively	η_x	—	Coefficient of passenger exchange
w	sps-km/h	Transportation work	φ	—	Fleet utilization factor
w_p	prs-km/h	Utilized work			
α	prs/sps	Load factor—capacity utilization coefficient			
γ	%	Terminal time coefficient			
δ	—	Scheduled line capacity utilization coefficient			

This chapter covers the basic elements, operations, and functioning of transit systems. Transportation offered to passengers is usually referred to as *service,* while *operations* covers system management, scheduling, and functioning from the operating agency's point of view. First, the basic elements of operations and passenger flows are presented; since terminology and concepts in this area often vary, special attention is paid to precise definitions. The next section gives a review of data-collection methods and surveys; it is followed by a description of passenger demand characteristics. Methodology of scheduling and run-cutting is presented in the last section.

Transit operations, scheduling, measurements of efficiency, and related transit line planning are covered in a variety of publications. The classical textbooks and handbooks with comprehensive definitions of concepts and operational procedures that have permanent value include the following, listed by order of publication: Rainville (1947); Lehner (1950, 1957, and 1978) (in German), Rüger (1974) (in German), Groche and Thiemer (1980) (in German), Banković (1982, 1984) (in Serbian), Levinson 1992, CUTA (1993), Molinero and Arellano (1997) (in Spanish), and Kittelson & Associates (2003). An excellent collection of exercises on this topic is Krstanoski (2000) (in Macedonian). Among many publications on specific issues in transit operations the reader may also find useful Furth (1980), Bowman and Turnquist (1981), Ceder and Stern (1981), and Ceder (1987).

1.1 BASIC OPERATING ELEMENTS

The vast majority of transit services are performed by vehicles or trains traveling along fixed lines according to predetermined schedules. Thus, a transit line represents the basic component of transit system operation. Its elements are defined here.

1.1.1 Line, Network, Stop, and Station

A *transit line* is the infrastructure and service provided on a fixed alignment by vehicles or trains operating on a predetermined schedule. The infrastructure may vary from simple stop designations along a street to a grade-separated, fully controlled right-of-way with stations. A *transit route* is often synonymous with transit line, but it usually designates street transit, often overlapping lines, rather than major metro or regional rail lines. A *transit network* is a set of transit lines that connect with or cross each other and that are coordinated for efficient operation and provision of integrated services in an area for the convenience of passengers and efficiency of operations.

Line length, usually expressed in kilometers (miles), is the one-way distance between the two terminals along the line alignment. *Network length* is the total length of all alignments served by one or more lines. *Total line* (*route*) *length* is the sum of all line lengths, regardless of whether they operate alone or overlap with other lines. Figure 1.1 illustrates these concepts.

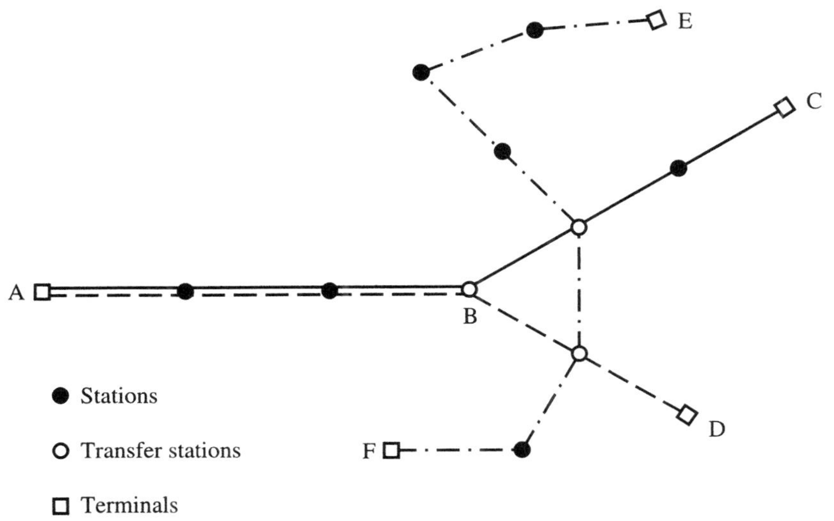

Figure 1.1 *Transit line, network, and station concepts. Assuming that each spacing between stations is 1 km long, the values in this network are:*

Line lengths: L_{AC} = *6 km,* L_{AD} = *5 km,* L_{EF} = *7 km*

Network length: $L_{AC} + L_{BD} + L_{EF}$ = *15 km*

Total line or route length: $L_{AC} + L_{AD} + L_{EF}$ = *18 km*

Transit right-of-way (*ROW*) is the strip of land on which a transit line operates; the term usually refers to the facility used exclusively by transit vehicles, but, in a broader sense, also refers to any of the numerous physical forms a transit line may follow, from street to fully controlled aerial (elevated) structure or tunnel. Among the different ROW features, the most important for transit operations, performance, and cost characteristics is the degree of its separation from other vehicles and pedestrians. With respect to this feature, transit rights-of-way are classified into three categories: *C*, *B*, and *A*.

Category C is a surface street with *mixed traffic;* transit vehicles share street space with other vehicles and pedestrians; they may have preferential treatment through visual devices (pavement markings, signs, or signals), but no physical separation from other traffic. Modes with this ROW category, such as buses, trolleybuses, and streetcars (tramways), represent the *street transit* generic class.

Category B includes a variety of ROW types that are longitudinally physically separated (by curbs, barriers, green strips, etc.) from other traffic, but with at-grade crossings for vehicles and pedestrians, including regular street intersections. This ROW category defines the generic class of *semirapid transit modes,* and its most common representative is light rail transit (LRT).

Category A is a fully controlled ROW without grade crossings or any legal access by other vehicles or persons. Also called grade separated, private, or exclusive, this ROW can be in a tunnel or a cut, on the surface, on an embankment, or on aerial structures. This ROW category is the main element defining the *rapid transit* generic class of modes, among which metro or rail rapid transit is the dominant representative. All automated guided transit (AGT) systems must have exclusive ROW, while regional rail often has some grade crossings with full signal override. This is still considered to be category A, since such crossings usually have no impact on line performance.

Capacity, reliability, speed, and other performance elements of transit lines on ROW category C and, to a much lesser degree, on category B are dependent on traffic conditions along the line, while the lines with ROW category A are fully controlled and therefore have highly reliable performance.[1]

A *transit stop* is a location along a line at which transit vehicles stop to pick up or drop off passengers; its equipment may include signs, information, a bench, and shelter. A *transit station* is a special structure and facility for passenger boarding/alighting, waiting, and transfer. It may have facilities for passengers, such as boarding platforms, mezzanines, stairways, and fare-collection equipment, and for vehicle operation, such as turnback and storage tracks. For the purpose of transit line operation analysis, the terms *stop* and *station* will be used interchangeably and will encompass all physical forms, from bus stops at curbs to major rail transit stations.

Transfer stations are joint stations for two or more lines at which passengers can transfer between lines. *Terminals* are, strictly defined, end stations on a transit line, but sometimes the term is also used for major transfer stations.

Different types of stations in a network are shown in Figure 1.1. Detailed description and analysis of transit lines and stations is given in Chapters 4 and 5, respectively.

1.1.2 Vehicles, Transit Units, and Fleet Size

Highway transit modes operate single vehicles only, with very rare exceptions (bus and trolleybus trailers are operated in Switzerland, Russia, and a few other countries). Guided modes, on the other hand, in most cases utilize their capability to couple vehicles into trains to achieve greater capacity and economy of operations. The scheduling of line operations is based on the concept of *transit unit* (TU)[2], which is defined as a set of n vehicles traveling physically coupled together. For single-vehicle operations such as buses, $n = 1$; for train operation, $n > 1$ (in rail transit terminology this n is also known as *train consist*). Thus, TU is the common concept for single vehicles and trains, whichever is used on a transit line. Frequency of operation on a line f is expressed in TU/h, so that the number of vehicles past a fixed point during one hour is $f \cdot n$.

Transit vehicles, bus and rail, are referred to collectively as *fleet*. For rail modes the specialized terms *rail cars* and *rolling stock* are often used.

Fleet size N_f is the total number of vehicles needed for operation of a line, or of an entire network. The vehicle fleet consists of the vehicles required for regular service N (determined by the peak hour operation), vehicles needed for reserve N_r, and vehicles on maintenance and repair N_m:

$$N_f = N + N_r + N_m \tag{1.1}$$

The utilization of a fleet, which depends on its physical condition and schedule efficiency, is measured by the fleet utilization factor φ, defined as percent of fleet available for service:

$$\varphi = \frac{N + N_r}{N_f} \qquad \left| \frac{\varphi}{-} \middle| \frac{N}{\text{veh}} \right| \tag{1.2}$$

1.1.3 Usage of Service: Passenger Flow and Volume

Transit service must be based on the demand for travel by transit along the line, as well as the required level of offered service. The demand is expressed as passenger flow (prs/time) or passenger volume p (prs/h). These two concepts represent the rate of passenger travel, or the number of passengers traveling past a fixed point in one direction per unit time. However, flow refers more to the continuous process of passenger travel in general, while volume designates the number of passengers traveling during a specified time interval, usually one hour.

Passenger travel on a transit line can be shown by a sequence of theoretical diagrams. First, assuming that

[1]For a detailed definition and description of ROW types and other physical components of transit systems, see Vuchic (1981), particularly Chapter 2.

[2]For definitions and descriptions of TUs see also Vuchic (1981), Sections 2.1.3 and 7.2.1.

passengers can board and alight TUs at any point along the line (distance s), boardings are plotted as a continuous density function $b(s)$, and alightings as $a(s)$; both are in persons per unit distance (usually kilometers, km) per unit time (hour, h). These functions are shown in Figure 1.2(a). From this diagram the cumulative (distribution) functions of boarding and alighting, $B(s)$ and $A(s)$, respectively, are derived and plotted in Figure 1.2(b). Naturally, $B(L) = A(L)$.

$$B(s) = \int_0^s b(s)\,ds \quad \text{and} \quad A(s) = \int_0^s a(s)\,ds$$

$$\left|\frac{B, A}{\text{prs/h}}\right|\frac{s}{\text{km}}\left|\frac{b, a}{\text{prs/km/h}}\right| \tag{1.3}$$

A passenger volume diagram along the line is then derived mathematically as:

$$P(s) = \int_0^s b(s)\,ds - \int_0^s a(s)\,ds = B(s) - A(s)$$

$$\left|\frac{P}{\text{prs/h}}\right|\frac{s}{\text{km}}\left|\frac{b, a}{\text{prs/km/h}}\right|\frac{B, A}{\text{prs/h}}\right| \tag{1.4}$$

and it is shown graphically, as in Figure 1.2(b) and (c). The extremes (maximum and minimum points) of the function $p(s)$ occur wherever

$$\frac{dp}{ds} = 0, \text{ so that } b(s) = a(s) \tag{1.5}$$

i.e., wherever the boarding and alighting functions intersect. If there are several such points, as in Figure 1.2(c), the values of all maximum points must be compared to find the maximum load for the line P_{max}, which is of critical importance for operations planning and scheduling.

Translating this theoretical approach to practice, in most cases passengers can board and alight only at transit stops or stations, i.e., at discrete points along the line. Therefore, passenger demand is computed and plotted for a set of existing (or planned) stations. A set of three diagrams with such discrete values, corresponding to the diagrams in Figure 1.2, is shown in Figure 1.3. Boarding and alighting passenger volumes at any station i are now discrete values shown as b_i and a_i bars, respectively; their cumulative functions are B_i and A_i, and the passenger volume (actual or forecasted) on a given section k is

$$P_k = B_k - A_k = \sum_{i=1}^{k} b_i - \sum_{i=2}^{k} a_i$$

$$\left|\frac{P}{\text{prs/h}}\right|\frac{B, A}{\text{prs/h}}\left|\frac{b, a}{\text{prs/h}}\right| \tag{1.6}$$

as shown in Figure 1.3(b) and (c). Naturally, total numbers of boardings, alightings and passengers using the line per unit time must be equal: $B_L = A_L = P_L$.

The maximum passenger volume for any section along the line, P_{max}, is the most important factor for planning the required transporting capacity on the line. The section on which that volume is found is called the *maximum load section* (MLS).

Temporal variations of demand and their handling in the scheduling process will be discussed in Section 1.3.

1.1.4 Operating Elements: Headway and Frequency

The basic process of transportation can be defined as the movement of objects u over a distance s during an interval of time t. As described in Vuchic (1981), various relationships of these three elements constitute most of the basic operating elements of transportation systems. The diagram in Figure 1.4 shows these relationships.

In transit operations, the most commonly used among these elements are *headway, frequency, capacity, work, productivity, travel time* and *speed*. Headway and frequency, which define movement of TUs on transit line, are defined here; the other elements are defined in the following sections.

Headway h is the time interval between the moments two successive TUs pass a fixed point on a transit line in the same direction. In scheduling it is expressed in minutes, while in capacity analysis seconds are often used.

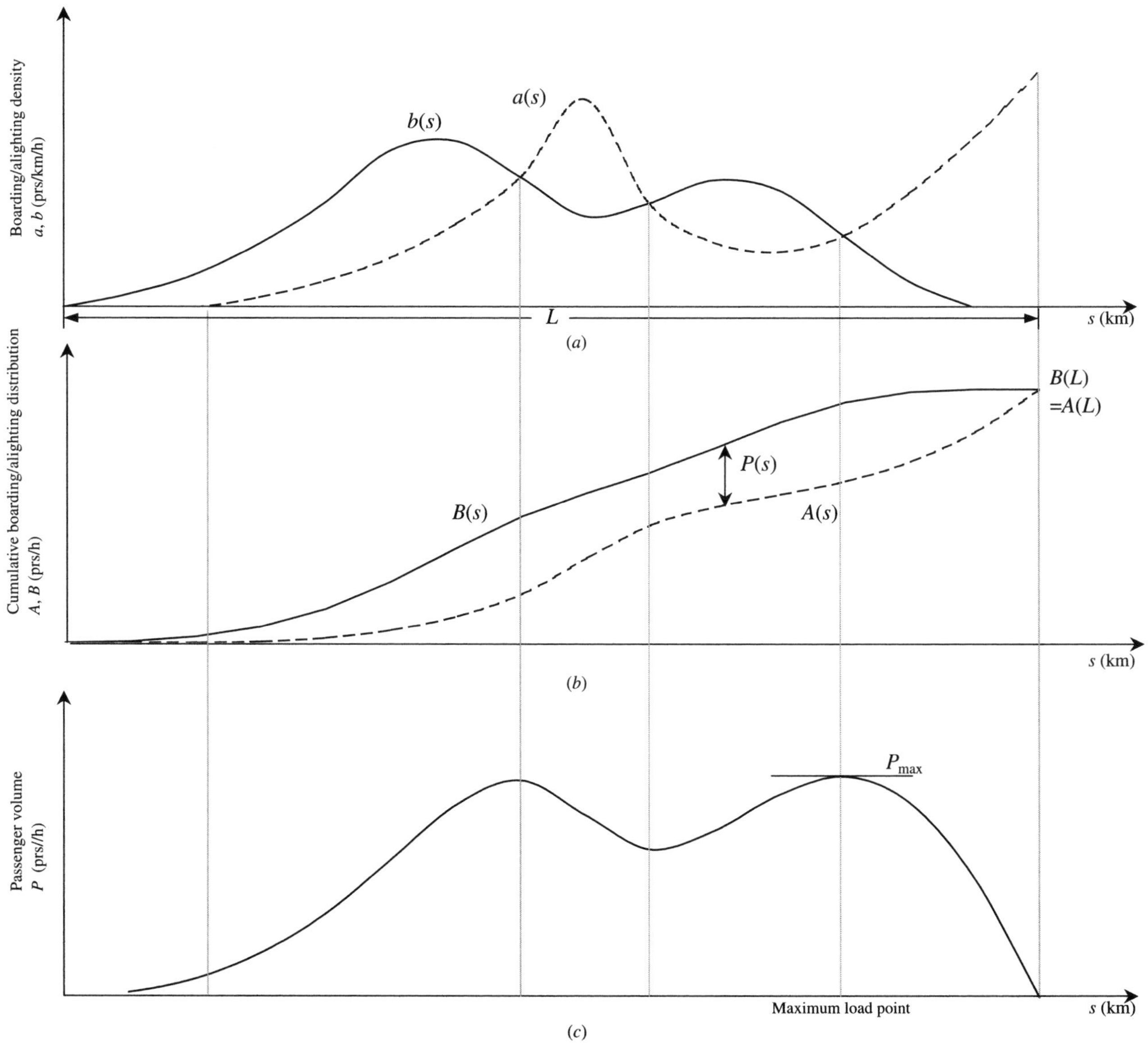

Figure 1.2 *Passenger boarding, alighting, and volume along an idealized transit line:* (*a*) *boarding and alighting distributions;* (*b*) *cumulative functions of boarding and alighting;* (*c*) *passenger volume profile.*

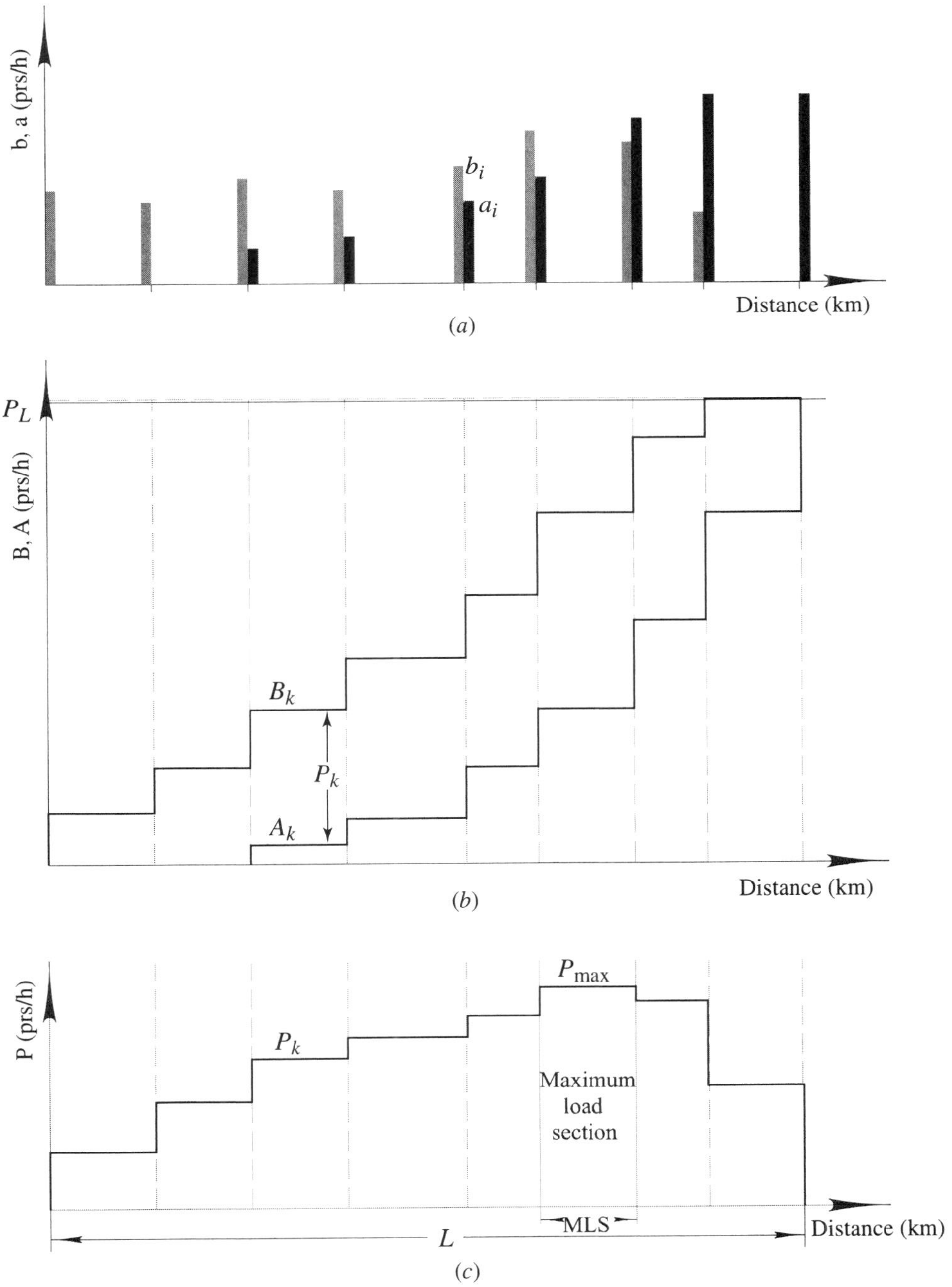

Figure 1.3 *Passenger boarding, alighting, and volume diagram for a transit line: (a) boarding and alighting volumes at stations; (b) cumulative boardings and alightings along the line; (c) passenger volumes transported along the line.*

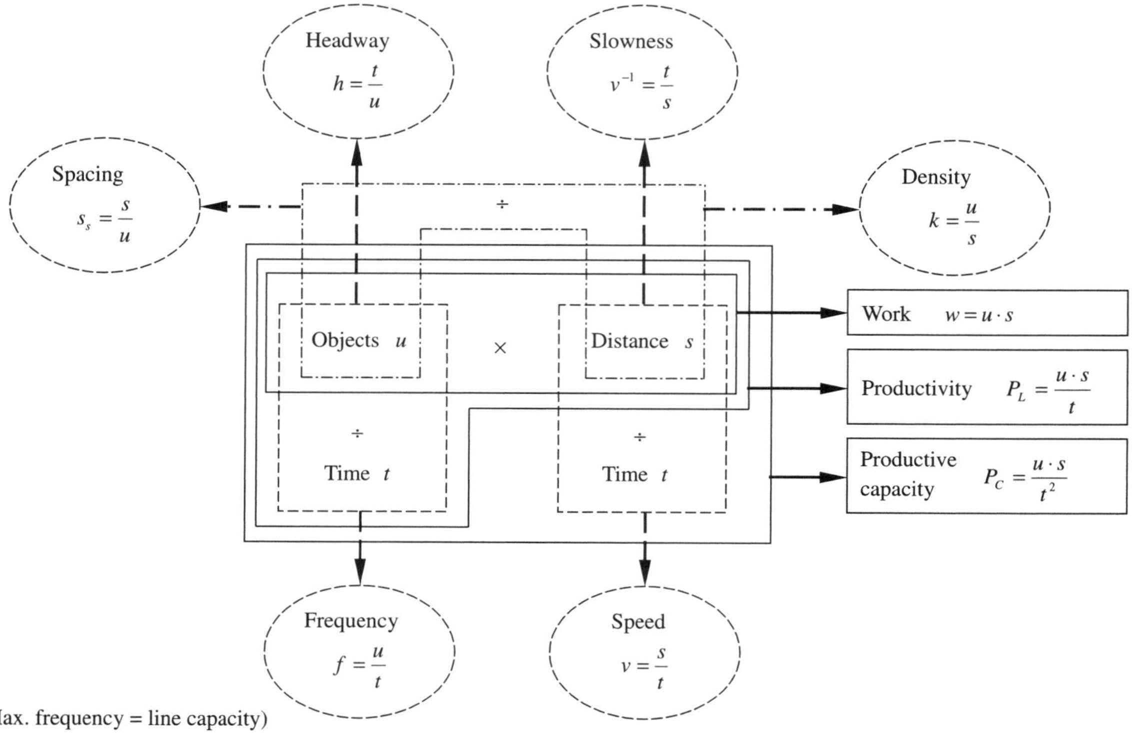

Figure 1.4 *Basic transit system operating elements and performance measures.*

Passengers are interested in having short headways to minimize their waiting time. However, since for any given passenger volume it is cheaper to operate a small number of large TUs than a large number of small TUs, a transit agency is usually interested in operating long headways. Consequently, headways are usually determined as a compromise between passenger travel time and convenience, and the cost of operation.

Whenever passenger demand is reasonably stable, uniform (constant) headways provide the most efficient operation (even loading of TUs and maximum schedule stability), and they are most attractive for passengers (simple, reliable, waiting time minimized). When headways are longer than 6 minutes, it is desirable to use only values divisible into 60 (7.5, 10, 12, 15, 20, 30 or 60 minutes), known as *clock headways,* because with them departure times at any one stop fall on the same minutes in each hour, so that passengers can easily memorize the schedule.

The longest headway scheduled for a line is determined by the minimum level of service considered acceptable for that line; it is therefore called *policy headway* h_p.

The *minimum headway* $h_{\min}$ on a line is, on the other hand, determined by the physical characteristics of the system (technology, method of driving and control, degree of safety required) and station operations (rate of boarding/alighting, departure control, etc.). The former factors influence the shortest headway achievable on open line between stops, *way headway* $h_{w\,\min}$, and the latter determine the shortest headway that can be operated at stations, *station headway* $h_{s\,\min}$. The greater of the two represents the minimum headway for the line:

Table 1.1 Capacity, work, and utilization concepts related to transit line operation

Category	Definition	Derivation Formula	Units: Offered	Units: Utilized	Utilization Coefficient
Static capacity	Vehicle capacity (seats only or seats + standing spaces)	C_v (given)	C_v (sps/veh)	P_v (prs)	$\alpha_v = P_v/C_v$
Dynamic capacity	Maximum frequency	f_{max}	f_{max} (TU/h)	f (TU/h)	$\delta_f = f/f_{max}$
	Vehicle line capacity	$f_{max} \cdot n$	c (veh/h)	$f \cdot n$ (veh/h)	$\delta_v = f \cdot n/c$
	Line capacity, min (C_w, C_s)*	$f_{max} \cdot n \cdot C_v$	C (sps/h)	P (prs/h)	$\alpha = P/C$
	Scheduled capacity	$f \cdot n \cdot C_v$	C_o (sps/h)	P (prs/h)	$\alpha = P/C$ $\delta = C_o/C$
Transportation work	Work on a line during one hour	$f \cdot n \cdot C_v \cdot L$	w (sp-km/h)	w_p (prs-km/h)	$\bar{\alpha} = w_p/w$
Productive capacity	Product of capacity and speed	$f \cdot n \cdot C_v \cdot V_o$	P_c (sp-km/h^2)	(not used)	(not used)

* Way capacity C_w and station capacity C_s are defined in Section 2.1.

$$h_{\min} = \text{Max}\ (h_{w\min}, h_{s\min}) \qquad (1.7)$$

In the vast majority of cases, $h_{s\min} >> h_{w\min}$, and the minimum headway at the station with the heaviest exchange (boarding and alighting) of passengers, i.e., the longest minimum headway among all the stations on the line, represents the minimum headway that can be achieved on the line.

The number of TUs passing a point on a transit line in one direction during one hour (or some other time interval) represents *frequency of service f*. Thus, frequency is inverse of the headway:

$$f = \frac{60}{h} \qquad \left| \frac{f}{\text{TU/h}} \middle| \frac{h}{\text{min/TU}} \right| \qquad (1.8)$$

and it is one of the components of quantity of offered service.

The concepts of headway and frequency are often confused not only by the lay public, but by transit operators. For example, many schedules state that a transit line has a "Frequency: 10 minutes." Actually, 10 minutes is the headway. Frequency is its inverse, and in this case it is six departures per hour.

1.1.5 Capacity, Work, and Utilization

Capacity of a system or facility in a broad sense refers to its maximum ability to perform under prevailing conditions. For public transport systems two different capacities are particularly important: *vehicle capacity,* expressed in spaces per vehicle, and *transit line capacity,* with the dimension of spaces per hour. With these dimensions they represent *offered capacities C* (sps/h). Maximum number of passengers, or flow of passengers per hour that are actually transported, represents the demand or *utilized capacity P* (prs/h). The ratio of utilized to offered capacity is the *capacity utilization coefficient* α, also known as *load factor,* with the dimension of persons transported divided by spaces offered:

$$\alpha = \frac{P}{C} \qquad \left| \frac{\alpha}{\text{prs/sp}} \middle| \frac{P}{\text{prs/h}} \middle| \frac{C}{\text{sps/h}} \right| \qquad (1.9)$$

These different capacity concepts, as well as transportation work and productive capacity of a transit line, are listed with their formulae, units, and utilization coefficients in Table 1.1. Each will be briefly defined here in sequence.

1.1.5.1 Vehicle Capacity. Vehicle capacity C_v is the maximum number of spaces for passengers a vehicle (or TU) can accommodate. This "static capacity" is expressed in spaces, and it can be computed in three different ways: *seats plus standing spaces, seats only,* and *ratio of seats to standing spaces.*

Seats plus standing spaces is the capacity definition used for high-volume rail and bus services. This capacity depends on the standard used for floor area per standee (for detailed discussion of passenger level of service as a function of area per standee, see Fruin, 1971). For capacity computations in different transit systems and adopted comfort levels, any one of the values shown in Table 1.2 can be used. In most cities in industrialized countries, the standard of 0.25 or 0.20 m^2/standee is used. In a few very large cities (Mexico, Moscow, New York, Tokyo) and in most cities in developing countries, heavy crowding of vehicles is common, so that the value of 0.15 m^2/standee might be more applicable.

With this definition of capacity, the load factor α cannot exceed 1.00 (unless assumed standing comfort standards are exceeded), and its value clearly shows how close the volume is to capacity.

Seats only; this definition, known as an exact number, is sometimes used for transit services designed to provide seats for all passengers, such as many regional rail and high-quality bus systems. However, if these systems often carry loads exceeding seating capacity, this definition yields load factors greater than 1, which give no indication how full the vehicles actually are and how much reserve capacity is available. For example, a load factor of 2.0 on the San Francisco Bay Area Rapid Transit (BART) cars, which have large standing areas, results in a higher comfort and has more reserve capacity than load factor of 1.5 on the regional rail cars with high-density seating, such as the Regional Rail cars in Philadelphia.

The ratio of seats to standing spaces (e.g., 40:60) is sometimes used as a goal for comfort standard in initial design of a vehicle interior. This measure is inconvenient because it does not describe the comfort (density) of standees and cannot be controlled in operations.

In summary, on transit systems which sometimes carry standees, total vehicle capacity rather than seats only should be used.

Average number of passengers per vehicle is used as a measure of the capacity of semipublic and private modes (such as van pools and private automobiles, respectively) and for the public mode used individually—the taxi. Utilization of these vehicles is determined by their owners or users and is independent of demand for travel on a given facility in the short run. Thus, their physical capacity is often much greater than the number of persons they carry, but the empty seats cannot be utilized in private transportation modes.

TU capacity C_{TU}, is simply $n \times C_v$, where C_v is the respective vehicle capacity from the three definitions given above.

1.1.5.2 Line Capacity. Line (offered) capacity is the maximum number of spaces that can be transported past a fixed point in one direction during one hour. This "dynamic capacity" is expressed in units per hour, and several different units can be used:

Transit units per hour represents the *maximum frequency* f_{max} that can be physically achieved on a line under given conditions (speed, safety, station operations, signal system, etc.).

Table 1.2 Passenger comfort as a function of floor area per standee

Density of Persons	*Area per Standee*		
prs/m²	*m²/prs*	*ft²/prs*	*Standing Passengers' Condition*
<1	>1.00	>10.8	Independent standing, easy circulation
2–3	0.50–0.33	5.4–3.6	Some body contacts, circulation disturbing others
4	0.25	2.7	Extensive body contacts, difficult movements
5	0.20	2.2	Pressed standing, extremely difficult movements
6.7	0.15	1.6	Crash loads, possible injuries, forced movements

Area conversion: 1 m^2 = 10.76 ft^2.

Vehicles per hour, designated *vehicle line capacity* c, defines the maximum number of vehicles that can pass a fixed point on a line in one direction during one hour, regardless of whether as single units or in trains: $c = n \times f_{\max}$. Obviously, for single-vehicle operation $c = f_{\max}$, but for train operation $n > 1$, so that $c > f_{\max}$.

Spaces per hour is the line capacity C, most frequently used in transit mode description, planning, and scheduling. It represents the actual capacity offered to passengers in spaces transported past a fixed point in one direction during one hour.

All three capacities are interrelated and are functions of a line's minimum headway:

$$C = C_v \cdot c = C_v \cdot n \cdot f_{\max} = \frac{60 C_v \cdot n}{h_{\min}} \quad \left|\frac{C}{\text{sps/h}}\right|\frac{C_v}{\text{sps/veh}}\left|\frac{c}{\text{veh/h}}\right|\frac{n}{\text{veh/TU}}\left|\frac{f}{\text{TU/h}}\right|\frac{h}{\text{min/TU}}\right| \tag{1.10}$$

Scheduled line capacity C_o is the number of spaces transported past a fixed point in one direction during one hour under a given operating schedule. Ratios of scheduled to absolute capacities of a line represent utilization coefficients of service frequency δ_f, vehicle line capacity δ_v, and scheduled line capacity δ, respectively:

$$\delta_f = f/f_{\max};\ \delta_v = n \cdot f/c; \quad \text{and} \quad \delta = C_o/C \quad \left|\frac{\delta}{-}\right|\frac{f}{\text{TU/h}}\left|\frac{n}{\text{veh/TU}}\right|\frac{c}{\text{veh/h}}\left|\frac{C}{\text{sps/h}}\right| \tag{1.11}$$

These coefficients are particularly useful for analyses of heavily used transit lines.

Transportation work w performed on a transit line represents its output or quantity of offered or utilized service. The offered work w_o can be expressed in train-, vehicle-, or space-km. When all TUs operate on the entire length L of a line, the work performed during one hour is:

$$w_o = C \cdot L = f \cdot n \cdot C_v \cdot L \quad \left|\frac{w}{\text{sp-km(/h)}}\right| \quad \left|\frac{C}{\text{sps/h}}\right|\frac{L}{\text{km}}\left|\frac{f}{\text{TU}/h}\right|\frac{n}{\text{veh/TU}}\left|\frac{C_v}{\text{sps/veh}}\right| \tag{1.12}$$

Passenger-km traveled on the line represent *utilized work* w_p; it is computed for one hour as:

$$w_p = \sum_i p_i \cdot S_i \quad \left|\frac{w_p}{\text{prs-km(/h)}}\right|\frac{p}{\text{prs/h}}\left|\frac{S}{\text{km}}\right| \tag{1.13}$$

where p_i is passenger volume on section i of the line, and S_i length of that section.

Work utilization coefficient $\bar{\alpha}$ is the ratio of utilized to offered work:

$$\bar{\alpha} = \frac{w_p}{w_o} = \frac{\sum_i p_i \cdot S_i}{C \cdot L} \quad \left|\frac{\bar{\alpha}}{\text{prs-km/sp-km}}\right| \quad \left|\frac{p}{\text{prs/h}}\right|\frac{S}{\text{km}}\left|\frac{C}{\text{sps/h}}\right|\frac{L}{\text{km}}\right| \tag{1.14}$$

This ratio represents the average utilization of offered capacity along the line, or the average value of the load factor weighted by the passenger volume.

Figure 1.5 shows some of the above-defined concepts: a transit line with passenger volume (as shaded area) and offered capacity. The load factor α varies along the line, reaching its maximum value on the maximum load section (MLS), where it is $P_{\max}/C$. The coefficient $\bar{\alpha}$ cannot be plotted numerically, but only visually: it represents the ratio of the shaded area to the area under the C line.

Productive capacity P_c, the product of line capacity and its operating speed, is one of the most useful quantitative indicators of line performance, since it incorporates capacity, which concerns the operator, and speed, which affects both passengers and the operator. Its detailed description and use for comparison of

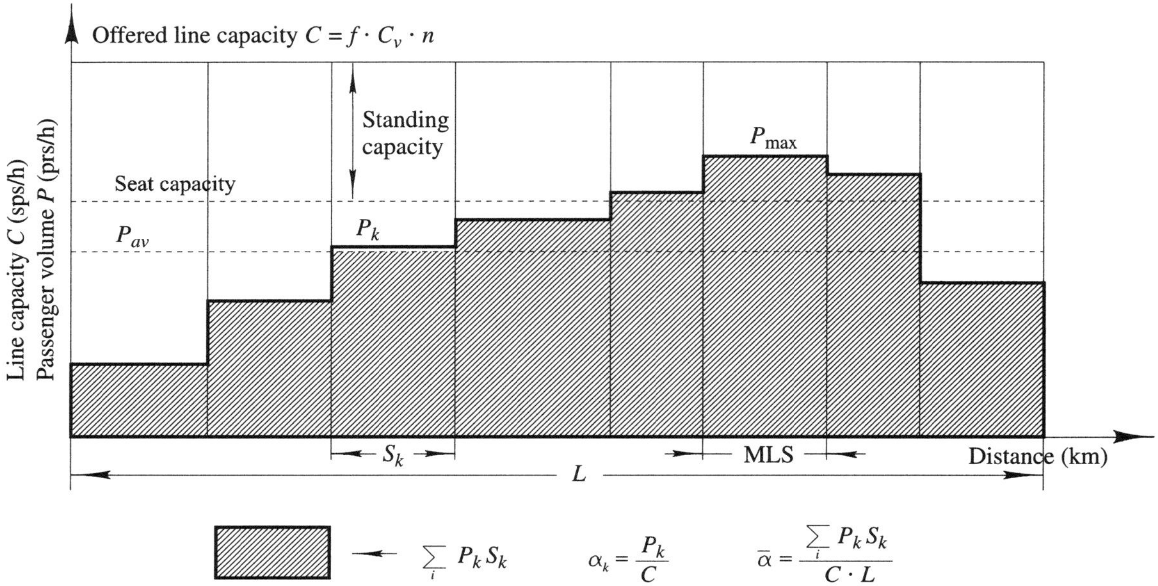

Figure 1.5 *Line capacity, passenger volume, and utilization factors.*

different modes is given in Vuchic (1981), Sections 2.3 and 7.6.

1.1.6 Travel Times

The durations of individual time intervals in transit system operation or in passenger travel are referred to as *travel times*. Many different intervals can be defined. Individual travel time components will be designated by *t*, the intervals consisting of several elements by *T*, while subscripts will indicate specific intervals. Transit travel times are usually expressed in minutes.

1.1.6.1 On-Line Travel Times. Several different travel times of TUs on a line can be defined.

Running time t_r is the time interval between a TU's starting from one station and stopping at the next one, i.e., the net travel time between stations.

Station standing (or dwell) time t_s is the duration of a TU's standing at a station for the purpose of boarding and alighting of passengers. Definitions of elements influencing t_s and its values for different modes are given in Vuchic (1981), Section 7.4.4.2.

Station-to-station travel time T_s is the time interval between a TU's departures from two adjacent stations, i.e., it is equal to running time t_r plus station standing time t_s on any spacing *i*:

$$T_{si} = t_{ri} + t_{si} \qquad \left| \frac{T, t}{\text{min or s}} \right| \tag{1.15}$$

This time interval is thus the basic module of TU travel along the line.

On lines where there are no geometric limitations or interferences by other traffic, such as on some rapid transit lines, station-to-station travel time is expressed by two different equations (derived in Vuchic, 1981, Section 3.6.3). For station spacings S', so short that TUs cannot reach their maximum speed, travel consists of only acceleration, braking and standing intervals t_a, t_b, and t_s, respectively. Travel time T_s' is expressed by the following formula:

$$T'_s = t_a + t_b + t_s = \sqrt{\frac{2(\bar{\alpha} + \bar{b})\, S'}{\bar{a}\bar{b}}} + t_s$$

$$\left|\frac{T,\ t}{\text{s}}\right|\frac{a,\ b}{\text{m/s}^2}\left|\frac{S}{\text{m}}\right| \qquad (1.16)$$

where $\bar{a}$ and $\bar{b}$ are average acceleration and deceleration rates, respectively. For spacings on which the maximum speed of TUs, $V_{\max}$, can be reached, the equation is:

$$T_s = \frac{3.6S}{V_{\max}} + T_\ell \qquad \left|\frac{T}{\text{s}}\right|\frac{S}{\text{m}}\left|\frac{V}{\text{km/h}}\right| \qquad (1.17)$$

where T_ℓ is the incremental time loss per stopping at one station, computed as:

$$T_\ell = \frac{V_{\max}}{7.2}\left(\frac{1}{\bar{a}} + \frac{1}{\bar{b}}\right) + t_s \qquad \left|\frac{T,\ t}{\text{s}}\right|\frac{V}{\text{km/h}}\left|\frac{a,\ b}{\text{m/s}^2}\right| \qquad (1.18)$$

using the approximation that both acceleration and deceleration rates are constant.

Operating (*or travel*) *time* T_o is the scheduled time interval between departure of a TU from one terminal and its arrival at the other terminal on the line. The operating time is therefore the sum of station-to-station travel times for all i interstation spacings between terminals:

$$T_o = \sum_i T_{si} = \sum_i (t_{ri} + t_{si}) \qquad \left|\frac{T,\ t}{\text{min or s}}\right| \qquad (1.19)$$

Strictly speaking, there should be $(i - 1)\ t_s$ time intervals if standing at terminals is included in terminal times (see below). However, the definition given by Eq. (1.19) is simpler, and yet correct, if it is assumed that a portion of terminal times at each terminal equivalent to $t_s/2$ is included in operating time. Since terminal times are usually much longer than t_s (several minutes versus 20–30 seconds), they are practically unaffected by this definition. In the cases where passenger boarding (or alighting) time at a terminal is very long, these definitions may have to be modified.

Terminal time t_t is the time a TU spends at a line terminal (strictly, after a t_s time or two $t_s/2$ times is subtracted). This time is provided for some or all of the following purposes:

- Vehicle turning or change of driver's cab
- Resting of the crew
- Adjustment in schedule (e.g., to maintain uniform headway)
- Recovery of delays incurred in travel

Transit agency work standards, and usually labor contracts, stipulate that a certain percentage of working time must be provided for a driver's rest or meal break. The need for schedule recovery is also related to line length, since the probability of delays and their propagation increases with the length. Total terminal time (at both terminals, $t'_t + t''_t$) is therefore usually determined as a percent of operating time on the line, γ:

$$\gamma = \frac{t'_t + t''_t}{2T_o} \times 100\% \qquad \left|\frac{\gamma}{\%}\right|\frac{T,\ t}{\text{min}}\right| \qquad (1.20)$$

The value of γ usually varies between 10 and 30% (i.e., the sum of the two terminal times represents 10–30% of total, two-way operating time on the line), the typical average being around 15%. Exceptions to this relationship exist in the following cases.

Modes with high reliability of operation, such as rail rapid transit and regional rail, can have fixed terminal times regardless of line length. If that time is short, drivers are exchanged: each driver falls back, i.e., leaves his train and takes over a later one, after his break for rest. The extreme case of this is found on circle lines, which often operate without any terminal time.

On lines with long uniform headways, cycle times (defined below), being integer multiple of headways, must sometimes include long terminal times, resulting in $\gamma > 30\%$.

Bus and other street transit lines are often scheduled with longer terminal times during peaks than off-peaks, to provide reserve times for schedule recoveries from delays caused by frequent peak hour traffic congestion.

Sometimes terminal times are related to cycle time T, defined below, giving factor γ':

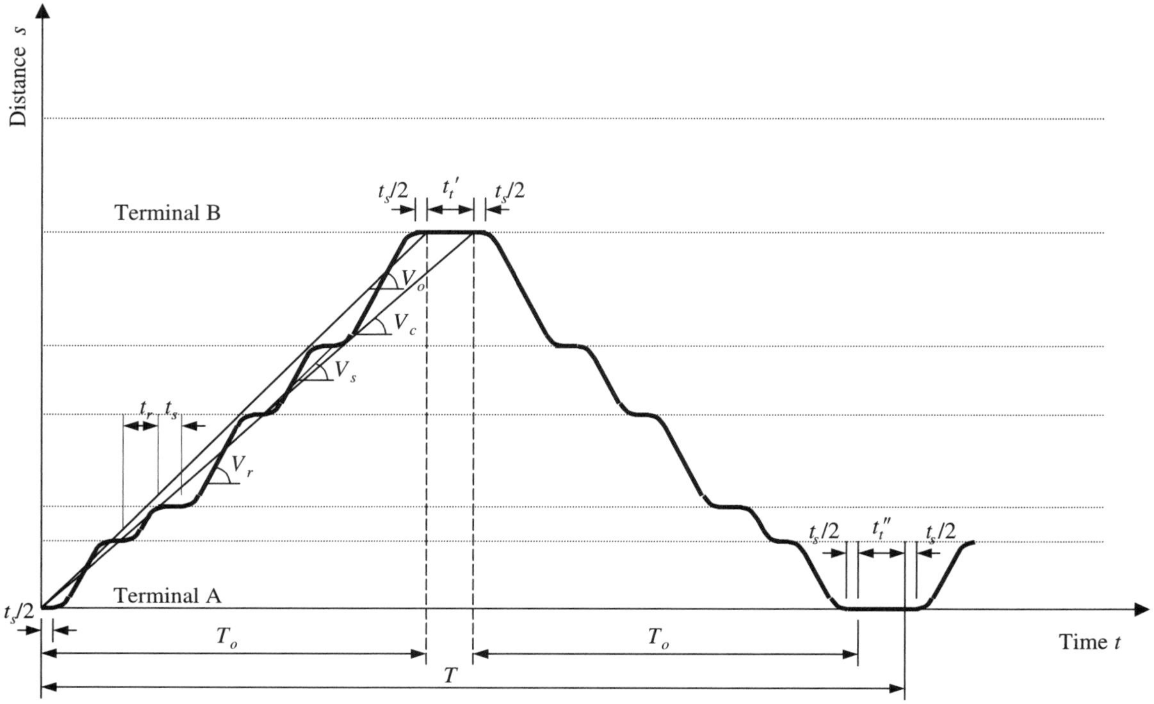

Figure 1.6 *Travel times and speeds on a transit line.*

$$\gamma' = \frac{t_t' + t_t''}{T} \times 100\% \qquad \left|\frac{\gamma}{\%}\middle|\frac{T,\ t}{\text{min}}\right| \qquad (1.21)$$

Naturally, $\gamma' < \gamma$, except on lines without terminal times, such as most circle lines, where $\gamma' = \gamma = 0$.

Cycle time T is the total round trip time on a line, or the interval between the two consecutive times a TU in regular service leaves the same terminal. It consists of operating times for the two directions, T_o' and T_o'', and terminal times:

$$T = T_o' + T_o'' + t_t' + t_t'' = 2(T_o + t_t) \qquad \left|\frac{T,\ t}{\text{min}}\right| \qquad (1.22)$$

The latter expression is for the case when operating times in both directions and terminal times at both terminals are equal, respectively. Cycle time is the basic time unit for scheduling transit service, and it strongly influences investment and operating costs.

All of the preceding travel times are shown graphically on the time-distance diagram of TU travel in Figure 1.6.

Deadhead time t_d is the portion of TU travel time during which the TU is not in passenger (revenue) service. It includes travel from depot to the line and back, or between lines when a TU is reassigned. Deadhead time is not directly productive, and it is therefore desirable to minimize it.

Platform time T_p is the total time a TU is in operation. When a TU makes k round trips on a line and has both deadhead times equal, its platform time is:

$$T_p = kT + 2t_d \qquad \left|\frac{T,\ t}{\text{min}}\middle|\frac{k}{-}\right| \qquad (1.23)$$

Sometimes platform time is also affected by the waiting times at terminals for schedule adjustments, or, if the TU is scheduled to operate on more than one line, by cycle times on different lines. Platform time represents the net on-duty time for computing labor requirements for transit line operation. To compute total labor time, check-in, check-out, and idle times of drivers must be added to T_p.

1.1.6.2 Passenger Travel Times. The travel of a transit passenger on an origin-destination path, including his/her approach to a transit stop or station, travel on the line, a transfer between lines, and departure from a stop to his/her destination, is shown on the diagram in Figure 1.7. This type of diagram is often used for analysis of passenger travel on individual lines and comparison of passenger travel times by alternative modes. The diagram shows that the passenger has the following time components.

Access time t_a is the time which an individual passenger requires for approach to a transit stop or departure from a stop to his destination for a given trip.

Waiting time t_w is the time between passenger arrival at a stop and the time of TU departure. For frequent transit service the average waiting time is equal to half of the headway. For longer headways (usually $h > 6$ or 10 minutes) passengers begin to use a timetable and adjust their arrivals to the scheduled TU departures, so that the average waiting time becomes somewhat shorter than for random passenger arrivals and remains approximately constant for longer headways, as the diagram in Figure 1.8 shows. An excellent analysis of t_w is given in Bowman and Turnquist (1981).

On-line travel time t_o is the duration of passenger travel in a TU for a given trip.

Transfer time t_f is the time used for transferring between different lines or modes, i.e., the interval between alighting one TU and boarding another. Transfer time depends on the walking time between the two line platforms, on headway of the second line, and on schedule coordination between the lines.

Origin-destination travel time T_{od} is the total passenger's travel time from his point of origin (o) to his

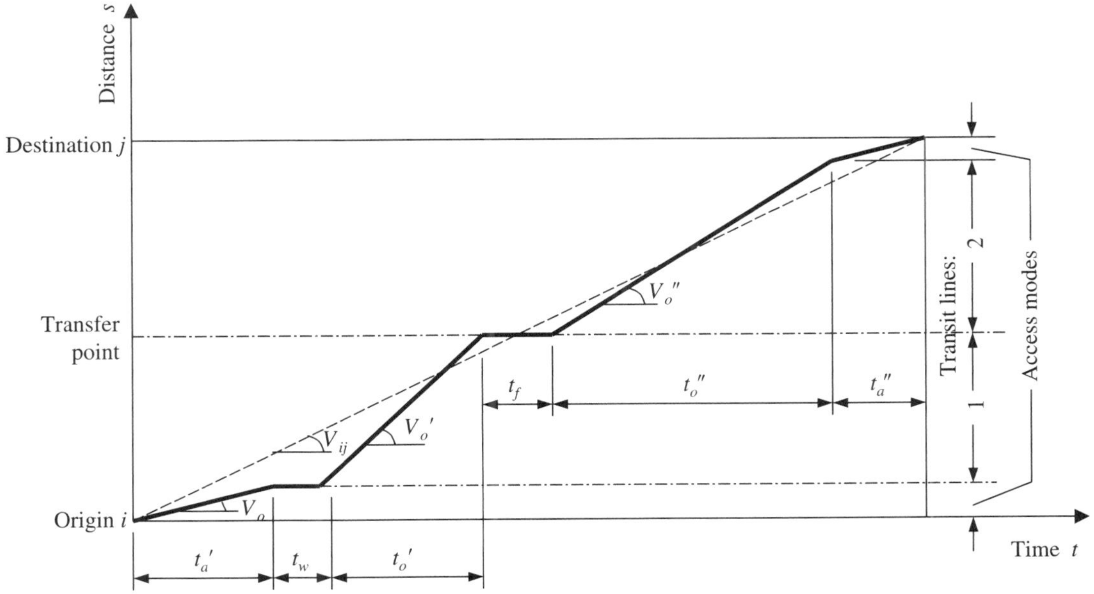

Figure 1.7 *Passenger travel times and speeds.*

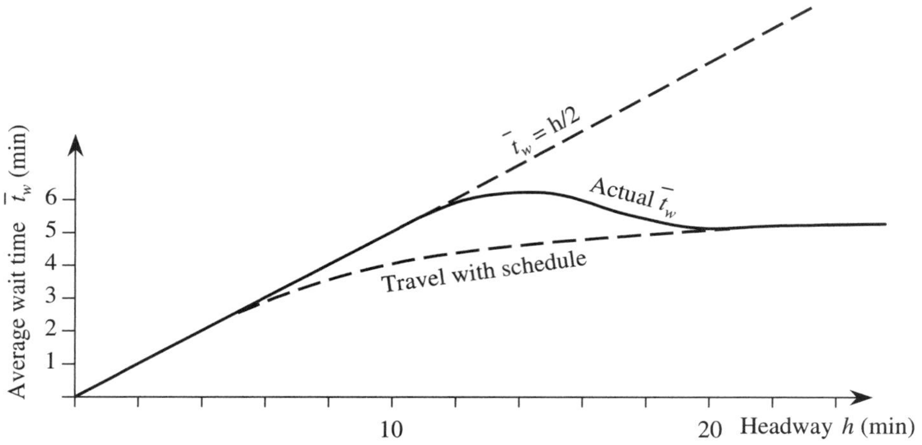

Figure 1.8 *Average passenger waiting time as a function of headway.*

point of destination (d). This time may consist of some or all of the following components:

$$T_{od} = t_a + t_w + t_o + t_f \qquad \left|\frac{T, t}{\text{min}}\right| \qquad (1.24)$$

1.1.7 Speeds

A number of different speed definitions are used in transit operations and analyses. They are listed in Table 1.3, classified in four categories. In the following sections each speed is defined and most of them are shown graphically on diagrams. Three types of diagrams are particularly useful for illustrating speeds:

- A time-distance diagram is used for plotting movement of a single TU along a transit line, for schedules, and for street traffic movement with signal control.
- A time-speed diagram conveniently shows vehicle performance, travel regimes, energy consumption, and travel time for station-to-station movement.
- A distance-speed diagram is best suited for plotting of speed profiles along a transit line and analyses of its different sections, braking and stopping distances, signal blocks and their locations, etc.

1.1.7.1 Vehicle Speeds. The variable of *actual vehicle speed* on a transit line is designated as V (km/h) or v (m/s).

Maximum technical speed V_{max} is the highest speed a transit vehicle is physically capable of achieving on a straight horizontal way under normal weather conditions when its maximum power is applied and acceleration has gradually ceased.

High technical speed of vehicles is usually desirable for achieving better performance of the line; however, speed of transit vehicle travel is often limited by traffic conditions and spacings between stops, so these should be carefully analyzed in determining the maximum speed vehicles should have for a given type of service.

1.1.7.2 Alignment Speeds. In connection with design and operation of transit lines, the following three speeds can be defined.

Line design speed V_d is the maximum speed transit vehicles can achieve on a given section of line with adequate comfort and safety when physical conditions govern. It may vary among line sections and directions, depending on geometry of alignment, gradient and other factors. Average line design speed for an entire line is:

Table 1.3 Transit speeds and corresponding travel times

Category	*Speed Designation*	*Symbol*	*Corresponding Travel Time*
Vehicle speed	1. Actual vehicle	V, v	—
	2. Maximum technical	V_{max}	—
Alignment speeds	3. Line design	V_d	—
	4. Legal	V_ℓ	—
	5. Programmed	V_g	—
Vehicle-on-line speeds	6. Running	V_r	t_r
	7. Station-to-station	V_s	T_s
	8. Operating	V_o	T_o
	9. Cycle	V_c	T
	10. Platform	V_p	T_p
Passenger speeds	11. Access	V_a	t_a
	12. Travel on line	V_o	T_o
	13. Origin-destination	V_{od}	T_{od}

$$\overline{V}_d = \frac{\sum_i S_i}{\sum_i \frac{S_i}{V_i}} = \frac{L}{\sum_i \frac{S_i}{V_i}} \qquad \left| \frac{V}{\text{km/h}} \middle| \frac{S, L}{\text{km}} \right| \tag{1.25}$$

where S_i is length of section i and V_i design speed on it.

Legal speed V_ℓ is the maximum speed transit vehicles can legally operate on a given section of line. Since the legal speed is determined on the basis of various limitations (design speed, traffic and environmental conditions, etc.), it can be equal to or smaller than the design speed. Legal speed is less permanent than design speed: it can be changed with traffic conditions, between day and night, with improved control devices, etc.

Programmed speed V_g is the speed transit vehicles can operate meeting given standards of safety, comfort, economy, and vehicle performance. Average programmed speed for a line cannot be computed as precisely as design and legal speeds because its changes among sections depend on vehicle performance and various influences, while the former ones change instantly. Programmed speed is the speed that transit vehicles can actually achieve: rapid transit systems with automatic train operation (ATO) can operate it precisely; manually driven TUs can follow it as closely as drivers' skills allow.

None of the speeds defined so far include transit stops. They refer to physical conditions of TUs running along the line, not considering requirements to stop and related time delays. All these speeds are shown on a distance-speed diagram in Figure 1.9, which illustrates the following facts:

- V_{max}, as an element of vehicle performance, is affected by gradient i, but not by V_d and V_ℓ.
- $V_g \leq V_\ell \leq V_d$.
- Alignment gradient, as well as different street conditions on directionally split lines, may result in different V_g's for the two directions.

1.1.7.3 Vehicle-on-Line Speeds. Actual speeds achieved on transit lines are influenced by vehicle and alignment speeds, as well as by stoppings at passenger stops and general traffic conditions (rapid transit is not affected by the latter ones). Five different speeds can be defined for transit line operations.

Running speed V_r is the average speed TUs achieve from leaving one station (stop) to arriving at the next one. On a spacing between stations S_i, this speed is:

$$V_{ri} = \frac{60 S_i}{t_{ri}} \qquad \left| \frac{V}{\text{km/h}} \middle| \frac{S}{\text{km}} \middle| \frac{t}{\text{min}} \right| \tag{1.26}$$

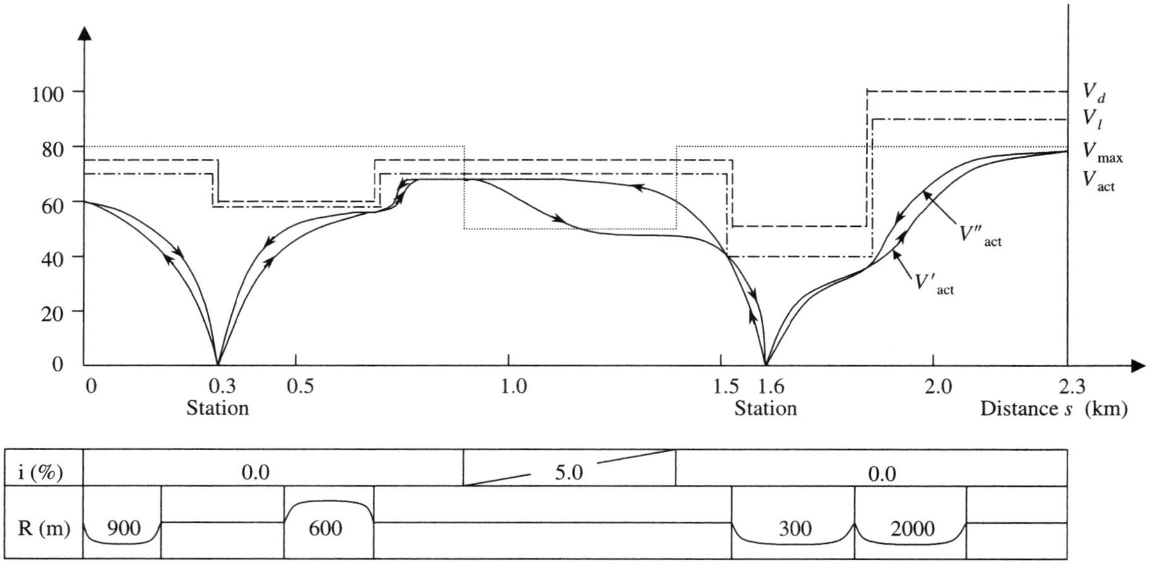

i— gradient; *R*—horizontal curve radius

Different types of speed given in km/h

V_{max}	80				50		80			
V_d	75	60		75				50		100
V_l	70	60		70				40		90
V_{act}	70	60	60	70	70	50	40		40	80

Figure 1.9 *An illustration of relationships among vehicle and alignment speeds on a transit line.*

Thus, running speed varies among spacings and is usually analyzed for individual spacings, rather than as an average for line.

Station-to-station speed V_s is the average speed of travel between moments a TU leaves two adjacent stations; it includes running and one station dwell time:

$$V_{si} = \frac{60S_i}{t_{ri} + t_{si}} = \frac{60S_i}{T_{si}} \qquad \left|\frac{V}{\text{km/h}}\right|\frac{S}{\text{km}}\left|\frac{t,\ T}{\text{min}}\right| \tag{1.27}$$

Consequently, this is the speed for the basic module of travel along a line, movement along one-station spacing. When these speeds are averaged along the line, one obtains the next speed definition—operating speed.

Operating or *travel speed* V_o is the average speed of TU travel along transit line with *j* spacings (or on a section of it):

$$V_o = \frac{60\sum_{i=1}^{j} S_i}{\sum_{i=1}^{j} T_{si}} = \frac{60L}{T_o} = \frac{120L}{T'_o + T''_o}$$

$$\left|\frac{V}{\text{km/h}}\right|\frac{S,\ L}{\text{km}}\left|\frac{t,\ T}{\text{min}}\right| \tag{1.28}$$

Operating speed is the speed of travel offered to the public; it is therefore one of the basic elements of offered transit service performance.

Cycle speed V_c is the average speed of a TU for a complete round trip on a line:

$$V_c = \frac{60 \cdot 2L}{T} = \frac{120L}{T} \qquad \left| \frac{V}{\text{km/h}} \middle| \frac{L}{\text{km}} \middle| \frac{T}{\text{min}} \right| \tag{1.29}$$

This speed is the most important one for the operator, since it directly influences the number of TUs required for a given level of service, and thus the transit line's capital and operating costs.

All vehicle-on-line speeds defined here are shown on a time-distance diagram in Figure 1.6. Clearly, $V_{ri} > V_{si}$, for every spacing i, but both vary among spacings, while $V_o > V_c$ is true for all cases, except for lines without terminal times (such as circle lines), on which $V_o = V_c$.

Platform speed V_p is the average speed of TUs operating on a line from the time they leave depot (garage or yard) until they return to it, i.e., during the platform time. In the case of simple, regular scheduling, platform speed can be computed as:

$$V_p = \frac{120(k \cdot L + L_d)}{k \cdot T + 2t_d} = \frac{60\ L_p}{T_p} \qquad \left| \frac{k}{-} \middle| \frac{L}{\text{km}} \middle| \frac{T,\ t}{\text{min}} \right| \tag{1.30}$$

where k is again the number of round trips performed, L_d is deadhead distance, and L_p is the total distance the TU travels while absent from the depot. Platform speed thus includes two deadheading trips (between the depot and the line) and a number of round trips on the line, so that it is influenced by the location of depot in relation to the line and various schedule requirements. It is therefore used for measuring efficiency of vehicle deployment on individual lines.

1.1.7.4 Passenger Speeds. Transit passengers experience several different speeds while they travel.

Access speed V_a is the average speed of passenger travel to and from transit stops or stations. Computed as access distance divided by access time, this speed can vary greatly: from 4–5 km/h for walking to 30–50 km/h for access by automobile. The same trip may have different access speeds for travel to and from a transit line, depending on access mode used. Access speed is important for planning and analyses of transit network layout and spacings of stations.

Travel speed on line V_o that a passenger experiences is actually the operating speed on the line section that he/she utilizes.

Origin-destination speed V_{od} is the average speed of passenger travel along his/her path from origin to destination, including access, waiting, on-line travel, and transfers, if any. For a total distance past between origin and destination S_{od}, it is:

$$V_{od} = \frac{60\ S_{od}}{T_{od}} \qquad \left| \frac{V}{\text{km/h}} \middle| \frac{S}{\text{km}} \middle| \frac{T}{\text{min}} \right| \tag{1.31}$$

This speed is an element of mobility (or ease of travel) by transit, and it influences modal split.

With all travel times and speeds defined, it is now possible to present a convenient graphical method for computation of passenger origin-destination travel time. The diagram for this computation, presented in Figure 1.10, is obtained in the following way.

The upper right quadrant represents a distance-time diagram of transit line operation: distance is on the abscissa, time on the ordinate, and a family of lines is plotted with slopes representing decreasing operating speeds (40–20 km/h). The abscissa from the origin to the left represents travel time on the line plus waiting time. This time is plotted at the same scale as the ordinate, so that a line at a 45° slope represents travel time without waiting ($t_w = 0$). Lines representing travel with different waiting times are plotted parallel to it toward the left. Assuming that the average waiting time is half of the headway on the line for headways up to 10 minutes and then remain constant for all headways longer than that, the lines for h = 3, 6, and 10 minutes are plotted at a horizontal distance of 1.5, 3, and 5 minutes, respectively, from the first (45°) line.

Finally, the total travel time T_{od} is plotted on the ordinate downward from the origin, usually at a smaller scale than travel time on the abscissa. A straight line from the origin through the intersections of the same values on the abscissa and ordinate in the third quadrant is plotted for travel time without access

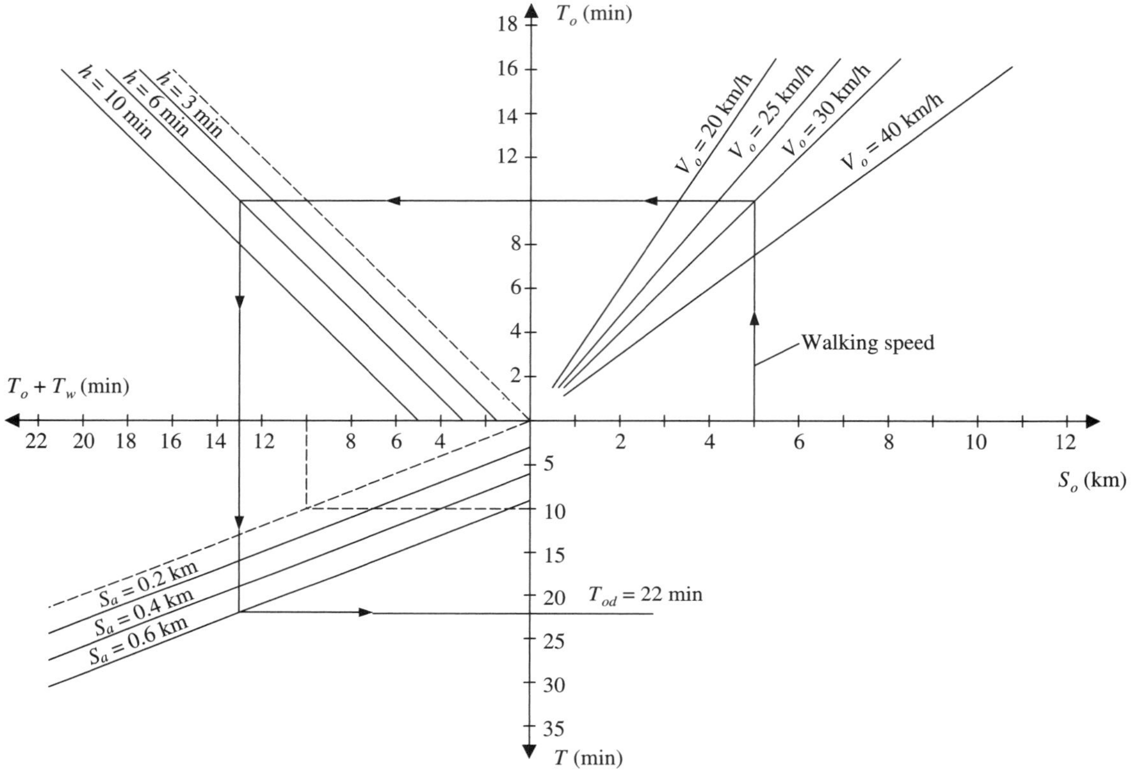

Figure 1.10 *Diagram for computation of passenger origin-destination travel time by transit.*

time ($t_a = 0$), and a family of parallel lines is plotted below it at a distance of travel time for various access distances (representing the sum of origin-to-station and station-to-destination distances) by a given mode. In Figure 1.10, the lines are plotted for access by walking speed of 4 km/h. If it is assumed that average bicycle and automobile access speeds are 20 and 40 km/h, these same lines will represent, respectively, 5 and 10 times greater distances than those shown.

The use of this diagram is very simple. As the plotted example shows, the derivation starts from the distances traveled on the transit line on the abscissa and then continues through the three quadrants for a given set of parameters, to the resulting origin-destination travel time on the downward ordinate. The values assumed in the plotted example are: S_o = 5 km, V_o = 30 km/h, h = 6 min, and S_a = 0.6 km, giving T_{od} = 22 min.

Naturally, the scales and values of parameters on the diagram may vary among different cases; the most convenient ones for a specific case should be chosen. Where appropriate, transfer times should be added to the computed times.

1.2 INFORMATION FILES AND DATA COLLECTION: SURVEYS AND COUNTS

Effective planning of transit operations, scheduling and analysis of efficiency, and sound management in general require accurate data about operations and usage of transit lines and networks. Transit agencies must therefore organize, maintain, and regularly update systematic files of information and data. These files should include:

- *Physical inventories of lines and facilities,* such as locations and equipment of bus stops on streets; designs and dimensions of rail transit rights-of-way, track layouts, signals, stations, yards and workshops, etc.
- *Vehicle data,* including body dimensions, design and performance data, age, condition, breakdowns, etc.
- *Operating conditions on lines:* traffic regulations, relationship of transit vehicles to other traffic, their speeds, standing times and reliability of service; rail transit ROW conditions, speed limits, signal operations, etc.
- *Types of services provided and schedules* for all modes and lines.
- *Usage of services:* passenger boardings, alightings and volumes along the lines, their time fluctuations and their trip lengths; trip generation of different major activity areas, such as stadiums, university campuses, medical complexes, and others.
- *Miscellaneous information on events in operations, fares, and passenger attitudes,* such as accidents, crime incidents, methods of fare collection and types of fares, train-dispatching efficiency, passenger preferences with respect to schedules, riding comfort, and other service parameters.

Much of this information (e.g., transit ROW and station facilities, vehicle characteristics) should exist on permanent plans and records. Data on variable phenomena must be obtained through field surveys: observations, counts, measurements, and interviews. Transit agencies must therefore perform a number of surveys, some on a systematic (regular) basis, others randomly or as special needs occur. The most common transit surveys are described here.

1.2.1 Organization of Surveys

Good planning and continuity in data collection are necessary to obtain and maintain the current data base. The frequency and comprehensiveness of field surveys must be determined based on a compromise between the need for accurate information and the cost of surveys. A good practice is to organize major, detailed surveys at longer intervals and supplement them by minor ones, often on a sampling basis, within these intervals. In other words, a comprehensive cross-sectional set of data, describing entire operations at one time, is combined with longitudinal or time data on key elements, such as passenger volumes on maximum load sections, the busiest (or most typical for the line) stations, or peak-hour flows into the CBD.

For example, every five years the transit agency may organize passenger counts on all routes on one weekday, and on several routes for a whole week. Selected routes are counted every month of that year to find seasonal variations of travel. This count provides data on passenger demand by line: its distribution along the line and its hourly variations. Then every year several lines selected as typical of the whole system are counted along their entire lengths or on their maximum load sections. This may be done during all hours, or during peaks only, depending on the variations that may have occurred and on desired accuracy of data.

Surveys of transit operations often focus on TU running conditions: operating speeds should be measured regularly since they must be known for the purpose of scheduling. However, analyses of traffic conditions, reasons for delays, etc. are usually done on an ad hoc basis, when a new service is planned, or on existing lines when conditions that affect transit operations change or operational improvements are contemplated.

Each type of manual field survey requires a special field sheet and at least one summary sheet. These sheets are usually specially designed for each transit system and type of survey to suit the specifics of the location and type of survey and/or analysis. Typical forms that can be used for many surveys with minor adjustments are given in most manuals for transit and traffic operations, such as Vuchic et al. (1978), Hulbert (1982), and Robertson (1994).

1.2.2 Transit Speed-and-Delay Survey

The purpose of a speed-and-delay survey on street transit lines is to find the distribution of time a TU spends in travel, classified as running, dwells at passenger stops, and several categories of delays. It is also intended to record the locations, durations, and causes of the delays. The measured travel times and delays are used to compute operating speed and reliability of service during different periods of day. This information is important for planning possible operational improvements.

A speed-and-delay survey is usually performed by an observer riding the transit vehicle in a front seat so as to observe reasons for slowdowns and stops. To obtain representative results, several runs are made on one day and repeated on other weekdays. Their exact number depends on statistical variability and the desired accuracy of results. The observer is given detailed instructions on how to define stopped times, how to classify causes of delays, and what codes to use on the field sheet. The observer's equipment may consist of the field sheet clipboard, pen, watch, and stopwatch for measuring delays (in seconds), or a portable electronic recorder with function keys for standing at passenger stops, red signals, and other delay categories. Use of such a recorder eliminates the need to count seconds for various time components.

Speed-and-delay data can also be collected by an on-board computer connected to the vehicle's power supply, drive shaft, and door-opening mechanism. It records the locations and times of each vehicle start and stop, as well as door openings. Its diskette is then given to a main computer for summary and analysis of the data. This type of survey provides more exact and detailed data with far less personnel, although with fewer personal observations about special events.

Manual speed-and-delay surveys have been replaced in many cities, such as Zürich, Oslo, and Helsinki, by equipment monitoring travel of TUs on lines: their passing of fixed points, global positioning system (GPS), or automatic vehicle location (AVL) systems. The information obtained is automatically transmitted to the control center, where computers convert it into graphical time-distance diagrams with the lines of scheduled and actual travel. Deviations from scheduled times are shown in different colors when they are within a certain time interval, such as two minutes. They are shown in different colors if the vehicle is late or if it runs ahead of schedule. This allows the controller on-line monitoring of operations and intervention when irregularities occur, thus increasing service reliability.

Finally, transit systems with automatic driving, control, or monitoring of TUs on the lines usually have automated recording systems for travel times and other operational data. This is the case with metro systems that have automated train operation (ATO) and LRT systems that have signalized lines, as well as with rail vehicles and buses that have various types of AVL systems. These systems continuously monitor locations of all vehicles, and the entire travel process is automatically recorded.

The data from speed-and-delay surveys on streets are summarized and statistically analyzed. Travel times are classified by line section, time period of day, and category: running times, dwells at stops, and standing for other reasons, such as traffic congestion, signals, parking maneuvers, turning movements, etc. Average travel times are then computed for each line section with deviations indicating the reliability of service.

Travel time breakdown on different categories can be presented numerically on a summary sheet, or, more clearly, graphically as a bar chart or a pie chart with segments showing percentage distribution. These are used for analysis of the efficiency of operations: comparison of running times with dwell times at passenger stops, and standing for other reasons. These data are essential in planning new routings, operational improvements, particularly the introduction of various transit priority measures, one-way street regulation changes, elimination of parking, etc. For such changes before-and-after speed-and-delay studies can be performed. Sometimes monitoring of individual driver's performance can also be done.

Speed-and-delay studies are usually not performed on rapid transit systems, since they do not have random delays caused by numerous external causes. However, very precise surveys are sometimes undertaken to measure station standing times and travel regimes between stations so that the extremely high level of service reliability required on rapid transit systems can be achieved.

1.2.3 Passenger Volume and Load Count

The purpose of passenger counts is to determine the passenger volumes on TUs over different sections of a line, the maximum TU load and the section on which that occurs, variations of volumes in time, analysis of service quality—i.e., all elements needed for scheduling of operations.

A detailed survey should include TU passenger load counts at several points along each line, particularly focusing on the sections with heavy loads to find the maximum load section (MLS). Minor annual surveys may be limited to MLSs and one or more additional sections on each line to verify the changes recorded on the MLS.

Personnel for the load count consist of observers at each counting location. For single vehicles operating on streets, one observer is usually sufficient at each location. If the route is heavily loaded or has simultaneous loading of several vehicles, two observers may be needed. Larger teams are needed at busy rail transit stations and bus terminals.

The observer must have adequate training in the method of counting. Very often, exact counts cannot be made because of short standing time of the TUs, so the observer must know how to make a fast estimate of number of people in each group or seating section. The observer should be trained in this through fast estimates at locations where exact counts can be made to check his/her estimates so that he/she can correct any tendency to over- or undercount. In addition, the observer must know the exact seating and total capacities of vehicles, so that he/she can accurately estimate the number of passengers in a full vehicle; or, when there are a few standees, their number can be added to the seating capacity of the vehicle.

It is often possible for the observer to count the load on the arriving vehicle as well as alighting and boarding passengers. These data then provide counts for both sections before and after that station.

Each observer must have an electronic recorder with the appropriate program, or a specially designed field sheet, clipboard, pen, and watch. The following information should be recorded:

- Data about the count: line, location, TU capacity, date and day, time period of count, weather conditions, name of observer, and remarks
- Count records in columns: run number, scheduled and actual arrival times, and passenger counts: on arriving vehicles, alighting and boarding (if possible), and on departing vehicles (this can be computed later)

Counts are summarized after the field survey on a single summary sheet for the 15- or 20-minute peak, 30- or 60-minute off-peak periods, and average TU loads for each period. These numbers are then ready for use for line scheduling and various analyses.

1.2.4 Passenger Boarding and Alighting Counts

The most detailed information on passenger volumes on a transit line is obtained through the counts of boarding and alighting passengers at each stop or station along its entire length. This count provides the data on the number of passengers using each station, as well as TU loads at all points along the line. One can also compute the distribution of passenger trip lengths and total work of the line in passenger-km for any hour or day. This represents virtually all the information needed for scheduling, analyses of operations, extending or shortening of the line, addition or abandonment of some stops, etc.

The decision on what time periods to make this survey is again dependent on the cost of the survey and the need for accurate data. Whenever the operator can afford its higher cost, a boarding/alighting survey should be undertaken instead of a passenger load count, since it provides much more detailed information.

Personnel are usually best utilized on street transit if one observer travels on each vehicle and counts passenger boardings and alightings at up to two doors. Thus, one person can cover a regular bus, but on long LRT vehicles two persons may be necessary. At very busy stops, additional observers may be placed to assist. For high-frequency service, particularly on rapid transit lines, it is more efficient to place one or more counting persons at each station.

Personnel must be given similar training to that for load counts. Boarding/alighting is usually simpler to count than vehicle occupancy, provided the observer can be positioned inside or outside the vehicle so that he/she can easily observe all doors.

The equipment required again consists of an electronic recorder or a special field sheet, clipboard, pen, and watch. It is convenient to use one sheet for each one-way or for each round-trip run. After the survey, recordings from all TUs (or stations) are summarized by time periods, usually 15 minutes for peak and 60 minutes for off-peak hours. A simplified example of a summary sheet is given in Table 1.4. These data are used for development of various passenger volume diagrams, such as shown in Figure 1.11, for computations of trip length distribution on the line, boarding density (prs/km), travel density (prs-km/km), and other characteristics of line usage.

There are a number of automatic passenger counter (APC) devices, from hand recorders and portable computers to automatic scanners and vehicle step pressure detectors, recording boardings and alightings of passengers. They can be utilized in different ways to decrease manpower required for counts, increase the volume and accuracy of data, and simplify their recording. For example, the transit agency in Oslo has developed a counter for its LRT vehicles that is installed behind the driver and connected with an axle to record its revolutions and thus compute vehicle location along the line. The counter records passage of every person entering through the doors. The disc from this counter is then taken to a main computer in the office, where its record is translated into passenger count summaries by station along the line.

The most complete and reliable automated passenger counts can be obtained on the modern rapid transit systems with fully controlled stations, i.e., with computer-controlled fare gates at which every passenger inserts a magnetic card—either a sliding or a contactless "smart card"—for entry and exit (e.g., BART, Washington Metro, many systems in Japan, Korea, and Hong Kong). These gates provide continuous counts

Table 1.4 Summary of passenger counts and computation of person-kilometers

Route no. *41* | Direction *NB* (Departure from South terminal) | Time period *16:30–17:30*

	Number of Passengers				
Transit Stop	*Boarded*	*Alighted*	*On Vehicle*	*Km Between Stops*	*Person-km*
1 (South terminal)	48	—			
			48	2.2	105.6
2	35	12			
			71	2.8	198.8
3	54	30			
			95	3.2	304.0
4	29	13			
			111	3.4	377.4
5	16	46			
			81	2.0	162.0
6 (North terminal)	—	81			
Totals	182	182	—	13.6	1147.8

Average trip length: 1147.8 : 182 = 6.3 km
Average boardings/km: 182 : 13.6 = 13.4 prs/km
Average passenger volume/km: 1147.8 : 13.6 = 84 prs-km/km

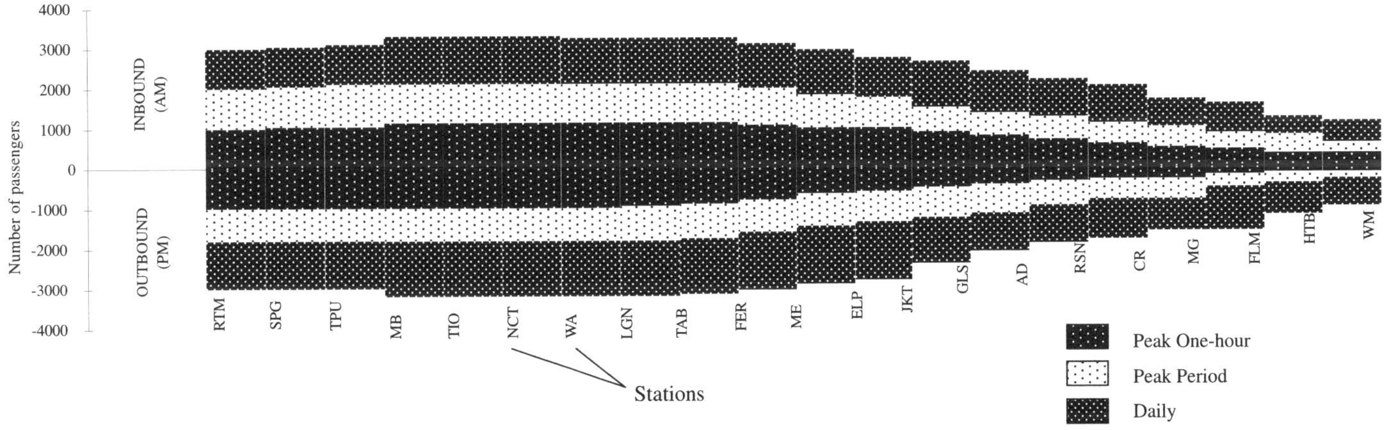

Figure 1.11 *Passenger volume profile by direction for different time periods (Warminster Line, Philadelphia Regional Rail System).*

of entering and exiting passenger volumes for each station, so that all data for station usage and a loading profile of the line are obtained. Slide-through magnetic tickets or smart cards that provide passenger count data are also increasingly used on street transit vehicles. When electronic fare collection is used, personnel are used only for special surveys of passenger attitudes, rates of boarding, etc.

1.2.5 Other Types of Surveys

In addition to the above-described rather standard surveys, many other surveys can be undertaken to obtain various types of data. Several examples of such surveys are described here.

- *Transfer counts*. Volumes of passengers transferring between two or more lines at transfer stations must usually be manually counted. However, on some transit systems transferring passengers are required to use magnetic or smart cards that record their path and automatically collect the statistical information about transfers. Results of such counts at major transfer stations are shown in Figure 1.12. Information about transferring is necessary to distinguish the number of passengers who use two or more lines from those traveling on individual lines only ("linked" from "unlinked" trips); it is also useful for planning the types of fares and methods of their collection.
- *Fare usage*. Counts of passengers by the type of fare they pay on street transit vehicles also must sometimes be performed manually when self-service fare collection is used. This is often the case with LRT systems that have open stations and simultaneous boarding and alighting on many doors. Their results are used for analyses of fare structure and collection and control methods, and sometimes to make possible passenger counts through fare revenues. Automated fare collection on closed systems, such as metros, records all transactions and on some systems, such as the San Francisco BART, provides origin-destination records for every passenger. Thus, it is possible to obtain entire origin-destination matrices for any time period.
- *Passenger travel information*. In addition to information on passenger origin and destination stations along a transit line, it is desirable to know actual origins and destinations of their entire trips, including lengths and modes of access to/from stations, i.e., so-called submodal split. This survey usually requires interviews with passengers or distribution of questionnaires, which they complete and drop off at their destination stations or mail back.
- *Attitudinal and modal split surveys*. Preferences of passengers with respect to service parameters, such as headway length, speed, reliability, fare level, and others, and the importance of these for their choice of travel mode, are also found through interviews or questionnaires.
- *Use of timetables*. Surveys of passenger arrival time distributions in relation to scheduled TU departures can show for what services passengers use schedules and how that use is influenced by headway lengths, reliability of service, and convenience of obtaining schedule information.

As this description shows, field data collection is performed both by automatic devices and by personnel in the field. Specialized surveys, particularly for information on operational details and passenger attitudes, are usually performed manually. In all cases, handling of collected data and their storage, retrieval, and presentation are most efficiently done by use of computerized data storage. Permanent numerical and other data files allow systematic organization, storage, and fast retrieval, as well as printing of various tables or graphical displays such as diagrams, bar charts, passenger volumes along lines, or on entire networks.

1.3 TRANSIT TRAVEL CHARACTERISTICS

Transit travel volume and characteristics must be known for short-range planning of operations (scheduling), as well as for travel forecasting, which repre-

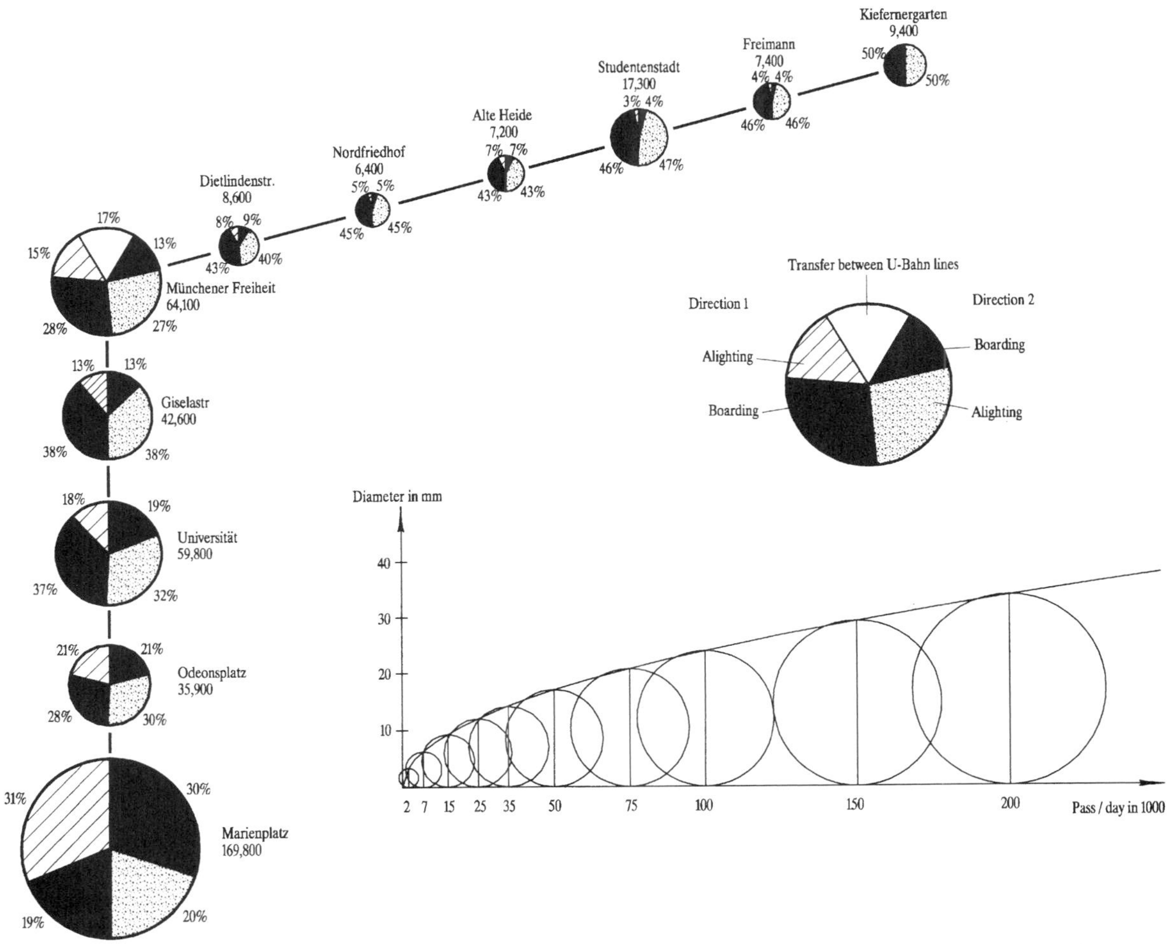

Figure 1.12 Graphical station passenger count summary: boardings, alightings, and transfers at the Munich U-Bahn stations. (Source: *Münchner Verkehrsverbund Annual Reports.*).

sents a basis for long-range planning. Although travel characteristics vary among cities, many general features are common. They are described here.

Transit travel, or use of transit services, is often referred to as *transit demand.* More precisely, this term may have two definitions. When it is defined as the number of passengers who want to use a given service and pay its price, the concept of demand is the same as passenger travel volume on a transit system. Alternatively, transit demand is defined as volume of travel that would take place when service is very good and price (fare) low or moderate. Since demand for travel is elastic with respect to level of service and its price, demand by the second definition, which should more precisely be referred to as *potential transit demand,* is at least equal to, but usually greater than, passenger volume on a given transit system. The difference between the two represents unrealized travel and is known as *latent demand.* Therefore, large-scale planning may refer to transit demand in a global sense, while in descriptions of actual passenger travel that occurs on transit systems the same concept is more precisely designated as *passenger volume,* or the number of passengers transported per hour.

Consequently, transit travel refers to actual passenger volumes under given conditions, and its magnitude varies with size, form, and character of the city, among transit modes, lines, and times of day. It is important to bear in mind that level of service (LOS) consisting of many factors affecting passengers, such as speed, reliability, comfort, and others, influences the ability of transit to attract passengers. Improved LOS, particularly if it is initially very low, can significantly increase transit travel. On the other hand, service deterioration or an increase in transit fares may decrease passenger volume.

1.3.1 Factors Influencing Transit Travel

Generally, the volume of transit passengers on a line or in a network depends on the total volume of passenger travel in a city (or in a specific area) and on the LOS/price package that transit offers as compared to the corresponding packages of competing modes, such as the private automobile, taxi, or bicycle.

The relationship between transit and auto LOS and price (or cost) of travel varies with many factors, but mostly with the concentration of potential travel demand. In cities with well-planned, coordinated intermodal transportation systems, transit is a superior mode for travel in the central city area and along heavily traveled corridors, even if auto ownership is high. In suburban areas with dispersed, low-density travel, travel by auto is more convenient, although transit systems can be organized to offer a reasonably good basic service. This is done by an integrated rail network with bus feeders, or with a timed transfer system (see Section 4.5) type of service. Transportation policies, particularly unlimited versus controlled parking, can strongly influence this balance between transit and automobile travel, referred to as *modal split.*

In cities that have concentrated all their efforts on highway construction, even major travel corridors are dominated by auto travel. Lack of acceptable transit service suppresses transit use and mobility in general. It also causes many travelers to shift to the automobile if they can own and drive one. Frequent traffic congestion then causes higher total cost and lower average quality of travel in the corridor by both modes, auto and transit.

The inherent advantage of transit over automobile in handling high passenger volumes explains the fact that transit share is highest in dense areas such as central business districts (CBDs) and other major activity centers; higher during peaks than at other times of day; and generally higher for work trips than for other purpose trips, because that travel is most concentrated in space and time. Work trips are followed by school and shopping trips, while social-recreational travel gives transit the lowest shares. Thus, in a city with good transit service, such as Munich, Paris, or San Francisco, typical shares of transit trips in central city by trip purpose are in the following ranges: work 60–90%; shopping 40–60%; school 50–70%; social-recreational 20–40%; and other purposes 10–30%.

Trip length distribution also varies among cities, depending on their density and form, types of transit and street networks, fare structure, and other factors. Several typical trip length distributions are shown in Figure 1.13. Cities with dense central areas (e.g., London, Milan, San Francisco) have very skewed distributions, due to large volumes of short (1–4 km) trips. Extensive commuting to the CBD from outlying areas (and to some extent among these areas) results in high volumes of long trips: this is found in most U.S. and Canadian and some West European cities, as exemplified by Hamburg in Figure 1.13.

Trip length distributions for individual modes depend on the role each mode plays, the structure of its network, and the type of service it provides. Street transit modes—streetcars, trolleybuses and buses—typically carry the shortest trips; average lengths are usually 4–8 km. Rapid transit systems carry longer urban trips, typically averaging 6–12 km. Only in cities that have extensive network coverage of central areas, such as Madrid, New York, and Paris, does rapid transit also serve many short trips. Suburban bus lines often carry rather long trips, unless they are local within individual areas and feed long bus or rail lines into central city. Finally, regional rail systems typically serve very long trips: average trip length on this mode for all U.S. cities is about 22 km. On the San Francisco BART, which has a network regional in character, the

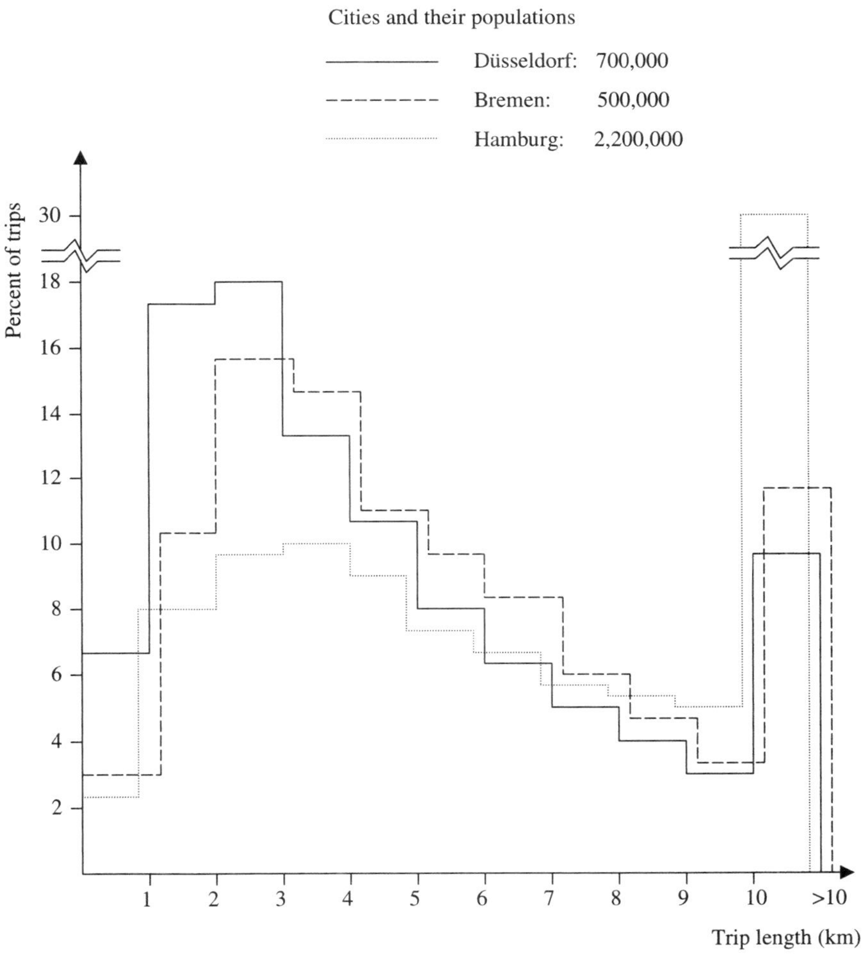

Figure 1.13 *Transit trip length distributions in several cities.*

average trip length is 21 km. A diagram of typical cumulative distributions of trip lengths for different modes is shown in Figure 1.14.

1.3.2 Spatial Distribution of Transit Travel

Distribution of travel demand by area and direction is a function of city form and land use in it. The highest travel volumes are always concentrated on radial directions toward the area of most intensive activities—the CBD. These directions therefore have the highest level of transit service. A typical transit travel volume diagram for a small city is shown in Figure 1.15.

Regional subcenters (or major activity centers) in suburban areas are also generators of large travel volumes. The transit share of these volumes depends greatly on the type of planning and design of these centers and their relationships to transportation networks. If the planning of the two is coordinated and transit services are carefully incorporated with the center's design, as has been the case in various new developments in Frankfurt, Gothenburg, Montreal, Stockholm, and many other cities, the transit share of

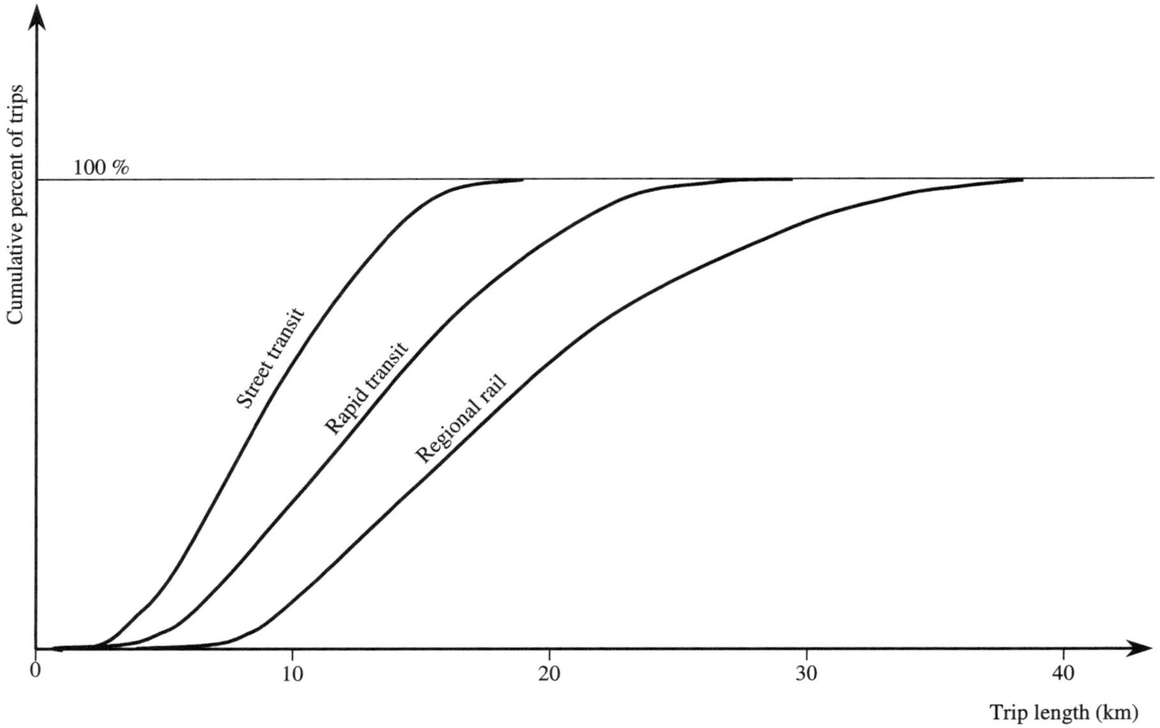

Figure 1.14 *Typical cumulative distributions of trip lengths for different modes in a large metropolitan area.*

travel is quite significant. The most important element that makes transit attractive and competitive with auto travel is the provision of separate transit ROW category B or A, which is easy during the initial design but very difficult to build later.

The other extreme, when transit is disregarded in the planning and design of major suburban activity centers (as was done in some auto-oriented cities not only in the United States but also in some countries in which transit has a very important role), it may be difficult to provide a reasonably good transit service economically. Lack of transit service may cause serious problems of excessive congestion and lack of mobility for persons who do not own or do not drive cars. Sometimes this condition leads to various citizen-organized actions, such as use of rented buses, group taxis, or other forms of semipublic paratransit. These actions can ameliorate but not solve the problems created by the absence of adequate public transit services, because semipublic modes do not offer all-day services, nor are they available to the general public.

There is sometimes a tendency to believe that if in a certain urban area or corridor there is little transit travel, there is not much demand for it. However, a major reason for low usage may be very poor transit service. The fact that people cannot use transit service if it is not offered, or that they do not want to use it if it is slow, unreliable, or expensive, should never be overlooked. Actual transit travel in such cases may be much lower than potential transit demand.

Cordon counts around CBDs represent a good indicator of the quality of transit service and its role in the city. The percentage of trips into the CBD by transit is highest during peak (commuting) hours, and it is generally higher for larger than for smaller cities. In small auto-oriented cities, transit often carries only 20–

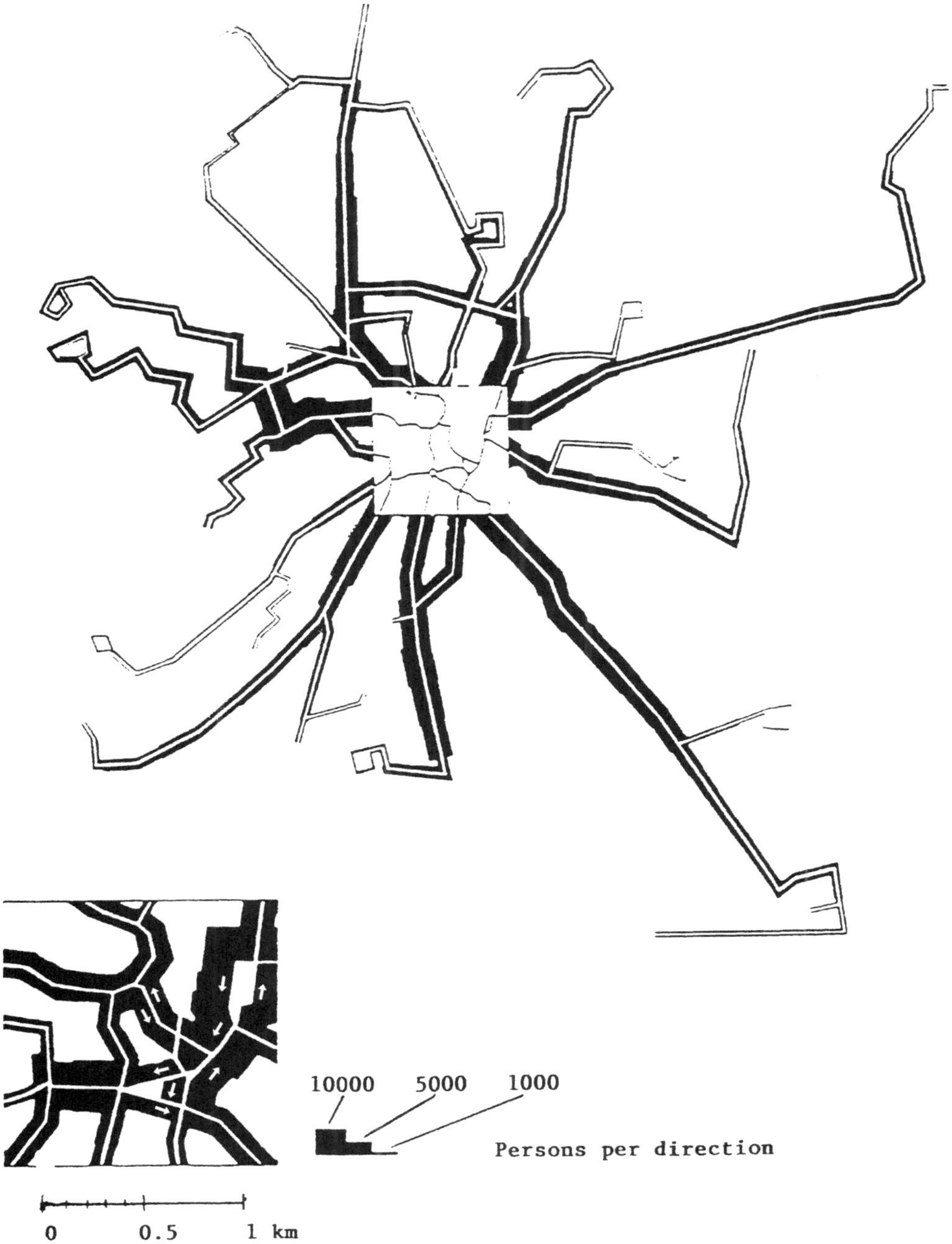

Figure 1.15 *Typical transit flow pattern.*

30% of trips during peak hours, and a lower percentage for the whole day. In medium-sized cities (400,000–1 million populations) this percentage is about 40–60% when transit service is good. The highest percentage is found in very large cities with extensive rapid transit systems, such as New York, Paris, and Tokyo, where transit carries up to 95% of peak-hour trips into their CBD areas.

Some cities, such as Chicago, London, and New York, have precisely defined central cities and make cordon counts regularly. Such counts show intermodal shifts of travel over time. During the 1950–1980 period, with increasing auto ownership and road construction there was a general trend of increasing use of automobiles as well as rapid and regional rail transit, mostly at the expense of buses, which were seriously affected by street congestion. This trend showed, however, the importance of the level of transit service. For example, a drastic improvement of bus speeds and service reliability achieved in Paris through the introduction of many reserved bus lanes during the 1970s resulted in an increase in the share of bus passengers; deterioration of rapid transit services in New York in the early 1980s resulted in corresponding passenger losses; subsequent service improvements were followed by increases in ridership. In the late 1990s, New York also presented an interesting demonstration of elasticity of demand to transit price. Introduction of the electronic "Metrocard" was used to allow free transfer between subway and buses. Compared to the preceding separate fares on each mode, this represented a reduction of fare by 50% for the transferring passengers. This resulted in a network-wide increase of passenger volumes by over 30% on buses, as well as a significant increase of subway ridership.

In general, the role of transit in CBD-oriented travel, particularly in large cities, remains very significant or dominant even with saturation-level auto ownership.

1.3.3 Temporal Variations of Transit Travel

Most of the temporal variations in transit travel are caused by the differences in travel patterns and levels of service by competing modes of transportation during different time periods. With higher travel concentrations in peak periods, the relative advantages of transit increase, so that modal split is higher for transit during peaks than for total travel throughout the day.

Transit travel varies with seasons, as its plot by month for several years in Figure 1.16(*a*) shows. The diagram shows considerable regularity: the summer period (June–August) is by far the lowest, late fall (October–December) the highest. May and September tend to be the closest to the average monthly volume.

These variations, however, are not typical for all transit systems. Variations are less pronounced in areas with diversified activities in large cities, while transit service in vacation resorts or other cities dominated by a single activity may have extremely pronounced and quite different variations. For example, transit ridership in such resort cities as Atlantic City (U.S.), Cannes (France), or Blackpool (England) is several times higher during the summer months than during the winter period. Special events, such as fairs, Olympic Games, or other sport events, create all-time peak demands, disrupting all regular patterns. Naturally, such peaks cannot be adopted as design values for permanent facilities, unless these are also justified for regular services after the event (construction of initial lines of Metro systems in Montreal, Mexico, and Munich were speeded up by the World's Fair in 1967 and Olympic Games in 1968 and 1972, respectively).

Daily variations within a week, shown in Figure 1.16(*b*), are relatively minor among workdays (Monday and Friday being usually the highest), but they are very pronounced for Saturday and Sunday. Holidays are usually similar to Sundays. The difference in passenger volumes between workdays and weekend days increases greatly when weekend trips are shifted heavily to private automobiles. However, there are some notable exceptions to this pattern. In some cities that have attractive central areas and good transit service, such as Portland, Oregon, and Newcastle, England, the highest transit ridership occurs on Saturdays. In cities with low automobile ownership and heavy reliance on transit, weekend ridership is also much more similar to weekday volumes.

Hourly variations of demand are usually very pronounced on workdays, as Figure 1.17 illustrates, mostly because of the peaking character of commuter

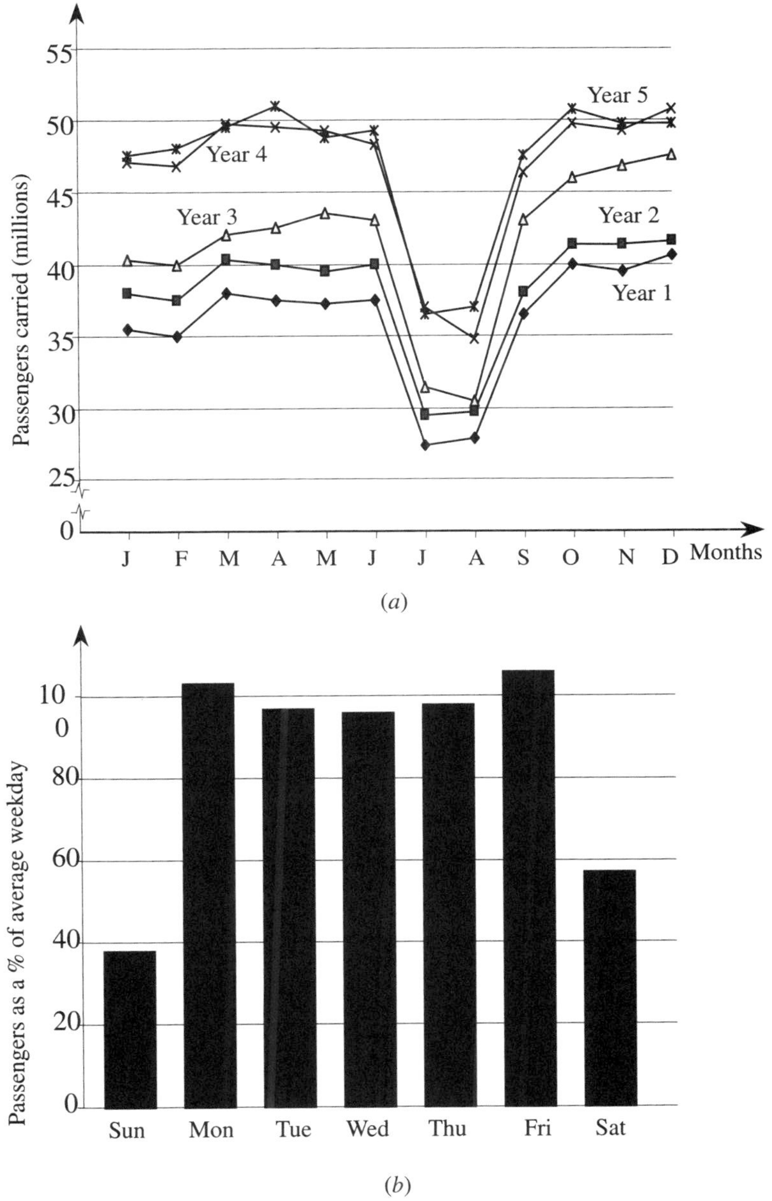

Figure 1.16 *Monthly and daily variations of transit travel: (a) monthly variations for consecutive years (Belgrade)* (source: *Banković, 1982*); *(b) typical daily variations.*

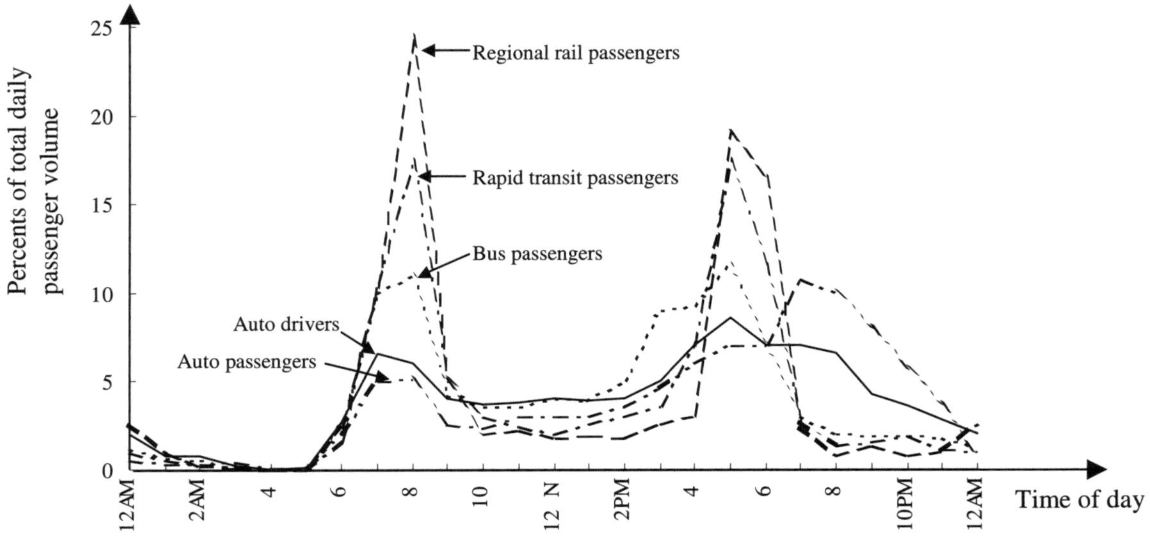

(*a*)

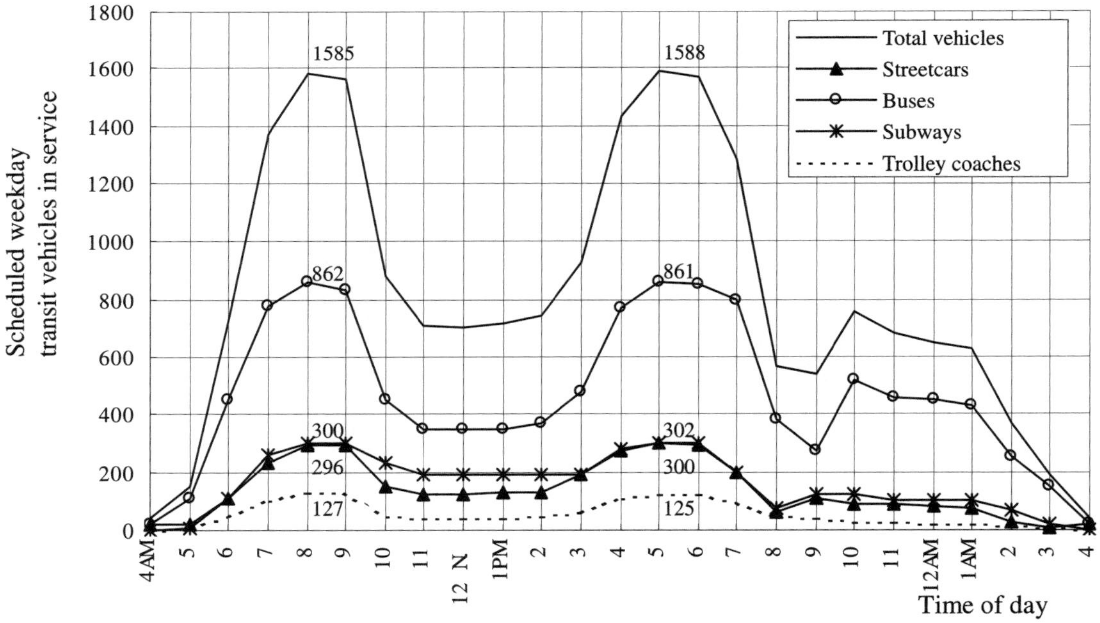

(*b*)

Figure 1.17 *Hourly variations in transit: peaking patterns of passengers and rolling stocks:* (*a*) *variation in passenger volume by mode in Chicago* (source: *Chicago Area Transportation Study*); (*b*) *variation in fleet requirements by mode in Toronto* (Source: *Toronto Transit Commission*).

travel. Passenger counts summed up by hours are usually accurate enough for most periods of day, but not for peak hours when sharp variations may occur within an hour. Peak-hour volumes are discussed further in Section 1.4.

The peaking characteristics of passenger volume can be measured by the ratio of the highest hourly volume to the average off-peak hourly volume, such as between 9 and 12 a.m. This *peak-to-base ratio* usually varies among modes. As the diagram in Figure 1.17 shows for Chicago, the highest peaking occurs on regional rail lines, followed by rapid transit and bus passengers. Auto drivers and auto passengers are less peaked than most transit passenger volumes.

Compared to the variations and peaking of passenger volume, variations of the number of vehicles in service are less pronounced because the load factor α is greater during peaks than off-peaks. The variations are even lower for the number of TUs on guided systems, which use larger TU consists (longer trains) during peaks than off-peaks.

Variations in the required personnel to provide transit services depend on vehicle capacity, train consist, and degree of automation. Consequently, personnel requirement varies directly with the number of buses (one driver per bus); it is somewhat lower for LRT and metro systems if different train lengths are used during peaks and off-peaks. Fully automated modes, such as AGT and several metros, have even lower operating personnel ratios, which gives them an advantage in economic efficiency.

With respect to travel categories, peak-to-base ratio is the highest for radial travel dominated by commuting, which is the case on some regional rail and express bus systems. It is the lowest for services within the central city area, as well as on tangential and circumferential lines, where the diversity of trip purposes is the greatest.

Similar to daily variations of transit travel, hourly variations are also less pronounced (i.e., the peak-to-base ratio is smaller) in cities with heavy, multipurpose use of transit, which fills the troughs between the morning and evening commuting peaks. Examples of such transit use are found in Hong Kong, Mexico, Moscow, and New York.

1.3.4 Passenger Volume Analysis and Service Capacity Determination

The volume of passengers who want to travel along a transit line from a suburban terminal to the CBD terminal during one day (assuming a certain level of service) can be presented on a three-dimensional diagram with distance and time representing a horizontal plane and passenger volume the vertical topography over the plane, as Figure 1.18 shows.

A TU traveling on a line is plotted on the diagram in Figure 1.19 as a sloped line representing its distance-time path on the diagram. Since station-to-station speeds vary with distances between stations, the TU travel line has different slopes among spacings. Capacity of the TU is plotted vertically. It is constant along the line, while passenger load on it varies as a function of boardings/alightings, and it reaches a maximum value P_{max} on the MLS. Service on the line can thus be imagined as a series of slices through the topography of passenger volume, as illustrated in Figure 1.20. Each slice represents a TU with its passenger load shown as the shaded area. For clarity, variations of speed along the line are not shown on this diagram.

Naturally, if constant service were provided throughout the day, it might not be sufficient during peak hours, particularly on the MLS, while its utilization during off-peak hours would be very low. Referring to the diagram in Figure 1.20, fitting a constant-height box over the passenger volume topography could result in protruding peaks and empty space toward all side walls of the box.

To satisfy the variable demand and achieve good utilization of offered capacity, service must be tailored to cover the demand as closely as possible. This can be done in three different ways:

1. Changing headways—shortening them (higher frequency) during the peaks
2. Using different TU sizes/capacities, such as different train consists (number of cars)
3. Operating TUs on certain sections of the line only ("short-turn trains"), so that the protruding topography peaks are covered by small additional TU capacity boxes.

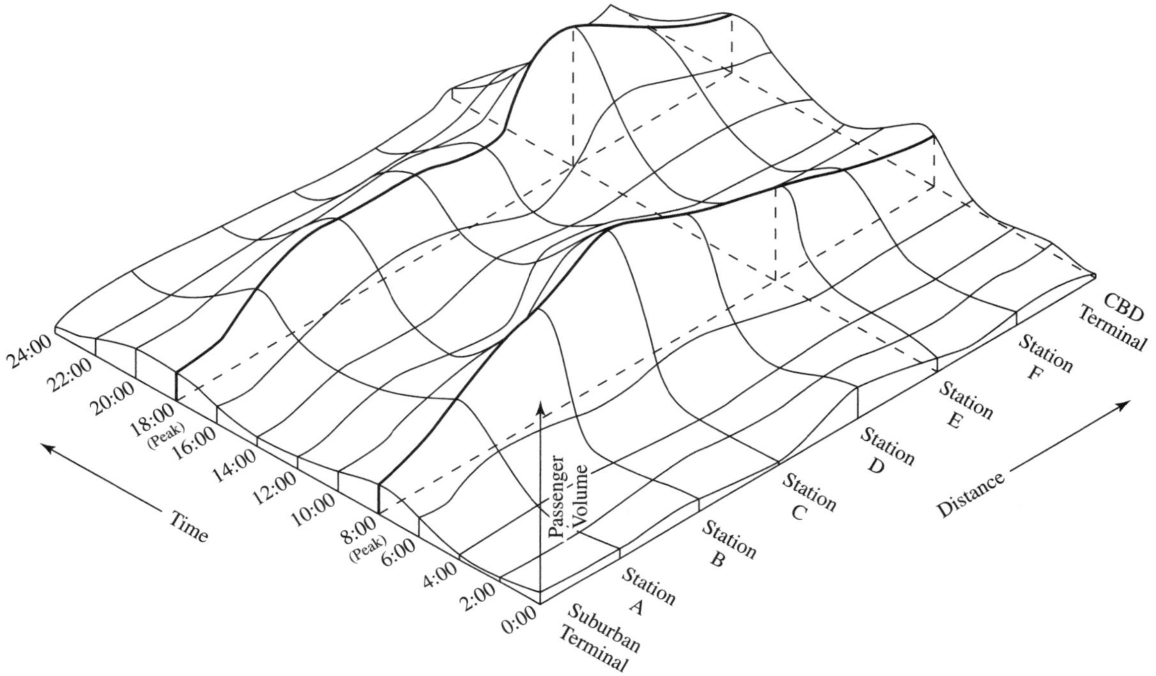

Figure 1.18 Three-dimensional presentation of passenger volume distribution in distance and time along the line from a suburban to the CBD terminal. (Design: *Thor Haatveit.*)

Another way to reduce offered capacity as the line proceeds from the CBD and volume decreases is to divide the line into two or more branches. That condition would be presented on the diagram in the same way as case 3—short-turning of trains at a point along the line.

The three capacity adjustments are illustrated in Figure 1.21.

Each of these three types of capacity adjustments is utilized in actual transit system operations, but not all are practical in every situation. The first, headway adjustments, is used in virtually all services, except when minimum headways, limited by line capacity, single track sections, etc., are already operated. The second, changing TU capacity, is used on many rail systems, since cars can be coupled or uncoupled from the train at terminals or points where storage tracks exist. With buses such adjustments (use of different buses during peak and off-peak periods) are extremely rare. The third type of capacity adjustment, addition of short-turning trains on some line sections, can be used only where intermediate turning points are available, good joint scheduling is possible, etc.

The preceding three-dimensional presentations of passenger volume and service capacity, Figures 1.18–1.20, are useful for conceptual clarity, but actual scheduling is usually performed with two-dimensional diagrams of passenger volume and capacity as functions of time and of distance, which will be presented in Section 1.4.

1.3.5 Characteristics of Travel on a Transit Line

Passenger travel on a transit line can be analyzed through various indicators. Several important ones are defined here.

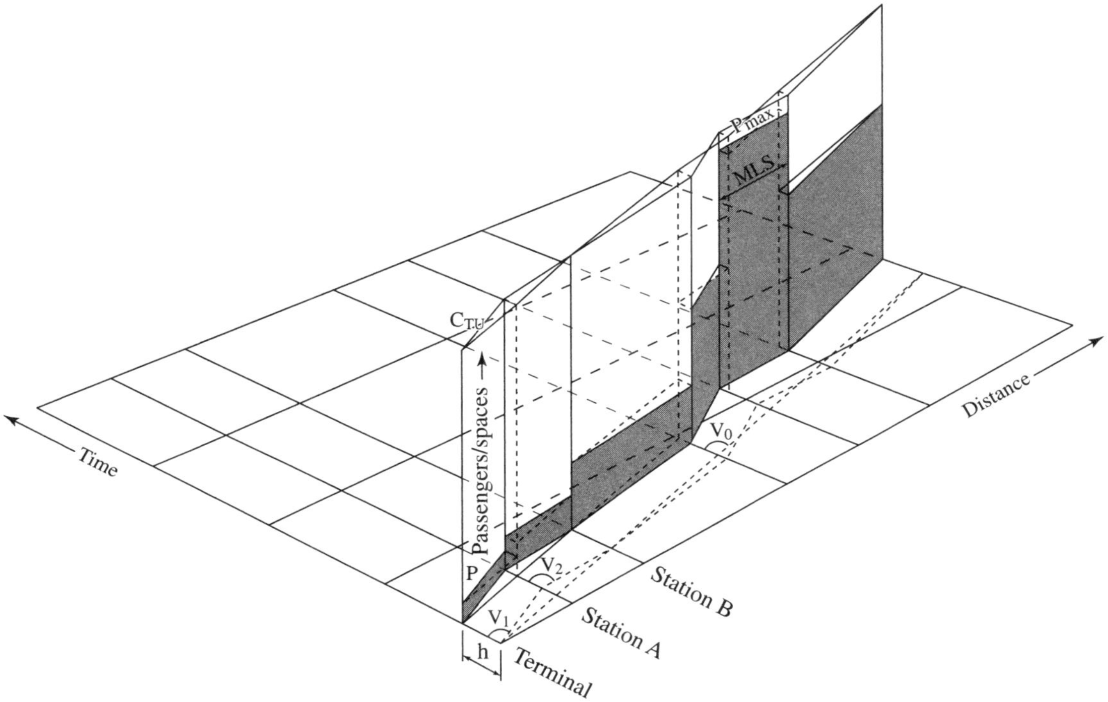

Figure 1.19 *Travel of a TU in its passenger load profile along the line.*

Average passenger trip length, or the average distance passengers travel on a line l_{av}, is obtained when the total passenger-km are divided by the number of passengers (both being for the same time period, such as hour or day):

$$l_{av} = \frac{\sum_{i=1}^{n} p_i \cdot l_i}{\sum_{i=1}^{n} b_i} = \frac{1}{P_t} \sum_{i=1}^{n} p_i \cdot l_i \qquad \left| \frac{l}{\text{km}} \middle| \frac{p, P, b}{\text{prs/h}} \right| \tag{1.32}$$

where p is the number of passengers, l is interstation distance or spacing, and P is total number of passengers boarding along the line during the same time period.

Average passenger volume P_{av} is computed as the total passenger-km on the line divided by its length L:

$$P_{av} = \frac{\sum_{i=1}^{n} p_i \cdot l_i}{L} \qquad \left| \frac{P}{\text{prs/h}} \middle| \frac{l, L}{\text{km}} \right| \tag{1.33}$$

Coefficient of flow variations η_f expresses the degree to which passenger volume peaks along the line. It is the ratio of the maximum volume (on MLS) P_{max} and the average volume P_{av}:

$$\eta_f = \frac{P_{max}}{P_{av}} = \frac{L \cdot P_{max}}{\sum_{i=1}^{n} p_i \cdot l_i} \qquad \left| \frac{\eta}{-} \middle| \frac{P, p}{\text{prs/h}} \middle| \frac{L, l}{\text{km}} \right| \tag{1.34}$$

The lowest possible value of η_f is 1, and it is found on lines with constant passenger load along their entire length, such as those which operate between two points only (for example, some CBD–airport expresses). The greater the value of η_f, the lower the average load

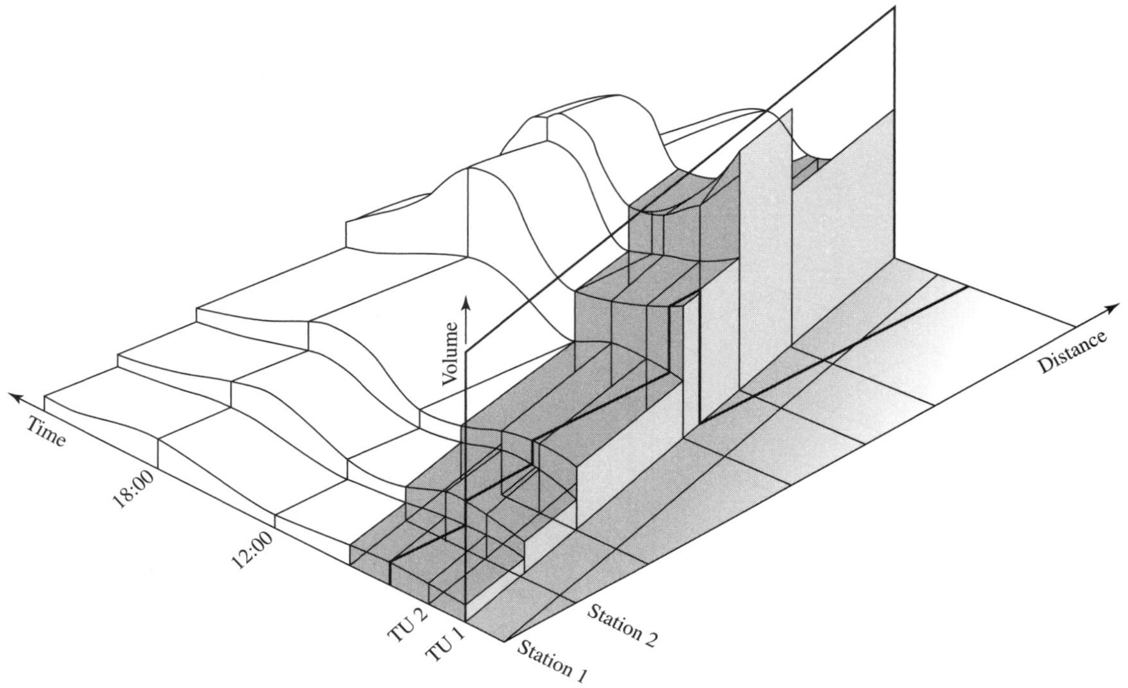

Figure 1.20 *Graphical presentation of TUs transporting passengers along a line.*

factor α_{av}, and the more desirable are adjustments of offered service to passenger volume. In the case when a line would consist of n equal station spacings and all travel would take place on one spacing only, η_f would assume its highest value:

$$\eta_f = \frac{P_{max}}{\frac{P_{max}}{n}} = n \qquad \left|\frac{\eta}{-}\right|\frac{P}{\text{prs/h}}\left|\frac{n}{-}\right| \tag{1.35}$$

This is, of course, a hypothetical situation only.

Coefficient of passenger exchange η_x indicates what portion of passengers are exchanged along a line, i.e., their turnover rate. It is defined as the ratio of total passengers who board along a line to those who did not replace the alighting passengers. Referring to Figure 1.22(a), η_x represents the ratio of the total B_L area to the portion of B_L area that does not overlap with A_L, i.e.,

$$\eta_x = \frac{B_L}{B_L - P_x} \qquad \left|\frac{\eta}{-}\right|\frac{B, P}{\text{prs/h}}\right| \tag{1.36}$$

where P_x is the area of overlap of B_L and A_L. Thus, η_x indicates the intensity of passenger exchange and therefore has significance for operations; it is useful to know for vehicle design, for scheduling of station dwell times, as well as in selecting fare structure.

In a general case, the boarding and alighting functions may increase and decrease several times along the line, intersecting themselves. Each intersecting point represents an extreme value in passenger flow—a local minimum or maximum. Figure 1.22(b) shows such a case, typical for a diametrical line that has two maxima, one at each end of central city.

For an actual transit line the boarding and alighting functions may follow the same general form, but with a discrete shape, as shown in Figure 1.22(c). The formula for the coefficient of passenger exchange is then:

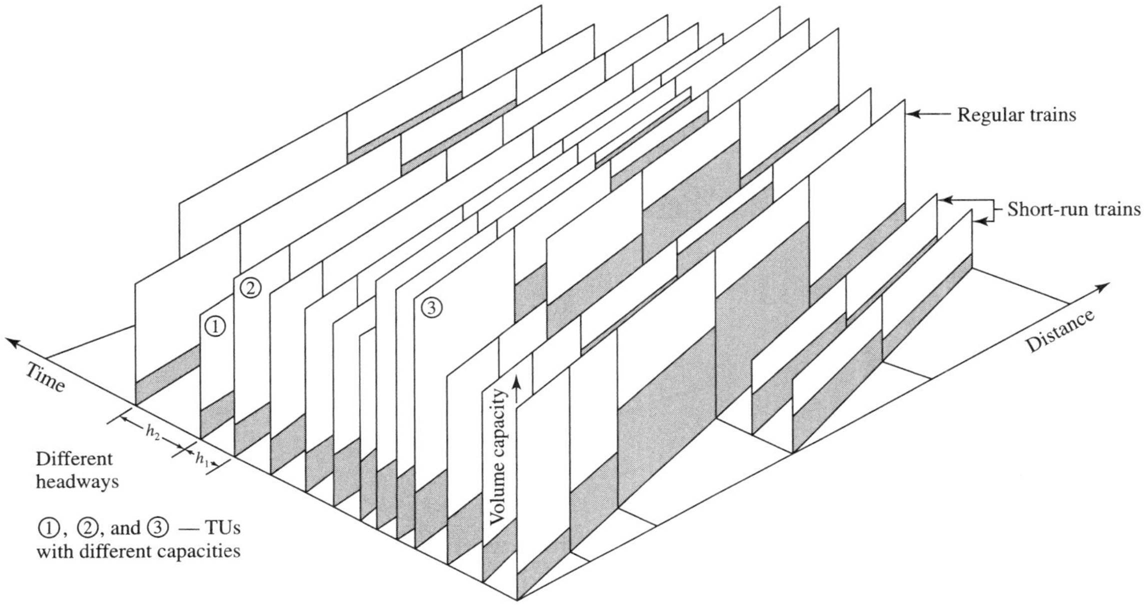

Figure 1.21 *Three methods of adjustments of service capacity to passenger volume: variation in headways, TU consists, and length of run.*

$$\eta_x = \frac{P_t}{P_t - \sum_{i=1}^{n} |b_i - a_i|} \qquad \left| \frac{\eta}{-} \middle| \frac{P, b, a}{\text{prs/h}} \right| \tag{1.37}$$

Descriptively, the sum in the denominator consists of absolute values of differences between boarding and alighting volumes at all stations at which both occur.

For the lines that have only one station (k) at which the boarding and alighting curves intersect, and therefore there is only one maximum (MLS with P_{max}), as shown in Figure 1.22(d), the expression for η_x can be simplified:

$$\eta_x = \frac{\sum_{i=1}^{n-1} b_i}{\sum_{i=1}^{n-1} b_i - \left(\sum_{i=2}^{k} a_i + \sum_{k=1}^{n-1} b_i\right)} = \frac{P_t}{\sum_{i=1}^{k} (b_i - a_i)}$$

$$= \frac{P_t}{P_{max}} \qquad \left| \frac{\eta}{-} \middle| \frac{b, a, P}{\text{prs/h}} \right| \tag{1.38}$$

Introducing P_{max} from Eq. (1.34) into Eq. (1.38) gives:

$$\eta_x = \frac{L \cdot P_t}{\eta_f \sum_{i=1}^{n} p_i \cdot l_i} \qquad \left| \frac{\eta}{-} \middle| \frac{L, l}{\text{km}} \middle| \frac{P, p}{\text{prs/h}} \right| \tag{1.39}$$

Further, replacing this sum with its relationship in Eq. (1.32) gives the coefficient of passenger exchange as a function of line length, average travel distance, and the coefficient of flow variations:

$$\eta_x = \frac{L}{l_{av} \cdot \eta_f} \qquad \left| \frac{\eta}{-} \middle| \frac{L, l}{\text{km}} \right| \tag{1.40}$$

Equation (1.40) shows that the lowest value of η_x occurs when all passengers travel along the entire line: there is no passenger exchange, and $\eta_x = 1$. If, theoretically, all passengers are exchanged at each of the n

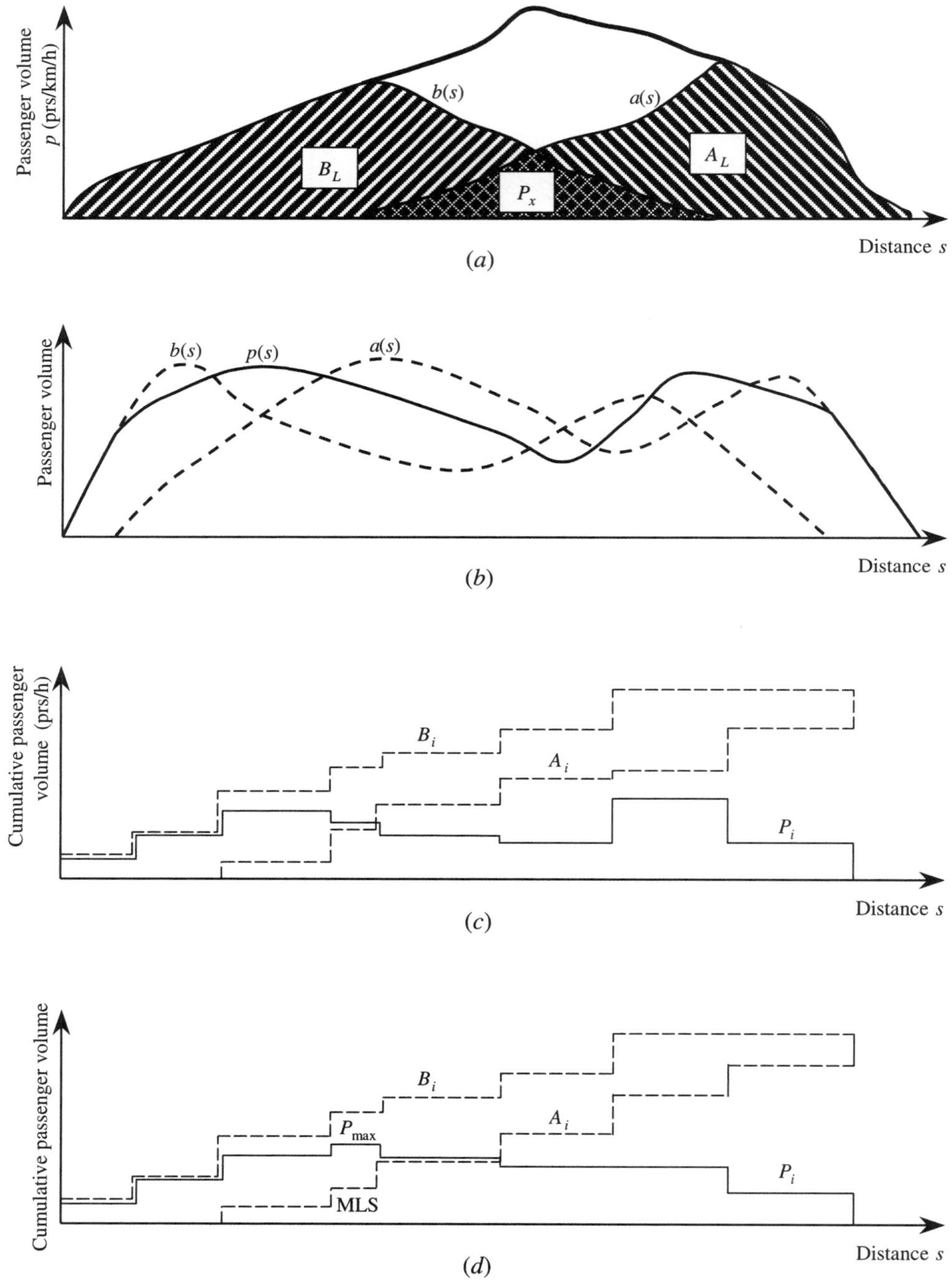

Figure 1.22 *Definition of the passenger exchange coefficient* η_x*: (a) the concept of passenger exchange coefficient; (b) general case of boarding, alighting, and flow functions; (c) discrete boarding and alighting functions; (d) discrete boarding, alighting, and load functions;* P_{max} *defines the MLS.*

stations of a line, the coefficient assumes its highest value, $\eta_x = L/\ell_{av} = n$.

1.3.6 Indicators of Transit Usage

Two indicators are most commonly used to express the absolute and relative magnitude of transit travel in a city; they reflect the role a transit system plays in the area and its relationship with other modes.

Riding habit, defined as the ratio of annual transit rides to population of the served area, indicates how much the population utilizes the transit system. Riding habit is generally greater for large than for small cities.

When transit was practically the only mode of travel for medium and long trips, riding habit in large cities was generally over 200, sometimes even close to 400 rides per capita per year (Chicago in 1926 recorded 374). After the widespread introduction of the private automobile and subsequent deterioration of transit services in U.S. cities, that number dropped to a range between 120 and 220 for large cities with rapid transit systems, frequent service, and low fare (New York, San Francisco, and New Orleans have the highest figures) and considerably below that range for smaller cities and cities that developed rapidly since 1950, generally without high-quality transit. Already in 1959 Cincinnati had only 70, Detroit 65, Minneapolis 52, and Kansas City 50 rides per capita.

This trend in transit use indicates that high-quality transit is a significant factor contributing to high transit riding habit. This correlation is corroborated by a comparison between West European and U.S. cities. Prior to the wide use of the private automobile, U.S. cities tended to have more extensive transit systems than their West European counterparts, and their riding habits were higher. In recent decades, however, as West European cities improved their transit systems to a higher level of service than most U.S. cities, riding habit in European cities has become higher: cities with populations of 600,000 to 2.5 million, such as Rotterdam, Düsseldorf and Hamburg, have about 150–200 annual rides per capita. Brussels (population 1.1 million) had 148 rides per capita in 1970; following construction of its Metro and LRT lines and other improvements, in 1984 its riding habit increased to 176. The highest riding habits are found in Oslo and Zürich: with populations of only about 500,000 but very extensive transit systems, these two cities maintained extremely high riding habits into the late 1990s: about 200 and over 300 annual rides per capita, respectively.

In medium-sized and small North American cities, reliance on the automobile is very high, so that riding habit is quite low. For example, Pennsylvania cities with populations of about 300,000, such as Wilkes-Barre and Harrisburg, have riding habits of only 15–30, while of the two large cities, Pittsburgh (1.8 million) has about 50 and Philadelphia (4.0 million) 90 annual rides per capita. These numbers are approximate because of the difficulty of defining the limits and population of the city—how many suburbs are included. This factor, very extensive dispersal of population in metropolitan areas, actually makes the riding habit numbers for U.S. cities considerably lower than in most other countries. For example, riding habits computed for transit travel in central cities of Philadelphia, San Francisco, New York, and other major cities are much higher than when they are computed for their metropolitan areas. This should be considered in comparing the computed riding habits for different cities.

It is important to note that cities with a consistent policy of transit improvement, such as was pursued in Toronto and Edmonton during certain time periods, increased their riding habits in spite of their low population densities and high auto ownership rates. The influence of good transit service on ridership attraction is also obvious from a comparison of several cities in Western Canada and the United States (1991 figures), which shows better performance of Canadian cities: Vancouver (population in thousands, 1,603), 129 annual rides per capita; Calgary (754), 103; Portland (1,172), 50; Seattle (2,348), 46; Houston (2,902), 28; and Phoenix (2,006), the most auto-oriented city, 14.

Transit travel as percent of total travel and relationship of transit to automobile travel (modal split) are indicators showing the relative role that transit plays in a city. Since these indicators vary greatly, they are usually analyzed by areas of the city, individual corridors, or travel purposes, rather than as global values for entire cities. The values given for entire cities, usually showing extremely low percentage for transit (5–15%), have been often used by laymen to downplay the importance of transit. However, these global numbers are deceptive for many reasons. First, transit does not provide ubiquitous coverage in the entire metropolitan area, as cars do. Calculating percentage of transit trips using as a base many trips that transit is not intended to serve is incorrect. Second, travel by auto and by transit is often functionally different: a person commuting by transit to work may make only two recorded trips per day; however, that person may make stops along the way or undertake, for example, another couple of walking trips during the day. To perform the same functions—get to work, go to lunch, go to a bank and a store—an auto commuter in the same city, but particularly in the suburbs, may make four auto trips in addition to the trips to/from work. Thus, while both persons have performed the same number of functions or trip purposes, the count of trips gives a false impression that auto served three times more trips and travel functions than transit.

Finally, another bias in the global modal split analysis is caused by neglect of (and difficulty in counting) walking trips. Proper land use planning and transportation policies lead to situations where walking and use of transit can serve many functions conveniently, while auto-oriented development requires a far greater number of trips and vehicle-km of travel in order to perform the same functions. Percentages of trips in an entire urbanized area is therefore an inappropriate measure of the importance of individual modes.

In some U.S. cities, another distorted statistic is also used to downplay the role of transit: the number of daily transit trips is divided by two, claiming that that figure represents the number of persons using transit (presumably, all transit riders are commuters, making two trips per day). By the same logic, if car users make, on the average, 3.2 (or a similar number) trips, traffic volume on every highway should be divided by 3.2 to show the "number of affected persons." Both such figures are incorrect, and the exact number of trips should be used for each mode, rather than divided by some imaginary number of commuting trips while ignoring other trips.

1.4 SCHEDULING OF SERVICE

Transit scheduling is the process of computing the frequency of service, the number of vehicles required, the timing of their travel, and other related operating elements. The products of scheduling include graphical and numerical schedules for operators and supervisors (also known as paddles, picking lists, dispatchers' lists, etc.), timetables for the public, as well as operating data for a line.

Public transportation services with low frequencies, such as some long-distance and suburban routes, or commuter transit, which operates during peak hours only, sometimes have variable headways, determined by demand, cycle times, crew requirements, and other constraints. For regular transit lines, however, uniform headways during each schedule period represent the optimum operation for several reasons. First, for random passenger arrivals, uniform headways minimize waiting times; second, they minimize probability of delay propagation, thus resulting in higher capacity and reliability of service. Moreover, use of clock headways allows simple information and represents an important convenience for both regular and incidental transit users. For these reasons, virtually all well-planned and operated urban transit services are scheduled with uniform clock headways for each scheduling period of the day.

1.4.1 Components of the Scheduling Process

As the flowchart in Figure 1.23 shows, the entire scheduling process can be divided into three phases.

I. *Input,* or preparation of data needed for scheduling, includes various line characteristics, schedules of lines that meet and have transfers, passenger volumes, service standards and considerations, characteristics of vehicles and train consists, operational factors and practices for each line, and work rules and standards. Clearly, these data include fixed numbers (e.g.,

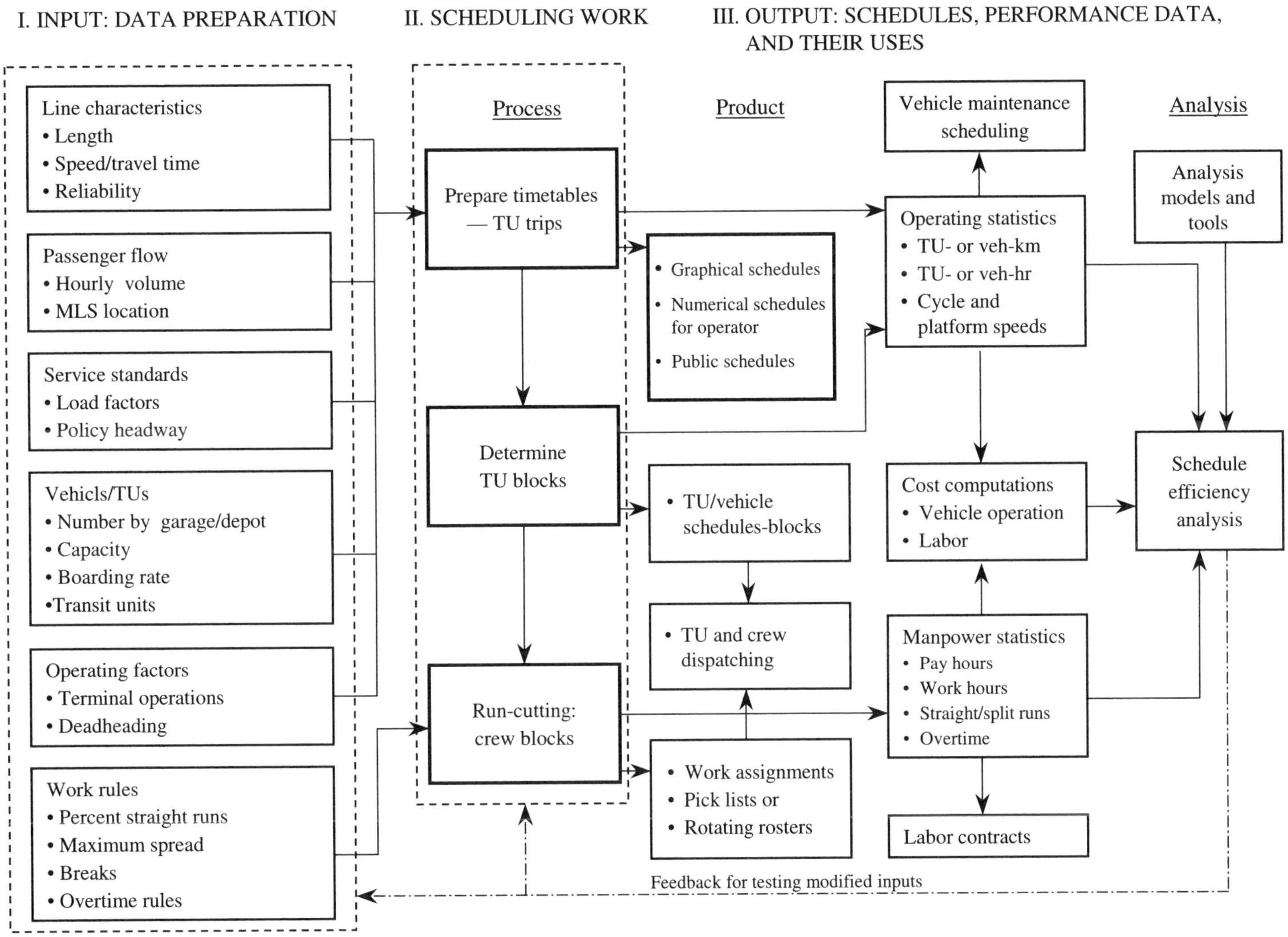

Figure 1.23 *Flowchart of transit scheduling.*

line length), data that must be periodically updated (passenger volumes), as well as various characteristics and standards, which experienced schedulers introduce in the course of their work or program preparation.

II. *Scheduling work* represents the central component of the process. In most cases it is divided into three major elements:

- *Preparation of timetables or trip building,* which determines headways, terminal times, and other elements. Its products are graphical schedules (string charts) and numerical schedules for operating personnel (headway sheets) and for the public (timetables).
- *Determination of blocks* or *block building* assigns TUs to all trips specified in the timetable. The product of this element are blocks, or work schedules for each TU for a day.
- *Run-cutting* or determination of work duties for individual drivers during the day. This process produces work assignments that are put together into pick lists or rosters from which each driver selects a specific run (which may be straight, split, include overtime, etc.).

In European transit systems, development of rotating duty rosters is sometimes treated as a separate element.

III. *Output* of the scheduling process, in addition to the direct products (schedules, blocks, runs, etc.), consists of various performance data, such as TU- and vehicle-km, pay-hours and work-hours, etc. These data are used for cost computations, various reports on transit operations, and, particularly important, analysis of schedule efficiency.

This scheduling process is shown in Figure 1.23 as a sequence of steps, from the input of information to the output of vehicle schedules and rosters. Actually, the process is often more complicated because many of the input elements, such as policy headways, types of vehicles or TU sizes, load factors, and others, may be varied to some extent, allowing testing of alternate schedules and improvements of efficiency of the initially developed schedule. The procedure therefore often has a feedback step, which allows testing of possible changes in parameters and their impact on the final schedule, so-called what-if analysis. This kind of testing is particularly common with computerized scheduling procedures because testing of many different situations is easy and fast.

It actually often happens that the last step in the process, schedule efficiency analysis, leads to the conclusion that the solutions are not satisfactory and calls for modifications in the input or scheduling process. Thus, as the dashed line in the flowchart in Figure 1.23 shows, there is a feedback from the last step, schedule efficiency analysis, to Input and Process, representing reiterative scheduling procedure.

1.4.2 Determination of Service Requirements

The schedule for every transit line must satisfy two basic requirements: it must provide adequate transporting capacity for passenger volume, and it must offer a certain minimum frequency of service (maximum acceptable headways) required from a level-of-service point of view.

During peak hours and on heavily traveled lines at all times, the former requirement is critical: the operator must provide adequate capacity on the line; if this is done, the minimum frequency is usually automatically met and exceeded. However, during low travel demand periods and on lightly traveled lines, if service is based on capacity requirements, frequency may be unacceptably low. For example, if there are 70 potential passengers per hour on MLS of a bus line, on the basis of capacity, only one or two bus departures would be provided, resulting in headways of 60 or 30 minutes. But such a schedule would be unacceptable for many travelers. Therefore headways of, say, 15 minutes should be adopted even though utilization ratio of the offered service would be rather low.

1.4.2.1 Passenger Volume Distribution in Distance and Time. The basic information needed for scheduling a transit line is its expected volume of passengers and its distribution in distance and time. Two types of diagrams are developed for this purpose.

A *passenger load profile diagram* is used to show passenger volumes P_i on each station spacing i along the line during a period of time, usually one hour, or the entire day. As shown in Figure 1.5, such a diagram shows the highest volume, P_{max}, which defines the MLS. The volume on MLS is critical for determination of the required scheduled capacity.

On the diagram, three horizontal lines are plotted: average passenger volume for the entire line P_{av}, offered line capacity per hour in seats, and in total spaces C. The shaded area shows person-km traveled on the line, the P_i/C ratio represents the utilization coefficient or load factor α on that spacing (with its maximum value reached on the MLS) and the ratio of the total passenger-km traveled to the total space-km offered $\overline{\alpha}$.

This diagram is analyzed to determine the basic type of service. If the volume is much greater on one portion of line than on others, it may be appropriate to operate some short-turn services or to divide the line into two or more branches at the point where volume decreases (this is common for radial routes from city center). An example of such a profile and service with alternating short- and through-routed TUs, which reduces the offered capacity on the outer section to one half, is shown in Figure 1.24.

A *temporal variations diagram* shows detailed variations of passenger volume on the MLS of a line for each hour, half hour or sometimes, during the peaks, 15-minute intervals, covering the entire period when transit service is operated. This diagram, shown in Figure 1.25, is used to determine scheduling periods—time intervals with rather uniform travel volumes, during which a fixed schedule is operated. While in some cases one schedule may be appropriate for the entire day (e.g., a line serving many different trip purposes on a Saturday), in most cases there are three, four, or as many as six scheduling periods. The diagram in Figure 1.25 shows a typical weekday passenger demand with a.m. peak, midday, p.m. peak, evening and night ("owl") scheduling periods.

1.4.2.2 Scheduling Periods and Design Passenger Volumes. Based on the temporal distribution of passenger volumes shown in Figure 1.25 for one line, the number and durations of different scheduling periods are determined, usually for the entire network of transit lines. Then, for each scheduling period, a diagram of passenger volume along the line, such as Figure 1.18, is developed and used for scheduling.

The general procedure of transit line scheduling will be described in the following text. However, since

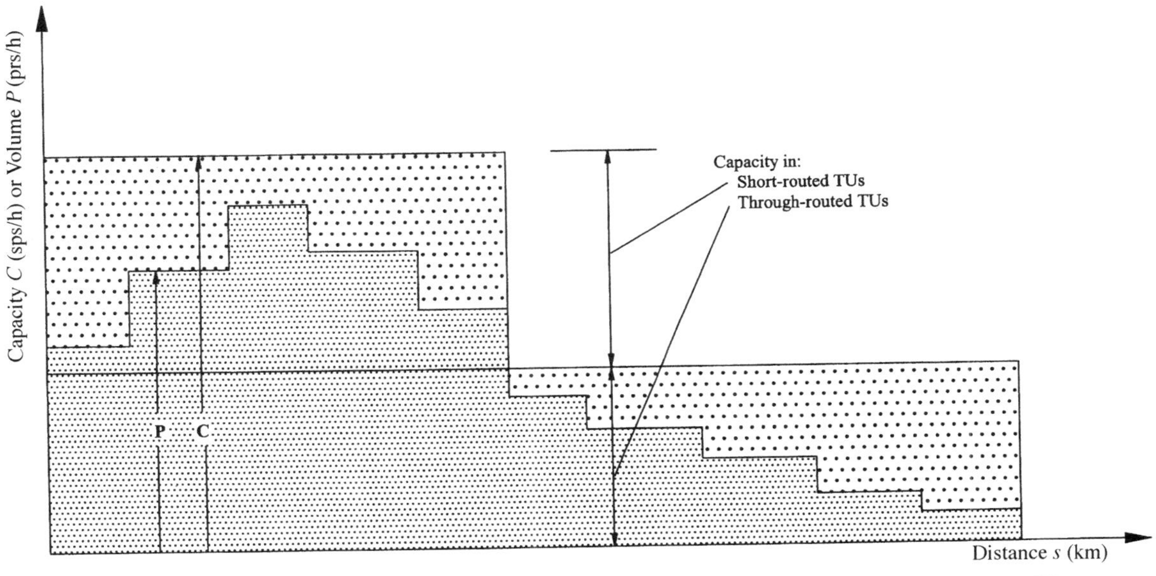

Figure 1.24 *Offered capacity and its utilization on a transit line with through- and short-turn TUs.*

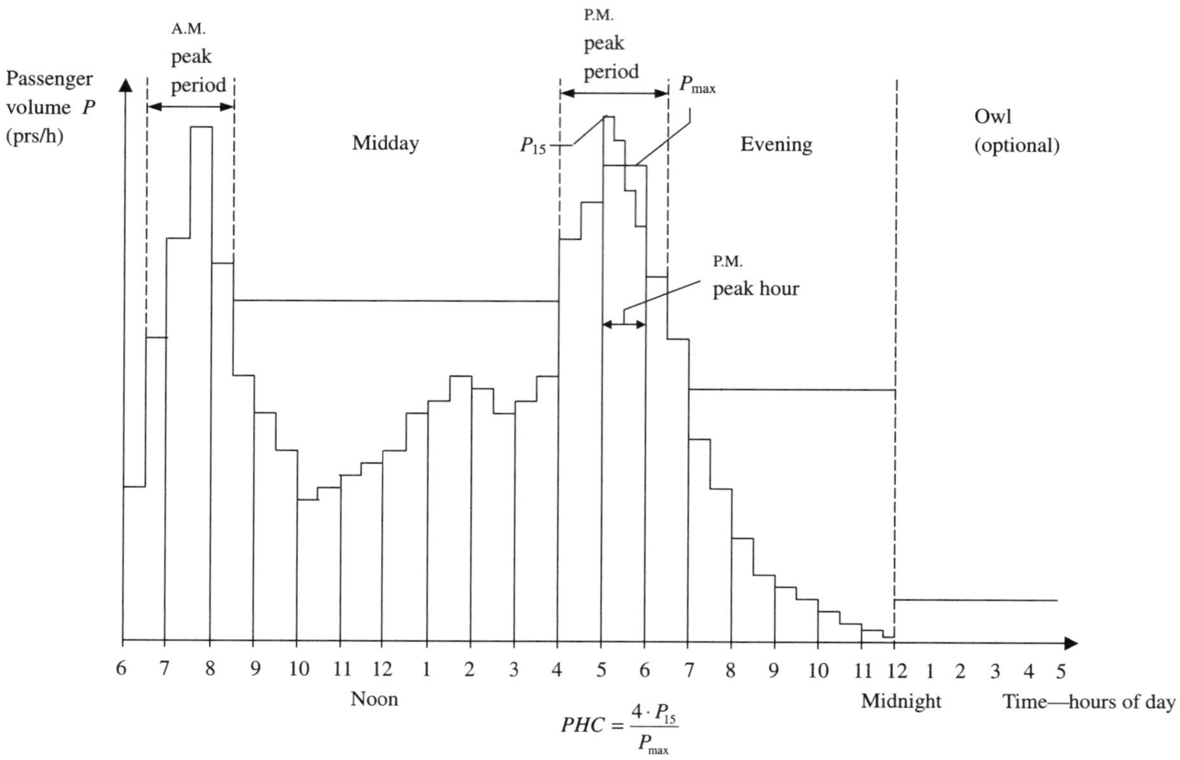

Figure 1.25 *Hourly variations of passenger volume on a line and its scheduling periods.*

scheduling is very important for providing good service on the passenger side, and achieving maximum operational efficiency on the transit operator side, rather sophisticated analyses are often performed. Two such refinements in scheduling are described here.

In some cases the MLS is not always the same station spacing during all scheduling periods. For example, a line may carry heavy commuter travel in the morning and afternoon, creating the MLS at the periphery of the CBD; midday shopper and lunch hour travel may make the MLS inside the CBD; and in the early afternoon the MLS may occur at a school in a suburban area. For such a case it is useful to make a tabulation or diagram of the line showing P_{max} for each hour by location along the line and its direction. The same table or diagram should indicate the offered capacity that is based on policy headways (see Sections 1.1.4 and 1.4.3). For every scheduling period, the higher of the two required capacities—the one determined by passenger volume and the other dictated by the policy-based level of service—should be used for developing the schedule.

Another refinement in analyzing demand is consideration of detailed variations in time. During the peak hours variations of passenger volumes are sometimes considerable, so that scheduling based on the total hourly volume may lead to both excessive crowding and oversupply of vehicles at different times within the same hour. To avoid this, passenger counts during the peaks must be made for 15-minute or even shorter time periods. Variations among these periods, as shown in Figure 1.24, are used to compute the peak hour coefficient (PHC), defined as the ratio of the highest 15-minute volume multiplied by four, and the total counted hourly volume on the MLS:

$$PHC = \frac{4\,p_{15}}{P_{max}} \qquad \left|\frac{PHC}{-}\right|\frac{p}{\text{prs/15 min}}\left|\frac{P}{\text{prs/h}}\right| \tag{1.41}$$

When the actual peak hour volume is multiplied by this coefficient, the hourly volume equivalent to the peak 15-minute volume is obtained, which should be used as the design volume:

$$P_d = P_{\max} \cdot PHC \qquad \left| \frac{P}{\text{prs/h}} \right| \frac{PHC}{-} \right| \qquad (1.42)$$

Theoretically, values of PHC can vary from 1 to 4. PHC is the inverse of the peak hour factor (PHF), used for highway volume variations, which can have values between 0.25 and 1.

In some cases, particularly on heavily traveled lines operating with short headways, use of P_d obtained this way for scheduling the whole peak hour results in uneconomical operation. Instead, scheduling is made for each 15-minute (or even shorter) period based on its actual passenger volume.

Scheduling must be done for each scheduling period separately. Lengths of individual periods need not always coincide with hours. Transitions between scheduling periods should always be gradual, fitted to the change in demand. In the beginning of peak periods, capacity can be increased in two ways: by shortening headways or by increasing the capacity of TUs. A combination of the two may also be used.

The simplest way to shorten headways is to insert additional TUs between regularly scheduled ones. Thus, the headway is cut in half. This regularity is convenient for scheduling and provides simple operation, but when all TUs have equal capacities, it represents an abrupt transition, providing double capacity rather than incremental increases. Reduction of headways to other than one-half is therefore often made. For example, a basic headway of 10 minutes is reduced to 7.5 or 6 minutes during the peak. The inconvenience to passengers of having to memorize a different schedule is not serious when headways are short. For long headways (>10 minutes) such a change must be clearly indicated in timetables.

The second technique for transition to higher capacity is often used on rail systems. A rapid transit line operating with four-car trains throughout the day can be changed into an operation with six- or eight-car trains during the peak hours by coupling additional cars at terminals or intermediate stations. With buses and trolleybuses this technique cannot be employed, except in special cases when standard buses are used for regular service, but larger-capacity vehicles (e.g., articulated) are used during the peak hours. The changes of fleet involve additional deadheading, however.

1.4.2.3 The Required Capacity. On lightly traveled lines and during off-peak hours, the capacity offered on a line is usually dictated by the policy headway, i.e., the volume of offered service is determined by the required service headway, rather than by passenger volumes. However, when passenger volume is heavy, capacity offered must cover passenger load profile, as shown in Figure 1.24. The level of offered capacity is determined by selection of the maximum value for the load factor α, i.e., by capacity utilization on MLS. This selection is based on the trade-off between the two basic requirements for transit operations:

- For passenger comfort and convenience, lower values of α result in less crowding, higher availability of seats, and more frequent service.
- Cost of operation is lower when a higher value of α is adopted, because smaller number of vehicles is required to transport a given passenger volume.

Several other factors may influence the selection of the maximum value for the load factor on the MLS, $P_{\max}$. For example:

- Rather uniform passenger volume along the line suggests lower α; uneven distribution allows higher α, because the maximum volume is found only on a short section of the line.
- High ratio of seated to standing passengers foreseen by vehicle design requires lower α (areas with seats cannot accept any overcrowding, only standing areas can).
- Long average trip length (e.g., express bus, regional rail) implies low passenger turnover rate; therefore, higher comfort should be provided—α should be lower.
- High percentage of senior citizens or shoppers with packages dictates lower α, larger volumes of school children, higher α.

- For transporting large crowds of passengers to special events, such as sport stadia, holiday events in the city, major concerts, etc., the maximum achievable value for α can be used.

As this review of different operating factors shows, the value of the load factor α is a result of the trade-off between two mutually conflicting requirements: passenger comfort, requiring lower α, and lower operating costs, dictating higher α. Other mentioned conditions may influence the selection of the load factor's value in this trade-off.

1.4.2.4 Selection of TU Size, Frequency, and Load Factor. Figure 1.26 presents a diagram that can be used to select the optimal combination of TU size n, service frequency f or headway h, and load factor α for any scheduling period of day. The diagram is based on the equation

$$C = n \cdot C_v \cdot \alpha \cdot f = \frac{60 \cdot n \cdot C_v \cdot \alpha}{h}$$

$$\left| \frac{C}{\text{sps/h}} \right| \frac{n}{\text{veh/TU}} \left| \frac{C_v}{\text{sps/veh}} \right| \frac{\alpha}{\text{prs/sp}}$$

$$\left| \frac{f}{\text{TU/h}} \right| \frac{h}{\text{min/TU}} \Bigg| \qquad (1.43)$$

The equation shows four different train consists: TU sizes of two, four, six, and eight cars, and the line capacity they can provide with operations at different

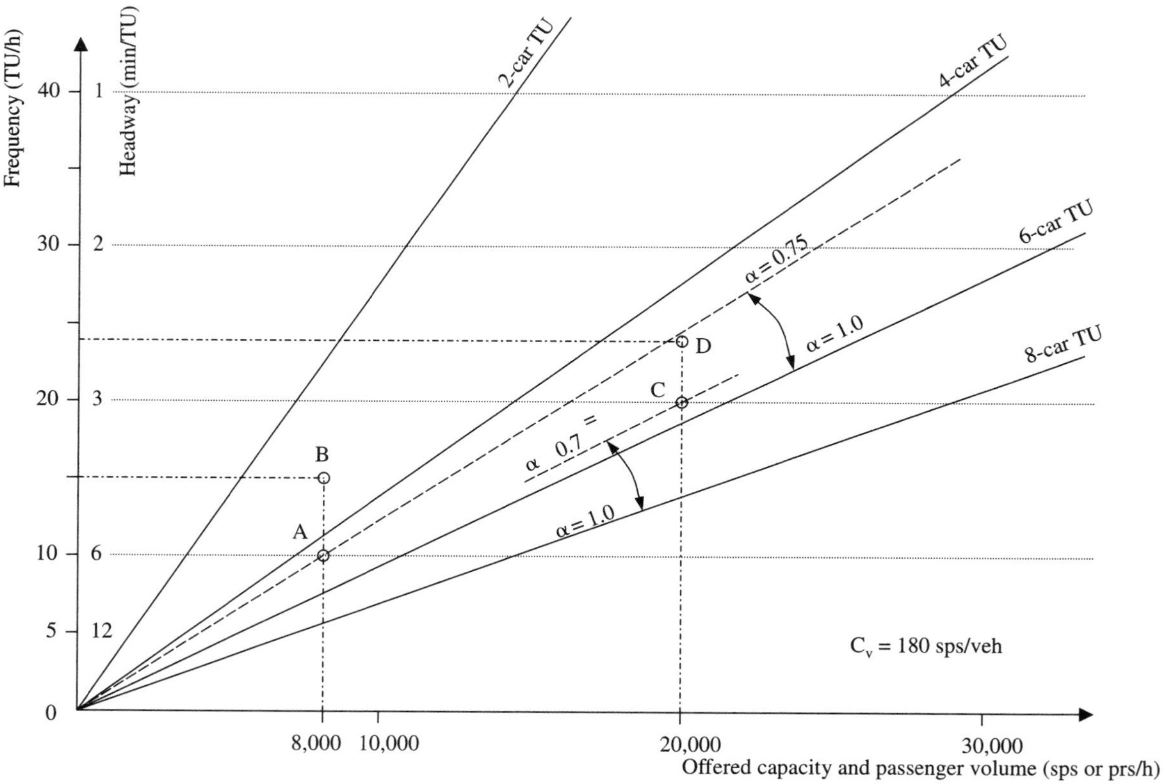

Figure 1.26 *Diagram for selection of TU size and headway for different scheduling periods.*

frequencies/headways and with different values of the load factor α. Each sloped line shows the values for a given TU size at full occupancy, $\alpha = 1.00$. A dashed line shows the capacities offered by six-car TUs with $\alpha = 0.75$.

This diagram can be used as follows. If during midday period $P_{max} = 8{,}000$ prs/h, reasonable choices would be to operate six-car TUs at $h = 6$ min with $\alpha = 0.75$—point A on the diagram; or four-car TUs at $h = 4$ min with $\alpha = 0.75$—point B. Suppose that the peak period has $P_{max} = 20{,}000$ prs/h; then the choices may be to operate eight-car TUs at $h = 3$ min and $\alpha = 0.70$—point C, or six-car TUs at $h = 2.5$ min and $\alpha = 0.77$—point D on the diagram.

1.4.3 Scheduling Procedure

There are many variations in the scheduling procedure, depending on the time period (peak, off-peak, owl, etc.), methods of bringing vehicles into and out of service, crew practices, etc. The basic physical relationships among operational elements are valid in all cases, however; scheduling for a regular case is presented here, with practical examples of the computational sequence.

The design hour volume P_d divided by the average number of passengers a TU will carry on MLS gives the required frequency:

$$f = \frac{P_d}{\alpha \cdot n \cdot C_v} \qquad \left|\frac{f}{\text{TU/h}}\right|\frac{P_d}{\text{prs/h}}\left|\frac{\alpha}{\text{prs/sp}}\right| \qquad \left|\frac{n}{\text{veh/TU}}\right|\frac{C_v}{\text{sps/veh}}\right| \tag{1.44}$$

and its inverse, the headway:

$$h = \frac{60}{f} = \frac{60\alpha \cdot n \cdot C_v}{P_d} \qquad \left|\frac{h}{\text{min/TU}}\right|\frac{f}{\text{TU/h}}\right| \qquad \left|\frac{\alpha}{\text{prs/sp}}\right|\frac{n}{\text{veh/TU}}\left|\frac{C_v}{\text{sps/veh}}\right|\frac{P_d}{\text{prs/h}}\right| \tag{1.45}$$

When $h \leq 6$ minutes, passengers are not particularly concerned about exact departure times, so that any value of h is acceptable, including, on precisely operated lines, such values as 1½ or 2½ minutes. However, if the computed value of $h > 6$ minutes, it should be rounded down to the nearest clock headway (see Section 1.1.4). This h should then be compared with the adopted policy headway h_p, and the smaller of the two should be used.

The actual values of f (which is an integer when $h > 6$ minutes) and α, after the rounding of h, are computed from Eq. (1.45):

$$f = \frac{60}{h} \quad \text{and} \quad \alpha = \frac{P_d \cdot h}{60\, n \cdot C_v}$$

To compute the number of vehicles needed to provide service with a given headway, cycle time T must be found by computing its components. *Operating time* T_o for each direction must be computed by field surveys. *Terminal time* t_t at each terminal depends on four factors, as explained in Section 1.1.6.1. For street transit systems, crew rest and delay recovery usually govern, while rapid transit has a constant terminal time (usually 5–10 minutes) irrespective of line length. Alternatively, instead of t_t, terminal time coefficient γ, expressing t_t as a percent of T_o, can be used for this preliminary computation of T:

$$T = 2(T_o + t_t) = 2\,T_o(1 + \gamma) \qquad \left|\frac{T,\ t}{\text{min}}\right|\frac{\gamma}{-}\right| \tag{1.46}$$

here γ can be given a value of 0.1–0.3.

The number of TUs ($N_{TU} = N/n$) on the line is obtained as the ratio of the cycle time and headway:

$$N_{\text{TU}} = \left[\frac{T}{h}\right]^+ = \left[\frac{f \cdot T}{60}\right]^+ \qquad \left|\frac{N_{\text{TU}}}{\text{TU}}\right|\frac{T}{\text{min}}\left|\frac{h}{\text{min/TU}}\right|\frac{f}{\text{TU}/h}\right| \tag{1.47}$$

The brackets indicate integer value equal to or greater than the computed value. In the following equations it will be assumed, for simplicity, that single vehicles are operated, $n = 1$, so that $N_{TU} = N$.

Since h is fixed, T must be adjusted to be

$$T = h \cdot N_{TU} \qquad \left|\frac{T}{\text{min}}\right|\frac{h}{\text{min/TU}}\left|\frac{N_{\text{TU}}}{\text{TU}}\right| \tag{1.48}$$

by increasing t_t (at one or both terminals). The adjusted value of T represents the final cycle time. The actual value of γ is recomputed with the new t_t by Eq. (1.20).

If it is desirable to minimize the number of TUs in service, terminal times can be kept at their minimum values by using two different headways, usually varying by one minute, satisfying the following equation:

$$k_1 \cdot h_1 + k_2 \cdot h_2 = T \tag{1.49}$$

where k's represent the numbers of the two respective headways in a cycle.

For example, the required headway is computed to be 5.4 minutes and the initial cycle time is 81 min. Ideally, a uniform headway of 5 minutes should be used, T should be rounded up to a multiple of h, i.e., to 85 minutes, and $N = 85/5 = 17$ TUs. However, it is possible to use nine 5-minute and six 6-minute headways, adding up to the cycle time of 81 minutes (Eq. 1.49): $9 \times 5 + 6 \times 6 = 81$.

The number of TUs is $9 + 6 = 15$, i.e., two TUs less than if uniform headways were used, and the average headway is exactly $81/15 = 5.4$ minutes.

Cycle speed V_c can also be related to operating speed via the terminal time coefficient γ:

$$V_c = \frac{V_o}{1 + \gamma} \qquad \left|\frac{V}{\text{km/h}}\right|\frac{\gamma}{-}\right| \tag{1.50}$$

but its most important relationship is with the number of TUs on the line, because that determines the cost of providing the service. From Eqs. (1.29) and (1.48), one obtains:

$$V_c = \frac{120L}{T} = \frac{120L}{h \cdot N_{\text{TU}}}$$

$$\left|\frac{V}{\text{km/h}}\right|\frac{L}{\text{km}}\left|\frac{T}{\text{min}}\right|\frac{h}{\text{min/TU}}\left|\frac{N_{\text{TU}}}{\text{TU}}\right| \tag{1.51}$$

From this equation and Eqs. (1.44) and (1.50) N_{TU} can be expressed as:

$$N_{\text{TU}} = \frac{120\ L}{h \cdot V_c} = \frac{2\ L \cdot P_d(1 + \gamma)}{\alpha \cdot n \cdot C_v \cdot V_c}$$

$$\left|\frac{N_{\text{TU}}}{\text{TU}}\right|\frac{L}{\text{km}}\left|\frac{h}{\text{min/TU}}\right|\frac{V}{\text{km/h}}\left|\frac{V}{\text{km/h}}\right|\frac{P}{\text{prs/h}}\left|\frac{\gamma}{-}\right.$$

$$\left|\frac{\alpha}{\text{prs/sp}}\right|\frac{n}{\text{veh/TU}}\left|\frac{C_v}{\text{sps/veh}}\right| \tag{1.52}$$

The first of these expressions, N_{TU} as a function of V_c, is plotted in Figure 1.27. The diagram shows that cycle speed strongly influences the required number of TUs, particularly at low speeds and for short headways. The second expression shows that, as is intuitively clear, the fleet size is directly proportional to the line length and passenger volume, while it is inversely proportional to TU capacity and operating speed. These elements, together with coefficients α and γ, therefore have a direct bearing on the cost of operation.

1.4.4 Procedure Summary, Examples, and Numerical Schedules

The preceding computational procedure is summarized here into five steps and illustrated by examples. It is assumed that the data on passenger volumes have been collected and analyzed and uniform scheduling periods determined. Four different cases are illustrated numerically:

a. A street transit route operated by 45-seat buses for peak periods

b. The same route for base periods

c. A rapid transit line operated by vehicles with total (seats plus standing) capacity of 140, in trains of up to six cars, for peak hours

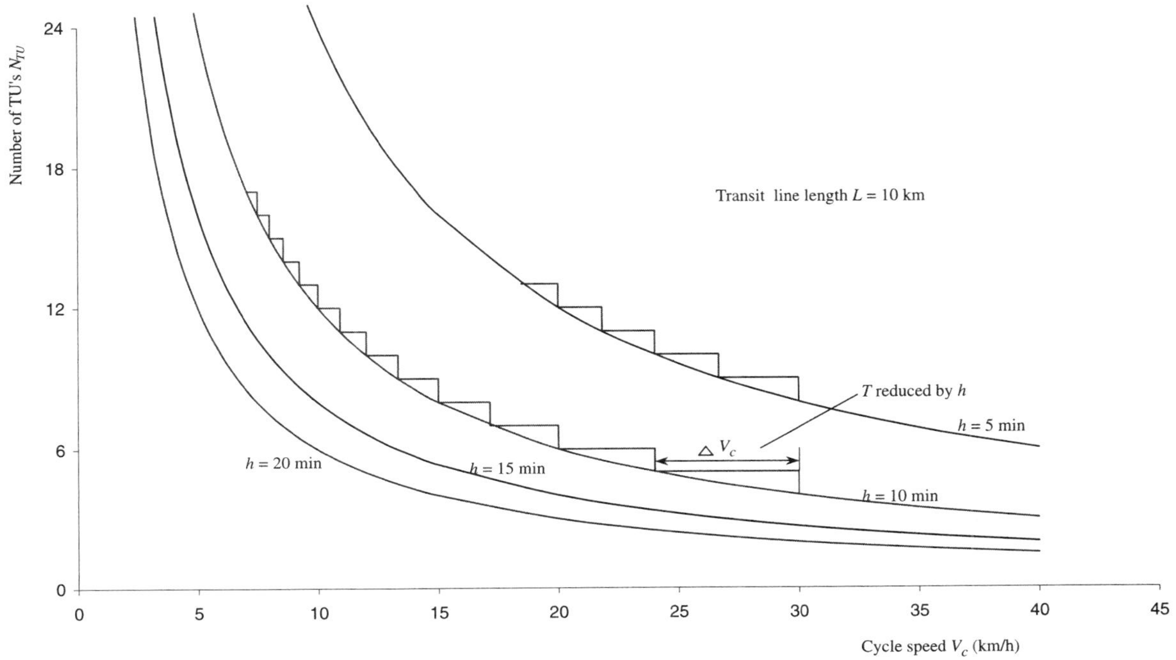

Figure 1.27 *Number of TUs on a line as a function of cycle speed and headway.*

d. The same line during base periods, minimum train consisting of a married pair

Scheduling steps are defined in a brief form. The data and computed values for the example cases a–d are presented in Table 1.5.

Step 1: Prepare data and determine factors.

- Line length: L (km)
- One-way operating time: T_0 (min)
- TU capacity: C_v (sps/veh) and n (veh/TU)
- Policy headway: h_p (min/TU)
- Load factor (on MLS): α (prs/sp)
- Design volume: P_d (prs/h) (includes PHC)
- Minimum terminal time: t_t (min) or minimum value of γ

Step 2: Compute headway and frequency. Headway is obtained directly from Eq. (1.45):

$$h = \frac{60 \cdot \alpha \cdot n \cdot C_v}{P_d}$$

if the computed $h > 6$ minites, round it down to the nearest smaller of the following numbers: 6, 7.5, 10, 12, 15, 20, 30, or 60 (sometimes 40 minutes is also used, repeating the times every two hours).

Compare the obtained headway with h_p and adopt the shorter one.

Compute frequency and the actual value of α using expressions based on Eq. (1.45):

$$f = \frac{60}{h} \quad \text{and} \quad \alpha = \frac{P_d \cdot h}{60 \cdot n \cdot C_v}$$

Step 3: Determine fleet size. Introduce the known T_o and t_t or γ in the first or second of the following expressions, Eq. (1.46):

$$T' = 2\,(T_o + t_t) = 2\,T_o(1 + \gamma)$$

Table 1.5 Examples of scheduling computations

					Cases			
Step	*Item*	*Symbol*	*Dimension*	*Source, Eq. ( )*	*a* *Bus-peak*	*b* *Bus-base*	*c* *RT-peak*	*d* *RT-base*
1	Line length	L	km	given	8	8	12	12
	Vehicle capacity	C_v	sps/veh	given	45	45	840	280
	Max. passenger volume	P	prs/h	given	980	160	10,000	1500
	Operating time	T_o	min	given	40	30	24	24
	Load factor	α	prs/sp	given	1.1	0.9	0.8	0.6
	Policy headway	h_p	min/TU	given	5	12	5	10
	Terminal time coefficient	γ, (t_t)	–, (min)	given	0.18	0.15	(6)	(6)
2	Headway	h	min/TU	(1.44)	3	15	4	6
	Min (h, h_p)	h	min/TU		3	12	4	6
	Frequency	f	TU/h	(1.45)	20	5	15	10
	Actual load factor	α	prs/sp	(1.45)	1.09	0.71	0.79	0.54
3	Approx. cycle time	T'	min	(1.46)	95	69	62	62
	Fleet size	N	veh	(1.47)	35	6	16×6	11×2
4	Cycle time	T	min	(1.48)	96	72	64	66
	Terminal time	t_t	min	(1.52)	8	6	8	9
5	Cycle speed	V_c	km/h	(1.50)	10.0	13.3	22.5	21.8
	Summary data:							
	h		min/TU		3	12	4	6
	α		prs/sp		1.09	0.71	0.79	0.54
	N		veh		32	6	96	22
	T		min		96	72	64	66
	V_c		km/h		10	13.3	22.5	21.8

to obtain an approximate value of T, called T'. With that T', compute fleet size using Eq. (1.47):

$$N_{TU} = \left[\frac{T'}{h}\right]^+$$

rounding the obtained N to the next higher integer.

Step 4: Compute cycle and terminal times by Eq. (1.48): $T = h\ N_{TU}$, using the rounded N_{TU}, and

$$t_t = \frac{T - 2\ T_o}{2} \tag{1.53}$$

The obtained t_t represents the average value of terminal times at the two terminals, which may not be equal.

Step 5: Compute cycle speed, using Eq. (1.29):

$$V_c = \frac{120\ L}{T}$$

Computations in Table 1.5 follow this sequence of steps.

The final products of the scheduling process are numerical and graphical presentations of transit line operation that are made for use by both the operator and the public. The numerical schedule for operation of each run, i.e., running of a TU from its leaving the depot to its return, is used by individual drivers and dispatchers (see Section 1.4.6). A tabulation of times for all TU runs is given to the line supervisors, while various public timetables are prepared in appropriate formats.

1.4.5 Graphical Presentations of Transit Operations

Graphical presentations, such as time-distance, time-speed, distance-speed, and time-energy consumption diagrams, can be used very effectively in planning, operations (scheduling), and analyses of transit systems.

They offer, for example, a much better overview (such as regularity of headways), as well as details (meeting locations) of a line operation, or impacts of schedule changes, than a numerical table can provide.

While there were speculations that use of computers in scheduling and various transit operations analyses would make graphical presentations obsolete, the opposite has happened: with computer plotter capabilities it has become easier to generate various diagrams and figures. Innovative types of plots and diagrams offer clearer understanding of operations than either mathematical equations or numerical tables. Thus, computers make complex types of graphical presentations easy to plot and make them very useful in operations.

1.4.5.1 Graphical Schedules for Single Lines. There are several types of graphical schedules, from the conventional time-distance diagrams (sometimes known in the transit industry as string charts) to presentations of schedules on complex multiline networks. They can also vary in the manner of plotting and degree of detail.

The basic graphical schedule is a *time-distance diagram,* which has the line (distance) plotted on the ordinate and time on the abscissa. As shown in Figure 1.28, the line is divided in sections with uniform speeds. The plot of every run of a TU, designated by a number, shows all schedule elements—travel times, speeds, etc.—for that TU on each section and at each terminal. The abscissa also shows headways as distances between subsequent TU runs, and cycle times as distances between two subsequent departures of the same run from the terminal.

The entire diagram shows TU arrivals/departures at each reference point along the line, terminal times, as well as locations and times where TUs meet. Short turns, those operating on a section of the line, can be plotted and their fitting in regular headways in both directions can be checked.

To avoid different slopes on the diagram, simplify the plotting, and improve readability, it is common to use a *real-time/operating-time diagram,* which has operating time between the two terminals instead of distance plotted on the ordinate. This diagram can be understood as one on which distance scale is stretched on slower sections until a straight line is obtained for the entire line; thus, the ordinate presents the line section lengths at different scales, depending on their speeds, or it presents operating time at a constant rate (therefore the designation time-time diagram). Figure 1.29 shows the schedule from Figure 1.28 as a time-time diagram.

With respect to scale and plotted details, graphical schedules may vary greatly. For example, the schedule for a light rail line in Figure 1.30 shows track layout with single and double track sections, pull-outs and pull-ins of TUs from depots for operations on some line sections, different terminal times, etc. Some diagrams, such as those for highly controlled rapid transit systems, particularly the automated ones, show precise times for stopping at every station. Large-scale schedules, which may show long sections of lines as the basic blocks only, are used for plotting networks with overlapping and branch lines.

1.4.5.2 Trunk Lines with Branches. For complex transit lines, such as a trunk line with branches, graphical schedules can be even more useful than for single lines. For example, for a line consisting of a trunk with two branches, each branch must be scheduled by itself, and then the joint schedule of the two lines on the trunk section must be checked with respect to service regularity as well as the total offered capacity. This relationship can be seen very clearly on a graphical schedule. For further details on trunk-and-branch scheduling, see Sections 2.5.3 and 4.3.3.

A graphical schedule for a line with two branches, A and B, is shown in Figure 1.31. It consists of a diagram of the trunk TX, consisting of three sections with different speeds, which continues directly into one branch, XA. Vertically separated, but plotted to the same real-time axis, is the diagram for the second branch, XB. Plots for TUs on line A continue from the trunk directly into that branch, while those to branch B are vertically transferred from point X to the XB diagram.

This plot shows TU movements and headways on all three sections, the trunk and the two branches; it also shows how the terminal times on the two branches must be coordinated if the trunk is to have uniform headways.

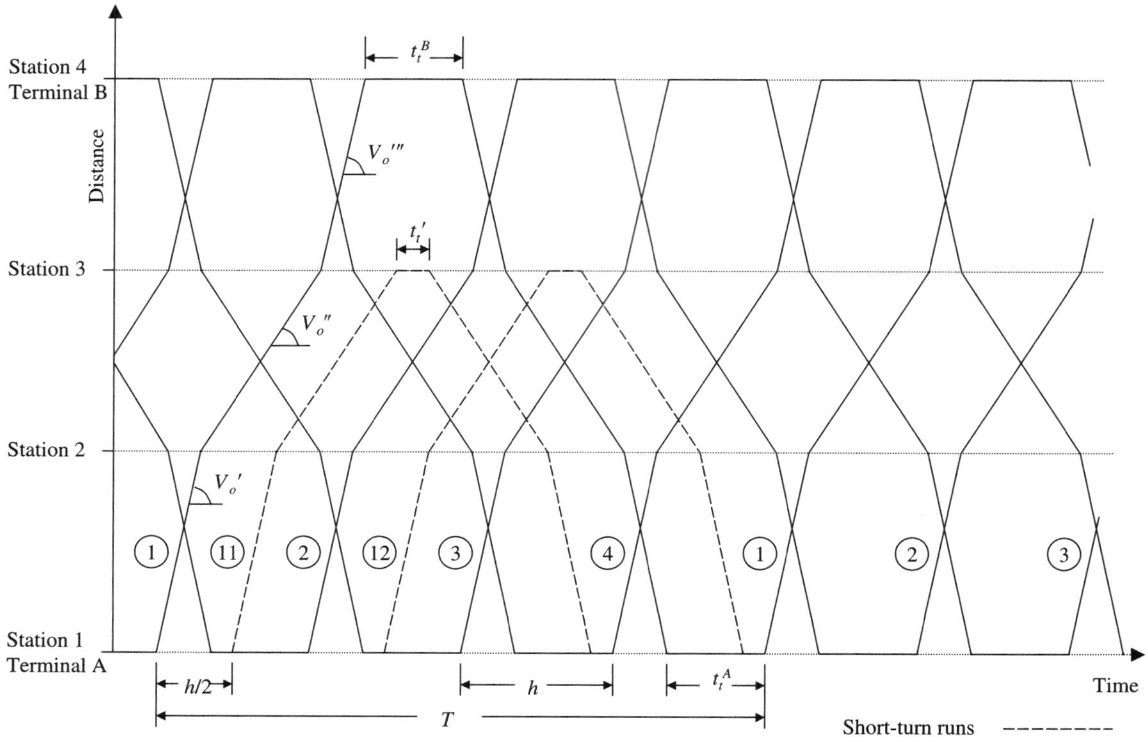

Figure 1.28 *Graphical schedule for a transit line with regular and short-turn runs.*

The same method of graphical schedule presentation can be used for a set of diametrical lines with a joint trunk section, typical for regional rail lines (Munich, Philadelphia). Such a diagram is shown in Figure 1.32. The basic purpose of such a plot is to present the relationship of TUs from different lines on the trunk section. Therefore, the branches can be shown only schematically, at a highly reduced scale, plotting only the entire travel time on each one.

This diagram should then be complemented by separate larger-scale diagrams of individual lines (such as the one in Figure 1.29), which have all the necessary details of sections, stations, terminal times, or short-turn runs.

1.4.5.3 Interconnected and Overlapping Lines. As the complexity of transit network increases, graphical schedules may become the only way to show the relationships of schedules clearly on joint sections as well as on individual branches.

An excellent example of the use of graphical schedule for a complex interconnected network is the San Francisco Bay Area Rapid Transit (BART) system. Figure 1.33 shows this network in 1996, when it consisted of a trunk line—M (Market Street)—and three branches—R (Richmond), C (Concord), and F (Fremont). The network was operated as three trunk/branch lines—MR, MC, and MF—and a fourth line between two of the branches, RF. The core point in the network is the Y junction in Oakland, where the three branches merge and the RF line crosses in the north-south direction, forming the triangular, Y connection. The MR and MC lines proceed jointly to the north, branching out at MacArthur Station (MA). Subsequently, in the late 1990s, several extensions as well as another branch, to Dublin/Pleasanton, opened, making the operations even more complex. The diagrams

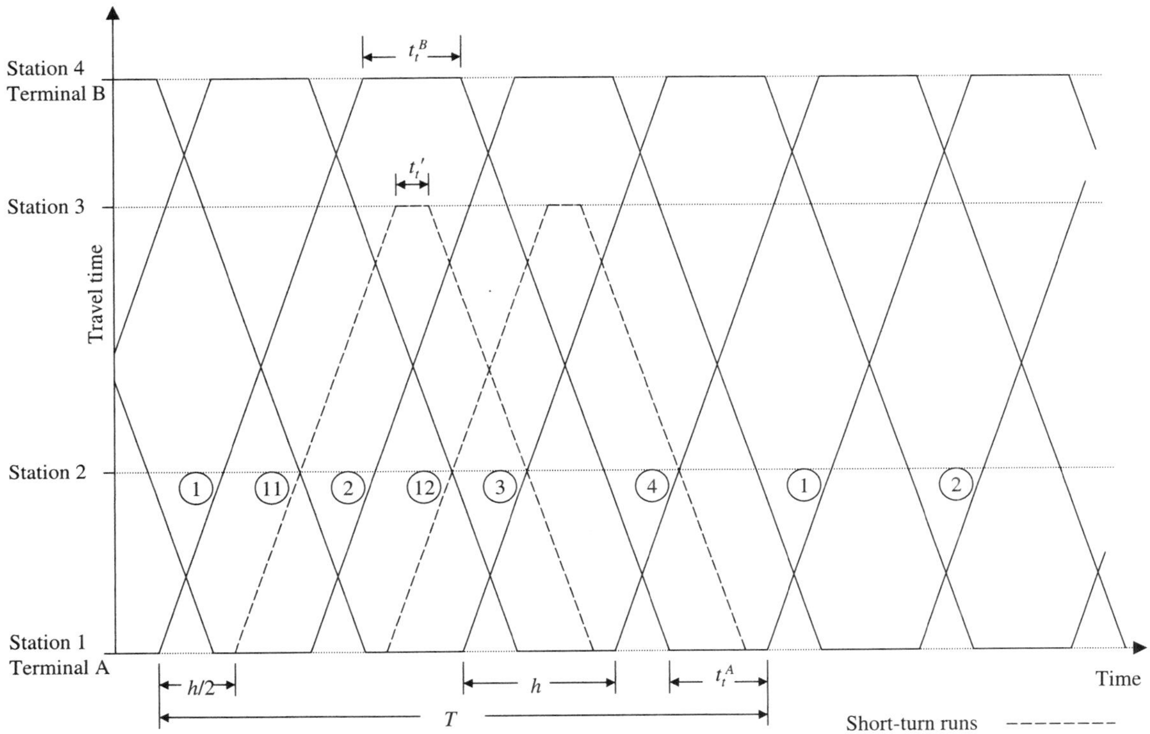

Figure 1.29 *Real-time/travel time schedule diagram for the line from 1.28.*

shown here are kept for the network prior to these extensions to avoid excessive schedule density that does not introduce any new concepts.

Scheduling of the BART network is quite complex because uniform headways, which are desirable on the trunk, impose certain relationships of headways on the branches, which then constrain the schedule for the cross-branch RF line. If the trunk headway is h_t and all branches should have equal headways, each one must have a headway of $3h_t$, and the sequence of trains on the trunk, or the basic module of the schedule, must be MR-MC-MF. If one line (in this case, MC) is more loaded than the others, another option is to operate the trunk with the basic module MC-MF-MC-MR. Then line MC has headways $2h_t$, while lines MR and MF have headways equal to $4h_t$. When combined with the cross-branch line RF, the average joint headway on each branch, R and F, also equals $2h_t$.

The headways on the branches, determined in this manner by the trunk operation, must also be synchronized with the RF line schedule. This must be done by an analysis of train arrivals/departures at the Y junction, because these determine the headways on all three directions radiating from it.

The relationships of headways on different lines and their regularity depend on the geometric setup of lines, headways that must be offered on each one, and other elements. It is therefore not always possible to provide uniform headways on all sections of the network.

Having determined the optimal relationship of schedules on all four lines, it is now necessary to schedule each line for its complete cycle, to determine terminal times and the required number of TUs. For this, a time-time diagram is used, again showing one branch as continuation of the trunk and the other branches on separate sections (Figure 1.34).

A special feature in this case is the plot of the RF line trains, i.e., those between two branches: inbound trains from R are shown as parallel to the inbound RM trains to the Y, but then their plot continues on the F

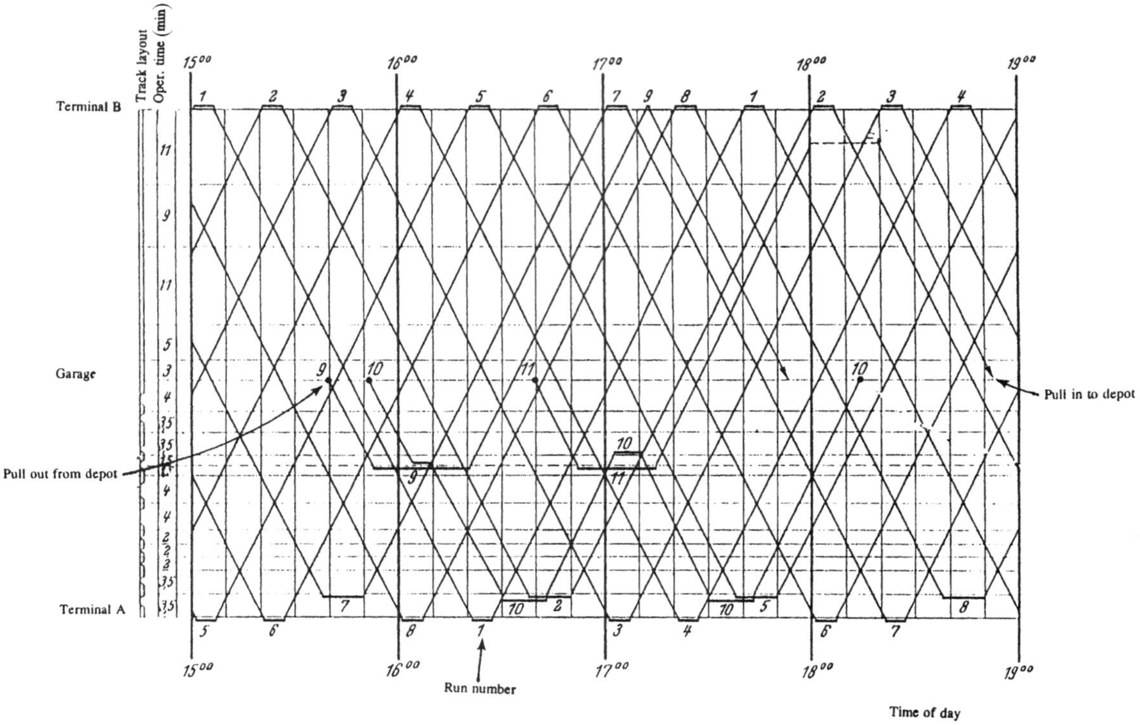

Figure 1.30 *An actual time-time graphical schedule with single track sections and operational details.*

branch as outgoing trains, i.e., with the opposite slope parallel to the MF trains. Thus, triangular interconnected lines can also be presented on a graphical schedule, and such a diagram gives an excellent overview of services on all network sections and their relationships, regularity of headways, terminal times, meeting points of trains, points critical for capacity, etc.

1.4.5.4 Other Types of Graphical Presentations.

A number of other types of diagrams can be used effectively in analyses of transit operations. The most important ones are described here.

A *time-distance diagram* for different stopping regimes allows analysis of such operations as local (all-stop), skip-stop, express, and zonal, presented in Section 2.4. Plotting of different stopping schedules by subsequent TUs allows an easy overview of headways, travel times, meetings, transfer possibilities and overtaking points along a line, etc.

A *speed-time diagram* is the most useful for analysis of driving regimes and travel times along a line. This is presented in considerable great detail in Vuchic (1981), Chapter 3.

A *speed-distance diagram* is useful in planning and analysis of TU travel along a line because it shows speed at any one point. The speed profile is used for determining signal blocks on rail lines, positions of different signals, braking distances, etc.

An *energy consumption-time diagram* shows consumption of energy for traction during different regimes and for a station-to-station travel cycle. As explained in Vuchic (1981), Chapter 3, this diagram can be used to further develop a diagram of travel time versus energy consumption on a given station spacing. This relationship is important in planning operating re-

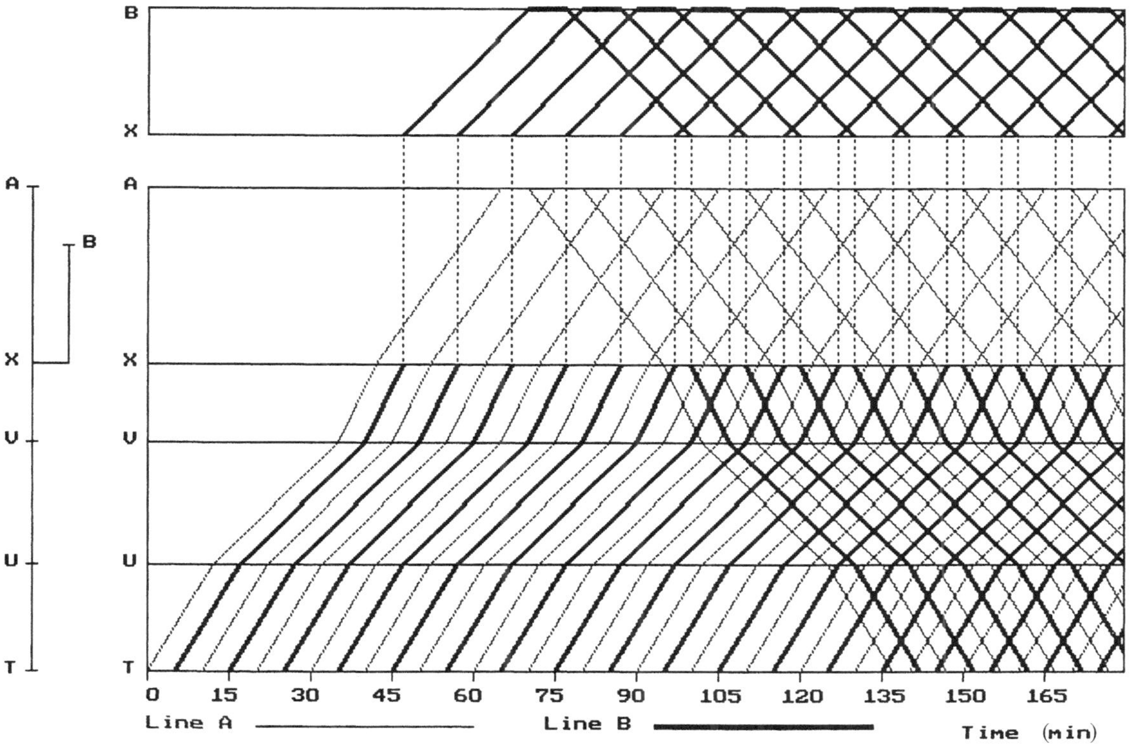

Figure 1.31 *A time-distance graphical schedule for a trunk line with two branches, A and B.*

gimes for electrically powered guided systems. Some rapid transit systems (e.g., BART and Washington Metro) have computer-controlled train operation which optimizes travel time-energy consumption relationship for different conditions.

1.4.6 Crew Scheduling or Run-Cutting

Assigning drivers or train crews to scheduled sequence of operations or TU runs is commonly known in the transit industry as *crew scheduling* or *run-cutting.* The task in run-cutting is to assign personnel to a given schedule of TU operations in such a manner that the total expenditure for wages is minimal, while satisfying the various operating and work rules set up by the agency or specified in its agreement with employee labor union.

Transit is a service that usually must be offered every day of the year for 16, 20, or even 24 hours. Since the quantity of service offered varies greatly among different days and hours of day, transit operating personnel must work during irregular hours, for periods of varying length, sometimes during nights, on weekends, and on holidays.

1.4.6.1 The Basic Concepts. The sharp peaking of passenger volumes that typically occurs on transit systems each working day in the morning and afternoon hours, about nine hours apart, requires different vehicle schedules for several periods during the day, as described in Section 1.4.2.2. The run-cutting—preparing schedules for drivers—is usually even more complex because sets of work assignments must consider workers' needs and meet established work rules and contractual standards.

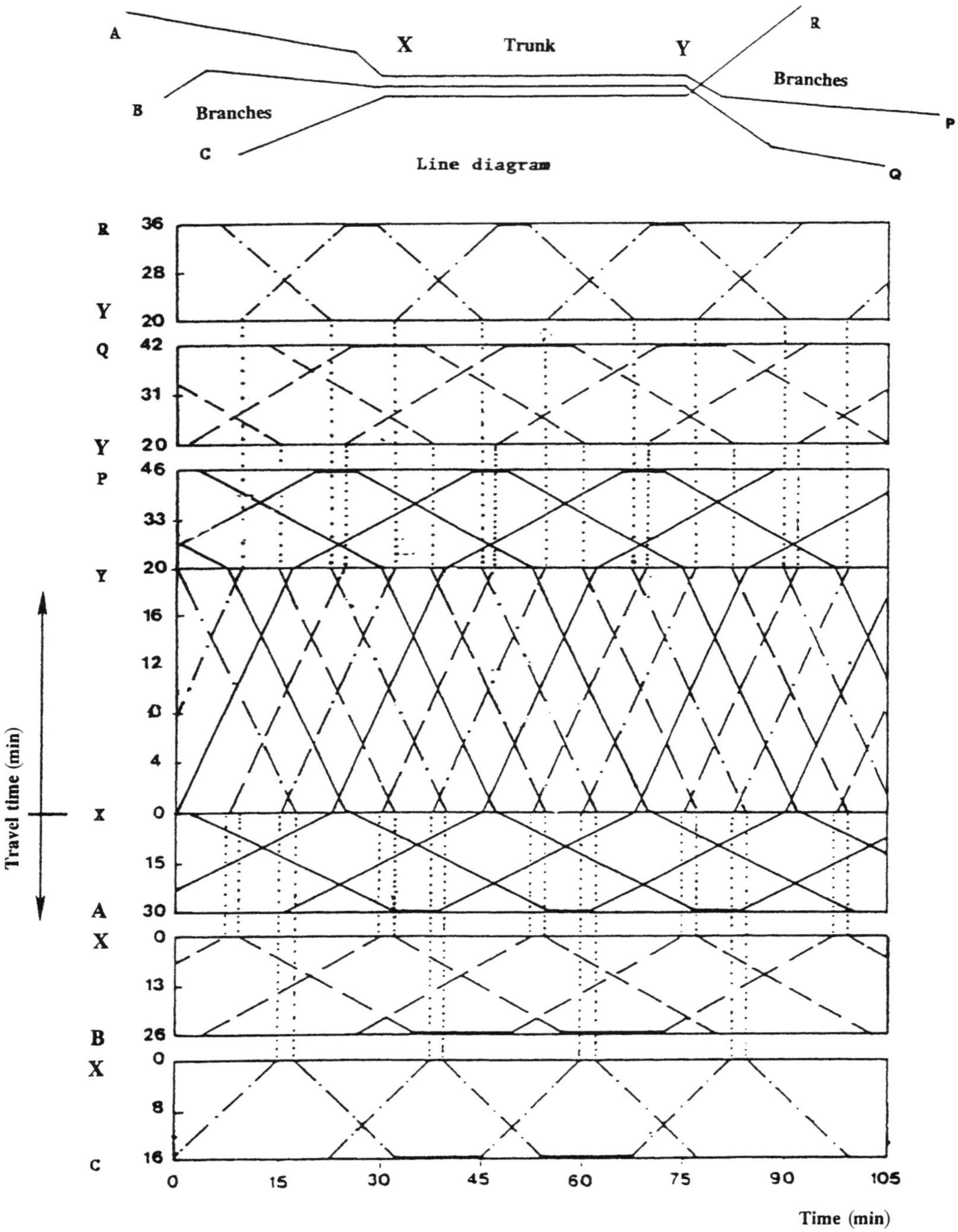

Figure 1.32 *A graphical schedule for diametrical lines with a joint trunk.*

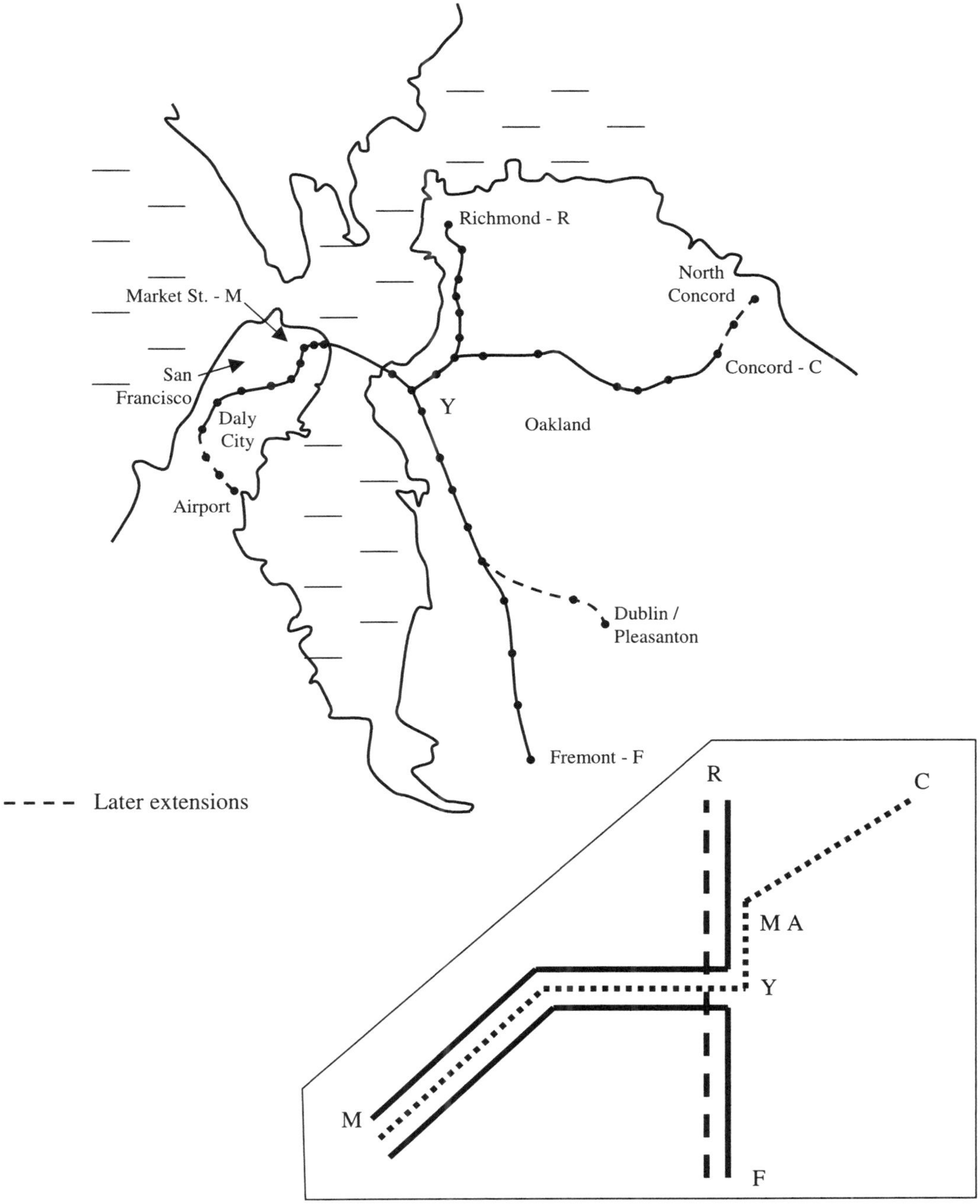

Figure 1.33 *San Francisco BART network and schematic presentation of its lines.* Note: *The scheduling is shown for the network from 1995, with later extensions as dashed lines.*

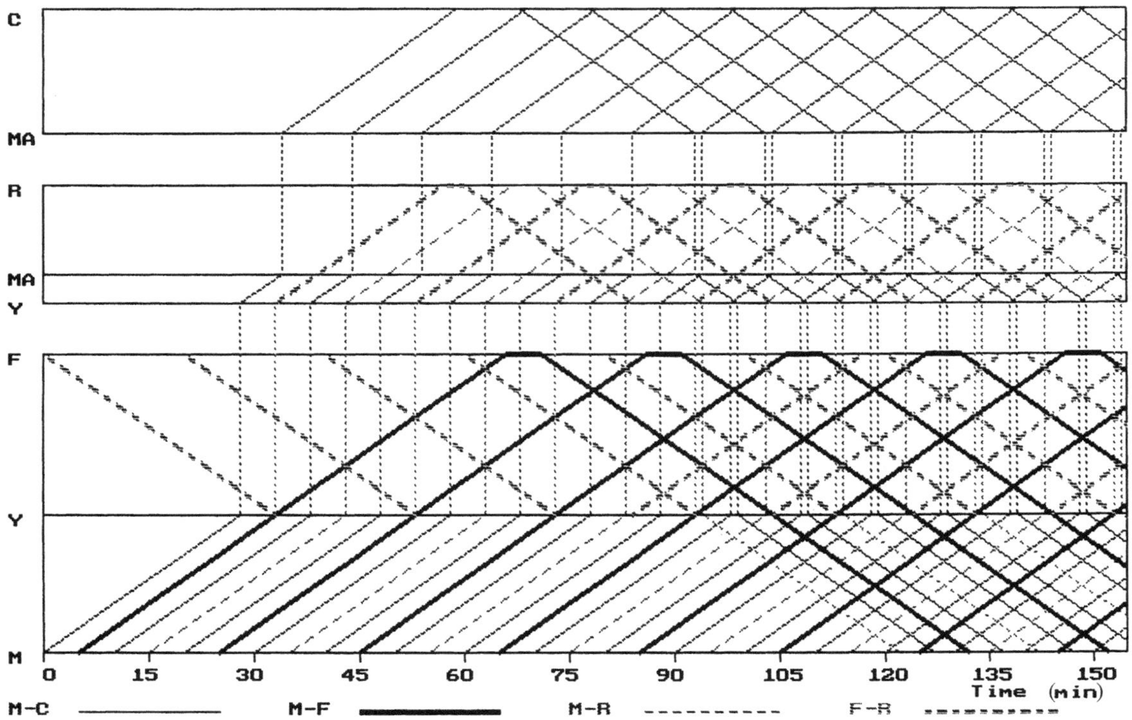

Figure 1.34 *Graphical schedule diagram for a trunk with three branches and the cross-branch line (San Francisco BART network).*

To cover the uneven demand for drivers during the daily hours on weekdays and weekends, working times are tailored in different forms. The basic and most desirable work shift or run is a continuous work shift, usually of eight hours. That is called a *straight run.* Then, since many more drivers are needed during peak than during off-peak periods, many drivers must be scheduled to work during the two peaks with several unproductive (paid or unpaid) hours between. Such a work shift is called a *split run.* The time interval between the beginning of work in the morning and its termination in the afternoon is referred to as *spread time.* Short time segments of work are called *extras* or *trippers.*

Work rules that the schedule must satisfy are determined in different ways. In some cases they are specified by the transit agency as a set of operating practices or standards, or spelled out in the contract between the agency and the labor union of its employees; further, there may be recommended or required practices developed by an association of agencies for the entire transit industry; finally, there are national government rules and regulations for maximum duration of driving between rests. Germany, for example, has extensive, carefully developed guidelines and regulations for work rules (as well as for safety, vehicle maintenance, and many other aspects of transit operations).

Consequently, there are wide variations in work rules and payment standards. For U.S. and European transit agencies, it is rather common for regular pay to be given for 8 hours of work, which are either continuous or within a 12-hour-spread time interval. Higher wage rates are usually paid for some or all of the following conditions:

- Overtime
- Longer spread time than prescribed
- Night shift
- Work on holidays

Alternatively, the contract may stipulate the frequency with which shifts involving night, holiday, or overtime work can be given to any one employee, after which a higher-than-regular wage rate must be paid. For example, the maximum percent of split runs, or ratio of split to straight runs, is often stipulated. When that is exceeded, an employee gets extra compensation.

Transit agencies usually have a scheduling department that produces schedule blocks for vehicles/trains and for crews. Specific assignments of individual workers to run-cutting blocks, which are then placed into work-assignment sheets, are usually done by individual rail yards and bus garages. This decentralization is more efficient because assignments can be more personalized at the local levels, to the extent that in some agencies drivers are always given the same vehicles, so that they tend to take better care of them. Moreover, work rules sometimes vary among yards and garages, particularly in such large agencies as those in London and New York, where there may be 10–20 rail yards and 50–100 bus garages.

1.4.6.2 Run-Cutting Procedure. The procedure of work assignments to employees for individual runs vary among agencies. In most U.S. transit systems, the runs are selected (picked) in the sequence of employees' seniority, so that workers with longest tenure can choose the most convenient runs. In most European transit systems, runs are programmed in a rotational manner, so that every worker gets generally the same overall schedule within a period of several weeks. An extremely systematic and detailed description of the methodology for making such rotating ("turnus") schedules, as well as analysis of work-assignment efficiencies, is given by Lehner (1978).

Various practices and requirements, together with many local conditions in each agency, make the task of achieving the minimum-cost driver assignments a very complicated one. There is no exact sequence of mathematical formulae or procedures for finding the optimal solution, but the basic procedure for run-cutting, developed through experience, consists of the following sequence of steps:

1. Develop as many straight runs as possible.
2. Form split runs within spread time.
3. Divide some straight runs into two or three segments (extras) and combine these segments with the extras left over from step 2 to form additional split runs.
4. Analyze the efficiency (basically, total expenditure) of the obtained solution. If it is not satisfactory, investigate possibilities for pairing with other routes so that TUs alternate between them, and repeat the procedure.

Figure 1.35(*a*) shows a typical diagram of operating hours that have to be served on a transit line. Actual numbered runs for such a service are shown in Figure 1.35(*b*), while the results of that run-cut are listed in Table 1.6.

If the peak-to-base ratio, i.e., the ratio of the number of TUs in operation during the peak hour to their number during midday period, is very large and the two peaks are spread far apart, it may be particularly advantageous to employ part-time workers. They can be used to cover one of the peaks, avoiding the runs with excessive spread times (>12 hours), which are inconvenient for workers and can be costly for the agency. Hiring part-time workers has other advantages and disadvantages (lower benefit payments but extra training costs) and is usually subject to agreement with the labor union.

Run-cutting procedures are also presented by Lehner (1978), Homburger (1982), Hoffstadt (1981), and Rainville (1982).

1.4.7 Use of Computers in Scheduling

The entire process of scheduling, including computations of headways, TUs, cycle times, etc., and then

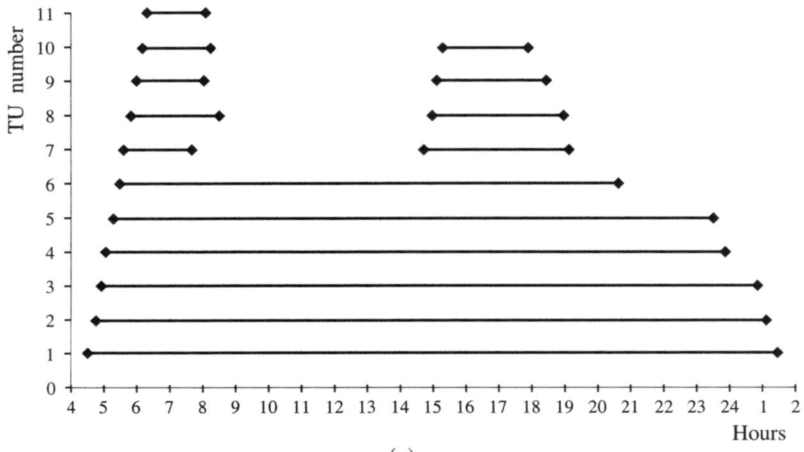

(*a*)

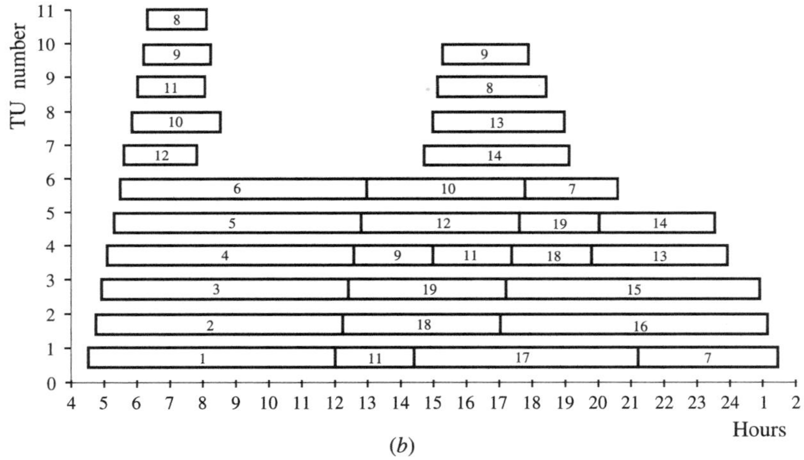

(*b*)

Figure 1.35 *Diagram of duty hours and run-cuts:* (*a*) *a typical diagram of duty hours;* (*b*) *run assignments for duty hours from* (*a*). (Source: *Lehner, 1978.*)

determination of specific TU blocks and driver runs, is very complex even for a single line. Scheduling for many lines is correspondingly more difficult. Moreover, the variations in schedules for different periods of day and among days of week, and considerations of various requirements and wage rates for labor, add to the computational complexity.

Efficient scheduling is, however, extremely important because it determines expenditure for what is usually the largest item of agency's operating expenditures: operating costs of vehicle-km operated and labor wages for hours worked. In large agencies, even small increases in schedule efficiency can translate into large annual funds. To facilitate this process, operations research algorithms with computer programs have been developed that can investigate hundreds of schedule and run-cutting variations. These may include line-pairing combinations, driver switches to other TUs after breaks in their work, variations in reserve TUs, and personnel shift changes at different locations and

Table 1.6 Summary of runs from the diagrams in Figure 1.35

Run	Duration (min)	Gaps (min)	Work Length (min)	Breaks (min)	Total Work Spread (min)
1	450	—	450	—	450
2	450	—	450	—	450
3	450	—	450	—	450
4	450	—	450	—	450
5	450	—	450	—	450
6	450	—	450	—	450
7	168 + 255 = 423	—	423	39	462
8	108 + 198 = 306	—	306	228	534
9	120 + 144 + 156 = 420	24	444	246	690
10	162 + 288 = 450	—	450	252	702
11	122 + 144 + 144 = 410	36	446	220	666
12	132 + 288 = 420	—	420	282	702
13	240 + 246 = 486	—	486	48	534
14	264 + 210 = 474	—	474	48	522
15	462	—	462	—	462
16	486	—	486	—	486
17	411	—	411	—	411
18	288 + 144 = 432	—	456	—	456
19	288 + 144 = 432	—	456	—	456
Totals	8312	108	8420	1363	9783

Characteristic elements:
1. Longest, shortest, and average work durations: 486, 411, and 443 minutes, respectively.
2. Longest, shortest, and average spread: 702, 411, and 515 minutes, respectively.

times of day, depending on the probability of demand variations, personnel absenteeism, etc.

Computer programs can produce schedules for several lines in a fraction of the time that a scheduler would take through manual computations. Even greater benefit is the ability to test variations in scheduling. When an initial schedule is produced, its total cost and various coefficients of utilization are computed; changes in the initial inputs, work assignments to individuals, wage rates, etc., can be assumed and immediately tested by the computer program. This testing of variations, shown by a dashed line as feedback in Figure 1.23, allows considerable improvements from an initial computation toward an optimal schedule.

Despite their advantages, however, computer applications for this purpose are not always straightforward because of numerous local details and procedural peculiarities that require the personal attention of a person familiar with the system. For example, travel times on some line sections may vary between peak and off-peak hours; acceptability of marginal values of load factor α must be subjectively decided; and labor contracts sometimes vary among divisions. All these details must be handled by experienced schedulers. For this reason, the introduction of computerized scheduling by transit agencies was historically a slow, often gradual process. For example, initial attempts to use computerized scheduling for buses of London Transport during the 1970s proved impractical because each one of its over 70 bus garages had a specific procedure of vehicle scheduling and different labor agreements. In many agencies, it took considerable efforts to develop computerized programs that could produce as efficient schedules as the experienced schedulers could prepare by traditional manual method.

With major efforts in the development of sophisticated and yet practical computerized scheduling programs, their applications have now become widely adopted and the size of agency in terms of TUs and operating personnel at which computer-based programs become more efficient than manual computations has decreased from hundreds to as few as 20–30 TUs, the exact number depending on local conditions and practices.

1.4.7.1 Structure and Contents of Packages. The best-known program for transit scheduling in the United States has been the Run Cutting and Scheduling (RUCUS) package. Developed under the sponsorship of UMTA/FTA, RUCUS was designed as a modular parametric package suitable for staged construction of schedules in any agency. Other programs were developed by several consulting firms, as well as by major transit agencies such as HHA in Hamburg, RATP in Paris, and SNCB in Brussels. Thus, there are several highly refined, comprehensive, but expensive packages sold internationally. Most packages consist of different modules: timetables, vehicle blocks, crew blocks, and various analytical tools used for revising inputs, as shown in Figure 1.23. In addition, there are numerous less expensive packages that perform fewer tasks but may be sufficient for small and medium-sized transit agencies.

The RUCUS package consists of three distinct modules: trips, blocks, and runs, corresponding to the three scheduling work elements in Figure 1.23. In addition, the program performs filing and preparing of passenger counts and travel time data for input into scheduling, and preparing and printing of results of scheduling—summary data for various reports. European transit operators utilize the convenient graphical schedules much more extensively than the U.S. transit agencies. Many European agencies also use different types of run-cuts than North America due to different labor practices. They use rotating duty rosters which give *equal time assignment cycle to each worker.*

Among these components, run-cutting is most frequently computerized, followed by determination of vehicle schedules (blocks) and timetables (trips). Summary data used for management reporting and regulation compliance are also frequently handled by computerized program packages.

Even with full computerization of the scheduling process, the work commonly remains modular because of the need for schedulers to intervene at different stages of the work. Actually, one of the reasons for the initial difficulties in introducing computers was that the first programs tended to be of the "black box" type, excluding the schedulers from the computations. The current trend is toward use of interactive programs that keep the scheduler informed and, more importantly, in charge of the computational process. The entire operation with computerized scheduling and information system requires considerable computer expertise, but that is not sufficient. A person thoroughly familiar with transit operations and experience in scheduling vehicles and personnel is always required to supervise the entire process.

1.4.7.2 Evaluation of Computer Applications. There are a number of benefits to as well as some costs of using computers for scheduling; there are also certain conditions that influence the benefits and costs. These should be evaluated to determine feasibility of computerizing that process for each transit agency.

Briefly stated, the benefits of computer-assisted scheduling are:

- Reduced manpower requirements for scheduling
- Increased probability of finding the optimal solution, i.e., minimal vehicle- and driver-hours for a given service level
- Ability to test quickly many alternatives and variations, including interlining, driver fallback after a break, and others, which is very useful for such actions as modifying schedules when passenger volume changes, for labor negotiations, cost considerations of different level-of-service policies, etc.
- Increased accuracy of computations
- Much greater ability to automate production of data and results of operations, such as timetables and payroll data
- Information produced that can be used for computer aided dispatching (CAD), automatic passenger counting (APC), real-time schedule adherence information, travel time data collection, scheduling of vehicle maintenance, and other computer-based operational functions

The costs or negative aspects of computer-assisted scheduling are:

- Implementation requires a major effort to upgrade and reorient the entire scheduling department information technology infrastructure and retrain its personnel.
- Agency must obtain the appropriate hardware and software.
- Agency must acquire competent programmers who understand transit operations.

Consequently, the benefits can be highly significant; the costs and effort are also major, but most of them are one-time investments. A few other factors should be borne in mind in evaluating the possible use of computers. First, for small agencies the use of computers is not appropriate because the benefits are so small (if any) that they cannot outweigh the efforts in preparing and adjusting the programs; benefits increase with the size of the agency. Hoffstadt (1981) estimated that a rationalizing effect is to be expected only when at least 30 vehicles are operated. Roy Lave in California also found 30 to be the threshold size for demand-responsive transit. And second, even the best computer programs cannot fully replace experienced schedulers. They replace the bulk of routine computations and person-hours of work, but good understanding of the process and experience in transit operations remain fundamental for efficient scheduling and optimal results.

1.4.8 Measures of Operating Efficiency

Strictly speaking, directly productive time of vehicles and personnel is only that when TUs actually travel and perform service on transit lines. Other times, such as terminal and deadhead, or sick leave for personnel, although necessary, should be minimized within the limits of various operating requirements and work standards. A very fundamental, systematic definition of concepts and analysis of schedules, run cuts, and their efficiencies was given by Lehner (1978). Based on his work, the following efficiency and utilization coefficients are defined.

1.4.8.1 Schedule Efficiency. The basic measure of efficiency of a transit line schedule is coefficient η_t, reflecting terminal time losses. It is defined as the ratio of the sum of operating times (in two directions) to cycle time:

$$\eta_t = \frac{T_o' + T_o''}{T} \qquad \left| \frac{\eta}{-} \middle| \frac{T}{\text{min}} \right| \tag{1.54}$$

Utilizing Eqs. (1.20) and (1.21), and assuming that $T_o' = T_o''$, η_t can be related to terminal time coefficients γ and γ':

$$\eta_t = \frac{T_o' + T_o''}{T_o' + T_o'' + t_t' + t_t''} = \frac{2 \cdot T_o}{T}$$
$$= \frac{1}{1+\gamma} = 1 - \gamma' \qquad \left| \frac{\eta, \gamma}{-} \middle| \frac{T,\ t}{\text{min}} \right| \tag{1.55}$$

If T and T_o are extracted from Eqs. (1.28) and (1.29) and introduced into Eq. (1.54), η_t can also be expressed as the ratio of cycle speed to operating speed:

$$\eta_t = \frac{V_c}{V_o} \qquad \left| \frac{\eta_t}{-} \middle| \frac{V}{\text{km/h}} \right| \tag{1.56}$$

As an example, if cycle time on a line is $T = 80$ min and operating times are $T'_o = 33$ min. and $T''_o = 35$ min, then terminal time coefficients are, from Eqs. (1.20) and (1.21), respectively:

$$\gamma = \frac{12}{68} = 0.18 \quad \text{and} \quad \gamma' = \frac{12}{80} = 0.15$$

while coefficient of schedule efficiency η_t is from Eq. (1.54):

$$\eta_t = \frac{68}{80} = 0.85$$

If it is assumed that $T_o' = T''_o$, which is true in most cases, and T is replaced by $N_{TU} \cdot h$ based on Eq. (1.48), then η_t can be expressed as:

$$\eta_t = \frac{2T_o}{N_{TU} \cdot h} = 1 - \frac{2t_t}{N_{TU} \cdot h}$$

$$\left| \frac{\eta}{-} \middle| \frac{T,\ t}{\text{min}} \middle| \frac{N}{\text{TU}} \middle| \frac{h}{\text{min/TU}} \right| \qquad (1.57)$$

Since T_o is a function of operating speed V_o and line length L, this equation shows that η_t depends directly or indirectly on several operating elements. The relationships are not always simple, however, because the cycle time can only be a discrete number, an integer multiple of h. Thus, most relationships have a sawtooth shape, as shown in Figure 1.36, with sudden increases of η_t when T is shortened by h (min), so that N can be decreased by 1 for a given h.

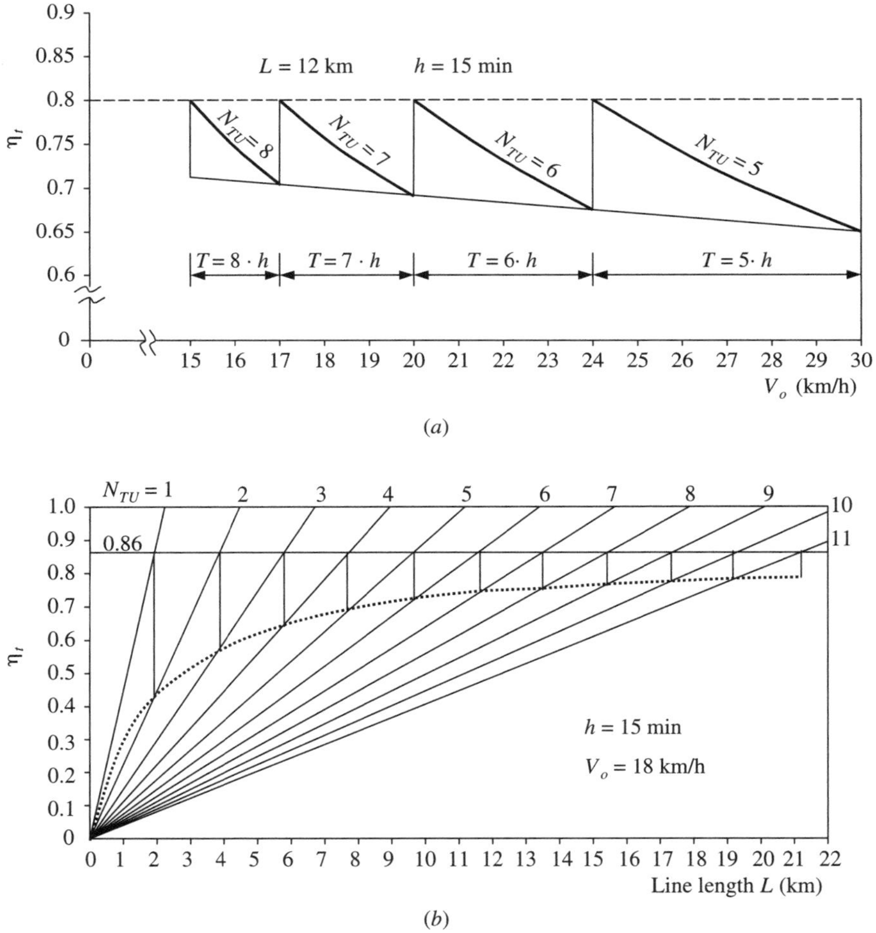

Figure 1.36 *Schedule efficiency coefficient* η_t *as a function of operating speed* V_o *and line length* L*: (a)* η_t *as a function of* V_o *for given* h *and* L*; (b)* η_t *as a function of* L *for given* h *and* V_o. (Source: *Lehner, 1978.*)

Table 1.7 Influence of operating speed on the schedule efficiency coefficient η_t

Operating Elements	Dimension	Case i	Case ii	Case iii	Case iv
V_o	km/h	18.0	19.5	20.0	21.5
T_o	min	40	37	36	34
T	min	80 + 25 = 105	74 + 31 = 105	72 + 18 = 90	68 + 22 = 90
η_t	—	0.76	0.70	0.80	0.76
V_c	km/h	13.7	13.7	16.0	16.0
N_{TU}	—	7	7	6	6

Line parameters: L = 12 km, h = 15 min, Min $(t_t' + t_t'')$ = 18 min.

The diagram in Figure 1.36(*a*) shows η_t as a function of V_o for fixed values of L and h. As V_o is increased, η_t decreases as long as T remains constant. When V_o is so high that T can be reduced by h [min], η_t jumps to the highest value possible for the minimum length of t_t, as determined by the crew rest and other operating requirements.

The diagram in Figure 1.36(*a*) shows that η_t generally tends to decrease as V_o increases. Lehner has shown that because of the discontinuous nature of the cycle time, impacts of other operating elements on η_t are also discontinuous; however, for a given line, η_t has a general tendency to increase with line length L (Figure 1.36(*b*)), but to decrease as headway increases.

The relationship from Figure 1.36(*a*) is also illustrated in Table 1.7 by schedule computations for four cases. As V_o increases from case i to ii, η_t decreases; then, from ii to iii T can be shortened, so that η_t increases and N decreases from 7 to 6. In case iv T remains the same as in iii, so that η_t decreases again.

1.4.8.2 Operating Personnel Efficiency. An important measure of labor force utilization is the ratio of net productive working hours on transit line to the total number of paid hours, designated as operating personnel efficiency coefficient η. That coefficient consists of the following elements.

- η_a, attendance coefficient, represents the ratio of reported hours (time) t^r, to paid hours t^p, accounting for losses due to vacation, illness, and other absences.
- η_s, coefficient of run-cutting, or the ratio of hours on transit line t^ℓ to reported hours t^r. It includes losses caused by split shifts, work preparation, deadheading, etc.
- η_t, coefficient of schedule efficiency, defined as Eqs. (1.54) and (1.55). Coefficient η is the product of the preceding three coefficients:

$$\eta = \eta_a \cdot \eta_s \cdot \eta_t = \frac{t^r}{t^p} \cdot \frac{t^l}{t^r} \cdot \frac{2 \cdot T_o}{T} = \frac{t^l}{t^p(1+\gamma)} \qquad \left| \frac{\eta}{-} \right| \frac{t,\ T}{\text{min}} \left| \frac{\gamma}{-} \right| \tag{1.58}$$

It is difficult to give typical values for these coefficients because they depend on many local factors. For example, rapid transit systems usually have higher values of η_s than buses and regional rail systems; labor contracts or work laws affect η_a, while fully automated systems have a different structure of their personnel (no drivers, but more supervisors). Although their absolute values vary, these coefficients can be very useful for comparisons of alternative vehicle and personnel schedules (run cuts) for an agency, or for comparison of operating efficiencies of similar transit services.

EXERCISES

1.1. A bus line has stops at every intersection, 200 m apart. The speed of traffic and of the buses except for their stopping at bus stops is 30 km/h, but for every stopping, buses lose a total of $t_s = 20$ s. Passengers are uniformly distributed along the line and walk to/from its stops at a speed of $V_w = 4.8$ km/h. How would a passenger's total travel time, including access to/from stops and travel on buses, change if every other stop was abandoned, i.e., the spacing between stops was increased from 200 to 400 m, if the average trip length on the line is: a) 4 km; b) 6 km; c) 8 km? For what passenger trip length L_p will the total travel time remain the same?

1.2. A heavily traveled bus line has a length of 12 km. It has stops at every intersection and there are 5 blocks per kilometer. The speed of traffic and of the buses except for their stopping is 34 km/h. Terminal time is 9 min at each end. Assume that the buses stop at every stop, losing 20 s for deceleration and acceleration plus standing (dwell) time of 15 s. The average passenger trip length on the line is 4.8 km. Service headway is 5 min.

You want to propose a reduction of stop density to one per two blocks. In that case, time loss for each deceleration/acceleration would not change, but standing times would increase to 25 s due to increased number of passengers boarding and alighting per stop. The company management believes that the existing operation is desirable for passengers (short access) and that the changes would not make much difference in operating costs anyway. To convince the management that the change would be useful, compute and systematically present the following consequences of the change in stop density, assuming uniform distribution of origins-destinations along the whole line:

a. Additional average walking distance per passenger along the line, which includes access and egress (in m/prs).

b. Additional average access and egress time (in min/prs), if the speed of walking is 75 m/min.

c. Reduction of the average passenger travel time on buses (in min/prs).

d. Change in the passenger total travel time.

e. Break-even passenger trip length L_a between the two stopping regimes. HINT: Compute the travel distance on the line on which passenger saves as much time as he/she loses for longer walking to/from stops.

f. How many buses can be saved due to higher operating speed. HINT: Compute operating time $T_r = 60L/V_r$ rounded in min; then add the present additional time for stopping (converted in minutes) to obtain T and then N. Do the same for the new stopping times and find the difference ΔN.

1.3. Operating (travel) speed on an 8-km-long bus line with $h = 12$ min is $V_o = 16$ km/h. Terminal time at each end is 6 min ($\Sigma t_t = 12$ min). Bus preferential treatments are considered to improve the operation, but they would require a certain level of investment. The company will make the investment if it can reduce the number of buses by one due to these improvements. What is the minimum value of V_o that would allow saving one bus if the total terminal time at both ends should not be less than $T_t = 0.15\ T$, and the headway remains 12 min?

1.4. A bus line with a length $L = 2430$ m has 6 stations, including terminals. Interstation distances have the following lengths: 520, 280, 680, 450, 500 m. Running speed on the line is $V_r = 32$ km/h, headway is 4 min, and terminal times at each end are 5 min. Draw a general form of a graphical schedule for two buses operating on this line at headway h: plot a diagram with 1500 s on the abscissa and 2500 m on the ordinate. Show on the diagram straight lines of bus travel between stops and time lost per stopping of 30 s. Show also the following elements: h, T_o, t_t, T, V_o and V_c, assuming T_o and t_t are the same in each direction.

1.5. A 14-km (one-way) long trolleybus line has an operating speed $V_o = 12$ km/h; terminal time at each end t_t is at least 6 min.

a. How many trolleybuses are required for operation with a 10-min headway? What will be the cycle speed V_c for the line with that schedule?

b. What new operating speed V_o' should be achieved in order to reduce the number of trolleybuses by two, while maintaining the same headway and increasing the minimum required terminal time at one terminal to 8 min? Compute also the new cycle speed V_c'.

c. What will be the offered line capacity if $C_v = 80$ sps? Compute α_{max} if the line carries 300 prs/h on its maximum load section.

1.6. A rapid transit line is 14.2 km long (one way) and has the following passenger volumes boarding and alighting at individual stations in the peak direction during the peak hour:

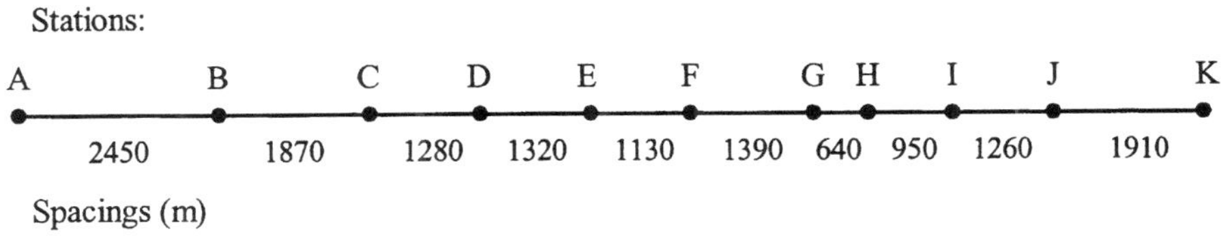

Station	*A*	*B*	*C*	*D*	*E*	*F*	*G*	*H*	*I*	*J*	*K*
Boardings (prs/h)	3300	1700	1900	3200	2900	1300	1600	600	400	700	0
Alightings (prs/h)	0	0	500	700	2100	700	4000	2700	2200	1700	3000

Capacity of each train is 590 spaces, α_{max} is selected to be 0.9, travel speed on the line is $V_o = 36$ km/h. The labor union contract requires that terminal times at both terminals together should be at least 15% of the two-way operating time, but no less than 10 min.

a. Compute the number of trains required for the service, excluding reserves. Using that number for N_{TU}, round the cycle time up to the next multiple of the headway, then compute the value of γ.

b. Assuming that there is no lower limit on terminal time, how much should the terminal time be reduced to permit withdrawal of one train, without changing the headway? What will be the new value of γ?

c. If terminal times remain as found under part a, how much should the travel speed be increased to reduce the number of trains on the line by one without changing the headway?

d. Draw a diagram of this line with distance and stations on the abscissa, passenger volume, and offered capacity on the ordinate (see Figure 1.5). Show the MLS.

e. Compute the work in person-km/h on this line and its work utilization coefficient α.

1.7. A 16.48-km-long regional rail line from center city to a suburb has the following characteristics during the afternoon peak hour:

The line has the following stations, station spacings, and boardings and alightings:

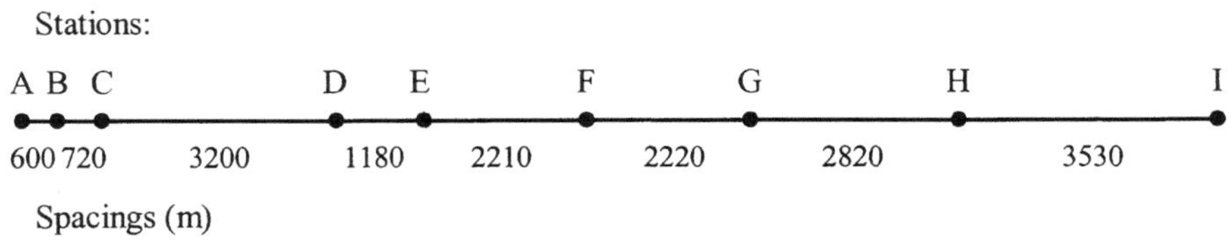

Station	*A*	*B*	*C*	*D*	*E*	*F*	*G*	*H*	*I*
Boardings (prs/h)	2640	1968	3156	288	160	20	48	12	0
Alightings (prs/h)	0	96	144	1032	1240	3020	1032	528	1200

Train consists are 5 cars, and each car has 94 seats and a total capacity of 145 spaces. Operating speed on the line is 42 km/h. Terminal times cannot be shorter than 5 min at each end.

a. Compute and draw the load diagram (profile of passengers, offered seated, and total capacities). Compute and show the MLS.

b. Compute schedule I for the line adopting the maximum load factor $\alpha = 0.65$ on the MLS.

c. Compute two other schedules for a more economical operation when alternative trains would turn back at station F and at terminal I: schedule II would again have load factor $\alpha = 0.65$, while schedule III should be based on a higher load factor $\alpha = 0.87$.

d. Compare the three schedules by:

- The number of trains and cars required
- Work utilization coefficients

e. Draw a graphical schedule showing operation III. Note that the headway on the joint section must be uniform in both directions.

f. List other factors that should be considered in selecting between schedule I and schedule II or III with short-turn trains.

1.8. Five bus lines merge to a joint section on a trunk street. If their headways are 2, 5, 7.5, 15, and 20 min, what will be the average headway h_{av} on the common section, including buses from all five lines?

The MLS, which is on the common section, has a volume of 2480 prs/h. The capacity of each bus is 68 spaces. What is the average load factor α?

1.9. In a situation similar to Boston, Philadelphia, and San Francisco, three radial LRT lines merge into one 4.6-km-long trunk line leading to the city center. Operating speed on the trunk is 19.6 km/h.

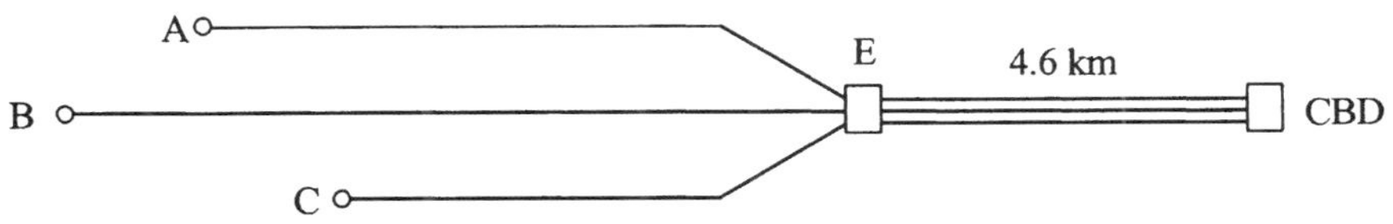

The branches have the following characteristics and passenger volumes during the peak hour:

Line	*Length* (km)*	*Operating Speed on Branches (km/h)*	*Passenger Volume (prs/h/direction)***
A	8.1	18.6	3520
B	10.1	24.0	2640
C	6.0	18.2	705

*From outer terminals to point E, one way.
**East of point E, which is the maximum load section for all lines.

Each LRT vehicle has a capacity of 156 spaces, including 58 seats; average maximum occupancy of each vehicle should be $\alpha_{max} = 66\%$ (all seats occupied plus 45 persons standing).

a. Compute the headway for each line using single cars. If single-car operation would result in a headway shorter than 6 min, use 2-car TUs. The policy headway for all the lines should be 7.5 min.

b. Find average headway of service on the trunk section (including vehicles from all lines).

c. Compute the total fleet required for this service, assuming 10% reserves.

d. Suppose that operating speed on the trunk is reduced to 15.1 km/h. What is the required increase in fleet size in order to maintain the same headway on all three lines?

e. If the operating speed on the trunk remains 19.6 km/h, what operating speed should be achieved on the outer section of line B to reduce the number of TUs used by one?

Note: Round off headway on individual lines to the closest half minute and make them divisible in 60. Average headway should be exact, in seconds. Total terminal times for each line should be at least 0.15 T.

1.10. Two LRT lines, shown below as the Blue Line and the Red Line, should be scheduled. Data about their individual sections are given in the table. Service on the joint section, BC, should be with uniform headway. The lines are served by two-car trains, and each car has a capacity of 160 spaces; $\alpha_{max} = 0.80$.

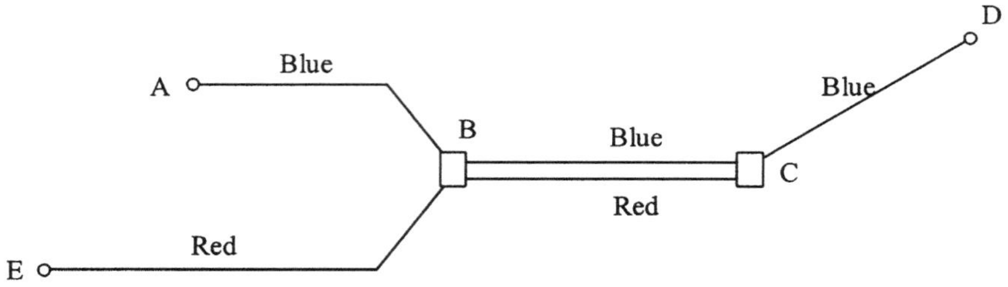

Line Section	*Length (km)*	V_o *(km/h)*	P_{max} *(prs/h)*
AB	3.6	18	1210
BC	4.2	15	2500
CD	2.8	16	1150
EB	6.4	18	1490

a. Begin by computing the headway for each line separately, assuming that the passengers on the joint BC section are distributed equally between the two lines. Then adjust these to meet the uniform headway requirement.

b. Determine the required fleet size, assuming that terminal times at both terminals together amount to at least 10% of the cycle time, and that 12% of all vehicles are in repair of reserve (i.e., the cars on the lines should represent 88% of the fleet).

c. Draw a graphical schedule with the derived values for a period including at least one cycle time of each line. The ordinate of the diagram should start from terminal D, then have CBA at given distances. From point A leave a 1-cm gap and then plot section BE, similar to Figure 1.31.

1.11 A network of three bus lines, AA′, BB′, and CC′, is shown here:

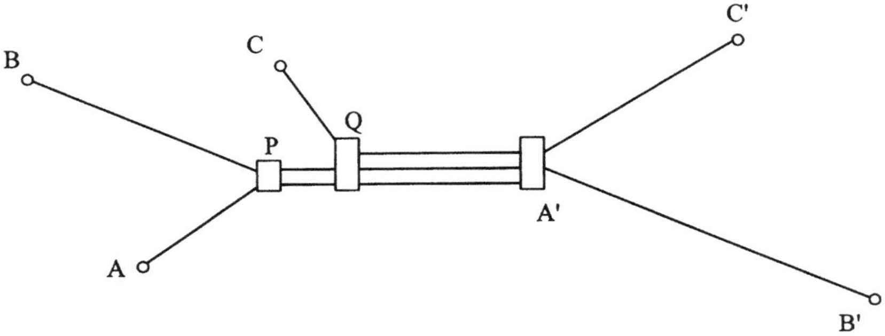

Each section of lines has the following characteristics:

Section	*Length (km)*	V_o *(km/h)*
AP	3	22
BP	6	16
PQ	1.5	13
CQ	1.5	14
QA′	5	13
A′B′	8	16
A′C′	6.5	22

Terminal time (at both ends together) is approximately $t_t = 0.15T$. Passenger volumes on the maximum load section, in persons/hour/direction, are: A-A′ = 980, B-B′ = 1515, and C-C′ = 500.

Two types of buses are available for service:

1. Standard buses with capacity of 53 seats for which the load factor is $\alpha = 1.0$

2. Articulated buses with 80 seats, with $\alpha = 0.9$

The maximum frequency of buses at any point without major delays is 60 veh/h.

a. Compute h for each line for each type of bus.

b. Determine which type of bus would be better for each line on the following basis. For headways longer than 6 min, select the type that provides the shorter headway; for headways shorter than 6 min, passenger waiting time is not a major factor, so that the bus type having the lower cost per seat-km (articulated) should be adopted.

c. Compute T, N, T', and V_c for each line for the selected vehicle type (1 or 2).

d. Compute the average headway on sections PQ and QA′.

e. Suppose that a number of improvements (such as reserved lanes, signal actuation by buses, presale of tickets, etc.) are introduced on section PA′, so that the operating speed is raised from 13 to 18 km/h. Compute how many buses each line would need.

1.12 A network shown below consisting of a trunk line MY with three branches, R, C, and F, and a cross line between two branches RF (similar to the initial San Francisco BART network) has to be scheduled. One-way travel times for each line section are given on the diagram. Compute the schedules for all the lines based on the following headways:

Line MR: 15 min
Line MC: 7.5 min
Line MF: 15 min
Line RF: 15 min

making sure that the headway on each line and on the joint section from Y to M are uniform.

Draw the computed schedules on a diagram using the patterns from Figures 1.33 and 1.34. Show all the trains leaving terminals from 12:00 noon, cutting off the diagram at 15:00. Note that the distance scales vary among the diagrams, so that the slopes of the scheduled lines will be different (but mutually parallel) on each section.

Compute the number of trains required for the service, assuming that terminal times at each terminal must be at least 5 minutes.

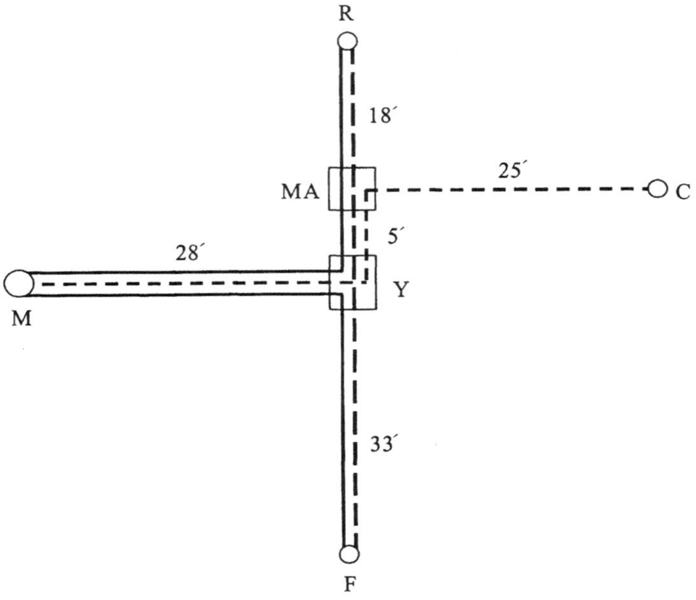

1.13. Suppose that, similar to the network of Green Lines in Philadelphia, four radial streetcar lines, AF, BF, CF, and DF, merge into a 4.6-km-long trunk line leading to the city center:

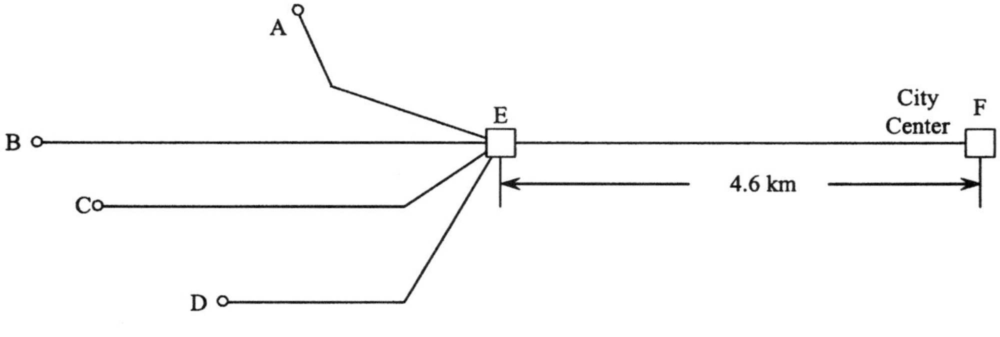

The lines have the following characteristics:

Line	*Branch Length from Outer Terminal to E (km)*	*Maximum Load on Each Branch (prs/h/direction)*
AF	4.8	1920
BF	7.6	1450
CF	8.2	2550
DF	5.4	1200

Each streetcar has a capacity of 125 spaces. Maximum occupancy of vehicles should be 80%.

a. Find headways for each line using only values divisible in 60.

b. Find the average headway on the trunk section (including vehicles from all four lines);

c. If the cycle speed on all lines is 25 km/h, compute cycle times T and find the minimum number of vehicles required for the derived headways on each line.

REFERENCES

Banković, R. 1982. *Javni gradski putnički prevoz* (Urban Public Transportation). Belgrade: Gradjevinska Knjiga.

Bowman, L. A. and A. M. Turnquist. 1981. Service frequency, schedule reliability and passenger wait times at transit stops. *Transportation Research* 15A(6):465–471.

Canadian Urban Transit Association (CUTA). 1993. *The Canadian Transit Handbook.* Toronto: CUTA.

Ceder, A. 1987. Methods for creating bus timetables. *Transportation Research* 21A(1): 59–83.

Ceder, A., and H. Stern. 1981. Deficit function bus scheduling with deadheading trip insertions for fleet size reduction. *Transportation Science* 15(4):339–363.

Fruin, J. J. 1971. *Pedestrian Planning and Design.* New York: Metropolitan Association for Urban Designers and Planners, Inc.

Furth, P. G. 1980. *Setting Frequencies on Bus Routes.* Research Report CTS-RAMP-80-2. Cambridge, MA: Massachusetts Institute of Technology.

Groche, G., and E. Thiemer. 1980. *Handbuch für den Öffentlichen Personen-Nahverkehr-ÖPNV* (Urban Transit Handbook). Darmstadt: Otto Elsner Verlag.

Hoffstadt, J. 1981. Computerized vehicle and driver scheduling for the Hamburger Hochbahn AG. In *Computer Scheduling of Public Transport,* ed. A. Wren, 35–52. New York: Elsevier.

Homburger, W. S. 1982. Urban transit. In *Transportation and Traffic Engineering Handbook,* 2nd ed., ed. W. S. Homburger, Chap. 7. Englewood Cliffs, NJ: ITE/Prentice Hall.

Hulbert, S. 1982. Human factors in transportation. In *Transportation and Traffic Engineering Handbook,* 2nd ed., ed. W. S. Homburger, Chap. 8. Englewood Cliffs, NJ: ITE/Prentice Hall.

Kittelson & Associates. 2003. *Transit Capacity and Quality of Service Manual.* TCRP 100. Washington, DC: Transportation Research Board.

Krstanoski, N. 2000. *Zbirka Rešeni Zadaci po Javen Gradski Patnički Prevoz* (Collection of Solved Problems in Urban Transit). Technical University Bitola, Macedonia.

Lehner, F. 1950. Menge, Arbeit, Leistung and Wirkungsgrad im Verkehr. *Verkehr und Technik* 1 and 2.

———. 1957. *Die Verkehrsbedieunung im Personennahverkehr.* In *Der öffentliche Personennahverkehr,* Risch und Lademann, Chap. 3, 44–143. Berlin: Springer.

———. 1978. (originally published in 1953), *Der maximale Wirkungsgrad des Personaleinsatzes.* Düsseldorf: Alba Buchverlag.

Levinson, H. S. 1992. Urban mass transit systems. In *Transportation Planning Handbook,* ed. J. O. Edwards, Chap. 5. Englewood Cliffs, NJ: Prentice-Hall.

Molinero, A. R. M., and L. I. S. Arellano 1997. *Transporte Publico: Planeacion, Diseno, Operation y Administracion.* Universidad Autonoma del Estado de Mexico.

Rainville, W. S. 1982. (originally published in 1947), *Bus Scheduling Manual: Traffic Checking and Schedule Preparation.* Washington, DC: U.S. DOT and American Public Transport Association.

Robertson, D. ed. 1994. *Manual of Transportation Engineering Studies,* 1st ed. Washington, DC: ITE.

Rüger, S. 1974. *Betriebstechnologie städtischer öffentlicher Personenverkehr.* Berlin: Transpress VEB Verlag für Verkehrswesen.

Vuchic, V. R. 1981. *Urban Public Transportation Systems and Technology.* Englewood Cliffs, NJ: Prentice-Hall.

Vuchic, V. R., et al. 1978. *Transit Operating Manual,* 2nd ed. Harrisburg: Pennsylvania DOT.

2

CAPACITY, SPEED, ACCELERATED AND SPECIAL OPERATIONS

List of Symbols

Symbol	Unit	Definition	Symbol	Unit	Definition
a	m/s^2	Acceleration rate	P_{AB}	prs	Number of passengers who travel between stations AB during a time interval
A	m^2/veh	Standing area in a vehicle	PT, PT_a, PT_t	prs-h	Passenger total, access, and travel time on line
b	m/s^2	Braking rate	R	%	Reliability
b_e	m/s^2	Emergency braking rate	S_c	km	Critical distance on which maximum speed can be reached
b_n	m/s^2	Normal braking rate	S_{to}	km	Distance between passing sidings
h_A, h_{AB}, h_B	min/TU	Headways at A, AB, and B stations, respectively	t_o	min/TU	Time lost prior to opening and after closing of doors
$h_{z\,min}$	min/TU	Minimum headway for zonal operation	t_r	min	Driver's reaction time
n	veh/TU	Number of vehicles in a transit unit	t_{to}	min	Travel time between passing sidings
n_d, n_f, n_o	—	Number of demand, fixed, and on-call stoppings	t_v	min	Travel time at a constant speed
N	—	Number of stops along a line	α_{serv}	prs/sps	Average capacity utilization factor
N_d, N_f, N_o	—	Number of demand, fixed, and on-call stops	σ	—	Assumed area per standee
O_{ij}	prs	Number of passengers originating from station i going to station j	τ_a	s/prs	Alighting time per person
P	prs/km	Number of passengers per kilometer	τ_b	s/prs	Boarding time per person
$P_{A,B}$	prs	Number of passengers who travel between stations A or B during a time interval			

This chapter presents descriptions, analyses, and planning methodology for a number of transit system operational aspects. A discussion of line capacity in the first section is followed by a review of many methods for speed increase and their evaluation. A brief description of different stopping practices then leads to a rather detailed discussion of accelerated operations obtained by different scheduled stopping patterns along a line. The fifth section covers scheduling of special types of transit lines, such as single-track lines, circle lines, and trunk lines with branches.

2.1 TRANSIT LINE CAPACITY

Capacity represents the capability of a transit system to transport a given passenger volume on a transit line under prevailing conditions. It also affects the level of service that the line can offer, because many other performance characteristics, such as speed, reliability, and comfort, depend on the ratio of volume to capacity (either persons in vehicles or vehicles on line). Capacity is therefore one of the basic characteristics of transit systems.

The definitions of capacity and related concepts are given in Section 1.1.5. A detailed analysis of theoretical and practical capacities is presented in Vuchic (1981), Chapter 7. An overview of transit systems capacity is given by Robinson et al. (1992). A very systematic and detailed report on rail transit capacity is given by Parkinson (1996). Kittelson & Associates (2003) present extensive material on capacity of all transit modes.

This section presents a review of the basic capacity concepts, followed by an analysis of transit systems capacity elements and a practical methodology for capacity computations, illustrated by several examples. Finally, a set of tables with theoretical and practical capacities of different transit modes is given.

The discussion of theoretical and practical aspects of capacity is presented to avoid the rather common mistake of treating capacity as a single, fixed number, independent of other transit system performance elements. The presented methodology for capacity computation can be applied to analyses and improvements of existing, and planning of new transit lines for different modes, such as street transit, semirapid, or rapid transit.

The basic concepts relating to transit line capacity, some of which were defined in Chapter 1, are briefly reviewed here for convenience. *Way capacity* is the maximum number of passenger spaces that can be transported in vehicles past a point in one direction per hour without stopping. *Station capacity* is the corresponding number of spaces in vehicles stopping at a station.

Offered capacity is the maximum number of spaces (capacity) in vehicles that can be transported past a point in one direction per hour, while *utilized capacity* is the maximum number of persons who are actually transported per direction per hour. The difference between these two concepts is caused by the fact that usually not all spaces in all vehicles are utilized, due to uneven passenger distribution, excessively long times which would be required for filling all spaces, etc. The ratio of utilized to offered capacity is load factor α, used in many other sections throughout this book.

The capacity computed utilizing minimum headway $h_{\min}$ for a line is *theoretical capacity;* in practice, theoretical $h_{\min}$ cannot be sustained for an extended period of time, so that the average achievable $h_{\min}$ during one hour should be used to compute *practical capacity*. The difference between theoretical and practical capacities depends largely on the regularity of operations and control of transit system: with rapid transit and AGT systems it is not great; with street systems, such as buses or tramways in congested traffic, practical capacity can vary greatly with street conditions.

Each one of these capacity definitions is useful for certain types of analyses, but for most cases in planning, scheduling, and analysis of operations, *practical utilized station capacity* is the most relevant transit line capacity concept.

The basic expression for *offered transit line capacity* is:

$$C = C_v \cdot n \cdot f_{\max} = \frac{60 C_v \cdot n}{h_{\min}}$$

$$\left| \frac{C}{\text{sps/h}} \right| \frac{C_v}{\text{sps/veh}} \left| \frac{n}{\text{veh/TU}} \right| \frac{f}{\text{TU/h}} \left| \frac{h}{\text{min/TU}} \right| \tag{2.1}$$

Descriptively, line capacity is the product of vehicle capacity C_v, the number of vehicles per TU n, and the maximum number of TUs that can pass a point on the line per direction during a unit of time, usually one hour, i.e., the maximum frequency of service $f_{\max}$.

2.1.1 Elements of Line Capacity

Offered transit line capacity can be expressed through its two major components: capacity of a TU (in spaces or persons), given by the numerator in Eq. (2.1), and headway between TUs, given by its denominator. Utilized capacity also includes the load factor.

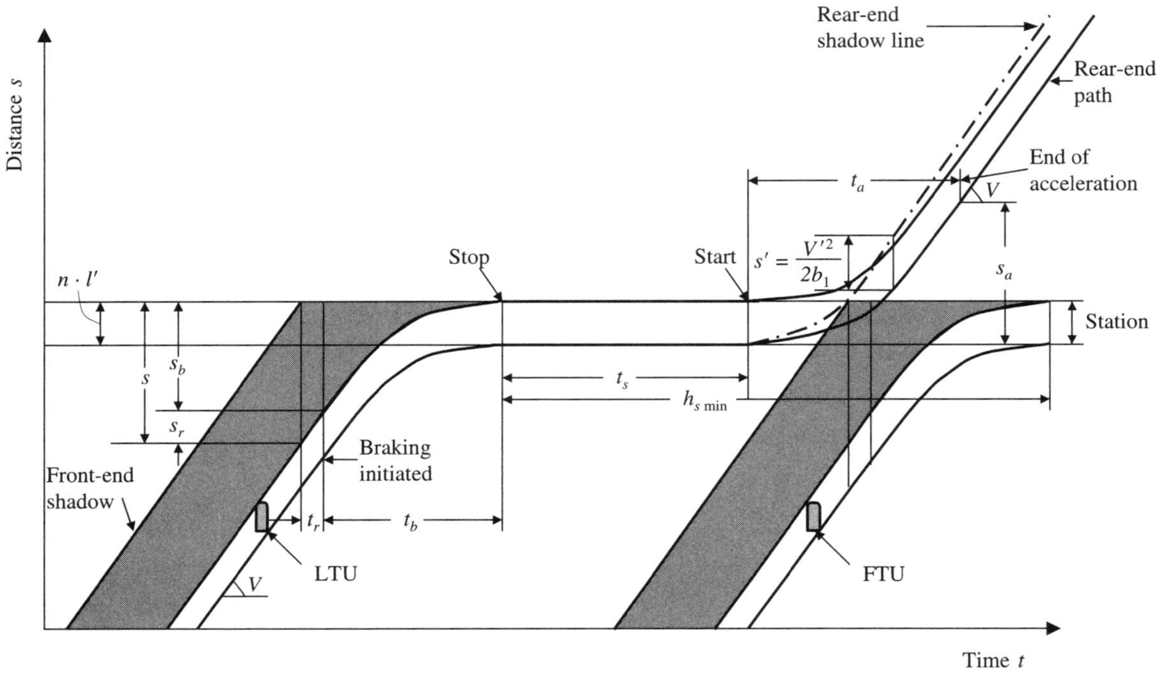

Figure 2.1 *Time-distance elements of departure/arrival sequence and headway between consecutive TUs at a station.*

2.1.1.1 TU Capacity. Three elements comprise this capacity: vehicle capacity, number of cars per TU, and, for utilized capacity, load factor.

Vehicle capacity C_v, as discussed in Section 1.1.5, depends on vehicle dimensions and layout, seat/standing ratio, and the standard adopted for the minimum floor area per person.

Load factor α can reach values close to 1.00 in individual vehicles under crowded conditions, but its maximum average hourly values are lower because the capacity conditions usually last less than one hour. Also, if a line operated with α averaging 1.00 for one hour, that condition would present extremely poor comfort, as well as delays to the passengers who could not board the first arriving TU.

Number of vehicles per TU n, depends on the technology of a transit mode. For highway modes $n = 1$ (with rare exceptions where bus trailers are used). For guided modes n varies from 1 to 10 (exceptionally even more on some regional rail systems), depending on ROW separation, station length, and, of course, passenger volumes, i.e., demand for long trains.

2.1.1.2 Minimum Headway $\mathbf{h}_{min}$. The minimum headway, which determines the maximum frequency of TUs on the line, is by far the most complex element influencing line capacity. On signal-controlled guided transit systems (e.g., metro), that headway is determined by the relationship of two TUs following each other at a station, shown in Figure 2.1. The critical situation is when the leading TU is accelerating from the station while the following TU is entering it and braking, so that its stopping shadow, or its movable stopping distance, just touches the path of the rear end of the leading TU (for a more detailed explanation and derivation of the following equation see Vuchic, 1981).

The minimum station headway (usually the critical one) $h_{s\,\min}$ as a function of speed v, based on this situation, is:

$$h_{s\min} = t_s + t_r + \sqrt{\frac{2 \cdot n \cdot l'}{a} + \frac{n \cdot l'}{v} + \frac{v(K+1)}{2b}} \quad \left|\frac{h,\ t}{\text{s/TU}}\right|\frac{l}{\text{m/veh}}\left|\frac{a,\ b}{\text{m/s}^2}\right|\frac{v}{\text{m/s}}\left|\frac{k}{-}\right| \tag{2.2}$$

This equation shows that the critical headway along the line depends on the following elements:

- Standing or dwell time at station t_s
- Type of control and required safety regime, which affect reaction time t_r and safety factor K (≥ 1); this factor is commonly used in rail transit operations to take into account the conditions when normal braking rates cannot be achieved due to variations in braking rates, etc.
- Length of TU $n \cdot \ell'$, which affects station clearing time of the train
- Acceleration and braking rates, a and b, respectively, and speed of the arriving TUs, v

The expression for offered station capacity C_s under these conditions is obtained when $h_{s\,\min}$ from Eq. (2.2) is introduced into Eq. (2.1) as $h_{\min}$. The plot of this equation with values for elements typical for several different transit modes is given in Figure 2.2. Table 2.1 presents a summary of physical and operational elements typical for the three basic transit modes: bus, light rail, and rapid transit used to compute their respective curves in Figure 2.2. The computed theoretical way and station capacities and the practical station capacities of these three modes are also given. These are only individual samples for each mode; actual capacities of different modes are discussed further in Chapter 12. Operating safety regimes vary among the modes: regime c is driver's control for buses, while regime a for rail systems provides a fail-safe signal system control.

The numbers in Table 2.1 indicate that way capacities are several times greater than station capacities of the same modes. This is because the stations impose much longer $h_{s\,\min}$ than are required between moving TUs, $h_{w\,\min}$. Practical capacities of the three modes, on the other hand, usually amount to about 50% of theoretical station capacities. However, practical capacities vary considerably with local physical and operating conditions, so that it is very difficult, and may be deceptive, to quote a single figure as capacity of a given transit mode.

In practice, headway computations are usually simpler, based on average values of operating parameters measured for existing or projected for new transit lines. Station standing time, the dominant element, consists of the time lost prior to opening and after closing of doors t_o (and for buses in mixed traffic, reentering travel lane), and the time required for boarding/alighting of passengers at the most heavily used (busiest) doors. Thus, if vehicles have boarding and alighting via different doors (e.g., most regular buses with fare control at the entry), the expression is:

$$t_s = t_o + \text{Max}(b' \cdot \tau_b,\ a' \cdot \tau_a) \quad \left|\frac{t}{\text{s}}\right|\frac{b',\ a'}{\text{prs}}\left|\frac{\tau}{\text{s/prs}}\right| \tag{2.3}$$

where b' is the number of boarding persons on the most heavily used entering doors, a' is the number of alighting persons on the most heavily used alighting doors, and τ_b and τ_a are, respectively, boarding and alighting times per person. On the systems with boarding and alighting on all doors, the formula is:

$$t_s = t_o + b' \cdot \tau_b + a' \cdot \tau_a \quad \left|\frac{t}{\text{s}}\right|\frac{b',\ a'}{\text{prs}}\left|\frac{\tau}{\text{s/prs}}\right| \tag{2.4}$$

In this case b' and a' are the numbers of, respectively, boarding and alighting passengers using the busiest door, i.e., the door involving the longest passenger exchange.

The values of b' and a' depend on the volume of boarding and alighting passengers at the station, b_i and

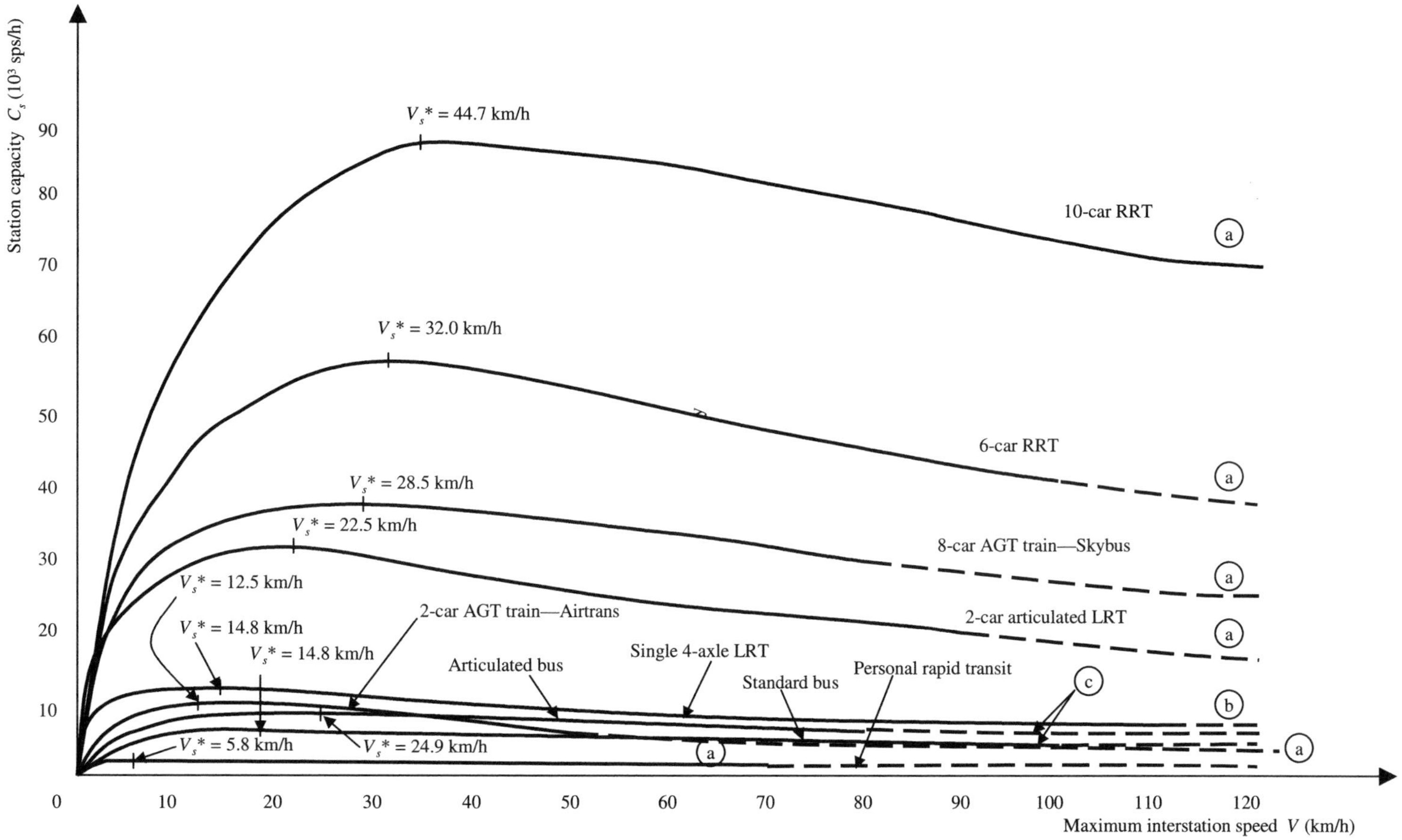

Figure 2.2 *Station capacities of different modes with their typical safety regimes.*

Table 2.1 Theoretical and practical capacities of typical transit modes

Item	*Symbol (dimension)*	*Standard bus*	*Light rail*	*Rapid transit*
Vehicles/TU	n (veh/TU)	1	2	10
Vehicle length	ℓ' (m)	12	24	21
Vehicle capacity	C_v (sps/veh)	53	189	175
Safety distance	s_o (m)	1	2	2
Reaction time	t_r (s)	1	1	0
Normal braking rate	b_n (m/s^2)	1.4	1.2	1.1
Emergency braking rate	b_e (m/s^2)	4.0	3.0	1.8
Maximum speed	V_{max} (km/h)	90	90	120
Operating safety regime	—	c	a	a
Speed for maximum way capacity	$V_w{}^*$ (km/h)	37	39	78
Maximum way capacity	$C_{vw}{}^*$ (sps/h)	53,754	134,350	320,888
Speed for maximum station capacity	$V_s{}^*$ (km/h)	29.7	22.5	44.7
Maximum station capacity	$C_{vs}{}^*$ (sps/h)	6373	30,758	89,950
Practical capacity (see note below)	C (sps/h)	3180	15,120	52,500

Note: The values shown assume average $h_{s\ min}$ of 1, 1.5, and 2.0 minutes for the three modes, respectively. They are realistic, except that at busy stops buses often have two lanes, allowing overtaking, thus averaging $h_{s\ min} < 1$ min and C up to 6,000 sps/h and in special cases much higher.

a_i, on the number of door channels available for boarding and alighting n', and the distribution of passengers among them. They can be computed by:

$$b' = \frac{b_i}{n'} \cdot \xi_b \quad \text{and} \quad a' = \frac{a_i}{n'}\,\xi_a \qquad \left|\frac{b,\ a}{\text{prs}}\middle|\frac{n,\ \xi}{-}\right| \qquad (2.5)$$

where ξ_b and ξ_a are coefficients of passenger distribution among doors, defined as the ratios of the maximum to average number of boarding and alighting passengers per door, respectively. Note that all b's and a's have dimensions of persons only; strictly, they represent numbers of boarding and alighting persons per TU, or those accumulated during one headway.

The values of boarding/alighting times τ and the numbers of boarding/alighting passengers b' and a' are thus critical for the standing times. They vary considerably among transit modes since they depend on level and physical design of station platform (or sidewalk at bus stops), vehicles (floor height, door width, steps, area in vehicle around doors, etc.), operations (one- or two-directional passenger flows and fare collection), volume of passengers on-board, their behavior and control: unorganized pushing, for example, takes longer time than disciplined alighting followed by boarding.

Table 2.2 presents an overview of major factors influencing operations and dwell times at stops/stations for different transit modes and vehicle types. The included factors are:

- *Vehicle floor height and platform height:* conventional bus and streetcar vehicles with high floors and low platforms require longer boarding/alighting times than the modern low-floor vehicles. The fastest boarding is on metro systems with high car floors and high platforms.
- *Number of boarding/alighting channels* n' is computed as $n' = n_{TU} \cdot m \cdot$ (1 or 2 channels per door), where n_{TU} is the number of vehicles per TU and m is the number of doors per vehicle.
- *Fare type and fare collection* range from the traditional manual payment of cash fares when boarding on one door to faster electronic processing, validation, and, finally, no fare collection on platform or in vehicles (metro systems).

The subsequent columns in the table give typical values of τ_b and τ_a for six cases for different modes,

Table 2.2 Physical and operating elements affecting TU standing times

Transit mode and TU sketch	*Vehicle floor/ platform height*	*Cars × doors × channels = TU total (board + alight)*	*Fare type/ collection*	$t_o(s)$	*Boarding time* $\tau_b(s/prs)$	*Alighting time* τ_a *(s/prs)*	Difference from preceding cases
i. Bus 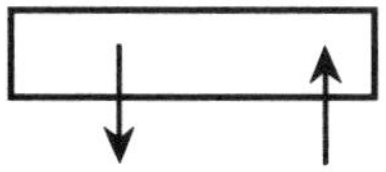	*H/L*	1 × 2 × 1 = 2 (1 + 1)	Graduated/at boarding	1.0	3.0–5.0	1.5–3.0	—
ii. Bus/streetcar 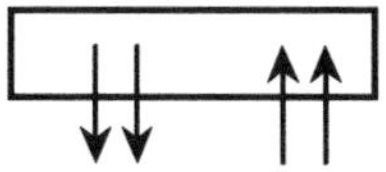	*H/L*	1 × 2 × 2 = 4 (2 + 2)	Flat, passes, validation	1.0	2.0–3.0	1.5–3.0	More channels, faster fare collection
iii. Bus/streetcar 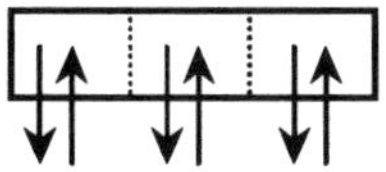	*L/L*	1 × 2 × 2 = 4 1 × 3 × 2 = 6	Self-service, validation	2.0	1.2–2.2	1.0–1.6	Low floor, faster fare control
iv. Articulated bus, single LRV 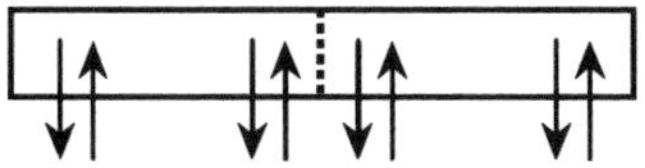	*H/L*	1 × 4 × 2 = 8	Self-service, validation	2.0	1.6–2.6	1.5–3.0	More channels but high floor
v. LRT train 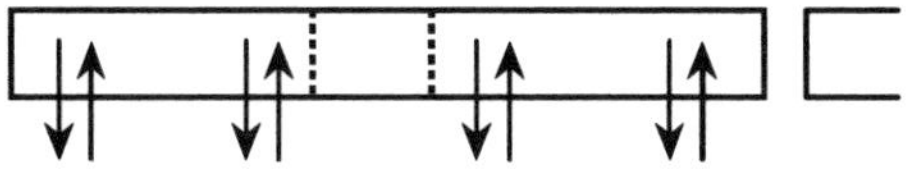	*L/L* *H/H*	2 × 4 × 2 = 16 3 × 4 × 2 = 24	Any type, prepaid	3.0	0.8–1.6	0.6–1.4	More channels, low floor; no fare collection
vi. Metro 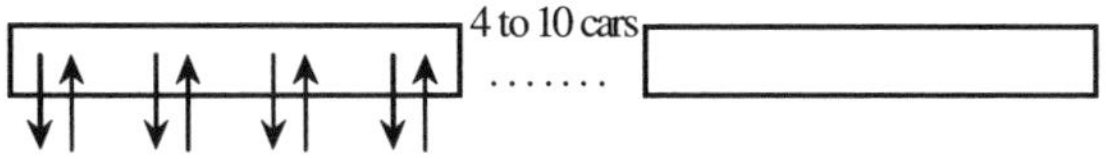	*H/H*	4 × 3 × 2 = 24 10 × 4 × 2 = 80	Any type, prepaid	3.0	0.6–1.4	0.5–1.2	More channels

Note: Bus includes trolleybus.

vehicles, fare collection, and other elements. The given average values of τ vary from as much as 5.0 s/prs for a bus with a single channel door and collection of graduated fares during the boarding, to 0.5 s/prs for boarding or alighting on rapid transit trains with high platforms, multiple two-way double-channel doors, and no fare collection or ticket validation on trains.

It is emphasized that the numbers in Table 2.2 are not exact ranges that can be applied directly to any given situation. They are given to illustrate influences of various factors and the method to use such numbers in planning transit system operations and analyzing alternative designs and operational elements. For example, it may be of interest what impact on dwell times would be achieved by introduction of low-floor LRT vehicles, introduction of self-service fare collection, or boarding/alighting through many double-channel doors along a train.

Local conditions always have a great role in transit operations. It is therefore necessary to acquire data regarding such factors as types of vehicles and their doors, fare collection, supervision of departure, etc., to obtain the values which are then used in planning or testing operational alternatives.

For illustration of the use of factors listed in Table 2.2, two examples are presented. In the examples, medium values from the ranges given in the table for all time intervals are used.

EXAMPLE 2.1

Planning data for a bus line show that stop number 17 will have the highest average passenger volumes per bus: b_{17} = 12 prs, a_{17} = 19 prs. To determine line capacity, it is necessary to predict t_s for this stop. Buses have two single-channel doors, graduated fares are collected at boarding, and there are no prepaid tickets. Alighting is slow because it requires pushing the door by passengers while stepping down. A significant portion of passengers at the stop are elderly.

Using Eq. (2.3) and maximum values from the ranges for τ's in case i, Table 2.2, because of described conditions, yields the following result:

$$t_s = 1 + \text{Max}(12 \cdot 5.0,\ 19 \cdot 3.0) = 61 \text{ s}$$

EXAMPLE 2.2

Station 7, which is critical for the capacity of a streetcar line, has low platforms. Single high-floor cars with two double-channel doors are operated; some passengers pay or validate tickets on board (case ii). Measurements have shown that the distribution of passengers between the two doors is equal, $\zeta = 1.0$. Average boarding and alighting volumes per TU are b_7 = 52 prs and a_7 = 63 prs, respectively, and presently the average standing time is t_s = 65 s.

A major office complex is being planned in the immediate vicinity of station 7. It is expected that passenger volumes during the peak hours will increase by 60%. The coefficient of distribution among doors for both boarding and alighting will be $\zeta_a = \zeta_b = 1.1$. Several measures for improving station operations are considered, so that three cases of station operation should be analyzed:

- **a.** Continued present operation—case ii in Table 2.2
- **b.** Fares are collected before boarding, TUs consist of single articulated high-floor cars, with four double-channel doors each—case iv in Table 2.2
- **c.** Same as b, but low-floor cars and two-car TUs (case v)

What would be the value of t_s for these cases?

First the information about present operations should be used to determine values of individual time parameters. Since the present conditions correspond to case ii in Table 2.2, average values for τ's will be taken from it and placed in Eq. (2.5):

$$b' = \frac{52}{2} = 26 \text{ prs} \quad \text{and} \quad a' = \frac{63}{2} = 31.5 \text{ prs}$$

introducing these values and t_o = 1 s into Eq. (2.3) yields:

$$t_s = 1 + \text{Max}(26 \cdot 2.5,\ 31.5 \cdot 2.2) = 70.3 \text{ s}$$

which is within 10% of the actually measured t_s; the selected values of parameters are thus considered valid. If there were a major discrepancy between the measured and computed values, different parameters would have to be used. With this validation of the equations, t_s will be computed for the three cases.

a. Continued present operation—still case ii. The same computation as above, including factor of 1.6 for 60% more passengers, gives:

$$b' = \frac{52 \cdot 1.6}{2} = 41.6 \text{ prs} \quad \text{and}$$

$$a' = \frac{63 \cdot 1.6}{2} = 39.2 \text{ prs}$$

$$t_s = 1 + \text{Max}(41.6 \cdot 2.5, 39.2 \cdot 2.2)$$
$$= 111.9 \text{ s}$$

b. Case iv: The number of door channels increases from four to eight, and each channel is used for both alighting and boarding. Applying Eq. (2.5),

$$b' = \frac{52 \cdot 1.6}{4 \cdot 2} \cdot 1.1 = 11.44 \text{ prs} \quad \text{and}$$

$$a' = \frac{63 \cdot 1.6}{4 \cdot 2} \cdot 1.1 = 13.86 \text{ prs}$$

Introducing these values and $t_s = 2$ s into Eq. (2.4),

$$t_s = 2 + 11.44 \cdot 2.1 + 13.86 \cdot 2.2 = 56.5 \text{ s}$$

c. Case v: The number of passengers per door channel decreases because the number of channels is $2 \cdot 4 \cdot 2 = 16$, instead of 8. Thus, $b' = 5.72$ prs, and $a' = 6.93$ prs. Standing time is thus:

$$t_s = 2 + 5.72 \cdot 1.2 + 6.93 \cdot 1.0 = 15.8 \text{ s}$$

Consequently, if the present system is retained, the 60% increase in passenger volumes would increase t_s from 71 s to 112 s. Elimination of all delays due to fare collection or ticket validation at boarding and introduction of articulated car with a total of eight instead of four door channels would result in a drastic shortening of t_s to 57 s. Finally, the last option foresees the same type of operation as in case b, but with low-floor cars and two-car TUs, allowing at level, faster boarding/alighting. This change would bring standing time down to 16 s.

It is emphasized again that due to numerous randomly distributed influences on t_s, including control system and organizational habits of drivers and passengers, the numbers given here are not exact. They should be used judiciously, with careful evaluation of local conditions and, whenever appropriate, organization of actual observations and measurements. However, even this simple example with approximate numbers shows that car design and type of operation at stations can have major influence on TU dwell times and thus on the operating speed and capacity of the line.

2.1.1.3 Review of the Capacity Elements. Since station standing time is, in some ways, less productive than travel time, it is always desirable to minimize it. Lightly traveled bus lines can operate with a single door for boarding, with τ_b = 2–4 s/prs, due to small boarding volumes. Thus, for $b' = b_i$ = 3–5 prs, standing times can be only t_s = 12–30 s. However, because of large τ's, t_s rapidly increases with b'; for large passenger volumes such bus lines, with traditional single-channel boarding and manual fare collection, are unattractive and uneconomical because of their low speeds and low reliability.

For modes serving large volumes—articulated bus, light rail, and, ultimately, metro—boarding conditions are progressively more efficient and the number of doors must be many times greater. Thus, at the upper limit, metro trains with 10 cars can have up to 40 double-channel doors and their boarding/alighting times may be as short as $\tau_b = \tau_a = 0.5$ s/prs, so that even if only half of the doors are fully used because of uneven passenger distribution or heavy alighting volume, a total of up to 80 prs/s can be boarded.

Therefore, dwell times can be maintained at 25–40 s even for very high passenger volumes.

The preceding analysis of the influencing elements shows that among many factors influencing capacity, several have a particularly important role. If an increase in capacity is desired, it can most effectively be achieved by changes in the following three elements.

First, an *increase in TU capacity* (C_v and/or n) is most effective because it increases line capacity nearly linearly. The greater TU length has little influence on the length of h_{min}, so that f_{max} is usually unaffected and C increases as a direct function of C_{TU}. Thus, replacement of standard buses by articulated buses can increase line capacity by about 50% without changing frequency; the capacity of a rail line with two-car trains can be virtually doubled if four-car trains are introduced. Actually, very high capacity of transit lines is most effectively achieved by usage of long trains of very large cars: rail transit systems that operate up to 10-car trains with cars up to 23 m long and 3.20 m wide provide by far the highest line capacity of all modes. Moreover, such operation of electrically powered signal controlled trains is greatly superior in performance, economic efficiency, reliability, and safety to other modes with more frequent operation of smaller units.

Second, *shortening of headway* h_{min} by decreasing t_s through such measures as an increase of door-to-capacity ratio on the cars, improvements of platforms, fare-collection procedure, door supervision, and dispatching of TUs from stations. These measures are particularly effective on high-capacity rail modes.

Third, when smaller vehicles, such as buses, must provide high capacity, their throughput at stops along the line can be greatly increased by such operations as:

- *Multiple stops for simultaneous stopping* without overtaking. This operation, usually applied on bus and trolleybus, but sometimes also on rail lines, increases capacity, but it may lower operating speeds because the TUs with longest dwell times dictate the length of dwell times at each stop.
- *Leapfrogging,* or mutual overtaking of vehicles that stand at stops. This operation requires two wide lanes per direction and extra long stop frontages.
- Some TUs (buses or even trains) operate as *expresses,* bypassing those stopped at stations that have extra lanes or tracks for that purpose.
- *Platooning,* or convoy operation of buses. Vehicles from different lines are collected in a special staging area into groups of two to five buses and then travel and stop following each other closely.

2.1.2 Capacity Computations

There is often a strong tendency, even in professional circles, to quote capacities of transit modes as single numbers. That is, however, a simplistic and seldom accurate approach for two reasons. First, since many elements influence transit system performance, as was shown above, line capacity may vary greatly, depending, for example, on vehicle characteristics (bus type, size, seating capacity), type of stations, fare collection, etc. It is therefore impossible to give a single number for capacity of a bus line. Second, even for a given line and all vehicle, operating, and other characteristics, capacity can vary substantially depending on other performance measures, such as comfort, safety, and reliability.

2.1.2.1 Relationship between Capacity and Other Performance Elements. For an accurate computation of line capacity, utilizing Eq. (2.1), a careful analysis of the parameters C_v, α, and h_{min} (f_{max}) and various factors that influence them must be made. It is through these parameters that capacity interrelates with other level-of-service and performance elements.

The adopted minimum area per standee σ represents the comfort standard and is traded off against capacity C_v (and C): greater values of σ result in higher comfort but lower capacity, and vice versa. Load factor α expresses the difference between offered and utilized capacity and thus also reflects passenger comfort and convenience. This factor depends on variations of passenger volume within the peak hour, distribution of boardings at the critical station, acceptable delays to passengers, and other factors. For example, tolerating

Table 2.3 Comparison of capacities and related performance elements on a transit line*

Performance Elements \ Transit Systems		*i. 1920s Streetcar—Typical*	*ii. 2000s LRT—Typical*	*iii. 2000s LRT—Maximum Performance*
Comfort standard	σ (m²/sps)	0.15	0.25	0.15
TU capacity	$C_{TU} = C_v \times n$ (sps)	$120 \times 2 = 240$	$190 \times 3 = 570$	$271 \times 3 = 810$
Offered line capacity	$C = C_{TU} \times f_{max}$ (sps/h)	$240 \times 100 = 24{,}000$	$570 \times 45 = 25{,}650$	$810 \times 45 = 36{,}450$
Load factor	α (prs/h)	0.92	0.86	0.90
Utilized capacity	$C_p = \alpha \times C$ (prs/h)	22,080	22,059	32,802
Reliability	R (%)	80	94	90
Operating speed	V_o (km/h)	12	24	20
Utilized productive capacity	$P_c = C_p \times V_o$ (prs-km/h²)	264,960	529,416	656,100

*Values in this table are typical for the first two cases (1920s and 2000s); the third case is estimated through extrapolation of conditions for the preceding two cases.

that some passengers may not be able to board the first TU increases α.

The assumed (or measured) value of h_{min} is even more complex: it may depend on operating speed, adopted safety standards, TU length, and terminal operations with TUs. Moreover, there is a direct trade-off between h_{min} and service reliability R, defined as percent of TU departures within a certain time interval (usually four to five minutes) after the scheduled times. It is through h_{min} that the trade-offs among line capacity and speed, reliability of service, safety, and operating cost are introduced. Consequently, to present a complete description of conditions under which the capacity is achieved, the following set of elements should be given:

- Offered capacity C (sps/h)
- Operating speed V_o (km/h)
- Comfort standard σ (m²/sp)
- Reliability R (%)
- Load factor α (prs/sp)

The need for this greater precision of capacity characteristics can be illustrated by the following example. Records show that during the 1920s a typical streetcar line in Boston, Frankfurt, or Vienna could operate with approximately the following performance elements: frequency of 100 TU/h, each TU offering 240 spaces with $\sigma = 0.15$ m²/sp, $\alpha = 0.92$, and $V_o = 12$ km/h. Utilized line capacity was thus very high, 22,080 prs/h, but reliability of service was estimated at only 80%. After major upgrading of such lines, today's LRT lines in these cities typically operate with f_{max} of only 45 TU/h, but their TU capacities are $3 \cdot 190 = 570$ sps; each car has about 70 seats and 120 standing spaces, assuming upgraded comfort standards of $\sigma = 0.25$ m²/sp. If $\alpha = 0.86$ prs/sp, their utilized capacity is 22,059 prs/h.

Since the capacities of the two systems are very similar, their direct comparison alone would not show any impact of streetcar upgrading into LRT. However, a comprehensive comparison through all the performance elements gives a very different picture. To allow such a comparison, Table 2.3 presents operational data for three cases: i, 1920s streetcar operation; ii, typical contemporary (2000s) LRT operation; and iii, the maximum performance that the LRT could produce if a need for such capacity would arise. The table shows two very important changes between the streetcar and LRT systems:

1. Comparison of cases i and ii shows that modern LRT lines offer a service that has a much higher comfort (σ is 67% greater), higher reliability (separate ROW and less congested vehicles have fewer delays), and twice the speed of their 1920s predecessors.
2. Case iii shows that if different conditions, such as increased ridership, special events, or emergencies, require, service standards σ and α can be brought closer to the old levels, yielding much higher capacity and productivity values. Utilized capacity of 32,805 prs/h could be achieved. Thus, the LRT service has not only much higher quality, but also a 49% reserve capacity that could be utilized if service standards were lowered to the levels from the 1920s. Speed and reliability of service under such load would be lower than those of LRT under present conditions, and yet considerably higher than those of the old streetcars due to higher vehicle performance and separate ROW (B and A instead of C). Productive capacity of the line would be about 2.5 times greater than in the 1920s.

2.1.2.2 Methodology for Capacity Computations with Examples. As an important performance element, transit line capacity is computed in different cases, such as (a) for an existing line and given conditions; (b) for an existing line with some planned changes; (c) for a planned line. The methodology among these cases varies in that some factors can be measured in case (a), measured and modified in (b), or estimated/predicted in case (c). Yet the procedure for all the cases consists of the same basic steps and sequence of computations, outlined on the next page.

Application of this methodology to compute transit line capacity and related performance elements is illustrated here by two examples.

EXAMPLE 2.3

A portion of a bus network is to be reorganized so that several lines will converge into a joint section in the CBD. They are operated by standard buses with two single-channel doors, 35 seats, and 12 m^2 of standing area. The busiest stop on the joint section will have a single stopping location and serve during the peak 15-minute period 25 boarding and 30 alighting passengers per bus. Measurements on similar lines in the same city show that on the average buses have a 2-s delay for opening and closing the doors; boarding is via front door and takes 2.0 s/prs, while alighting via rear door requires 1.6 s/prs. Average operating speed of buses is 14 km/h and reliability has been recorded to be 75% when the average headway between departure of a bus and arrival of the next one to a stop is 18 s, increasing to 82% if this headway averages 32 s. Compute capacities and related performance elements on this joint section adopting $\sigma = 0.25$ m^2/sp, $\alpha_{max} = 0.85$ prs/sp, and $R = 80\%$.

Solution

Step 1: The data are: $C_s = 35$ seats; $A = 12$ m^2; $b = 25$ prs; $a = 30$ prs

Step 2: $\tau_a = 1.6$ s/prs; $\tau_b = 2.0$ s/prs; $V_o = 14$ km/h; $R_{18} = 75\%$; $R_{32} = 82\%$

Standing time at the busiest stop is computed by Eq. (2.3):

$$t_s = 2 + \text{Max}\,(25 \cdot 2.0,\ 30 \cdot 1.6) = 52 \text{ s}$$

Assuming linear change of R with h_{min} within the observed range, interpolation gives that for $R = 80\%$ a headway between buses of 28 s should be planned for; thus, $h_{min} = 52 + 28 = 70$ s.

Step 3: $\sigma = 0.25$ m^2/prs; $\alpha_{max} = 0.85$

Step 4: Vehicle capacity is: $C_v = 35 + 12/0.25 = 83$ sps

Step 5: Line capacities are

$$C = \frac{3600 \cdot 83 \cdot 1}{70} = 4269 \text{ sps/h and}$$

$$C_p = 0.85 \cdot 4269 = 3629 \text{ prs/h}$$

Step 6: Performance of the joint line section is thus:

Procedure for computing transit line capacity:

- *Step 1:* Collect the following data:
  - Vehicles: seating capacity C_s (sps/veh), standing area A (m^2/veh), and number of door channels n' (chan/veh)
  - Maximum number of cars per TU n_{max} (veh/TU);
  - Maximum passenger volume P_{max} and peak hour coefficient PHC
- *Step 2:* Observe, measure, compute, or predict:
  - Boarding and alighting unit times τ_b, τ_a (s/prs)
  - Minimum average headways at all potentially critical points, such as major stations (use Eqs. (2.3–2.5)), intersections, terminals; select the longest one Max h_{min} (s/TU)
  - Maximum line frequency $f_{max} \left[\frac{TU}{h}\right] = \frac{3600}{\text{Max } h_{min}}$
  - Operating speed V_o (km/h)
  - Reliability of service R (%).
- *Step 3:* Adopt standards for comfort σ (m^2/sp) and utilization α_{max} (prs/sp)
- *Step 4:* Compute vehicle and TU capacities:

$$C_v \left[\frac{\text{sps}}{\text{veh}}\right] = C_s + \frac{A}{\sigma}$$

- *Step 5:* Compute offered and utilized capacities:

$$C \left[\frac{\text{sps}}{\text{hr}}\right] = f_{max} \cdot C_{TU} = \frac{3600 \cdot C_v \cdot n_{max}}{\text{Max } h_{min}} \quad \text{and} \quad C_p \left[\frac{\text{prs}}{\text{h}}\right] = \alpha_{max} \cdot C$$

- *Step 6:* Summarize the capacity data: C, C_p, σ, α_{max}, V_o, and R.

C = 4,269 sps/h; C_p = 3,629 prs/h; σ = 0.25 m^2/sp
α_{max} = 0.85 prs/sp; V_o = 14 km/h; R = 80%

EXAMPLE 2.4

A planned metro line is projected to have peak 15-minute boarding and alighting passenger volumes at Monument, the most heavily used station, respectively, $b^M{}_{15}$ = 2580 prs and $a^M{}_{15}$ = 1850 prs. The line will be operated by six-car trains, each car with three double-channel doors and C_v = 180 sps. Boarding and alighting times will be $\tau_b = \tau_a$ = 1.1 s/prs, coefficient of passenger distribution is ξ = 1.3, t_o = 3 s, and the minimum interval between starting of one train from the station and stopping of the next one is 60 s. At the terminal stations the minimum headway for turning trains back is $h^T{}_{min}$ = 110 s. What will be the offered line capacity and what will be its utilized capacity if on the MLS α = 0.90?

Solution

Step 1: The data are: C_v = 180 sps/car; n_{max} = 6 veh/TU; m = 3 doors/car, n' = 2 chan/door; $b^M{}_{15}$ = 2580 prs/15 min, and $a^M{}_{15}$ = 1850 prs/15 min

Step 2: $\tau_b = \tau_a = 1.1$ s/prs; $\xi = 1.3$; $t_o = 2.0$ s; $h^T{}_{\min} = 110$ s

To find which location along the line is critical for capacity, use the minimum headway at the terminal $h^T{}_{\min}$ to compute passenger volumes/TU at the Monument station, b_{TU} and a_{TU}, and then the minimum headway at that station, $h^M{}_{\min}$. This computation can be most conveniently made via $f_{\max}$ for 15-minute period:

$$f^T_{\max} = \frac{3600}{4h^T_{\min}} = \frac{3600}{4 \cdot 110}$$

$$= 8.2 \approx 8 \text{ TU/15 min}$$

Since $\tau_b = \tau_a$, computations can be made for b_i and a_i together:

$$b_{TU} + a_{TU} = \frac{2580 + 1850}{8}$$

$$= 554 \text{ prs/15 min/TU}$$

With $n = 6$ and $m = 3$, Eqs. (2.5) and (2.4) give the critical volume per door as:

$$b' + a' = \frac{b_{TU} + a_{TU}}{n \cdot m \cdot n'} \cdot \xi$$

$$= \frac{554}{6 \cdot 3 \cdot 2} \cdot 1.3 = 20 \text{ prs}$$

The critical standing time and the minimum headway are then, respectively:

$t_s = t_o + (b' + a') \cdot \tau_b = 3 + 20 \cdot 1.1 = 25$ s and $h^M{}_{\min} = t_s + 60 = 85$ s

Since $h^M{}_{\min} < h^T{}_{\min}$, the latter governs; thus, $h^T{}_{\min}$ is critical.

Step 3: $\alpha_{\max} = 0.9$

Step 4: $C_{TU} = 6 \cdot 180 = 1080$ sps/TU

Step 5: The offered and utilized line capacities (based on Eq. (2.1)) are, respectively:

$$C = C_v \cdot n \cdot 4f^{15}{}_{\max} = 180 \cdot 6 \cdot 4 \cdot 8.2$$

$$= 35{,}345 \text{ sps/h}$$

$$C_p = \alpha_{\max} \cdot C = 0.9 \cdot 35{,}345 = 31{,}810 \text{ prs/h}$$

Step 6: $C = 35{,}345$ sps/h; $C_p = 31{,}810$ prs/h; $\alpha_{\max} = 0.9$

σ, V_o, and R are not included in this question.

2.1.3 Systems Approach to Transit Line Capacity

If a metro line is observed as a system, its ability to transport passengers is actually determined by two types of capacity:

- Line transporting capacity in TUs, spaces, or passengers carried per hour
- Passenger entrance/exit capacities at stations along the line

These two capacity concepts, shown in Figure 2.3, are computed in different ways. While the line capacity is analyzed as a continuous system for the entire line, each station must have sufficient capacity to accommodate its passenger volumes, and these capacities vary among stations.

The line transporting capacity should always be the critical one because the line—its tracks, stations, and trains—performs the transportation work. Therefore, the capacity discussed until now refers to line capacity and its critical locations, usually the station with the largest exchange of passengers, causing the longest minimum headway. This "busiest station" may be one in center city, an outlying terminal, or a station with transfer to other metro lines. In some cases, one of the terminal stations may be critical if it has only two tracks and its reversing maneuver takes longer time than station dwells at any other station along the line. However, to ensure that the line capacity is achieved, the station capacities must be designed so that they can handle their entering/exiting passenger volumes to serve the trains without delays.

A model for station capacity analysis is shown in Figure 2.4. It consists of a series of passenger flows,

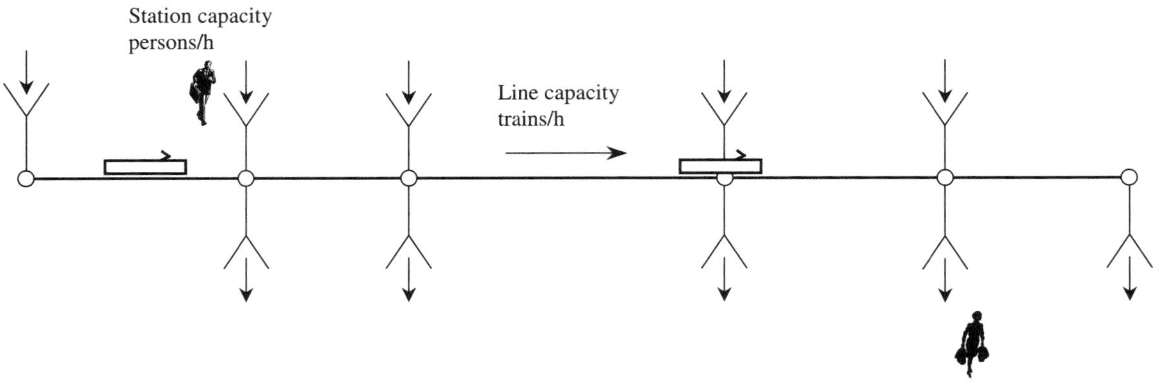

Figure 2.3 *Concepts of line capacity and station capacity.*

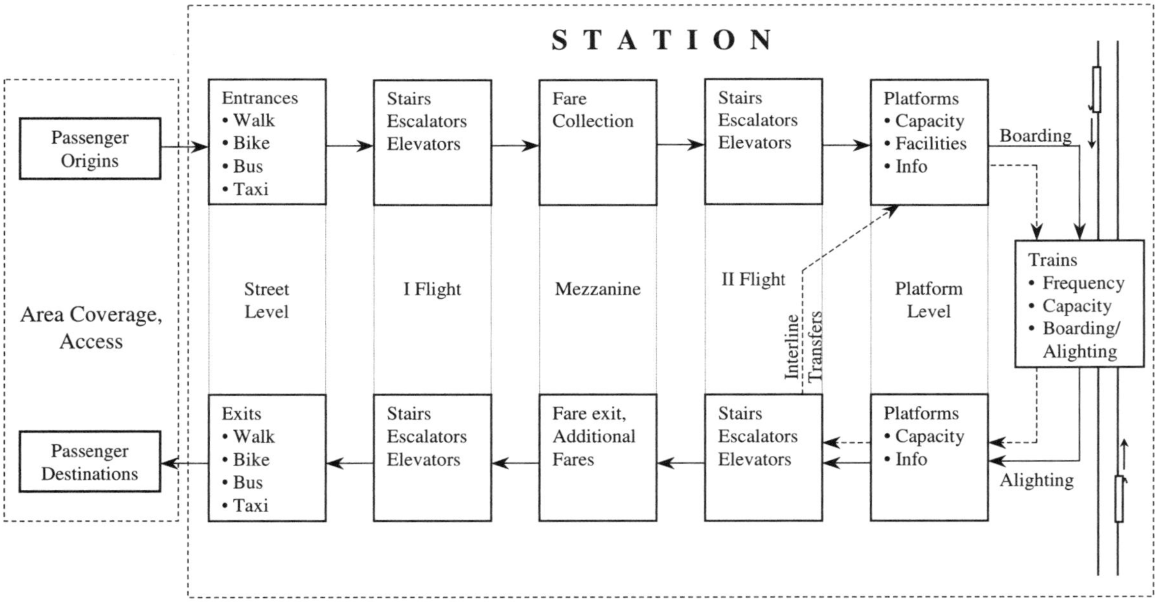

Figure 2.4 *Analysis of station capacity—passenger throughput.*

from the station access at the street through entrances, stairs, and escalators to the mezzanine and then platform level, where this capacity interacts with the line capacity that trains provide. At transfer stations a special analysis must be made for passenger volumes transferring between the lines, whether at different levels (as shown) or across the platform.

This type of station capacity analysis can be used for testing system capacity on existing lines when it is expected that passenger volumes will increase. Actually, the specific model in Figure 2.4 was developed by this author for application in the core capacity studies that the Boards of the Washington Metro and San Francisco BART initiated to examine the capability of the existing lines in central cities to handle the increasing passenger volumes as radial lines are extended further in the suburbs. One of the findings was that in Washington the bottlenecks on the lines were the stations where the lines intersect due to the large volumes of transferring passengers.

2.1.4 Capacities of Different Modes

It is often convenient to use round figures as capacities typical for each transit mode. Yet, as mentioned in Section 2.1.1, capacities may vary considerably. For example, it is often said that rapid transit has capacity of 40,000 prs/h, but the fact is that some metro lines (Boston, Lyon) have capacities of less than 20,000, while others (New York, Hong Kong) exceed 70,000 sps/h. It is therefore important to understand the factors that determine capacity, to know the method for its analysis, and to include all the other relevant performance elements in the capacity analysis.

A summary of all major elements relevant to capacity, as well as ranges of capacities for several sets of characteristics of all main transit modes, is given in Table 2.4. The table presents single values or their ranges, as appropriate, for the following elements:

- Vehicle dimensions: length L and width W, and TU capacity C_{TU}
- Minimum headway $h_{\min}$ (or maximum frequency $f_{\max}$)
- Maximum offered line capacities C (rounded to 100s) and their ratios to the capacity of a standard bus line, used as a base value. The ratios are between the middle values of the respective capacity ranges.
- Operating speed V_o and offered productive capacity P_c. The lower value of each P_c range is computed as the product of the lower values of V_o and C, increased by 20%; the upper values represent the product of the upper values of V_o and C, decreased by 20%. These 20% adjustments were made to avoid the unlikely coincidence of the extreme values on any one transit system
- Productive capacity ratios of modes to that of the standard bus line, computed in the same way as the ratios of line capacities

The numbers given in Table 2.4 are for typical vehicle dimensions, TU sizes, and well-organized operation. Local conditions in any one case may differ and result in lower or higher capacities than those given in the table. Major deviations in specific cases should be computed as follows.

- Offered line capacity increases linearly with vehicle dimensions and TU size. For example, LRT line operated by single 25-m-long articulated cars offer approximately 25% higher capacity than if 20-m long cars are used. Operation of two-car trains results in nearly 100% higher offered capacity than operation with single cars.
- Most vehicles in the table are assumed to have seating/standing ratios typical for the respective modes. For minimum seating (longitudinal) capacities can be considerably greater, for maximum seating somewhat lower than most given numbers.
- Wherever sufficient data were available, vehicle capacities have been based on 5 sps/m^2 ($\sigma = 0.20$ m^2/sp). For conditions where lower comfort standards can be used, such as 6 or 7 sps/m^2, C_v should be recomputed accordingly.
- Capacities can also be greater than shown in the table where special operational arrangements are made to decrease $h_{\min}$, such as operations of buses in convoys, very efficient departure control for LRT and metro trains at stations, signals that allow reduction of $h_{\min}$ in metro stations, etc.

Capacity of each major transit mode is briefly described here.

Standard buses and trolleybuses (nonarticulated) traveling in a single lane and having single stop locations (case 1 in the table) can operate at minimum headways as short as 40 s, but at heavily used stops, which are critical for $f_{\max}$ on the line, they usually average at least 50 s; including the clearance interval between buses, it is seldom possible to achieve less than $h_{\min}$ = 60 s, i.e., $f_{\max}$ = 60 veh/h. With C_v = 75–90 sps, depending on seating arrangement and

Table 2.4 Maximum offered line capacities of transit modes

Mode	Vehicle dimensions $L \times W$ (m)	TU capacity $n\,C_v = C_{TU}$ (sps)	Minimum headway h_{min} (s)	Maximum frequency f_{max} (TU/h)	Max offered line capacity C (sps/h)	Ratio to basic bus capacity C_{av}/C_{av}^{bus}	Operating speed at capacity V_o (km/h)	Offered productive capacity P_c (10^3 sp-km/h²)	Ratio to basic bus productive capacity P_c/P_c^{bus}
1. Standard bus, single stops	12.00 × 2.50	1 × 75 = 75	70–50	51–72	3800–5400	1.0	8–12	36–52	1.0
2. Articulated bus, single stops	18.00 × 2.50	1 × 120 = 120	80–60	45–60	5400–7200	1.4	7–11	45–63	1.2
3. 50% standard, 50% articulated bus multiple stops	12.00 × 2.50 18.00 × 2.50	1 × 75 = 75 1 × 120 = 120	40–20	90–180	8800–17,500	2.9	10–16	105–225	3.7
4. High-capacity bus	12.00 × 2.50 22.00 × 2.50	1 × 75 = 75 1 × 160 = 160	30–12	120–300	9000–30,000	4.2	16–20	173–360	6.1
5. Streetcar ROW C, multiple stops	14.00 × 2.20	2 × 110 = 220	60–30	60–120	13,200–26,400	4.3	8–14	158–296	5.2
6. Light rail transit, ROW B, multiple stops	24.00 × 2.65	3 × 170 = 510 2 × 280 = 560	150–75	24–48	12,200–26,900	4.2	18–30	280–622	10.2
7. AGT—VAL, Westinghouse	12.50 × 2.05	4 × 86 = 344	90–75	40–48	13,800–16,500	3.3	20–36	330–475	9.1
8. AGT—rail	12.70 × 2.50	6 × 100 = 600	100–90	36–40	21,600–24,000	4.9	20–36	518–690	13.7
9. Rapid transit	18.00 × 2.90	8 × 180 = 1440	100–90	36–40	51,800–57,600	8.6	22–36	988–1200	24.9
10. Rapid transit	21.00 × 3.15	10 × 240 = 2400	150–120	24–30	67,200–72,000	15.1	24–40	1935–2304	48.2
11. Regional rail, diesel	26.00 × 3.20	10 × 200 = 2000	240–200	15–18	30,000–36,000	7.2	30–40	1080–1152	25.4
12. Regional rail, electric	26.00 × 3.20	10 × 200 = 2000	180–120	20–30	40,000–60,000	10.9	35–55	1680–2640	49.1

comfort standard, a line averaging these values can achieve an offered capacity of C = 3800–5400 sps/h, in rounded figures. In many cases, however, particularly where signal and traffic interferences exist, this capacity is achieved only with rather low operating speed and service reliability.

Articulated buses/trolleybuses (case 2) can achieve nearly the same h_{min} when they have at least three double-channel doors and predominantly self-service fare collection. With vehicle capacities approximately 50% greater than standard buses, lines with articulated buses, single lane, and little interference from traffic signals can offer capacities of 5400–7200 sps/h.

High-capacity bus lines (cases 3 and 4) represent bus/trolleybus lines designed for maximum capacity, such as those in São Paulo and Curitiba, Brazil, and Transmillennio in Bogotá, Colombia. Each one of these systems has used some special features, such as multiple stops, overtaking at stops (Bogotá), bus convoys and their leapfrogging (São Paulo), and double-articulated buses with high-platform boardings and off-board fare collection (Curitiba). Many of these features require some special conditions: long and wide station areas, sometimes with grade-separated access; large areas for convoy formation and dispatching (São Paulo), four-lane roadways, etc. However, each one of these features has been proven operational where conditions allowed it.

The common feature for all these systems has been careful planning and design and very strict operations control and enforcement of regulations by police. Particularly important, all these cities have had very strong policies favoring transit and prioritizing transit over other vehicles.

Multiple stops along bus lines (case 3) allow simultaneous stopping of two or more buses. As a result, the average headway between buses is much shorter than at stations with single stop locations, f_{max} is greater. On streets with heavy traffic and irregular bus travel, the resulting frequency increase is limited due to the mutual impedance of buses to 90–120 veh/h, the upper limit being achieved where buses travel in pairs, or leapfrog at stops so that simultaneous standing times are possible, minimizing mutual interference. In Portland, Oregon, a two-lane one-way street is used exclusively by buses. With multiple stops staggered along the street and extensive overtaking, the facility allows maximum frequency of some 180 buses/h, offering a capacity of up to 17,500 sps/h.

Bus convoys, double-articulated buses, four-lane stations, and other capacity-increasing measures were introduced in the São Paulo, Curitiba, and Bogotá high-capacity bus corridors, achieving a total throughput of about 300 buses/h. In São Paulo, a capacity of 24,000 and in Bogotá over 30,000 prs/h have been claimed. Some of these figures (São Paulo) were given in the literature as theoretical computations. However, as in many cases of operations at maximum-capacity level, some of the quoted figures cannot be easily corroborated in actual operation.

With increasing frequency of bus services, operating speed and service reliability decrease. However, these Brazilian cities and Hong Kong have shown that the quoted very high bus frequencies can be operated with acceptable level of service if the city has policies favoring transit: competent traffic engineering, enforcement of regulations, and well-organized transit operating agencies. No city without these conditions has achieved such high frequencies.

The highest capacity of a bus facility in the United States is the contraflow lane on I-495, the New Jersey approach to the Lincoln Tunnel. Several lanes converge into this lane, which has no stations, so that a steady flow of over 500 buses per hour offering about 25,000 sps/h is achieved. After these buses leave the Lincoln Tunnel, they enter the Port Authority Bus Terminal, one of the largest in the world, which has over 150 berths for bus stops. Consequently, this is a unique case where way capacity governs because this section has no stops. On regular transit lines with on-line stops, the capacity would be much lower.

Streetcar lines (case 5) with traditional operations in streets or street medians can operate with up to f_{max} = 60 TU/h on lines with single stops, offering C = 6600 sps/h when single, 14-m long cars with C_v = 110 sps are used, and C = 13,200 sps/h with car/trailer or articulated cars. However, it is common on heavily used lines to organize double stops. Then, with good organization of boardings/alightings, frequencies of up to 120 TU/h can be achieved. With

TUs with capacities of 220 spaces, offered capacity exceeded the rate of 26,000 sps/h.

There are no exact records about average operating speeds and reliability of these record-volume lines. However, they generally tend to be rather low. It is a characteristic of all transit modes operating on streets (ROW C), without full system control, that when line capacity is stretched to its maximum, other performance elements, particularly passenger comfort, operating speed, and service reliability, are considerably lower than under normal operations.

Light rail transit (LRT) (case 6) differs from streetcars in many performance elements. As described in Section 2.1.2.1, LRT operating larger TUs at lower frequencies offers capacities similar to those of streetcars, but at substantially higher operating speeds and service reliability.

Although in most U.S. and West European cities passenger volumes on transit lines are lower now than they were several decades ago, a number of LRT systems have been designed to offer similar capacities to those of streetcars, but at much higher operating speed, reliability, and comfort. Here are several examples.

Cologne has upgraded its streetcar into an LRT network. Short signal blocks in the trunk tunnel section allow headways of 80 seconds, or frequency of 45 TU/h. Some trains consist of two articulated cars (model B), each with 72 seats and 210 spaces of total capacity. With these trains it is possible to offer a transporting capacity of 6480 seats/h or 18,900 sps/h.

Calgary operates trains of three articulated cars with 190 spaces, so that the total TU capacity is 570 sps. With headways of two minutes, the line can offer 17,100 sps/h. Operating speeds of most recently built LRT lines are in the range of 26–35 km/h. For example, the Blue Line in Portland, Oregon operates at 33.6 km/h. These are much higher than the speeds of 10–16 km/h that were typical for streetcars.

San Francisco's Muni Metro, an LRT system with trunk line in the tunnel, was capable of operating four-car trains at two-minute headways. With TU capacity of 744 sps, line capacity can reach 22,320 sps/h with much higher comfort, speed, and reliability than buses or streetcars offer. That particular operation was replaced about the year 2000 by a fully automated train operation (ATO), which can handle two-car trains at headways of one minute, offering the same line capacity at a higher frequency that is required for handling trains from five converging lines.

Automated guided transit (AGT) (cases 7 and 8) as a mode includes many systems with different technologies and operating characteristics. Consequently, their capacities also vary. The four major AGT technologies in operation in the 1990s have the following capabilities:

- The Westinghouse C-100 system (used in many airports and in Miami CBD) operates trains with up to three 100-space cars at h_{min} = 1.5 min, offering line capacity of 12,000 sps/h.
- The Kobe New Transit system, operating six-car trains with a TU capacity of 372 spaces, can offer at h_{min} = 2.0 min a line capacity of 11,160 sps/h. Several other Japanese AGT systems (Osaka, Yokohama) have similar characteristics, except that the Osaka system provides less capacity because it operates four-car trains.
- The VAL system, operated in Lille, Toulouse, and several lines in other cities, was designed with rather low-capacity cars: its TU is a married pair with a capacity of only 172 spaces. To obtain higher line capacity, the system is designed to operate at h_{min} = 1 min, or, in practice, it can achieve f_{max} = 48 TU/h. Thus, the capacity of these VAL systems is about 8256 sps/h. A VAL line in Taipei, Taiwan has four-car TUs, offering line capacity of about 16,000 sps/h.
- The latest but most promising AGT system is the one using the oldest guided technology—conventional rail. Automated LRT, such as the Vancouver ALRT, operates TUs of up to six cars, each with a capacity of 100 sps. With f_{max} = 48 TU/h, its offered line capacity is 28,800 sps/h.

While the above-described four AGT systems dominate this mode, its development is still very lively. New features and new systems are likely to appear in the future. However, it is likely that the capacity domain of the AGT mode will stabilize in the range of 12,000–18,000 sps/h for rubber-tired and somewhat higher, up to 28,000 sps/h, for rail AGT systems.

Rapid transit or metro systems (cases 9 and 10) have a great diversity of features determining line capacity. Dimensions of metro cars vary from 15.00 · 2.5 m in Paris to 21.00 · 3.23 m in Toronto and Hong Kong; maximum train lengths go from 4 in Boston and 5 in Paris to 10 in New York, Tokyo, and San Francisco; London has operated f_{max} = 38 TU/h, while some systems reach only f_{max} = 24 TU/h.

Performance, including capacity of rapid transit systems, changes with technology and operating conditions. For example, many systems can operate h_{min} = 90 s and even approach this value as their hourly average. While Moscow uses midplatform signals to reduce h_{min} to as little as 70 s, some other heavily traveled systems, such as Tokyo, intentionally operate well above the shortest possible headway in order to maintain very high reliability. While automation is increasingly used, some of its elements result in longer h_{min} than manual driving can provide. New metro systems generally utilize larger cars (at least 18 m long), but their capacities vary greatly between the systems that want to maximize comfort by providing maximum and spacious seating (BART) and the systems that must carry maximum passenger volumes, such as Hong Kong, Moscow, São Paulo and Tokyo.

Case 9 in Table 2.4 represents a system typical for a large city in a developed country, such as Brussels or Munich. Capacities of such systems are typically in the vicinity of 40,000 sps/h; the range of their operating speeds is wide because it encompasses old system with short station spacings (Berlin, Madrid, Paris) and new ones with longer suburban high-speed running (Baltimore).

Case 10 is typical for the systems recently built for maximum transporting capacity. These systems, exemplified by metros in Caracas, Hong Kong, São Paulo, Tokyo, and large-profile lines in New York, offer capacities actually counted in excess of 70,000 prs/h! They represent by their capacities and productive capacities by far the highest performance transit systems of any mode.

Regional rail systems (RGR) (cases 11 and 12) vary in their characteristics even more than metro systems, so that it is not possible to define one system as typical for this mode.

There are two basic types of RGR systems: diesel-powered, usually with diesel locomotive pulling or pushing trailer cars (Boston, Chicago, Toronto); and electrically powered, mostly with MU (multiple-unit or self-powered) cars (New York City, Philadelphia, most European RGR systems). As the two cases in Table 2.4 show, both types usually have very large cars and trains with up to 10 cars, but diesel trains operate at longer h_{min} values (due to lower acceleration/deceleration capabilities) and at lower average speeds than electric trains.

Diesel-powered lines seldom approach h_{min} = 3 min. With very long trains or double-decker cars, they can reach capacities of 30,000–36,000 sps/h. Electrically powered lines, such as the Long Island Rail Road in New York, many SNCF lines serving Paris suburbs, and the Munich S-Bahn, operate at headways as short as 2.5 minutes, so that they can offer capacities of 40,000–48,000 sps/h.

Many reviews of capacities of different modes, particularly common in European professional literature, show RGR as the highest-capacity mode. Actually, this is incorrect. RGR systems typically have lower capacity than metro systems for two reasons. First, RGR systems, serving much longer lines, operate at higher technical speeds than metros and therefore must have longer minimum headways. And second, most RGR systems provide considerably greater comfort (higher seats/standing spaces ratio) than metro systems, which decreases their car capacities.

Exceptions to the rule that RGR systems that have capacities matching those of metro systems are found in the cases where RGR systems are technically, operationally, and by comfort standards identical to those of metro systems. One example of this is found in Tokyo, where JR (RGR) lines have virtually the same rolling stock and operations as those of the Teito Rapid Transit Authority. The JR Yamanote Line operates trains with 10 cars, each with capacity exceeding 225 spaces at h_{min} = 2 min, offering over 67,500 sps/h,

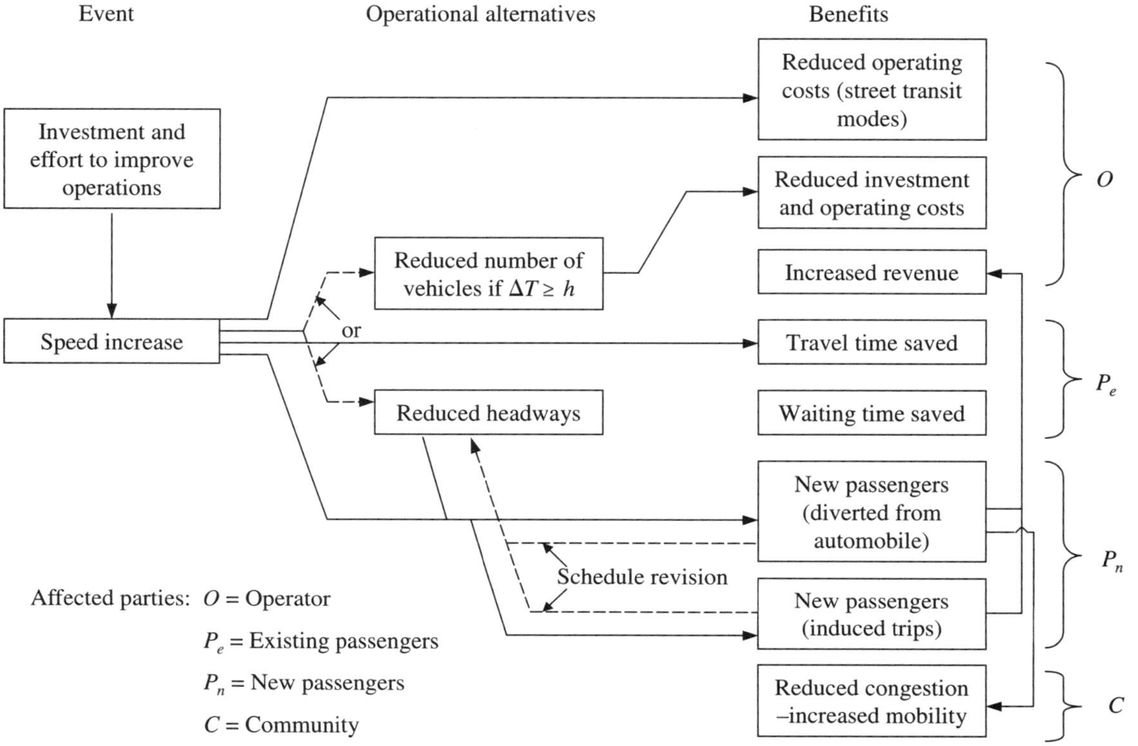

Figure 2.5 *Evaluation of transit speed increase.*

making this one of the highest line capacities in the world.

Another example is the San Francisco BART system, which was designed for high comfort and speed typical for RGR, but with metro-type operation. Its 10-car trains have a total of 700 seats and operate at h_{min} = 2.25 min. Reaching technical speeds of 120–130 km/h and average operating speed of nearly 60 km/h, BART offers 18,700 very comfortable seats per hour and a total capacity of approximately 37,400 sps/h. This capacity is moderate in comparison to other metro systems, but its comfort and speed are among the highest in the world, comparable to the best RGR systems.

2.2 INCREASE OF TRANSIT SPEED

Operating speed of transit has a major impact on its performance, costs, and role in the city. Increasing speed on an existing line results in shorter passenger travel times and usually in decreased investment and operating costs for the agency. Even more significant is the impact such an increase may have on attraction of new transit passengers. On the other hand, decrease in transit speed, such as traffic congestion often causes, results in losses to passengers, increased operator's costs, and diminished significance of transit services for the city. Therefore, the importance of speed in transit service should not be underestimated and the operator should make every effort to achieve as high a speed as operational and economic conditions allow.

2.2.1 Desirability of Speed Increase

A conceptual structure of the consequences of an increase in transit speed, including benefits accrued by various affected groups, is shown in Figure 2.5. The diagram starts with an investment (financial and implementation effort) into a measure increasing transit speed (e.g., introduction of transit-actuated signals; higher-performance vehicles; construction of exclusive ROW, etc.) and traces the consequences of the resulting higher speed through various operational changes to the final benefits accrued by the three major affected groups—passengers, operator, and community—shown in the last column's boxes. Some of the items in the diagram can be exactly defined and computed; others are qualitative or cannot be predicted with accuracy in most cases. It is important, however, that, in evaluating a measure for transit speed increase, all items in these boxes be included. The same structure of analysis can be used in estimating losses due to a speed decrease.

There is a considerable polarization of opinions among transit operators in U.S. cities with respect to transit speed. Many agencies that operate surface routes have made little effort to improve the speed of their vehicles; for example, the obsolete policy of stopping for passengers at every corner, still practiced in many cities, results in such low speed that few persons who have other modes of travel available can be attracted to transit (see Section 2.3.3). Another group, the operators who plan and operate rapid transit and regional rail systems, has often opted for very high speed by planning rail lines with extremely long station spacings (2–3 km in inner cities, 3–5 km in suburban areas) and very high technical speeds (up to 130 km/h) at the expense of area coverage.

One can therefore expect that with improving attitudes toward transit in cities and with progressive managements in transit agencies, more attention will be given to increasing the speed of urban street transit, while the trend toward very high speeds of regional rail systems may slow down due to the need for better area coverage and better integration of transit networks. In both cases, evaluation procedures presented in this section should be applied to ensure inclusion of all important items affected by the operating speed on the line.

2.2.2 Possible Measures for Speed Increase

Passengers are interested in high operating speed, while the operator's costs are directly related to the cycle speed, which influences the fleet size, as was shown in Figure 1.27. Usually, an increase in operating speed leads to an increase in the cycle speed, so that the focus of efforts to increase speed should be on measures affecting vehicle running, as well as on minimizing terminal times.

A great number of physical and operational changes may result in increased transit speed. The operator should consider which ones may be applicable to his system. This section presents potential changes to surface transit modes—bus, trolleybus, and LRT. They are classified into several categories. The sequence of items is not necessarily related to their significance, since applicability of individual measures and their effects depend heavily on local conditions.

2.2.2.1 Vehicle Design and Performance Characteristics. Several changes in vehicle design may lead to increased travel speed. Most of them should be planned for new vehicles since only limited modifications can be made on the existing ones.

Interior circulation may cause delays at stops if the aisle is narrow and passenger exchange is heavy. In such cases elimination of one row of seats, at least between the front and center door, should be considered.

Double-channel doors substantially decrease boarding-alighting times: they should be considered for all transit vehicles except those for longer suburban lines with low passenger exchanges, where advantages of faster boarding/alighting are outweighed by the requirement for maximum seating capacity.

Low-floor vehicles offer easier and faster boarding/alighting. Introduced in the 1980s, low-floor LRT, trolleybus, and bus vehicles have become standard designs for urban transit services, first in West European and then in many other countries.

Dynamic characteristics of vehicles must be carefully determined in connection with route layout and type of operation. For operation on steep streets, electrically powered vehicles or buses with stronger motors or special gearing are needed to provide reasonably high speeds. Buses operating on freeways and rail systems on long regional rail lines require high maximum speed, while street operation benefits more from high acceleration/deceleration than from high maximum speed. Thus, while electric traction offers advantages in the dynamic performance (as well as many others, such as absence of exhaust and noise) on any transit system, those advantages are particularly significant on lines with frequent acceleration and braking and on those on hilly streets. In such situations, they can provide shorter travel times.

2.2.2.2 Intersection and Street Design. In most cities, street layout and traffic regulation were traditionally designed without giving much attention to transit vehicles. For example, in determining design capacities or signal timing, buses and streetcars are included with other vehicles without any special considerations except for their different space occupancy. Physical size and dynamic characteristics of transit vehicles are only expressed through their passenger car equivalent (p.c.e.) factors. Since the basic objects of transportation are persons rather than vehicles (freight is seldom a governing factor on urban streets), the relative priorities given to different vehicle classes should be proportional to the number of persons they carry. Thus, a bus that carries 20–80 persons and an LRT train carrying 50–200 persons should be given strong priorities over private cars carrying on the average 1.3 persons. If, in addition, the vital role transit plays in the city is taken into account, one can easily see that introduction of transit priorities for increase of speed and reliability can be justified not only for heavily traveled, but for most major transit lines. The most important such improvements are described here.

Intersection design that facilitates transit operation should have precise channelization, conveniently located transit stops, turning lanes, pedestrian control and protection, etc.

Construction of bus stop bulb-outs or sidewalk extensions through a parking lane allows buses to stop without maneuvering into a parking lane to make a stop and then reentering the travel lane, losing time and decreasing riding comfort.

Elimination of curb parking on transit streets, with proper enforcement, greatly enhances transit operation by providing for more efficient use of the full width of street for moving vehicles. At the same time, it allows faster travel for transit vehicles due to decreased street congestion and vehicular friction. This action also simplifies the bus maneuver to approach the curb and then return into the traffic lane and increases passenger safety by allowing boarding at the curb.

Elimination of parking often creates objections from merchant associations and residents. Several actions can alleviate this problem. The eliminated parking may sometimes be compensated for by providing additional parking on close-by off-street areas, vacant lots, etc. or by provision of additional parking on adjacent streets. Alternatively, demand for parking on adjacent streets can be reduced by increasing parking rates or reducing allowed parking duration. In addition, various transit improvements may be helpful to attract some of the affected auto users to transit. For example, when vehicle travel time is decreased by the elimination of parking and this is coupled with other traffic improvements, a more frequent and faster transit service may become feasible without any fleet size increase. Another possible improvement is introduction of shopper-special service with discount fares, or a transit validation system (reimbursement of fares) in cooperation with merchants.

Introduction of transit lanes or exclusive transit ROW category B for rail lines represents the most significant measure in increasing transit speed and reliability. An exclusive transit lane may be obtained by:

- Converting an existing traffic lane by prohibiting all other vehicles from it
- Eliminating curb parking and use of that lane for transit or for other traffic and construction of physically separated category B ROW in the center of the street
- Placing transit lanes or tracks in wide unused medians
- Building transit rights-of-way independent of street alignments

Various design options for exclusive bus and light rail rights-of-way are described, respectively, in Vuchic (1981), Chapters 4 and 5.

Introduction of transit/pedestrian malls gives transit a very strong identity and relates it closely to pedestrian-oriented urban centers. By provision of such malls, transit obtains a distinct advantage over cars for access to areas of very concentrated activities. Bus/pedestrian malls have been well received in some cities, such as Minneapolis, Vancouver, and Ottawa, while a few were later abandoned (Philadelphia). Pedestrian malls and precincts with tramway and LRT lines have been extremely successful in cities in many European countries (Germany, Sweden, France, Switzerland), in U.S. and Canadian cities, and in Melbourne. In most of these cities, transit malls have become the most attractive pedestrian areas and symbols of the vital role of transit in the city.

2.2.2.3 Traffic Signal Priorities for Transit. Delays caused by traffic signals at intersections often cause considerable delays and affect reliability of transit services. There are, however, a number of different methods to decrease or eliminate TU delays at signalized intersections (see ITS America, 2003). It must be borne in mind, however, that transit signal priorities are fully effective only where transit has ROW category B, separate tracks, or exclusive lanes. In mixed traffic, category C, transit vehicles cannot utilize priority if they are behind other vehicles on intersection approaches.

Technological solutions for detection of TU positions along a line, for the logic in signal control, and for signal actuation vary considerably. Detection of approaching TUs can be achieved by a contact on the overhead catenary. For buses, a loop in the pavement may be used. Increasingly, this detector may become a component of an AVL system, which may use radio signals, scanning devices registering passage of buses along their line, or a GPS that senses exact location of each TU at all times.

Priority measures are defined here in sequence from minor adjustments to signal timing to absolute override when transit vehicles have absolute right of passage at any time.

1. *Passive priority measures* are those introduced to favor traffic flow through an intersection on the street that carries transit. They include longer green phase, special signal phase for transit turning movements, etc. These measures cannot be used at intersections where transit lines intersect, however.
2. *Active priority measures* include:
   - Early green—when a TU arrives during the red phase, the beginning of the green phase is started up to 10 s earlier than programmed.
   - Extended green—when a TU arrives at the end of a green phase, it is extended up to 10 s to allow the TU passage.
   - Changed phase sequence (sometimes known as phase rotation)—arrival of a TU activates the green phase it needs even if is not next in sequence by the basic signal timing program.
   - Activation of a special phase or a special turning movement that is programmed to be used only on-call by TUs.
3. *Signal preemption* is absolute priority for TUs by initiating its green phase whenever it arrives, regardless of the timing program, only ensuring that all conflicting movements have been safely stopped. In its ultimate application, this override is timed with such time advance that TUs can pass through the intersection or crossing at any speed without slowing down. This system is used quite often on regional rail lines when they cross streets or go through intersections, sometimes at speeds exceeding 100 km/h.

Priority levels may vary in some installations. As early as the 1980s, the LRT system in Hannover, Germany, and some other cities was equipped with a program capable of selecting among three levels of priority. LRT TUs had a programmed priority over other traffic at intersections for their scheduled operations. If a TU arrived later than scheduled, it received a higher degree of signal preemption. If a TU was running ahead of its schedule, it did not get any priority at all.

Priority levels can be determined taking into account several other factors, such as:

- The impact that transit priority will have on other street traffic. This is particularly considered when transit service is frequent, such as buses on a heavily traveled street. Priority is then limited to prevent excessive green phase reductions for crossing traffic.

- When two transit streets intersect, the priorities between them must be shared.

Studies have been conducted on how to include TU occupancy and give it priority on the basis of the number of passengers. This is not appropriate, however, because a transit line with such control would have variable reliability depending on its volume of passengers, which would reduce regularity and attractiveness of transit services.

2.2.2.4 Transit Stops. Improvements in the design and location of stops may have a major impact on transit speed. One possible improvement is increased spacings between stops to the extent which area coverage allows. Determination of optimal spacings is described in detail in Section 2.3.

Another possible improvement is alternating near-side and far-side stop locations along a street with synchronized signalization. Most common transit stop locations along a street are at its intersections, either prior to crossing a street (near-side, NS) or after crossing (far-side, FS). The selection between the two depends on the following factors: convenience of passenger access, geometry of bus access to the curb, traffic conditions (turning lanes), and minimization of delays due to signals along the street. The last factor is seldom considered, but it can have a major impact on transit travel times.

When signals along a street are fully synchronized, as is often the case on one-way streets or in the peak-direction on two-way streets, stopping of a TU for passengers often delays it sufficiently to lose the signal through-band. The TU must then wait through a red phase. Figure 2.6 shows the path of a TU with NS stops and another TU with FS stops. Time lost at stops for passenger boarding/alighting is assumed to be the same for both runs. The diagram shows that both runs have equal travel times over the plotted section, but the all-FS policy involves more stoppings. If the NS and FS locations are alternated, however, the solid line on the diagram shows that transit vehicle delay is shortened considerably because each time the vehicle starts from a NS stop, it progresses through two signals without delay.

In some situations, there may be a sequence of NS and FS stops even better than regular alternation, since vehicle delays depend on the frequency and duration of stopping for passengers and the signal-synchronization pattern. However, one basic rule is always valid: when an NS stop is followed by two or more progressively coordinated signals, the following stop should be FS. It is also a fact that regardless of signal timing and length of delays at stops, alternating NS and FS stops result, on the average, in considerably shorter travel times than either all-NS or all-FS stops. In the theoretically worst case, travel times are equal.

There is a slight inconvenience to passengers when stops are not at the same location at each intersection, but this is easily overcome by adequate signing.

2.2.2.5 Elements of Transit Operations. Several operational elements that are largely under the transit operator's control can often be modified to increase transit operating speed.

Faster fare collection: manual payments by passengers when boarding a street transit vehicle, which is still common on bus services in many cities, is the slowest fare-collection method. Speed, efficiency, and convenience of fare collection have been increased through several stages of technical and operational innovations over several decades. These developments can be grouped in three major stages.

The first stage, introduced many decades ago, was the replacement of cash by more convenient payment forms, such as tokens, multiple-ride tickets, and weekly or monthly passes.

The second stage was the concept of self-service or presale of tickets prior to boarding with validation prior to or immediately after boarding. It was first introduced in several European countries during the 1960s, followed by North America about 20 years later.

The third stage of fare-collection development was started with innovations in electronics and computers: payments by sliding magnetic cards at the entry and exit of transit stations. In recent years, use of contactless "smart cards" has allowed their use on street tran-

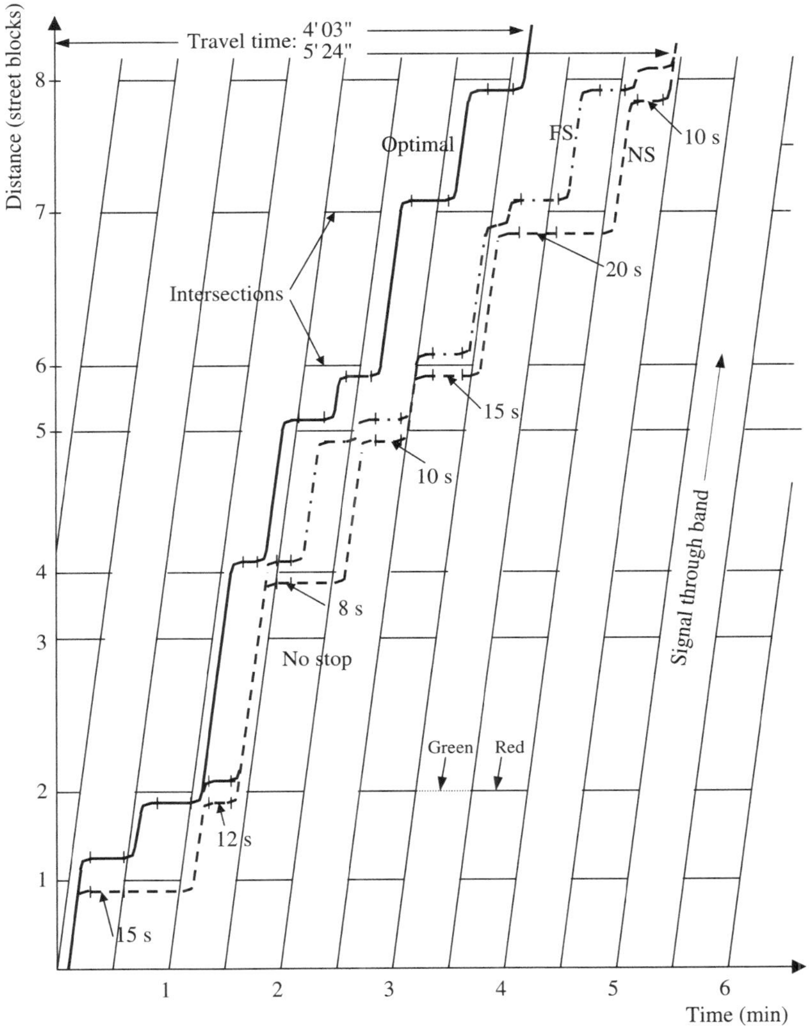

Figure 2.6 *Transit unit travel time for all-NS, all-FS, and alternative stop locations.*

sit vehicles also, further decreasing processing time per passenger and reducing travel time on transit lines.

For more details on fares, see Chapter 7.

Changes in route layout for elimination of excessive turns or indirect routings. It is a common practice, particularly in smaller cities and suburban areas, to operate very complicated transit routes to cover as much area as possible by each route. Many turning movements on a route not only increase vehicle travel time, but also decrease reliability of service and passenger comfort. Also, complicated route layouts decrease the simplicity and image of the system and create passenger confusion, resulting in lower patronage. A partial solution to this problem may be to divide routes that have many turns into several more direct routes. Since the vehicle speed on each route is then increased, it may be possible to operate direct routes with higher frequency without changing the fleet size.

Introduction of express, skip-stop, and zonal operations is described in Section 2.4.

2.2.3 Sensitivity of Transit Speeds to Elements of Cycle Time

Each measure for transit speed increase discussed in Section 2.2.2 reduces one or more elements of cycle time: station standing or dwell time t_s and acceleration and braking times t_a and t_b; or it lengthens the interval of travel at constant speed t_v, etc. In order to evaluate the impact of changes in individual time elements on the changes in system's operating speed V_o and cycle speed V_c, it is necessary first to examine sensitivity of speeds to each element individually and then to all of the time elements together.

The most important time intervals for analysis are operating time T_o and cycle time T, corresponding to the operating speed V_o and cycle speed V_c, respectively. For their analysis the following assumptions are made (for detailed derivation of the following equations see Vuchic, 1981, Chapter 3):

1. The length of any interstation spacing i, S_i, is sufficiently great that the vehicle can reach its maximum speed, i.e., $S_i \geq S_c$, where S_c is the shortest distance on which maximum speed can be reached.
2. The average acceleration rate $\bar{a}$ and deceleration rate $\bar{b}$ are constant;
3. Standing time in each station is constant, t_s, while standing time at each terminal exclusive of terminal layover time is $t_s/2$.
4. There are no influences of way alignment or grade on vehicle travel.
5. No coasting is applied.

These assumptions are reasonably realistic and do not significantly affect the results of this analysis. In the cases in which some of the assumptions are not satisfied (e.g. assumption 1 for a line with very frequent stops), this fact must be taken into account and the results must be treated with appropriate limitations.

The operating and cycle times can be expressed, respectively, as:

$$T_o = n(t_a + t_b + t_s) + \sum_{i=1}^{n} t_{vi} = \frac{n\,V}{432}\left(\frac{1}{\bar{a}} + \frac{1}{\bar{b}}\right) + n\,t_s + \frac{60n\,S}{V} = \frac{60L}{V} + n\,T_\ell \qquad (2.6)$$

$$\left|\frac{T,\,t}{\text{min}}\right|\frac{a,\,b}{\text{m/s}^2}\left|\frac{V}{\text{km/h}}\right|\frac{S,\,L}{\text{km}}\left|\frac{n}{-}\right|$$

where

$$T_\ell = \frac{V_{\max}}{7.2}\left(\frac{1}{\bar{a}} + \frac{1}{\bar{b}}\right) + t_s \qquad \left|\frac{T,\,t}{\text{s}}\right|\frac{V}{\text{km/h}}\left|\frac{a,\,b}{\text{m/s}^2}\right| \qquad (1.18)$$

and

$$T = 2(T_o + t_t) \qquad (2.7)$$

where S is average station spacing, n is the number of spacings, and L is line length; 432 reconciles the units of a, b, V, and T.

Sensitivity of T_o and T to individual time elements can be explored by marginal analysis of their expressions. The change of T_o and T due to unit changes in any one of the time elements t_s, and t_t, and parameters S, n, V, $\bar{a}$, and $\bar{b}$ (which determine t_{vi}, t_a, and t_b), can be obtained through partial derivatives. If x designates any one of these independent variables, the partial derivative of T can be expressed in the general form:

$$\frac{\partial T}{\partial x} = 2\left(\frac{\partial T_o}{\partial x} + \frac{\partial t_t}{\partial x}\right) \qquad (2.8)$$

If x represents any variable except t_t, one obtains

$$\frac{\partial T}{\partial x} = 2\,\frac{\partial T_o}{\partial x} \qquad (2.9)$$

This equation indicates that all independent variables except t_t affect both T_o and T, which is intuitively clear. Marginal analysis with respect to individual variables will therefore be performed for T_o only. The same expressions, multiplied by two, are valid for T.

***2.2.3.1 Sensitivity of* $\mathbf{T_o}$ *to* a, b *and* V.** The relationship between T_o and the dynamic characteristics of a transit vehicle, $\bar{a}$, $\bar{b}$, and V, is given by Eq. (2.6). T_o as a function of acceleration (or deceleration) rate $\bar{a}$ can be obtained from that equation, assuming all other parameters are constant:

$$T_o = \frac{60L}{V} + \frac{n \cdot V}{432\bar{a}} \qquad \left|\frac{T}{\text{min}}\right|\frac{L}{\text{km}}\left|\frac{V}{\text{km/h}}\right|\frac{n}{-}\left|\frac{a}{\text{m/s}^2}\right| \tag{2.10}$$

Sensitivity of T_o to changes in $\bar{a}$ (or $\bar{b}$) is expressed by its partial derivative with respect to $\bar{a}$:

$$\frac{\partial T_o}{\partial \bar{a}} = -\frac{n\,V}{432\bar{a}^2} \qquad \left|\frac{T}{\text{min}}\right|\frac{a}{\text{m/s}^2}\left|\frac{n}{-}\right|\frac{V}{\text{km/h}}\right| \tag{2.11}$$

These two equations are plotted in Figure 2.7. The diagram shows a family of curves using different values of n in Eq. (2.10). The curves decrease with increasing $\bar{a}$, asymptotically approaching the value $T_o = 60L/V$. The sensitivity of T_o, expressed by Eq. (2.11), appears on the negative side of the ordinate T_o.

The curves indicate that with increasing acceleration rate, sensitivity of travel time to acceleration rate decreases. Consequently, it is particularly useful to increase acceleration rates when they are low. The diagram also shows that sensitivity is higher when the number of stations is greater.

To examine the influence of the maximum speed V, T_o will be expressed and plotted as a function of V and as a partial derivative with respect to V. In this case, S will be used as an independent parameter rather than n (the two are equivalent since for a given line $n \cdot S = L =$ const.) because the possibility of reaching V on an interstation spacing depends on its length, S. The two expressions plotted in Figure 2.8 are Eq. (2.6) and:

$$\frac{\partial T_o}{\partial V} = \frac{n}{432}\left(\frac{1}{a} + \frac{1}{b}\right) - \frac{60\,L}{V^2}$$

$$\left|\frac{T}{\text{min}}\right|\frac{V}{\text{km/h}}\left|\frac{n}{-}\right|\frac{a,\,b}{\text{m/s}^2}\left|\frac{L}{\text{km}}\right| \tag{2.12}$$

It can be seen on the diagram that for any given value of S, T_o depends on V only to a certain point, S_c, since beyond that point the given spacing is not sufficiently long for the vehicle to attain that maximum speed V. The maximum speed achieved on a given spacing S, V^*, can be obtained by setting Eq. (2.12) equal to 0 and solving it for V:

$$V^* = 36\sqrt{\frac{5\bar{a}\,\bar{b}\,S}{\bar{a} + \bar{b}}} \qquad \left|\frac{V}{\text{km/h}}\right|\frac{a,\,b}{\text{m/s}^2}\left|\frac{S}{\text{km}}\right| \tag{2.13}$$

As expected, the reduction in T_o from a unit increase in V is the largest when the initial value of V is small.

***2.2.3.2 Sensitivity of* $\mathbf{T_o}$ *to* $\mathbf{t_s}$, n, *and* $\mathbf{t_t}$.** Equation (2.6) shows that T_o is very sensitive to t_s and to the entire marginal time loss due to an additional stopping, T_ℓ, because these time increments are multiplied by n. A unit change in the time required for passenger boarding and alighting results in a change of n units in T_o and of $2n$ units in the cycle time T. Consequently, the number of stops along a line n, and duration of TUs standing at the stops, have a direct influence on T_o regardless of other variables.

Terminal time t_t does not affect T_o, but it directly influences T (Eq. (2.7)).

2.2.3.3 Application of Findings. The preceding analysis has examined the relative sensitivity of the transit line travel time to changes in the various parameters. An example based on ranges of values of parameters typical for urban bus lines will illustrate use of this analysis in a practical situation.

A model transit line is assumed to be 9 km long, and it is analyzed for a wide range of seven parameters listed in Table 2.5 with low and high values of the range. The set of low values has been adopted as the initial basis on which all changes of parameters are examined. For the low values one obtains cycle time $T = 81$ min, cycle speed $V_c = 13.3$ km/h, and operating speed $V_o = 15.7$ km/h.

Figure 2.9 presents a diagram of travel time on the ordinate as a function of different sets of parameters. The values of each parameter are plotted below the abscissa to such scale that its assumed range fits the

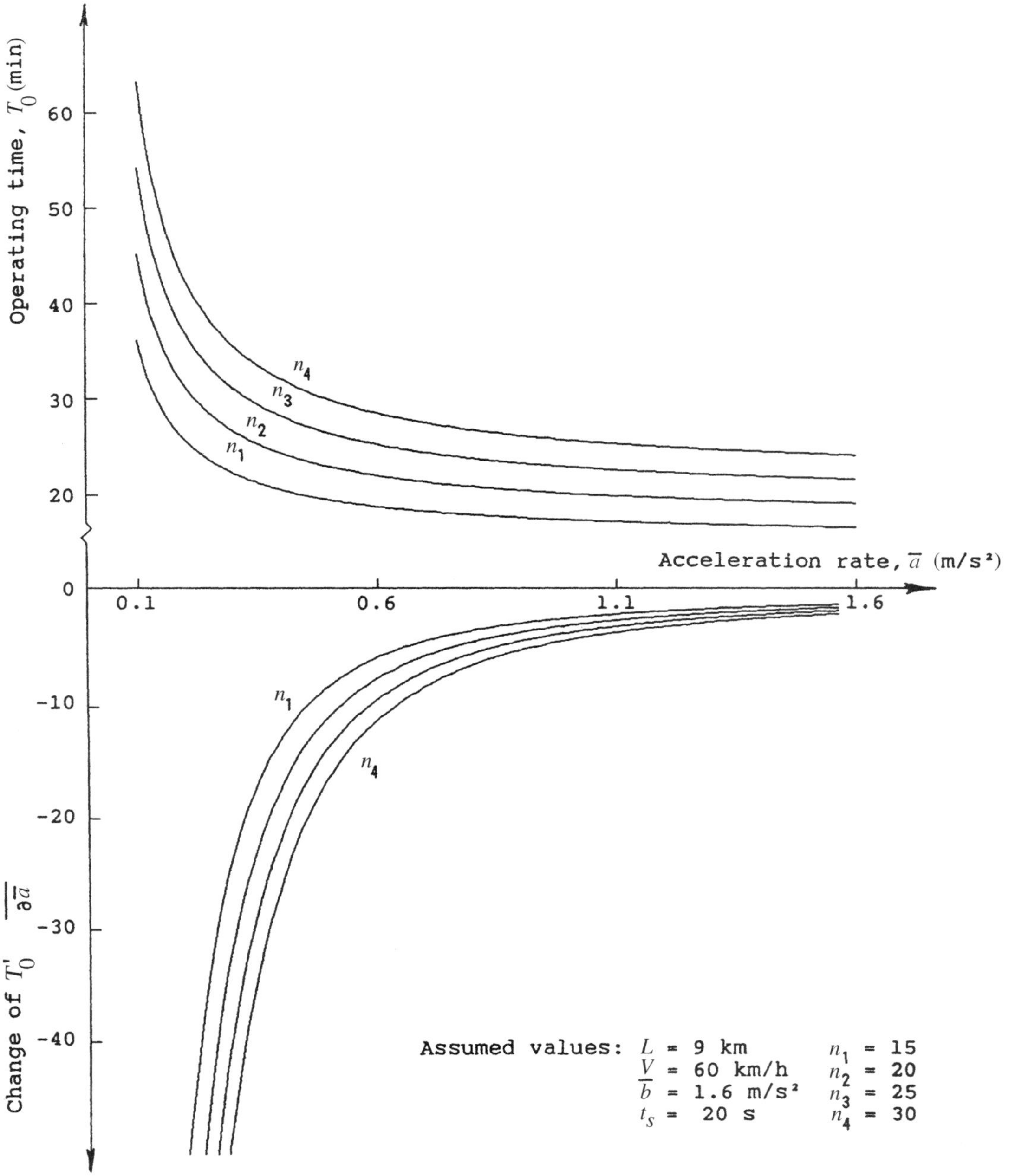

Figure 2.7 *Operating time as a function of acceleration rate* a *for different number of stations* n.

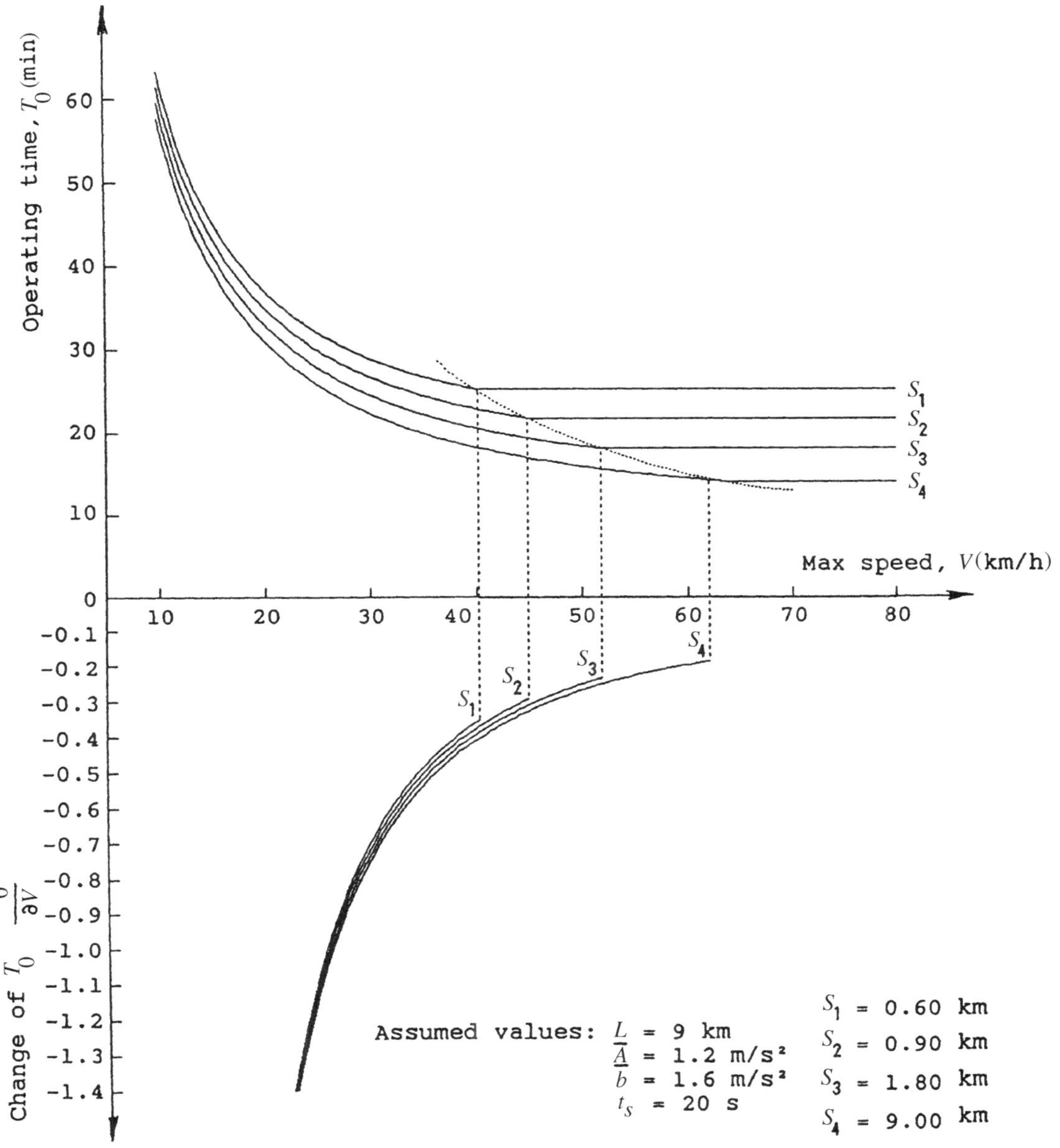

Figure 2.8 *Operating time as a function of maximum speed* v *for different station spacings* s.

Table 2.5 Operating values for an example of a bus line

Parameter	*Symbol*	*Dimension*	*Range of Values*	
			Low	*High*
Number of interstation spacings	n	—	30	1
Stop spacing	S	km	0.30	9.00
Standing time per stop	t_s	s	30	10
Terminal time at one end	t_t	min	6	0
Maximum speed	V	km/h	40	80
Acceleration rate	$\bar{a}$	m/s^2	0.8	1.6
Deceleration rate	$\bar{b}$	m/s^2	1.2	2.0

shown diagram. The initial value of T, along the ordinate, represents cycle time for all low values of parameters (80.6 minutes), designated as case A. From that point of low values, travel time is computed for the initial values of all parameters (case A) except for one, which has the values plotted for it under the abscissa. The obtained curve is designated by the parameter that has changing values. Thus, on the diagram, the V curve represents cycle length T for initial values of all parameters except V, which increases from 40 to 80 km/h along the abscissa. The line designated as $\bar{a}$ is for all initial or low values of parameters except for the acceleration rate a, which increases from 0.8 to 1.6 m/s^2, as shown on the corresponding line under the abscissa.

The composite curve represents T when all parameter values change as plotted along the abscissa. This curve goes from $T = 80.6$ min, for the A case with all low values to $T = 14.3$ min, for the B case with all high values. The latter case is the upper boundary case for speed, which would represent a line which would have a single spacing S and only the two terminals.

The diagram in Figure 2.9 can be used in planning improvements for an existing transit service, as well as in selecting operational characteristics in planning new systems. An operator who wants to increase speed of transit services can develop a diagram like this one with the existing value of travel time plotted along the ordinate as low value. Along the abscissa all parameters that could possibly be changed should be plotted.

The diagram then shows what reduction in cycle time can be achieved by any given change in acceleration rate, speed, or other parameters. For example, in the case shown in Figure 2.9, a reduction in cycle time T from the initial 81.6 to 78.0 minutes, i.e., by 2.6 minutes, could be achieved by any one of the following changes: reduction of the number of stops from 30 to 28, reduction of average standing time at 30 stops from 30 to 27.4 seconds, reduction of terminal times at each end by 78 seconds, or increase of acceleration rate from 0.80 to 1.28 m/s^2 (in the real world, these values would be rounded, of course).

2.2.4. Evaluation of Measures for Speed Increase

To decide whether the introduction of a measure for speed increase is justified, the operator must make an analysis of all its benefits and costs in the broad sense, i.e., not only through travel time or monetary aspects. In some cases, particularly with respect to changes that the operator can introduce alone (e.g., a new fare-collection method, faster vehicles, etc.), this analysis is rather simple; in others, particularly those involving introduction of transit lanes, signal actuation, or reconstruction of streets, which are under the jurisdiction of other bodies and affect not only transit users but also auto drivers and other parties, it is necessary to make a comprehensive analysis. Results of such analyses should be used in justifying the effort required for speed increase and determining who should participate in it and to what extent.

Due to the great number of direct and indirect consequences of any significant speed change, it is useful to have a graphical presentation such as a flowchart of the entire process. Two examples of such diagrams and

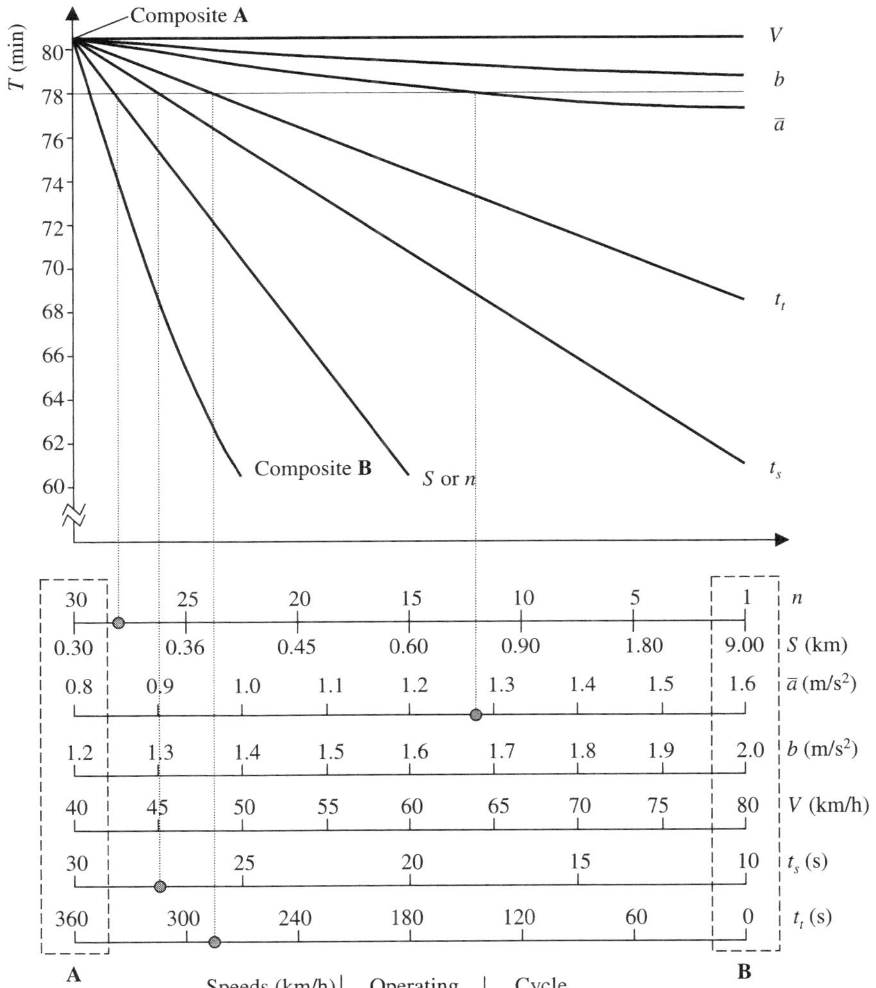

Speeds (km/h)	Operating	Cycle
Composite A	15.75	13.40
Composite B	75.79	75.79
$L = 9$ km		

Figure 2.9 *Cycle time as a function of operating parameters.*

computations based on them are presented here. One analyzes the consequences of a measure that increases speed of fare collection, such as introduction of self-service fare collection with prepaid tickets. The second example defines the steps in evaluating the benefits and costs of introducing an exclusive transit lane. Design details and warrants for exclusive transit lane are presented in Vuchic (1981), Chapter 4.

EXAMPLE 2.5: Fare-Collection Improvement

A transit agency is considering an improvement in fare collection on a bus line with the following characteristics:

$$L = 9 \text{ km}$$
$$T_o = 36 \text{ min}$$

$\ell_a = 4.5$ km
$t_t = 6$ min
$h = 6$ min
$\tau_b = 3$ s/prs
$P_L = 180$ prs/TU

where ℓ_a is average passenger trip length on the line and P_L is the total number of boarding passengers per one-way trip. The fare-collection innovation will result in a reduction of boarding times from 3.0 to 2.0 s/prs.

For the examination of the feasibility of this improvement, major items shown in Figure 2.10 are analyzed. Actions required for introduction of the new fare-collection method are self-explanatory. Major benefits are computed and described here.

1. *Increased vehicle operating speed.* The self-service fare-collection method will cause a time saving at bus stops amounting to 180 (3 − 2) = 180 s = 3 min per one-way trip. Operating speeds before and after the improvement, respectively, are computed by Eq. (1.28):

$$V_o = \frac{60\ L}{T_o} = \frac{60 \cdot 9}{36} = 15.0 \text{ km/h and}$$

$$V'_o = \frac{60 \cdot 9}{36 - 3} = 16.3 \text{ km/h}$$

Consequently, operating speed is increased by 1.3 km/h.

2. *Reduced user travel time.* An immediate effect of vehicle operating speed increase is the sav-

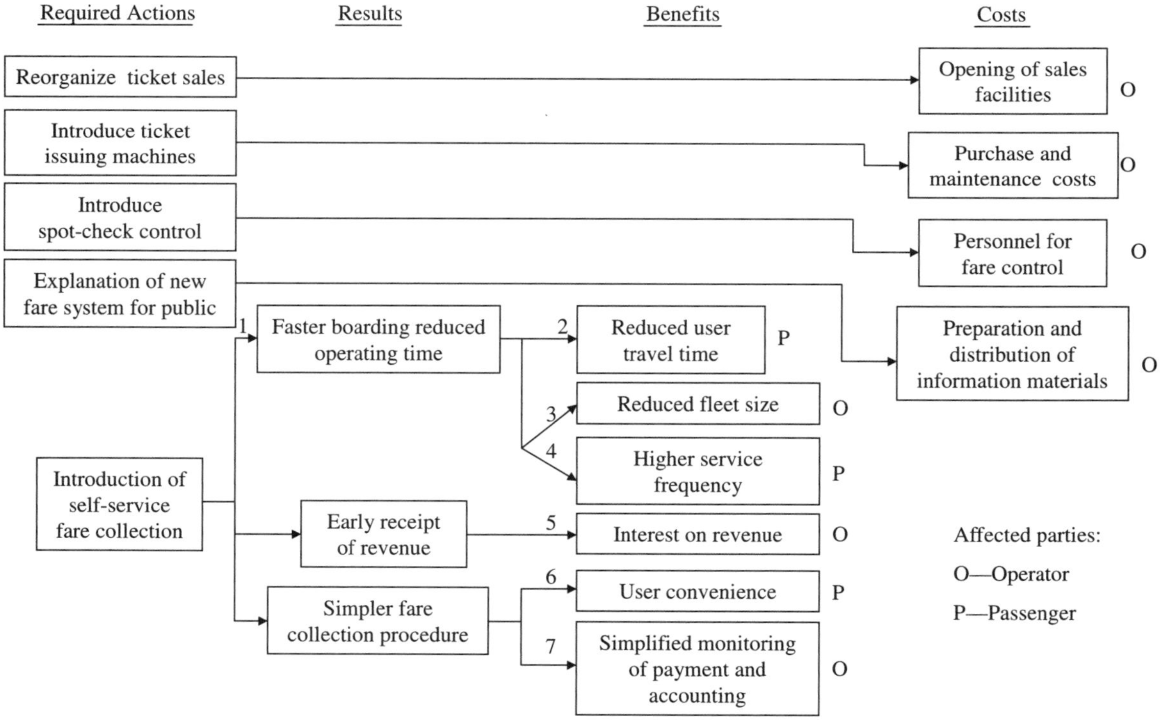

Figure 2.10 *Example 2.5: Evaluation of self-service fare-collection system (numbers correspond to those in the text).*

ing in user travel time. For a 4.5-km average trip length, travel time changes from:

$$t_o = \frac{60\ \ell_a}{V_o} = \frac{60 \cdot 4.5}{15.0} = 18.0 \text{ min to}$$

$$t_o' = \frac{60\ \ell_a}{V_o} = \frac{60 \cdot 4.5}{16.3} = 16.6 \text{ min}$$

Thus, an average saving of $t_o - t_o' = 1.4$ min per passenger is achieved.

3. *Reduced fleet size.* With the increase in vehicle operating speed, the operator has two options:

- Decrease fleet size while maintaining the existing headway
- Decrease headway while maintaining the existing fleet size

The first option reduces operating costs (depreciation, driver, fuel, and maintenance costs). This option, however, can only be taken when the difference between the cycle times before and after the improvement $(T - T')$ is at least equal to one headway, so that the same service frequency can be maintained if one TU is withdrawn. In this case, the cycle time, cycle speed and fleet size requirements before and after the improvements are as given in Table 2.6. They show that one TU is saved under the assumed numerical values.

4. *Higher service frequency.* If the same fleet size is maintained, the operator can try to reduce headway as a result of cycle speed increase. The headway of six minutes could become, after the improvement:

$$h' = \frac{T'}{N_{TU}} = \frac{78}{14} = 5.57 \text{ min}$$

Rounding this value to the higher integer, one obtains six minutes. Thus, in this case, decreasing the headway would not be operationally practical.

5. *Interest on revenue.* Another advantage if prepaid tickets or passes are introduced is that the operator receives revenue in advance and can collect interest on it.

6. *User convenience.* The introduction of the self-service system with prepaid tickets greatly increases user convenience: passengers do not have to prepare cash for each ride or to show transfers. Greater convenience, coupled with reduced travel time, may result in attraction of new riders.

7. *Simplified monitoring of payment and accounting.* The simpler fare-collection procedure resulting from the improvement eases the driver's supervision of fare payment and shortens dwell times. With less fare collection, accounting procedures at the office are simplified.

The cost items defined in Figure 2.10, which are self-explanatory, should be computed and presented in a summary form.

EXAMPLE 2.6: Transit Lane Introduction

This measure for speed increase is more complex than change in fare collection, since it concerns not only the transit operator and passengers, but also other traffic, police, residents along the streets, etc. Preparation for its introduction must therefore include cooperation with the traffic engineering department and other concerned parties. All these

Table 2.6 Changes in operating elements due to speed increase

Equation	*Unit*	*Before*	*After*	*Change*
$T = 2(T_o + t_t)$	min	$2(36 + 6) = 84$	$2(36 - 3 + 6) = 78$	−6
$V_o = \frac{120\ L}{T}$	km/h	$\frac{120 \cdot 9}{84} = 12.9$	$\frac{120 \cdot 9}{78} = 13.8$	+0.9
$N_{TU} = \frac{T}{h}$	TU	$\frac{84}{6} = 14$	$\frac{78}{6} = 13$	−1

organizational and often political actions should be based on a comprehensive analysis of the technical/economic factors involved in such a change, which is outlined here. The better the analysis is, the easier it is to refute subjective arguments of opposition groups.

The analysis of benefits and costs of introducing a reserved transit lane on an arterial street follows a procedure similar to the one in the preceding example. It is described in sequence of the numbered boxes in Figure 2.11.

1. *Required actions.* The investment cost, time of implementation, and disruption incurred by the introduction of an exclusive transit lane must be specified. They depend on the location and design of the facility to be provided. The lowest-cost alternative would be conversion of an existing parking or driving lane into a transit lane. With this option, some sort of visual control (such as markings, signs, and signals) or physical restraint devices (such as mountable curbs along lane lines) that either prohibit or discourage entry of nontransit vehicles into the lane must be provided. In addition, visual controls and special minor construction may be required at certain intersections and stop areas to ensure safety of operation and eliminate potential conflicts with other vehicular traffic. Other, more expensive measures for transit lane introduction are possible, including the construction of lanes or rail tracks in wide freeway medians or on unused railroad rights-of-way. The magnitude of operating benefits and operating costs for each alternative de-

Photo 2.1a *Bus caught in street congestion—Philadelphia (courtesy of Christopher Puchalsky).*

Photo 2.1b *Bus operation in reserved bus lane—Philadelphia.*

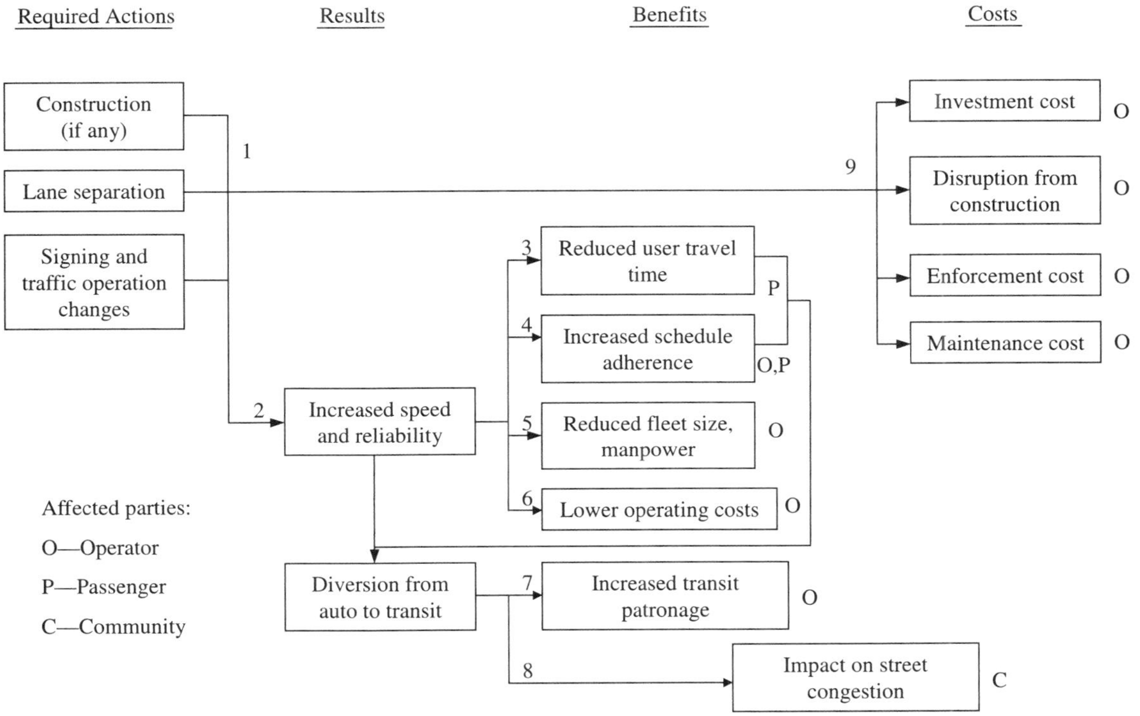

Figure 2.11 *Example 2.6: Evaluation of transit lane introduction* (*numbers correspond to those in the text*).

sign varies with the investment cost, but each has the same kind of effects that must be analyzed.

2. *Increased speed and reliability.* Provision of a transit lane or a fully separated ROW for transit significantly improves the average speed of vehicles since interference from other traffic is substantially reduced or eliminated. Likewise, reliability of service is improved. These improvements provide added benefits, which are described below.

3. *Reduced user travel time.* Calculated as in Example 1, the travel time savings for present transit users can be significant. Auto drivers who switch to transit will also experience reduced travel times or increased convenience (otherwise they would not have switched). Finally, auto drivers who continue to drive their cars may also save time, as explained under item 8 below.

4. *Increased schedule adherence.* Because of their greater independence from street traffic, transit vehicles traveling on their own exclusive lane have higher reliability and can therefore meet schedules more easily and consistently. The improved schedule adherence can be measured by observing the percent of on-time arrivals before and after the lane installation.

5. *Reduced fleet size and manpower.* The savings in fleet size and manpower resulting from the speed increase should be calculated in the same way as in Example 1.

6. *Lower operating costs.* With fewer delays in traffic and fewer stops and starts, the vehicle operating and maintenance costs are reduced. Savings are accrued from lower fuel consumption, maintenance, and replacement of vehicle parts, but these are usually difficult to quantify.

7. *Auto-to-transit diversion and increased transit patronage.* The combined improvements of transit service resulting from the introduction of

exclusive transit lanes usually result in some diversion of travel from auto to transit. The amount of diversion depends heavily on traffic conditions on the streets prior to operation of the transit lane and whether auto drivers are adversely affected by the transit lane introduction. For example, diversion from auto to transit would be higher where a lane on a heavily congested expressway is removed from use by auto traffic ("convert-a-lane") and assigned to transit vehicles than it would be if the transit lane were built in the median ("add-a-lane"). This is because highway travel may be substantially hurt by removal of one lane in the first alternative, whereas it would be improved by the second. This leads to the next point.

8. *Impact on street congestion.* The nature of the impact of a transit lane on street congestion varies with the right-of-way used (existing or new construction) and the amount of passenger diversion to transit. Obviously, if an entirely new facility is provided, present highway users will benefit from having transit vehicles removed from traffic. On the other hand, if a presently used street lane is taken for transit, auto drivers may be adversely affected to the point where so many of them switch to transit that original traffic conditions are again achieved or even improved. There are, however, cases in which conversion of an existing traffic lane to an exclusive transit lane results in increased speed of both due to lower friction between the two different types of vehicles.

Since the impact of a transit lane on congestion may be positive or negative, depending on conditions, its box in Figure 2.11 is placed between benefits and costs.

9. *Costs.* Investment costs and disruption due to construction vary with the type of facility used or constructed. There is always a trade-off between the level of investment cost and required enforcement cost. For example, on a physically separated exclusive transit ROW, particularly rail tracks, there is no opportunity for violators (nontransit vehicles) to enter upon the way; consequently, enforcement costs of patrolling or monitoring are virtually nil. An expressway or urban street lane that is easily accessible from other lanes by autos, on the other hand, must be monitored closely for potential abusers. The maintenance costs incurred by the transit lane introduction, such as the upkeep of signs, markings, and other traffic control devices, exist at any level of investment, although they are also lower for high-investment, high-quality options.

Fully controlled ROW (category A) generally requires the highest investment, but offer the highest service quality and lowest operating costs as compared to ROW categories B and C.

Results of this analysis, which should be presented in a summarized form for easy review, thus contain all significant monetary and other quantitative (e.g., travel times) aspects, as well as a number of qualitative aspects of the proposal for transit lane introduction.

2.3 STOPS AND STOPPING REGIMES

The frequency of transit unit stopping along a line, or distance between the locations where the TU stops for passengers, affects both passenger travel time and efficiency of operations. It is therefore important to plan stop locations and stopping regimes carefully for each line.

2.3.1 Definitions and Relationships

The basic concepts for analysis of transit line stops are:

- *Stops* are locations designated for passenger pick-up and drop-off by TUs; they include stops along streets and stations. Depending on the type or regime of operation, stops may be designated as demand, on-call or fixed, as defined below.
- *Stopping* is the event of a TU making a stop for passengers.
- *Stopping regime* represents the method of selecting stoppings, i.e., determination whether TUs make stoppings at all designated stops or only at those at which passengers request.

With respect to passengers, the main factor influencing selection of stops and stopping regime for a given transit line is passenger travel time and convenience; these depend on the ratio between the time for access to stops and on-vehicle travel time. Frequent stopping makes access time and distance short, while long spacings between stoppings result in high operating speed on the line. The optimum spacings for minimum passenger travel time can be derived from the trade-off between these two components.

However, that need not represent the optimum solution: if it results in low operating speed, transit service may not attract many passengers traveling longer distances. Moreover, both components of operator's costs, operating (energy, wear and tear) and investment (required fleet size) decrease with increased spacings. Consequently, the most desirable spacings are often greater than those resulting in minimum average passenger travel times.

2.3.2 Stopping Regimes and Stops

With respect to operating practice, there are three different regimes:

- *Demand stopping* is the regime when TUs stop at any location that passengers on board or along a street demand. Theoretically, lines with this stopping regime have an infinite number of stops and TUs stop twice for every passenger: once for boarding, once for alighting. In reality, stopping is limited by physical and traffic conditions along a line, yet their potential locations are numerous.
- *On-call stopping* is the operating practice in which there are fixed stop locations, but TUs stop only at stops where passengers demand.
- *Fixed stopping* is the regime in which all TUs stop at all stopping locations—stops or stations—regardless of whether any passengers will board or alight.

Stops are classified, correspondingly, into *demand, on-call,* and *fixed stops*. The number of fixed stops along a given line is smaller than the number of on-call or demand stops; therefore, their facilities (passenger protection, information, signing) can be more extensive than at on-call stops, while demand stops have no facilities at all.

Figure 2.12 shows schematically the three stopping regimes. When there are very few passengers along a line, demand stopping is obviously optimal: transit vehicles stop at any locations along the line where passengers want to board and alight.

As the number of passengers along the line increases, demand stopping results in ever slower operation and it becomes necessary to aggregate passenger boardings/alightings to a finite number of locations with on-call stopping regime: transit vehicles still stop only when passengers signal (call) for stoppings, but fewer times than with demand stopping.

For passenger convenience, there are usually a large number of on-call stops. Since any one TU stops at only a fraction of on-call stops, operating speed remains high. But when passenger volume becomes high, frequency of stopping increases. To maintain reasonable operating speed, it is logical to change the regime to fixed stopping with fewer fixed than on-call stops.

Table 2.7 gives the numbers of stops and stoppings for the three stopping regimes. The diagrams in Figure 2.13 show the number of stoppings and operating speeds as functions of passenger volume for the three stopping regimes. The two diagrams together show that with demand stopping regime the number of stoppings increases and operating speed decreases rather rapidly as passenger volume increases.

At a certain point the on-call regime offers an increase in operating speed sufficient to offset the additional walking of passengers to stops as compared to ubiquitous stopping in the demand regime. As increasing incidence of stoppings further decreases operating speed, it eventually becomes advantageous to change to a fixed stopping regime. This change again increases the average access distance of passengers compared to on-call operation, but it increases the operating speed of the line. The diagrams show schematically the optimal regions of the three stopping regimes.

The optimal spacing between fixed stops S can be mathematically derived as a function of average travel distance on the line ℓ_a, access speed to/from stops V_a,

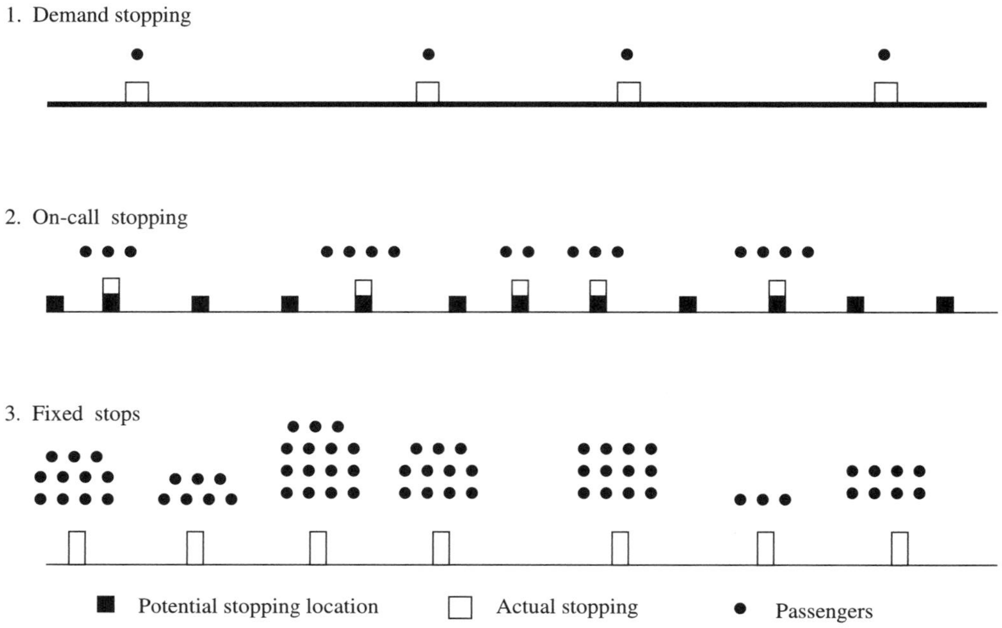

Figure 2.12 *The concept of the three stopping regimes along a transit line.*

Table 2.7 Numbers of stops along a line N and stoppings by a TU n for the three stopping regimes (subscripts indicate the regimes)

Stopping Regime	*Number of Stops*	*Number of Stoppings*		
		Relationship	*Min*	*Max*
1. Demand	$N_d = \infty$	$n_d = 2P_{TU}$	0	∞
2. On-call	N_o	$n_o \leq n_d$	0	N_o
3. Fixed	$N_f < N_o$	$n_f \leq n_o$	N_f	N_f

P_{TU} is the number of passengers boarding TU along the line.

and time lost per stopping T_ℓ. Similar to the derivations of optimal station spacings in Chapter 5, with passenger density p (prs/km) passenger travel times for access PT_a, and traveling on the line PT_t, are derived as, respectively:

$$PT_a = \frac{S \cdot p}{2V_a} \quad \text{and} \quad PT_t = p\left(\frac{\ell_a}{V} + \frac{\ell_a}{S} T_\ell\right) \qquad \left|\frac{PT}{\text{prs-h}}\right|\frac{S,\ell}{\text{km}}\left|\frac{p}{\text{prs/km}}\right|\frac{V}{\text{km/h}}\left|\frac{T}{\text{h}}\right| \tag{2.14}$$

Adding these two time components up, finding the derivative of that expression with respect to S, and setting it equal to 0 gives the following expression for the spacing resulting in the minimum total passenger travel time:

$$S^* = \sqrt{2\ell_a \cdot T\ell \cdot V_a} \qquad \left|\frac{S,\ell}{\text{km}}\right|\frac{T}{\text{h}}\left|\frac{V}{\text{km/h}}\right| \tag{2.15}$$

2.3.3 Practical Values of Stop Spacings

Determination of station spacings on rapid transit lines is discussed in Chapter 5. For surface transit, spacings

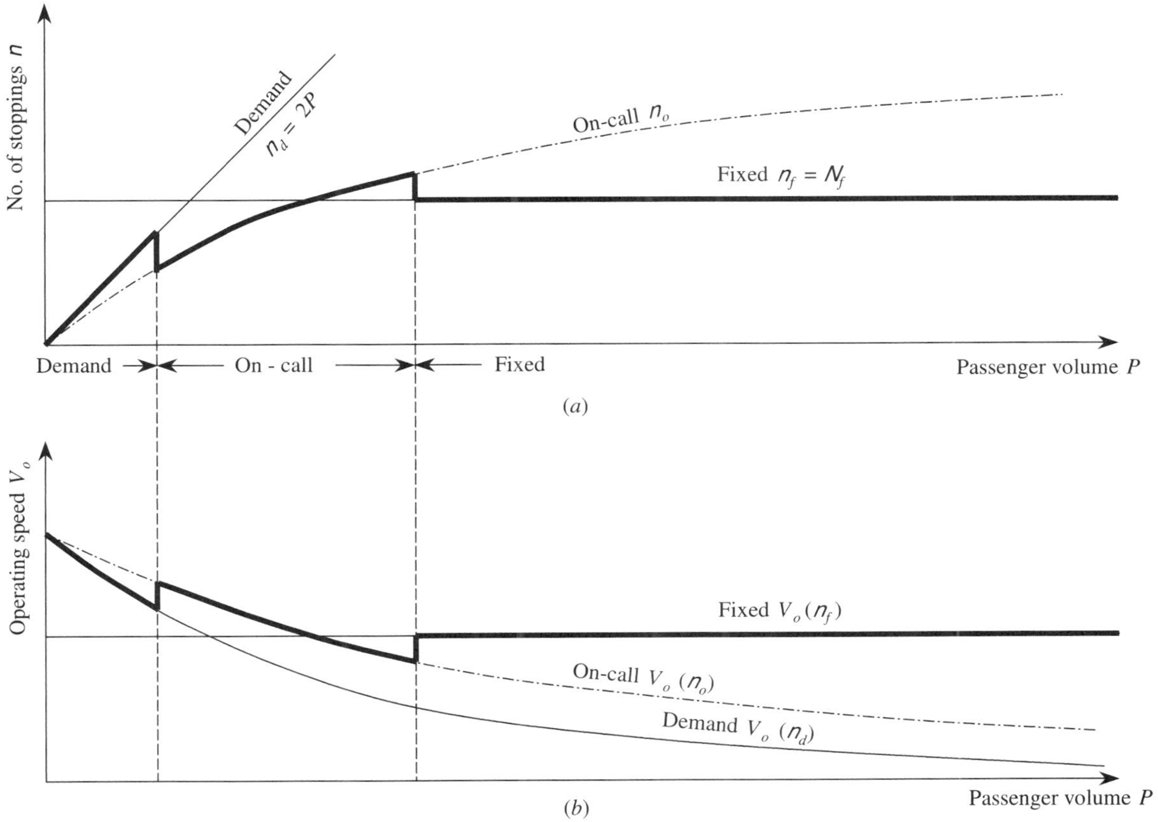

Figure 2.13 *Stopping regime domains as functions of passenger volume: (a) number of stoppings for different stopping regimes; (b) operating speeds for different stopping regimes.*

should be such that TUs stop on the average not closer than 300–500 m, with smaller cities at the lower and larger cities at the upper end of this range. With fixed stops this results in an average access distance along the line of 75–125 m and a maximum walking distance of 150–250 m. On lines with low passenger volume and on-call stops, considerably shorter spacings may be used. Since some of the stops are skipped, operating speed is not affected. If, however, short spacings are used and demand is high so that most stops are used, operating speed may be excessively lowered. For this reason, some cities use fixed stops in the above mentioned range with additional on-call stops between them. During peak hours, fixed stop operation is used, while in off-peak hours, when demand is low, operation is changed to on-call stopping.

The old-fashioned practice of on-call stops at every corner, originating from the times of horse-drawn vehicles, is still used in some cities. Inefficiency of this practice is shown by a specific example. A comprehensive comparison of a bus service with stop spacings of 320 m to an operation with 160-m spacings is presented in Table 2.8. The table shows that the only disadvantage of the longer spacing is somewhat longer average access distance for passengers to/from bus stops. The benefits for passengers, the transit agency, and other organizations and public are so numerous, however, that they easily outweigh the increase of ac-

Table 2.8 Comparison of a bus line with 320-m stop spacings with 160-m spacings

Directly Affected Party	*Advantages*	*Disadvantages*
Passengers	+ Higher travel speed: time saving + Higher comfort (reduced frequency of acceleration-deceleration) + More passengers attracted	– Walking to and from stops increased by an average or 40 m each, or a total of 80 m. This walk takes approximately one minute
Transit agency	+ Fewer vehicles may be required for a given schedule (if $\Delta T \geq h$), appreciable savings + Less energy consumption + Less vehicle wear and tear + Number of stops requiring signs, markings, benches reduced to a half + Improved stopping bays or parking prohibition possible	
Traffic engineering and police departments, general public	+ Less curb space taken by stops + Easier enforcement of no parking at stops + Reduced impedance of traffic by bus stopping in traveling lane + Less air pollution and noise	

cess distance of 80 m, or one minute of walking. The stop spacings of 300–400 m are clearly superior to those of 150–250 m.

2.4 ACCELERATED RAIL TRANSIT OPERATIONS WITH FIXED STOPPING SCHEDULES

Stopping at stops/stations represents a significant component of travel time on transit lines, particularly on rapid transit systems, which do not have delays caused by external elements, such as street congestion and traffic signals. Yet station spacings on metro lines in urban areas are usually determined by the area coverage requirement, amounting to 500–900 m. With these spacings, the average operating speeds TUs can achieve are in the range of 25–35 km/h. These speeds are satisfactory for medium trip lengths within cities, such as up to 8 or 10 km. However, for long lines and trips of 10–20 km, the service sometimes becomes too slow, particularly if a competing freeway parallels the line. This trade-off between area coverage and travel speed on the line and the demand for faster speeds has become an increasing problem with the spatial spread of cities.

If station locations (and thus the number of stations along a line) are fixed, the only way to increase speed is to introduce *accelerated operations,* in which some TUs do not stop at some of the stations, following a fixed predetermined schedule (demand stops are neither desirable nor possible on major transit lines). The three basic accelerated operations are *skip-stop, zonal,* and *express/local,* shown in Figure 2.14. Each is described here and compared with standard (all-stop) operation. Rail lines will be referred to as their most common users, but other guided modes or buses can also employ accelerated operations.

Accelerated operations are primarily applicable to rather long lines with many stations; they also require frequent service. Thus, they are applicable to many rapid transit lines, particularly during peak hours. Introduction of fully automated train operation (ATO) without crews on some metro lines may increase applications of accelerated services because it makes operation of short TUs at short headways economically feasible throughout the day.

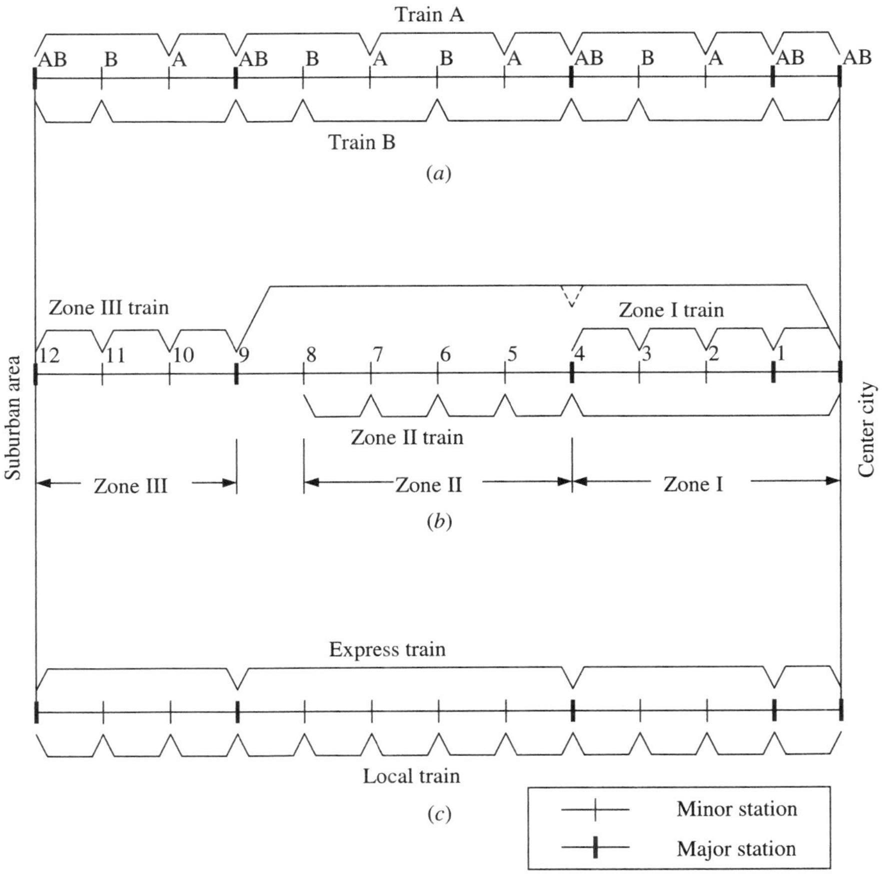

Figure 2.14 *Types of accelerated transit operations: (a) skip-stop operation; (b) zonal operation; (c) express/local operation.*

2.4.1 Skip-Stop Operation

Skip-stop is the only type of operation by which operating speed on transit lines with only two tracks can be increased while maintaining high frequency and high line capacity. Description and analysis of this operation was given in Vuchic, 1976.

2.4.1.1 Description and Characteristics. Skip-stop operation, shown schematically in Figure 2.14(*a*) and with standard operation on a time-distance diagram in Figure 2.15, is obtained by classifying stations along a line into three groups: A, B, and AB. TUs are then operated in two patterns: trains A stop at A and AB stations, while trains B stop at B and AB stations. As trains A and B alternate, A and B stations are served by every other TU, while at AB stations all TUs stop.

Considerations for the selection of A and B stations are:

- They should be the stations with the smallest numbers of passengers.
- The total number of passengers at A and those at B stations should be similar to maintain even loadings of A and B TUs.
- The number of A and B stations should be the same to maintain uniform headways at AB stations.

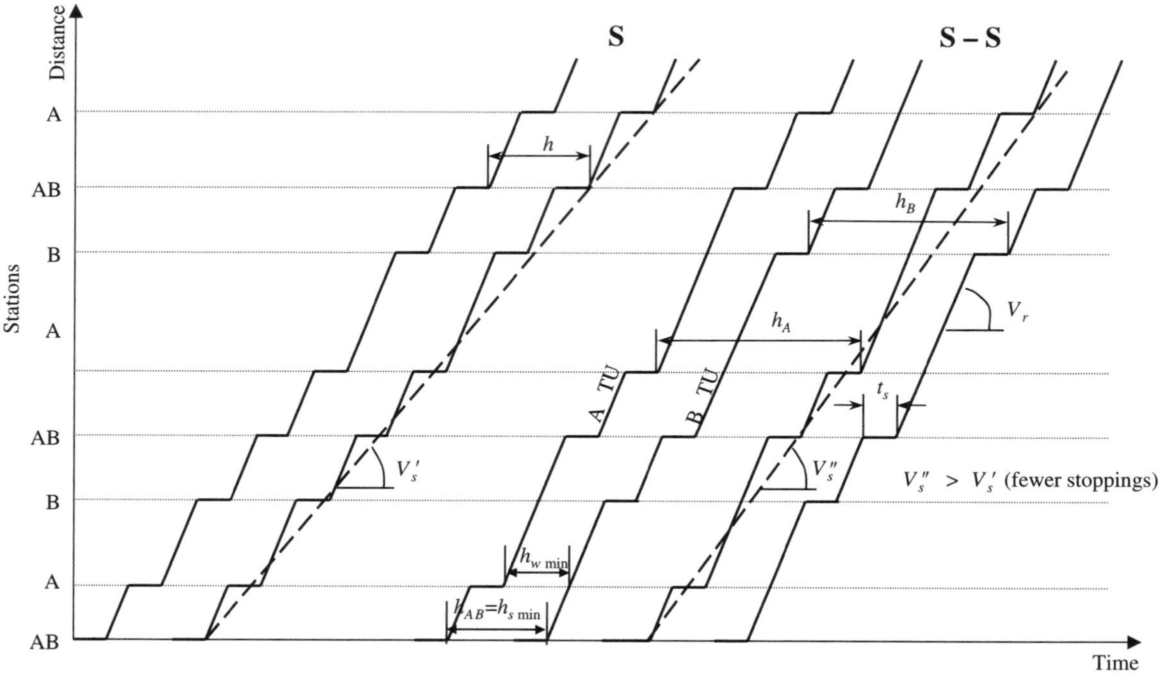

Figure 2.15 *Time-distance diagram for standard (S) and skip-stop (S-S) operations.*

- There should be as few consecutive A−B station pairs as possible, to minimize the number of links between A and B stations that cannot be traveled without reversing, i.e., traveling to an AB station beyond the destination and then taking another TU back.

The main characteristics of skip-stop operation are:

- It has higher operating speed and fewer stoppings along the line (see Figure 2.15). Passenger travel time is reduced and service is more attractive.
- For a constant number of passengers, total transportation work (prs-km) of the line remains the same. However, increased operating, and consequently cycle speed, of TUs results in their higher productivity (expressed in prs-km/veh-h). If this allows a decrease in the number of TUs, there are substantial cost savings.
- Capacity of the line is not affected. The common belief that skip-stop increases headways is in most cases not correct. As can be seen in Figure 2.15, if A and B trains leave an AB station at $h_{s\ min}$, train B catches up with train A at the following A station. However, train B can follow train A at a shorter way headway: $h_{w\ min} < h_{s\ min}$ to station B, where train B falls back to a $h_{s\ min}$, so that all trains arrive at the following AB station at uniform headways $h_{s\ min}$ again. Thus, heavily used stations, which are AB, continue to determine line capacity.
- Skip-stop operation involves somewhat more irregular movements of trains, so that a more rigorous control of train movements is required. If service is not reliable, line capacity may be negatively affected.
- Skip-stop requires practically no investment for introduction on existing or new lines, since the only preparation required is public information about types of stations (A, B, or AB) and TUs (A or B).

- Generally, skip-stop service can be introduced only on the lines—or during periods of day—for which headways are rather short (below five to six minutes) to avoid long waiting at A and B stations.

Following is a detailed comparison of skip-stop with standard operation, analysis of all differences between them, and a methodology for an exact, yet simple evaluation of alternative types of operation, standard and skip-stop, with different numbers of A–B station pairs.

2.4.1.2 Comparison with Standard Operation. The characteristics of skip-stop compared with standard operation and the resulting relative advantages and disadvantages are shown schematically in Figure 2.16. Rectangular boxes show operational differences; boxes with rounded sides contain advantages (*a*) and disadvantages (*d*) of skip-stop operation as they affect the two parties: passengers (P) or operator (O), i.e., the transit agency.

Difference 1 in Figure 2.16, higher operating speed, results in direct advantage a-1 in the figure (reduced passenger travel time on trains). In addition, the operator has two basic options of changing the operation to utilize the increased speed. First (I), the same number of TUs may be operated on the line, so that headways are reduced at AB stations. This is possible if $h_{AB} > h_{s\,\min}$. Two advantages are achieved: a_I-2 (waiting time) is shorter (P); and a_I-3 (line capacity) is increased (O,P). There is one disadvantage: d_I-1 (operating cost) is higher due to more TU-km. Alternatively, the second option (II) is to maintain the same

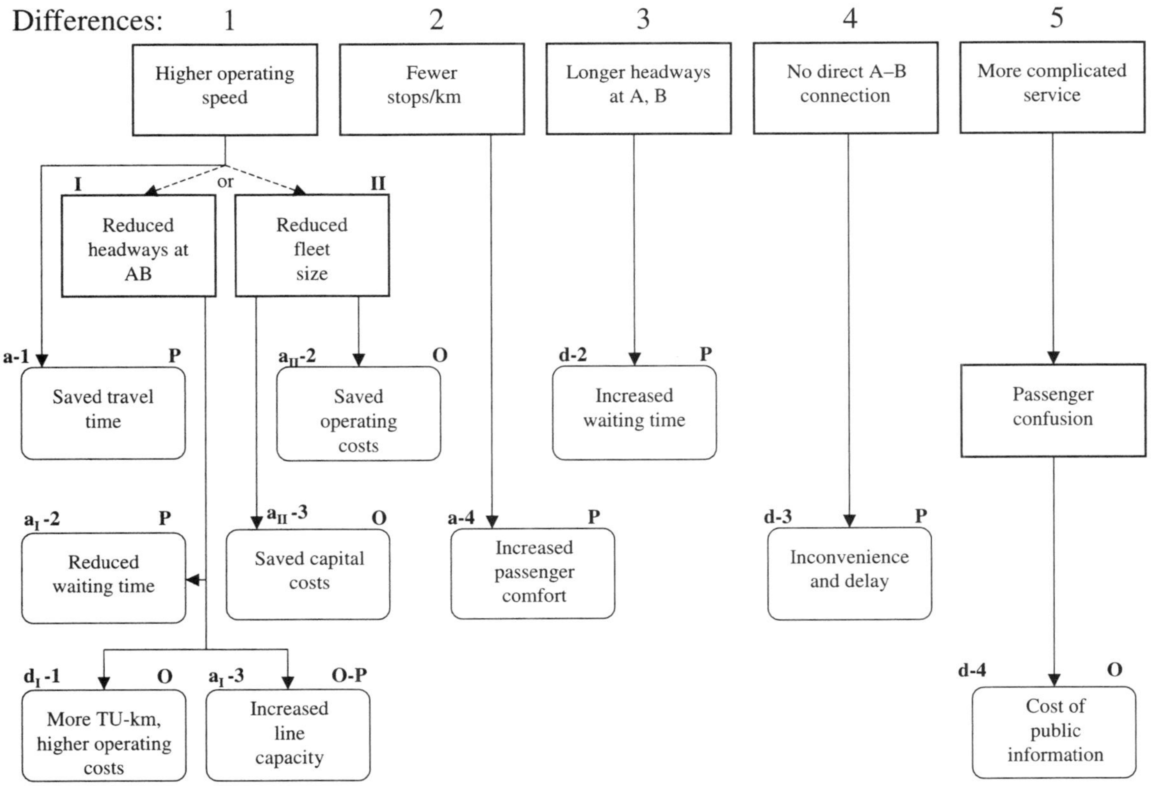

Figure 2.16 *Skip-stop compared with standard operation.*

headway and reduce the number of TUs in service. This brings two advantages: a_{II}-2 (operating cost) is reduced (O), and a_{II}-3 (capital costs) are reduced (O). This option is the most common. In cases, when capacity of the line is reached, it is the only feasible option.

Differences 2–5 result, respectively, in a-4 (increased passenger travel comfort) (P); d-2 (increased waiting times at A and B stations) (P); d-3 (inconvenience and delay due to transferring of some passengers) (P); and d-4 (cost of providing information to passengers) (O).

The above-listed advantages and disadvantages will now be analyzed for a line with a length L, ℓ spacings, and $\ell + 1$ stations. All factors that change when skip-stop operation is introduced (such as speed, headway, number of TUs, etc.) will be designated with a prime sign (′) added to the original symbol.

Since the attractiveness of skip-stop operation depends heavily on the length of headways (and thereby indirectly on passenger volumes), its relative advantages and disadvantages vary for different periods of the day. The analysis is therefore based on hourly values. The assumption is that the number of passengers boarding and alighting from the trains during that hour at each station is independent of the type of service, and that their arrivals are uniform. If this is not the case, the analysis should be done for a shorter period of time.

For standard operation, travel time from one terminal to the other is:

$$T_o = \frac{60L}{V_{max}} + \ell \cdot T_\ell \qquad \left|\frac{T}{\text{min}}\right|\frac{L}{\text{km}}\left|\frac{V}{\text{km/h}}\right|\frac{\ell}{-}\right| \qquad (2.16)$$

where T_ℓ is the time loss due to stopping at one station. It consists of incremental time losses in braking and acceleration, and dwell time, as derived from Eq. (1.18):

$$T_\ell = \frac{V_{max}}{432}\left(\frac{1}{a} + \frac{1}{b}\right) + t_s \qquad \left|\frac{T_\ell}{\text{min}}\right|\frac{V}{\text{km/h}}\left|\frac{a, b}{\text{m/s}^2}\right|$$

The computed value of T_ℓ can be easily checked experimentally on the transit line: train travel time between two points is measured with and without one stopping between them. T_ℓ is the difference between these two times. It will be assumed in this analysis of accelerated operations that T_ℓ is constant, although it is somewhat shorter when the train does not reach V_{max} on a spacing between two stations.

The operating speed is based on Eq. (1.28):

$$V_o = \frac{60L}{T_o} \qquad \left|\frac{V}{\text{km/h}}\right|\frac{L}{\text{km}}\left|\frac{T}{\text{min}}\right|$$

Using Eqs. (1.48), (1.46), (1.20), and (2.16), the headway can be expressed as:

$$h = \frac{2}{N_{TU}}\left(\frac{60L}{V} + \ell \cdot T_\ell + t_t\right) \qquad \left|\frac{h}{\text{min/TU}}\right|\frac{N_{TU}}{\text{TU}}\left|\frac{L}{\text{km}}\right|\frac{V}{\text{km/h}}\left|\frac{\ell}{-}\right|\frac{T, t}{\text{min}}\right| \qquad (2.17)$$

and transporting capacity of the line is:

$$C = C_v \cdot n \cdot f_{max} = \frac{60C_v \cdot n}{h_{min}} \qquad \left|\frac{C}{\text{sps/h}}\right|\frac{C_v}{\text{sps/veh}}\left|\frac{n}{\text{veh/TU}}\right|\frac{f}{\text{TU/h}}\left|\frac{h}{\text{min/TU}}\right| \qquad (2.1)$$

Speed Increase. With skip-stop operation, the travel time changes to:

$$T'_o = T_o - k \cdot T_\ell \qquad \left|\frac{T}{\text{min}}\right|\frac{k}{-}\right| \qquad (2.18)$$

where k is the number of A−B station pairs. The new scheduled speed, V'_o, is expressed by Eq. (1.28), with T'_o substituting for T_o. The new cycle time is:

$$T' = T - 2k \cdot T_\ell \qquad \left|\frac{T}{\text{min}}\right|\frac{k}{-}\right| \qquad (2.19)$$

and the headways at AB stations under policy I are reduced to:

$$h_{AB} = \frac{T'}{N_{TU}} = h - \frac{2k \cdot T_\ell}{N_{TU}} \qquad \left|\frac{h}{\text{min/TU}}\right|\frac{T}{\text{min}}\left|\frac{N_{TU}}{\text{TU}}\right|\frac{k}{-}\right| \tag{2.20}$$

With policy II, the headways at AB stations do not change. At A and B stations the headways, under both policies, become:

$$h_A = h_B = 2h_{AB} \qquad \left|\frac{h}{\text{min/TU}}\right| \tag{2.21}$$

All changes in passenger travel times can now be analyzed together.

Passenger travel time (a-1, a_f-2, d-2). Two aspects of passenger travel time are important: how does the total travel time of all passengers change, and how is the time saving/loss distributed among passengers?

The total change in passenger travel time on the line consists of the changes in the time on trains and the changes in the waiting time at stations:

$$\Delta PT = \Delta PT_t + \Delta PT_w \qquad \left|\frac{PT}{\text{prs-h}}\right| \tag{2.22}$$

Considering time saving as positive and increased time as negative, the change in the time on trains is:

$$\Delta PT_w = \frac{h - h'_{AB}}{120} \sum_{AB} P_{AB} - \frac{h'_{A,B} - h}{120} \sum_{A,B} P_{A,B} \qquad \left|\frac{PT}{\text{prs-h}}\right|\frac{T}{\text{min}}\left|\frac{R}{\text{prs}}\right| \tag{2.23}$$

where $P_{A,B}$ is the number of passengers on all trains passing through A and B stations during the analyzed period (hour). This number is divided by two because half of the passengers passing through, say, an A station will be on B trains (assuming even loading) and save T_ℓ. The other half, on A trains, will make the stop.

The waiting time in the stations changes by:

$$\Delta PT_w = -\frac{h}{120} \sum_{A,B} P_{A,B} \qquad \left|\frac{PT}{\text{prs-h/TU}}\right|\frac{h}{\text{min/TU}}\left|\frac{P}{\text{prs}}\right| \tag{2.24}$$

where P_{AB} is the number of passengers who travel between AB stations and $P_{A,B}$ is the number of passengers boarding at A and B and those boarding at AB but alighting at A or B. This equation holds for both policies, although for policy II, $h'_{AB} = h$ and $h'_{A,B} = 2h$, so that Eq. (2.24) simplifies to:

$$\Delta PT_t = \frac{T_\ell}{120} \sum_{A,B} P_{A,B} \qquad \left|\frac{PT}{\text{prs-h/TU}}\right|\frac{h}{\text{min/TU}}\left|\frac{P}{\text{prs}}\right| \tag{2.25}$$

In most cases for which skip-stop service would be considered, the aggregate time saving ΔPT would be considerable. However, that is not a sufficient reason for introduction of the service, since the distribution of the time savings may be quite uneven. Some passengers, suffering a considerable increase in travel time, might stop using the line. It is necessary, therefore, to analyze the time savings and losses for individual groups of passengers.

Passengers boarding at one and alighting at another AB station are clearly only gaining; they have either the same or decreased headways and increased travel speed on the line. The passengers boarding at A and at B stations, however, have waiting time changed (increased) by:

$$\Delta t_{w\,A,B} = -\frac{h'_{A,B} - h}{2} \qquad \left|\frac{t_w, h}{\text{min/TU}}\right| \tag{2.26}$$

while their train time saving depends on the distance they travel, or more precisely, on the number (j) of A–B pairs on the section of the line that they travel:

$$\Delta t_p = j \cdot T_\ell \qquad \left| \frac{t,\ T}{\text{min}} \middle| \frac{j}{-} \right| \tag{2.27}$$

The passengers realize an overall saving in travel time if $\Delta t_p > |\Delta t_w|$ (for their trips). Substituting Eqs. (2.27) and (2.26) and then (2.21) and (2.20), with policy I, passengers save time if

$$j > j_c = \frac{h}{2\ T_\ell} - \frac{2k}{N_{TU}} \qquad \left| \frac{j,\ k}{-} \middle| \frac{h}{\text{min/TU}} \middle| \frac{T}{\text{min}} \middle| \frac{N_{TU}}{\text{TU}} \right| \tag{2.28}$$

This expression is very simple to use for any given line. For example, if there are six pairs of A–B stations on a line, $T_\ell = 0.75$ min, and the number of trains in service is 20, the passengers at stations A and B would save time if they traveled over j A–B pairs, defined as:

$$j > \frac{k}{1.5} - 0.6$$

Thus, one can examine the distribution of time saving for any given headway for any period of day. All trips satisfying inequality (2.28) realize a gain. Note that h is the initial headway, i.e., for standard operation. For policy II the expressions for j_c is:

$$j_c = \frac{h}{2T_\ell} \qquad \left| \frac{j}{-} \middle| \frac{h}{\text{min/TU}} \middle| \frac{T}{\text{min}} \right| \tag{2.29}$$

Operating Costs (a_{II}*-2*). TU operating costs are lower for skip-stop operation in most cases, since the reduced number of stoppings per hour of operation reduces both the power requirements and the wear and tear on vehicles. Although no exact data on this item are available, an approximate value may be obtained by measuring the number of TUs that can be taken out of service when skip-stop operation is introduced (policy II). The savings in the operating costs are at least equal to the operating costs for the TUs taken out of service. Cost of additional information for skip-stop service is negligible.

Transporting Capacity (a_I*-3*). Under policy I, the number of TUs on the line is maintained constant, so that the capacity of the line increases due to the shorter headways by:

$$\Delta C = C' - C = 60n \cdot C_v \left(\frac{1}{h'_{AB}} - \frac{1}{h} \right)$$

$$\left| \frac{C}{\text{sps/h}} \middle| \frac{n}{\text{veh/TU}} \middle| \frac{C_v}{\text{sps/veh}} \middle| \frac{h}{\text{min/TU}} \right| \tag{2.30}$$

Reduced Fleet Size (a_{II}*-2 and 3*). Under policy II, the headways are retained without change for skip-stop operation, so that, due to the decreased cycle time, the number of TUs on the line can be reduced. The savings due to this reduction can be expressed as:

$$\Delta K = K \cdot \Delta N_{TU} = K\,\frac{T - T'}{h} = 2K\,\frac{k \cdot T_\ell}{h}$$

$$\left| \frac{\Delta K}{\$} \middle| \frac{K}{\$/\text{TU}} \middle| \frac{N_{TU}}{\text{TU}} \middle| \frac{T}{\text{min}} \middle| \frac{h}{\text{min/TU}} \middle| \frac{k}{-} \right| \tag{2.31}$$

where K is being the total (capital and operating) cost of a TU per unit of time (day or year). It will be shown later that this saving can be very substantial.

Fewer Stops: Increased Passenger Comfort (*a-4*). Stopping a transit vehicle is undesirable for the passengers not only because of the delay, but because it represents an interruption in their ride and affects them through deceleration-acceleration, opening and closing of doors, walking through cars, etc. The significance of this interruption is not possible to measure in quantitative terms, but it has been observed that some passengers do not take the first train if it is local, but rather wait for an express, even when the latter will bring them to their destination later than the local. When the train does not stop, the passenger's impression is that the saving is actually considerably greater than the 30–60 s reduction in travel time. The perceived benefits of not stopping are, therefore, an important advantage of the skip-stop operation.

Travel between A and B Stations: Inconvenience and Delay (d-3). With the introduction of skip-stop operation, there is no direct connection between A and B stations, so that the passengers traveling between such stations suffer an inconvenience. Those traveling between distant A and B stations have to transfer from an A to a B train at an intermediate AB station. This involves certain discomfort and a small loss of time (h'_{AB}) for the passengers.

The passengers traveling between adjacent A and B stations cannot make those trips, unless they would travel past their station to the first AB station and then backtrack to their destination. This is highly inconvenient, and if the AB station has side platforms, another fare payment may be necessary. However, in most cases either such trips do not exist or their number is quite negligible, since very few passengers travel very short distances on rapid transit. Yet this factor should be considered in selecting A and B stations: many consecutive A and B stations should be avoided.

Increased Service Complexity: Additional Information for Passengers (d-4). Skip-stop represents a somewhat more complicated service than standard operation. Passengers must pay more attention to which train they take. This is an item of inconvenience, but in most cases it can be solved by providing adequate information, particularly at the time such service is initially introduced. This involves very low cost. If skip-stop operation is used only during certain periods of the day, this information should be displayed more distinctly during those times.

2.4.1.3 Methodology for Skip-Stop Planning and Evaluation. The preceding analysis can be utilized for a systematic evaluation of standard versus skip-stop operation as well as for finding the optimal number of A−B station pairs. The methodology consists of five steps: data collection, planning decisions, data preparation performance computations, and evaluation of alternatives.

Data Collection. This calls for obtaining operating data of the analyzed line: L, T_o, or V, ℓ and T_ℓ, t_t (or T), h (or N_{TU}), C_v, and K for the time period considered for skip-stop operation. Trip matrix for the line—the number of trips from each to each station—for the studied period should be obtained. In metro systems with automated fare collection, complete tables of station-to-station trips for any time period are readily available. If complete counts are not available, the best possible estimate should be made from counts at individual stations, train loadings, etc.

Planning Decisions. The alternative skip-stop combinations that are to be analyzed (e.g., with two, four, and six A−B pairs—even numbers maintain uniform headways at AB stations) should be selected using the principles given in Section 2.4.1.1. Which stations would be A and B in each alternative should be determined, as should whether policy I or policy II will be used.

Data Preparation. The trip table, shown for a simple example in Table 2.9, should be prepared from the collected data. This table includes all required passenger volumes. Using the symbols listed below Table 2.9, one finds the following information.

The numbers of passenger trips among all stations are written in respective squares within the heavy lines, except those along the diagonal. For example, in this case there are 64 trips from station K to station N. Totals of all westbound origins from each station, in column O_w, are sums of rows to the left of the diagonal; those to the right of it represent eastbound origins O_e. Total trip origins for each station, O, are obtained by adding O_w and O_e. Corresponding summations of columns give sums of destinations in the three rows for D's.

The squares along the diagonal line show the number of passengers in trains leaving each station by direction. If the number of passengers leaving station M in the westbound direction is designated as S_{Mw}, it is computed as follows:

$$S_{Mw} = S_{Nw} - D_{Mw} + O_{Mw} = 343 - 36 + 188 = 495 \text{ prs}$$

Table 2.9 Trip table for computations of passenger volumes on a line: a numerical example

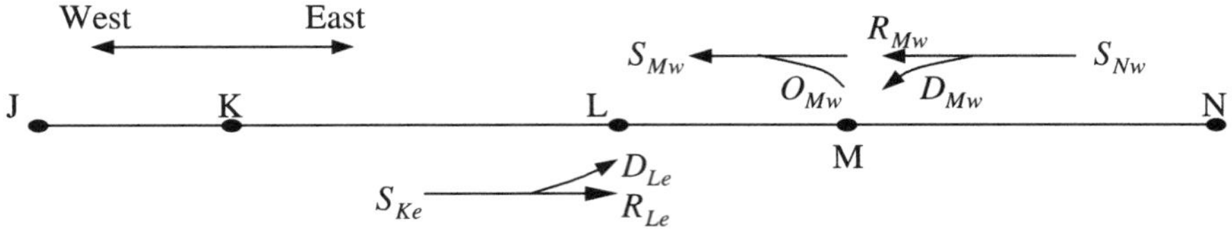

	To →	*J*	*K*	*L*	*M*	*N*		
From ↓	D_e → / ↓ O_w	*0*	*12*	*91*	*280*	*291*	O_e ↓	*O* ↓
J	0	242 / 0	12	50	128	52	242	242
K	17	17	438 / 240	41	103	64	208	225
L	46	38	8	526 / 367	49	130	179	225
M	188	107	55	26	291 / 495	45	45	233
N	343	78	81	148	36	0 / 343	0	343
	D_w →	240	144	174	36	0	↖ S_e ↖ S_w	1268
	D →	240	156	265	316	291	1268	—

J–N: Station designation
O, D: Originating and destination passengers, respectively
e, w: Eastbound and westbound, respectively
S, R: Passengers traveling on a section or through a station, respectively
Source: Vuchic 1976.

The number of passengers traveling through a station is not written in the table, but it can be computed easily. For example, the number of passengers passing through station L in the eastbound direction R_{Le}, is computed as follows:

$$R_{Le} = S_{Ke} - D_{Le} = 438 - 91 = 347 \text{ prs}$$

A performance summary is then prepared for review of evaluation computations. A numerical example of such a table, assuming that the more common policy II is considered, is shown in Table 2.10 (the numbers are not related to those in Table 2.9).

Performance Computations. The following should be computed for each studied alternative:

T_o by Eq. (2.16), T_o' by (2.18), V_o and V_o' by (1.28), T by (1.48), T' by (2.19); for policy I, h_{AB}' is computed by (2.20), and for policy II by (1.46). Also, j_c is computed by Eq. (2.28), ΔC by (2.1), and ΔK by (2.31).

Table 2.10 Performance summary for comparison of skip-stop with standard operation: rapid transit line, peak hour, policy II

Operating Element	Dimension	Standard Operation	Skip-stop Operation: 3 A–B Pairs	Skip-stop Operation: 6 A–B Pairs	Skip-stop Operation: 7 A–B Pairs
V_o	km/h	33.2	34.8	36.7	37.4
T	min	86	82	78	(77) 78
N	veh	43	41	39	39
$\Delta\ Pt_t$	h/day	—	354	765	870
$\Delta\ Pt_{wA,B}$	h/day	—	−18	−57	−67
$\Delta\ Pt$	h/day	—	336	708	803
$\Sigma\ P_{A,B}$	h/day	—	1077	3392	4036
j_c	—	—	1.7	1.7	1.7
$\Delta\ K$ operations	\$/yr	—	240,000	480,000	480,000
$\Delta\ K$ investment	\$/yr	—	200,000	200,000	200,000
$\Delta\ K$ total	\$/yr	—	680,000	680,000	680,000

$L = 21.0$ km
$T_l = 36$ s $= 0.6$ min
$t_t = 5$ min
$T_o = 38$ min
$h_{AB} = 2$ min
Source: numerical example adapted from Vuchic, 1976.

Passenger time is computed in three steps:

- Find ΔPT_t by Eq. (2.23); the value for $\Sigma_{A,B}\ R_{A,B}$ is obtained by computing R for all A and B stations as explained in the trip table, then summing them.
- Compute ΔPT_w by Eq. (2.25). The number of affected passengers is the total of all passengers boarding trains at A and B stations (in both directions), found in the last column of the trip table.
- Derive total time saved by subtracting the increased waiting from the reduced travel time, by Eq. (2.22).

Finally, it is necessary to compute how many passengers traveling between A and B stations are affected and how (h_{AB} added to their travel time due to changing trains, or impossible to make the trip). If significant, the time savings estimates should be corrected and the number of affected passengers added to the performance summary in Table 2.10.

Evaluation of Alternatives. All the major differences among the alternatives are consolidated in the performance summary. Because it is very difficult to bring the various items in the table to a common denominator (values of time and convenience would be particularly difficult to handle), and often many other local factors must also be included in considerations, the planning engineer must evaluate the alternatives by observing simultaneously the following items:

- V_o': in addition to the time savings, the increased speed makes the service more attractive for new passengers.
- ΔPT: total time savings of passengers, although relatively small for each individual, may represent a significant social benefit.
- j_c: if its value is high, the number of passengers negatively affected by skip-stop operation may be significant; in many practical cases, it is, however, quite negligible.
- ΔN_{TU} and ΔK: savings in the number of required TUs represent a very direct cost reduction for the

operator. If the analysis is made prior to a purchase of cars, the savings are not only in the operating, but also in the investment costs.

The number of passengers affected negatively in different ways should be analyzed with particular attention. This should be given a greater relative weight than such items as the amount of time lost, number of additional transfers, etc.

2.4.1.4 Potential Applications of Skip-Stop Operation. In summary, the main benefits from skip-stop operation are reduced passenger travel time, increased traveling comfort, enhanced attractiveness of service for potential travelers due to the increased speed, and savings to the operator, all of which can be quite substantial on some lines. The main problems of skip-stop operation are increased headways at A and B stations and some initial passenger confusion, which can be easily overcome. Physical capacity of line remains unchanged.

Skip-stop operation is particularly effective on lines with many stations, since it can then offer significant speed increase. However, it can be introduced only when headways are short, so that the double headway at stations A and B is still acceptable. In most cases, skip-stop could be readily introduced when headways are shorter than three to four minutes, since the new headways of up to six to eight minutes would not be excessive. In some cases, skip-stop may be desirable even for headways of five to six minutes if a significant number of skip-stop stations pairs can be introduced, so that the increased waiting time is more than offset by reduced travel time for most passengers. In the example following Eq. (2.28), for a headway of six minutes, passengers traveling through four or more AB pairs would realize greater saving in travel time than loss in waiting. For headways of three minutes, passengers traveling over only two A−B pairs would already realize a net time saving. Consequently, in many cities skip-stop service would result in benefits to users and operator far exceeding the inconvenience it would cause, particularly during the peak hours. It appears that skip-stop might be well suited to many metro lines in Paris, Montreal, Hamburg, Stockholm, Moscow, and other cities. Some cities with high-frequency light rail lines (e.g., San Francisco, Boston, Brussels, Essen) or bus lines (several British cities) could also benefit from skip-stop operation.

2.4.2 Zonal Operation

Transit lines in large cities predominantly serving commuters from large suburban areas into center city, such as regional rail lines in New York (LIRR, Metro North, and NJT), London (British Railways), and Paris (RER and SNCF), often have zonal operations (Eisele, 1978). Bus lines in some cities also utilize this type of operation.

2.4.2.1 Description and Characteristics. Zonal operation, as the diagrams in Figures 2.14 and 2.17 show, is obtained when a line is divided into two or more sections, referred to as zones, and each zone is served by a different set of trains. Then the first train leaving center city goes nonstop to the farthest zone and stops at all its stations; the next train goes nonstop to the next-closer zone and serves its stations, etc. Each train is turned back at the end of the zone it serves, usually making the same stops that it made traveling outbound.

Zonal operation can only be used on long radial lines on which a large majority of passengers travel from many suburban stations into one (or a few) stations in CBD (in the afternoon the travel pattern is reversed—it is one, or a few, to many stations). The line can have only two tracks, but there must be crossovers or, desirably, a third track for reversing at terminal stations of each zone.

Passenger volume on the line must be substantial, so that acceptable headways can be provided for each zone; but the line cannot operate close to its capacity, because zonal operation has lower maximum hourly frequency than regular operation.

To achieve greater regularity of service and allow interchangeability of trains among zones, it is necessary to make zones similar in numbers of stations and passenger volumes they serve. End stations for each zone should be major ones by passenger volumes and have an adequate turnback facility.

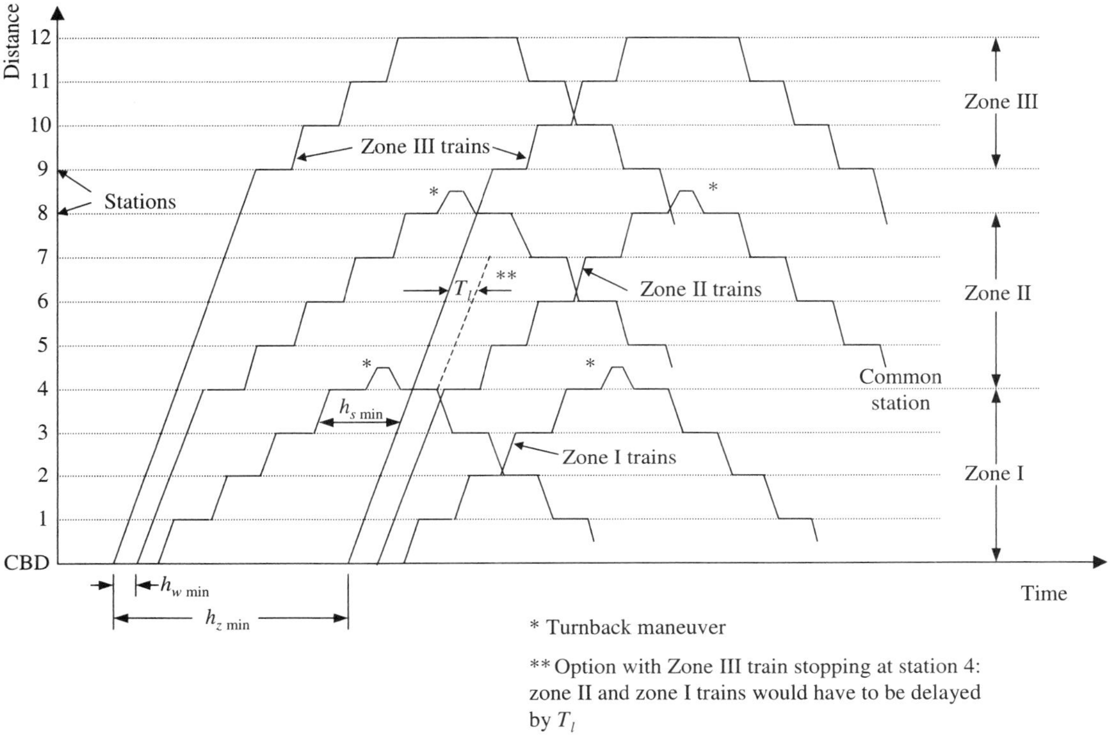

Figure 2.17 *Time-distance diagram for zonal operation.*

The operating pattern of trains with respect to zone terminals can be of several types:

1. There are no joint stations among zones, such as between zones II and III in Figures 2.14 and 2.17. All stations have the same service frequency, and travel between the two zones is practically impossible.
2. Trains of both adjacent zones serve the terminal station of each zone, as between zones I and II in Figures 2.14 and 2.17. These stations have double service frequency, and travel between adjacent zones is possible through one transfer between consecutive trains.
3. Each zone terminal is also served by trains serving all zones beyond it. In the example in Figures 2.14 and 2.17, station 4 would be served not only by the trains serving zones I and II, but also by those for zone III, as shown by dashed lines. This operation causes additional delays to trains for remote zones, but it allows travel among all zones with one transfer only.

As Figure 2.17 shows, zonal operation is characterized by platoon-type train dispatching (or arrivals) at the CBD end of the line. If the line has m zones, m trains leave the CBD station at short intervals. They can be as short as $h_{w\ min}$ if operations are as described under 1 and 2 above, while in the case of 3, the headways must be at least $h_{s\ min}$, except for the m-th train, which can still leave after $h_{w\ min}$ interval. Following the m-th train, there must be a headway at least $(\ell_I T_\ell + h_{w\ min})$ long, ℓ_I being the number of stations in zone I. Thus, the minimum headway between trains serving any one zone is:

Zonal operation has the following advantages (+) and disadvantages (−) compared to standard operation:

+ Faster and more comfortable travel between center city and stations beyond first zone
+ Greater number of shorter TUs usually results in lower fleet size requirement
+ Usually lower operating costs
− Inconvenient or impossible travel among zones
− Lower frequency of service at most stations
− Lower line capacity
− More complicated operation and uneven headways may cause passenger confusion and lower reliability of service

$$h_{z\min} = (m - 1)h_{w\min} + \ell_I \cdot T_\ell + h_{w\min}$$

$$= m \cdot h_{w\min} + \ell_I \cdot T_\ell \qquad \left|\frac{h,\,T}{\text{min}}\right|\frac{m,\,\ell}{-}\right| \qquad (2.32)$$

This equation shows that the minimum headway (determining line capacity) for zonal operation increases with the number of zones, number of stations in the first zone, and time lost per station for the first zone. Consequently, if line operation is close to capacity, zonal operation should have a short first zone.

Operating speed for zone I remains the same as for regular operation:

$$V_{oI} = \frac{60\,L_I}{60\,L_I/V + \ell_I \cdot T_\ell} \qquad \left|\frac{V}{\text{km/h}}\right|\frac{L}{\text{km}}\left|\frac{\ell}{-}\right|\frac{T}{\text{min}}\right| \qquad (2.33)$$

For each subsequent zone, there is a time saving in the amount of delays for all stations where the train did not stop. For zone m that is:

$$\Delta T_m = \sum_{i=1}^{m-1} \ell_i \cdot T_\ell \qquad \left|\frac{T}{\text{min}}\right|\frac{\ell}{-}\right| \qquad (2.34)$$

and operating speed to the end of zone m is:

$$V_o = \frac{60L}{60L/V + \ell_m \cdot T_\ell} \qquad \left|\frac{V}{\text{km/h}}\right|\frac{L}{\text{km}}\left|\frac{\ell}{-}\right|\frac{T}{\text{min}}\right| \qquad (2.35)$$

If zonal schedule is made with the same load factor α as the comparable standard operation, the zonal operation results in lower fleet size requirement, for two reasons. First, train capacities are better tailored to passenger volume in each zone (not all TUs go to the end of the line); and second, most trains operate with shorter cycle times.

On the negative side, there are also two major and several minor factors. The major ones are that service headways at most stations are increased and the number of trains required for operation is greater. Although shorter, the greater number of trains involves more personnel. The relationship between the additional labor costs and reduced fleet costs depends on local unit costs, but in most cases zonal operation involves a lower total cost than standard operation.

2.4.2.2 Comparison with Standard Operation. It is important to understand that zonal operation, unlike skip-stop, cannot be used on most transit lines; it is limited to lines with cumulative loadings, usually radials from center city into suburbs. Even if travel among different points along the line is low, inconvenience to those riders must be taken into account. Differences between zonal and standard (local) operations are summarized in the box.

The two major advantages of the zonal service, faster service and smaller fleet, are often very significant, outweighing the disadvantages. On very long lines with many stations, passengers traveling to or from outer zones may save substantial travel time. Also, reduction of cost of operation due to lower fleet requirements, similar to skip-stop operation, may amount to considerable savings.

Disadvantages of this operation may, however, also be important. Although frequency of service is somewhat less critical for commuters than for transit riders, it is still an important service feature. Many potential passengers are lost to commuter lines that offer service with long, sometimes irregular peak hour headways of over 30 or even 60 minutes.

The second major disadvantage is splitting the line into zones with poor or no service among them. Passengers making such trips may be lost. This can be a serious problem since some commuter lines serve considerable number of intrasuburban trips. Moreover, non-CBD-oriented trips represent economically desirable diversification of service that regional rail lines offer. Some reverse commuting, served well by such lines, may also be prevented by zonal operation.

2.4.2.3 Potential Applications of Zonal Operation. In conclusion, zonal operation has potential for application on the lines when the following conditions exist:

- They serve virtually exclusively cumulative travel from many points to one or several center city stations.
- The line operates considerably below its capacity.
- There are, or can be constructed, turnback track facilities at the stations that would be convenient as terminals for individual zones.

2.4.3 Express/Local Operation

The only way to provide regular service among all stations as well as higher-speed service stopping only at major stations is to operate both local and express services. Several types of such service exist on various rail rapid transit, and sometimes also on bus lines.

2.4.3.1 Types of Express/Local Operations. Depending on the number of tracks and station types on a transit line, there can be several types of express/local operations. The main types are shown in Figure 2.18 and described here. For these operations, stations may have two (exceptionally three) or four tracks, and they may be operated as local (L, for local trains only), or express (X, for both local and express trains) stations. The four-track stations allow simultaneous stopping of local and express trains, or express trains passing stopped local trains. Thus, designating the number of tracks by subscripts, there are four (six) types of stations: L_2, (L_3), L_4, X_2, (X_3), and X_4.

Because travel of local and express trains on all two-track lines is mutually dependent, their express stations must be located at distances with approximately equal travel times, because train overtakings must occur at equal headways. Their schedules must then be synchronized. On four-track lines, local and express trains run independently, but if simultaneous transfer between trains is to be provided at express stations, the schedules must be synchronized, as discussed below.

Case 2a: Two-Track Lines with Two-Track Stations. This operation, shown in Figure 2.19, allows only limited express operation: an express train departs from the terminal, followed after an interval $h_1 = h_{w\,min}$ by a local train. Another pair of express/local trains departs, as shown in Figure 2.19, after an interval $h/2$ that allows the preceding train to make local stops before the express train catches up with it:

$$h_2 = h_{s\,min} + \ell_\ell \cdot T_\ell \qquad \left| \frac{h,\ T}{\text{min}} \middle| \frac{\ell}{-} \right| \qquad (2.36)$$

where ℓ_ℓ is the number of L_2's at which the express train does not stop. Thus, for every L_2, the minimum headway between the pairs of express/local trains increases by a T_ℓ, and the scheduled line capacity, which, assuming equal number of local and express trains, can be expressed as:

$$C = f_{max} \cdot n \cdot C_v = \frac{120}{h_1 + h_2} n \cdot C_v$$

$$\left| \frac{C}{\text{sps/h}} \middle| \frac{f}{\text{TU/h}} \middle| \frac{n}{\text{veh/TU}} \middle| \frac{C_v}{\text{sps/veh}} \middle| \frac{h}{\text{min/TU}} \right| \qquad (2.37)$$

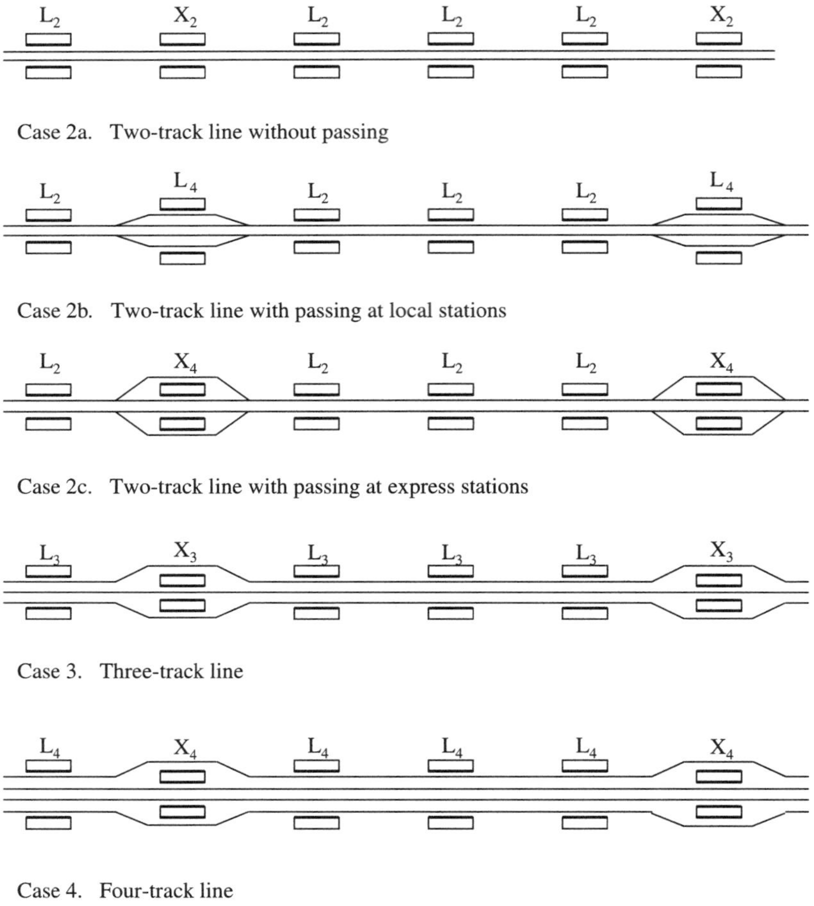

Figure 2.18 *Line and station design for different types of express/local operations.*

decreases correspondingly. On a typical rapid transit line, time elements have the following values: $h_{w\,\min}$ = 40–70 s, $h_{s\,\min}$ = 90–120 s, T_ℓ = 40–60 s. Thus, h_2 increases by 40–60 s for each L station, and scheduled line capacity decreases by about 0.25 $n \cdot C_v$ compared with local-only operation. Consequently, on two-track lines that operate considerably below capacity (examples are Cleveland, PATCO in Philadelphia, and some RGR lines), express/local operation is sometimes used with several (four to eight) local stations.

Cases 2b and 2c: Two-Track Lines with some Four-Track Stations. These allow express trains to overtake locals. The four-track stations may be of two types: first, L_4, in which case the express passes local train at full speed (case 2b); and second, X_4, which has platforms for all tracks, allowing simultaneous stopping of express and local trains (case 2c). The latter allows passenger transferring, so that local trains serve as collectors-distributors for expresses. However, the delay to local trains in case 2c is even greater than in

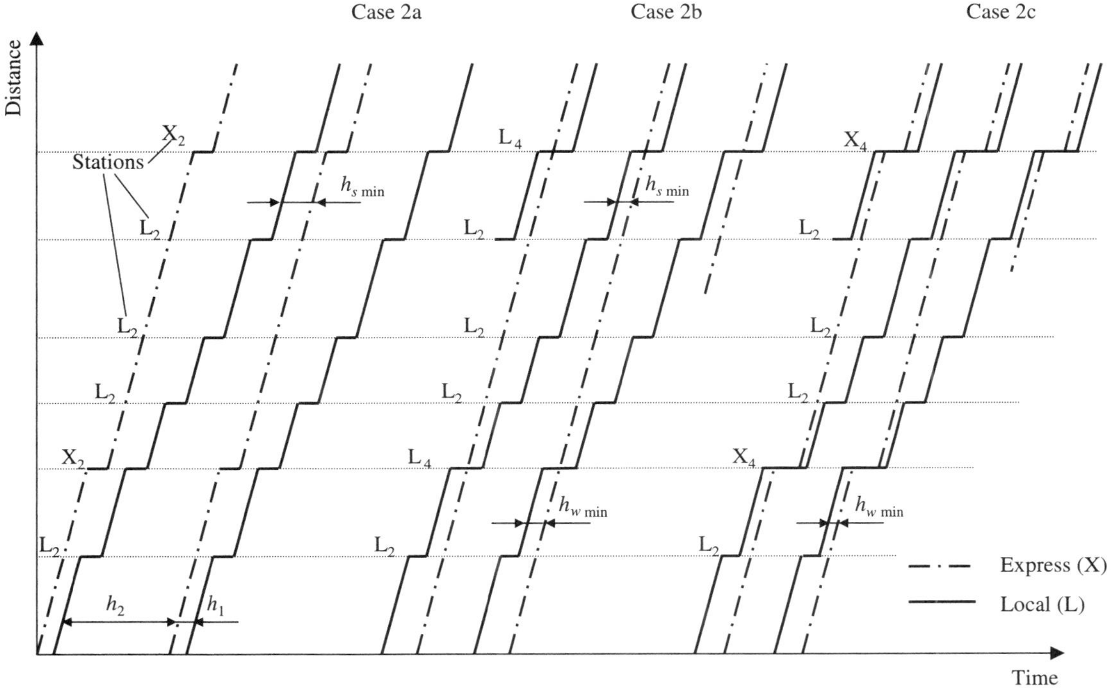

Figure 2.19 *Time-distance diagrams for express/local operations on two-track lines.*

case 2b. Time-distance diagrams for the two cases are shown in Figure 2.19.

These types of operations are possible only with precise scheduling and highly reliable operation. Examples of them are found on the Tozai Line of the TRTA system in Tokyo and on a number of regional rail lines in Tokyo (Odakiu), Kobe, and other Japanese cities.

Case 3: Three-Track Lines. These can be used for operation of two tracks in each peak direction, reversing the center track during the morning and evening peaks. With two tracks in the peak direction, local and express trains operate independently and therefore there are neither limits on headways nor on schedule coordination.

Three-track lines have the problem, however, that their operation is directionally unbalanced; they can therefore be used only on the lines with heavy directional peaking when either the single track opposite to peak direction has sufficient capacity to accept both sets of returning trains, local and express, or there is adequate storage track capacity at each terminus to accept some of the peak-direction trains. Due to these requirements, there are not many three-track lines.

Case 4: Four-Track Lines. These allow utilization of all advantages of express/local service without any operational complexities: the two services can be operated independently of each other, as shown in Figure 2.20(*a*). A common design is that the two express tracks have central position, with local tracks placed on the sides. Local stations have side platforms, so that track alignments are not affected. At express stations the local tracks are curved outward, providing space for two directional platforms between local and ex-

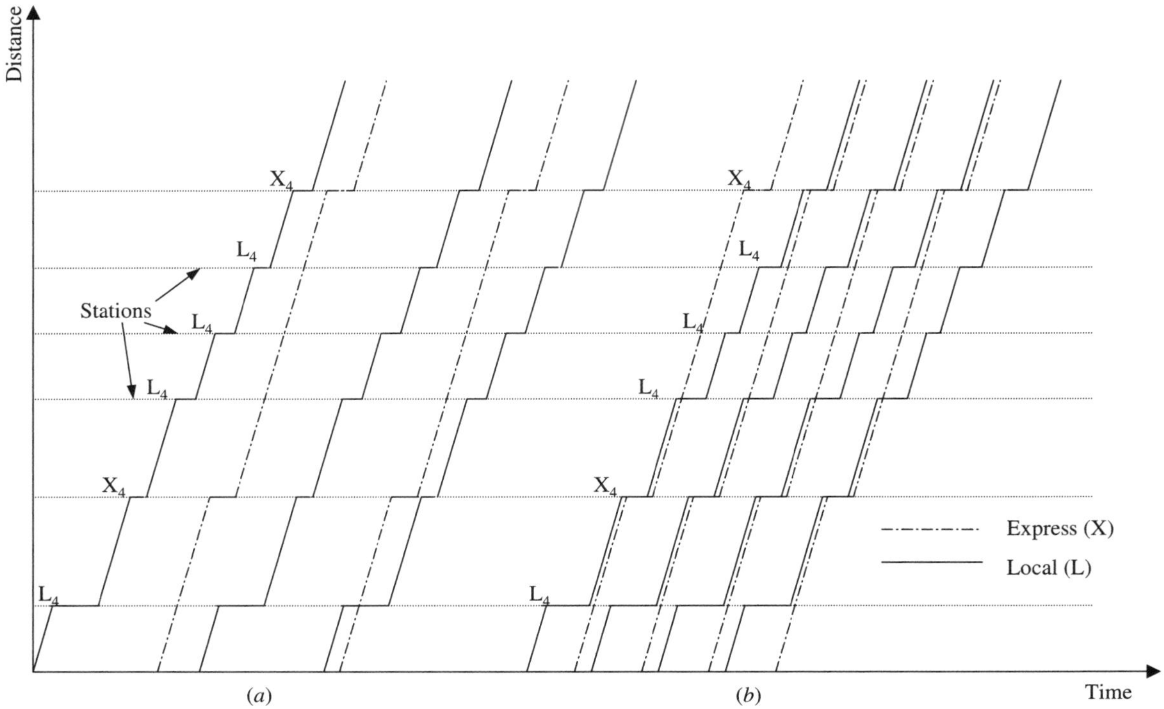

Figure 2.20 *Express/local operations on a four-track line (case 4): (a) independent running of X and L trains; (b) synchronized running of X and L trains for simultaneous transfers.*

press tracks. Both station layouts are shown in Figure 2.18.

In some cases, conditions permit synchronization of train schedules with travel times between X_4's such that local and express trains arrive simultaneously at them, allowing a two-way passenger transfer. Such operation, shown in Figure 2.20(*b*), can be achieved when the headways of both local and express trains are equal to the marginal time loss per station T_ℓ multiplied by the number of L_4's between each two X_4's, ℓ_ℓ:

$$h = \ell_\ell \cdot T_\ell \qquad \left| \frac{h,\ T}{\text{min}} \middle| \frac{\ell}{-} \right| \tag{2.38}$$

Thus, each express train catches up with one by one local train as it travels along the line. The transferring of passengers is similar to the case of two-track lines with four-track express station, except that the local train suffers no extra delay: it can arrive and depart simultaneously with the express.

2.4.3.2 Comparison with Standard Operation.

Only with two-track operation can express/local be considered as an alternative to standard (local only) operation. Their differences are summarized in the box.

With case 2b, several more elaborate stations (with four tracks) are required, but the benefits to passengers in reduced travel time and increased comfort are much greater.

Three- and four-track lines with express/local operation are not direct alternatives to standard operation because they require a substantially higher investment (additional tracks and more complex stations); they

Express/local on two-track lines (case 2a), compared with local operation only, results in:

+ Reduced travel times and increased comfort for some passengers
+ Better fleet utilization (more car-km of service operated)
− Longer waiting time for passengers at local stations
− More complex service, requiring high reliability

also offer two types of service (standard or local and express) and approximately double capacity. In other words, such express/local line represents a higher cost/better service package than standard operation.

2.4.3.3 Potential Applications of Express Operation. Four-track lines with express/local operations represent the highest-investment/highest-capacity transit facilities, so that their applications are mostly found on older rapid transit lines in large cities such as New York, Chicago, and Philadelphia. In recent years, several cities have built separate parallel lines to increase capacity of existing rapid transit lines; the new lines have fewer stations, thus providing higher speed. Examples are the RER Line A in Paris, paralleling Metro Line 1, and Hanzomon Line in Tokyo, paralleling a section of Ginza Line.

Multitrack express/local operation is also provided on a number of regional rail lines in major cities, such as London, Paris, New York, Philadelphia, and many Japanese cities.

Two-track lines with express/local operations are most extensively used on Japanese metro and regional rail systems, which are characterized by high reliability of operations.

With increasing automation and sophistication of train controls, conditions for wider applications of express operations on guided systems are growing. Their extensive applications, however, may not be imminent because many operators prefer to offer their passengers regular, simple, and reliable service rather than make efforts to operate more elaborate service patterns that require high reliability.

2.4.4 Comparison of All-Stop, Skip-Stop, Zonal, and Express/Local Operations

Basic operational and service differences among these four types of transit line operations are illustrated on a model of a simple radial transit line shown in Figures 2.21–2.24. To show the differences clearly, the model line has regular geometry: it is 12 km long with 13 equally spaced stations (including the two terminals); every third station is a major one with a passenger volume per direction of 2,000 prs/h, while every two intermediate stations have only 1,000 prs/h each. Thus, the total number of passengers boarding the trains along the line is 16,000 prs/h. It is assumed that all these passengers travel to the CBD. The four diagrams show the profiles of passengers boarding the trains along the line, as well as the offered capacities, for each type of operation.

Figure 2.21 shows that for local and for skip-stop operations the offered capacities are identical. The passenger volume boarding trains is divided between A and B skip-stop trains in equal numbers.

Figures 2.22 and 2.23 show local with zonal operations with two and four zones, respectively. The total number of passengers boarding each group of trains is equal, but the numbers of person-km carried by trains for each zone differ. The offered capacities are also tailored better to passenger volume, since capacity offered for each zone is limited only to the distance from CBD to the end of that zone. Thus, the unused offered capacities, shown on the diagrams, are less than in the preceding case and are reduced further as the number of zones increases; this is obvious from the comparison of the diagrams for two and four zones.

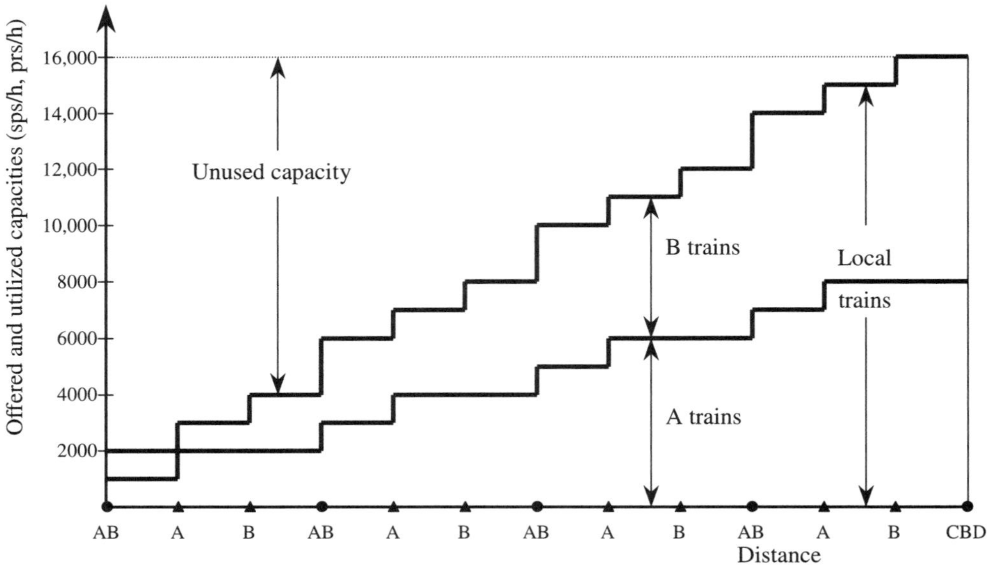

Figure 2.21 *Offered and utilized capacities for local and skip-stop operations.*

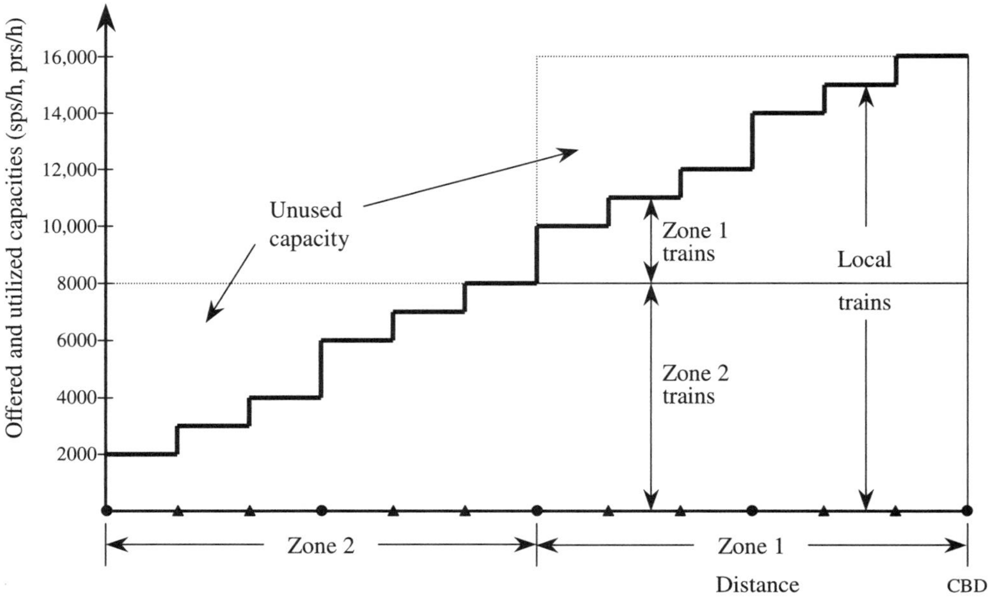

Figure 2.22 *Offered and utilized capacities for local and zonal (two-zone) operations.*

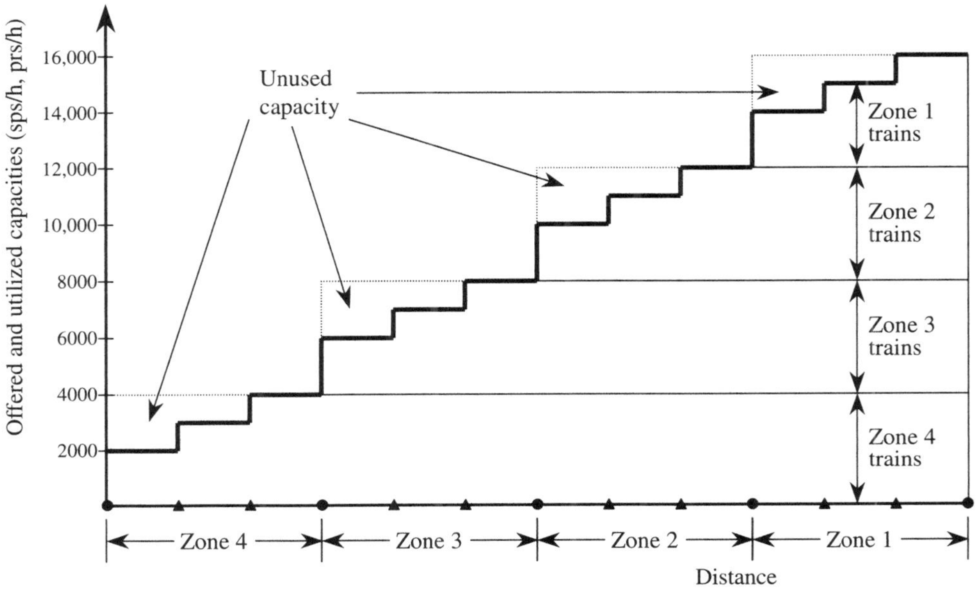

Figure 2.23 *Offered and utilized capacities for local and zonal (four-zone) operations.*

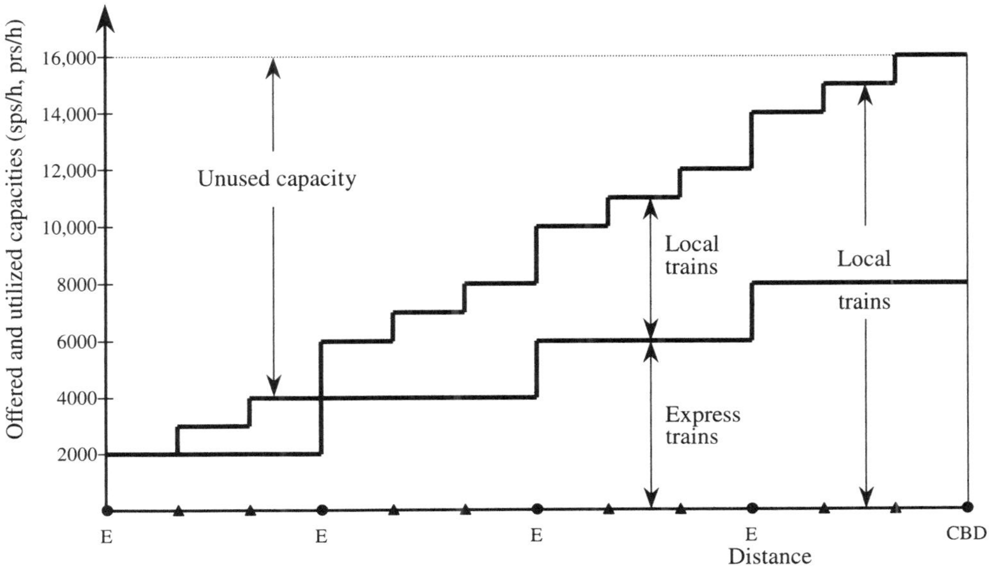

Figure 2.24 *Offered and utilized capacities for local and express operations.*

Figure 2.24 shows that, assuming that passengers at major (i.e., express) stations all take express trains, the local and express trains share the passengers again at 8,000 prs/h each. The offered capacity is still the same as in the case of local and skip-stop operations.

For a numerical comparison of the four types of operation, a set of computationally convenient operational values was assumed, and for each service schedules have been computed and presented in Table 2.11. This table shows a number of interesting comparative features. Some major ones will be pointed out, taking the local (all-stop) service as the basic type and comparing the others to it.

Skip-stop service compared to local has travel time shorter by four minutes, so that the average travel speed is increased from 30 to 36 km/h. Headways for local service are three minutes at all stations, for skip-stop they are three minutes at major and six minutes at minor stations. Although the offered capacities are identical in the two cases, the number of trains required is smaller by two for the skip-stop service because of its shorter cycle times. The fleet size is reduced correspondingly, from 160 to 128 cars, while the average capacity utilization factor for the entire service, $\overline{\alpha}_{serv}$, is the same—only 0.56.

Zonal service with two zones compared to local service shows equal speed for zone I, but significant increase of speed (from 30 to 40 km/h) for the trains for zone II. Headways at all stations are now six minutes, but cycle times for each group of trains are considerably shorter than for local operation. Consequently, the total number of trains required drops from 18 to 12 and the number of required cars drops correspondingly, from 160 to 96. Capacity utilization coefficient increases to 0.75.

It is interesting to compare four-zone to two-zone operation. The former one has increasing average speeds for each further zone, the same headways, but shorter cycle times. The number of trains increases to 18, but the total fleet size is the lowest of all types of operations, only 72. This is due to the best tailoring of offered capacity to passenger volume, which is also shown by the very high average utilization coefficient of 0.90.

Finally, comparing the express/local operation to the local-only operation shows that the speed of the express trains is 50% greater. Cycle time of these is correspondingly shorter and the number of required trains decreases to 16, fleet size from 160 to 128. The overall service utilization coefficient $\overline{\alpha}_{serv}$ is the same as for local service, 0.56.

2.4.5 Methodology for Selection of Accelerated Operations

The above comparison of local, skip-stop, zonal, and express/local operations shows that they differ in many performance and service elements. The comparison among them and selection of the optimal service type for a given line is therefore quite complex, but it can be performed as a sequence of steps that leads first to selection of the type of service and then to determination of zones or stations for skip-stop or express service. For a two-track line the procedure for selection is as follows.

Step 1. Analyze the number of stations along the line. If there are few stations, accelerated services should not be considered, since the potential for time saving is small and it is not worth the increased complexity for passengers. If the number of stations is considerable, accelerated services should be considered.

Step 2. Examine travel patterns. If most trips follow a "many-to-one" pattern, such as travel on radial line into center city, zonal service (as well as skip-stop and express/local) may be considered. If there is a significant number of trips among different stations along the line, zonal service should be eliminated as an option.

Step 3. Compare passenger volume on the line to its capacity. If the volume is considerably lower, all types of accelerated services can be used; if the line operates close to capacity, only skip-stop remains as an alternative to the local service.

Step 4. Analyze headway lengths. Short headways allow all accelerated services, subject to operational feasibility, described in Section 2.4 for zonal and express services. Long headways make accelerated services infeasible because it is not tolerable for passengers to wait long and then

Table 2.11 Comparison of operating data for different types of service

Regime	Train Type	L	T_o	V_o	P_{max}	f	n	h	T	V_c	N_t	ΣN_t	N_f	ΣN_f	$\bar{\alpha}$	$\bar{\alpha}_{serv}$
Local (all stop)		12	24	30	16,000	20	8	3	54	27	18	18	144	144	0.56	0.56
Skip-stop	A	12	20	36	8000	10	8	6/3	48	30	8	16	64	128	0.54	0.56
	B		20	36	8000	10	8	6/3	48	30	8		64		0.60	
Zonal—2	I	6	12	30	8000	10	8	6	30	24	5	12	40	96	0.63	0.75
	II	12	18	40	8000	10	8	6	42	34	7		56		0.81	
Zonal—4	I	3	6	30	4000	10	4	6	18	20	3	18	12	72	0.75	0.90
	II	6	9	40	4000	10	4	6	24	30	4		16		0.88	
	III	9	12	45	4000	10	4	6	30	36	5		20		0.92	
	IV	12	15	48	4000	10	4	6	36	40	6		24		0.94	
Local/ express	L	12	24	30	8000	10	8	6/3	54	27	9	16	72	128	0.50	0.56
	X	12	16	45	8000	10	8	6/3	42	34	7		56		0.63	

Assumed values: $\alpha\, C_v = 100$ prs/TU; $T_r = 1$ min/km; T = 1 min; min $\Sigma t_t = 6$ min.

have a train pass through their station without stopping.

The definition of short headway depends considerably on the type of transit line and level of transit service in a particular city. For example, short headway on a metro system is usually up to 7.5 minutes, while on a regional rail system headways up to 15 or 20 minutes are usually considered short.

These four steps that define general feasibility of different types of accelerated services are summarized in Table 2.12. The table shows that while local service can be used under all conditions, accelerated services generally require four conditions: many stations on a line, many-to-one travel pattern, passenger volume lower than line capacity, and rather short headways. Among the three accelerated services, skip-stop is the most diversified one because it can serve in some cases when trip patterns are many-to-many, and it is the only accelerated service that can be used at capacity conditions.

If the four steps of analysis show that one or more types of accelerated services are feasible, planning of operations should proceed with two additional steps.

Step 5. Design several possible alternatives for each feasible type of service (e.g., different sets of A and B stations for skip-stop, or different number of zones for the zonal service), compare them, and select the optimal one for each type.

Step 6. Compare local service with each other selected type of accelerated service and decide which one is optimal for the given line. This selection need not be for the entire day; in some cases a line may be operated as a local at all times except during the peaks, when one of the accelerated services is used.

2.5 SCHEDULING OF SINGLE-TRACK LINES, CIRCLE LINES, AND TRUNK LINES WITH BRANCHES

Planning and design of special geometric forms of transit lines, such as closed circles or major trunk lines with several branches, often impose special relationships between geometric elements and operating conditions on the lines. This section covers three types of lines with specific layout characteristics. First are transit lines that have some sections or entire length with single track (guideway) and turnouts (two-track sections for meeting and passing of TUs). These lines are encountered on some rail and other guided systems. Second are circle lines, which have certain operating characteristics regardless of whether they use guided or steered technology. And third are trunk lines with two or more branches, a very common form of lines on many transit systems.

2.5.1 Single-Track Lines

Single-track streetcar lines, common several decades ago, were discontinued when growing street conges-

Table 2.12 Transit line characteristics for different types of local and accelerated services

Characteristics	*Type of Service*	*Local*	*Skip-stop*	*Zonal*	*Express/local*
Stations	Many	+	+	+	+
	Few	+	−	−	−
Travel pattern	Many-to-one	+	+	+	+
	Many-to-many	+	+	−	−
Passenger volume	Moderate	+	+	+	+
	At capacity	+	+	−	−
Headways	Short	+	+	+	+
	Long	+	−	−	−

+ Condition allowing respective service.
− Condition not conducive to respective service.

tion prevented reliable operations, which single-track lines require. However, many rail lines on ROW categories A and B, particularly regional rail and light rail, utilize single tracks with turnouts in regular transit operations. Many regional rail lines, including some lines in Boston, London, and Los Angeles, have single tracks, and several new LRT systems, such as those in Sacramento and Baltimore, have been recently built as single-track lines, or double-track lines with some single-track sections. Single-track lines must be carefully designed, however, because the locations of turnouts and double-track sections determine which headways will be feasible to operate. Similar analysis must be done to determine locations of crossover tracks on two-track lines, which allow single track operations when one track must be closed for track-maintenance purposes.

Naturally, double-track lines are superior for operations because TU movements in opposite directions are independent of each other, while on single-track lines delay in one direction affects TUs traveling in the opposite direction. However, because single-track lines often involve much lower investment and maintenance costs, on lines with reliable operation single-track is in many cases fully acceptable. Actually, experience has shown that on single-track lines the requirement for precise timing of the TU meetings forces the drivers to maintain schedule more rigorously than on double-track lines. Single guideway with turnouts is also used on many specialized modes, such as funiculars, and on some AGT systems.

Transit units operating on a line with uniform headways always meet at fixed locations along the line. These locations are at distances on which travel times are equal to one half of the minimum line headway $h/2$. That is the shortest operational headway, but the line can also be operated at headways h or $2h$, when the TUs meet only at every other or every third

Photo 2.2 *Single track with passing siding—Philadelphia R5 Line (courtesy of Bob Vogel).*

meeting point, respectively. Since at all meeting points TUs must pass each other, the line must have either double track, or a turnout (short double-track section) at all those points. Thus, the schedule on the line must be
exactly coordinated with meeting locations—double-track sections or turnouts—along the line.

Figure 2.25 shows a sketch of a line with three sections between meeting points $S_{to}^{\,i}$ and two end sections S_e' and S_e''. Length of this line is in general:

$$L = \sum_i S_{to}^i + S_e' + S_e'' \qquad \left|\frac{L,\ S}{\text{km}}\right| \qquad (2.39)$$

In this case, $i = 3$.

Under the line sketch, there are three examples of double- and single-track designs of this line that have the same meeting locations and thus must all be operated with the same headway. The last one is the case when the entire line has single track and all meeting points are turnouts.

The time-distance diagram for operation of this line in Figure 2.26 shows its two terminals and four meeting points, with five sections and paths of five TUs. The general formula is that on a line with N sections there are $N - 1$ meeting points, $N - 2$ sections between meeting points, and two end sections, and for full schedule there must be N runs or TUs in service. In this case, $N = 5$.

The formulae for computations of these physical and operational elements are based on the fact that the length of sections (i.e., spacings) between meeting locations depends on operating speeds and headways. Assuming that the speed is uniform along the line, the spacings S_{to} can be determined as follows.

Travel time on a spacing between two turnouts, t_{to}, can be expressed by the distance between turnouts S_{to} and operating speed V_o. That time must be equal to one half of the line headway:

$$t_{to} = \frac{60 S_{to}}{V_o} = \frac{h}{2} \qquad \left|\frac{t,\ h}{\text{min}}\middle|\frac{S}{\text{km}}\middle|\frac{V}{\text{km/h}}\right| \qquad (2.40)$$

From these equations, one obtains the following relationships among S_{to}, h, and V_o:

$$S_{to} = \frac{h \cdot V_o}{120}; \quad \text{or} \quad h = 120\,\frac{S_{to}}{V_o} \qquad \left|\frac{S}{\text{km}}\middle|\frac{h}{\text{min}}\middle|\frac{V}{\text{km/h}}\right| \qquad (2.41)$$

The relationship expressed by these equations is plotted on a diagram in Figure 2.27. The diagram gives spacing between passing sidings as a function of operating speed (assumed to be uniform along the line) for different headways. Commonly used headways from 5 to 20 minutes are shown. The diagram can be used to find the required turnout spacings on a line with given operating speed and headway. The example shown on the diagram indicates that for $V_o = 18$ km/h and $h = 10$ minutes, the required spacing is $S_{to} = 1.50$ km; for the same speed and h = 20 minutes, one obtains $S_{to} = 3.00$ km.

The diagram can also be used to examine how much should operating speed be increased to allow operation with a shorter headway. For example, if V_o on a line with $S_{to} = 2.00$ km is increased from 16 to 20 km/h (point A to point B on the diagram), the headway can be decreased from 15 to 12 minutes. Such a case may occur when speed is increased by physical or operational improvements along a line or new rolling stock is introduced.

The two end sections of a line, between the last meeting points and its terminals, should have operating times shorter than the travel times on the middle sections between turnouts to allow for the terminal times. The length of an end section is computed as:

$$S_e' = \frac{h - t_t'}{120} \cdot V_o \qquad \left|\frac{S}{\text{km}}\middle|\frac{h,\ t}{\text{min}}\middle|\frac{V}{\text{km/h}}\right| \qquad (2.42)$$

An example will illustrate the procedure for computation of elements of lines with single tracks on their sections or entire length. Suppose that a line is planned that has the following elements:

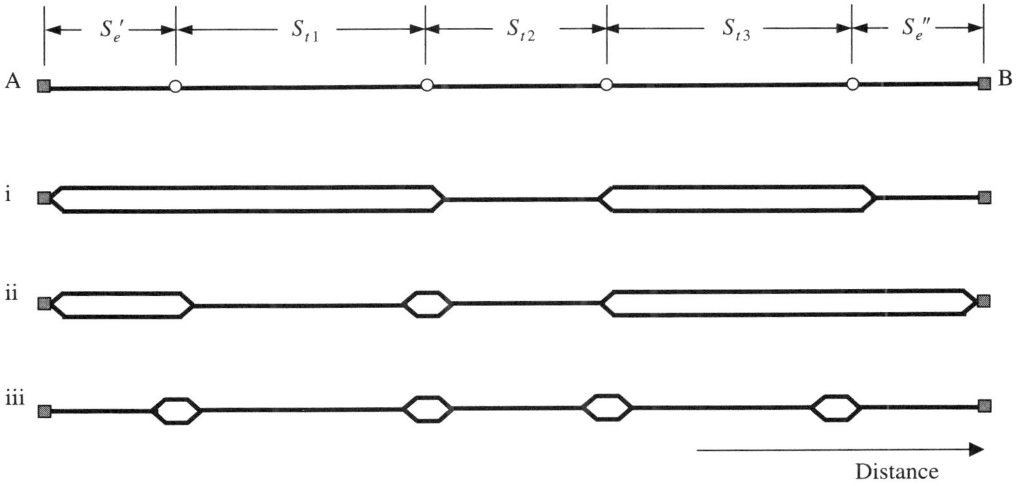

Figure 2.25 Line with single- and double-track sections and four meeting points.

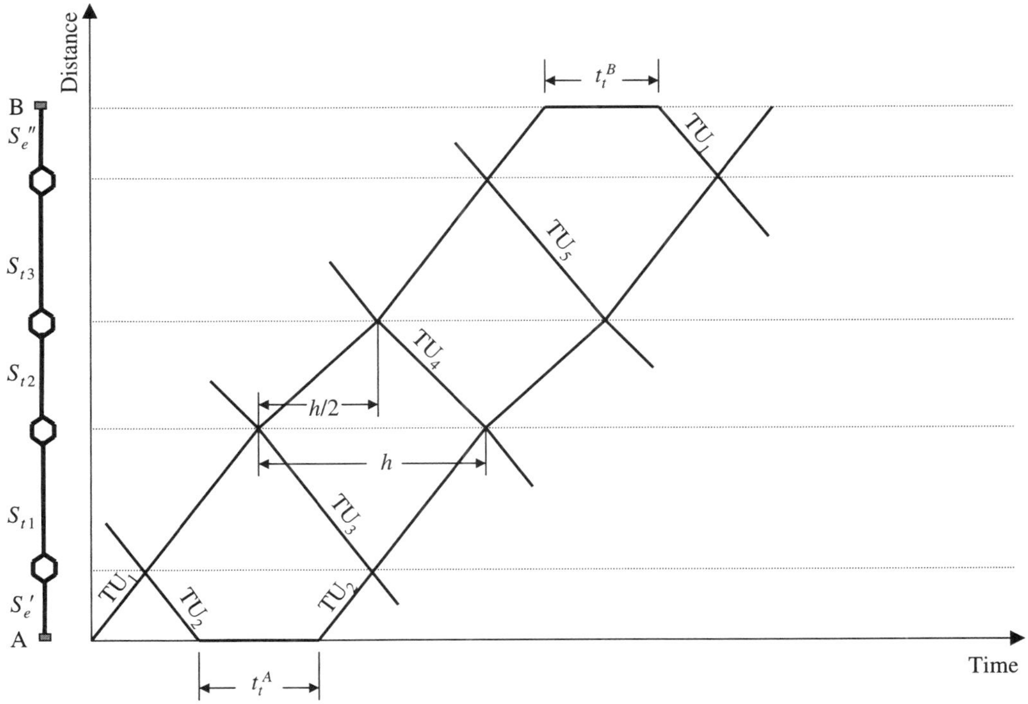

Figure 2.26 Time-distance diagram for the single track line iii from Figure 2.25 operating with minimum headway and five TUs.

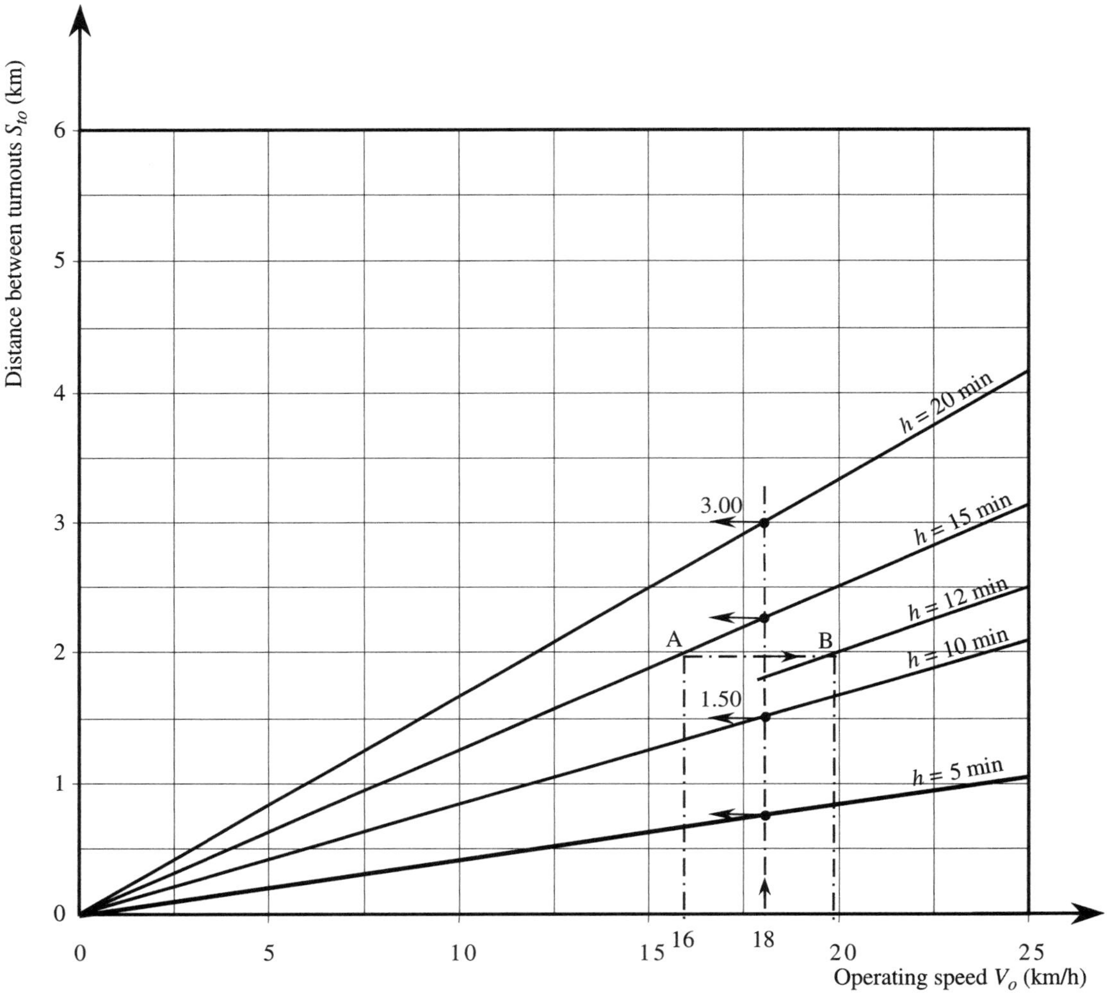

Figure 2.27 *Relationships among operating speed, headway, and distance between passing sidings on single-track lines.*

Line length:	$L = 10.850$ km
Operating speed (assumed constant along the line):	$V_o = 21$ km/h
Operating headway:	$h = 10$ min
Minimum terminal time at both terminals together:	$\Sigma t_t = 6$ min, with any extra time to be allocated to the second terminal, t_t''

Line schedule is now computed:

- Travel time, using Eq. (1.28):

$$T_o = \frac{60L}{V_o} = 31 \text{ min}$$

- Minimum cycle time:

$$T' = 2\,T_o + t_t' + t_t'' = 2 \cdot 31 + 6 = 68 \text{ min}$$

This is rounded up to the next multiple of the headway of 10 min, so that the final cycle time is $T = 70$ min and total terminal time is 8 min

- Individual terminal times are $t_t' = 3$ min and $t_t'' = 5$ min
- Number of TUs:

$$N = T/h = 70/10 = 7 \text{ TUs}$$

The spacing between meeting points is now computed by Eq. (2.41):

$$S_{to} = \frac{10 \cdot 21}{120} = 1.750 \text{ km}$$

and the lengths of the two end section are computed by Eq. (2.42):

$$S_e' = 0.00833\ V \cdot (h - t_t')$$
$$= 0.00833 \cdot 21\ (10 - 3) = 1.225 \text{ km}$$
$$S_e'' = 0.00833 \cdot 21 \cdot (10 - 5) = 0.875 \text{ km}$$

The line length consists of the following elements, from Eq. (2.39):

$$L = 5 \cdot 1.750 + 1.225 + 0.875 = 10.850 \text{ km}$$

For efficient operation, spacings between turnouts along a line should be designed so that operating times on all of them are equal. If they are not, the spacing with the longest travel time governs the headway and travel times on other spacings must be lengthened to match it.

2.5.2 Circle Lines

Transit lines in the form of a closed circle (also known as loop or ring lines—see Section 4.2) are unique due to the fact that they operate without terminal times. This feature gives circle lines several operational and service characteristics, which are discussed in Section 4.2.5. This section covers only the scheduling of the circle lines.

On circle lines, operating time around the circle T_o is equal to the cycle time T. Therefore, if uniform headways are operated, T_o can only have values that are integer multiples of the headway:

$$T_o = T = h \cdot N \qquad \left| \frac{T}{\text{min}} \middle| \frac{h}{\text{min}} \middle| \frac{N}{-} \right| \tag{2.43}$$

Operating speed V_o can be computed from the line length L and T_o:

$$V_o = \frac{60L}{T_o} \qquad \left| \frac{V}{\text{km/h}} \middle| \frac{L}{\text{km}} \middle| \frac{T}{\text{min}} \right| \tag{2.44}$$

The highest operating speed $V_{o\ max}$ (and the shortest travel time $T_{o\ min}$) that can be operationally realized on a given circle line can therefore be achieved only when $T_{o\ min}$ is an integer multiple of the headway. If this is not the case, operating time must be lengthened to the next integer multiple of h. To achieve this, operating speed must be deliberately lowered, usually by insertion of a short catch-up time, playing the role of terminal time, at one or two stations along the lightly used line sections.

The relationships among these operating elements are plotted on the diagram in Figure 2.28. Two sets of lines show the relationships for headways of 5 and 10 minutes, respectively, and each set shows lines for one, two, three, and four TUs in service. Operation can be performed only at points along these lines.

Possible uses of this diagram are illustrated by two examples. If the maximum physically feasible speed on a line is 27 km/h and the line length is $L = 6.75$ km, then service with $h = 5$ min can be provided by three TUs (point A on the diagram). If the line length is slightly greater, e.g., 7.00 km, an additional TU is needed, and for the same headway operating speed must be reduced to 21 km/h (point B). For further increases in line length, higher operating speeds would be required; when the length reaches $L = 9.00$ km, the maximum physically feasible speed of 27 km/h would again be achieved (point C).

Consequently, the full potential of a circle line—maximum operating speed for passenger attraction, as well as the best utilization of its rolling

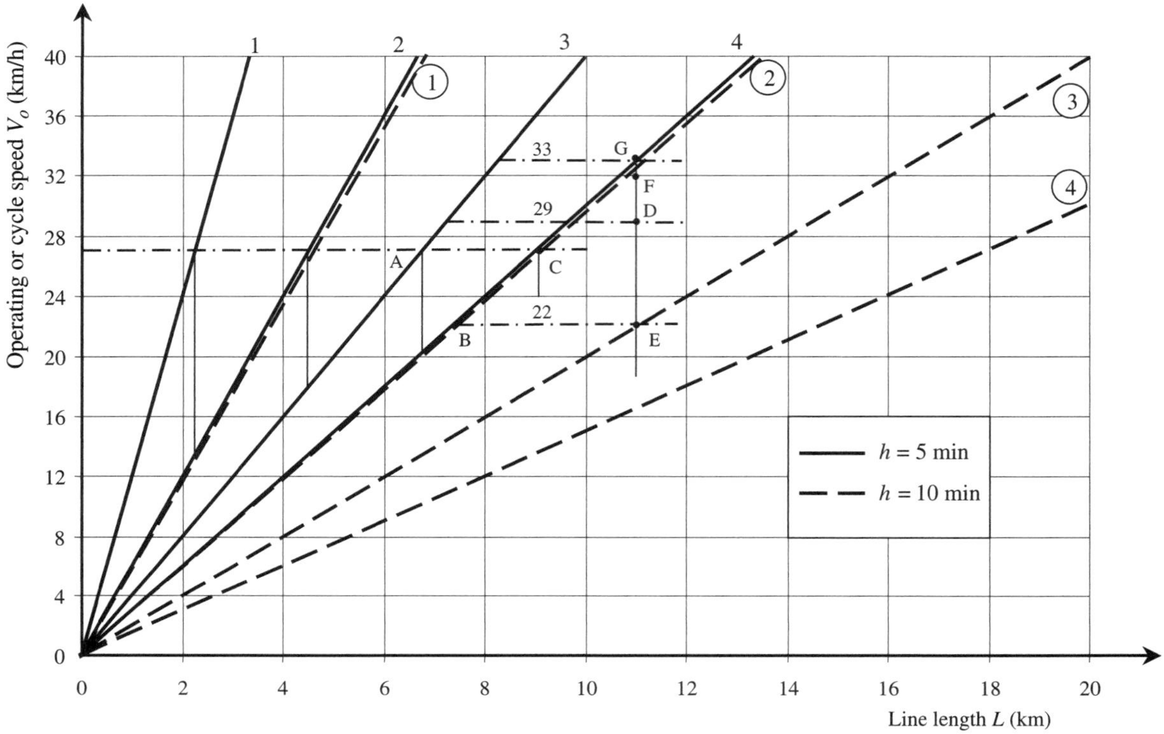

Figure 2.28 *Relationships among line length, headway, number of TUs, and operating speed on circle lines.*

stock—can be achieved only when the expression for V_o obtained from Eqs. (2.43) and (2.44) gives the value equal to the maximum speed achievable on the line:

$$V_o = \frac{60L}{h \cdot N} = V_{o\,\max} \qquad \left| \frac{V_o}{\text{km/h}} \middle| \frac{L}{\text{km}} \middle| \frac{h}{\text{min}} \middle| \frac{N}{-} \right| \qquad (2.45)$$

Therefore, in planning and design of circle lines, their projected operating speed V_o should be computed from its line length L, planned headway h, and an integer value of N. The obtained value of V_o should then be the basis for design speed, number of stations, rolling stock characteristics, and other elements affecting speed.

In another example, if the maximum possible operating speed on an 11-km-long metro circle line is $V_{o\,\max}$ = 29 km/h (point D) and h = 10 min, the diagram shows that three TUs would be required, traveling at only V_o = 22 km/h (point E). Suppose the new rolling stock is being designed. If, due to better dynamic characteristics, the new cars could achieve V_o = 32 km/h (F), that would bring no benefits because the line would still require three TUs with V_o = 22 km/h (E). However, if the new cars could achieve V_o = 33 km/h (G), the speed increase could be utilized and only two TUs would be needed for h = 10 min; alternatively, with four TUs and the same speed, headway could be reduced to 5 min (the same point, G).

2.5.3 Trunk Lines with Branches

Trunk lines with branches are very common in transit networks. Their scheduling involves the determination of frequencies, capacities, and headways offered on the trunk and branch sections. These operating elements should correspond to the design passenger volumes on the trunk and on each branch as much as that is possible with acceptable headways and operating reliability and within trunk line capacity limitations.

When trains or buses from two or more branches merge into a joint line section (trunk) without coupling short into longer trains, the trunk frequency is the sum of frequencies on branches:

$$f_t = \sum_{i=1}^{k} f_i \qquad \left| \frac{f}{\text{TU/h}} \right| \tag{2.46}$$

The average joint headway on the trunk section $\bar{h}_t$ can be computed from the headways on individual lines h_i via their frequencies:

$$\bar{h}_t = \frac{60}{f_t} = \frac{60}{\sum_i f_i} \qquad \left| \frac{h}{\text{min/TU}} \right| \frac{f}{\text{TU/min}} \right| \tag{2.47}$$

The average waiting time of passengers $\bar{t}_w$ on a joint section of two lines with headways h_1 and h_2 is given by the formula:

$$\bar{t}_w = \frac{h_1}{2}\left(1 - \frac{h_1}{2h_2}\right) \qquad \left| \frac{t,\, h}{\text{min}} \right| \tag{2.48}$$

where h_2 is the longer headway ($h_1 \leq h_2$). This equation is valid only when passengers arrive randomly, i.e., for lines with short headways ($\leq$10 min). For $h_2 = \infty$ (only one line operates), it gives $\bar{t}_w = h_1/2$, while for $h_1 = h_2$ (uniform merging), one gets $\bar{t}_w = h_1/4$. For all other values, $h_1 < h_2 < \infty$, the average waiting time is between these two values: $h_1/4 < \bar{t}_w < h_1/2$.

The relationship between capacities offered on the trunk, C_t, and on k branches ($C_1, C_2, \ldots, C_k$) can be expressed by branch frequencies and TU capacities:

$$C_t = \sum_{i=1}^{k} C_i = \sum_{i=1}^{k} f_i \cdot n_i \cdot C_{vi}$$

$$\left| \frac{C}{\text{sps/h}} \right| \frac{F}{\text{TU/h}} \left| \frac{n}{\text{veh/TU}} \right| \frac{C_v}{\text{sps/veh}} \right| \tag{2.49}$$

where n_i is the number of cars per TU and C_{vi} is car capacity on branch i.

2.5.3.1 Trunk/Branch Line Capacities and Headways. The ideal case with respect to operations is when all branches diverging from a trunk line have the same critical passenger volumes. In that case, TUs with equal capacities go to different branches at equal headways, so that the trunk has service with uniform headways equal to the branch headway divided by the number of branches. However, when passenger volumes on branches are different, services provided on them may vary, so that often certain operational irregularities, such as different TU capacities or irregular headways, must be accepted either on branches or on the trunk, or on both. The range of possible service adjustments in response to uneven passenger volumes on different branches will be explained here using the simplest model: a trunk line T with two branches, A and B. The types of schedules described apply in principle to any number of branches.

Eight cases with different ratios of critical passenger volumes P_{max} on the two branches and different offered services are listed in Table 2.13; four of them, cases i, ii, iii, and vi, are illustrated in Figure 2.29. The diagrams in the figure show TU departures in the form of a clock; the number of circles around the clock represents hourly frequency of services. The number within each circle represents the number of cars per TU, n. The line diagram shows bands proportional to offered TU capacities, while their dark portions represent passenger volumes, or utilized capacities, on individual TUs. Since band widths do not reflect service frequencies, the differences between offered line capacities on the branches are in some cases greater than shown graphically. For example, in case iii offered capacities are 16 veh/h on branch A and 8 veh/h on branch B, or in proportion 2:1, as is the ratio of the passenger volumes shown on the diagram by the shaded bands. However, band widths of offered capacities are equal, because they represent TU capacities, which are equal on the two branches, $n_A = n_B = 4$, while TU frequencies are shown by the clock-diagrams.

Table 2.13 Different types of scheduling for trunk line T with branches A and B

Case	*P*	*n*	*f*	*h*	*Regular section*
i	$P_A \approx P_B$	$n_A = n_B$	$f_A = f_B$	$h_T = ½\, h_A = ½\, h_B$	T, A, B
ii	$P_A > P_B$	$n_A = n_B$	$f_A > f_B$	h_T, h_B uniform h_A irregular	T, B
iii	$P_A > P_B$	$n_A = n_B$	$f_A > f_B$	h_T irregular h_A, h_B uniform	A, B
iv	$P_A > P_B$	$n_A > n_B$	$f_A = f_B$	$h_T = ½\, h_A = ½\, h_B$	A, B
v	$P_A \gg P_B$	$n_A > n_B$	$f_A > f_B$	h_T, h_B uniform h_A irregular	B
vi	$P_A \gg P_B$	$n_A > n_B$	$f_A > f_B$	h_T irregular h_A, h_B uniform	A, B
			With coupling of branch trains at the merging point, $n_T = n_A + n_B$		
vii	$P_A \approx P_B$	$n_A = n_B$	$f_A = f_B$	$h_T = h_A = h_B$	T, A, B
viii	$P_A > P_B$	$n_A > n_B$	$f_A = f_B$	$h_T = h_A = h_B$	T, A, B

Case i is the regular one: the trunk line divides into two branches with equal (or similar) critical passenger volumes. The trunk served by TUs of the same length running at regular headways alternatively to the two branches. Being regular in all respects, this service allows high reliability.

In all other cases, branch A has larger passenger volume than branch B. The difference between the two volumes in cases ii–iv is accommodated by different frequencies or by different TU sizes. Case ii is applicable where the trunk line is much more important than the branches: it has regular headways and uniform TU sizes. Branch B gets only every third TU (but at regular headways), while branch A has two out of three TUs, but at irregular headways. Case iii is very similar, but it applies where branches are more important than the trunk: both branches have uniform TU sizes and regular headways, but the trunk headways are irregular. Case iv (not illustrated) provides regular headways on all three sections (trunk and branches), but nonuniform TU sizes cause uneven loadings and station standing times on the trunk.

Cases v (not illustrated) and vi (Figure 2.29) apply to the lines on which trunk section is not important and branches have very uneven volumes. They have uneven TU sizes and different frequencies. Case v has regular headways on the trunk, while case vi has regular operations on the branches, but irregular headways on the trunk. These two schedules are the same as the schedules in cases ii and iii, but with different branch TU capacities.

There are many examples of transit lines with branches utilizing each one of the above described types of operation. Generally, the relative importance of trunk versus branches depends on their geometric and usage characteristics: on a long line with two branches in suburbs, as is found on the Boston's Red Line, the capacity and regularity of service on the trunk dictate the entire line operation. In such cases, equally loaded branches can be operated efficiently; however, if the loads differ, the lower capacity utilization on the lighter branch is accepted to maintain the trunk regular. However, when one branch carries substantially more than half of the trunk load and it is necessary to offer regular services on the two branches, the lighter traveled branch is sometimes separated and operated as a shuttle. An example of this situation is the Paris Metro Lines 3 and 7 (north), on which the trunk has been extended into the heavier branch, while the lighter ones have been cut off and operated as independent feeders.

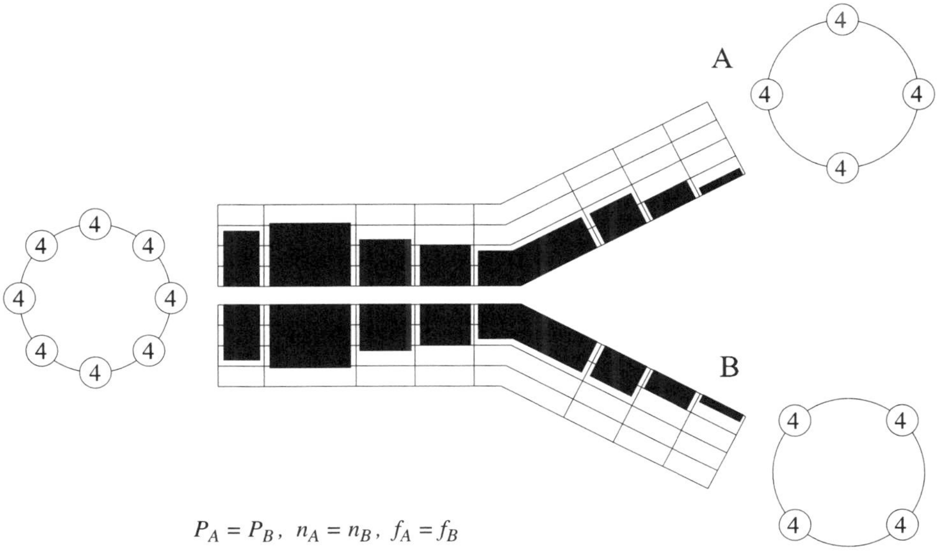

Case i: Branch volumes equal, 4-car train consists, equal headways

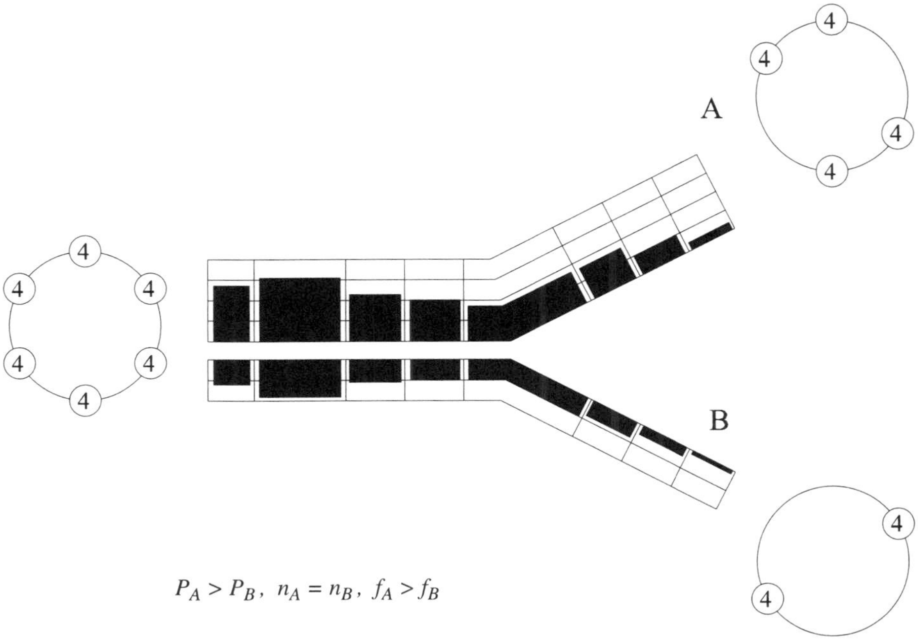

Case ii: Branch A volume heavier, 4-car train consists, uneven headways on A

Figure 2.29 *Selected cases of different trunk-branch volumes, train consists, and headways.*

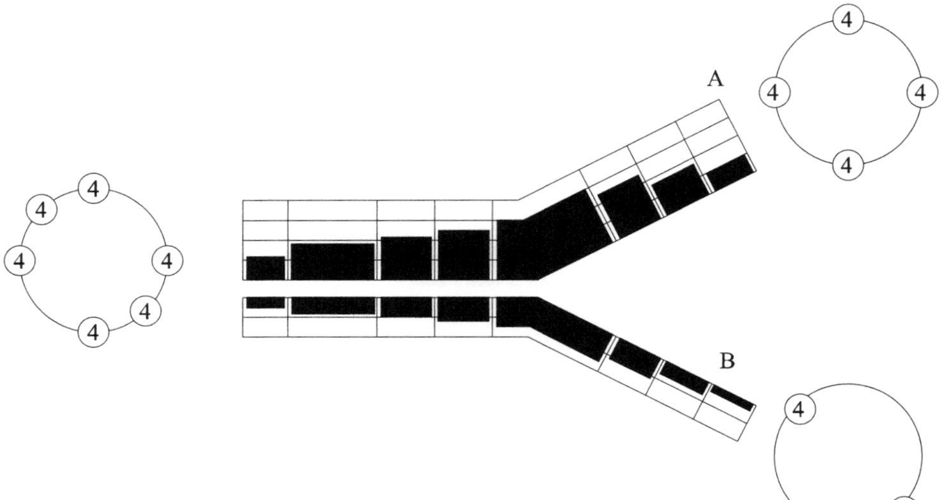

$P_A > P_B,\ n_A = n_B,\ f_A > f_B$

Case iii: Volumes the same as in case ii, but headways on branches uniform, on the trunk irregular

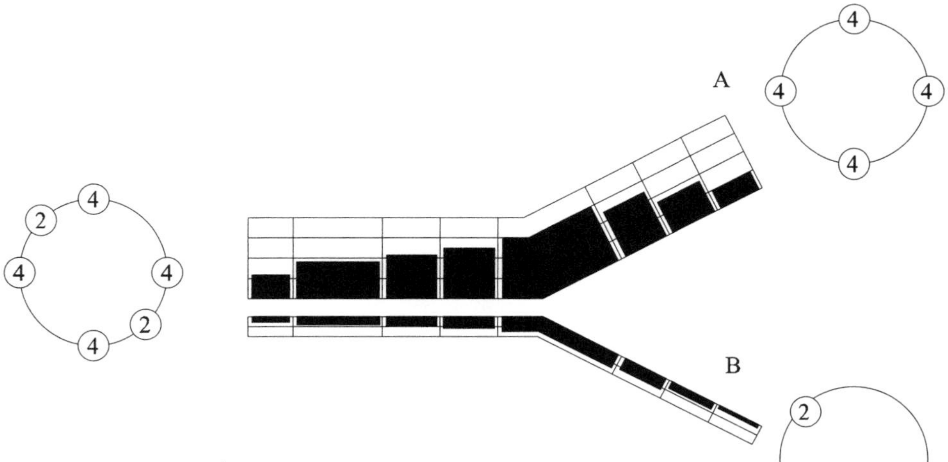

$P_A > P_B,\ n_A > n_B,\ f_A > f_B$

Case vi: Branch A volume much heavier, different train consists and headways, trunk service irregular

Figure 2.29 *(continued)*

The two major branches of the main north-south line in Hong Kong are also operated as a trunk with feeder, rather than as a trunk with two branches.

On the other hand, when numerous bus lines converge on a joint corridor in CBD, a situation found in many cities, the branch sections represent most of the lines and dictate all their operating elements. The trunk section then unavoidably has irregular headways. Similar situations occur on many regional rail systems: lines from many suburban areas converge into a short trunk line through center city (Munich, Philadelphia). Then the trunk has not only uneven headways, but also different TU lengths (case vi); however, this does not cause unnecessary irregular train loadings and operational difficulties, since few passengers travel on the trunk only. Most of them select one of the branches, so that TU loads correspond to the capacity and frequency offered on each particular line.

Accuracy of operations and adherence to scheduled headways on the trunk and branches depend mostly on the transit mode. Fully controlled systems, such as rapid transit and regional rail, maintain very precise schedules and they are not suited to any operations that inherently create disturbances. Therefore, they are often operated with type of service from case i even when the branches are not completely balanced; types ii to iv are often acceptable, but v and vi only in the cases where trunk section is short and most trips take place between the trunk and branches. Buses merging from lines on different streets, on the other hand, operate with much less precision and they are therefore less sensitive to schedule disturbances. They can use any one of the six types of service.

2.5.3.2 Coupling of TUs. On lines with many branches, it is often difficult to provide sufficiently short headways on each branch without creating excessive frequencies (i.e., less than the physically possible h_{min}) on the trunk. This problem can be solved on rail (and other guided) modes by coupling of short TUs into longer ones at one or more convergence points.

In addition to reliable automatic couplers, this operation on regional rail and rapid transit systems requires precise schedule adherence and special signal system design that permits visual approach of trains to the coupling location with short delays only. On LRT systems with incidental delays, irregularities in arrivals can be solved by forming TUs of different sizes, as was the case in San Francisco, which is described below.

Operation with coupling is more complex than simple merging of TUs, but it allows service with short headways on branches and longer than minimum headways on the trunk, which allows higher speed and reliability. In addition, personnel utilization increases: drivers from nonleading TUs on the coupled trains can either disembark and drive an outbound TU or perform other duties.

There have been examples of operations with coupling/uncoupling at convergence points or adding cars along the line on a number or rail systems, such as the Munich S-Bahn and the London Suburban Railways. A particularly interesting case has been the San Francisco Muni Metro network, shown schematically in Figure 2.30. When five converging streetcar lines operated with headways of 5–7.5 minutes, operation on the trunk with headways often falling below 1 minute was extremely slow, involving many delays. When the BART system was built from 1965 to 1972, it included construction of two tracks for the Muni Metro trunk line. To upgrade streetcars into LRT while maintaining the same headways on the branch lines, however, longer headways were obtained by coupling of one- and two-car trains at their convergence points

As Figure 2.30 shows, one-car trains from lines K, L, and M couple at the West Portal station into three-car trains, while two-car trains from line N couple with one J car at the Duboce portal. Thus-formed three-car trains (which can have four cars if one had been delayed) from the two tunnel portals merge into the common tunnel trunk section under full signal control at longer headways that allow reliable, full-speed operation. In the outbound direction, changeable signs at stations inform passengers at what location to board cars for individual lines. Uncoupling of TUs in the outbound direction is very simple: the driver on a nonleading car pushes a button that uncouples his coupler and separates the long TU into two (or more) short ones. This operation was maintained for over two

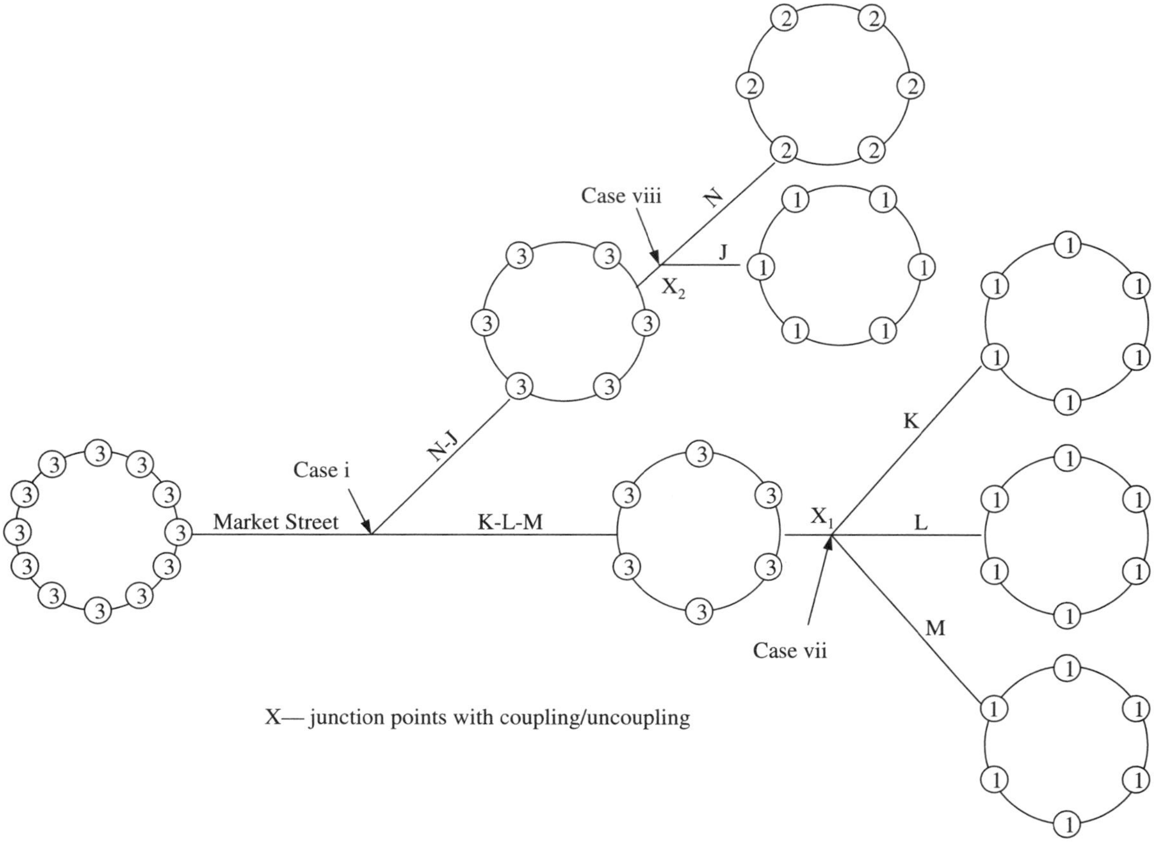

Figure 2.30 San Francisco LRT network operation: coupling/uncoupling of TUs at junction points.

decades. During the 1990s, fully automated signal and train control was introduced that allows high-speed operation at short headways, so that now the San Francisco Muni Metro system is a unique LRT system that includes automatic train operation (ATO) on its trunk section and manual driving elsewhere.

As can be seen in Figure 2.30, in the operation with coupling, each branch line can have a headway that is only two times longer than the trunk headway, instead of five times. Delays caused by street traffic can be absorbed by coupling the delayed car into the subsequent train, since the cars are designed for multiple unit operation of up to four vehicles.

Although desirable in many networks, operation with on-line coupling/uncoupling has not been frequently used due to unreliable operation of some couplers and problems with reliability, communications, and operational supervision. With improved coupler designs and advanced track signal and communication systems, it should be possible to increase use of coupling/uncoupling operations with its excellent service advantages of operating shorter headways on branches and lower operating cost due to higher load factors.

2.5.3.3 Synchronization of Branch Schedules. In addition to determining capacities and frequencies on trunk and branch lines, discussed above, schedules for all branch lines must be coordinated so that a desired sequence of diverging and merging of trains is

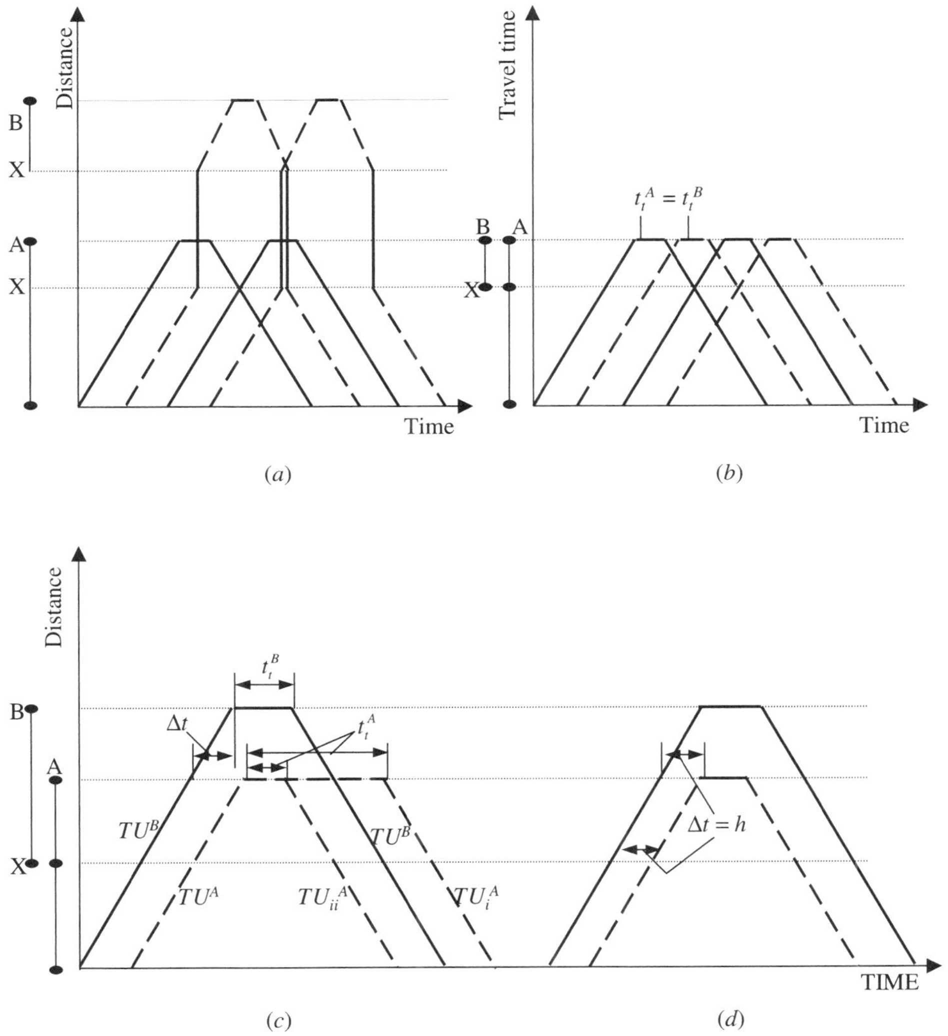

Figure 2.31 *Trunk with branches line schedules: (a) separate presentation at branches; (b) joint diagram for branches; (c) operations with the same and inversed TU sequences; (d) case with* Δt = h.

achieved. This is particularly the case when uniform headways are required on the trunk section. In that case, trains diverging at uniform headways go to different branches and when returning must merge on the trunk at uniform headways again. That is done by coordinating terminal times on each branch with the differences in travel times between the branches. On lines with short headways, such synchronization can be achieved with small adjustments of terminal times, but when long headways are operated, special analysis is required to ensure efficiency, i.e., minimal time losses for trains.

Coordinated scheduling of branches will be explained on an example of a trunk line OX with two branches, designated as XA and XB. They are shown in Figure 2.31 on time-distance diagrams for different schedules. Figure 2.31(a) shows a line with different lengths of branches and different speeds on them. The two branches are plotted separately: XA as an extension of the trunk and XB separately above it, as explained in Section 1.4.5.2 and shown in Figure 1.31. Operating speed on XA is shown as greater than on XB.

Schedules for two branches can also be shown on the same diagram, with branches XA and XB overlapping on the upper section. To illustrate this type of diagram, the simplest case, when branches have equal lengths and speeds, and therefore also equal terminal times ($t_t^A = t_t^B$), is shown in Figure 2.31(b). This diagram shows that when branches A and B have equal travel times from the diverging point X to their respective terminals and back, trains diverge and later merge in the same sequence and with the same headways; there are no time losses. But when travel times on the two branches are different, as shown in Figure 2.31(a), adjustments in terminal times are needed to provide uniform headways between trains returning from A and from B. This operation with diverging, merging and terminal times of trains on them can be seen clearly when the two branches are shown together as overlapping plots. Figure 2.31(c) shows such a case: the longer branch XA and shorter XB are plotted together above point X.

With different travel times on the two branches, there are two different types of schedule.

Case i. Trains merge in the same sequence as they had diverged (Figure 2.31(c)). Because trains on line B have shorter travel time, they have to wait on returning A trains by extending their terminal times to be:

$$t_t^B = t_t^A + 2\Delta t \qquad (2.50)$$

where Δt is the difference between one-way travel times on respective branches. If terminal times are to be minimized for higher schedule efficiency, t_t^A should be minimized because it also dictates the minimum value of t_t^B.

Case ii. A train to the longer branch, A, is followed by a train going to a shorter branch, B; when returning, train B precedes train A, i.e., they merge in different sequence than they diverge (Figure 2.31(d)). Travel time on line A is so much longer, that train on line B can return to the trunk before train A. In that case terminal time on branch A is related to the terminal time on B as:

$$t_t^A = t_t^B + 2|h - \Delta t| \qquad \left|\frac{t,\ h}{\text{min}}\right| \qquad (2.51)$$

Thus, in this case greater efficiency and line capacity require that t_t^B be minimized, because that determines also the t_t^A.

EXERCISES

2.1. Describe which elements of a rapid transit line typically determine its maximum passenger capacity. If a line begins to develop frequent congestion and delays, describe several measures that could be undertaken to increase its capacity and reduce delays.

2.2. An LRT line is operated with two-car TUs, each car with capacity of $C_v = 164$ spaces. The headway is $h = 3$ min and cycle time $T = 48$ min. The maximum load factor is $\alpha_{max} = 0.88$, and the line is operating at capacity. Compute the present ridership on the line. Ridership is growing and offered capacity must be increased by at least 25%. Make the necessary computations to evaluate the following measures:

a. Increase train consists from two to three cars

b. Improve signal system to allow operation of 2 min headway

c. Decrease station dwell and terminal times so that the cycle time is reduced to 42 min. Check whether each one of these alternative measures can meet the required increase in capacity.

Discuss the following related aspects of these capacity increasing measures:

- Investment cost as a function of the number of cars (N_C)
- Operating cost as affected by the number of drivers (N_{TH})
- Impacts on passenger capacity through headway lengths

2.3. In many U.S. cities, the standard practice is still that buses operating with general traffic have stops at every intersection and all passengers board passing by the driver and pay in the farebox. Describe several measures that could increase operating speed and reliability of bus services. Classify the measures by the investments they would require and impacts they would have on the service.

2.4. Operating speed on a long two-track rail transit line should be increased without construction of additional tracks. You are asked to determine whether this is feasible by introduction of an accelerated service.

a. List and briefly define alternative methods for accelerating the service that you would consider. For each type of service, list the elements which would favor or work against its application.

b. Describe factors that you would use to make a comparative analysis and to select the optimal accelerated service. Classify factors into quantitative and qualitative ones.

2.5. Describe skip-stop service and compare it with regular (all-stop) service. Draw a sketch and present a list of items that differentiate the two services to make the comparison clear. Based on that description and comparison, define characteristics of a rail line that make it a likely candidate for skip-stop operation.

2.6. In selecting how many pairs of A and B stations there should be on a line with skip-stop operation, what major criteria would you use to choose among alternative sets (e.g., among two-pair, five-pair and eight-pair alternatives)?

2.7. An all-stop local service is replaced by a skip-stop service (with A, B, and AB stations) with the same number of trains. Discuss the impacts of this change on:

a. Train operating speed

b. Waiting time for different groups of passengers

c. Required fleet size

d. Line capacity

2.8. Give a detailed and clear description of zonal service of transit line operation and define its advantages and disadvantages in comparison with regular, all-stop operation. Use sketches or lists of items to make the description clear.

2.9. Service on a long two-track transit line with many stations should be accelerated through either zonal or skip-stop service. Compare the two types of service and list factors that would favor one or the other. Describe a line that would clearly be suited only to zonal, and one that could be accelerated only by skip-stop service. Sketch each service on a simple time-distance diagram. Show on the diagram which sequences of trains determine line capacity, i.e., at which points the minimum headways occur.

2.10. Draw a time-distance of a regular, all-stop train on a two-track line (one track per direction) with 10 stations, i–x. Plot to the right of that train zonal service: a sequence of paths of trains serving two zones: i–v and vi–x. Plot two cycles of the zonal trains and show the minimum headways at critical points. Plot the train paths from station i outbound, one-way, with a turnaround at their respective terminals.

2.11. A line with operating speed V_o = 23 km/h has the following layout of single and double-track sections (distances are in meters):

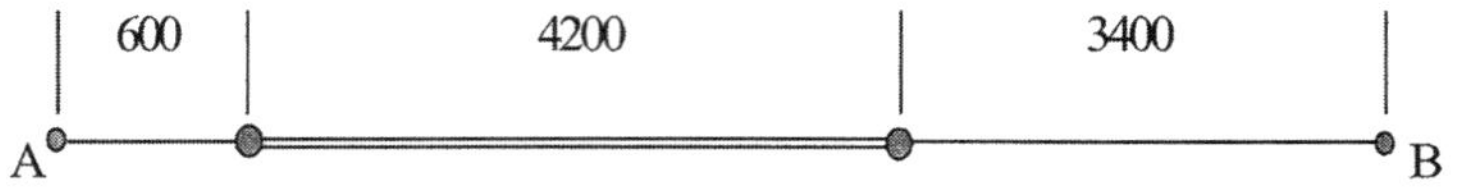

The terminal times must be at least two min at both ends. Compute the cycle time and number of TUs needed, then determine the locations of passing sidings (meeting locations) on the single-track sections and indicate them in the graph above for operation with:

a. Headways of 10 min

b. Headways of 15 min

2.12. An 18.8-km-long metro circle (ring) line has 20 stations. It operates with a cycle time T = 44 min and headways of 4 min. Average station standing time is t_s = 31 s.

a. It is planned to reduce the standing times so that the number of TUs can be reduced by one. What should be the new average t_s to allow operation of the same schedule by one less TU?

b. Compute the operating speed on the line now and with the change under *a*.

c. New rolling stock is being designed. Due to its higher performance operating speed on the line could be raised by 3.2 km/h over the speed under *a*. Assuming the headway remains the same, compute the new operating speed V_o', cycle time T', and number of TUs N'.

d. For the same change as in c, what would be the new headway h' if the number of TUs remained the same as in the old schedule?

2.13. (True-False) Zonal compared to local service

Offers higher travel speed to some passengers	T	F
Improves travel convenience among all stations along the line	T	F
Increases capacity	T	F
Has simpler scheduling and operations	T	F
Requires smaller car fleet for comparable service	T	F

2.14. Two lines, M and N, have a common trunk section and separate branches, as shown below. Operating speed on all sections is 32.5 km/h. Terminal time at terminal A is 6 min. Headways on each line are 6 min so that on the trunk the joint headway is 3 min. Their schedules should be computed.

a. Compute the cycle time T_M and number of TUs on line M, N_M, if its terminal time at M is at least 8 min (the computed cycle time has to be the smallest multiple of h that allows that terminal time at M is at least 8 min).

b. Compute the cycle time and number of TUs on line N, T_N and N_N, providing that t_N is at least 6 min, for two conditions:

i. Every train M followed by train N on the trunk has such terminal time at terminal A that when returning from outlying terminals M train is again followed by the same train N.

ii. Train M followed by train N on the outbound trip returns to the trunk going inbound before the train M.

Make sure that in both cases, i and ii, cycle times are integer multiples of headways. Round all computed travel times to the nearest integer.

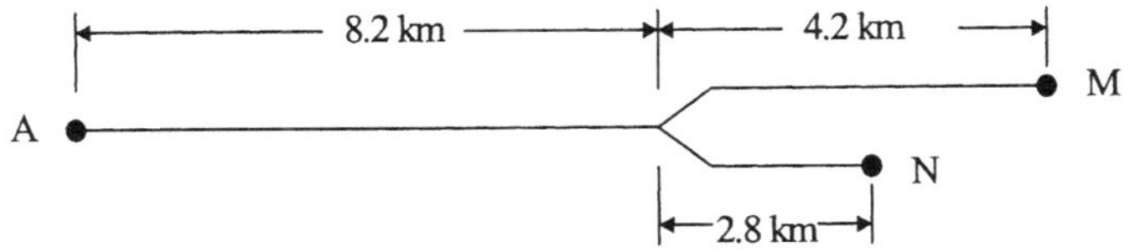

REFERENCES

Transportation Planning Handbook, ed. John D. Edwards, chap. 12. Englewood Cliffs, NJ: Prentice-Hall.

Eisele, D. 1978. Zone theory of suburban rail transit operations: revisited. *Traffic Quarterly* January, 32:5–22.

ITS America. 2003. *An Overview of Transit Signal Priority*. Washington, DC: ITS America.

Kittelson & Associates. 2003. *Transit Capacity and Quality of Service Manual*. TCRP 100. Washington, DC: Transportation Research Board.

Parkinson, T. and I. Fisher 1996. *Rail Transit Capacity*. TCRP Report 13. Washington, DC: Transportation Research Board.

Robinson, C, H. S. Levinson, and L. Goodman. 1992. Capacity in Transportation Planning. In, ed. John D. Edwards, chap. 12.

Vuchic, V. R. 1976. Skip-stop operation: high speed with good area coverage. (In French, English, and German.) *Revue de l'UITP* 2, 105–128.

Vuchic, V. R. 1981. *Urban Public Transportation Systems and Technology*. Englewood Cliffs, NJ: Prentice-Hall.

3

MODELING AND OPTIMIZATION IN TRANSIT SYSTEMS ANALYSIS

List of Symbols

Symbol	Unit	Definition	Symbol	Unit	Definition
C	\$/h	Total cost	S_j	\$	Vehicle salvage value in year j
C_d	\$	Annual vehicle depreciation	β	—	Ratio of passenger arrival rate to the difference between boarding and arrival rates
c_o	\$/TU-h	Unit TU operating cost			
C_o	\$/h	Operating cost			
c_p	\$/prs-h	Average value of passenger time	η_p	—	Ratio of passenger volume on MLS to the volume on the entire line
C_p	\$/h	Passenger cost			
C_r	\$	Vehicle purchase cost	λ	prs/h	Passenger arriving rate at a station
P_L	prs/h	Total passengers boarding along the transit line per hour	μ	prs/h	Passenger boarding rate
			τ	min	Initial delay of a train along the line
ΔT_{mn}	min	Delay of train m at station n			

Planners, managers, and operators in transit agencies and companies often face complex problems that require expertise in different aspects of transit and various types of systems analysis. Traditional approaches to problem solving by individual professionals, such as civil or mechanical engineers, economists, and managers, have therefore been increasingly replaced by more comprehensive approaches and cooperation among different professionals, teams of experts, or multiple organizations.

Solving many complex problems requires consideration of multiple goals and objectives, definition of requirements, development of alternative solutions, qualitative and quantitative impacts, etc. Most important, however, is a thorough understanding of the problem. Many new tools and methods have been developed for this purpose. In recent decades, there has been an intensive development of *systems approach, systems analysis,* and *systems engineering,* activities characterized by a broad view of complex systems, consideration of multiple relationships, and involvement of different disciplines and often different organizations. More precisely, these concepts and activities can be defined as follows.

- Systems approach represents a broad view of an object, system, or process with attention to a thorough understanding of its characteristics and functions.
- Systems analysis is a methodology of work involving a comprehensive analysis of systems and their components, as well as a review of alternative solutions and trade-offs among them.
- Systems engineering is a field of engineering dealing with planning, analysis, design, and operation of complex systems, using engineering methods but often also including economic, social, environmental, and other factors.

Two other major tools often applied to transit systems are defined here.

- *Operations research* (*OR*), a major component of systems analysis, is the application of mathematical and logical methods or models to plan, design, analyze, or operate actual systems or processes.

- *Models* is a broad term that includes physical, conceptual, mathematical, and other kinds of models. Transit agencies sometimes use physical models, such as vehicle-control models for driver training. The emphasis here will be on conceptual, analytical, and mathematical models, which are a major tool in systems analysis and OR work.

This chapter provides an introduction to the use and applications of systems analysis, systems engineering, and OR in urban transit. An introduction to general methodology is presented and illustrated by examples.

3.1 APPLICATIONS OF SYSTEMS ANALYSIS IN TRANSIT

Urban transit systems are by their nature conducive to many applications of system analysis for several reasons. First, large transit agencies, which are typically organizations with complex transportation systems, diverse goals, sizable financial resources, and numerous constraints, cannot be operated efficiently only by routine day-to-day managerial decisions. Second, the trade-off between economically efficient operation and adequate service for the public is very complex and requires sophisticated analyses and decisions. Third, since transit systems represent a public service, investment decisions must not only be based on single (usually quantitative) objectives (such as minimum cost or maximum revenue), but must also include a number of qualitative factors and impacts. Finally, various models and other results of systems analyses and OR can be effectively implemented in transit design and operations because transit systems, particularly rapid transit, are centrally controlled and therefore have much more deterministic operation than general traffic on streets and highways.

3.1.1 Introduction of Methodology

Systems methodology for planning and design of complex projects can be defined as a sequence of phases or steps shown as a flowchart in Figure 3.1. This sequence varies somewhat among types of projects, local conditions, etc., but the major phases are common in most cases. Briefly described, the methodology consists of the following steps.

Problem definition is very important to determine the direction of work, and it leads logically to the statements of general project goals and more specific, shorter-term objectives. The objectives are then used to establish the requirements that the project should meet and technical standards it must follow. Then the purpose and scope of the project are determined, considering local conditions and constraints. A common constraint is the available budget, which may influence both the project purpose and scope.

With the framework, purpose, and direction of the project defined, technical work starts with collection of general information (e.g., present transit services, conditions in the planning area) and specific data (present and projected ridership, construction or operating costs, etc.). Then comes the technical and creative phase: analysis of conditions, modeling of the present and alternative future systems, and design of several possible new facilities or services, referred to as alternatives.

When several alternative solutions are developed, they must be evaluated and compared. This evaluation is performed using a set of criteria based on project objectives, requirements, conditions, and constraints. The process of project evaluation can be complicated and sensitive, and many different methodologies can be used. In some cases, a single criterion may be appropriate to use, such as maximum ridership increase on a new transit line, minimum investment (or life-cycle) costs, etc. However, with respect to transit, several objectives must typically be considered. Three types of evaluation measures may be used: monetary units; other quantitative measures, such as the number of passengers, passenger hours, operational efficiency, safety, etc.; and qualitative criteria, such as system image and positive impacts on urban design and livability, which are listed with the party each one affects. These are usually passengers, operator (agency), and city or society. Short- and long-term effects may be included.

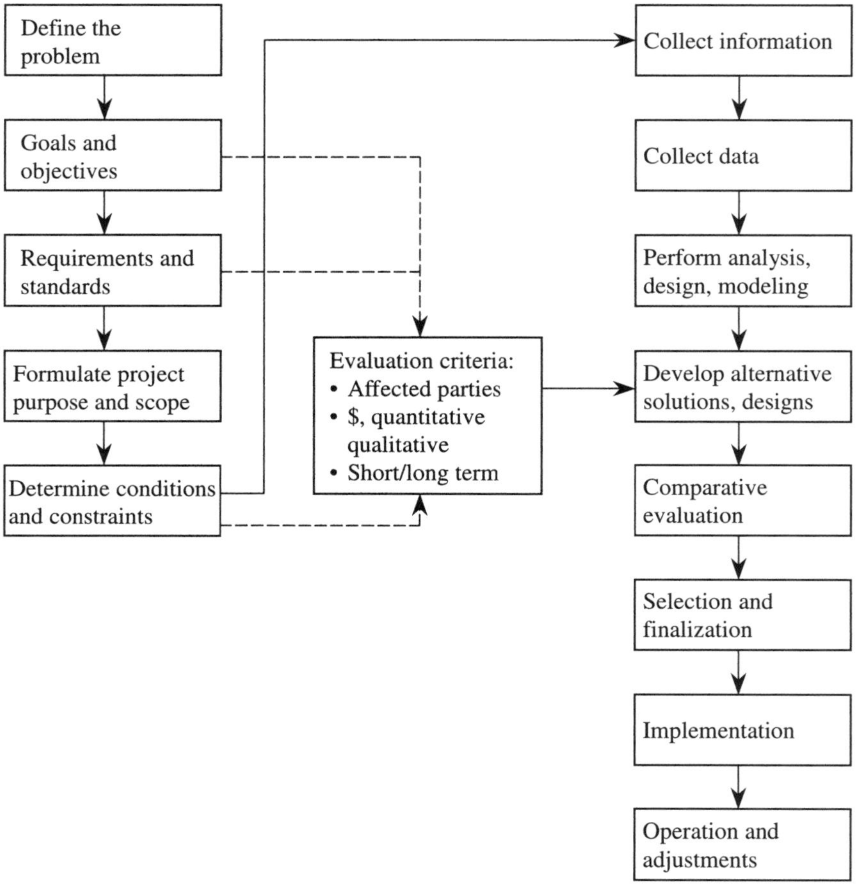

Figure 3.1 *Systems methodology for project planning and design.*

When the evaluation is completed and the optimal alternative selected, the project can be finalized and implemented. In some cases, its operation is observed and adjusted if necessary.

The process of modeling is often the main phase of systems analysis, design and testing of various possible solutions. The model enables systematic evaluation and assists in creativity, testing, and optimization. Therefore, the modeling phase deserves special attention.

The process of modeling, which is a part of the above-described systems methodology, is used for analysis, design, and testing of different systems or operations. A mathematical model is a formal representation of real-world systems or phenomena, using mathematical equations, diagrams, and other formulations. Models are made for the better understanding of complex systems and their components and to examine interactions among the components, their sensitivities to individual changes, etc.

The modeling process is shown in Figure 3.2. Starting from an existing system that should be analyzed or planned, an analyst makes assumptions that simplify the real-world system. For example, an assumption may be that a transit vehicle travels in a deterministic manner, disregarding irregularities that may occur; or that passengers arrive to a station at a certain uniform rate, such as eight persons per minute. These assumptions are made to simplify the real-world system so

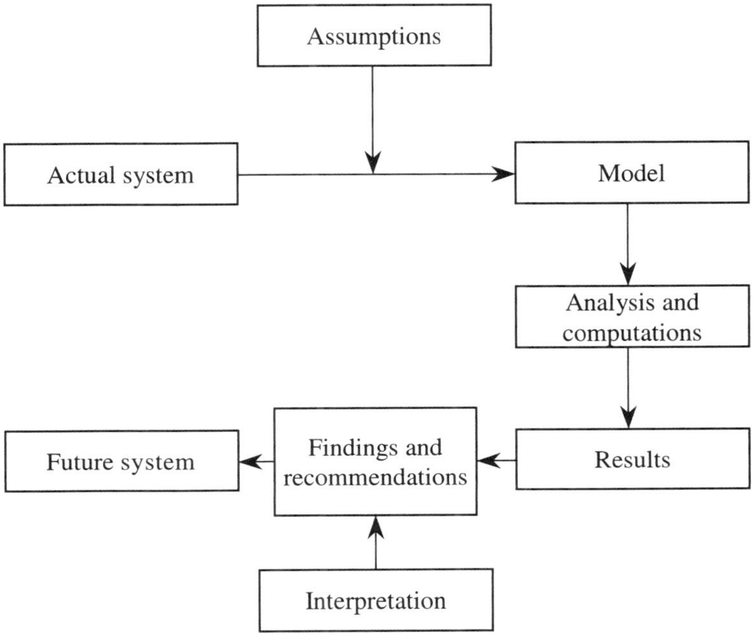

Figure 3.2 *The procedure for theoretical modeling.*

that its functioning can be described by mathematical equations. The number and type of assumptions is therefore a compromise between the simplification of the real system and the ability to handle mathematical expressions of the model. It is usually convenient to make many assumptions initially in order to make the basic model that is easy to understand, solve mathematically, and use. Then, with the formulated and successfully used basic model, some assumptions may be gradually relaxed (canceled), making the model more complex and more realistic and powerful for analysis.

With the operational model, various analyses, tests, and design computations can be performed, as the diagram shows. This phase of work produces the most important results of the analysis or design process. The obtained results are then analyzed and translated back to the real-world system.

The formulation of the model, its interpretation, and the application of results to the real-world system are often difficult because they require two different types of backgrounds and skills, which are not often found in the same person. One person must be thoroughly familiar with the transit system and its operations; another must have mathematical and other theoretical skills to develop and analyze the model. A certain apprehension by transit experts toward theoretical models on one side, and a lack of practical experience of operations researchers on the other, often create a major obstacle in systems analysis and OR studies applications to transit systems.

Experiences in many countries and cities, in transit agencies, and among academics and other theoreticians have led to the development of the following eight points summarizing conditions required for successful development of theoretical models and their applications to transit systems operations and planning (see the box).

3.1.2 Classification of Applications

The great variety of systems analysis and OR applications in urban transit systems can be classified by their subject areas into three major categories:

Eight Rules for Successful Transit Systems Modeling:

1. The analyst should be familiar with and understand the transit system and its functioning and impacts.
2. The system, function, or problem to be analyzed should be defined and set up using the systems approach.
3. The relationships among system components and functions should be established and used as the basis for the model.
4. The goals and scope of modeling should be defined, such as:
   - Develop a permanent analytical tool
   - Compute specific values
   - Examine sensitivities of the system to individual elements and operating conditions
   - Simulate system's operation under different conditions and influences
   - Test new system designs, operations, or their modifications
5. Assumptions for the model should be based on an appropriate trade-off between the model's reality and analytical simplicity.
6. Analytical methodology (e.g., conceptual analysis, mathematical or numerical computation) should be selected so that model's objectives can best be achieved, rather than for simpler solution or for mathematical elegance.
7. Mathematical results should be translated back to the real system to realize the benefits of the systems analysis effort.
8. The model's findings should be presented with a clear definition of their applicability and its limitations due to the assumptions the model is based on.

- *Planning of transit systems* includes such problems as the optimum network configuration and extensiveness, selection of optimum modes for different lines, intermodal integration, trade-offs among alternative investments, and others.
- *Transit system design* includes more specific decisions about transit line alignment, station locations, size and design of terminals, optimum vehicle capacity and TU size, etc.
- *Transit system analysis and operations* can benefit from such OR studies as various scheduling programs and schedule evaluations, reliability of service (minimization of delays), impacts of transit priority schemes on transit level of service and on street traffic, and many others.

In Sections 3.2–3.4, applications are presented of the three major OR methodologies—conceptual models, mathematical modeling, and simulation.

3.2 CONCEPTUAL MODELS

To analyze a complex systems problem in urban transportation, it is necessary to understand fully the relationships of its components, its function and method of operation and its interactions with other systems. Only with such understanding can the goals and objectives of the analysis be defined and, if appropriate, a mathematical model developed. Preparation of a conceptual model in such cases is very useful as a basis for the analysis and modeling.

3.2.1 The Conceptual Modeling Methodology

Most commonly, a conceptual model is presented as a chart with system components, or sequence of steps in an analysis, or flow of events. Relationships among different events, directions of influences, incidence of benefits and costs, etc. are included. Often an initial event or input into a system flows into a series of activities and consequences and, eventually, in one or more final results or outputs.

Some of the boxes in the flowchart are descriptive items or events, while others may be relationships conducive to mathematical expressions. Thus, a mathematical model may be a portion of the conceptual model, or it may encompass the entire system that is presented by the conceptual model. The model is a guideline for planning and organizing the entire analysis process. It can also be used as a basis for presentation of results in the final report.

3.2.2 Representative Applications of Conceptual Models

Systems analysis and development of conceptual or functional models of different processes and phenomena in transit systems are presented in many sections of this book. The following are references to such models presented in other chapters, as well as several additional systems analysis models useful in transit systems planning and operations.

EXAMPLE 3.1: Evaluation of Transit Speed Increase and Its Impacts

It is intuitively known that increase of speed in transit services is desirable for passengers and also usually brings benefits to the operator. Section 2.2 reviews various methods for speed increases. Systems analysis is used in developing a methodology to evaluate the implementation of a speed-increase measure: the necessary investment costs and required actions, as compared to the direct benefits and broader impacts. The entire evaluation procedure is presented in Figure 2.5 as a conceptual flowchart. Overall evaluation of the results of a project to increase transit speed—its feasibility, benefits and costs—can then be made by reviewing all final results, including and comparing the positive and negative ones. The same model can also be used to evaluate the consequences of a decrease of speed.

EXAMPLE 3.2: Evaluation of Fare-Collection Improvements

Decisions about various measures that speed up fare collection often require an evaluation of costs and benefits of such an innovation. Typically, new fare-collection equipment requires considerable initial investment and organizational effort, while the benefits from faster fare collection include a number of qualitative and quantitative elements. These costs and benefits must therefore be systematically defined and evaluated, as discussed in Section 2.2.4, Example 2.5. A conceptual model for this evaluation is shown as a flowchart in Figure 2.10.

EXAMPLE 3.3: Introduction of Exclusive Transit Lanes

This is another measure for increase of transit speed and reliability that requires a systems analysis. That analysis is presented in Section 2.2.4, Example 2.6, and a conceptual model diagram for analysis and evaluation of such a project is shown in Figure 2.11. Again, the benefits and costs are clearly shown in the last two columns.

EXAMPLE 3.4: Conversion of a Busway into a High-Occupancy Vehicle (HOV) Roadway

Efficient busways and bus lanes in many cities, particularly in the United States, have been converted into HOV facilities. The highway authorities that introduced this change justified it by the argument that it increased vehicle-km operated on the facility. That evaluation of this action is, however, too narrow because it does not identify the affected groups and ignores its intermodal impacts. If a systems analysis is applied that includes impacts on different modes and groups of travelers, shown in Figure 3.3, it shows that auto and HOV users are beneficiaries, but that bus passengers have only losses. Decreased

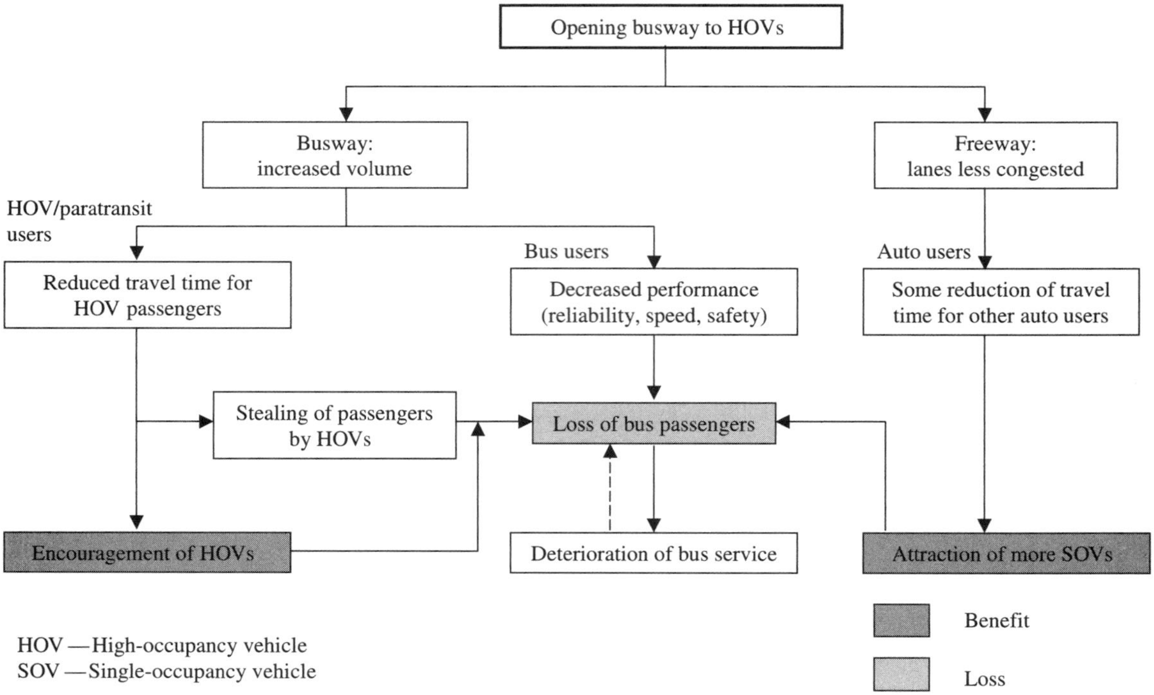

Figure 3.3 *Consequences of busway conversion into a HOV roadway.*

level of service and image result in a diversion of bus passengers to autos and HOVs. This loss may lead to a reduction of bus service, resulting in further passenger losses.

In this case, a systems analysis with a conceptual diagram can be used not to advance transit, but to prevent its deterioration due to an action that would directly harm it. The diagram can be effectively used to demonstrate that from the point of view of transit services, every conversion of exclusive bus facilities into HOV facilities is damaging. Moreover, with respect to rational intermodal policy that encourages a shift of travel from cars to transit this action is directly counterproductive.

EXAMPLE 3.5: Estimating Value of Reliability in Transit Services

It is known that reliability is one of the most important service features for passengers. It also affects costs of operations: unreliable service requires more TUs and drivers to provide a given service than if operation is very reliable. In making various decisions about investments that would increase reliability (such as modernization of communications, modernization of the operations control center, improved vehicle maintenance, etc.), it is sometimes necessary to define the value of reliability systematically through its results and impacts. A conceptual flow diagram showing the major results from increased reliability is shown in Figure 3.4. The diagram includes both quantifiable and qualitative elements and indicates who are the affected parties, i.e., which results benefit which of the three major parties. (An analysis of elements critical for service reliability is presented in Section 6.4.4.) The diagram shows that all consequences of reliability improvements are positive, so that the decision about the feasibility of such a project is rather clear: it

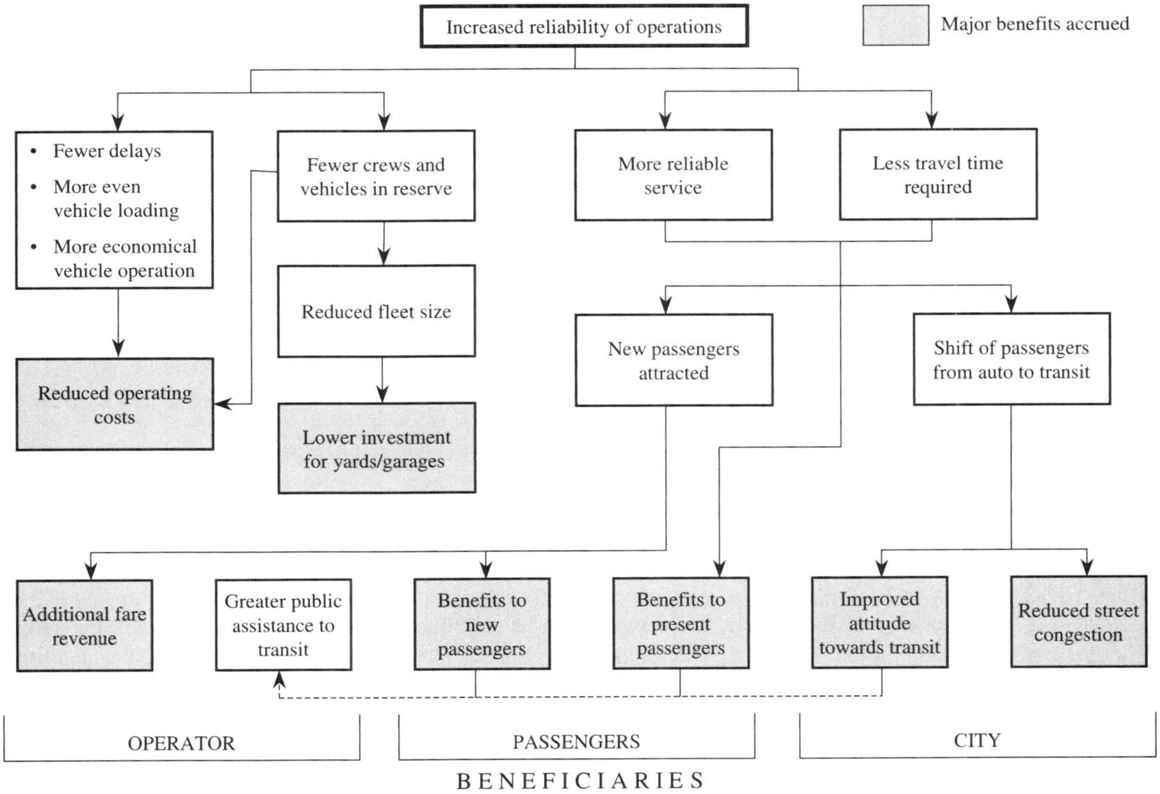

Figure 3.4 *Estimation of the value of reliability increase in transit services.*

depends on the evaluation whether the benefits justify the expected investment.

While this model discusses overall reliability of system operations, a conceptual model focusing specifically on rolling stock reliability is shown in Figure 6.2. The diagram shows major elements that contribute to high reliability of vehicles, and the benefits accrued by the operator and passengers from the reliable rolling stock.

EXAMPLE 3.6: Determination of Optimum Rolling Stock Life

Minimum average annual cost per vehicle can be adopted as the main objective in determining optimum rolling stock life. This economic analysis usually includes two major cost elements and derives the minimum value of their total for the entire vehicle life. One element is operating and maintenance cost averaged over the vehicle life C_m. Since this cost, particularly the maintenance, increases with vehicle age, the average cost increases with vehicle life. The other element is the average annual vehicle depreciation cost C_d, which, if straight-line depreciation is assumed, decreases when vehicle life becomes longer, as shown in Figure 3.5. Salvage (resale of rolling stock) value, if any, decreases with life length, so that its specific value for each life length should be subtracted from the depreciation value. The optimum is then found as the minimum total of these two cost elements $C_{\min}$. It should be noted that the costs plotted on the diagram for maintenance and depreciation are average

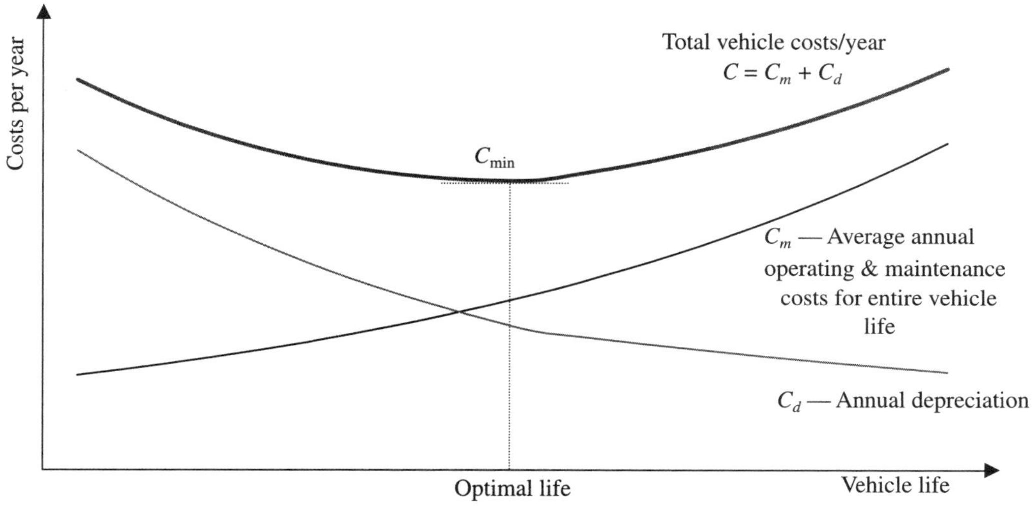

Figure 3.5 *Optimum vehicle life for minimum direct costs.*

costs per year for any given vehicle life length, rather than annual costs during the vehicle life. Thus, annual depreciation C_d for year j is:

$$C_d = \frac{C_r - S_j}{j} \qquad \left|\frac{C,\ S}{\$}\right|\frac{j}{-}\right| \qquad (3.1)$$

where C_r is the vehicle purchase cost and S_j is the vehicle salvage value in year j.

If the systems approach is used, however, many other factors are considered, which may change this analysis and result in different decisions on vehicle life. For example, in some cases capital and operating funds may come from different sources. If capital funds are more easily available than operating funds, the optimum is shifted toward shorter vehicle life. If, to the contrary, financing for purchase of new vehicles is not easy to obtain and depreciation is not regularly collected, vehicle fleets are maintained longer and extensive overhauls are performed.

An even broader view is presented in Section 6.4.4.2. First, it is pointed out that in addition to economic vehicle life, based on minimum life-cycle costs, the concepts of physical life and functional life include influences of such non-monetary factors as quality of service, technological innovations, and image. The shortest of the three lives is usually the critical one. Typical life lengths for rolling stock of different transit modes, which are a result of all economic, service, and other factors discussed, are presented in Table 6.3.

This example shows that some computational models of quantitative elements, such as costs, can be used for analyses of individual aspects of complex systems, but their results should then be combined with other, often qualitative, factors to derive optimum solutions for specific local conditions.

3.3 MATHEMATICAL MODELING PROCEDURE AND APPLICATIONS

The general approach to mathematical modeling is presented here with several simple examples of model development and applications in transit systems planning and operations. Additional uses of models and other OR techniques can be found in other chapters.

3.3.1 The Mathematical Modeling Procedure

As already discussed for conceptual models, when building mathematical models, the analyst must first carefully define the goals of the analysis and then decide how the model should be formulated. This decision requires that he/she become thoroughly acquainted with the analyzed system or process. The next set of decisions is the choice of assumptions, including the compromise between simplicity of the model and its realistic value. Once the assumptions are made, a set of mathematical equations is developed that defines the relationships of variables and parameters of the analyzed system. These equations represent a mathematical model of the analyzed system. It is a powerful quantitative tool for computations of system components, operations, and characteristics.

Systems analysis and computations performed by a mathematical model can take different forms. If a single objective for the modeled system can be established (such as minimum passenger time, minimum operating cost, maximum utilization of vehicles, etc.), the analysis has a form of an optimization problem. If alternative solutions should be examined, the analysis may provide insight into trade-offs among variables that are needed for the comparison and selection of the optimal solution. Also, sensitivity of results to individual variables and parameters is often useful to examine. For example, a mathematical model of TU travel along a line computes travel time as a function of TU acceleration, speed, braking rates, and station standing times. It can be used to compute sensitivity of travel time to the acceleration rate, station standing times, or other parameters.

The results of computations may be a range of values for different conditions, or a single numerical value. They can also be presented as a set of equations and diagrams.

The power of mathematical models is that they allow analyses in parametric form, i.e., utilizing symbols for all variables and parameters. This permits introduction of numerical values in the model for various sets of conditions. However, when the model is so complicated that it does not lend itself to an understandable analytical solution, numerical values may be used to derive specific solutions for individual cases only. Finally, if a highly irregular phenomenon is analyzed, particularly when it involves stochastic processes, computer simulation may be the best or only possible technique to obtain a solution to the problem.

Understanding the results of an analysis and drawing conclusions from them is the final step in modeling. The results of the model usually lead to a number of relationships, some of which are known, but others of which may not be intuitively clear. They give new insights, describing relationships, causalities of phenomena, and sensitivities of a system to individual variables and components introduced as parameters. As already mentioned, the analyst should be particularly careful about the limitations of the results imposed by the types of assumptions and detail of accuracy in the data used for the model equations.

Applicability of a model is limited only to the situations for which the assumptions made are valid. Thus, if in analyzing an optimum schedule the assumption was made that one bus-hour costs $65 and a passenger-hour has a value of $9, the derived optimal solution may not be applicable to the situation where a bus-hour costs $58 and average passenger-hour is valued at $5.50.

In conclusion, the analyst who possesses adequate understanding of real-world systems and sufficient theoretical background to formulate and analyze mathematical models can draw a number of interesting observations, conclusions, and numerical values from such models if he/she is fully aware of their strengths and limitations. To illustrate the process of mathematical modeling and its results, several simple cases of transit models are presented here.

3.3.2 Models for Optimization of Rolling Stock

Selection of the size or type of transit vehicle or TU is one of the areas that are conducive to mathematical modeling. Two examples of such models are presented here.

EXAMPLE 3.7: Optimum TU Capacity

On a transit line with a given P_{max}, the operator is interested in providing a required transporting capacity by a smaller number of large TUs because of their lower cost per unit capacity. However, passengers require shorter headways, i.e., higher frequency of service, which dictates the use of a greater number of smaller TUs.

To find the optimal balance between the interests of the two parties considering costs only, a model including the operating costs C_o (\$/h) and passenger costs C_p (\$/h) should be developed and used to derive the expression for the total cost, $C = C_o + C_p$. With mathematical expressions for these costs, the TU capacity that results in the minimum total cost C should then be derived. This is illustrated here by a rather simple model of a transit line with length L (km), headway h (min), cycle time T (min), and load factor $\alpha_{\max}$ (prs/sp). The TU (bus or train) capacity is C_{TU} (sps/TU).

Simplifying assumptions for this model are:

- Total hourly cost of operation of the line C_o is equal to the product of unit vehicle operating cost c_o (\$/TU-h) and the number of TUs on the line N_{TU}: $C_o = c_o \cdot N_{TU}$.
- Passenger demand (design hour volume $P_{\max}$) is constant and is related to the total number of passengers boarding along the entire line, P_L, by a coefficient: $P_{\max} = \eta_p \, P_L$.
- Cycle speed V_c on the line is independent of TU size and of the number of passengers.
- Average passenger waiting time is equal to half of the headway, $h/2$.
- Average value of passenger time is constant: c_p (\$/prs-h).

First, the number of TUs on the line will be expressed by the given line parameters. Using Eqs. (1.29) and (1.45), one obtains the equations for T and h:

$$T = \frac{120L}{V_c} \quad \text{and} \quad h = \frac{60\alpha_{\max} \cdot C_{TU}}{P_{\max}}$$

$$\left| \frac{T,\ h}{\text{min}} \right| \frac{L}{\text{km}} \left| \frac{V}{\text{km/h}} \right| \frac{\alpha}{\text{prs/sp}} \left| \frac{C_{TU}}{\text{sps/TU}} \right| \frac{P}{\text{prs/h}} \right| \tag{3.2}$$

N_{TU} can be expressed by these operating elements as:

$$N_{TU} = \frac{T}{h} = \frac{2L \cdot P_{\max}}{\alpha_{\max} \cdot C_{TU} \cdot V_c}$$

$$\left| \frac{N}{\text{TU}} \right| \frac{T,\ h}{\text{min}} \left| \frac{L}{\text{km}} \right| \frac{P}{\text{prs/h}} \left| \frac{\alpha}{\text{prs/sp}} \right| \quad \left| \frac{C_{TU}}{\text{sps/TU}} \right| \frac{V}{\text{km/h}} \right| \tag{3.3}$$

Operating cost can now be expressed, using the first assumption above, as:

$$C_o = N_{TU} \cdot c_o = \frac{2L \cdot P_{\max}}{\alpha_{\max} \cdot C_{TU} \cdot V_c} \cdot c_o$$

$$\left| \frac{C}{\text{\$/h}} \right| \frac{L}{\text{km}} \left| \frac{P}{\text{prs/h}} \right| \frac{\alpha}{\text{prs/sp}} \quad \left| \frac{C_{TU}}{\text{sps/TU}} \right| \frac{V}{\text{km/h}} \left| \frac{c_0}{\text{\$/TU-h}} \right| \tag{3.4}$$

Passenger cost for all passengers on the line, P_L, is:

$$C_p = P_L \cdot \frac{h}{120} \cdot c_p = \frac{P_L \cdot \alpha_{\max} \cdot C_{TU}}{2P_{\max}} \cdot c_p$$

$$\left| \frac{C}{\text{\$/h}} \right| \frac{P}{\text{prs/h}} \left| \frac{h}{\text{min}} \right. \quad \left| \frac{c_p}{\text{\$/(prs-h)}} \right| \frac{\alpha}{\text{prs/sp}} \left| \frac{C_{TU}}{\text{sps/TU}} \right| \tag{3.5}$$

Total cost, consisting of the operation and passenger costs, is obtained by summing up Eqs. (3.4) and (3.5):

$$C = \frac{2L \cdot P_{\max}}{\alpha_{\max} \cdot C_{TU} \cdot V_c} \cdot c_o + \frac{P_L \cdot \alpha_{\max} \cdot C_{TU}}{2P_{\max}} \cdot c_p$$

$$\left| \frac{C}{\text{\$/h}} \right| \frac{L}{\text{km}} \left| \frac{P}{\text{prs/h}} \right| \frac{\alpha}{\text{prs/sp}} \quad \left| \frac{C_{TU}}{\text{sps/TU}} \right| \frac{V}{\text{km/h}} \left| \frac{c_0}{\text{\$/veh-h}} \right| \frac{c_p}{\text{\$/prs-h}} \right| \tag{3.6}$$

To find the optimum value of vehicle capacity that will minimize total coast, the derivative of Eq. (3.6) with respect to C_{TU} is set equal to 0:

$$\frac{\partial C}{\partial C_{TU}} = -\frac{2L \cdot P_{\max}}{\alpha_{\max} \cdot V_c \cdot C_{TU}^2} \cdot c_o + \frac{P_L \cdot \alpha_{\max}}{2P_{\max}} \cdot c_p = 0 \quad \text{(Units as above)} \tag{3.7}$$

from which the optimum vehicle capacity, C^*_{TU}, is:

$$C^*_{TU} = \frac{2P_{\max}}{\alpha_{\max}} \sqrt{\frac{L}{P_L \cdot V_c} \cdot \frac{c_o}{c_p}} \quad \text{(Units as above)} \tag{3.8}$$

Figure 3.6 shows a diagram of C_o, C_p, and C as functions of C_{TU}, as well as the position of C^*_{TU}.

The solution given by Eq. (3.8) shows that the optimal TU capacity is proportional to the critical passenger volume $P_{\max}$ but is inversely proportional to the square root of the total number of passengers using the line (their waiting time is shorter if TUs are smaller and therefore more frequent). One can assume, however, that in most cases these two passenger volume values, $P_{\max}$ and P_L, are related; then the two passenger volumes can be treated together through the coefficient η_p. Replacing $P_{\max}$ in Eq. (3.7) by $\eta_p \cdot P_L$ gives the optimal capacity as proportional to the square root of the total number of passengers on the line P_L:

$$C^*_{TU} = \frac{2\eta_p}{\alpha_{\max}} \sqrt{\frac{P_L \cdot L}{V_c} \cdot \frac{c_o}{c_p}} \quad \text{(Units as above)} \tag{3.9}$$

This equation is more realistic than Eq. (3.8) because the impact of the passenger volume on the economy of larger TUs is usually greater than its impact on passenger waiting times via shorter headways on the line, particularly when headways are already short, such as < 6 minutes. Thus, it is logical that C_{TU} *increases* with passenger volume on the line.

In interpreting the derived optimal solution, one must consider the assumptions. For example, in many cases the cost of line operation cannot be expressed just by TU cost multiplied by the number of TUs, since the amount of fixed costs (e.g., the way, terminals, and overhead expenses) per unit decreases as fleet size increases. Further, passenger demand is constant only for certain periods of time. Because of demand variations, different TU capacities are optimal at different times of day, and the optimum capacity should be determined taking all periods of day into account.

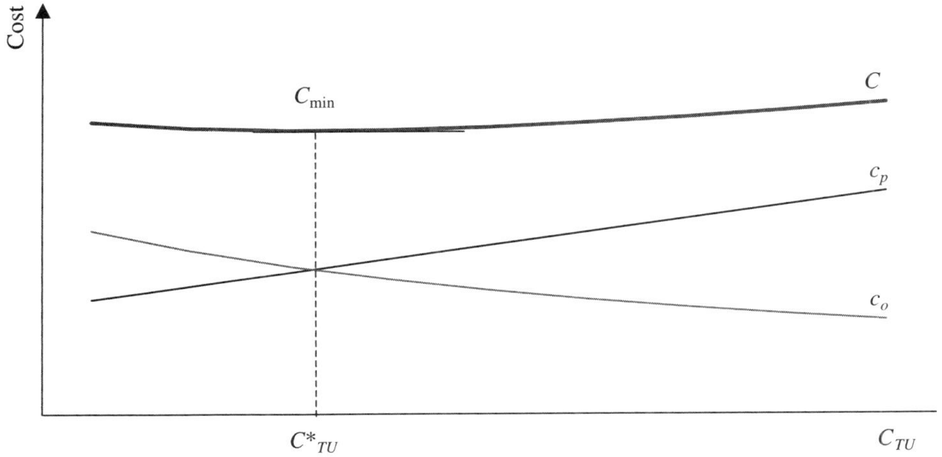

Figure 3.6 *Operator and passenger costs as functions of transit unit capacity.*

Consequently, it is not likely that all assumptions will be satisfied for any one realistic case, so that Eq. (3.8) can probably never be used alone in this form for derivation of an actual definitive solution. What, then, is the value of this analysis?

Despite these limitations, the value of this type of general OR model and expression for optimal solution is considerable. First, the model clearly shows the basic relationships among system parameters. Each one of these relationships is correct, *ceteris paribus*. Some of the relationships are intuitively clear. One would expect, for example, that the optimum TU capacity increases with the increase of peak demand and unit operating cost. But it is somewhat more revealing to see that the optimal TU capacity increases with line length. Second, the expression indicates sensitivity: C^*_{TU} is linearly proportional to P, but only as a square root to L and c_o, etc. A third value of the model is that it can serve as a basis for further, either more sophisticated or simpler analyses. Simply by relaxing the assumptions or introducing different features, the model can be made more realistic for any given set of conditions.

Direct application of this theoretical case to real-world situations is rather limited for several reasons. Actual numerical values for individual elements are often difficult to establish. Also, the question is which passenger volume should be used for vehicle selection: if peak hour volume is used, the selected vehicle will not be optimal for off-peak operation, and vice versa. Finally, there is no possibility of obtaining vehicles with any exact capacity: the choice is usually available from a certain number of vehicles and TUs, so that there is a limited set of values to choose from.

An illustration of selecting an optimal vehicle in a real-world situation is the choice between a standard and an articulated bus, considering two sets of conditions, for peak and for off-peak hours. This analysis is given in the next example, which relies more on the systems approach and conceptual model than on the mathematical model used in Example 3.7.

EXAMPLE 3.8: Selection of the Optimal Bus Type

A transit agency is planning service on a bus line that has a heavy load during peak hours: P_{max} is 2040 prs/h, and load factor $\alpha = 0.85$ can be reached; during off-peak hours, the volume is only 450 prs/h, but to offer seats for most passengers, load factor of only $\alpha = 0.65$ is adopted. To provide an attractive service, policy headway $h_p = 6$ min is established. The agency is considering for its fleet two types of buses: standard buses with a capacity of 80 sps and articulated buses with a capacity of 120 sps. Articulated buses involve higher investment and operating costs per vehicle, but lower total cost per space (or space-km) than standard buses. The problem is to select which type, or a combination of the two types, should be used for this line. To find the optimal solution, different operational scenarios must be developed and compared with respect to the established objectives and criteria.

To begin, the required offered capacity on the line for peak and off-peak periods are computed from passenger volumes and load factors. They are:

$$C_1 = P_{max}/\alpha = 2040/0.85$$
$$= 2400 \text{ sps/h for peak}$$
$$C_2 = 450/0.65$$
$$= 692 \sim 700 \text{ sps/h for off-peak period}$$

For development and an easy overview of different scenarios, a diagram of frequency versus line capacity, $C = f \cdot C_v$, is drawn in Figure 3.7. For convenience, headways are also shown on the diagram: they are inverse values of the frequency: $h = 60/f$. Correlation between frequency and line capacity for each bus type is shown by a sloped line. The standard bus has a steeper slope because more buses are needed to provide any given capacity than with articulated buses. For example, the required peak hour capacity of 2400 sps/h is achieved by articulated buses with frequency of 20 veh/h ($h = 3$ min) and by standard buses operating at a frequency of 30 veh/h ($h = 2$ min).

Photo 3.1 *Minibus operating on a shuttle line—Philadelphia (courtesy of Mario Semmler).*

A systematic analysis of the two vehicle types leads to formulation of five different scenarios: the initial case with standard buses only, three scenarios with articulated buses and different combinations of headways and capacities, and a combination of standard and articulated buses. These are listed as scenarios I–V in Table 3.1 and plotted graphically in Figure 3.7. On the diagram, the required peak hour capacity is at or above the $C_{\mathrm{I}} = 2400$ sps/h vertical line, off-peak capacity is $C_{\mathrm{II}} = 700$ sps/h, and the required frequencies are on the $f = 10$ veh/h line (policy headway of six minutes), except in scenario III. The five scenarios are described and evaluated here.

I. *Standard buses only.* During peaks and off-peaks these operate, respectively, at headways of two and six minutes, providing the required capacities for the peak as well as capacity and policy headway for off-peak.

II. *Articulated buses only, the same headways as standard buses in scenario I.* This scenario exceeds the required comfort (load factor is lower than required), but it involves higher cost because the same number of buses must be operated.

III. *Articulated buses only, offering the required capacities.* Services during both periods meet the capacity requirements with longer headways than standard buses in scenario I. Thus, total costs are lower because of lower cost per space. Headways are still adequate for peaks, but they exceed the policy headway during off-peaks. Therefore, looking strictly, this scenario must be eliminated.

Photo 3.2 *Regular bus—Philadelphia (courtesy of Mario Semmler).*

IV. *Articulated buses only, meeting the capacity required for peaks and policy headway during off-peaks.* This scenario is more economical than scenario II and meets the requirements better than III. Compared to I, scenario IV involves lower cost during peaks (fewer buses) but higher cost in off-peaks (the same number of articulated as standard buses). Overall, it is likely to be more economical than I because the difference in the number of buses saved during peaks is very large (frequency of 20 instead of 30 veh/h results in about 33% fewer buses).

V. *Articulated buses in peaks, standard during off-peaks.* This is a combination of most economical operation during both peak and off-peak periods, and it meets all requirements. This economic advantage must be evaluated against the significant disadvantage of this scenario, the complexity of owning two different vehicle types and changing the vehicles between peaks and off-peaks.

These descriptions of the scenarios and their review in the table allow the following comparative analysis and selection of the best scenario. Compared to scenario I, which can be considered as basic, scenario II provides comfort higher than required at a higher cost than scenario I. It is rejected because the additional cost cannot be justified. Scenario III is rejected because off-peak headway exceeds the requirement. Scenario IV satisfies both capacity and headway requirements and involves lower cost than I. Finally, scenario V also satisfies all requirements and has even lower costs than IV, but it has the serious disadvantage of re-

Photo 3.3 *High-capacity double-deck bus—Berlin.*

quiring two different fleets for peak and off-peak periods.

In conclusion, based on the stated service requirements, scenarios IV and V are superior to others, and the decision between them depends on the complexity of using two types of buses in scenario V versus the lower cost of scenario IV.

EXAMPLE 3.9: Optimum Station Spacings on Rapid Transit Lines
Determination of station spacings and locations along rapid transit lines, which has a major impact on passenger attraction, operations, and costs, is discussed in considerable detail in Chapter 5. A model for computing the optimal locations for minimum passenger travel time is first presented. Then, other factors such as area coverage, costs, and speed are added in a conceptual model with multiobjectives. This theoretical part is followed by a review and discussion of actual station spacing values on different modes and cities.

3.3.3 Models for Analysis of Operations

Many aspects of transit operations, such as vehicle travel, signal control, and handling of passengers, can be analyzed through modeling, as the following examples illustrate.

EXAMPLE 3.10: Analysis of Delay Propagation on a Transit Line
When a TU experiences delay on a line, there is a tendency for the delay to increase as the TU progresses along the line, while the following TU begins to get ahead of its schedule. Eventually, the

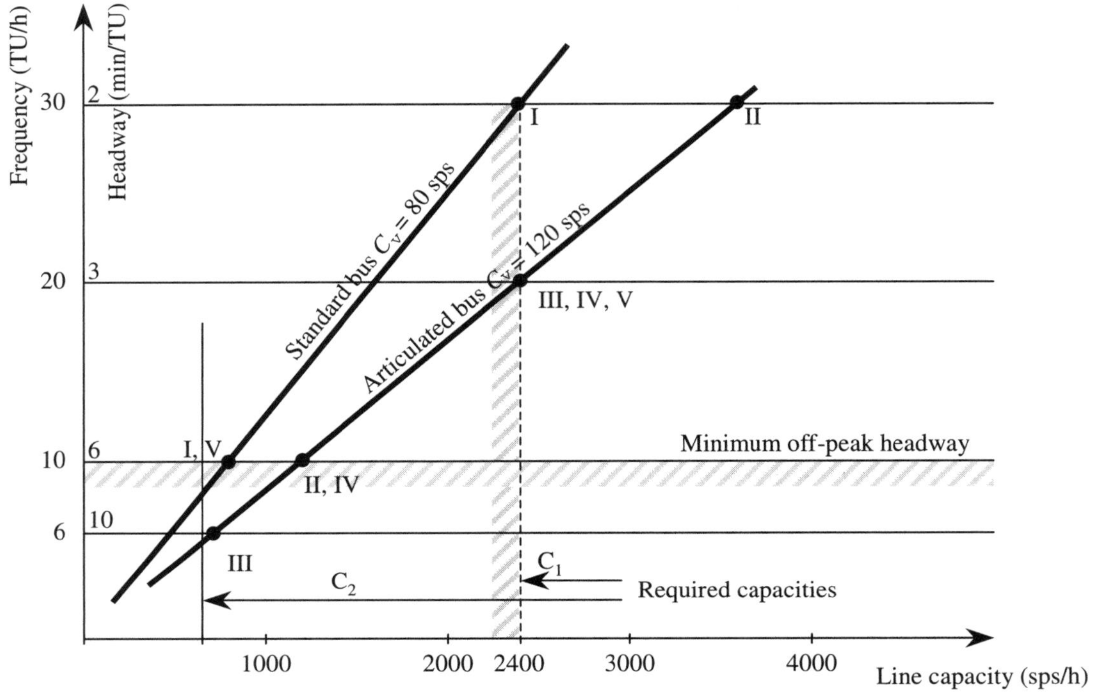

Figure 3.7 *Analysis for the selection of the optimal bus type.*

Table 3.1 Alternative bus fleet compositions for peak and off-peak hours

Period	*Peak*				*Off-peak*			
Scenario	*Bus type*	*Criterion*	*Headway*	*Capacity*	*Bus type*	*Criterion*	*Headway*	*Capacity*
I	Standard	Capacity	2	2400	Standard	Capacity	6	800
II	Articulated	Service	2	3600	Articulated	Service	6	1200
III	Articulated	Capacity	3	2400	Articulated	Capacity	10	720
IV	Articulated	Capacity	3	2400	Articulated	Service	6	1200
V	Articulated	Capacity	3	2400	Standard	Capacity	6	800

two TUs begin to travel as a pair, the next two TUs also form a pair, etc. With buses, when short headways are operated, this pairing may develop into bunching of many buses. A theoretical model may be helpful in the analysis of causes of the pairing phenomenon and exploration of influencing factors.

Since it is known that changed durations of passenger boarding at stations cause this phenomenon, external influences on vehicle travel can be eliminated without reducing the realistic value of the model. A rapid transit mode is adopted for the model with the following assumptions:

- Trains initially travel at uniform headways h.
- Passengers arrive at stations at a uniform rate λ.
- The rate of vehicle boarding μ ($>\lambda$) in prs/h is also constant.
- Each train collects all waiting passengers at each station and departs as soon as they have boarded (standing time is t_s).

• Train travel time between stations is constant.

The diagram of the number of passengers versus time is used to show operations graphically at a station (passenger accumulation, train arrival, boarding, departure, etc.). Figure 3.8 shows such diagrams for three subsequent stations with all origins representing, for convenience, departure time of the train at time and location 0; thus, the abscissa for each subsequent station is actually moved to the left by the travel time of the train between that and the preceding station, so that their beginnings are vertically aligned. The solid lines at all three stations depict accumulation of passengers at a uniform rate until the moment the train arrives. From that point the accumulation of passengers on the platform begins to dissipate at a rate $\mu - \lambda$ (a negative rate of accumulation) since passengers continue to arrive at rate λ while boarding the trains at rate μ. At the moment all passengers have boarded the train, it departs and the accumulation of passengers at rate μ starts again, starting the next triangle on the diagram.

To observe the phenomenon occurring after an external delay has occurred, the arrival of train 1 at station 1 is shown on the diagram delayed by an interval τ. It is seen in the diagram that the delay resulted in a slightly higher accumulation of passengers. From that point on, the situation caused by this delay is obtained by plotting parallel lines to the solid ones representing regular operation. The operation with delay is represented by the dashed lines. It can be clearly seen that the delay of train 1 results in a slight advance of train 2 in leaving station 1, which then causes a slight delay in the departure of train 3. Thus the vehicles begin to oscillate around their scheduled times.

The delayed departure of train 1 from station 1 results in delayed arrival of that train at station 2 by the same interval of time, so that a similar phenomenon of delay propagation takes place at that station as well as at station 3 and the subsequent ones. It is seen, however, that the delay τ in the arrival resulted in a longer delay in the departure of the same train from the station, which then led to a further increase of this delay at station 2, etc.

This diagram can be used to develop analytical expressions for the delays, as explained in Foulkes et al. (1954) and Vuchic (1969). The main result of that analysis is that the delay of train m at station n (ΔT_{mn}) is defined by the expression:

$$\Delta T_{mn} = \tau \frac{(m + n - 2)!}{(m - 1)!(n - 1)!} (-\beta^{m-1})(1 + \beta)^n \qquad \left|\frac{T,\ \tau}{\text{min}}\right|\frac{m,\ n,\ \beta}{-}\right| \tag{3.10}$$

where β represents the following relationships:

$$\beta = \frac{\lambda}{\mu - \lambda} = \frac{t_s}{h - t_s} \qquad \left|\frac{\beta}{-}\right|\frac{\lambda,\ \mu}{\text{prs/min}}\left|\frac{t_s,\ h}{\text{min}}\right| \tag{3.11}$$

Equation (3.10) shows that the behavior of delays depends greatly on factor β. Looking at the definition of this factor in Eq. (3.11), one can conclude that the stability of operations depends on the ratio of the rate of passenger arrival at stations to the rate of dissipation of the accumulated passengers during t_s, or on the ratio of the standing time t_s to the time between departure of one train and arrival of the following train, $(h - t_s)$. Transferring this conclusion to practical system operation, it is clear that for stability of operation the rate of boarding vehicles must be very high in relation to the passenger arrival rate.

It is interesting to illustrate this finding by a couple of numerical examples. If in one case there is a bus stop to which passengers arrive at a rate of $\lambda = 150$ prs/h and boarding, due to a very complicated fare-collection method, takes 6 s/prs, amounting to a rate of only $\mu = 600$ prs/h. In another case, $\lambda = 3600$ prs/h arrive at a rapid transit station and, boarding simultaneously on many doors, achieve a rate of $\mu = 4$ prs/s $= 1440$ prs/h. It may appear that the heavily used rapid transit line will have a much greater tendency than lightly loaded buses to propagate delays when they occur. To find the propensity for delay propagation, the factor β which influences it, is computed by Eq. (3.11). The

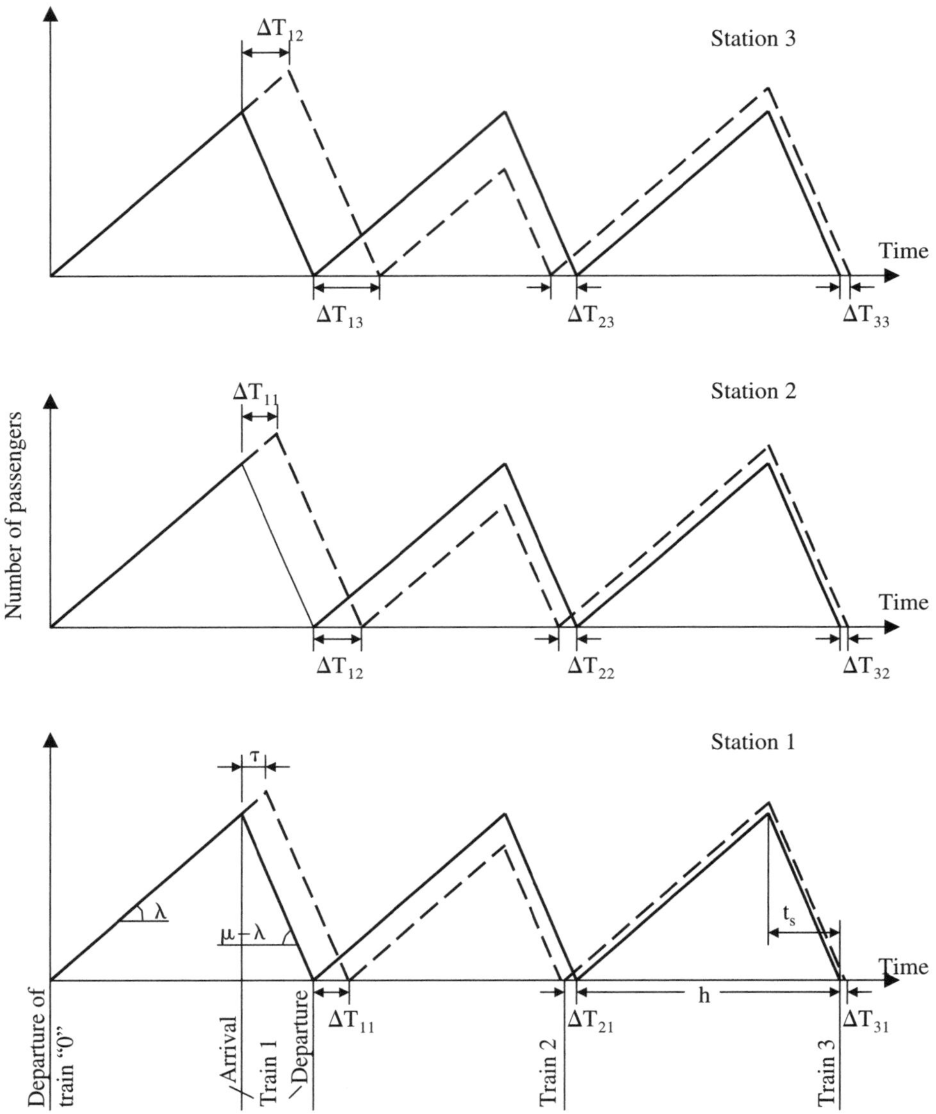

Figure 3.8 *Transit unit delay propagation over time and distance.* (Source: *Vuchic, 1969.*)

computations for the bus and rail lines give, respectively:

$$\beta_B = 150/(600 - 150) = 0.333 \quad \text{and}$$

$$\beta_R = 3600/(14{,}400 - 3600) = 0.333$$

Thus, the two systems with drastically different passenger volumes actually have identical tendency to propagate delays. The reason for this is that the bus line, although it has much lighter passenger volume, has extremely slow boarding. If it suffers from instability of service that causes pairing more frequently than rapid transit (since it is exposed to external delays), it will obviously be logical to correct the problem through acceleration of the boarding process. That will decrease its propensity to increase delays and thus improve reliability.

In evaluating this model, it must be borne in mind that it is based on the assumption of uniform passenger arrivals, so that it is valid only for time periods when this condition is approximately true. In addition, the phenomenon may look somewhat more serious in the model than in practice because of the assumption that the travel time and the boarding rates are constant; in actual operation, the driver of the delayed vehicle attempts to catch up with his schedule and the passengers board the train at a faster average rate if there are more of them (unless congestion occurs), thus somewhat dampening the oscillations around the schedule. However, despite these limitations of the model, it basically does represent the characteristics of the studied phenomenon, and the conclusion of this analysis clearly shows its most important causes and relationships. The model also allows testing of different measures (such as speeding up the boarding process, departing without collecting all passengers, etc.) with respect to their effectiveness in reducing delay propagation.

EXAMPLE 3.11: Skip-Stop versus Standard Operation

Another example of using a systems approach and modeling of transit operations is the comparison of skip-stop with standard operation, presented in Section 2.4.1. A conceptual flow diagram was developed to show all differences between the two types of operations, relationships among elements, and individual results to different parties. Some of the results, benefits and costs, are quantifiable and equations are presented for their computations. Finally, a method for systematic evaluation, including quantitative and qualitative elements, is presented and its application is demonstrated on a specific case.

EXAMPLE 3.12: Computation of the Optimal Headway

The optimal headway on a transit line h^* is derived from the trade-off between agency's operating cost C_o, which depends on the number of TUs on the line, and passengers' costs C_p in terms of waiting time which is assumed to amount to one-half of the headway. These two costs are added, differentiated with respect to the headway h, set equal to zero, and solved for h:

$$C = C_o + C_p = N \cdot c_o + P_L \frac{h}{2 \cdot 60} c_p$$

$$= \frac{T}{h} c_o + \frac{P_L \cdot h}{120} c_p \qquad (3.12)$$

$$h^* = \sqrt{\frac{120T}{P_L} \frac{c_o}{c_p}} \quad \left| \frac{C}{\$/\text{h}} \right| \frac{N}{\text{TU}} \left| \frac{c_o}{\$/(\text{TU-h})} \right| \frac{c_p}{\$/(\text{prs-h})} \left| \frac{P_L}{\text{prs/h}} \right| \frac{h,\ T}{\text{min}} \right| \qquad (3.13)$$

As in Example 7, c_o and c_p are hourly costs of TU operation and passenger time, respectively, while P_L represents the number of passengers per hour traveling on the line. This expression shows that the optimal headway increases with cycle time T, while it decreases as passenger demand increases. As is intuitively clear, headway should increase with vehicle operating cost and decrease with increased value of passenger time. Headway is related to the square root of all these parameters.

Fleet size N for this minimum total cost headway is obtained when h in Eq. (3.13) is replaced by T/N and solved for N:

$$N = \sqrt{\frac{P_L \cdot T}{120} \frac{c_p}{c_o}} \quad \text{(Units as above)} \qquad (3.14)$$

The relationships in this equation are intuitively clear.

3.4 APPLICATIONS OF SIMULATION METHODOLOGY

The preceding sections have shown that in the analysis of complex systems and processes, conceptual models are very important in organizing the approach, procedure and results of analysis. Mathematical models formulate quantitative relationships and produce precise results, but when systems are very complex or involve stochastic processes, they become too complicated to set up or solve. In such cases, models using computer simulation become a useful and powerful tool for analysis (Banks et al., 2001; Hillier and Lieberman, 1995; Law and Kelton, 2000).

In transit systems, there are numerous situations where computer simulation can be useful. For example, operation of a rapid transit line with regular alignment can be expressed or modeled by deterministic equations for the basic case. However, if the line has geometric irregularities or if some nonuniform passenger arrivals cause irregularities in train dwell times in stations, simulation by computer models becomes necessary. Moreover, a simulation model allows examination of influences of different components, such as acceleration and braking rates, passenger handling at stations, etc., on system performance. Once developed, the simulation model can also be used to predict the effects of possible system or operational changes.

A computer-based simulation model allows introduction of a much greater number and variety of influencing factors. For example, TU station dwell times can be analyzed assuming random passenger arrivals, different boarding and alighting rates as functions of passenger volume, certain adjustments of TU travel regime when a delay occurs, etc. In developing solutions to problems, a simulation model allows easy testing of many different modifications, new procedures, and multiple scenarios.

Thus, simulation models can be used for analyses of complex processes that mathematical models cannot solve. However, simulation models are numerical, so they must be adjusted to every application and their results can apply only to the specific simulated process.

The procedure in using simulation models has general steps similar to any modeling. It first requires a clear definition of the problem and objectives of the simulation. In developing the model, it is necessary to define the inputs, processes, and desired outputs. In addition, a model's output validity has to be tested using measures of effectiveness specified for every case.

Simulation models may utilize physical designs, such as networks or facilities, and often require collection of data, sometimes in considerable detail, and their translation in a form that can be put into mathematical formulae and relationships of the program. Once the model is set up and numerical data for the specific case put into it, the model should be verified: does it operate and give results as it is designed to do? Basically, the correctness of the model's setup and mathematical structure has to be tested. Then model operation with numerical data has to be validated, i.e., its numerical results should be tested for their real-world accuracy. This can be done by simulating a simple case with known parameters and results. Once verified and validated, the model is prepared and analyses can be performed.

The situations and changes that are to be simulated depend on the desired results: testing an existing operation, different relationships in it, influences of different changes, operation of the modified system, etc. Model outputs may be results of different changes to the system, modified operations, costs of input actions and value of its outputs, etc.

The level of complexity depends on the type of system or process that is simulated, the desired details of outputs, and the level of effort and funds available for the simulation. While some simple transit models can be made internally with a limited effort of the planning or managerial staff, others may require elaborate models that are commercially provided and require fitting, or calibration, for every specific transit system. The commercially available simulation programs are usually more elaborate and more powerful for analyzing details and present results than those most transit agencies develop internally. Their providers usually organize training for transit agency personnel.

This situation may cause serious problems, however. Many commercially provided simulations, as well as other computer-based programs, are of a "black box" type, that is, not understandable to the user. This is a very serious shortcoming because transit experts cannot use them knowledgeably. In some cases, the computer program owner keeps control over it and transit operators become its passive users. This creates a situation where transit professionals become much less knowledgeable about the transit system and its operations than they would be without the computer-based programs. This has been the case with many simulation and other analytical and planning models. For example, many scheduling and run-cutting programs are designed so that their users lose understanding of the process and ability to make knowledgeable analyses of it.

In many transit agencies, particularly large ones, simulation is often applied in the design of transit systems and analysis of their operations. For example, in the 1960s the Hamburg Transit Agency (HHA) performed extensive analog and later digital simulations of its U-Bahn train operations to measure the correlation between travel times and electric power consumption, to find the optimal driving regimes (maximum speeds and braking rates) to maintain scheduled times if travel time adjustments are necessary (Mies, 1974). Designers of the San Francisco BART and the Washington Metropolitan Area Transportation Authority (WMATA) built into their systems automatic train operation (ATO), which optimizes driving regime by using one of several (six and eight, respectively) performance levels with different acceleration rates and cruising speeds. The Paris Metro has used computer-based driving programs extensively, culminating in fully automated trains, without personnel, on its Meteor Line, opened in 1998 (Schwebel, 2000). The Toronto Transit Commission provided extensive survey data for a very elaborate simulation of station boarding/alighting and dispatching process and service reliability under different conditions (Krstanoski, 1996).

Simulation is also appropriate for transit operations on streets because their travel conditions are much more random than is the case with rapid transit on fully controlled ROW. Passenger flows, fare collection, and many other processes in transit system operations are also subjects for simulation, as are many predictive models for passenger demand estimation.

With continuous development of automation in various phases of transit systems design and operations, it is expected that simulation models will be increasingly used in transit systems for many different purposes. It is therefore important that transit agencies acquire personnel with expertise in these operations.

3.5 EVALUATION OF SYSTEMS ANALYSIS AND OPERATIONS RESEARCH IN TRANSIT

Transit systems present numerous opportunities for applications of systems analysis and OR studies. Present uses of these methodologies and tools are extensive in some agencies, particularly large ones. For example, transit agencies in Hamburg, London, Paris, Singapore, Tokyo, and Zürich have used them for many years, and some of them have special systems analysis or operational research departments. They have developed diverse models and procedures and use them extensively in their planning and analytical tasks. Unfortunately, many of these studies are not known widely because of insufficient exchanges of experiences among transit agencies or failures of the transit experts to publish their work.

There is a rather voluminous academically oriented literature on systems analysis, modeling, and OR studies of various transit functions, including design, operations, and analyses of efficiency. However, few of these articles reach actual applications, for several reasons. First, some of these publications are so theoretical that their results cannot be applied in real-world transit systems. Some of the studies even pursue incorrectly formulated objectives. Second, some researchers choose analytical techniques that are the most convenient for solving equations, or the most elegant, rather than those that lead to useful results. Third, these studies tend to be published in academic journals that are not read by experts in transit agencies. Finally, some transit experts in charge of actual systems are suspicious of theoretical studies and tend to avoid them entirely.

Consequently, while there is a considerable need and potential for fruitful applications of systems analysis and OR to transit system planning, operations, and analyses, many efforts in this field are underutilized because of the gap between the transit industry and theoreticians. The examples of successful systems analysis activities in transit systems planning and operations show that reducing this division is highly desirable and can be beneficial to both groups and, most importantly, to transit users and the society. To achieve a more productive cooperation, transit operators should make a greater effort to utilize the numerous advances in systems methodologies and techniques. On the other side, theoreticians wishing to analyze transit systems should fulfill the eight rules for transit system modeling given in Section 3.1.

To stimulate the exchange of knowledge between transit experts and academics interested in transit systems, several leading journals in this field, applied and academic, are listed in the References below.

EXERCISES

3.1. Discuss applications of operations research in transit systems:

- **a.** Which characteristics of transit systems make them conducive to modeling and systems analysis?
- **b.** Give three examples of using a systems analysis methodology for decision making in the following major categories of transit applications: planning, design, operations, economic analyses, network improvements, or expansions.

3.2. Describe and show by a flowchart major steps in systems analysis of complex transit projects. Then answer the following questions:

- **a.** Why is it important to define the problem or goal initially?
- **b.** Define a transit project that can be planned using systems analysis and give examples of goals and objectives for it. Which one, goals or objectives, logically precedes the other and why?
- **c.** Describe what alternatives are in transit system design and explain why is it necessary to examine them in the process of transit system design.

3.3. Why is the definition of objectives and evaluation criteria important in systems planning or, more specifically, in selecting the solution?

3.4. What are the three general categories of system characteristics that are used in project evaluation? What are the advantages and what are the disadvantages of converting all system characteristics used as decision variables into monetary values?

3.5. Models and modeling:

- **a.** What models are applied in different areas of engineering? What is their purpose?
- **b.** In addition to physical models, often used in labs, how can other types of models be classified?
- **c.** What are parameters, independent variables, and dependent variables (or input and output variables, respectively) in mathematical models?
- **d.** What are assumptions in formulating models? Why are they made? What is the major trade-off in making assumptions in a theoretical model?

e. In interpreting the results of system modeling, what limitations must be kept in mind? What is the danger of ignoring these limitations? Give an example.

3.6. Describe the three basic types of models: conceptual, mathematical, and computer simulation. Define types of optimal applications for each one of them and give one example for each type.

3.7. Which kinds of problems in systems engineering applied to transit are suited best for computer simulation? How do simulation models compare with mathematical models with respect to:

- Required information and data?
- Model calibration?
- Ability to derive results for many system variations?
- Ability to test system sensitivity to changes in individual components?

3.8. A city is considering the introduction of special signals for buses on a street with a bus lane, which would increase their travel speed. The transit company, looking only at its own direct financial interests, is willing to contribute toward the cost of the signals if it can save a corresponding amount within two years by reducing the number of buses needed for the line operation. Savings amount to \$40,000/bus per year. The line is $L = 5.33$ km long (one-way). Present operating (travel) speed is $V_o = 8$ km/h. Terminal time (which cannot be reduced) is 8 min at each end, and the headway is 12 min.

a. What travel speed should the signals provide so that the company would be willing to contribute: (a) \$50,000; and (b) \$90,000?

b. What other benefits in addition to the financial ones should the company consider?

3.9. Explain what causes the phenomenon of pairing of buses (or trains) on a transit line. Assuming that one bus was delayed by τ min, illustrate by a diagram of passenger accumulation at a station versus time what happens with bus operation at that station. Then describe the formation of bus pairs.

Describe two operational measures that can be introduced in the short term that would reduce the tendency for bus pairing to develop when one bus gets delayed.

3.10. A metro line operates with a headway of 4 min. Draw a diagram of the number of passengers P accumulated at a station platform on this line as a function of time t while two consecutive trains arrive, board passengers, and leave, i.e., for the duration of two headways. Use the following methodology:

Starting from an empty platform at time $t = 0$ s, show the passenger accumulation assuming a uniform arrival rate of $\lambda = 81$ prs/min. Next, compute the dwell time assuming passengers board the train at a uniform rate $\mu = 1200$ prs/min (note $\mu >> \lambda$). Show on the plot the change in passengers P during boarding. Repeat the passenger volumes for the next train's arrival, boarding, and departure under normal operation (i.e., no delay).

Then draw on the same diagram, by dashed lines, the number of passengers when the first train gets a delay $\tau = 30$ s prior to its arrival to that station. Show what happens with the following (second) train that arrives at this station on schedule.

a. Show graphically the schedule disturbance caused by the delay of the first train.

b. How could one use this diagram to examine whether the total passenger waiting time is increased by this disturbance or remains constant?

c. Do all delays (including the negative ones, i.e., running ahead of schedule) increase at other stations along the line (not shown on your diagram) for any relationship of λ and μ?

d. Which measures can be undertaken in train operations and platform control to reduce the impact of a delay on heavily loaded lines?

3.11. In providing transit services, there is a trade-off between regular TUs (bus or short metro train) operating at short headways and large TUs (articulated bus or long metro train) providing service at longer headways. There is therefore a need to find the optimal TU size for a given line, with operating speed V_o and maximum passenger volume P_{max}.

a. Define the two major objectives that are traded off in finding the optimum TU size and explain which of the major parties (passengers, operator, city) is interested in what element in the trade-off.

b. Draw a diagram of costs or disutilities of the two major elements and their total, indicating the optimal solution. Give the dimensions of the two variables.

c. Show on the diagram by dashed lines the changed shapes of the plotted curves that would be obtained if the assumed value of passenger time were increased.

3.12. Transit agencies often try to determine the optimum life for their bus fleet or rail rolling stock. Considering only annual depreciation and annual operating and maintenance (O&M) costs, explain how these change with respect to vehicle life. Plot a diagram that shows general shapes of these two costs using the vehicle life as an independent variable. Show the total cost and the optimal life based on these two factors only.

How would the shown curves change if: (a) the transit agency obtained public funding for a sizable share of vehicle purchase costs, and (b) the transit agency had to issue bonds to purchase vehicles and the interest on bonds increased significantly?

3.13. Passenger travel time on a transit line can be divided into access to station time PT_a and on-line travel time PT_t. Draw a diagram of these two time elements as functions of the density of stations, and their sum, total passenger travel time PT. Show the optimum station density with respect to the total passenger travel time.

Assuming that the access time was for walking, show the same curves for the case when passengers use bicycles for access (as is common in many Japanese cities). Mark these new curves as B for each respective PT curve. Then draw the same curves assuming that all passengers use cars to access stations (as found in many U.S. suburban areas). Mark these curves by C. Discuss the effects of these two changes on the optimum station density.

3.14. In determining the optimal interval for vehicle maintenance (usually defined by vehicle-km traveled), there is a trade-off between the cost of maintenance and vehicle performance (more frequent maintenance results in higher vehicle efficiency and longer mean distance between failures MDBF). Draw a diagram of general shapes of these two costs and, assuming that such curves could be obtained computationally, show how one can define the optimum maintenance interval from such a diagram (as a point on the total cost curve, which is a function of the maintenance interval).

3.15. Suppose that a detailed study of an actual investment in a new bus fleet by a transit company was performed. Extensive theoretical modeling was done and observations of many aspects of vehicle performance (quantified into monetary values), investment conditions, maintenance conditions and costs, and other monetary and operational factors were included. The study concluded that the "optimum investment was exceeded by 24.3%." You are asked to evaluate this study and its result.

a. List items that you would examine in the report on this study. In other words, which aspects of the analysis and components of the model would you examine?

b. From your general knowledge of analysis of public investments, such as the trade-off between capital and operating costs, benefit/cost analysis, can you point out some aspects of the analysis you would particularly suspect?

c. How would you examine the reliability of the numerical finding (24.3%) of this claimed overinvestment?

3.16. Develop a model for computing the optimal vehicle capacity C_v. Make the following assumptions:

- Cost of vehicle operation is c_o (\$/veh-h), and it is independent of vehicle capacity.
- Total hourly operation cost of all vehicles C_o is equal to the product of c_o and the number of vehicles on the line N.
- Passenger demand (P_{max}) is constant, independent of service frequency.
- Cycle speed V_c is independent of C_v and P_{max}.
- Average passenger waiting time is $h/2$.
- The maximum load factor is α_{max}, and the average value of passenger time is c_p (\$/prs-h).

Use the following procedure:

a. Compute total operating cost C_o in terms of line length L, P_{max}, C_v, V_c, and c_o.

b. Compute passenger costs C_p in terms of the total number of passengers on line P_L, c_p, and other parameters.

c. Compute total cost, including C_o and C_p, take its derivative with respect to vehicle capacity, set it equal to 0, and solve for C_v.
Draw a diagram of costs C_o, C_p, and C as functions of C_v. Discuss the derived expression.

3.17. Compute the optimal train length for a metro line that can be operated by 2-, 4-, 6- or 8-car train lengths (consists) and has the following values:

$$L = 16.4 \text{ km}$$

$$P_L = 14{,}600 \text{ prs/h}$$

$$\eta = 0.72$$

$$\alpha_{max} = 0.75$$

$$C_v = 180 \text{ sps/veh}$$

$$V_c = 32 \text{ km/h}$$

$$c_o = \$24/\text{veh-h}$$

$$c_p = \$6/\text{prs-h}$$

Note that instead of one optimal value for C_{TU}, you should compute specific values of costs for operations of four different train lengths. Therefore, use Eq. (3.6), expressing in it C_{TU} as $n \cdot C_v$, for values $n = 2$, 4, 6, and 8. Your answer will be the lowest of the four values you obtained for the total cost C by Eq. (3.6).

REFERENCES

Banks, J., J. S. Carson, II, B. L. Nelson, and D. M. Nicol. 2001. *Discrete-Event System Simulation.* Upper Saddle River, NJ: Prentice-Hall.

Foulkes, J. D., W. Prager, and W. H. Waner. 1954, On bus schedules. *Management Science.* 1.

Hillier, Frederick S., and Gerald J. Lieberman. 1995. *Introduction to Mathematical Programming,* 2nd ed., New York: McGraw-Hill.

Krstanoski, Nikola. 1996. *Rapid transit line performance analysis: a stochastic approach.* Ph.D. dissertation, University of Pennsylvania, Philadelphia.

Law, Averill M., and W. David Kelton. 2000. *Simulation Modeling and Analysis,* 3rd ed. New York: McGraw-Hill.

Mies, Arnold. 1974. Underground railway automation of the Hamburger Hochbahn AG. *Revue l'UITP* 1, 45–48.

Schwebel, Henry. 2000. Les métros automatiques dans le monde. In *Internationales Forum zum automatischen U-Bahn-Betrieb,* 27–50. Berlin: Berliner Verkehrsbetriebe.

Vuchic, Vukan R. 1969. Propagation of schedule disturbances in line-haul passenger transportation. *Revue de l'UITP* 4, 281–284.

FURTHER READING

Byrne, B. F., and V. R. Vuchic. 1972. Public transportation line positions and headways for minimum cost. In *Traffic Flow and Transportation,* ed. G. F. Newell. New York: American Elsevier, pp. 347–360.

Furth, P. G., and N. H. M. Wilson. 1983. Setting frequencies on bus routes: theory and practice. *TR Record 818,* 1–7.

Kikuchi, S., and V. R. Vuchic. 1982. Transit vehicle stopping regimes and spacings. *Transportation Science* 16(3):311–331.

Kraft, W. H., and H. Deutschman. 1977. Bus passenger service time distributions. *TRB Record 625,* 37–43.

Sheffi, Yosef. 1984. *Urban Transportation Networks: Equilibrium Analysis with Mathematical Programming Methods.* Englewood Cliffs, NJ: Prentice-Hall.

Vuchic, Vukan R. 1994. Erfahrungen mit Busspuren in USA. *Der Nahverkehr* 11, 61–64.

Vuchic, V. R., S. Kikuchi, N. Krstanoski, and Y. E. Shin. 1995. Negative impacts of busway and bus lane conversions into high-occupancy vehicle facilities. *TR Record 1496,* 75–86.

REFERENCE JOURNALS

Applied:

Public Transport International. Brussels: International Association of Public Transport (UITP)

Railway Gazette International. London.

Urban Transport International. Paris.

Verkehr und Technik. Bielefeld, Germany: Erich Schmidt Verlag.

Applied/theoretical

Der Nahverkehr. Düsseldorf: Alba Fachverlag.

Journal of Public Transportation. Tampa, FL: Center for Urban Transportation Research.

Journal of Transportation Engineering. Reston, VA: American Society of Civil Engineers.

Transportation Quarterly. Washington, DC: Eno Foundation.

Theoretical

Transportation Research. Amsterdam: Elsevier.

Transportation Science. Linthicum, MD: Institute for Operations Research and the Management Sciences.

Transportation Planning and Technology. London: Taylor & Francis.

4

TRANSIT LINES AND NETWORKS

List of Symbols

Symbol	Unit	Definition
a_i	—	Number of interstation spacings on line i
A	—	Number of station spacings
C	sps/h	Highest design line capacity in a network
f_w	TU/h	Frequency of service during peaks weighted by stations
h_p	min/TU	Pulse headway
k	—	Number of transfers
l_i	km	Length of line i
L	km	Length of network
L_a	km/km^2	Density of metro network
L_p	km/10^6 prs	Network extensiveness per 10^6 population
n_l	—	Number of local stations between express stations
$n_m{}^k$	—	Number of k multiple stations
N	—	Number of stations in a network
N_a	%	Area coverage (percent of area in station sheds)
N_e	—	Number of terminating (ending) lines
N_t	—	Number of through lines
OD	—	Number of station-to-station travel paths
OD_d	—	Number of direct paths
OD_t	—	Number of paths that include transfers
q	—	Number of lines in a network
R	trips/prs	Riding habit (annual passenger trips/population)
S	km	Average interstation spacing length
W_p, W_s	prs-km, sps-km	Utilized and offered work, respectively
α	prs-km/ sps-km	Work utilization factor
β	—	Network complexity indicator
δ	—	Directness of service
η_a, η_t	%	Auto access and street transit integration ratio, respectively
λ	—	Line overlapping index

The network of lines with their stops/stations (defined in Section 1.1.1) represents the principal infrastructure component of each transit system. The characteristics of lines vary greatly with their right-of-way (ROW) category, technology, and mode. Bus lines with ROW category C involve negligible investment and can be modified or relocated rather easily. Rail lines with exclusive ROW, i.e., category A, involve substantial investments, are permanent, provide much higher level of service, and have a strong potential for impact on urban development and form. Consequently, lines with ROW A must be planned and designed with particular care. The lines with ROW category B are between these two (C and A) with respect to all characteristics: investment, level of service, permanence, etc.

This chapter covers networks of all modes, but for the reasons explained above, the emphasis is on rail, particularly metro and regional rail modes, because they have ROW category A. A review of theoretical principles and analysis of geometric elements is followed by a systematic description of lines and networks. The function, design, and operation of different types of transfers are covered in a separate section. The last section presents a methodology for analysis and evaluation of metro networks. Examples illustrating different types of lines and networks are given throughout the text.

4.1 PLANNING OBJECTIVES, PRINCIPLES, AND CONSIDERATIONS

Objectives in planning a transit network can be grouped into three major categories:

- *Perform maximum transportation work:* This is expressed by the number of passenger-trips or passenger-km. This implies provision of high travel speed, passenger convenience, and other elements that attract passengers. In planning of a given network, if average trip lengths do not vary much among alternative designs, these two objectives may have similar values; however, in comparing an urban transit system with a regional one, both passenger-trips and passenger-km values should be considered because they may be quite different among alternative designs.
- *Achieve maximum operating efficiency:* This objective can ultimately be expressed as the minimum total system cost for a required performance level.
- *Create positive impacts:* These include a variety of effects, from short-range ones, such as reduction of highway congestion, to the long-range goals, such as achievement of high population mobility, desirable land use patterns, sustainability, and high quality of life.

These three categories of objectives generally correspond, respectively, to the requirements of the three interested parties considered in transit system evaluation: passengers, operator, and community (see Chapter 11). The objectives discussed here are very specific: they refer to the characteristics that can be affected by geometric design of lines and networks.

4.1.1 Passenger Attraction

The greater the number of passenger trips a transit system carries, the better it serves citizens' and a city's needs. The more person-km a transit network carries, the more economically it operates (its unit costs decrease) and the more it reduces less socially desirable automobile travel. Consequently, the dominant goal in designing most transit systems is to attract as many passengers as possible. The main network design features that affect transit service quality and thereby influence passenger attraction are defined here.

Area coverage, defined as percent of the total urban area that is within 5-minute (primary) and 10-minute (secondary) walking distance of transit stations, represents the basic element of transit system availability. For planning purposes, it can be considered that most potential transit users within a 5-minute (400-m) walking distance will use the system. Between that distance and a 10-minute (800-m) walking distance the potential ridership gradually decreases practically to zero, as Figure 4.1 shows. Willingness of potential passengers to walk or use feeder modes, from bicycles and buses to personal cars, depends greatly on the quality of rail transit service, trip lengths, and conditions of travel by alternative modes. Thus, the decrease in ridership with access distance is more rapid for street transit modes than for metro and regional rail, due to the higher level of service the latter modes offer (Bandi et al., 1974). For stations with extensive use of kiss-and-ride (K+R) and, particularly, park-and-ride (P+R) access, these standards do not apply. In many cases, substantial numbers of users come to stations by car from distances of many kilometers, including intercity travelers.

A separate analysis of area coverage should be made for street transit lines and rapid transit networks. In small and medium-sized cities, the goal should be to achieve high area coverage (80–100%) by street transit. In large cities, such as Boston, Montreal, Paris, and Moscow, trip lengths are greater, so that metro systems are the basic carriers. Therefore, high area coverage by rail transit systems, metro, light, and regional rail should be one of the basic planning goals. An analysis of such coverage for planning purposes is illustrated by the example of Boston's rapid transit and LRT lines in Figure 4.2. Bus lines are supplemental, serving predominantly shorter trips.

Operating speed is mostly a function of ROW category and station spacings. In street operations (ROW category C), the speed depends on street and traffic conditions; in rapid transit (ROW category A), station spacing is usually the determinant factor. Station plan-

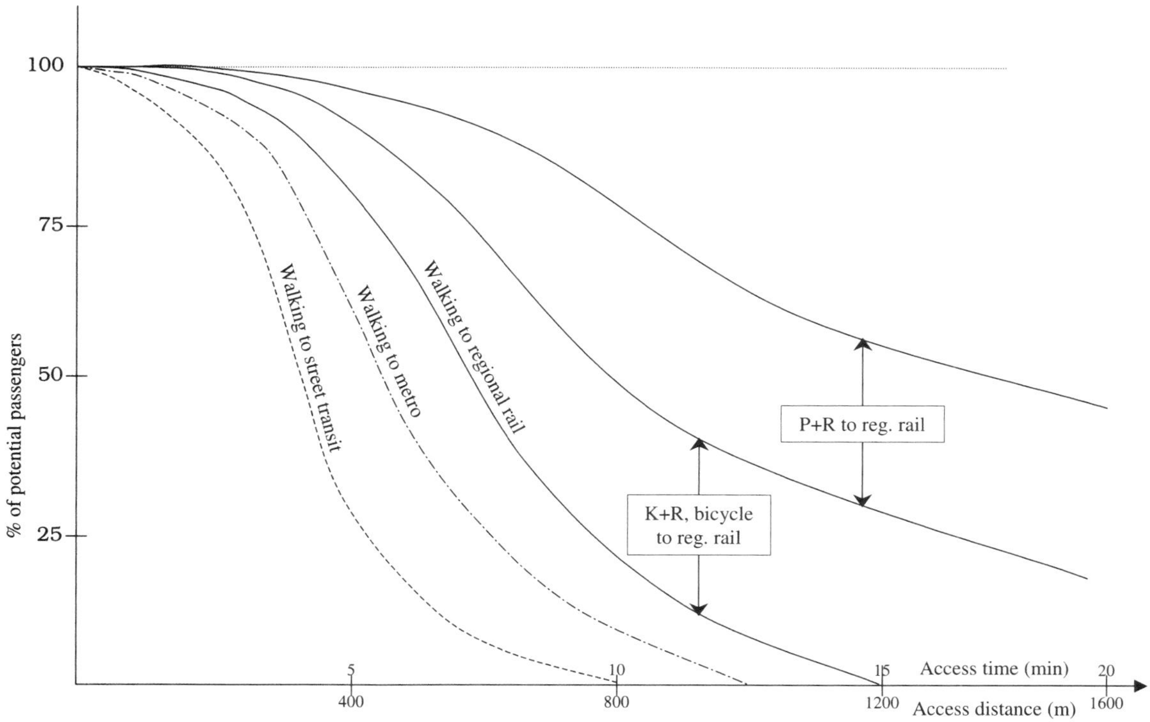

Figure 4.1 *Acceptance of access distances for different transit services by mode of access.* Note: *Walking distance curves are approximately scaled; K+R and P+R vary greatly with local conditions.*

ning is therefore a major element influencing travel speeds.

Travel desire lines: transit lines should be designed to follow as closely as possible major origin-destination (O-D) patterns of travel in order to attract and serve efficiently the maximum number of passengers.

Directness of travel, measured as the ratio of passenger-km traveled on the network to passenger-km along straight O-D lines, expresses an efficiency of the network design. For a given travel demand, the network configuration that results in the smallest ratio is most efficient.

Simplicity, connectivity, and easy transfers must be considered in design because they are important for passenger convenience and attraction. The greatest operational simplicity is achieved by operating lines independently of each other with easy passenger transfers among them. Excellent examples of this are the Paris Metro (Figure 4.3) and Tokyo Subway. Their extensive networks consist mostly of independent lines, i.e., each line operating by itself, without branches, merges, or overlaps with other lines. Only a few lines have two branches toward outlying areas. Each line intersects several other lines, allowing a majority of trips between any two stations in the network to be made with no more than one transfer. Such a network is operationally simple, but its connectivity involves large volumes of transferring passengers among the lines (transfers are discussed in considerable detail in Sections 4.4 and 4.5). Networks with integrated (overlapping) lines, such as the New York City Subway (Figure 4.8 in Sec. 4.2) and Washington Metro (Figure 4.38 in Section 4.7), are more convenient for passengers because they allow more trips to be made without transferring. A comprehensive comparison of

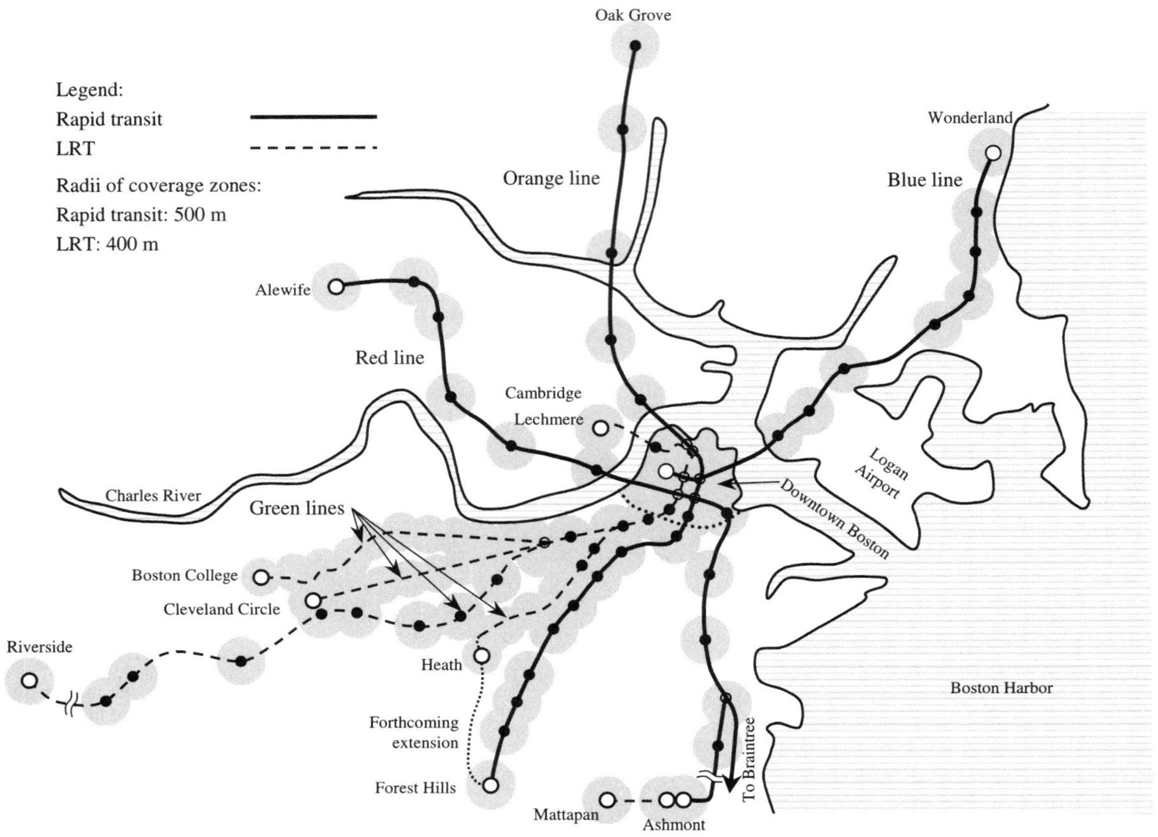

Figure 4.2 *Area coverage by rapid transit and LRT networks in Boston.*

independent and integrated lines is given in Section 4.2.4.

4.1.2 Network Operating Efficiency

Operating efficiency, representing performance and cost of transit operations, is a major concern of the second interested party—the operator or transit agency (Rockwell, 1959, 1967; Robbins and Pampel, 1975; Musso and Vuchic, 1988; Vuchic and Musso, 1991). There are five major determinants of efficiency.

Continuity and balancing of lines, or provision of direct services among areas with heavy travel demand, is desirable for both passenger convenience (direct service) and operating efficiency (no transfer delays, better fleet utilization). Continuity should be planned so that each line has reasonably well-balanced capacity requirements on all of its major sections.

Operating flexibility is increased through provision of connections between lines at which TUs can be switched among lines for scheduled or unforeseen changes of service. Greater extensiveness of networks and track connections among lines always increase operating flexibility. Turnback tracks, usually center dead-end tracks connected with both through tracks, allow short-turn operations, i.e., use of intermediate stations as terminals. Storage or pocket tracks can be used for holding reserve trains, storing of disabled trains and maintenance equipment, etc. Frequent cross-

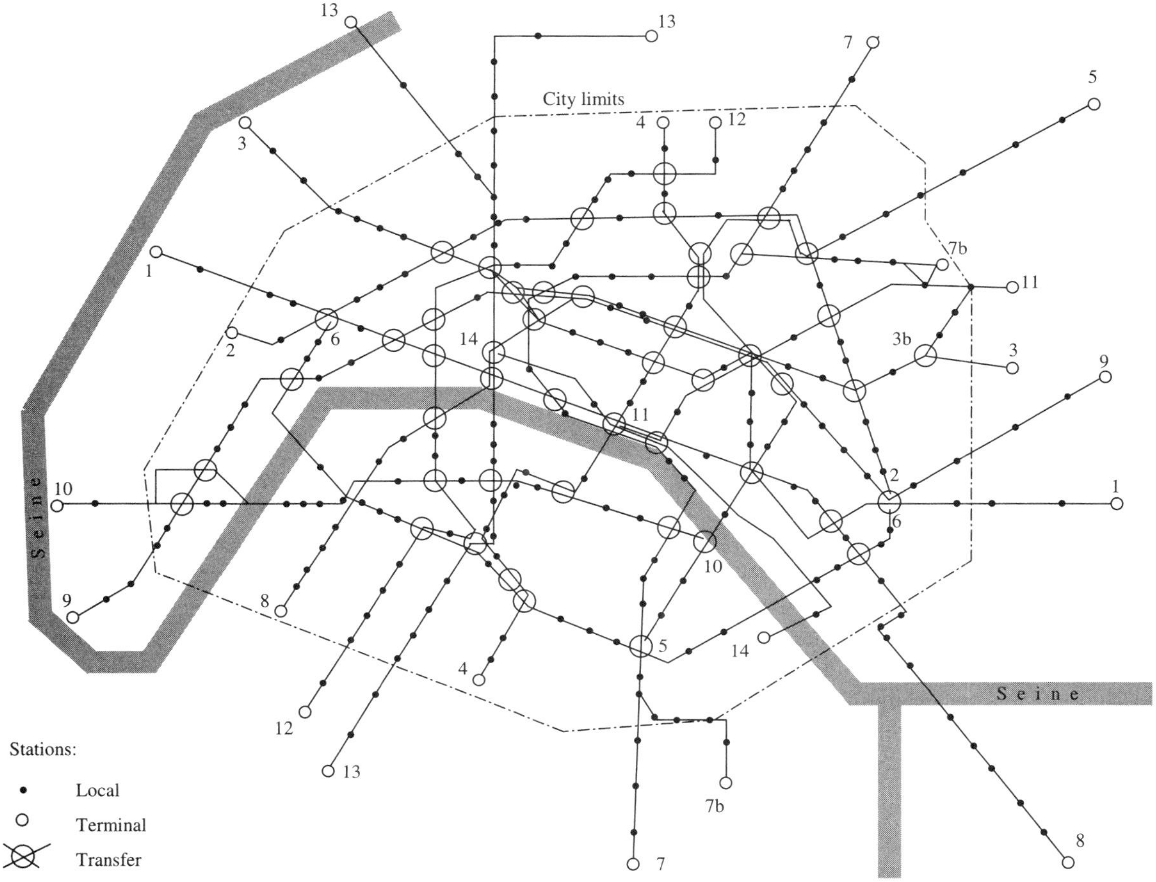

Figure 4.3 *The Paris Metro network: 14 independent lines.*

over tracks between the two main tracks make single-track operations easy, facilitating maintenance. All these types of tracks increase reliability and efficiency of network operations.

Integration with other modes ensures that a network provides a service in an area greater than immediate sheds of its stations. It allows efficient utilization of joint rail/highway corridors. Most important are intermodal stations, which permit easy interchange of passengers between different modes such as bus feeders to rail trunk line, P+R facilities, etc.

Terminals, depots, and yards should be placed at such locations that deadheading, i.e., travel between depots and the lines, is minimized. Also, because these facilities occupy large areas, they should be located where cost of land is not high. Both requirements make outlying sites better suited for depot and yard locations than central city areas. Terminals should be placed near major streets to allow good accessibility by walking and vehicles of different access modes.

Cost of the system is, of course, often the most important single factor in network design. Investment cost depends mostly on ROW category and alignment, its type (at-grade, elevated, tunnel, or other), and stations (their size, complexity, equipment, storage facilities, etc.).

Life-cycle costs include the investment as well as operating costs of the system. Although the impact of line and network geometry on operating costs is complex and often difficult to compute, it should be considered at least in general terms. There is often a trade-off between the investment and operating cost, and the optimum solution between them should be derived through cost models and analyses. For example, a more expensive alignment allows higher operating speed; or construction of turnback and crossover tracks may not be included in the line design to reduce the investment cost, but the lower operational flexibility may be very costly in system operation, involving extra TU-km of travel daily.

Generally, higher investment should provide either lower operating cost or higher service quality, or a combination of the two, thus attracting more passengers. It is a frequent mistake that investment cost is given excessive relative weight. Although it may be a large sum, investment cost often has only a moderate influence on the total annual system cost, because the long life of facilities makes annual capital cost low. High investment can also be justified by increased passenger attraction, which has a permanent effect on the urban area. This benefit should always be carefully considered.

4.1.3 Network-City Interactions

To ensure that the transit system will have positive impacts and minimal undesirable effects on the community (the third interested party), the following major factors must be considered in its design.

Land use should always be a major factor in planning a transit network, particularly metro and other rail systems. In many cities, the main objective of building a metro system is to enable and stimulate development of certain land use patterns. The Shaker Heights area in Cleveland is an example of the many streetcar suburbs built during the 1920s. In recent decades, good coordination between city and rail transit planning has been done in many cities, including Amsterdam, Frankfurt, the San Francisco Bay Area, Stockholm, Toronto, and the Washington metropolitan area. In Hamburg, transportation planners in the 1970s developed a model for density coordination between land use and transportation planning. Numerous major residential complexes and commercial centers were developed around rail transit stations. Actually, an aerial photograph of Toronto shows locations of its rapid transit stations by the density of land use developments around them. Full coordination between urban development and design of major transit networks is often the only way to achieve high efficiency of both land use activities and transit system operation.

Topography and environment usually impose stringent requirements and constraints on a transit system. Hilly terrain restricts the possibilities of the network design. The existence of valleys, rivers, or lakes often gives natural corridors for transit facilities, such as in New York City and Caracas. Aesthetics (positive and negative), noise, etc. may have varying importance in different types of areas.

Major geographic constraints, such as hills, valleys, rivers, or bays requiring major tunnels or bridges for crossing, always make construction of high-capacity transit lines far more efficient than construction of roads because two track transit lines have capacities equivalent to 10 or more highway lanes per direction. Examples of logical locations for transit crossings through tunnels and bridges have been the San Francisco BART tube under the Bay and through Berkeley Hills, Rotterdam Metro under the Maas River, and, for long-distance travel, the English Channel Tunnel.

The existing transportation network must be taken into account since it may be utilized for location of transit facilities (bus lanes, light rail medians, metro tunnels, or railway corridors), or it may sometimes create obstacles, such as difficult at-grade crossings. Consideration should be given to the possible use of sections of freeway medians, major arterials and streets, and railway rights-of-way for location of metro and light rail lines or track sharing.

As mentioned, the influence of most of these factors on transit network design cannot be defined precisely; however, it is important that the designer be familiar with them and consider them carefully. In most cases, network design represents a compromise among such conflicting requirements as overlapping lines for passenger convenience versus operating simplicity of independent lines; or speed, capacity and area coverage versus investment cost. It is therefore very important that persons with extensive experience in transit systems operations participate in the design of lines and networks, from general line alignments to the details of track layouts, station designs, and operational plans.

4.2 GEOMETRY OF TRANSIT LINES

During the early stages of transit development, when service was provided by many different companies, numerous predominantly short lines were operated on streets. In cities with grid street patterns, typical for North American cities such as Philadelphia, Chicago, and Toronto, transit lines were often operated on each parallel street in the central area, often as one-way pairs. Although such dense lines required minimum walking distances for access, their services created considerable duplication, resulting in low economic efficiency and long headways on many lines.

Urban growth, increasing automobile ownership, integration of transit companies, and other related developments have caused major changes in transit line layouts. Increasing attention has been given to the coordination of different services and the formation of integrated networks. The trend has been to reduce duplication while increasing both operating efficiency and service quality. Transit networks designed by contemporary principles have fewer lines than was typical for transit in the period between 1900 and 1950. The spacings among the lines are greater, but more of them have higher types of ROW A, B or preferential treatment for transit vehicles and better organized transfers and therefore offer higher speed and reliability and stronger image. Passengers tend to prefer frequent, fast services over infrequent, slower ones and accept longer walking access to such lines.

In designing individual transit lines or routes and coordinating them into networks, factors influencing their geometric elements must be analyzed. The major ones—spacing, length, alignment, and interconnections among lines—are presented here.

4.2.1 Spacing of Parallel Lines

The distances between parallel lines serving a given corridor area depend on the trade-off between the walking access distances to the lines, and their frequency of service, as a simple model in Figure 4.4 illustrates. The model is based on assumptions that the total passenger volume in the corridor with a width W requires transit services with a total frequency of F vehicles per hour; population is evenly distributed and the number of transit riders does not depend on the positions of lines. Line positions in each case minimize total access distances.

If the required service is provided by n lines, their optimal locations for minimum access distance are shown in Figure 4.4 for $n = 1, 2$, and 3. It is expressed for each i-th line as W_i distance from the left boundary of the corridor:

$$W_i = \frac{W}{2n}(2i - 1) \tag{4.1}$$

Thus, for two lines, the second line is at a distance $3W/4$ from the left boundary. For $n = 3$, the third line, $i = 3$, is $5W/6$ from the left boundary. Frequency of service on each line is $f = F/n$, while the average and maximum access distances to the lines, $\bar{L}_a$ and $L_{a\,\max}$, respectively, are:

$$\bar{L}_a = \frac{W}{4n} \quad \text{and} \quad L_{a\,\max} = \frac{W}{2n} \tag{4.2}$$

Thus, if a corridor is $W = 1200$ m wide and the required total service frequency is $F = 12$ TU/h, the alternative line designs, assuming that they could be placed at any lateral location in the corridor, would be as given in Table 4.1.

Of course, when planning transit lines in real-world situations, this model must be enhanced to take into account several additional factors. First, passenger

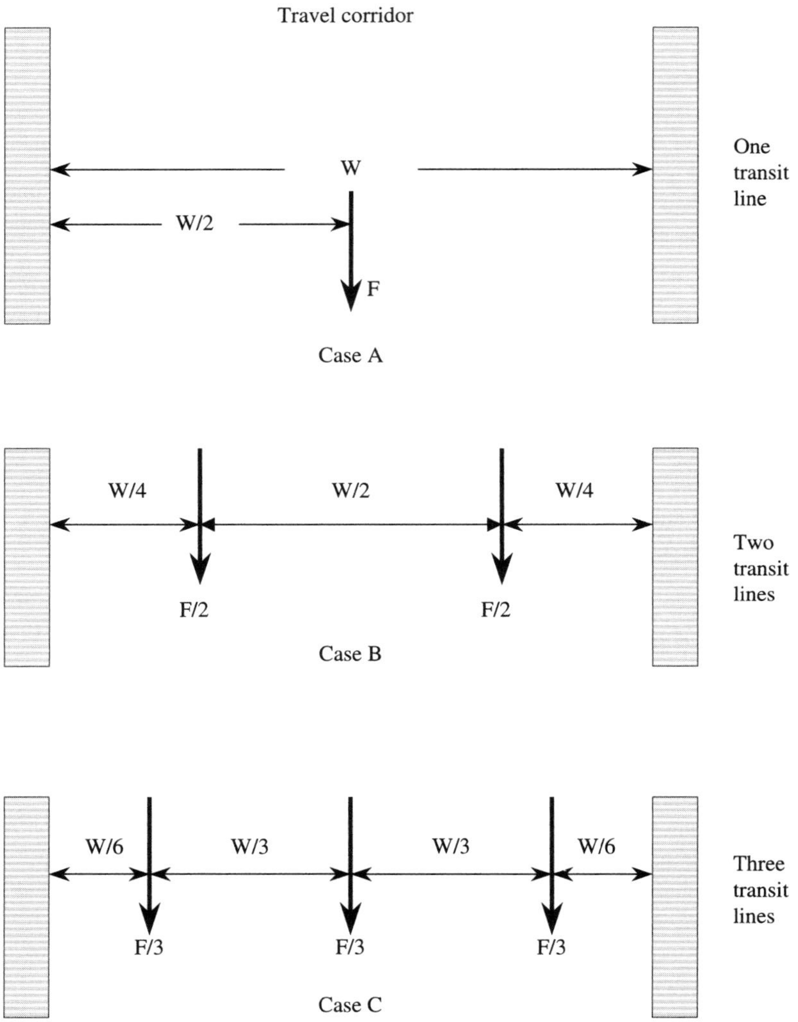

Figure 4.4 *Relationship between line spacing and frequency of service for a fixed total number of bus trips* F *in a corridor.*

Table 4.1 Relationship between access distances and service frequency

Case	*Number of lines* n	*Access distance (m)* Average $\bar{L}_a$	Maximum $L_{a\,max}$	*Frequency* f(*TU*/*h*)	*Headway* h(*min*/*TU*)
A	1	300	600	12	5
B	2	150	300	6	10
C	3	100	200	4	15

preferences between access distance and frequency of service, a major factor in selection among the above options, depend on such local factors as passenger trip lengths, walking conditions, and quality of transit service. Second, the concentration of passenger volume on fewer lines, and ultimately on one line, increases the feasibility of upgrading that line to improved operating conditions or introduction of a higher-performance mode. Such a change, in turn, attracts passengers from a greater distance. Thus, these additional factors tend to favor the solution of fewer, higher-quality lines over many lines with short access distances but low service quality. This situation may result in justification of the construction of a rail line that can serve a wide urban corridor.

In the real world, specific values may be as follows. In low-density areas, rail lines can be 800 m apart, requiring a maximum walking distance of 400 m or walking time of five minutes (this is only lateral access to the line; longitudinal walk to the transit stop is additional). In higher-density areas, shorter access is desirable, but spacing of lines should not be so small that frequency during off-peak hours drops below 6 TU/h (i.e., headways should not exceed 10 minutes). Transit service on every street, still operated in some U.S. cities, in many cases creates this problem. It gives excessive relative weight to minimization of walking at the expense of frequency, which is a strong element of passenger attraction.

While parallel metro lines at a distance of 800 m are considered to provide the basic area coverage, in cities with large populations and high density the demand for quality of service as well as high passenger volumes often justify construction of lines under parallel avenues at a distance of 300–500 m. Examples of such dense metro networks are central areas of Paris, Tokyo, Berlin, and New York City (Manhattan and Brooklyn). In Manhattan, closely parallel lines even have four tracks and operate local and express services.

4.2.2 Line Lengths

Spatial size of the urban area and the type of line are the major factors influencing the length of transit lines. Long lines have two major advantages. First, they serve more trips directly than short lines: a line with k stations offers direct service for $k(k-1)/2$ station-to-station two-way links; if it is extended by one station, k additional links are served; another station adds $(k+1)$ links, etc. For example, a line with 10 stations serves $10(10-1)/2 = 45$ bidirectional links. If it is extended by one station, it serves 55 bidirectional links, or 10 more. If a line with 20 stations is extended by one station, its number of links increases from 190 to 210, or 20 more. Thus, the number of direct point-to-point services increases with the extension of a line, and each additional station increases the number of links by a greater number on long than on short lines. Second, long lines have a smaller proportion of terminal time (factor γ is lower) because terminal maneuvers (changing ends of the train or turning vehicles or trains around a loop) are almost independent of line length and are performed after longer travel times, thus allowing better utilization of vehicles and labor.

On the negative side, long lines may result in less efficient scheduling and run-cutting because long cycle times are difficult to fit into shifts. Long surface lines may also have a problem of major delay propagation (see Section 3.3.3, Example 3.10).

In planning radial metro lines from center city, the question arises how far they should extend into a suburb (Black, 1962). Typically, a radial rail line complemented by feeder bus routes and/or P+R facilities serves a segment of the metropolitan area's circle. If the line is short, a large portion of its riders must transfer to feeder bus routes, as Figure 4.5(a) shows, and a major section of the trunk is still served by less efficient buses. At the other extreme, the rail line is extended in the alignment with the heaviest transit ridership so far into the suburb, that no feeders are needed in that direction (Figure 4.5(b)). A solution between these two extremes, where the trunk divides in two branches and still has some feeders, is shown in Figure 4.5(c).

The optimal length of radial rail lines in suburbs should be found by comparing the benefits of operating a longer line in terms of greater passenger convenience, higher ridership generation, and reduction of traffic volume on parallel highways versus the costs of

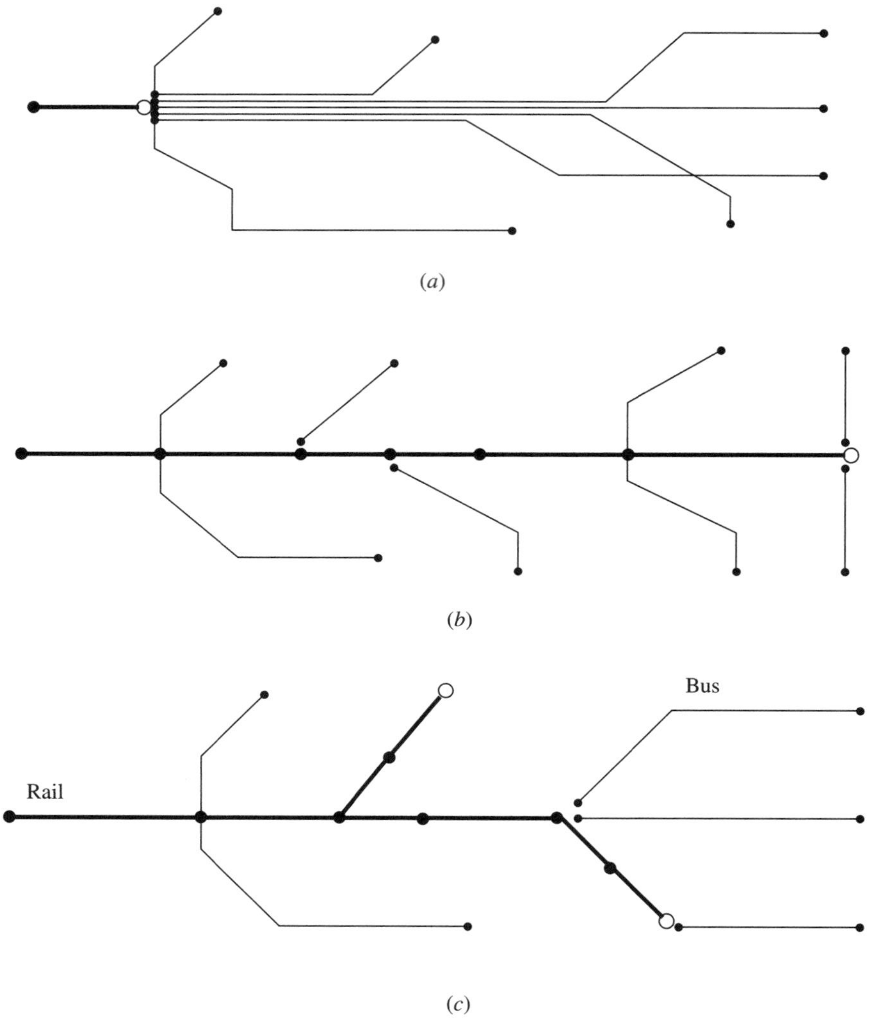

Figure 4.5 *Alternative radial line lengths in suburbs: (a) minimum length trunk with extensive feeders; (b) maximum length trunk with lateral feeders; (c) trunk incorporates two branches; feeders complement them.*

constructing the additional line segment. In addition, the difference in operating costs for different line lengths should be computed. Using the very short line with extensive feeders as a base case, longer trunk line involves additional investment, but it results in reduced operating costs because the metro replaces a large number of labor-intensive bus operations; as the line extends farther and ridership density decreases, this advantage diminishes and may become negative when the rail line reaches the area with very low ridership.

A typical one-way length of bus and other street transit lines is 5–12 km; in large cities 15–20-km-long lines can be found. However, when an extensive metro system exists, bus line lengths are shorter because street transit serves more as a feeder than as the principal carrier. Many recently built LRT lines in the U.S.

(e.g., Sacramento, Baltimore, San Diego, St. Louis), as well as several older intercity LRT lines in Europe (Cologne-Bonn, Belgian Coast), are up to 30–40 km long. On metro lines no upper limit can be defined, particularly for those with regional character, such as the London Underground, San Francisco BART, and Washington Metro. Similarly, regional rail and suburban bus lines practically have no limits that can be defined, because some of them cannot be distinguished from intercity lines. In Tokyo and several other Japanese cities, the rapid transit and regional rail lines are in many cases overlapping or have joint trains, providing so-called through services between networks of the two modes. The maximum cycle times on all modes are about two, but sometimes even three or four hours.

4.2.3 Line Alignments

Each transit line must perform two basic functions: collection/distribution of passengers, and their transport (haul) over a distance. To satisfy the first function ideally, a TU should theoretically go from origin to destination of each passenger's trip. To satisfy the second function, it should provide travel without stopping along the direction of the highest volume of travel desire lines.

The compromise between these two functions—good area service and high travel speed—depends on the number and distribution of passengers along the line and their trip lengths. The greater the number of passengers and the longer the distances they travel, the more direct the line alignment should be. Thus, the range of services is theoretically from very few short trips, which are best performed by walking, shuttle lines, or taxis, to high passenger volume with long trips, which are optimally served by a rail transit line that follows a direct alignment without any deviations to individual passengers' origins/destinations. In their collection/distribution, such lines are complemented by various feeder modes, including walking, bicycles, bus, and auto.

The trade-off between the two functions of transit lines is illustrated in Figure 4.6. Street transit often follows circuitous routing for collection/distribution in low-density suburban residential areas (Figure 4.6(*a*)). Such routing is, however, acceptable only on the end sections of lines, where there are no through passengers. It should not be used at midroute, as Figure 4.6(*b*) shows, since such alignment creates delays that are often unacceptable for passengers from outer areas. When passenger volume is substantial, this service should be provided by two separate lines (Figure 4.6(*c*)), one performing local collection/distribution and the other running express over the inner section to center city.

The alignment of heavily traveled lines, such as rapid transit, should follow, as much as possible, the directions of major passenger travel desire lines. Excessive geometric irregularities and indirectness should be avoided to ensure satisfactory operational efficiency as well as high operating speed, competitive with automobile travel.

4.2.4 Independent versus Integrated Lines

With respect to their relationship to other lines in a network, there are two basic types of transit lines: independent and integrated lines. These two types of lines are different with respect to networks they form, the service they provide, and the operations they perform. Each one has its advantages and disadvantages, and their use is related to the modes: bus, LRT, metro, and regional rail. They are defined here, their basic geometric and operational features are explained on a simple model, and their applications for different modes are then discussed.

Independent lines have independent alignments, without overlaps with other lines. These lines have a strong, clear image to the point that in metro networks of some cities (Tokyo, Boston) they have specific names. They are simple to operate, but their utilization of offered capacity is often rather low because all TUs operate the entire length, while their load factor may decrease as passenger volumes drop toward the ends of the line. Sometimes short-turn TUs are used to reduce unused capacities.

Integrated lines are those that have branches, converge, diverge, and mutually overlap on portions of

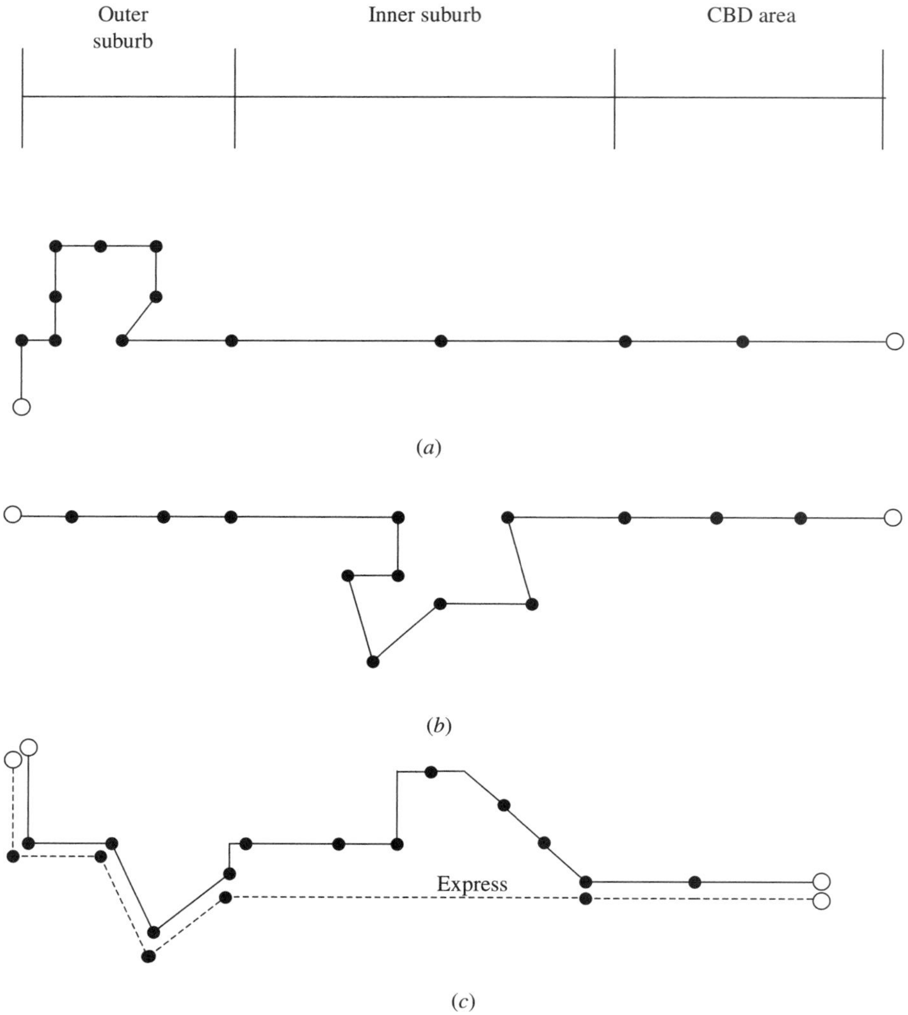

Figure 4.6 *Types of bus routings: (a) circuitous routing at the end of the route: acceptable; (b) circuitous routing at midroute: unacceptable; (c) express complementing local route.*

their alignments. Sometimes also referred to as overlapping, these lines can be scheduled to fit the passenger volumes better than independent lines. Networks with integrated lines offer a more diversified service and more station-to-station services among lines, branches, and sections, and the number of passenger transfers is lower.

Excessive overlapping, typical for many bus networks, leads to weak image and passenger confusion, which prevents many potential passengers from using the system. However, with moderate overlapping, such as is found on some metro networks (London, Washington), individual designations are possible and image is very strong.

The operational complexity of integrated lines is greater than on independent lines: a delay on one line may be transferred to other lines. They are more difficult to schedule and supervise. However, if major delays do occur, these networks often can offer rerouting of trains and revised services which are not possible on independent lines. Their recovery may be easier.

The basic characteristics of the two types of lines, independent and integrated, are illustrated on a simple network model of three intersecting lines, connecting six outlying terminal points, A–F, as shown in Figure 4.7. This model resembles the metro networks in Prague, St. Petersburg, and Tashkent.

If the three lines are operated independently, as is done in the mentioned cities (sketch (a)), with a frequency on each line of, for example, 15 TU/h (four-minute headway), passengers can travel between any two stations of the network with at most one transfer, and their total average waiting time is four minutes ($h/2$ at their origin and $h/2$ at the transfer station).

If the same network is operated as fully integrated, i.e., direct service is provided among all six terminals (sketch (b)), 15 lines are operated instead of 3. Since there are direct lines between all network sections and all stations, no passenger has to transfer. However, the 15 TU/h are now distributed to five lines instead of one, so that the frequency of service on each line is $15/5 = 3$ TU/h. With $h = 20$ minutes, the average waiting time is 10 minutes, or two and a half times longer than with independent line operation, in spite of the transferring in the former case.

Table 4.2 presents comparative values of service elements in parametric and in numerical forms for the networks with n terminals and line frequencies f operated with independent and integrated lines, such as in Figure 4.7(a) and (b).

Consequently, one extreme concept in rail transit network design and operation is to have only independent lines, often without any branches. The networks of the Moscow Metropolitan (Figure 4.11), Paris Metro (Figure 4.3), Osaka and Tokyo subways consist of lines that not only operate independently, but in most cases have separate yards and different rolling stock. This results in very simple and reliable services on individual lines, but no ability to shift cars among lines if the demand changes, special events have to be served, etc. The network actually does not represent a physically integrated system, but a collection of independent lines

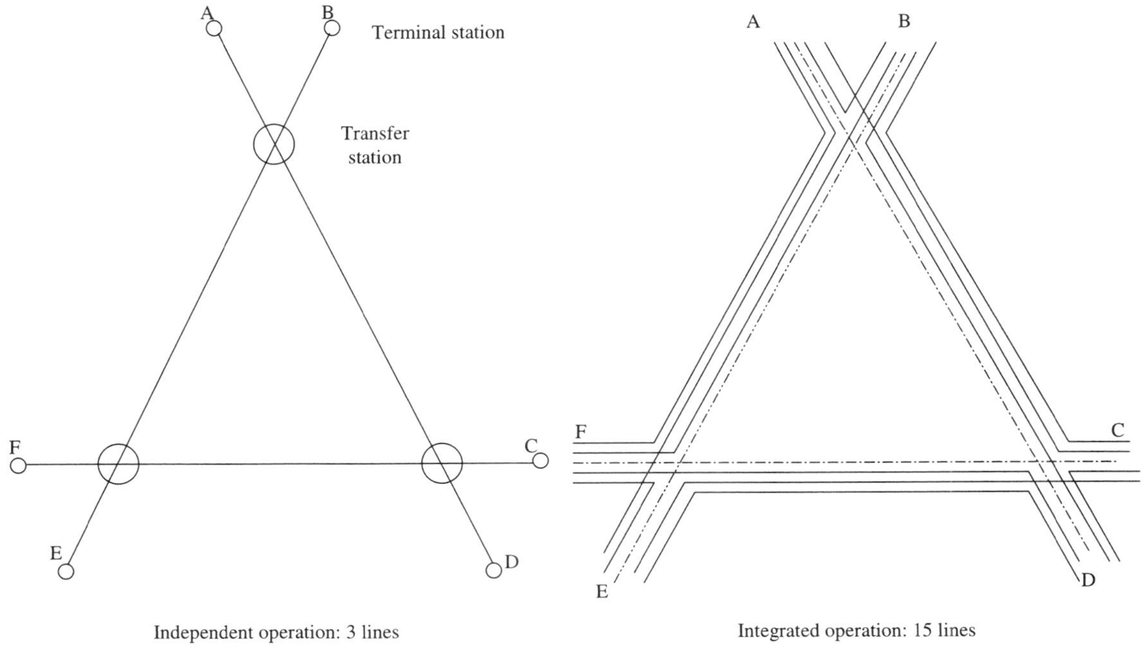

Figure 4.7 A theoretical transit network with independent and with integrated line operations.

Table 4.2 Characteristics of networks connecting n terminals with independent and with integrated lines

Characteristics / *Type of Operation*	*Independent*		*Integrated*	
	General	*Example*	*General*	*Example*
Number of lines/routes	$n/2$	3	$n(n-1)/2$	15
Frequency on each route	f	15	$f/(n-1)$	3
Transfers	Many		None	
Network operation	Simple		Complex	
Rolling stock utilization	Independent among lines		Good network-wide	

with transfer stations. Passengers have frequent services, but they have to make transfers for many trips. Extensive transferring is facilitated through careful design of stations with escalators and other amenities and elaborate information systems. Yet stations with heavy transfers often represent capacity bottlenecks on metro lines, as is the case in the Washington Metro (see Section 2.1.2.2).

Integrated line operation, as compared to independent operation, has the benefit of eliminating transfers for many passengers and providing a greater flexibility of travel through the network, but at the expense of longer waiting times for individual lines. Operationally, scheduling of interconnected lines is more complex than scheduling of independent lines (see Section 2.5), but utilization of offered capacity is usually higher. Integrated networks are more sensitive to delays, but they sometimes have a greater network-wide flexibility for rerouting of trains. How significant these advantages and disadvantages of integrated as compared to independent lines are depends on the network characteristics and travel patterns within network.

Some metro system planners and operators have categorical opinions that independent line operation is always optimal for rail transit networks. While this is correct for some cities, it is not correct for many others. Integrated lines are operated successfully on many metro networks, including Chicago, New York, San Francisco BART, and Washington Metro. The London Underground, Oslo, and Stockholm also have many branches radiating from individual trunk lines.

The New York City Subway system is the best example of a very complex integrated metro network that operates efficiently and reliably. Each of its two divisions, A and B, has not only interconnected lines, but three- and four-track lines with local/express and skip-stop operations (see Section 2.4). The smaller of the two, Division A, is shown in Figure 4.8 with the following special features:

- Three parallel lines with multiple branches at all three outward sections
- Two of the trunk lines (lines 1–3, 9, and 4–6) with converging branches on both sides (similar to the split trunk on the Northern Line of the London Tube network)
- Multiple tracks on many sections of the lines, with different lines using separate pairs
- Skip-stop operation on some sections (north end of lines 1 and 9)
- Local-express operations on several long line sections

None of these features exist in the independent-line networks, such as in Hamburg, Paris, or Moscow.

A review of different modes with regard to the type of lines and networks leads to interesting and important observations. Networks with independent lines are used mostly on very heavily used metro networks, as well as on the modes that are mechanically less conducive to switching and crossing of guideways than rail systems, such as monorails and various AGT modes.

Systems operating considerably below capacity levels, when extreme reliability is not absolutely essential, tend to have overlapping lines. Most LRT systems have networks with considerable overlapping of lines, as the example of Stuttgart in Figure 4.9 shows. With their strong image due to the presence of separate rights-of-way and tracks and short headways, LRT networks

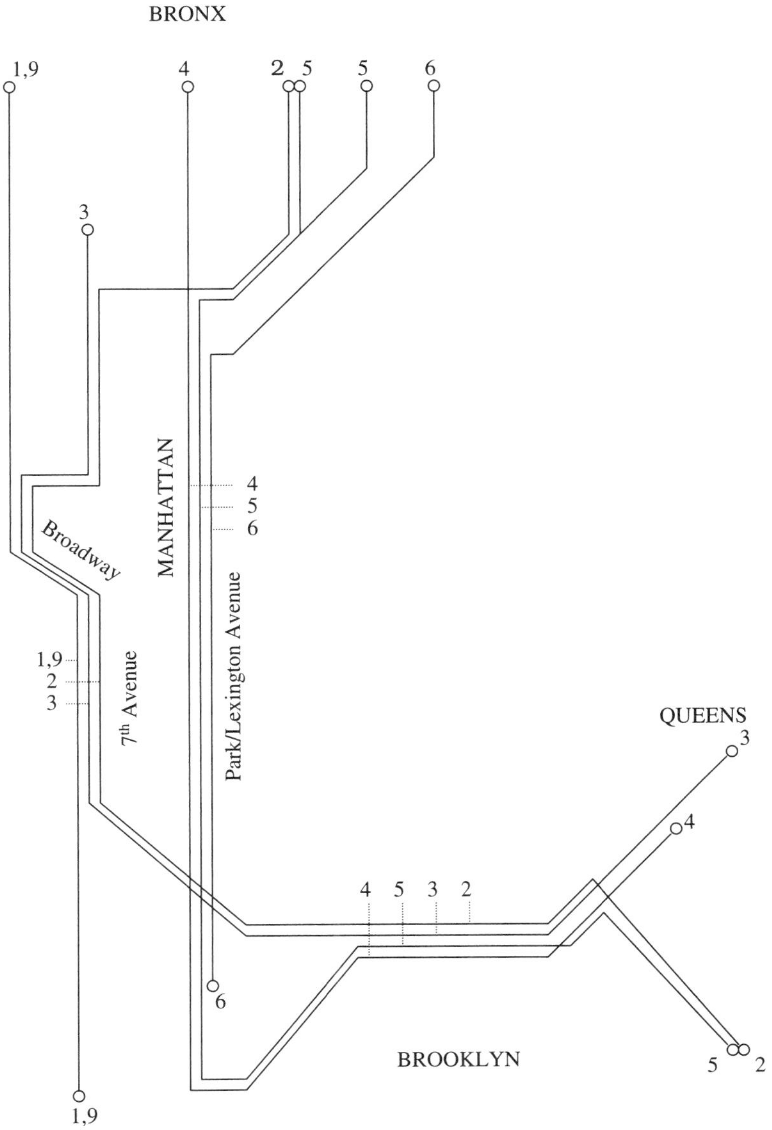

Figure 4.8 *Schematic layout of the contiguous Division A network in New York City.*

with three to five or even more overlapping lines offer numerous direct connections as well as easy recognition and strong image.

Regional rail networks are particularly characterized by large numbers of branches (see Section 4.3.3). Actually, regional rail networks in Berlin and Hamburg (S-Bahn), Paris (RER), and Tokyo (JR) are more similar by their complexity to the New York Subway than to the Paris Metro and Tokyo Subway. Several factors may play a role in the use of integrated lines on regional rail networks. First, their network geometry, with heavy orientation toward center city, requires merging/diverging patterns. Second, interruption of long suburban trips by transferring is less acceptable

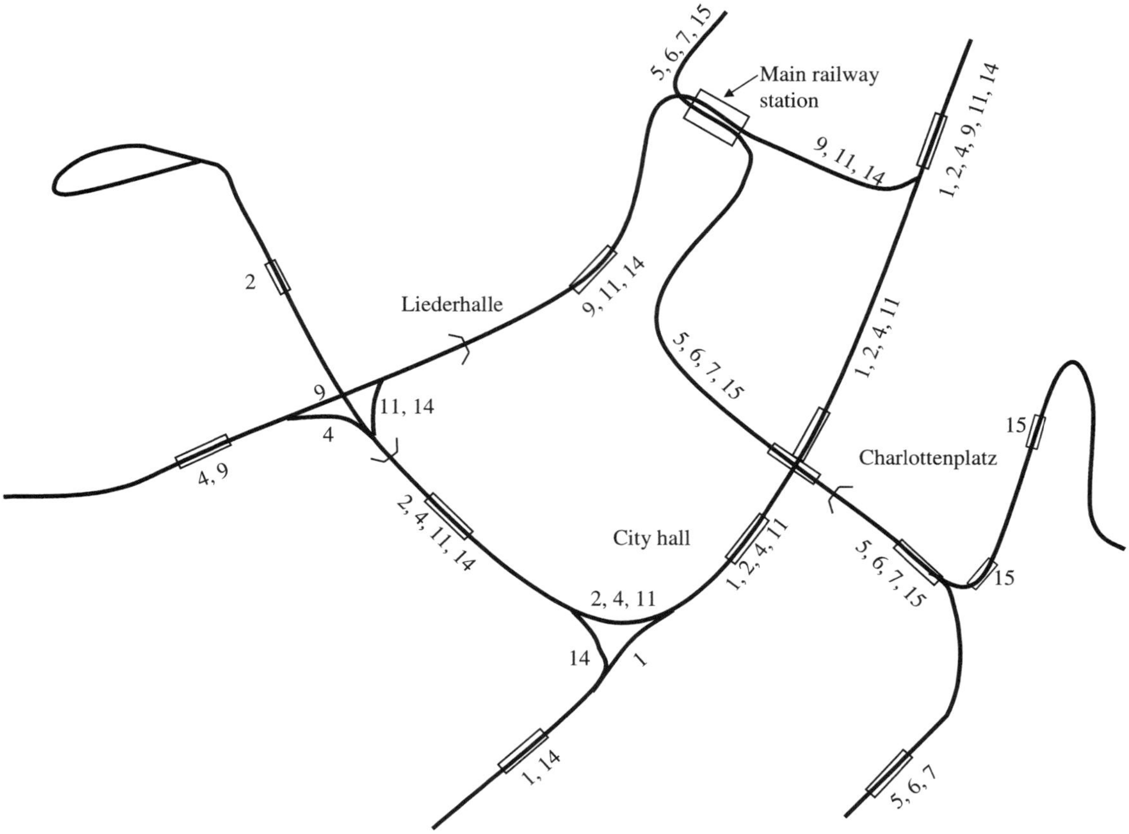

Figure 4.9 *Typical light rail transit integrated network in center city: Stuttgart.*

than transfers within integrated metro networks. And third, regional rail lines usually operate with longer headways than metros, which makes transferring more difficult.

As buses can operate at very short headways, there is no physical restriction to merging of many bus lines, except that their service speed and reliability deteriorate when buses begin to catch up with each other. Because of this ability to merge lines easily, many cities have developed bus networks with very complex sets of overlapping lines. Such networks do minimize the need for transferring, but they often have an image of excessive complexity that is confusing to passengers and discourages occasional transit users from considering travel by transit. This is actually a large group, including not only city residents who are incidental transit users, but also city visitors, tourists, and many others. The entire role of transit in the city thus deteriorates from the basic mode of travel (which many rail systems maintain) to an auxiliary travel mode used mostly by commuters and captive transit riders.

Many cities that have buses as their main or only transit mode have extremely complicated networks with overlapping lines and weak image. This weakness of such complex networks is demonstrated when some cities want to upgrade some bus lines into a semirapid transit mode, such as BRT. An important element in such upgrading is always reorganizing the bus network from many bus lines with weak image into one or a few with distinct image and reliable and frequent service. This change results in substantial ridership increases. Thus, the fact that buses can easily be used on many different lines and have no problems overlapping does not mean that this feature should be used to such an extreme that the lines have low performance and a confusing, unattractive image.

In summary, characteristics of transit lines and networks depend strongly on transit mode and relationships of lines: independent or integrated. The modes and types of lines are interrelated in the following manner.

Metro networks with independent lines with well designed transfer stations have the simplest, most reliable, and therefore highest possible line capacity of all transit modes. They operate well in large cities with high density of travel throughout a large central city area, as is the case in Paris, Tokyo, and Moscow. They also have some serious disadvantages, however. First, they involve extensive transferring, which becomes inconvenient when a passenger has to make two or more transfers on one trip. Second, their capacity utilization is low toward line terminals. And third, if lines are physically separate, rolling stock cannot be shared among lines.

Integrated operation of metro lines generally offers more direct and more diversified services and requires fewer transfers than independent operations. However, since the integrated networks are more sensitive to irregularities in operations, it is desirable that joint (trunk) lines in them operate below capacity, i.e., with longer than minimum operational headways. Integrated networks, such as the New York City Subway and San Francisco BART, can approach the minimum headways of independent lines when they have precise schedules and effective control centers.

Other rail modes, particularly LRT and regional rail, usually have integrated lines that provide direct connections throughout their networks while still maintaining the reliable services, network clarity, and strong image that public requires. Buses can operate integrated lines without any limits, but when this ability is used excessively, the convenience of minimized transfers is more than outweighed by unreliable and irregular service on the trunk sections and confusing services for passengers. Replacement of excessively complex interconnected networks by fewer lines with higher performance often results in better attraction of passengers.

4.3 TYPES OF TRANSIT LINES AND THEIR CHARACTERISTICS

Geometric form and alignment in the city give transit lines certain operational and functional characteristics. Their design must be based on a good understanding of geometric features of lines and experiences from different cities to ensure efficient operation and attractive service and avoid situations that create serious operational problems.

Although some lines have irregular forms, many can be classified into several basic types. These basic types, defined below, are presented schematically in Figure 4.10. They can be found in the metro networks of most cities. For example, the map of the Paris Metro (Figure 4.3) allows the following approximate classification of lines into different types:

Radial: 11;

Diametrical: 1, 3, 4, 7, 8, 9, 12, and 14

Tangential: 5 and 10

Circumferential: 2 and 6

Trunk with branches: 13

Trunk with a feeder: 3/3b, 7/7b

Loops: on lines 7b and 10

A typical circle line, as well as many diametrical ones, are found on the Moscow network, shown in Figure 4.11.

4.3.1 Radial and Diametrical Lines

The highest density of travel in urban areas is usually concentrated in radial directions converging on the city center area. Consequently, most of the heavily traveled lines lie between the central area and suburban centers or residential areas.

Radial lines, with one terminus in the city center, the other in the suburbs, tend to follow the major demand directions. They have decreasing passenger vol-

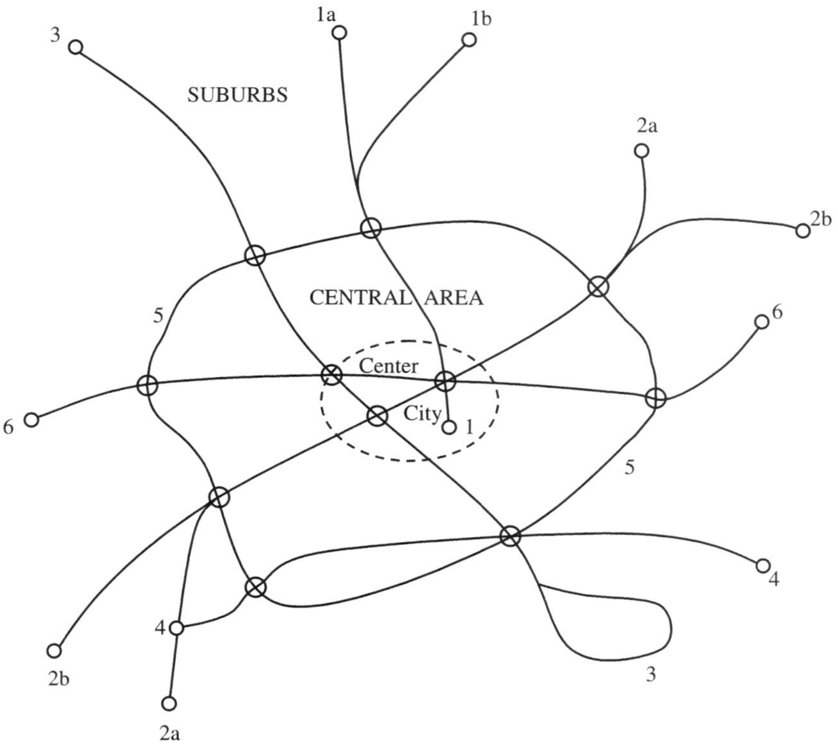

Line	Type
1	Radial with branches
2	Diametrical with branches
3	Diametrical with loop
4	Tangential (crosstown)
5	Circle or ring
6	Diametrical
⊗	Transfer station
○	Terminal station

Figure 4.10 *Geometric forms of transit lines.*

umes outward from the center. To fit offered capacity to the demand, short-turn service may be used: some TUs operate from the center only to intermediate points and then turn back (see Section 1.4.2.1). This decreasing capacity can also be obtained by splitting a radial line into several branch lines, thus also increasing area coverage in suburbs. Typically serving major volumes of commuter traffic, radial lines often have sharp peaking.

Radial lines also have two major disadvantages. First, they provide only limited distribution in the center, involving long access distances or transferring by most passengers. Drastic examples of this are the Caltrain Rail Line in San Francisco and North Station in Boston. The second disadvantage is that their terminals are located in high-density areas where land is most valuable and space least available. Typically, streets in the center are congested and tunnel construction is ex-

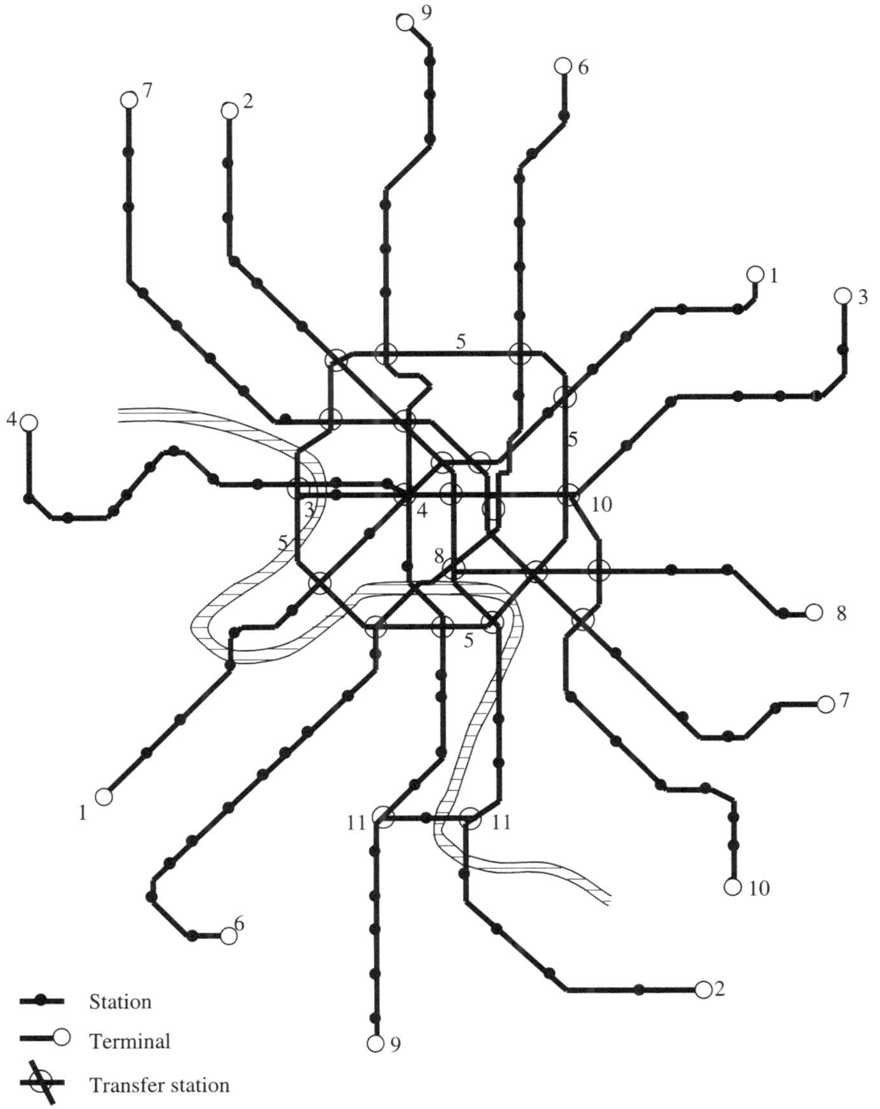

Figure 4.11 The Moscow Metropolitan network.

pensive, so that the storage of TUs for peak hour service is difficult to accommodate.

Radial lines are widely used. Many street transit and rapid transit lines, such as the Boston Blue Line and PATCO Line in Philadelphia, are radial. Many regional rail networks consist of radial lines only, since railroad stations were often built as stub terminals. Examples of the latter are all British Rail stations in London (except Blackfriars) and several of the railway stations in Paris, Moscow, and Chicago.

In suburban areas, many bus and sometimes LRT lines that are feeders to outlying terminals of rapid transit or regional rail lines are themselves radials. There are numerous examples of such transfer terminals in Philadelphia, Hamburg, Munich, São Paulo, and many other cities. These feeder radial lines are functionally very good, since neither disadvantage of radial lines applies to them: distribution at their destination is not necessary because most passengers transfer to the rail line at the intermodal transfer station, and the terminal can be designed to accommodate easy passenger transfers, terminal maneuvers, and storage of feeder lines vehicles.

Diametrical (also called *transverse* or *through*) *lines* connect suburbs on different sides of the city center. They follow radial directions, passing through the central area. To allow high load factors (capacity utilization), these lines must be designed so that their two sections on different sides of center city are balanced (similar) in terms of design passenger volumes. Diametrical lines are usually superior to radial lines because they serve more areas, offer better distribution in the central city, and have more opportunities for transfers with other lines. Moreover, they have no downtown terminals. One problem that diametrical street transit lines may have is the transfer of delays from their inbound to their outbound sections. With radial lines, delays from inbound travel may be recovered at downtown terminals. Semirapid and rapid transit lines do not have this problem because of their independence from street traffic congestion.

Diametrical lines can have several geometric forms. They may be straight or diagonal lines going through central city and connecting suburbs on its opposite sides by the shortest path. L-shaped lines, found in New York (due to geography) and Tokyo (by design), have a less direct alignment, but they usually provide more transfer points with other metro lines and better area coverage in the center than diagonals. Finally, U-shaped lines exist on some metro networks, such as in Hamburg (line U1), Toronto (Yonge-University-Spadina), and Washington (Red Line). They are formed in order to balance the volumes on two corridors on the same side of the central city, so that they are efficiently scheduled; however, this geometric shape has a serious drawback in that it provides such circuitous travel between two close suburbs that it cannot attract suburb-to-suburb travel from other modes, particularly the automobile. This eliminates one of major advantages of diametrical lines.

As already mentioned, with the growth of urban areas and the integration of transit services into coordinated networks, many radial transit lines were converted into diametrical ones. Most typical examples of this trend have taken place in recent decades on regional rail systems.

Most suburban or commuter railway lines were, in the past, radial, terminating on the fringes of central cities. With the spatial growth of metropolitan areas, it became necessary to improve these lines into a regional high-speed network. For that purpose, a number of cities have built connecting lines between railroad terminals on different sides of the central city. This conversion of radial into diametrical lines improved the distribution of passengers in the city and increased their connectivity with other transit lines.

There are many examples of this type of conversion from radial to diametrical regional rail networks. The construction of the Munich's Verbindungstunnel (connecting tunnel) converted 11 radial into 6 diametrical lines. The Paris RER lines A, B, and C have the same role. A similar solution was implemented in Philadelphia, where a central city tunnel was built to connect the former Penn-Central and Reading lines. Brussels, Frankfurt, Hamburg, and Oslo also built center city connections, while Manchester has built LRT-type lines to extend regional rail lines from all three of its railway terminals (Piccadilly, Victoria, and West Stations) through the city center.

A major exception to this trend toward diametrical lines are express bus routes, which are nearly exclusively radial; this is because they are designed to serve only commuters to the CBD. Thus, they represent special commuter transit lines, rather than a regular transit network.

4.3.2 Tangential, Circumferential, Circle, and Loop Lines

Lines serving noncentrally oriented travel can take several forms. *Tangential* or *crosstown lines,* following a tangential direction with respect to the city center, are common in cities with a grid street pattern, such as Chicago, Philadelphia, and Toronto. They often operate on streets with major commercial activities, schools, etc., and are spaced several blocks from each other.

Demand is usually lower, but less peaked, on tangential than on radial lines. Most tangential lines are operated by bus, although trolleybus, streetcar, and even rapid transit tangential lines exist in Philadelphia, San Francisco, Toronto, and other cities.

Circumferential lines are laid out around central city, such as Lines 2 and 6 of the Paris Metro. They intersect radial and diametrical lines and have transfers with them, and thus create an integrated network. Sometimes these lines go around the whole central area, representing bidirectional circle or ring lines that have no terminals.

Circle or ring lines usually have three main functions. First, they provide a direct connection among numerous medium-to-high density areas around city center (inner ring). Second, they can be effective distributors for radial metro and regional rail lines. For example, the Circle Line of the London Underground connects nine British Rail stations. And third, circumferential and circle lines enable many suburb-to-suburb trips to use a radial-circle-radial path. Although such trips require two transfers, they eliminate the need to follow circuitous travel via the city center.

With respect to operations, the circle lines have two disadvantages. First, since they have a fixed cycle time (without the flexible terminal times) for a given operating speed, they can operate only with headways that are integer fractions of the cycle time, as explained in Section 2.5.2. This is not a problem when short headways are operated, such as less than five minutes. However, with longer headways that becomes a problem that is resolved by operating a train for one circle and then sending it out onto a branching line. This is done on the London Circle Line and Berlin S-Bahn Ring Linie. The second disadvantage is that the circle lines are very vulnerable to delays because they can have at most a short (two- to four-minute) terminal time at one of the stations for schedule recovery.

Utilization of circle, as well as tangential and circumferential, lines is usually very high. They provide direct service links among a number of different activity areas (commercial, residential, schools, transfers from radial lines, etc.), so that they serve relatively short trips and their passenger turnover rates are high. With passenger volumes rather evenly spaced along the line and loads well distributed through most daily hours, average load factors on these lines tend to be high. This results in lower cost per passenger than on commuter-dominated radial lines.

Because of their potential operational problems, some cities have circle alignments operated as two separate lines. Paris Metro Lines 2 and 6 form a circle, and Hamburg reconstructed its Ring Linie into two independent lines, U2 and U3 (Figure 4.12). Although easier to operate, these lines fail to provide many important station-to-station links without transfers. To facilitate transfers, however, the Hamburg U-Bahn has four stations with simultaneous cross-platform transfers (as shown in Figure 4.16), which require very precise scheduling synchronization and reliable operation, particularly if this type of operation is used when headways are as short as 2.5 minutes.

The skepticism of Paris and Hamburg transit agencies toward circle lines is not shared in a number of other cities in which circle lines play very important roles. Milan has a trolleybus and Belgrade a tramway circle line that are heavily used. Metro circle lines exist in Beijing, Glasgow, London, Moscow (Figure 4.11), and Seoul (Figure 4.13), where it carries the highest passenger volume of all lines in the network. The circle Yamanote Line in Tokyo represents the basic element of its regional rail network, carrying over 10 million passengers per day. The Yamanote Line is operated very carefully to prevent delays: it has 10-car trains and handles extremely heavy passenger volumes (up to 60,000 prs/h per direction), but with headways and travel times that are longer than the minimum possible ones to allow some reserve time for delay recovery. When significant delays occur, there are reserve trains at three locations along the circle line that can be placed into service to fill the gaps. An interesting new concept is a new circle line in Madrid that goes through outer suburban areas, intersects many regional lines, and has a length of 50 km.

Circumferential and circle lines are less used in North American than in European and Asian cities for two reasons. First, grid street patterns common for American cities seldom provide rights-of-way for circumferential alignments. Second, American cities have highly concentrated CBDs surrounded by less active

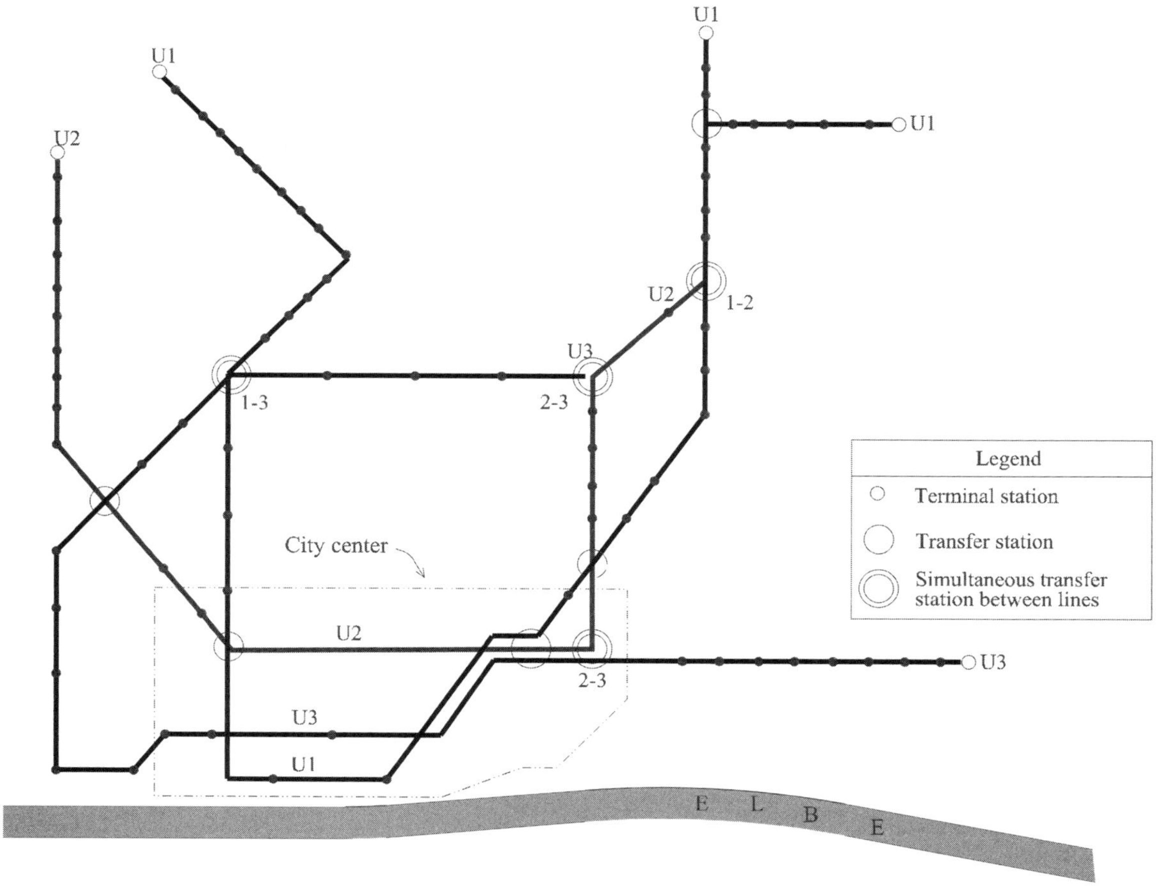

Figure 4.12 *U-Bahn (rapid transit) and simultaneous transfer stations in Hamburg.*

areas, while European and Asian cities tend to have larger central areas with a wide belt of many activities, which circumferential lines can serve very efficiently.

In conclusion, circle lines play a very significant role in transit networks of many major cities. Intersecting many (or all) radial/diametrical lines, circle lines tend to integrate the entire network. Due to their diverse ridership by trip directions and times, they usually have high capacity utilization and can therefore be operated very economically. However, they require special attention in operations and contingency preparations to ensure high reliability of service.

Loop lines are portions of regular two-way lines, which have a closed circle at one end of the line, usually with one-way operation. Thus, such lines may have only one terminal. Loop lines are often used for short-haul services within CBD areas, or for collection/distribution of rail passengers in the CBD or in the suburbs. The most famous example of this type of line is the Loop on the Chicago Rapid Transit System (Figure 4.28), which has become the symbol of the city's CBD.

4.3.3 Trunk Lines with Branches and Feeders

Due to the dominance of travel to and from center city, transit lines that follow radial directions usually have

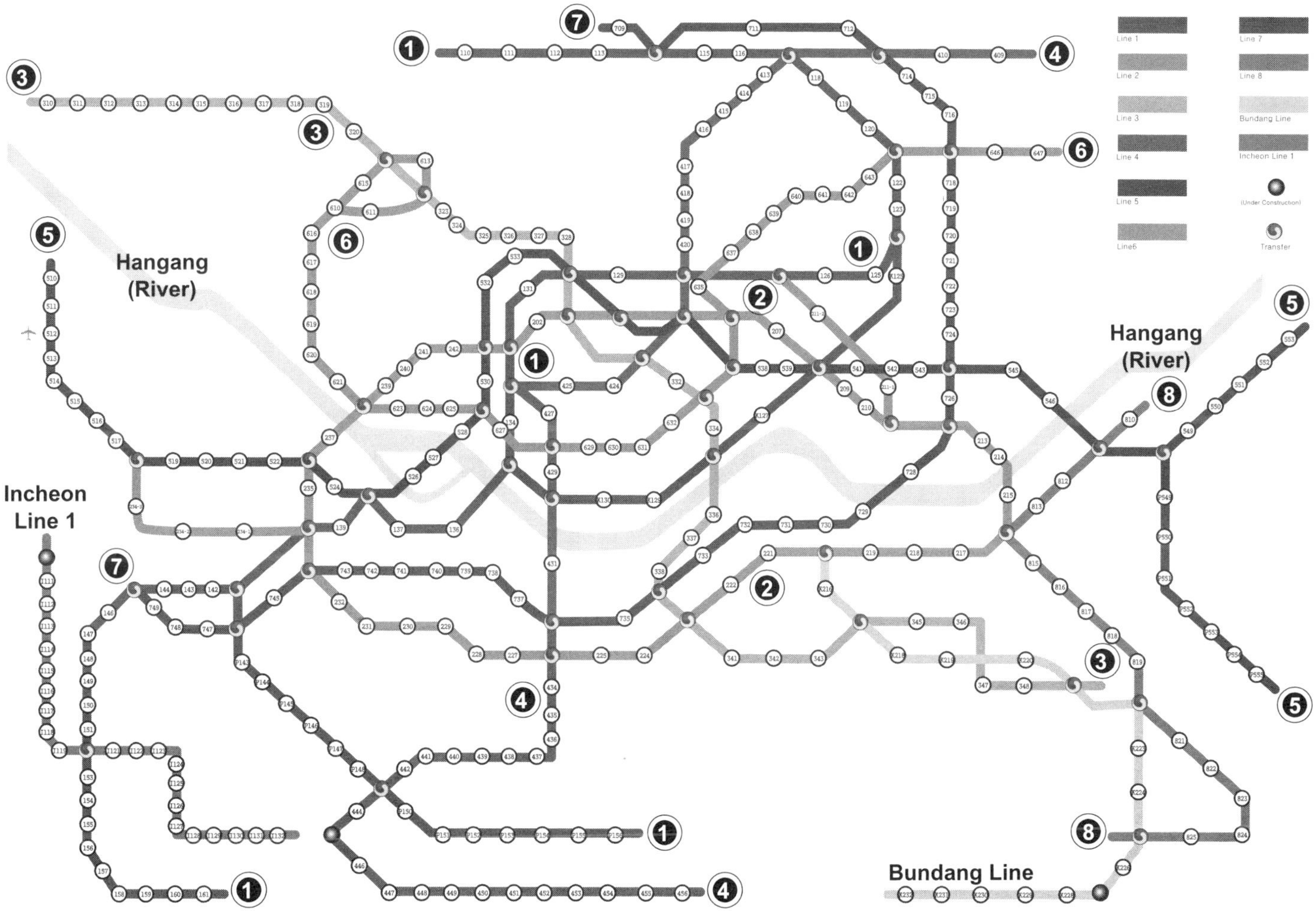

Figure 4.13 Seoul, South Korea, metro map.

the highest passenger volumes. Starting as heavily traveled *trunk lines* from the city center outward, they divide into a greater number of lines with lower loads serving large, often low-density suburban areas. By their design and type of operation, these distributed lines may be of two types: those that provide direct service from the trunk are *branches;* others, which operate independently from trunk line stations in outward directions, are *feeders*. Naturally, branches can be used only when the same transit mode and the same TUs are used on the trunk and on the branches. Change of mode results in feeder lines, although in some situations the same mode may also be operated as trunk and feeders, either because different vehicles (minibus/articulated bus) or different TUs (short and long trains, or electric and diesel lines) are used.

4.3.3.1 Functional Aspects. Figure 4.14 illustrates schematically trunk line with branches and with feeders. The advantages of each one with respect to the other are summarized in the box on page 210.

This analysis shows that each type of lines, with branches and with feeders, has certain functional advantages and disadvantages. The selection between them depends on the transit mode, number of branches or feeders, and other local conditions.

4.3.3.2 Operational Aspects. Physical and operational aspects of trunk/branch lines vary among transit modes. Buses and streetcars can merge into a trunk line without precise scheduling because they operate on visibility, short headways are possible, and simultaneous stopping at stations by several TUs can be organized. Metro and regional rail systems, on the other hand, must be scheduled with exact sequences because their headways are determined by station headways and signal blocks. Major operational aspects are reviewed here in general, then related to modes.

Capacity limitations on the trunk may limit the maximum frequency that can be offered on individual branches. Thus, if the minimum headway on the trunk is 2 minutes, and there are five branches with similar volumes, each branch cannot have headways shorter than 10 minutes. One way of handling the cases in which the sum of branch frequencies exceeds trunk line capacity is to couple TUs at converging points (see Section 2.5.3.2). However, this is seldom used because it involves some delay and requires high precision of

Photo 4.1 *Convenient transfers between trunk line and bus feeder—Cologne* (source: *G. Girnau, A. Müller-Hellmann, and F. Blennemann, eds.,* Sustainable Mobility: People in Motion. *Cologne: Verband Deutscher Verkehrsunternehmen, 1997. Reprinted by permission*).

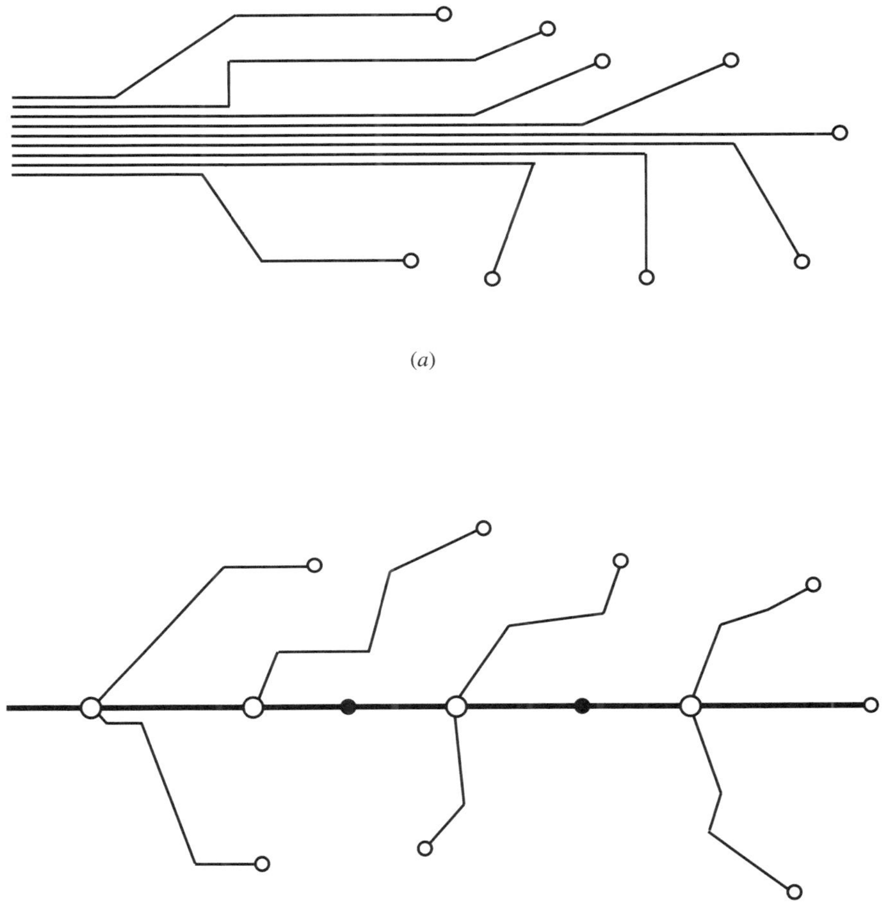

(*a*)

(*b*)

Figure 4.14 *Radial transit lines: trunk with branches and with feeders:* (*a*) *trunk with branches;* (*b*) *trunk with feeders.*

operations. If the trunk line does not have any stations, its capacity is greater, but that is the case only on commuter transit (express bus) lines or, theoretically, on lines with off-line stations, and the line must have a very large terminal. Regular transit cannot have this type of line and operation.

Uneven passenger volumes on different branches may create problems in scheduling individual branches. As shown in Section 2.5.3, when critical volumes vary considerably among branches, there must be either different TU lengths or uneven headways on the trunk or on some branches. These service irregularities can be avoided only by providing the same headways on all branches, i.e., providing the capacity required on the heaviest line on all branches. The higher operating costs on less loaded lines are then accepted in order to maintain regular service on the entire line.

Branches, compared to feeders, have the following advantages:

+ Provide for passengers a continuous service without transfers and delays between center city and suburban areas.
+ Being longer lines, trunk and branches have smaller percentage of cycle time as terminal times than trunk with feeders.
+ Transfer stations are not needed.

Feeders, on the other hand, have the following advantages over branches:

+ Each line section can be optimized with respect to mode and vehicle type (or TU) used, schedule, etc.; load factors are higher, fleet size smaller, and operating costs lower.
+ Use of higher-performance vehicles or mode (usually rail) on the trunk provides superior service at lower operating cost than smaller vehicles from branches (buses) can provide.
+ Regular headways can be operated on the trunk and on each feeder line.
+ More reliable service: delays are less likely to transfer between trunk and feeders.
+ Suburban terminals for trunks offer not only trunk/feeder transfers, but also transfers among feeder lines, providing greater network connectivity.

Rights-of-way of the trunk and branches must be coordinated. If the trunk line has a high-type ROW (category A or B), branches should also be free from frequent delays to prevent service degradation on the trunk.

Type of operation and control of branch convergence is very important for reliability and capacity of the lines. Any delays (positive, running behind schedule; or negative, running ahead of schedule) on one branch cause disturbance and loss of capacity on the trunk. This problem of delay propagation becomes increasingly critical when the trunk headways approach their physical minimum and passenger volumes are heavy, i.e., when maintaining service reliability is most important.

With respect to operations and control, there are three basic operating methods, each closely related to a transit mode.

Metro and Regional Rail Modes. The first method of operation is with predetermined sequencing of TUs entering the trunk from different branches and maintenance of precise schedule adherence, as described in Section 2.5.3. This operation is used on systems with rather short minimum headways, such as 90–180 seconds (1.5–3.0 minutes), typical for metro and regional rail systems. It offers high quality of service if two conditions are met: delays and disturbances are rare and the frequency on the trunk (i.e., the sum of branch frequencies) is considerably below the line capacity. For example, if the minimum operational headway is 90–120 seconds, the scheduled (average) headway should be at least 135 seconds for a very reliable operation, or 150–180 seconds if some irregularities frequently occur. Thus, there is a trade-off between these two conditions: the more reliable the operation is, the closer can the TUs run to the maximum possible frequency (capacity) of the trunk.

An example of this trade-off between headway and reliability (and thus also capacity) of service is found on many lines of the Tokyo Subway. They can operate at 90-second headways, but they are scheduled for 120 seconds, or a frequency of 30 trains/hour. Service reliability on these lines is very high.

Providing maximum reliability of service on metro and regional rail systems is the main reason that many of these systems were designed without branches (São Paulo, Mexico, Tokyo). However, many metro lines have two branches (Munich, Washington), while the London Underground Metropolitan Line has an exceptionally complex convergence of five branches. The Munich S-Bahn (Figure 4.27) has six long lines converging into a two-track trunk line. The lines are operated at 20-minute headways, so that the trunk operates with an average headway of 200 seconds, while its minimum technical headway is 120–150 seconds (2.0–2.5 minutes).

Light Rail Transit Mode. The second type of operation is with converging lines that operate with partial ROW control on streets and thus have lower schedule adherence than metros. Their TUs converge into a trunk line, which often operates with driving on visibility (or with advisory signals only) and at rather short headways. Typical examples are streetcar/LRT networks, which have up to five converging lines (Brussels, Essen, Philadelphia, Stuttgart—Figure 4.9). By far the most imaginative solution to multiline convergence by TU coupling was specially designed and operated for many years on the San Francisco Muni LRT system, as described in Section 2.5.3.2.

Bus Mode. The third type of operation is when vehicles from as many as 10 or more lines merge and operate at very short headways (averages may be as low as 20 seconds) and often overtake each other. This operation is typical for buses running on streets. Examples of such trunk-branch bus line operations are found in Ottawa, Portland, and São Paulo. An even more intensive convergence of lines is found on the New Jersey approaches to the Lincoln Tunnel under the Hudson River into New York City. The convergence of over 400 buses per hour from dozens of lines is greater than any street could handle, and it can be operated only because there are no stations on that section. All buses go into the Port Authority Terminal, where they are distributed into 233 bus berth locations on three levels. This is a very unusual situation, not typical for urban transit lines.

This type of bus line convergence on streets or on freeway ramps is much simpler to operate that convergence of rail lines. However, bus operations are less reliable and have significantly lower performance (speed, comfort, and particularly safety) than the first and second types of operation. Thus, while these heavily traveled corridors in São Paulo and other Brazilian cities, as well as in New Jersey/New York, are very impressive, their reconstruction into bus feeders to a high-capacity rail line would result in a major increase in the speed, reliability, and safety of the trunk line. This upgrading would in most cases involve a major investment in construction of a rail system, but it would result in a great reduction of operating costs.

4.3.3.3 Passenger Volumes. The dominant factor influencing the choice between branches and feeders for suburban distribution, as well as the selection of modes for the feeders, is the ratio of passenger volumes on the trunk and on the distribution sections. When most of the trunk volume is distributed into two directions with similar volumes, the same transit mode can be used to serve the trunk and branches. But when there are many distribution lines and they have much lower volumes than the trunk, use of the same mode for both sections would lead to excessive stretching of modes out of their optimal domains: the trunk mode may be too heavy for individual branches, while low-capacity modes that would be optimal for some branches could not satisfy the capacity of the trunk.

For example, suppose that a trunk has a critical volume of 18,500 prs/h and some outlying distribution lines have only 450–700 prs/h. Buses could not serve the trunk, and even LRT would be heavily loaded; by volume, metro would be the optimal mode for the trunk, but that mode would be poorly suited to serve some of the distribution lines as branches. A metro trunk line with bus feeders would represent the optimal modes for the two sections, providing both high service quality and good operating efficiency. These advantages would easily outweigh the inconvenience of passenger transfers.

4.3.3.4 Selection of Branches or Feeders. Considering the above-discussed factors, the selection between branches and feeders should be made in the following sequence of steps.

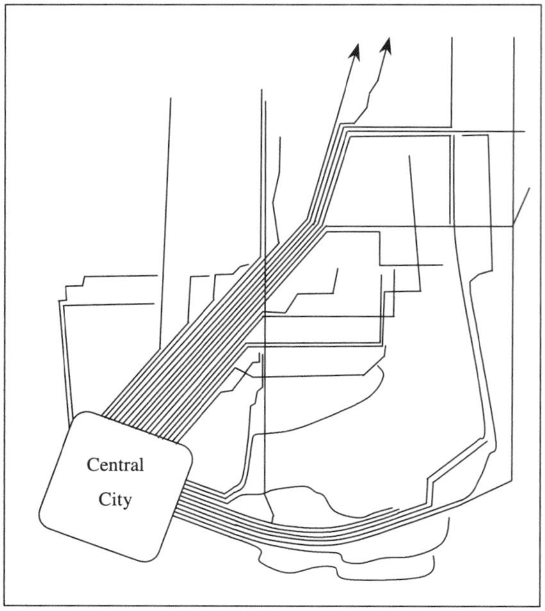

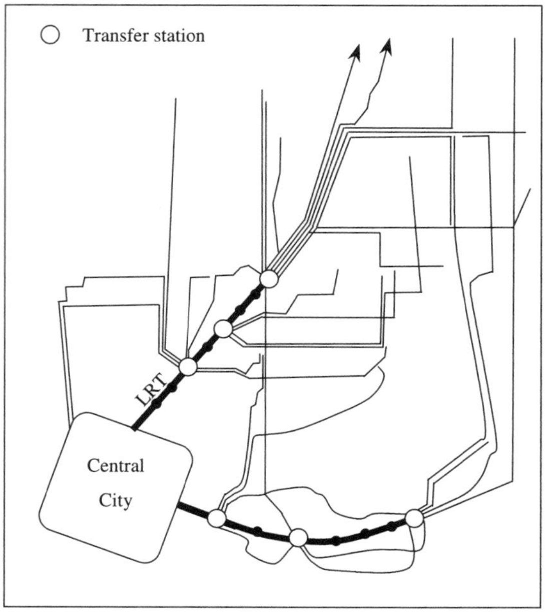

Figure 4.15 *Change from an extensive bus network into an intensive, more efficient and attractive LRT-bus network (Sacramento, CA).*

1. *Select the optimal mode for the trunk section.* This selection should be based on such factors as critical passenger volume, required level of service, costs, and impacts.
2. *Tentatively select of optimal modes for each distribution section (potential branch or feeder).* In addition to the factors considered in mode choice for the trunk, mode selection for these sections must be made considering the selection made for the trunk.
3. *Develop a tentative schedule* (TU capacities and operating headways in peak and off-peak hours) for the system of modes optimal for each section and test its operational feasibility:
   - Could the trunk schedule (TU capacities and headways) be reasonably split into schedules of two or more branches?
   - Alternatively, could the trunk be operated if TU frequencies from two or more branches are added up at convergence points?

If these questions are answered positively, branches can be operated; if not, outlying sections should be operated as feeders. In that case, one or more transfer facilities must be built and coordinated schedules must be planned.

In practice, many branches converging onto a trunk lead to very high operating frequencies, which decrease service reliability, limit capacity, and involve high operating costs. As explained in Section 4.2.2, it becomes logical to utilize a higher-capacity mode on the trunk and convert branches into feeders. This has been the case in many cities that have built rail lines. In Atlanta, several extensions of the Metro allowed shortening many bus lines with considerable reductions in operating costs. Similarly, opening of LRT lines in Edmonton, Calgary, Portland, and Sacramento allowed the same type of network restructuring with service improvements, ridership increases and reduction of operating costs.

Figure 4.15 shows the Sacramento transit network before and after the opening of the LRT line: a large number of bus lines converging on the two major corridors has been replaced by a single LRT line with service coordination through timed transfer stations (see Section 4.5). This replacement of buses by the LRT line has resulted in a considerable increase in ridership and decreases in operating costs. The LRT line now carries over 30% of the entire system's ridership but uses less than 20% of the total operating budget.

4.3.4 Rights-of-Way in Special Alignments

In recent decades, several new types of alignments, on freeways and urban streets, have been used for transit. Their characteristics are as follows.

4.3.4.1 Rapid Transit in Freeway Medians. The first rapid transit line in a freeway median was the Congress Line in the present Eisenhower Expressway in Chicago, opened in 1958–60. It was hailed as an excellent integration of modes: use of the freeway median reduced the investment for the rapid transit line. However, this type of alignment is far from ideal for rapid transit for several reasons.

First and basic, no freeway in North America was planned with considerations for the optimum alignment for a transit line. Actually, even if such consideration was given, in most cases not much could be done because freeway and transit alignments have inherently different optimal locations: freeways should go around or between urban subcenters or suburban major activity centers; transit should go through their axes and central areas to maximize convenience of pedestrian access. The two are therefore incompatible, so that transit lines in freeways are seldom in desirable alignments.

The second problem is that stations in freeway medians are by definition remote from any trip-generating objects. Walking distances to the nearest houses or stores are usually increased by 100–300 m by the presence of the freeway. In freeway interchanges, these access distances are even greater. This problem can be only partially ameliorated by utilization of bus feeders, K+R and P+R, as access modes.

Finally, stations in freeway medians are in an extremely non-human-friendly environment: intensive noise and air pollution are constantly produced by high-speed vehicular traffic on both sides, or even at several levels when stations are in the centers of interchanges.

Consequently, freeway medians can be used to obtain rail transit ROW at relatively low investment cost where both systems follow topographically constrained corridors; however, that is convenient mostly for sections of lines between central city and suburban areas, as has been successfully done by several lines of the San Francisco BART system (between Oakland and Walnut Creek), the Washington Metro (Orange Line to Vienna), and several others. Inside urbanized areas, rapid transit should preferably follow alignments with intensive generation of pedestrian traffic and potential transit riders.

4.3.4.2 Busways and High-Occupancy-Vehicle (HOV) Facilities. Busways can be built in freeway alignments at lower investment costs than in independent alignments. Many freeway corridors are good for commuter transit, but they are seldom functional for regular bus lines, for the same reasons as listed above for rapid transit.

HOV facilities, which exist virtually exclusively in the United States, often provide an improved operating regime for buses as compared to their operation in mixed traffic. However, their effectiveness is reduced whenever additional categories of vehicles are allowed to use it, such as the liberalization from three or more (3+) passengers to 2+ passengers vehicles (Vuchic et al., 1995). Because bus lines in HOV facilities usually operate only in one direction during peak hours on weekdays, they represent commuter transit, rather than regular transit. They do not have a distinct image of a high-performance transit line, but they can be very attractive to commuters and attract them from personal cars.

4.3.4.3 Transit Malls in CBD Areas. In recent decades, transit lines have been introduced in pedestrian precincts and malls in many European as well as North American cities. Most such installations, known as "transit malls," have been very successful in attracting considerable ridership and thereby making central cities more viable and human-oriented.

Photo 4.2 *Rapid transit alignment in freeway right-of-way—San Francisco BART.*

Experiences from many cities have shown that bus malls are usually successful: Minneapolis, Portland, and Denver bus transit malls are good examples. In some cases, enforcement of the prohibition of other vehicles is a serious problem; intensive noise and pollution also sometimes cause complaints. These problems and economic changes have led to partial reversals of bus malls into streets that also accommodate general traffic. Trolleybuses are environmentally much more friendly for mall operations than buses, and they fit well into pedestrian-oriented zones, such as the central core of Salzburg, Austria.

Light rail transit has become the most popular transit mode for pedestrian precincts and malls. In addition to their quiet and exhaust-free operation, their fixed travel paths and absence of street curbs are popular with pedestrians; they also greatly reduce the problem of encroachment by other vehicles. Thus, LRT tracks play a significant role: they visually give a distinctive image to transit, and illegal parking is easy to control because the required clearances for LRT cars are exact and unchangeable. The best-known LRT malls in Europe are those in Zürich, Karlsruhe, Mannheim, and Grenoble; in North America, Calgary's introduction of LRT/bus mall was followed successfully by Buffalo, Portland, Sacramento, Dallas, and several other cities. In Bremen, Germany, and Gothenburg, Sweden, LRT lines represent the dividers between traffic cells—sections of the central city core. Since vehicular traffic is prohibited from traveling among the zones, LRT rights-of-way are independent from general vehicular traffic, ensuring for it faster and more reliable service.

Photo 4.3 *Busway in freeway median—Lima, Peru.*

4.4 TRANSFERS IN TRANSIT NETWORKS

To attract passengers and compete with private automobiles throughout an urban area, transit must offer integrated service by area-wide networks. To achieve intramodal integration of different lines (e.g., within metro or bus networks), as well as intermodal integration, convenient and efficient transfers among lines must be provided. Network integration through transfers includes the functional design of lines, the optimal layout of transfer stations, coordinated scheduling, information, and, a particularly important factor, joint tariff.

Passenger transfers among transit lines and modes involve a certain resistance because they interrupt travel and require passengers to orient themselves and walk between vehicles on different lines; they also cause some delay. Therefore, it is sometimes believed that transfers should be avoided whenever possible. The fact is, however, that transit networks with many transfer opportunities offer passengers much greater selection of travel paths than networks with a large number of integrated lines that involve little or no transferring. In addition, the more transferring is performed, the greater is network efficiency, because each line can be designed optimally for its physical conditions, volume, and characteristics of demand.

Consequently, when transfers are planned correctly and resistance is minimized through good station design, convenient walking paths, station amenities, schedule coordination, etc., the inconvenience to passengers can be so small that it is easily outweighed by the benefits transfers bring with respect to the availa-

Photo 4.4 *Light rail in lively pedestrian area—Karlsruhe, Germany (courtesy of Mario Semmler).*

bility of different travel routings through the network, shorter headways on some lines, and, generally, better services. Passenger transfers among lines thus represent an important element of transit travel. This importance of transfers is clearly demonstrated by the fact that most successful large transit networks, such as Munich, Toronto, and Washington, rely on extensive intramodal and intermodal transfers.

This section presents classification of transfers by types of lines and their headways and analysis of some aspects of scheduling and of station layout that minimize transfer times and distances, major elements of resistance to transferring.

4.4.1 Classification of Transfers by Headway Length

Passenger transfers between lines occur where two or more transit lines intersect or terminate at one point. To analyze this phenomenon, a simple model of intersecting transit lines is presented. On each line TUs operate at fixed, uniform headways. Time delays caused by transferring depend on the lengths and relationships of headways on the intersecting lines.

For a general analysis, lines are classified into those with short headways (generally, $\leq$ 10 minutes) and lines with long headways ($>$10 minutes). Transfers among them can be classified by headways of originating and destination lines into four cases, A through D, as presented in Table 4.3. The cases have the following characteristics.

Cases A and B: short-to-short and long-to-short headway, respectively. Transferring from any line, with short or long headway, to a line with a short headway involves short transfer times. There is no need for any special schedule coordination at transfer points.

Case C: Short-to-long headway. This case is the reverse of that in case B; it happens, for example, with

Table 4.3 Transfer times between lines with short and long headways

Origin Line \ Destination Line	Short Headway	Long Headway
Short headway	*Case A* Always short, convenient	*Case C* Varies greatly Information about connecting runs required
Long headway	*Case B* Always short, convenient	*Case D* Variable depending on headways: 1. Equal and simultaneous—all transfers convenient (TTS) 2. Equal but not simultaneous—convenient in one direction 3. Different—impossible to coordinate; long transfer times

transfers from a trunk line, which has short headways, to feeder lines with long headways. Such transfers may involve waiting times that vary from very short ones to those close to the long headway on the line to which the passenger is transferring. Thus, the delay varies randomly. This uncertainty can be eliminated when schedules for all lines are provided to passengers, so that each passenger can plan his/her trip so as to take the TU on the trunk line that connects with his particular feeder departure with minimum delay.

Case D: Long-to-long headway. Transfers between two lines with long headways can be classified into three subcases based on the relationship of headways on the two lines.

Case D_1: Long-to-long, equal headways, synchronized, with overlapping station dwell times. TUs from the connecting lines arrive at the same times, in constant intervals (called pulse headway) and dwell for a few minutes to allow exchange of passengers among all TUs. This operation, called timed transfer system (TTS) (see Section 4.5), provides for easy and convenient transfers among all TUs, i.e., in all directions.

Case D_2: Long-to-long, equal headways, but no overlapping standing times. TU arrivals on different connecting lines are always in the same time sequence. It is possible to make convenient transfers from one line to another (e.g., line M to line N), but not in the opposite direction (N to M). For example, if line M arrives at 07, 27, and 47 minutes after each hour and line N arrives at 10, 30, and 50 minutes after each hour, the M to N transfer involves only a 3-minute wait, but the N to M transfer requires a 17-minute wait.

Case D_3: Long-to-long, different headways. No coordination is possible. Transfer times are random and can approach the headway length on the destination line.

Table 4.3 presents a summary review of characteristics of all these cases and subcases. It shows that transferring from any line to one with short headways is always convenient (cases A and B); transferring to a line with long headways varies from very convenient in case D_1 to very inconvenient with different headways, case D_3. Coordination of schedules is not necessary in cases A and B, but it is very important in cases D_1 and D_2.

4.4.2 Classification of Transfers by Type of Line

The number and character of transfers are influenced considerably by two aspects of the lines among which transfers take place. The first aspect is the relationship of each line to the transfer point: whether it terminates or passes through it. In this respect, lines will be referred to as *terminating* (t_e) and *through* (t_t) lines. The second aspect is whether all lines are of similar nature (frequency, capacity, which transit mode) or one of them is a dominant or *trunk line* with considerably higher performance than any other line, while the others, with low frequency, capacity, etc. represent its *feeders* with a collection/distribution function. This aspect of line type, similar or different, influences transferring patterns of passengers (i.e., many-to-many among similar lines, or many-to-one and one-to-many between different line categories, such as trunk and feeders).

Suburban bus lines meeting at different points often have a similar nature, while the feeder/trunk situation is typically found where suburban bus lines converge on a major radial line toward the central city. The trunk line has much higher performance (frequency, speed, capacity, comfort, reliability, and image) than its individual feeders; the trunk may be any mode, but it is typically a rail line.

Table 4.4 presents descriptions and characteristics of transfers classified by the above-defined aspects of transit line types. Three combinations of terminating and through lines, expressed in general terms (N routes), are given as cases 1–3; the other three cases, 4–6, are simple numerical examples.

All the cases are described for two types of situations: for similar lines and for a trunk-feeder relationship. Each case will be discussed here.

Case 1 is the simplest. When all of the N_e lines coming to a transfer station terminate, the total number of transfer permutations among the lines, k, is:

$$k = N_e\,(N_e - 1) \qquad (4.3)$$

This type of station is the prime case for application of TTS to minimize passenger delays and integrate the transit network. A typical example of this case is found where many suburban transit routes have a common terminal; they may be of similar nature (columns 5–7 in Table 4.4), or there can be the case of suburban low-frequency lines terminating at a trunk line terminal (columns 8–10). This case is found, for example, at most metro line terminals in Atlanta, Hamburg, and Philadelphia.

In the latter case, a trunk with feeders, the application of TTS again greatly facilitates transfers among the feeders, but it creates uneven loadings on the trunk line. Due to its much shorter headways, many of the trunk line's TUs would meet no feeders, while a few, meeting the pulse of feeder arrivals, would get passenger loads from all feeders. Therefore, TTS should be used only if transferring among feeders is substantial, and, whenever possible, feeders should be divided into two to three groups that meet in separate, staggered pulses. If the transferring among the feeders is negligible, TTS should not be used, but feeder arrivals/departures should be staggered as much as possible to provide even loading of the trunk line.

Case 4 represents the simplest numerical example of case 1 type of transfer: two lines with a common terminus. The number of possible transfers is two: one from line A to B, and one from line B to A.

Case 2 and case 5 represent transfer stations at which all N_t lines pass through. The total number of transfer permutations k is:

$$k = 4N_t\,(N_t - 1) \qquad (4.4)$$

This is a very large number. For only two intersecting lines (shown in case 5), there are already eight possible transfers. It is therefore highly desirable to organize a TTS among lines for which there is an appreciable number of transferring passengers and that can have the same headways. However, since this transfer point is in the middle of both lines, it causes delay to all through passengers. To minimize that delay and prevent eventual loss of through passengers, it is important to ensure short layover times through precise scheduling, reliable operation, and design of the common station with short transferring distances. This is achieved in metro stations with cross-platform passenger exchange, described in Section 4.4.3.1.

Case 3 represents the most general situation: N_e terminating lines meet N_t through lines at a joint terminal. The number of possible transfers k among these lines is given by the expression

$$k = (N_e + 2N_t)^2 - (N_e + 4N_t) \qquad (4.5)$$

When all lines are terminating or all go through, case 3 collapses into cases 1 and 2, respectively.

Case 6 in Table 4.4 shows the simplest situation of this type: a transfer between one terminating and one through line. There are only four possible transfers, and coordination is best achieved if TUs on the terminating line arrive before and leave after TUs on the through line pass through. This condition is similar to the one where feeders intersect a trunk line (cases 2

Table 4.4 Classification of transfers by types of lines

Case Number	Number of Lines: Terminating	Number of Lines: Through	Transfer Permutations	Similar Lines/Routes: Sketch	Similar Lines/Routes: Typical Case	Similar Lines/Routes: Comment	Trunk with Feeders: Sketch	Trunk with Feeders: Typical Case	Trunk with Feeders: Comment
1	N_e	0	$N_e (N_e - 1)$		Many suburban lines terminating	Coordination between lines needed. TTS can be applied successfully	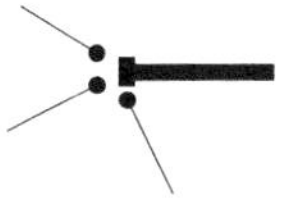	Trunk line terminating with many feeders	Requires coordination in case of transfers among feeders (TTS)
2	0	N_t	$4N_t (N_t - 1)$	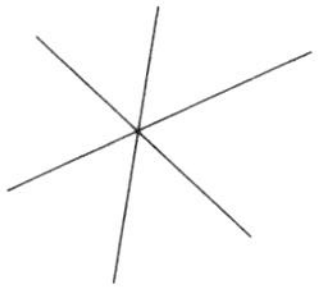	Any point with several intersecting transit lines	TTS desirable, but causes delay of through passengers	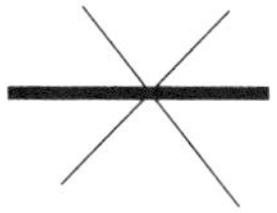	Trunk line station where feeders intersect and/or terminate	TTS possible. Feeders' arrivals and departures around trunk's stopping
3	N_e	N_t	$(N_e + 2N_t)^2 - (N_e + 4N_t)$	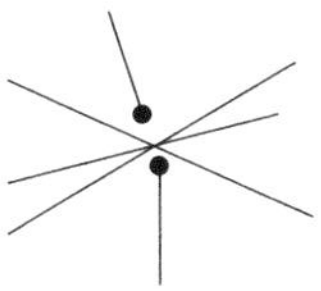	Many transit lines terminate or pass through	TTS desirable; easier to achieve than with case 2	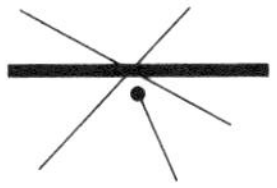	Trunk line passes through where feeders intersect and or terminate	
4	2	0	2		Terminal point of two suburban lines			Trunk line terminal with one feeder	
5	0	2	8	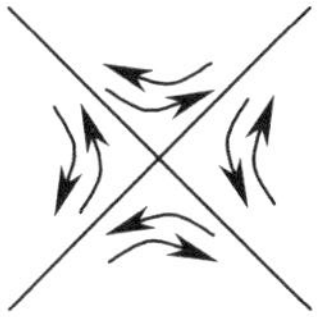	Intersecting point of any two transit lines			Trunk line station where a feeder intersects	
6	1	1	4	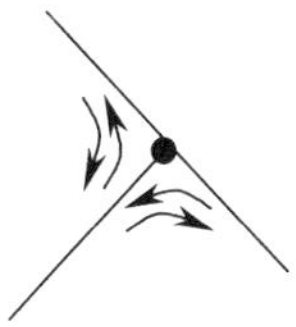	Point where one line terminates another passes through			Trunk line passes through and feeder terminates	

and 5, columns 8–10), except that no additional delay is caused on either line by this scheduling, because terminal time on the terminating line is used for overlap with the arrival on the through line.

4.4.3 Metro Station Layouts and Schedules for Simultaneous Transfers

Most transfer stations in metro systems have two or more lines crossing themselves at different levels. Since passengers must negotiate one or two flights of stairs, schedules of trains on different lines cannot be coordinated for simultaneous transfers. However, whenever geometry of line alignments allows, and particularly when large passenger volumes make fast and convenient transfers highly desirable, station layouts that provide for across-platform simultaneous two-way transfers between trains should be planned. Several such cases are described here.

4.4.3.1 Simultaneous Transfers between Weaving Lines. When the alignments of two metro lines can be brought to intersect under a flat angle, tracks should be woven as shown in Figure 4.16. The two trains in each direction are then scheduled to operate with equal and simultaneous headways. The passengers are thus given the ultimate type of transfer: across platform between two standing trains, thus requiring minimum distance and no waiting.

Very successful applications of this type of station design exist in Hamburg. As shown in Figure 4.12, three U-Bahn lines meet each other in pairs at four different stations of the network, shown as double circle junctions in the figure. This operation requires coordinated scheduling of all three lines, precise operations, and schedule corrections at stations approaching these junctions. Experience shows that operation with 5-minute headways is easily synchronized. When 2.5-minute headways are operated, synchronization of train arrivals is more difficult, but trains missing each other is not a major problem with such short headways anyway. Moscow Metro and several other transit systems also have this type of stations, but a particularly interesting design for transfers at two stations is found in Hong Kong, described next.

4.4.3.2 Station Design for Transfers at Y Junctions. If a Y-shaped radial network is operated with the trunk continuing into one branch while the other branch is disconnected, i.e., operates as a feeder, there may be heavy transfers at the convergence of the two lines. Such a case exists on the Hong Kong MTR: its main radial line from the CBD (Hong Kong Island on the south) (S), goes northbound to Kowloon and then diverges into two lines. The trunk line continues to

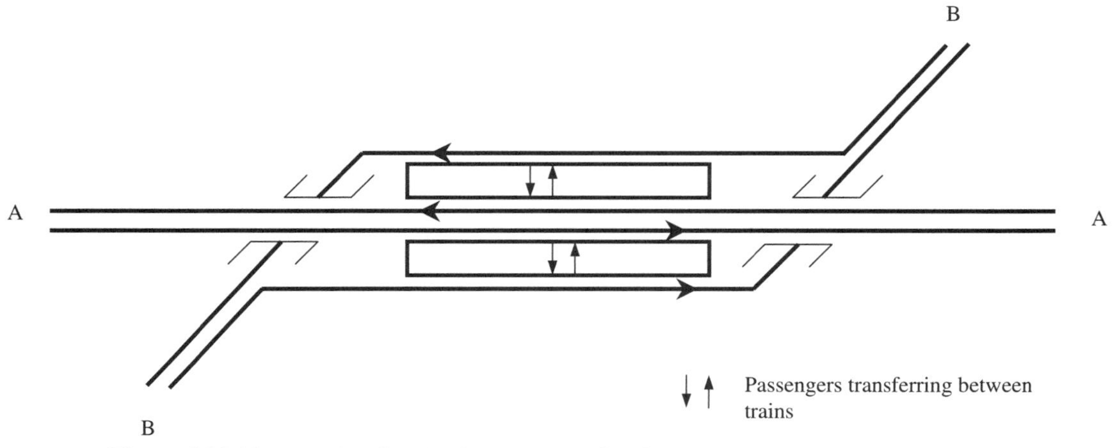

Figure 4.16 *Metro station for simultaneous transfers between trains on two weaving lines.*

Tsuen Wan on the north (N), while the line to Kwun Tong on the east (E) is operated as a feeder.

The majority of passengers from the E line travel toward the CBD, S, but a certain volume also travels N. Thus, there are heavy passenger transfers between the E line and both directions, N and S, of the trunk line at the Y junction. To ensure efficient operations, an imaginative station layout, shown in Figure 4.17(*a*), was designed.

The four directional transfers, E→N, N→E, E→S, and S→E, are divided into two stations, Prince Edward and Mong Kok. The lines run parallel between the two stations and have two platforms at different levels at each one of the two stations, as shown in Figure 4.17(*b*), to provide direct convenient transfers for all four transfers.

Logically, all transfers between E and N directions take place at the first joint station, Prince Edward, to minimize passenger travel. The two pairs of tracks for these two transfers, E→N and N→E, are woven around joint platforms on two levels. As shown in the figure, E→N transfer is performed on the upper, N→E on the lower level. Proceeding south, two of the tracks change levels, so that at the second joint station, Mong Kok, two northbound tracks are on the upper level providing for the S→E transfers, while the two southbound tracks on the lower level provide for E→S transfers. Thus, all four sets of transfers are organized in the most efficient manner: across-the-platform, one direction of transfer across each platform, two platforms at each station.

A similar convenient across-the-platform intermodal transfer arrangement exists at two stations between U-Bahn and S-Bahn lines under the main shopping street Zeil in Frankfurt am Main, Germany.

4.4.3.3 Transfers between Local and Express Trains. A few metro and many regional rail systems have lines that provide not only local, but also express

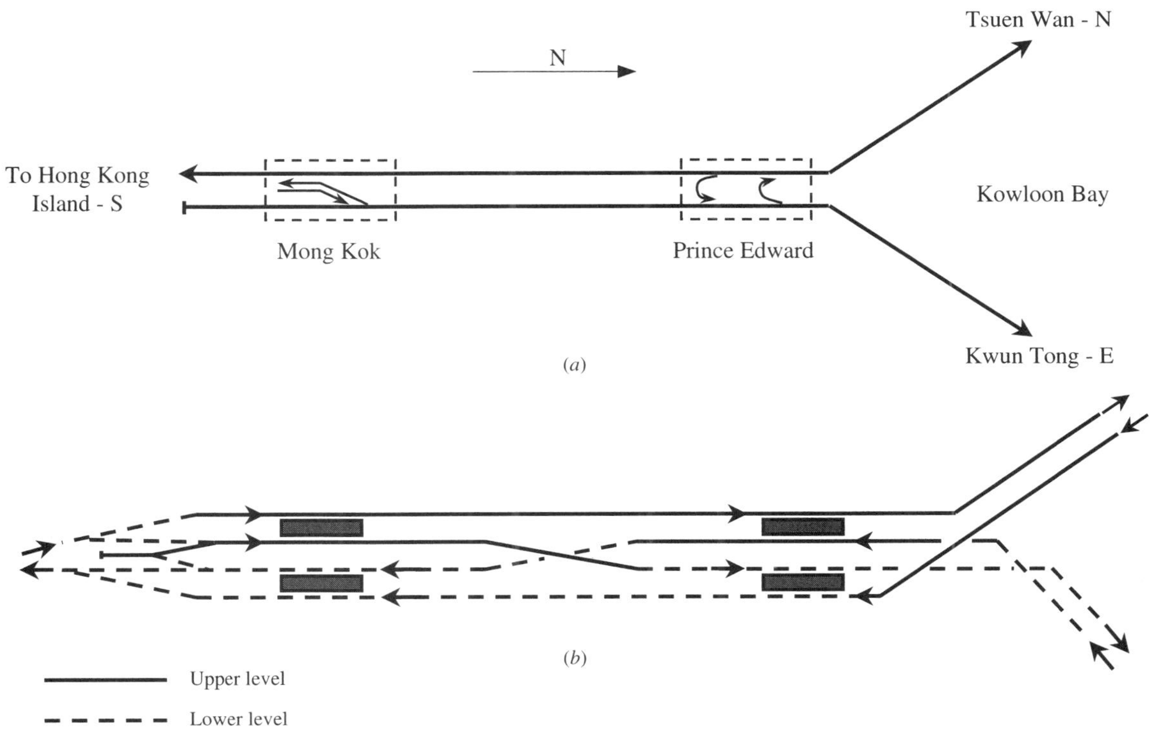

Figure 4.17 *Two-station Y junction with across-platform transfers for all movements (Hong Kong MTR—left-hand driving): (a) distribution of transfers; (b) layout of tracks and platforms.*

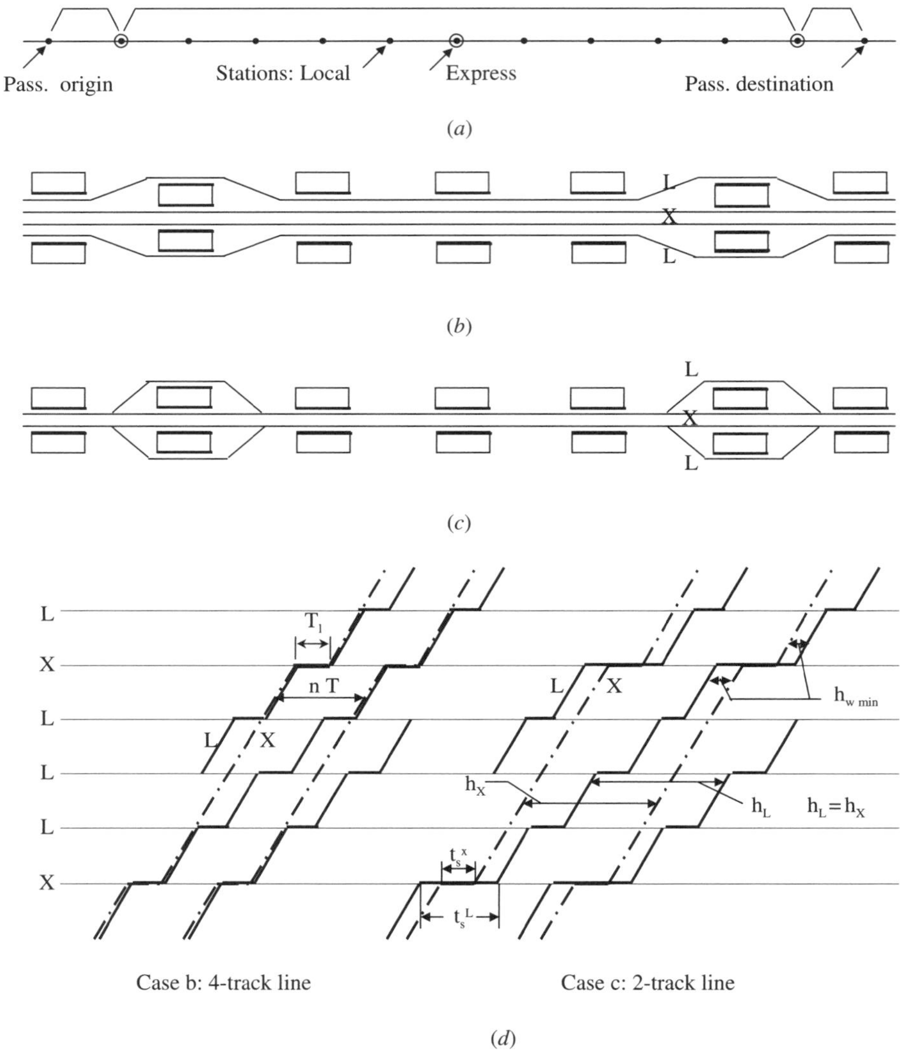

Figure 4.18 *Station layout and schedule for local/express services with simultaneous transfers:* (*a*) *line layout and travel pattern for a typical passenger's trip;* (*b*) *four-track with local* (*L*) *and express* (*X*) *operations;* (*c*) *two-track line with local* (*L*) *and express* (*X*) *operations;* (*d*) *graphical schedule of synchronized local/express operations.*

services (for definition of these services see Section 2.4.3). Such services exist on rapid transit systems in several U.S. cities: New York (by far the most extensive), Chicago, and Philadelphia have lines with these operations on four, exceptionally three or even two tracks. Several Japanese cities (Tokyo, Yokohama, Osaka, and Kobe) also have such operations on a few metro lines and, more commonly, on regional rail networks.

The two most typical track layouts and diagrams of local/express operation, one on four- and the other on two-track lines, are shown in Figure 4.18. Figure 4.18(*a*) shows the basic concept of this type of operation. A passenger who enters the line at a local station for a long trip can take a local train to the first express station, transfer to an express train for the largest portion of his/her trip, then again transfer to a local train to get to his/her destination at a local station.

Sketches in Figure 4.18(*b*) and 4.18(*c*) show track and station layouts for four- and two-track lines, respectively. Figure 4.18(*d*) shows time-distance diagrams for the two cases. On the four-track line the headways of local and express trains are independent of each other, but ideal operation, shown on the diagram, is when local trains are scheduled to stop simultaneously with expresses at every express station, so that passengers can transfer without waiting. On two-track lines the local and express trains must be scheduled so that they meet in express stations because those are the only locations where express trains can overtake the local ones. Thus, the two-track operation involves additional delays to local trains, because they must arrive before and leave after the expresses.

The difference between operations of the two cases is shown on the time-distance diagram in Figure 4.18(*d*). On the four-track line from Figure 4.18(*b*), locals and expresses can operate independently, but the ideal schedule is achieved when the headways h for both local and express lines are the same and equal to the sum of the times lost T_ℓ by local trains for stopping at the n_ℓ local stations between the express stations (or, an exact multiple of this headway, when locals and expresses meet only at every second or third express station):

$$h = n_\ell \cdot T_\ell \qquad \left| \frac{h}{s} \right| \frac{n}{-} \left| \frac{T}{s} \right. \tag{4.6}$$

The diagram for the two-track line from Figure 4.18(*c*) shows that the local train must arrive at the express station first. Then the express arrives, exchanges passengers across the platform, and departs, followed by the local. The time elements in this operation must be very precise, and the sequence of train operations is described here. After the local train's arrival, the express needs at least an $h_{w\,min}$ to enter the station; then follows the dwell time of the express train t_s, and another $h_{w\,min}$ interval after the express train's departure, when the local can proceed. Thus, the local train's standing time is longer than that of express train by at least two way headways: $t_s^L = t_s^X + 2h_{w\,min}$, as can be seen in Figure 4.18(*d*), case c.

This type of service is not only theoretical. It requires precise scheduling (in seconds rather than minutes) and strict control, and it is operated, for example, daily in the Tokyo region, on several metro and regional rail lines.

4.4.4 The Importance of Transfers

The quantity and quality of transfers in a network are closely related to the character of transit services in any given city. Generally, transit services without many transfers represent commuter transit, specialized only for travel to and from work. Examples of such systems are express bus services utilizing freeways and HOV facilities in many U.S. cities, such as Houston, Seattle, and Washington, DC (the Shirley corridor). They provide efficient peak hour travel to CBD and sometimes to other employment centers, but they cannot serve many-to-many type of travel and usually offer no service outside of the peak hours.

All transit systems offering all-day area-wide services rely heavily on passenger transfers among lines and modes. This is typical for extensive metro systems (New York, Paris, and Tokyo), all-bus networks (Canberra, Ottawa), and intermodally integrated systems (Atlanta, Calgary, Hannover, and São Paulo).

Inadequate transfer arrangements usually create major obstacles to present passengers and discourage potential passengers from using transit. The obstacles may be in various forms. Poor design of terminals, as is found in many developing countries, makes transfers tedious and physically difficult. Lack of coordinated schedules, joint fares, and integrated information, such as has been recently created by deregulation and legal prevention of multimodal companies in Great Britain, makes transfers confusing, long, and costly to passengers.

In recent decades, there have been many innovations in facilitating transfers, such as new designs of intermodal stations, increased use of escalators, better information, and joint fares. The problem of facilitating transfers in medium- and low-density cities and suburbs has been alleviated considerably in many North American cities, such as Edmonton, Portland, and Sacramento, by the development of the TTS networks, described in the following section.

4.5 TIMED TRANSFER SYSTEM NETWORKS

A *timed transfer system* (TTS) is a network consisting of transit lines and one or several *transit centers* (or focal points of the network) at which TUs from all intersecting lines arrive simultaneously, allowing easy passenger transfers in all directions. Services on the lines are scheduled for such arrivals and simultaneous dwell times, which are as long as is needed to allow passengers to transfer and then TUs to depart. For this type of operation, travel times between the transit centers and scheduled headways have a certain relationship. Therefore, the line network and the basic schedule of operations must be designed together in a coordinated manner (Sullivan, 1976; Vuchic et al., 1983).

The basic service concept of pulsing, where TUs from two or more lines meet simultaneously at a transit center or focal point, perform transferring, and then depart, is actually used in many different transportation systems, although with different times between the meets and their durations. For example, many airlines in North America use such operations. Each airline has one or several hub airports at which aircraft from many directions land during an interval of 30–60 minutes. About an hour later, there are a number of take-offs for different destinations. Known as hubbing operations, this allows convenient transferring of passengers. Some express freight services use the same type of operation with one or two meets during the day.

The timed transfer concept is used most effectively on systems that operate with precise schedules. The pulses or joint headways depend on the type of service. For example, intercity and regional rail systems in Germany, the Netherlands, and Switzerland have extensive networks with timed transfers every 60, 30, or 15 minutes. Transit systems in many cities have traditionally used TTS for night ("owl") services, which have long headways. Typically, TUs from all major lines in the city meet at hourly intervals at a central location—usually city hall, the main railway station or a central square. On the other hand, there are metro systems that have simultaneous meets for convenient transfers between lines at high frequencies: every 5 minutes or at even shorter headways. Such operations with synchronized TU arrivals for transfers between lines and between local and express trains on the same line are described in Sections 2.4 and 4.4.

The TTS concept in a somewhat different form has been introduced very successfully since the 1970s in the transit networks of a number of North American cities. With the growth of low-density, auto-oriented suburbs in many in North America and other industrialized countries, it became very difficult to serve these areas by conventional, independent transit lines. Their long headways made transferring among lines inconvenient, involving irregular, often long waiting times. Transit was therefore very inferior to auto travel for all suburban trips that do not follow a single transit line and require a transfer. To provide attractive transit services in such auto-oriented areas, cities like Edmonton, Canada, Portland, Oregon, and Sacramento, California, introduced TTS-based networks and types of operation. Strengthened by their new LRT systems, the TTS system of operations upgraded uncoordinated transit lines in these cities into integrated networks with easy transfers among lines. The design and scheduling of such networks are described in the next sections.

Three major actions must be undertaken to introduce a successful operation of a TTS network. First, lines and networks must be laid out and scheduled so that the system operates in a synchronized manner. The layout consists of the locations of transit centers and alignments of lines connecting them or radiating from them in different directions. Schedules must be prepared together with network design because the two are interdependent: travel times on the lines must correspond to pulse headways.

Second, the operations must be made as reliable as possible to minimize the probability of missed TU meets at transit centers. With reliable bus services in suburbs, the joint dwell times should be two to four minutes. If traffic congestion or other factors cause frequent delays, dwell times may be extended to four to six minutes. If longer times are used, the transfers lose the main advantage of TTS—convenience and very short transfer delays.

Third, a comprehensive effort for distribution of information and marketing of the services must precede and accompany the introduction of a TTS service to be understood and accepted by the public. Passengers get used to synchronized transfers and rely on them. They like the coordinated TU meets with convenient transferring at transit centers, but they are irritated when the system is not reliable, because every missed meet causes one headway-long wait, which may be 10–15 minutes or longer.

4.5.1 Timed Transfer System Scheduling

TTS is a system in which arrivals and departures of TUs from different lines at a transit center must be simultaneous, at time intervals referred to as the *pulse headway* h_p. Since TTS is usually operated with lines that carry light or moderate passenger loads, their pulse headways are long, typically between 10 and 30 minutes. To have schedules that are repetitive every hour, there is actually a choice between only two different basic pulse headways, each one with a set of headways that are their fractions or multiples:

- 15 minutes, which also allows use of 7.5, 30, 45, and 60 minutes on some lines;
- 20 minutes, which allows 10-, 40, and 60-minute headways.

The selection between the 15- and 20-minute pulse headway is made on the basis of:

- The desired level (frequency) of service
- Distances between transit centers, which influence efficiency of scheduling
- Investment (fleet size) and operating costs, considering frequencies and the required number of TUs for services on all lines

To design a TTS operation, the network form and size (length of lines) must be coordinated with schedules of TU (vehicle) travel. Computations of these network and operating elements will be shown using the simplest TTS model, a *unifocal network,* which is a set of radial and diametrical lines terminating or passing through a single transit center, as shown in Figure 4.19. To operate this unifocal TTS, all the lines terminating at the center, or on through lines each one of the two sections radiating from the transit center, L_s, must have the following relationships among T, N, and h_p:

$$j_1 \frac{T_1}{N_1} = j_2 \frac{T_2}{N_2} = \cdots = j_i \frac{T_i}{N_i} = h_p, \qquad \left| \frac{j}{-} \middle| \frac{T}{\text{min}} \middle| \frac{N}{\text{TU}} \middle| \frac{h}{\text{min/TU}} \right| \tag{4.7}$$

where j is an integer number: when TUs from a line arrive at every pulse, $j = 1$. If they come to only every second or third pulse, $j = 2$ and 3, respectively. Or, if a line has a cycle time $2T$ and it is operated with two TUs ($N = 2$), $j = 1$ because one TU arrives after each h_p.

This equation shows that when one TU operates on a line ($N = 1$), the pulse headway is equal to the cycle (round trip) time from the transit center to the end of each line i: $h_{pi} = T_i$. If the number of TUs on a line with cycle time T is increased from 1 to 2, the headway will be reduced to $h_p/2$. On the other hand, some lines may have longer cycle times, such as $2T$ or $3T$. If they are operated with one TU, they will have longer headways, such as $2h_p$ or $3h_p$, respectively. Then TUs from that line arrive only at every second or third pulse.

The cycle time T is now expressed through its time components and as a function of line length L and operating speed on each line V_o:

$$T = 2T_o + t_{t1} + t_{t2} = \frac{120L}{V_o} + t_{t1} + t_{t2} \qquad \left| \frac{T,\ t}{\text{min}} \middle| \frac{L}{\text{km}} \middle| \frac{V}{\text{km/h}} \right| \tag{4.8}$$

The two terminal times t_t are used to adjust cycle time to be equal to h_p, or to a multiple of h_p, satisfying Eq. (4.7).

Photo 4.5a *Intermodal timed transfer system: (a) regional rail trains; (b) bus transfer at same station—Zürich.*

The cycle time can also be expressed as a function of line length L and cycle speed V_c:

$$T = \frac{120\ L}{V_c} = h \cdot N$$

$$\left| \frac{T}{\text{min}} \middle| \frac{L}{\text{km}} \middle| \frac{V}{\text{km/h}} \middle| \frac{h}{\text{min/TU}} \middle| \frac{N}{\text{TU}} \right| \qquad (4.9)$$

The number of TUs on each line or its radiating section from the transit center L_s, N, can now be derived from Eqs. (4.8) and (4.9) if T is replaced by $N \cdot h$ and solved for N:

$$N_i = \frac{120L/V_o + \sum t_t}{h} = \frac{120L_{si}}{h_p \cdot V_{ci}}$$

$$\left| \frac{N}{\text{TU}} \middle| \frac{L}{\text{km}} \middle| \frac{V}{\text{km/h}} \middle| \frac{h}{\text{min/TU}} \middle| \frac{t}{\text{min}} \right| \qquad (4.10)$$

For diametrical lines, N's are computed for each line section from the center outward, L_s, and then the two N's added up for the entire line.

Using the preceding equations, planning of a TTS should consist of the following steps:

1. Determine for each line its T_o and the minimum duration of the two terminal times; their sum (Eq. 4.8) represents the minimum possible cycle time for that line.
2. Plot a diagram with the obtained headways and show their adjustments to the usable headways if h_p is based on 15 minutes, and the adjustments needed if h_p is 20 minutes, as shown in Figure 4.20.
3. Compare the two sets with respect to the level of service and number of TUs required (i.e., cost), and select the preferred one. That headway will be adopted for the entire TTS network.

Photo 4.5b *(continued)*

Then headways for all lines must be adjusted to one of the headways from the adopted set and terminal times adjusted so that T for each line is the same as, or a multiple of, h_p.

4. Compute the number of TUs for each line.

This planning procedure is illustrated by an example presented in Figure 4.20. The data and computed values for the five lines that converge on a transit center, A, B, C, D, and E, are presented on a diagram and in a table in two groups: for schedules based on h_p of 15 and 20 minutes, respectively. The minimum cycle times for each line, using the field measurements of T_o and operational considerations for terminal times, are computed by Eq. (4.8), shown graphically, and given in the table. The following two rows of data in the table give for each line the desired headway based on passenger volumes or policy headways, and the offered headway, which is the least smaller multiple of the respective h_p.

The minimum cycle times are then adjusted to the values compatible with the respective h_p using some of the following measures:

- Lengthening terminal times (most common)
- Increasing operating speed (by giving preferential treatment to transit vehicles, reduction of the number of stops, faster fare collection, shorter dwell times at stops, etc.)
- Changing the line length or alignment
- Changing the number of TUs

In the case of line D for 30-minute headway, it may be possible to use a cycle time of 45 instead of 60 minutes, but only if that line is cross-routed, or jointly scheduled, with another line on which the ratio of cycle time and headway would also yield $N = 1.5$.

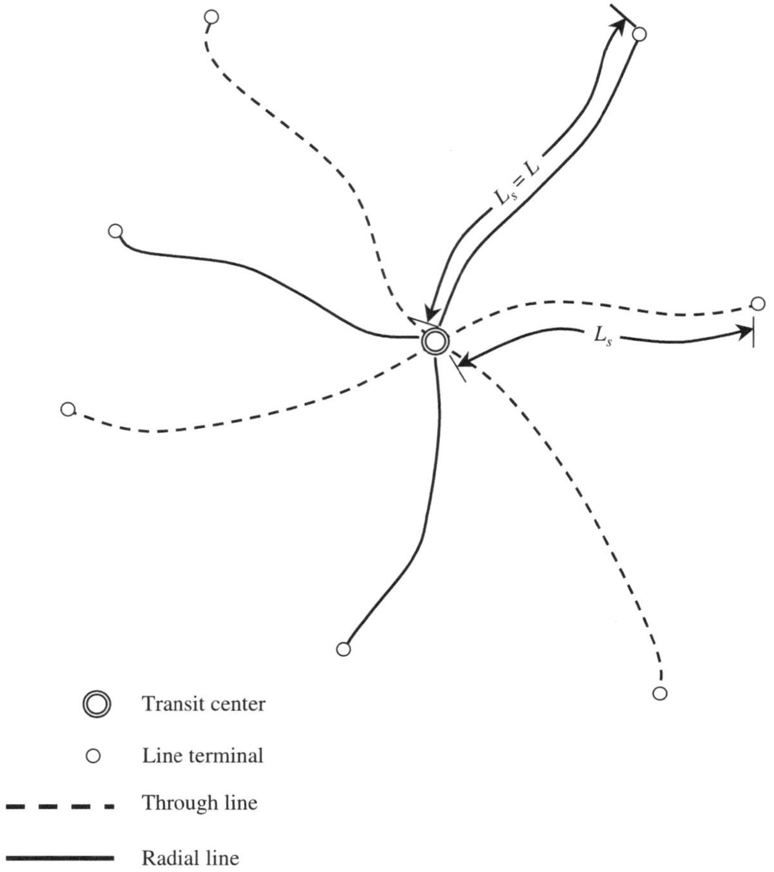

Figure 4.19 *Unifocal TTS network.*

The table then gives the additional cycle times ΔT, i.e., the difference between the minimum required and actual feasible T, required number of TUs N by line, and the total for all lines. Finally, the total TU-minutes due to cycle time adjustments by line, and their overall total, are given.

The diagram and table in Figure 4.20 give all elements needed for evaluation of the two cases. Compared to the 15-minute pulse headway operation, the 20-minute-based one offers shorter headways on lines A, B, and E, but longer ones on lines C and D. It involves lower ΔT (a total of 75 versus 99 minutes), but it requires one more bus, 10 instead of 9, as can also be seen by the drawings of buses at adopted headways on each line. The selection should be made by comparing these differences.

4.5.2 Multifocal Networks

The basic relationships among f or h, L, N, V_c, and T that exist for a unifocal network also apply when one or more additional focal points (transit centers) are added to the network. However, the interactions among focal points introduce some additional scheduling options.

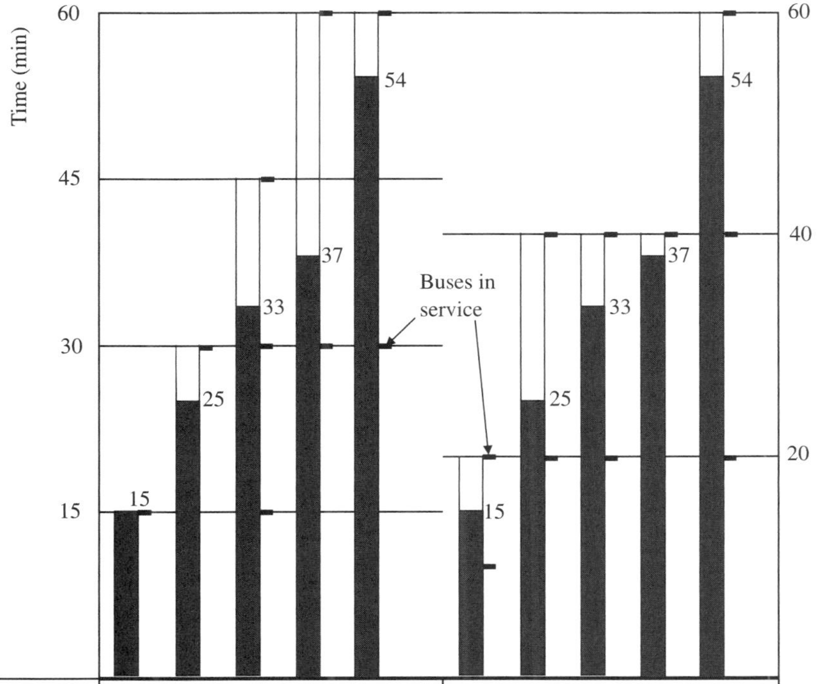

	h_p = 15 min						h_p = 20 min					
Line	A	B	C	D	E	Σ	A	B	C	D	E	Σ
T_{min}	15	25	33	37	54		15	25	33	37	54	
h: Desired	15	30	20	40	30		15	30	20	40	30	
Offered	15	30	15	30	30		10	20	20	40	20	
T	15	30	45	(45)60	60		20	40	40	40	60	
$\Delta T = T - T_{min}$	0	5	12	23	6	46	5	15	7	3	6	36
N	1	1	3	2	2	9	2	2	2	1	3	10
$N \cdot \Delta T$	0	5	36	46	12	99	10	30	14	3	18	75

Figure 4.20 *Planning a five-line TTS for pulse headways of 15 and 20 minutes.*

In a bifocal network, shown in Figure 4.21, each transit center has radial and diametrical lines, the same as unifocal network. But between the two focal points there may be two types of lines:

- *A direct connector line* is a fast line with the primary role to connect the two centers.
- *A collector-distributor connector line* operates between the two centers, but has the primary purpose to serve local trips in that corridor. This line may have a considerably longer cycle time than the former, so that a greater number of TUs in operation may be needed to have the same headway as the direct connector line.

Another bifocal TTS network is shown in Figure 4.22 by line layout (4.22(*a*)) and time-distance diagrams for all lines, the connector between A and B, three radials from A (11, 12, and 13), and two from B (21 and 22), each served by one TU: $N_r = 1$. Two cases

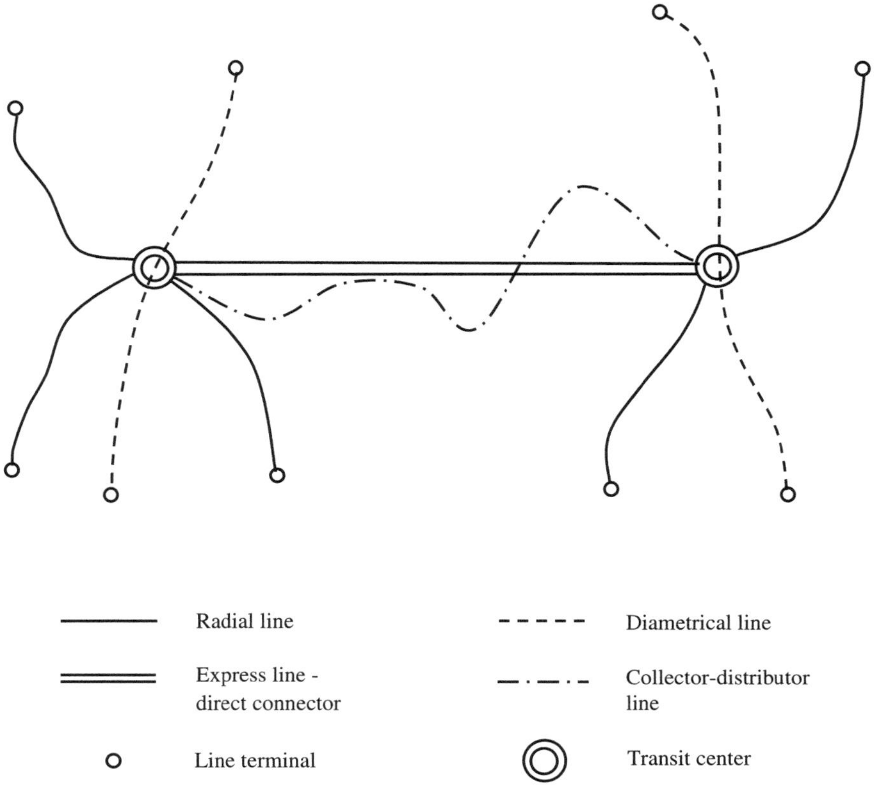

Figure 4.21 *Bifocal TTS network.*

are shown with different distances between A and B and correspondingly different schedules. In the first case, Figure 4.22(*b*), the connector line is served by one TU (N_d = 1) and a cycle time *T*. The schedule has *staggered pulsing,* i.e., the times when the TUs from the connector and the radials meet at the two centers take place alternatively, half a cycle (T/2) after each other. In the second case, Figure 4.22(*c*), the distance between A and B is approximately twice as great as in the first case, the cycle time is 2*T*, and two TUs are in service (N_d = 2) with *simultaneous pulses* at the two centers, A and B.

The schedules of the radiating lines do not change between the two cases. As the diagrams show, they all meet with the TU on the connector line and then follow their respective alignments. Each line goes to a different terminal, but they are plotted for convenience on the same time-distance diagram. Although they follow different alignments and have different line lengths, all the lines must have the same cycle time *T*. Therefore, longer lines must have higher operating speeds, as shown by the slopes of lines on the diagrams. For example, line 11 is longer than 12 and 13 is shorter, but 11 has a higher operating speed and 13 lower than line 12, so that cycle times and headways of all three lines are the same: they all must satisfy Eq. (4.8).

A triangular TTS network, shown in Figure 4.23, forms a closed geometric pattern, which places certain constraints on the connecting lines. In general, if all three connector routes have the same cycle *T*, a simultaneous pulse with two TUs per line must be utilized. Each radial line with cycle *T* must also have two TUs, while the radials with cycles of 0.5*T* need only one TU. The figure shows all these geometric and

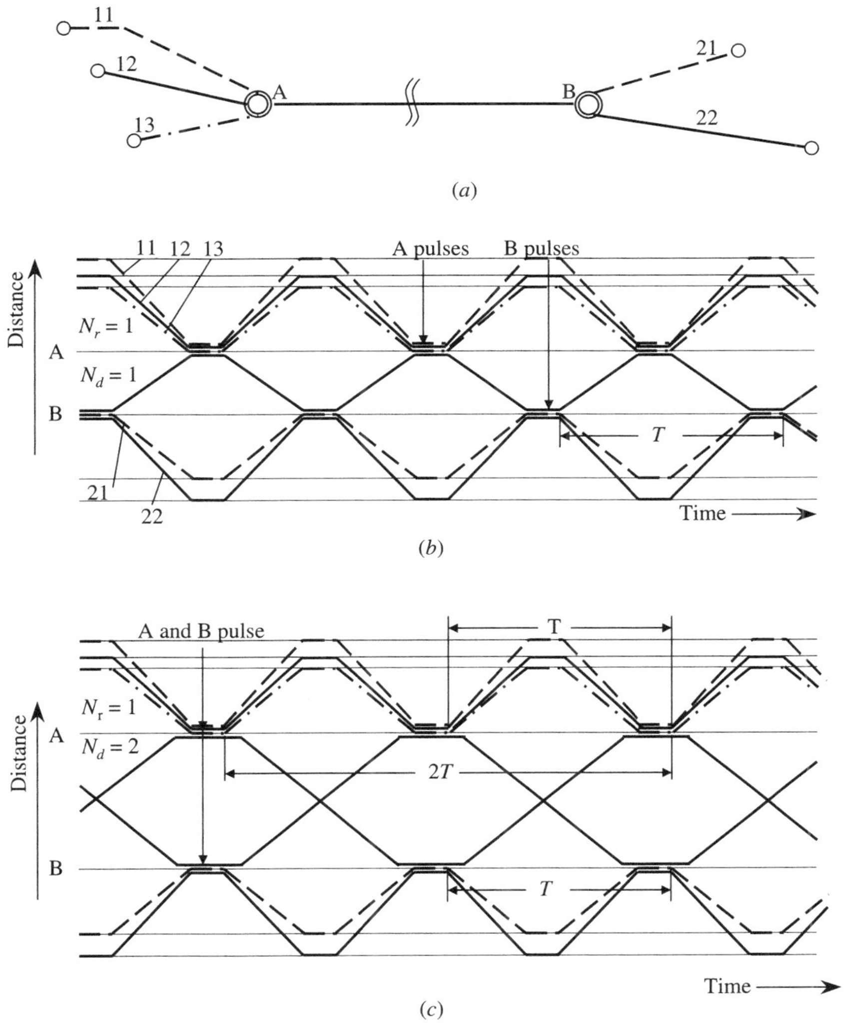

Figure 4.22 *Concepts of staggered and simultaneous pulsing in a bifocal network: (a) layout: connector and radials; (b) staggered pulsing between A and B focal points; (c) simultaneous pulsing at A and B focal points.*

operational relationships with pictorial presentation of buses (TUs) at a given moment of time.

Multifocal TTS networks are effective in large areas that have many subcenters and multidirectional travel. Independent transit lines in such situations often cannot provide extensive area coverage with good connectivity. They can effectively serve only trips along individual lines and, exceptionally, with one convenient transfer between the lines. When such services can be replaced by a well-designated TTS network, the city or suburban area obtains an area-wide integrated transit service. Such service provides convenient and fast network-wide travel, which attracts much greater ridership and represents a good alternative to driving.

Planning of a multifocal TTS should consist of the following major steps:

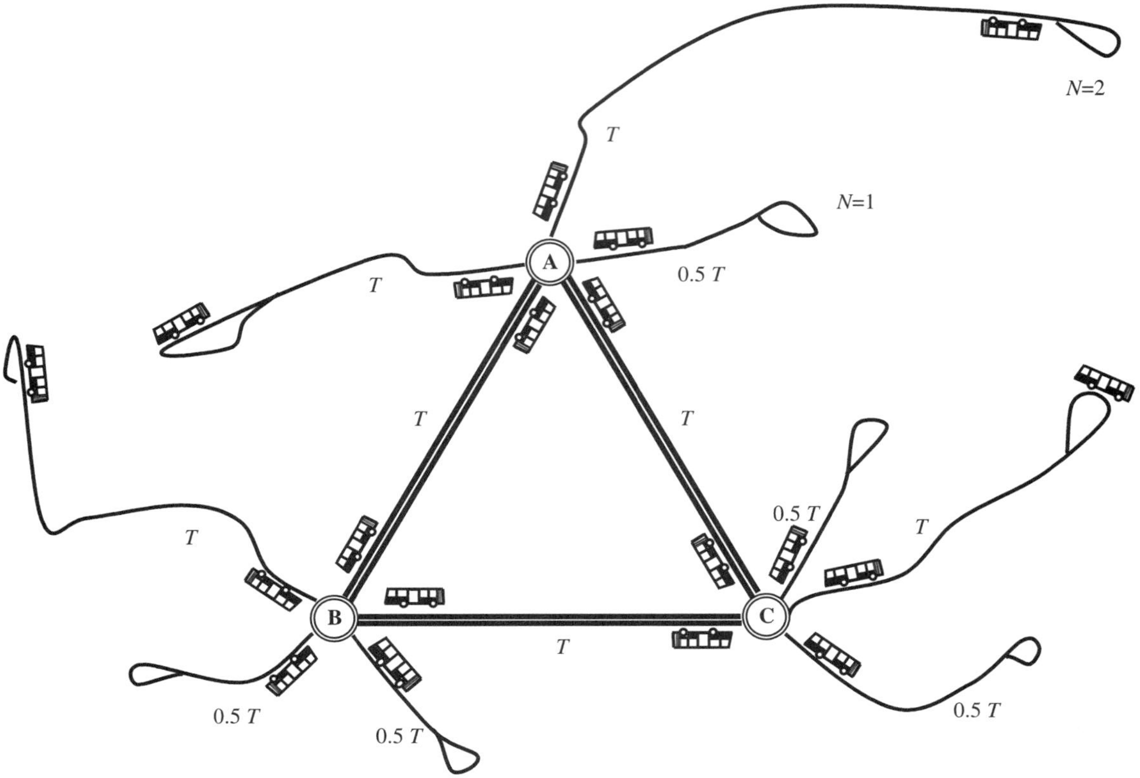

Figure 4.23 *Triangular TTS network with simultaneous pulsing.*

1. Define service area and collect information about its transit services and street network
2. Develop projected travel demand on all major sections of the street and transit line network.
3. Select a large number of potential locations for transit centers, such as centers of suburban towns, railway stations, university campuses, shopping centers, or other major activity centers.
4. Considering the potential travel demand and cost of providing the service, use the procedure shown in Figure 4.20 to decide whether the basic pulse headway of 15 or 20 minutes should be used. Check whether the selected pulse affects the demand projections in step 2, and revise steps 2 and 3 if necessary.
5. Estimate transit operating speeds on the streets where TUs will operate, then compute their travel times by sections between potential transit centers. Adding minimum or desirable terminal times, determine approximate distances between transit centers L that fit the schedule with the adopted pulse headway h_p with given operating conditions on the possible transit lines between potential transit centers. The expression for L as a function of h_p, V_o, N, and t_t is obtained when in the equation for cycle time consisting of operating and terminal times T is substituted by $h \cdot N$, and T_o by $60L/V_o$:

$$T = 2T_o + \sum t_t \qquad \left| \frac{T, t}{\min} \right| \qquad (1.22)$$

$$h \cdot N = \frac{120L}{V_o} + \sum t_t$$

$$\left|\frac{h}{\text{min/TU}}\right|\frac{t}{\text{min}}\left|\frac{N}{\text{TU}}\right|\frac{L}{\text{km}}\left|\frac{V}{\text{km/h}}\right|$$

This equation is now solved for L:

$$L = \frac{V_o}{120}\left(h \cdot N - \sum t_t\right)$$

$$\left|\frac{L}{\text{km}}\right|\frac{V}{\text{km/h}}\left|\frac{h}{\text{min/TU}}\right|\frac{N}{\text{TU}}\left|\frac{t}{\text{min}}\right| \quad (4.11)$$

For example, if operating speed of buses on a major arterial in a suburban area is 22 km/h and terminal times are 3 minutes at each transit center, then for a line that would operate at $h_p = 30$ minutes with one bus, maximum distance between transit centers would be:

$$L = \frac{12}{120}[30 \cdot 1 - (3 + 3)] = 4.4 \text{ km}$$

If two buses are used on this line, it could be operated with headways of 15 minutes ($h = 15$, $N = 2$). Or, if the headway of 30 minutes is retained, the line could be extended to a length of $L = 9.9$ km. Alternatively, if a line with the same operating speed and terminal times were operated with $h_p = 20$ minutes, its maximum length would be $L = 2.6$ km for operation with one bus, $L = 6.2$ km if two buses, or $L = 9.9$ km if three buses are used.

6. Select from the large number of potential transit center locations a set that are approximately at distances of the computed L or $2L$ apart from each other.
7. Design the network consisting of lines connecting the selected transit centers or radiating from them. Schedule services with selected headways h_p or its multiples, with $N = 1$, 2, or more TUs.

Multifocal networks serve more than one or two points and activity centers. They provide a network-wide service, which has a much stronger passenger attraction potential than independently designed and scheduled transit lines. As shown, however, they require more sophisticated area-wide transit network planning and well-organized and supervised operations to ensure the reliability and strong positive image of the service.

If a large area, sometimes most of a metropolitan area, is served by a TTS network, there may be a number of transit centers that are specially designed not only for transit transfers, but also with various stores and conveniences. Designing and scheduling of the network serving these centers as focal points basically uses the same procedure as described above for the unifocal and multifocal networks.

4.5.3 Graphical Presentation of Synchronized Schedules

Synchronization of schedules with clock-type headways at any transit center can be conveniently shown by a clock-type diagram: arrivals and departures of individual TUs for each line are shown on a separate concentric circle. With shaded areas showing dwell times for each line, this diagram presents a clear view of their time overlaps that allow passenger transfers.

Two such diagrams, developed for a terminal in a Philadelphia suburb served by two streetcar lines (11 and 13) and six bus lines, are illustrated in Figure 4.24. The two diagrams show schedules of all lines at a transit center prior to and after the introduction of a TTS operation. All transferring times can easily be seen. For example, prior to synchronization (diagram (*a*)), passengers could transfer from line 76 to 74 with only a 4-minute delay (arrivals on 76 at 26 and 56 minutes in each hour, leaving on 74 at 30 and 60 minutes). However, transferring from line 74 to line 76 would take 23 minutes (arrivals at 17 and 47 minutes, departures at 40 and 10 minutes after the hour).

With the TTS-type schedule (diagram (*b*)), transfers among lines 46 through 83 can be made in both direc-

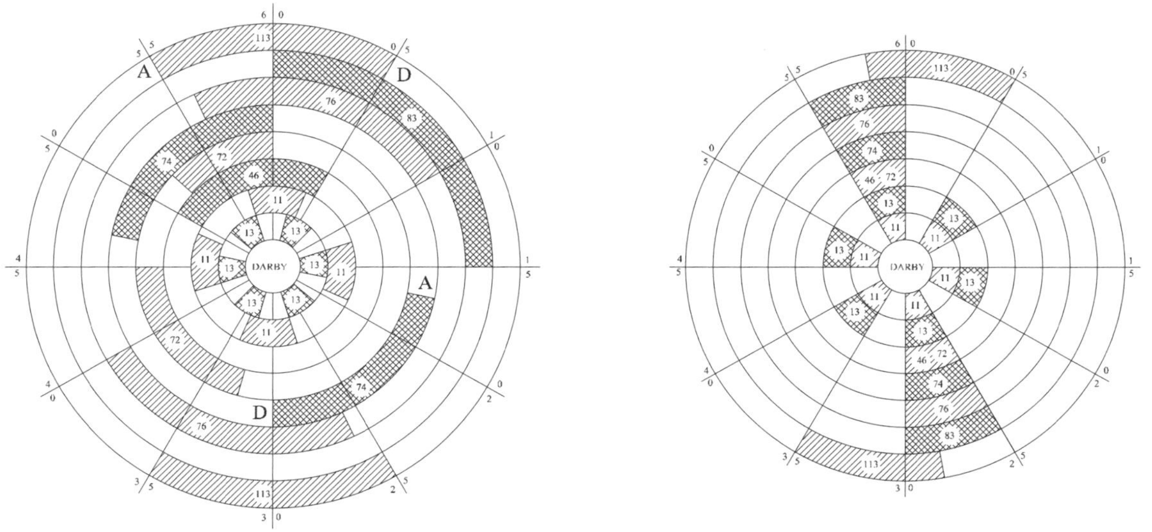

(*a*) (*b*)

Figure 4.24 *Clock-type diagrams showing two types of line schedules at a transit center:* (*a*) *independent line schedules;* (*b*) *TTS schedule.*

tions with only a 5-minute delay. Lines 11 and 13 operate with three times greater frequency (10- as compared to 30-minute headways), so that every third TU on these lines meets with the pulse of all other lines. Line 113, due to other constraints, slightly misses the pulse, but still has an overlap allowing reasonably good transfers. Lines 46 and 72 are shown in diagram (*b*) in the same slot: this can be done to simplify the diagram, if desired.

Rescheduling from independent line operations to a TTS-based schedule may require a greater number of TUs (usually on the order of 10–15%) because in coordinating cycle times for all the lines some cycle times have to be lengthened. Therefore, TTS may involve higher operating costs. However, as mentioned above, TTS provides not only much better service, allowing network-wide travel, but it also changes the image of transit services and generates greater ridership and revenue.

Finally, intermodal TTS centers are often organized at stations of regional rail lines. Feeder buses are scheduled to arrive a few minutes before the train and leave a few minutes after its departure from the station. TTS is not used for metro lines that operate with short headways because it is not necessary and would cause uneven loadings of trains.

4.6 TRANSIT NETWORK TYPES AND THEIR CHARACTERISTICS

Due to the great number of factors influencing network design and differences among local geographic, economic, and cultural conditions, geometric forms of urban transit networks vary greatly. However, general characteristics of networks, such as their geometric forms, relationship to the city, and types of component lines, are known. Also, a number of general service and operational features of network elements, particularly contact points between lines, can be defined and then used in network planning and analysis.

4.6.1 Networks of Modes with Different Right-of-Way Categories

Networks of transit lines utilizing ROW category C (street transit) differ considerably from networks with

ROW category A (rapid transit or metro), for three major reasons:

1. Street transit lines must follow urban streets; metro, although usually following major arterial streets, can also use alignments independent of the street pattern.
2. With street transit, area coverage and passenger collection/distribution are usually important functions, so that its network is denser and has closer stops than metro systems. That limits the ability of street transit to attract long trips. Metro networks must primarily provide high line-haul speed, capacity, and service quality, which require longer station spacings. This makes metro systems less attractive for short trips, but more attractive for medium and long trips.
3. Driving by sight makes street transit operation possible even with high vehicle frequency, but as headways become shorter, operating speed, reliability, and safety decrease. Rapid transit, operating with signal control, always maintains high speed, reliability, and safety, but usually it cannot exceed a frequency of 30–40 TU/h. To maintain maximum reliability of operations, some networks have been designed with independent lines only, as described in Section 4.2.4. Complex integrated networks, which have converging and diverging lines, trunks, and branches, can also be operated with short headways, but they require very precise schedule maintenance and on-line control of train movements, particularly prior to converging points.

Consequently, street transit networks follow street pattern and have higher density and more integrated lines than metro; metro offers much higher service quality on fewer lines, but with convenient transfers in stations.

Networks of lines with ROW category B, semirapid transit such as LRT, may have even greater variety of characteristics, covering their broad range between the features of street and metro modes. There are cases where unsignalized streetcar and even signalized LRT systems operate five or more converging lines where joint headways on the trunk are as short as 45 s/TU (Boston and Philadelphia during the 1920s, Essen), although with rather low operating speeds. On the other end of the spectrum, there are LRT systems operated with two- to four-car TUs on signalized tracks that have headways of 1.5 to 2.0 minutes and high speed, safety, and reliability on their trunk (Frankfurt, San Francisco).

Cities with extensive metro networks use street transit modes (often referred to as surface lines) mainly as supplementary or feeder lines. In suburban areas, feeder lines converge from a large area to metro or regional rail stations as focal points. Suburban feeders may have a form of radials, loops, or connectors to other metro stations. This type of intermodal network integration into a unified transit system for an area is shown schematically in Figure 4.25. Although street transit and metro lines together compose a total transit system network, metro has a particularly strong image because of its physical identity, simplicity, and high service quality. Similar to the advantages of freeways over streets, metro is distinct and has a much higher passenger attraction than street transit. For this reason, most cities publish separate metro maps in addition to standard transit system maps.

Another factor contributing to the strong image of metro systems is the distinct ROW with guideway and, especially, rail technology. That feature also gives LRT lines a significant advantage over buses, even when both operate on the same ROW category. Although not tested, indications are that this looseness and weak image of buses might be somewhat alleviated by striping a continuous distinct pavement line along bus routes. Passengers place considerable value on permanent, fixed, high-quality service because they can easily memorize its layout and rely on its presence. Due to this permanence, careful planning of metro and LRT networks is particularly important.

4.6.2 Rail Transit Network Types and Their Characteristics

The basic geometric forms of individual lines and their integration create networks that can be classified in several different geometric/functional categories.

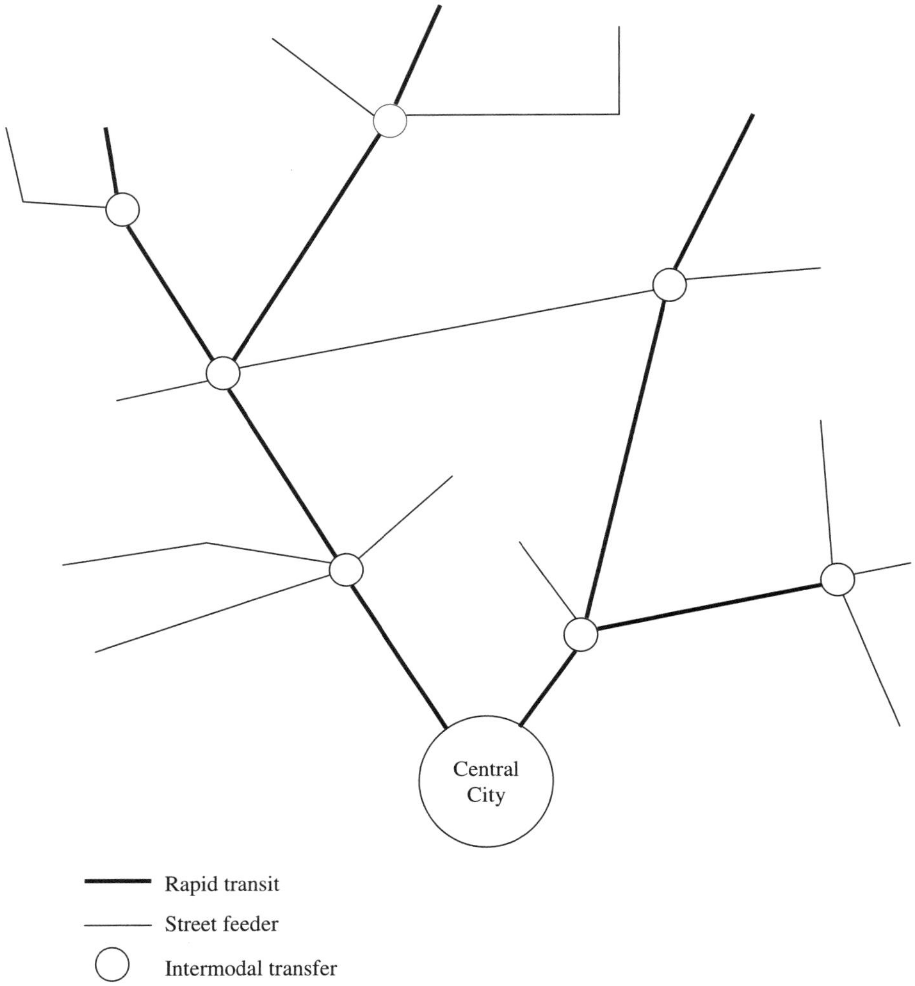

Figure 4.25 *Integrated suburban network: rapid transit with surface feeders.*

These categories and their characteristics are described here.

4.6.2.1 Radial Networks. Some rail transit networks consist mostly or entirely of radial and diametrical metro, LRT, or regional rail lines converging on one point or a small area in central city. Such networks exist in the cities built with a strong emphasis on the center. For example, Philadelphia has two diametrical metro lines intersecting under City Hall, five LRT/streetcar lines starting at the same point as radials, and seven diametrical regional rail lines passing through the same area.

Characteristics of radial transit networks are summarized in the box.

Most common radial networks are found on regional rail systems because they use railroad lines that typically radiate from one or more center city stations into several directions. A typical radial network is the Chicago Regional Rail system, Metra (Figure 4.26),

Purely radial networks compared with other network types have the following main positive (+) and negative (−) characteristics:

+ Lines follow major travel directions, and thus maximum ridership per kilometer of constructed line is achieved.
+ Most of the served trips require no transfer.
− They promote extreme concentration of activities in a small CBD area.
− They do not attract most non-CBD oriented trips, which would require indirect (V-shaped) travel via CBD with a transfer.
− Because of the high density of lines in a small CBD area station, sheds in that area may overlap, reducing total area coverage.

whose lines branch out of four center city stations. Regional rail systems in New York also radiate from two stations in Manhattan, one in Hoboken, New Jersey, and one in Brooklyn, while London has about a dozen railway stations with radial lines. Paris, Munich (Figure 4.27), Hamburg, and Oslo also had radial regional rail networks, but in recent decades they built new connections and lines so that now their networks are more diversified.

Many rapid transit networks, such as the San Francisco BART (Figure 1.33), Chicago (Figure 4.28), and Stockholm, are also basically radial. Similarly, numerous LRT networks have radial lines that carry the heaviest loads. Good examples are the San Francisco Muni-Metro and Boston Green Lines. In Philadelphia, five LRT Green Lines terminate at City Hall in center city. Two main rapid transit diametrical lines also cross this point, as do seven diametrical regional rail lines, created by construction of a connection between two formerly stub-end railroad stations. Thus, as Figure 4.29 shows, rail lines from 23 directions converge to an area of less than one km^2 around City Hall.

Convergence of many radial lines on a small area in the CBD of a major city, as well as high density of local trips in that area, are sometimes served by a special loop or similar shuttle transit line with high-frequency service to a number of stations. The best example of such networks is the Chicago elevated rapid transit Loop (see Figure 4.28), famous for its convenience, visibility, and symbolism of the center city area: buildings "inside the Loop" enjoy special prestige.

In Miami, the Metrorail line has two stations on the edge of the CBD where transfers are provided to an aerial Downtown People Mover (Figure 4.30), a fully automated guided transit (AGT) system consisting of three lines with a total of 15 stations, serving a fairly large center city area with frequent service. Detroit also has a Downtown People Mover, a rail AGT system, but its patronage is smaller because the city does not have a high-capacity metro or light rail system that would bring large passenger volumes that require distribution.

4.6.2.2 Radial/Circumferential Networks. In many large cities, transit networks consist of radial/diametrical lines as well as either a circle, circumferential or tangential lines (see Section 4.3.2). The latter types of lines usually intersect the radials at right angles and have transfer stations with them, forming an integrated network. Because circle and circumferential lines typically serve busy corridors and many commercial, educational, medical, and other subcenters in the areas around the central city, networks with these lines have more even loads throughout their networks as well as throughout daily hours than the more commuter-oriented radial lines only. Intersecting of lines creates contact points among them and allows the serving of a greater diversity of origins-destination pairs, particularly non-CBD-oriented ones.

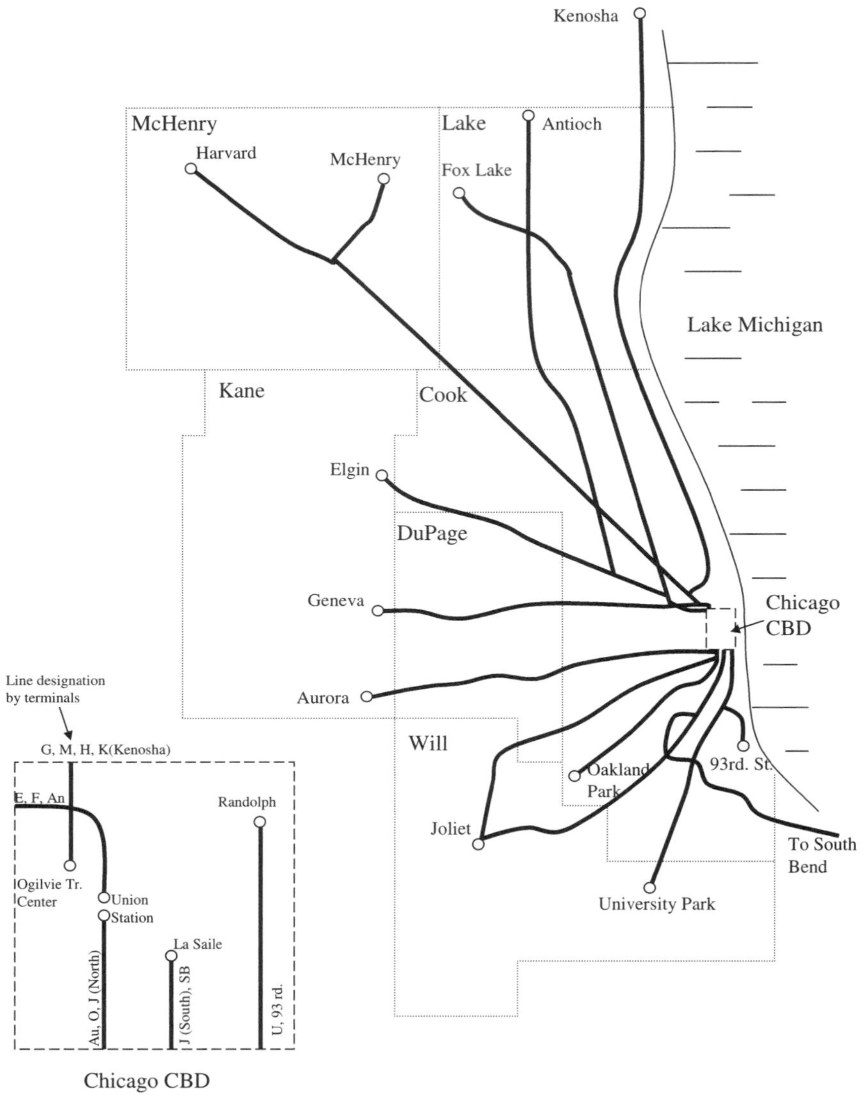

Figure 4.26 *Typical radial regional rail network: Metra in Chicago.*

The nonradial lines also serve for more direct collection/distribution of trips between these centers and radial lines than if they are routed through center city. This multiple function of circumferential or crosstown lines, their more diversified geometry, and greater area coverage result in attraction of substantially greater patronage than the networks consisting of radial lines only.

Radial/circumferential networks are found on metro and LRT systems in many medium and large cities. The London Circle Line connects most railway stations around center city, as well as all other, mostly diametrical lines of the Underground. The Moscow Metropolitan network (Figure 4.11) has a typical radial/circumferential geometry, including a circle line, and on the Seoul Subway System (Figure 4.13)

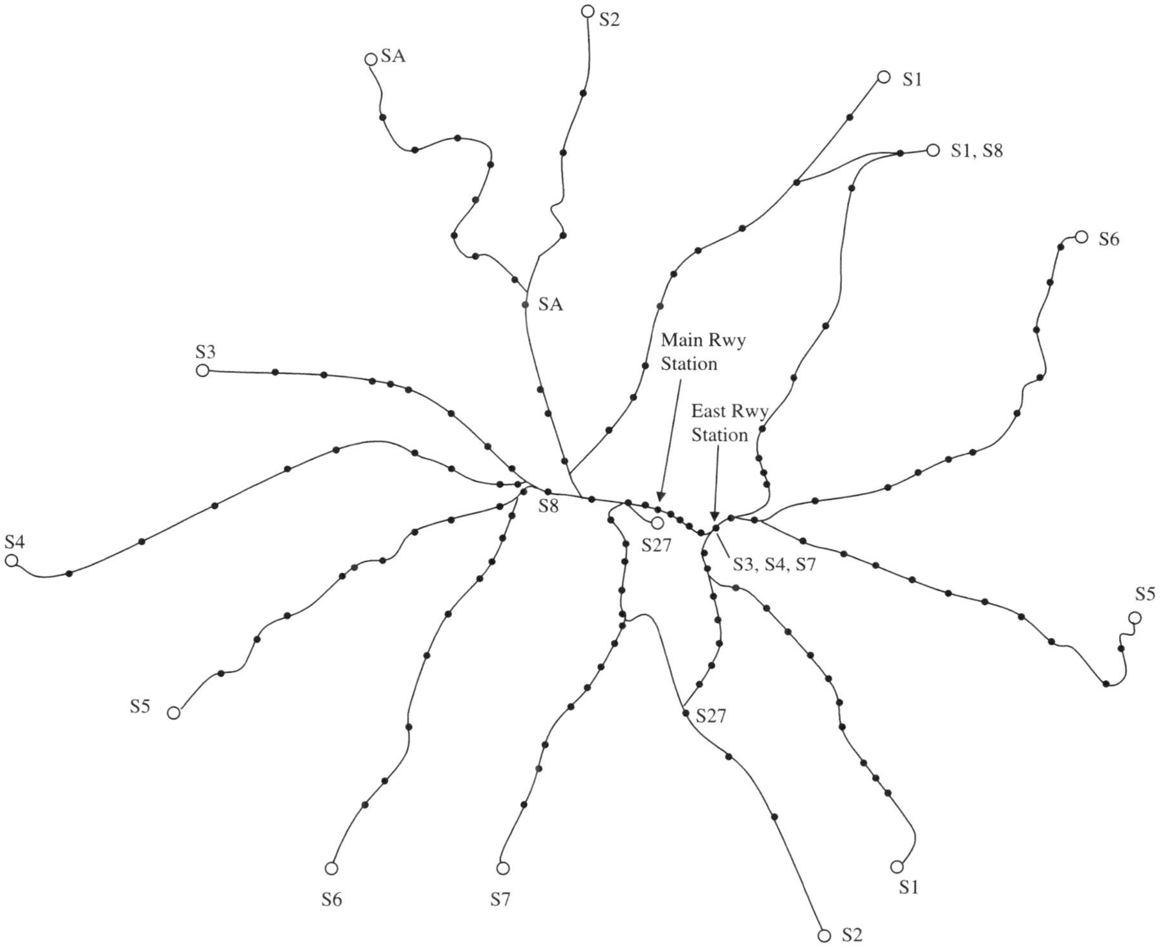

Figure 4.27 *The Munich S-Bahn (regional rail) network.*

the circle line represents the main carrier of the entire network.

The radial/circumferential networks are even more common among LRT systems because they are more extensive than metro networks. Cologne's LRT (Köln Stadtbahn) network, shown in Figure 4.31, and Vienna's tramway network have such patterns. Although the basic orientation of these networks is also radial, they offer much better area coverage and serve more diversified trips than the purely radial networks.

Although regional rail networks are usually radial, there are significant exceptions. The S-Bahn in Berlin has a circle and diametrical lines. The most interesting case, however, is the regional rail system in Tokyo. Carrying over 10 million passengers per day, it is the largest urban rail transit system in the world. Its Yamanote Regional Rail circle line delineates the areas served by the subway (metro) network from the outlying areas served primarily by the Japanese Railway Companies and private regional rail lines. The Yamanote line serves a great number of transfer stations with other JR lines, TRTA Subway lines, and a number of private lines, such as Shibuya, Shinjuku, Ikebukuro, Ueno, and Tokyo-Central, all of which are among the

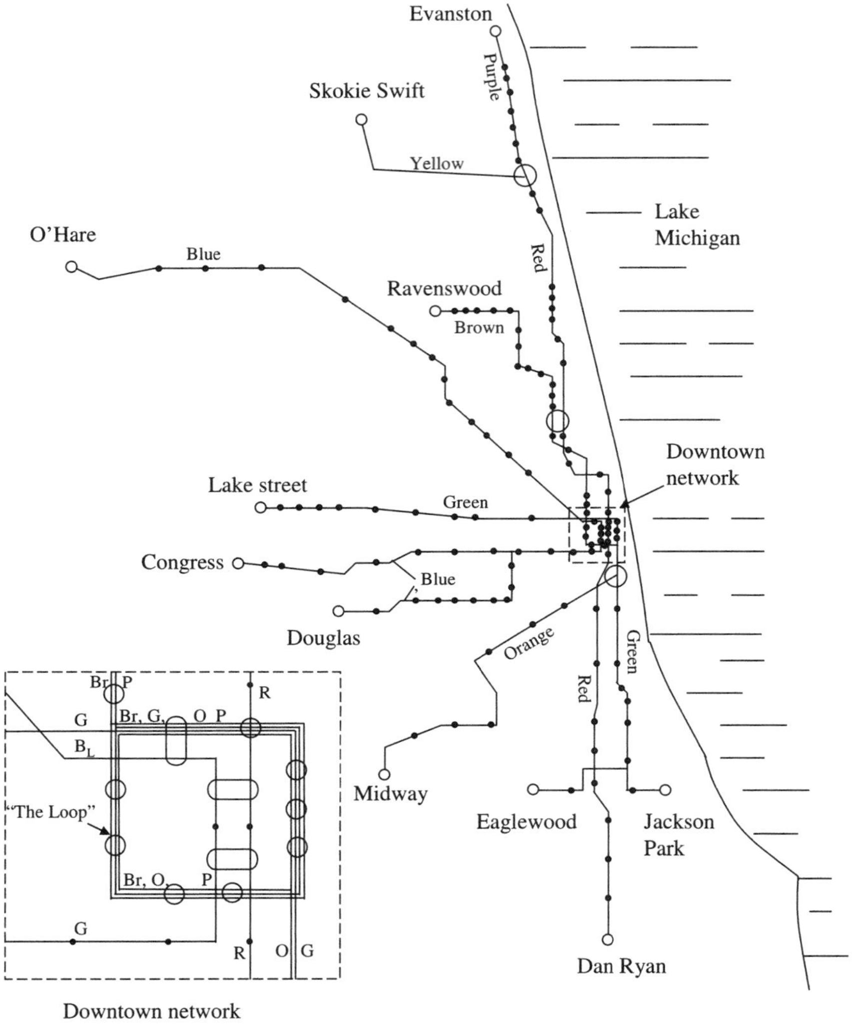

Figure 4.28 *A radial rapid transit network with CBD distribution loop: Chicago.*

most patronized transportation facilities in the world. The line itself has been a powerful element in shaping the city and forming its multipolarization from the opening of its first section in 1885 to its completion as a circle line in 1925. In addition to the above-mentioned transfer stations with other rail lines, Yamanote stations Shinagawa, Meguro, Ebisu, and others are major centers of office and commercial buildings and activities (Tiry, 1997).

4.6.2.3 Rectangular or Grid Networks. Some metro networks, such as Barcelona, Mexico City (Figure 4.32), New York/Manhattan, Osaka and Toronto, consist basically of lines following a rectangular geometric pattern. Some of the networks consist of two sets of parallel lines perpendicular to each other, while others also have L- or U-shaped lines (discussed in Section 4.3.1). Rectangular networks provide uniform area coverage, so that they are suited to the cities that

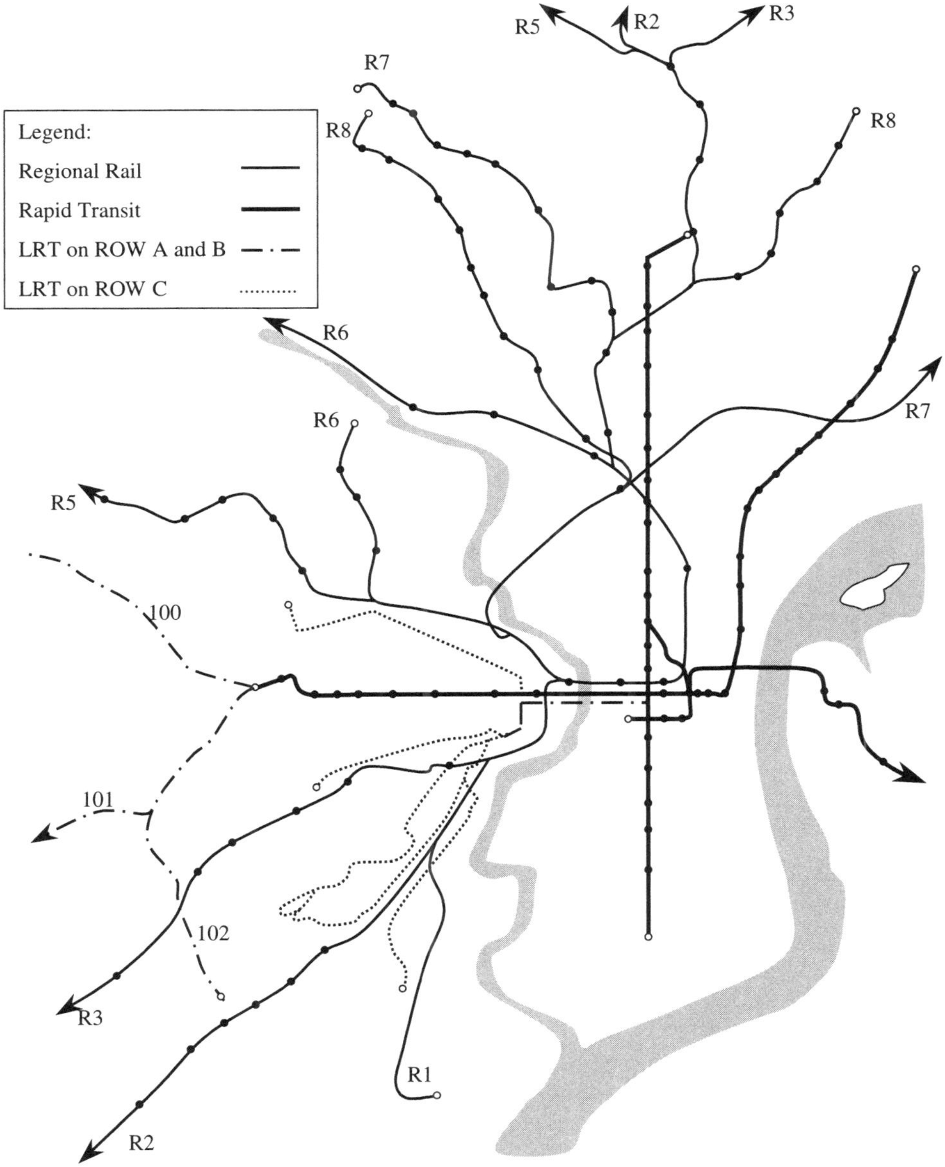

Figure 4.29 Light rail, rapid transit, and regional rail network in Philadelphia.

have a large area with relatively uniform density of activities. They do not fit well travel patterns that are predominantly oriented toward a small, highly concentrated central business district.

Rectangular networks rely heavily on passenger transfers at rectangular stations between metro lines or metro lines and feeders. Many trips require transferring, but very few passengers need to make more than

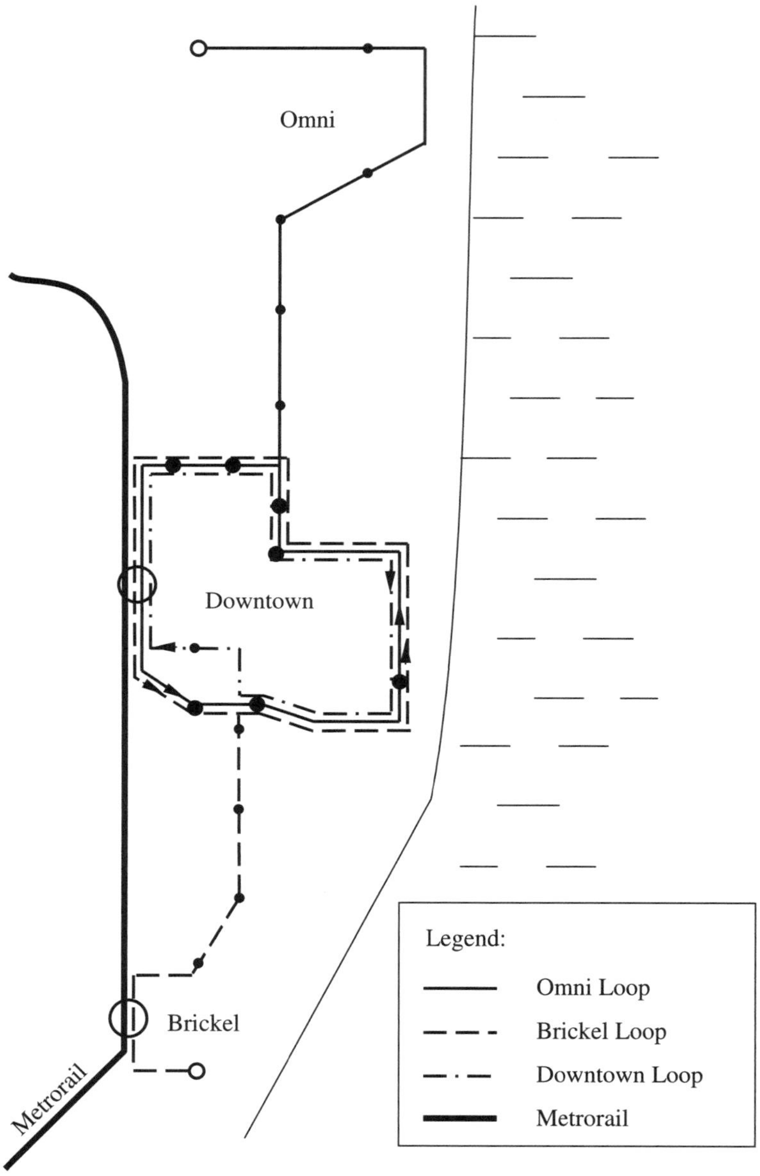

Figure 4.30 *Miami Downtown People Mover as a feeder to Metrorail.*

one transfer. Travel circuity is never excessive: at worst it is rectangular with respect to the hypotenuse of origin-destination lines. If the network consists of straight lines, travel between parallel lines is inconvenient because it requires two transfers. An example of this is the rectangle of rail lines in the Boston CBD, in which there is no direct transfer between the Blue and Red lines. This problem may be solved by con-

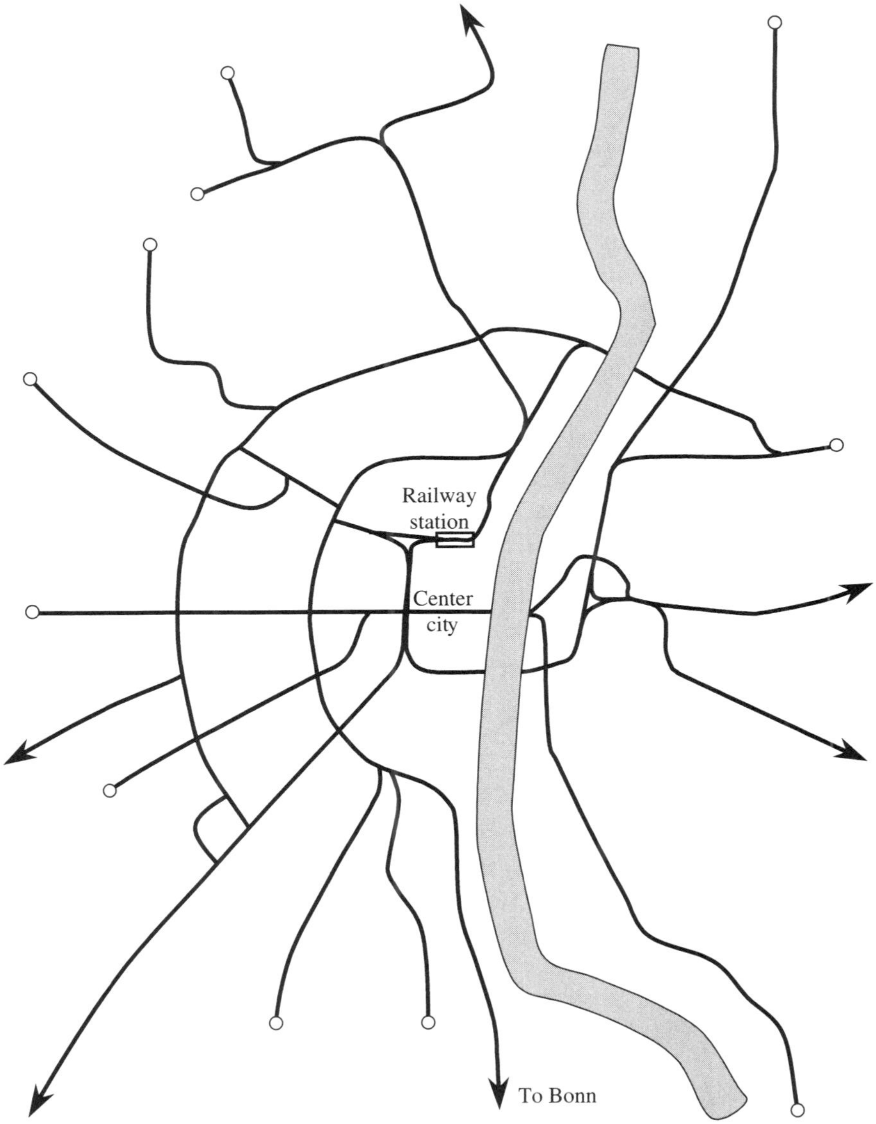

Figure 4.31 *Cologne (Köln) LRT: a radial-circumferential LRT network.*

struction of L-shaped lines that provide intersections among different perpendicular and parallel lines and thus increase connectivity. This is illustrated on the Tokyo Subway network.

U-shaped lines are diametrical by function, but they have a major disadvantage compared to other diametrical lines in a network in that they do not serve trips between suburbs through center city. It serves two relatively close suburbs by an excessively circuitous routing through center city. Examples of such lines are the Toronto Yonge-Spadina Line and Washington Red Line (see Figure 4.38).

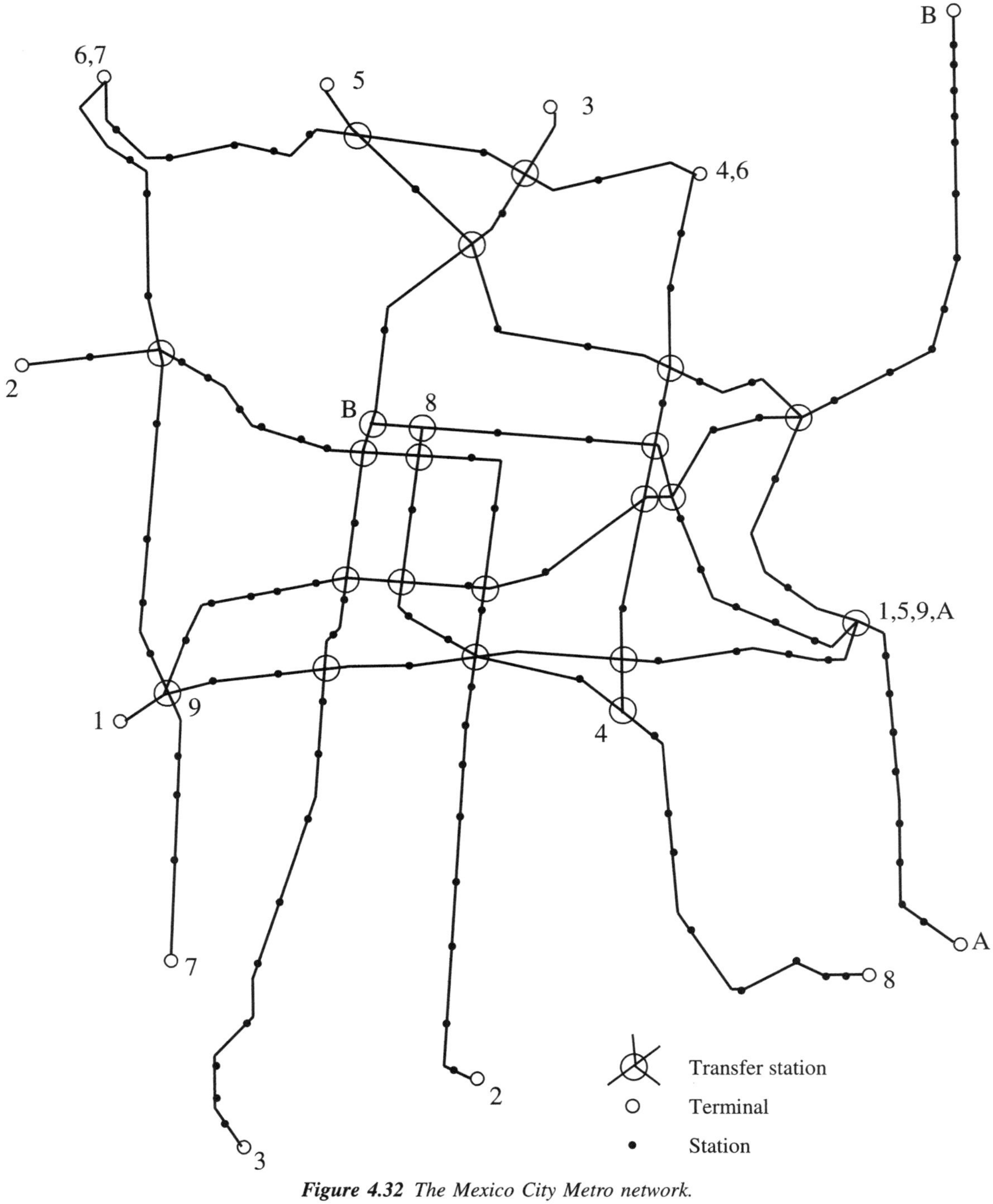

Figure 4.32 *The Mexico City Metro network.*

In some North American cities, which typically have grid street patterns, street transit lines also follow that geometric form. For example, Toronto has Yonge and Bloor Street subway lines crossed by many streetcar and bus lines feeding many of their stations. The well-designed transfer stations and extreme geometric simplicity of the network contribute to the simplicity and convenience of use, strong image, and popularity of transit services in that city.

4.6.2.4 Ubiquitous Networks. In contemporary rapid transit planning, there has been a trend to consider that the most advantageous form for many cities is a network providing extensive, ubiquitous coverage for a large central area while extending diametrical lines with branches into suburban areas. Thus, the network usually consists of diametrical lines that pass through the central area but do not focus on one point: their transfer stations are dispersed throughout that area, providing wider coverage than a radial network focusing on a small center city area. Toward suburbs most lines split into two and sometimes more branches.

Ubiquitous networks are usually designed to have the following features with respect to service and operations:

- Service in all heavily traveled corridors and provision of direct connections among major origin-destination pairs
- At least one station close to each major activity area or center
- Adequate coverage throughout the central area and close suburbs
- Different sections of each line reasonably well balanced by passenger volumes
- Good connectivity among lines, requiring not more than one transfer for all major travel directions
- Convenient transfers between lines, as well as to and from other modes
- Reasonably good connections also for non-CBD-oriented trips

It is obvious from these features that transit networks should be planned considering all objectives and requirements listed in Section 4.1. More specifically, in determining the network form and individual lines, the needs of passengers as well as operating aspects must be considered. In some cases, networks have been designed so that the operation proved to be inefficient, or even so that some expected services could not be provided. Some common problems in network design that create inefficiencies in operations may be:

- Radial lines without branches, leading to excessive unused capacity on outer extremities, where passenger volume is much lower than on the maximum load section in center city
- Many branches with different passenger volumes, which cannot give regular headways on the trunk
- Diametrical line with very different passenger loads on the two outlying sections
- Triangular (Y) connections, which are difficult to schedule with regular headways

A carefully designed ubiquitous network can usually avoid these operational problems and can incorporate most major advantages of both radial and grid networks.

The most typical example of a ubiquitous metro network is the Paris Metro (Figure 4.3), generally recognized as the most complete metro network of any city in the world. It covers the entire urban area well: no point within central city is more than 500 m from a metro station! Munich, Madrid, and, in some areas, New York also have many elements of the ubiquitous network pattern.

The discussed characteristics of network geometric forms are important to consider in network planning and design, but in many cases other factors, such as topography and existing street and transit facilities, have overriding importance. In such cases, no regular geometric form of network can be discerned. *Irregular networks* exist in a number of cities, such as Hong Kong, London, and Montreal.

4.6.3 Review of Transit Network Types

The above-discussed rail transit network forms, their characteristics and functional concepts, are summarized in several illustrations. Figure 4.33 shows two pairs of rail networks that have basically similar forms and extensiveness but some differences in individual features. The sketches on the right (ii) show the features that are usually more desirable than those on the left (i).

Case (*a*)-i shows two major trunk radial lines with numerous branches, typical for regional rail networks. In (*a*)-ii, the two sets of radial lines are converted into diametrical lines, which offer better connectivity and distribution in center city, as well as higher operating efficiency.

Cases (*b*)-i and (*b*)-ii depict a radial network with diametrical lines intersecting at one and at several points, respectively; the latter form results in better area coverage and more convenient transfers. A possible additional circle line can improve connectivity of the radial networks in both cases.

Figure 4.34 compares two other features of line relationships and forms. Diagrams (*a*)-i and (*a*)-ii show networks with, respectively, independent lines and integrated lines. The latter type of lines may be considered and referred to as overlapping lines when they have short joint sections, or trunk-branch lines when they divide into relatively short branches toward their outer terminals. Generally, the integrated lines are somewhat more complex in operation, but they offer better connectivity, i.e., fewer origin-destination trips require transfers than in networks consisting of independent lines only.

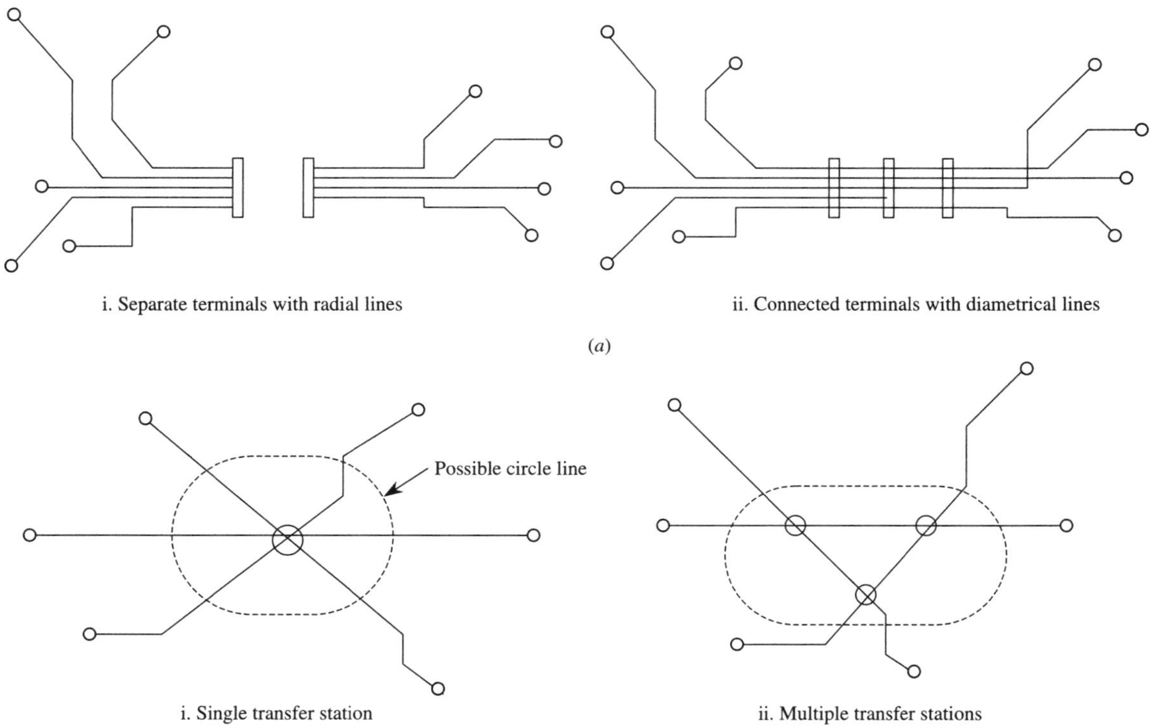

i. Separate terminals with radial lines

ii. Connected terminals with diametrical lines

(*a*)

i. Single transfer station

ii. Multiple transfer stations

(*b*)

Figure 4.33 *Radial rail transit networks: (a) typical regional rail networks; (b) radial metro networks with different layouts.*

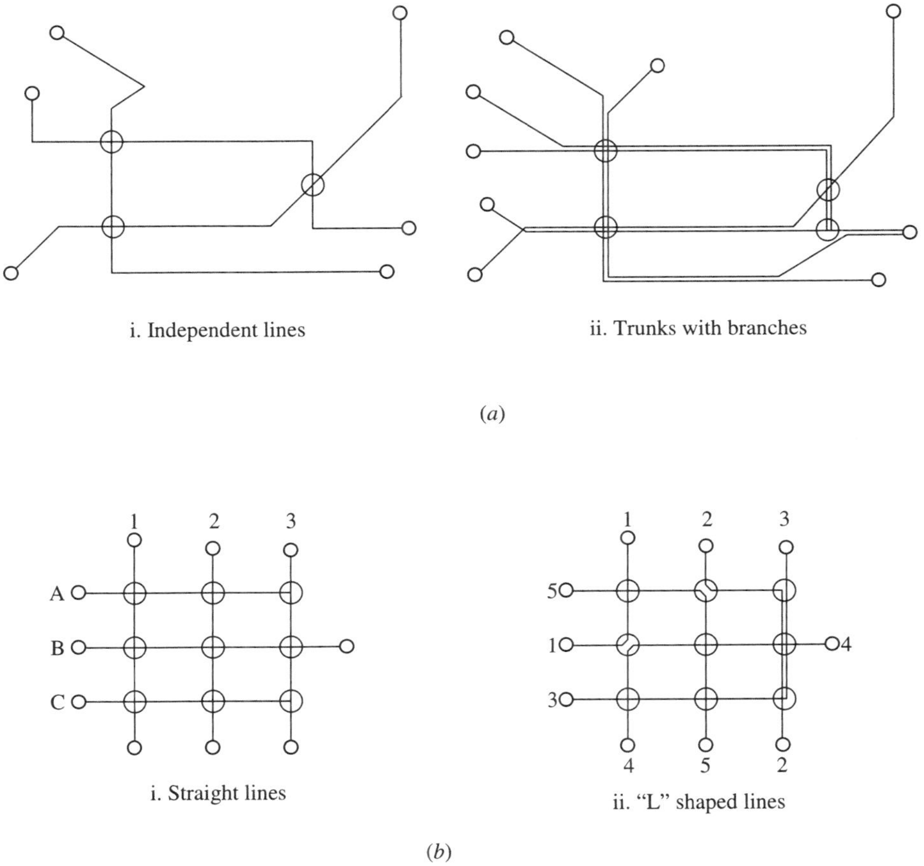

Figure 4.34 *Rail transit line relationships and forms:* (*a*) *network with independent and interconnected lines;* (*b*) *rectangular pattern rail transit networks.*

Case (*b*) illustrates two rectangular networks, one with two sets of parallel lines only, the other with some L-shaped lines. The latter offers more diverse alignments and interline transfers. For many diagonal trips across the city, it provides travel on a single line, i.e., without transfers. The L lines instead of parallel ones avoid the problem of requiring two transfers for travel on parallel lines, such as exists in Boston (Red and Blue Lines) and Mexico City (Lines 7, 3, and 4).

Figure 4.35 represents a network incorporating most of the desirable geometric/functional features of metro lines and networks. Naturally, in some cases some of these forms may not be optimal because of special local conditions. But these general descriptions of characteristics of different forms are useful for consideration in the planning and design of metro networks. The shown line and network forms include the following features:

- Diametrical lines across center city: straight, line 1; L, 2, 5–6, and 7–8, but not U-shaped
- Two overlapping lines dividing into branches in suburbs: 1–2, 3–4, 5–6, and 7–8

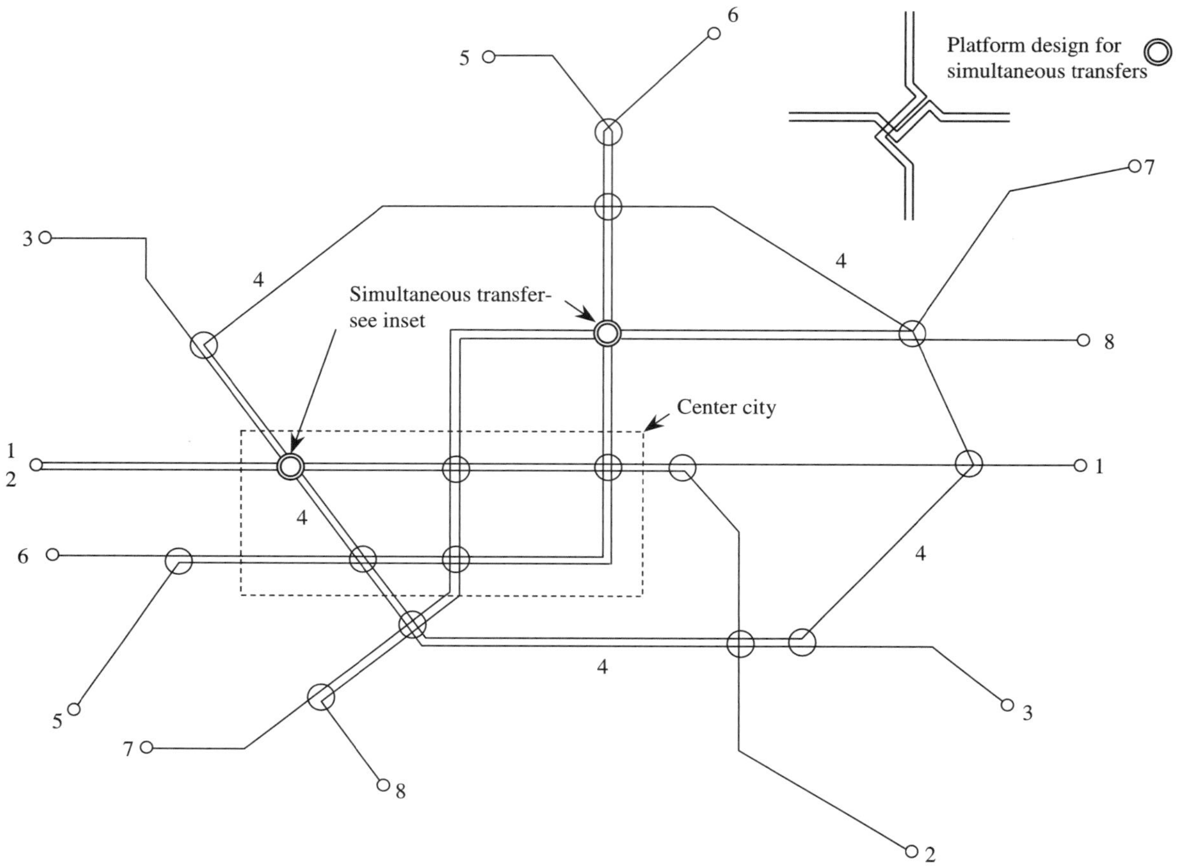

Figure 4.35 *Area-wide metro network of lines with operational efficiency.*

- A circle line, 4, intersecting all other lines and serving circumferential trips around center city
- Line-intersecting points (transfer stations) distributed throughout a large area
- Two stations with simultaneous across-platform transfers: lines 1–2 and 3–4, 5–6, and 7–8
- Additional transfer stations for transfers between branches in the suburban areas: 1–2 at one side, 3–4, 5–6, 7–8 at both diverging points
- Layout of lines for ubiquitous coverage, particularly in central city

In conclusion, geometric shapes, relationships of lines in a network, and their transfer stations have an important role in providing service attractiveness and efficiency of a metro network. It is therefore necessary that in the design of lines and network, and in its evaluation, all major requirements of the passengers, the transit system operator, and the city be carefully considered to achieve an optimal layout.

4.7 ANALYSIS OF METRO NETWORK GEOMETRIC FORMS

The planning of metro and other rail transit networks is usually performed empirically, without extensive analytical computations. Consideration of local conditions, such as urban form and topographic features, demand characteristics, existence of transportation corridors, requirements for certain station locations, and long distance terminals (bus, rail, and airports), tends to suppress the analyses of network topology and geometric characteristics. However, even when designers of metro lines and networks want to use some design guidelines, perform comparative analyses, or make use of experiences from network forms and operations in other cities, very few of these can be found in the professional literature (Berger, 1951; Rockwell, 1959, 1967; Franz, 1975; Vuchic and Musso, 1991).

4.7.1 Classification of Metro Network Measures and Indicators

Presented in this section is a systematic set of quantitative elements that defines the network characteristics of metro systems that can be used for their description, evaluation, and comparative analysis. Examples of such uses include planning of new or analysis of existing networks, comparison with networks of other cities, and comparison of alternative networks or their extensions.

The quantitative elements are grouped into five general categories:

a. Measures of network size and form
b. Indicators of network topology
c. Measures of relationship between network and city
d. Quantity and quality of offered service
e. Measures of service use

An application of these measures to metro networks in 10 cities, using many of the elements from categories a–c above, is given in Musso and Vuchic (1988) and Vuchic and Musso (1991).

Many different aspects of metro networks can be analyzed. The emphasis here is on the geometric form and the method of operation of transit lines and networks; that is, on categories a–c listed above. Categories d and e are given only as a guideline for analysis focusing on service quality and efficiency of operations.

In this section, theoretical concepts are introduced and used, particularly from graph theory (Garrison and Marble, 1974) with practical relevance to actual metro network planning. Data and experiences from many cities on metro network design, service, and operating characteristics are utilized.

4.7.2 Network Size and Form

The main elements that express the size and form of a metro network are defined here. Corresponding terms from graph theory are given in parentheses. Values for lines are designated by lower-case, for network by upper-case letters; the symbols in Eqs. (4.12)–(4.19) are nondimensional.

a-1 Number of stations (nodes) on a line i, n_i; this includes the two terminal stations.

a-2 Number of interstation spacings (arcs) on a line i, a_i, is related to n as follows:

$$a_i = n_i - 1 \qquad (4.12)$$

a-3 Length of line i, l_i.

a-4 Number of multiple stations (jointly used by two or more lines), $n_m{}^k$, in which k designates the number of lines using a station or a spacing. This superscript must be introduced to avoid double counting in computations of the number of stations, spacings, and origin-destination (OD) pairs. For these computations the number of multiple stations $n_m{}^k$ must be computed as one less than the number of lines it serves (i.e., as $k - 1$). This will be illustrated on cases following *a*-6, *a*-7, and *a*-8.

a-5 Number of lines in a network q.

a-6 Number of stations in a network N is computed as the sum of stations on individual lines minus the multiple stations as follows:

$$N = \sum_{i=1}^{q} n_i - \sum_{k=2}^{k_{\max}} (k - 1) \cdot n_m^k \qquad (4.13)$$

For example, the network in Figure 4.36 consists of three lines with eight, nine, and four stations, respectively. There are four stations served by two lines (three along the trunk and one terminal on the east side) and two stations served by three lines:

$$n_m^{\ 2} = 4 \quad \text{and} \quad n_m^{\ 3} = 2$$

Thus, the total number of stations (single and multiple) in the network is, by Eq. (4.13):

$$\begin{aligned} N &= 8 + 9 + 4 - (1 \times 4 + 2 \times 2) \\ &= 13 \text{ stations} \end{aligned}$$

These N stations include terminal stations and transfer stations.

a-7 Number of interstation spacings in a network A is computed as the sum of spacings on individual lines a_i, minus the multiple spacings, using an equation corresponding to Eq. (4.13):

$$A = \sum_{i=1}^{q} a_i - \sum_{k=2}^{k_{\max}} (k - 1) \cdot a_m^k \qquad (4.14)$$

In the case illustrated in Figure 4.36, the number of spacings (including single and multiple) computed by line and subtracting the duplications, by Eq. (4.14), is:

$$A = 7 + 8 + 3 - (1 \times 3 + 2 \times 1) = 13 \text{ spacings}$$

a-8 Length of network L is the sum of line lengths l_i minus lengths of multiple (duplicative) spacings l_m:

$$L = \sum_{i=1}^{q} l_i - \sum_{k=2}^{k_{\max}} (k - 1) \cdot l_m^k \qquad (4.15)$$

Using the lengths of spacings from the example in Figure 4.36, network length is:

$$\begin{aligned} L &= 12 + 14 + 5 \\ &\quad - (1 \times 1 + 1 \times 1 + 1 \times 2 + 2 \times 2) \\ &= 23 \text{ km} \end{aligned}$$

a-9 Number of station-to-station (origin-destination) travel paths OD consists of direct paths on a single line, OD_d, and paths that include one or more transfers between lines, OD_t. In a network with N stations, the number of all possible OD paths is:

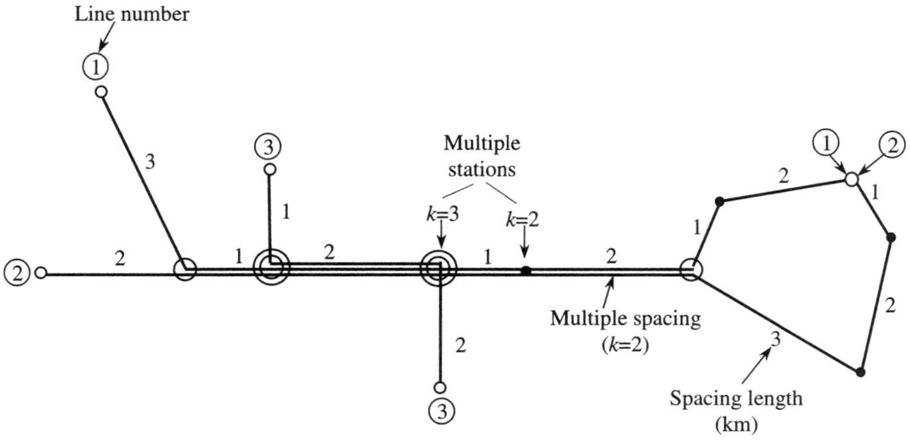

Figure 4.36 *Network schematics: stations and spacings (nodes and arcs).*

Table 4.5 Metro network size and form measures and indicators

Item code	*Definition*	*Symbol*	*Expression*
a-1	Number of stations on line *i*	n_i	
a-2	Number of interstation spacings (arcs) on line *i*	a_i	$n_i - 1$
a-3	Length of line *i*	l_i	
a-4	Number of multiple stations, spacing and their lengths, respectively	$n_m{}^k$, $a_m{}^k$, $l_m{}^k$	
a-5	Number of lines in network	q	
a-6	Number of stations in network	N	$\sum_{i=1}^{q} n_i - \sum_{k=2}^{k_{\max}} (k-1)n_m^k$
a-7	Number of interstation spacings	A	$\sum_{i=1}^{q} a_i - \sum_{k=2}^{k_{\max}} k \cdot a_m^k$
a-8	Length of network	L	$\sum_{i=1}^{q} l_i - \sum_{k=2}^{k_{\max}} (k-1)l_m^k$
a-9	Number of station-to-station travel paths:		
	Total	OD	$\frac{1}{2} N\,(N-1)$
	Direct	OD_d	$\frac{1}{2} \sum_{i=1}^{q} n_i(n_i - 1) - \sum_{j=1}^{q_m-1} n_{mj}(n_{mj} - 1)$
	With transfer	OD_t	$OD - OD_d$

$$OD = \tfrac{1}{2}\, N \cdot (N - 1) \qquad (4.16)$$

For a single line with *i* stations, the number of *OD* paths, all direct, is:

$$OD_i = OD_d = \tfrac{1}{2}\, N_i \cdot (N_i - 1) \qquad (4.17)$$

The total number of direct paths in a multiline network is equal to the sum of *OD* paths along each line minus the paths on each section that overlaps with any other line.

$$OD_d = \frac{1}{2}\left[\sum_{i=1}^{q} n_i(n_i - 1) - \sum_{j=1}^{q_m-1} n_{mj}(n_{mj} - 1)\right] \qquad (4.18)$$

where:

n_{mj} = the number of joint stations on each section that each line *j* shares with any already counted line

q_m = the number of lines that have joint sections with other lines

These additional direct *OD* paths must actually be computed among all line sections that such lines connect. The number of paths that require transfers, OD_t, is computed as the difference between all paths in the network and direct paths that involve no transfers:

$$OD_t = OD - OD_d = \frac{1}{2}\left[N(N-1) - \sum_{i=1}^{q} n_i(n_i - 1)\right] - \sum_{j=1}^{q_m} n_{mj}(n_{mj} = 1) \qquad (4.19)$$

The nine elements expressing size and form of metro network are listed in Table 4.5 with their mathematical expressions.

An illustration of the computations of *OD* values is given in Figure 4.37. The network shown by solid lines consists of two lines with one joint (transfer) station. Line 1 has six stations, and Line 2 has seven. To compute *OD* values, Eqs. (4.13) and (4.16) and the first member of Eq. (4.18) are used:

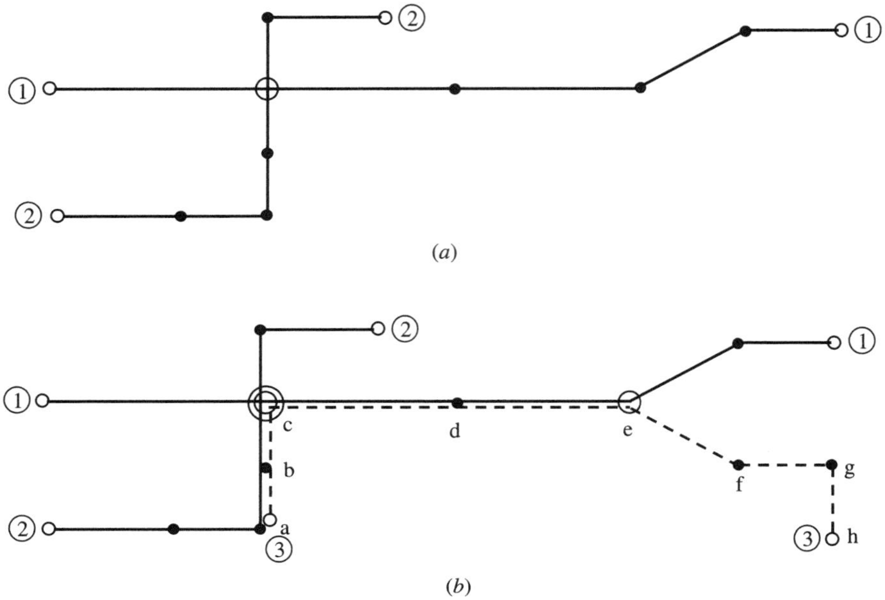

Figure 4.37 *Schematics of lines in a transit network: (a) network of two independent lines; (b) network with overlapping lines.*

$$N = 6 + 7 - 1 \times 1 = 12 \text{ stations}$$

$$OD = \tfrac{1}{2} \times 12\,(12 - 1) = 66 \text{ paths}$$

$$OD_d = \tfrac{1}{2}[6\,(6 - 1) + 7\,(7 - 1)] = 36 \text{ paths}$$

$$OD_t = 66 - 36 = 30 \text{ paths}$$

If line 3 with eight stations, shown by dashes in Figure 4.33, is added to this network, the numbers change as follows:

$$N = 6 + 7 + 8 - (4 \times 1 + 1 \times 2)$$
$$= 15 \text{ stations}$$

$$OD = \tfrac{1}{2} \times 15\,(15 - 1)$$
$$= 105 \text{ paths}$$

The computation of OD_d now becomes more complex. The first sum in Eq. (4.18) is:

$$\frac{1}{2}\sum_{i=1}^{q} n_i(n_i - 1)$$

$$= \frac{1}{2}(6 \times 5 + 7 \times 6 + 8 \times 7) = 64 \text{ paths}$$

The second sum from Eq. (4.18) must include ODs on the two overlapping sections of the three lines (q_m = 3): a-c and c-e, where line 3 overlaps with lines 2 and 1, respectively:

$$\frac{1}{2}\sum_{j=1}^{q_m-1} n_{mj}(n_{mj} - 1) = \frac{1}{2}(3 \times 2 + 3 \times 2) = 6 \text{ paths}$$

so that the number of direct paths is $OD_d = 64 - 6 = 58$ paths, and the number of ODs with transfers is $OD_t = OD - OD_d = 105 - 58 = 47$ paths.

4.7.3 Network Topology

Various ratios of the previously defined measures of network size can be used as quantitative indicators of

Table 4.6 Network topology indicators

b-1	Average interstation spacing	$\overline{S}$	$\frac{L}{A}$	
b-2	Line overlapping	λ	$1 + \frac{\sum_{k=2}^{k_{max}} l_m^k}{L}$	$\lambda \geq 1$
b-3	Network complexity	β	$\frac{A}{N}$	$\beta \geq 0.5$
b-4	Directness of service	δ	$\frac{OD_d}{OD}$	$\frac{2}{A+1} \leq \delta \leq 1$

network topology. Those particularly useful in metro network planning and analysis are selected here.

b-1 Average interstation spacing $\overline{S}$ can be computed for a network as

$$\overline{S} = \frac{L}{\sum_{i=1}^{q}(n_i - 1) - \sum_{k=2}^{k_{max}}(n_m^k - 1)} = \frac{L}{A} \qquad \left|\frac{S, L}{\text{km}}\middle|\frac{A}{-}\right| \tag{4.20}$$

Spacings between stations $\overline{S}$ are usually selected as a compromise between good area coverage (short spacings) and high operating speed (long spacings). Therefore, longer lines tend to have longer spacings: regional rail networks, such as San Francisco BART and the Munich S-Bahn, have average spacings of 1000–2500 m, compared with the spacings of 500–800 m on typical urban metro systems such as those in Paris, Philadelphia, and Mexico.

b-2 Line overlapping index λ is computed as:

$$\lambda = \frac{\sum_{i=1}^{q} l_i}{L} = 1 + \frac{\sum_{k=2}^{k_{max}} l_m^k}{L} \qquad \left|\frac{\lambda}{-}\middle|\frac{l, L}{\text{km}}\right| \tag{4.21}$$

Descriptively, this index is the ratio of the sum of line lengths to the network length. Therefore, a network consisting of independently operated lines (e.g., St. Petersburg, Mexico, or Toronto) has a value of 1.00. The more the lines are interconnected, sharing common sections with other lines and then branching, the greater is the value of λ.

b-3 Network complexity indicator β is the ratio of the number of spacings (arcs) and stations (nodes):

$$\beta = \frac{A}{N}; \quad \beta \geq 0.5 \tag{4.22}$$

This indicator reflects complexity in terms of the number of interstation spacings (arcs) related to the stations (nodes) of the network. Its minimum value of 0.5 is obtained on an elementary line with two stations; as the line is extended, adding more stations and spacings, β asymptotically approaches 1. On closed networks with cross connections, β can exceed the value of 1.

b-4 Directness of service δ is the indicator that reflects the proportion of OD paths that can be traveled without transfer:

$$\delta = \frac{OD_d}{OD_d + OD_t} = \frac{OD_d}{OD}; \; 0 > \delta \geq 1 \tag{4.23}$$

For a single line, $\delta = 1$; for more complex networks, δ tends to decrease, but operation of integrated lines increases it.

For an easy review, these indicators of topology are listed in Table 4.6 with their symbols, equations, and, where applicable, ranges of value.

In Table 4.7, eight different types of networks are given with their sizes and indicator values to illustrate their computations and the relative magnitudes of different network topologies. In all eight cases, network length is the same but other elements are changed one by one in subsequent cases to show how each one of the changes in measure influences different indicators.

Case 1 is a single line with variable interstation spacing lengths and a longer $\overline{S}$ than the others. Case 2 is similar to case 1, a single line, but with 12 equal spacings. The difference in b-1 also causes the changes in b-3. In case 3, the same network length is reorganized into three lines with one joint (transfer) stations. Compared with case 2, of all the indicators only b-4 changes. Case 4 is the same network as case 3, but with lines staggered to divide a single triple station into three double stations. Case 5, with four lines intersecting at four points, changes only b-4. Cases 6 and 7 have the same network topology but different line operations: in case 6 there is one line with a feeder, and in Case 7 there are two lines with a joint trunk. The differences are reflected in indicators λ (b-2) and δ (b-4). Finally, case 8 is the same basic network as case 4 but has line overlaps on the entire network; λ (b-2) is doubled and δ (b-4) is increased, reflecting a greater portion of travel paths without transfers. It should be noted, however, that this increase in direct service is traded for decreased frequency of service on each line. If one branch requires, for example, a service frequency of 12 TUs/h, or headways of 5 minutes, operation from case 4 will offer such headways, whereas case 8 will offer only a 10-minute headway for each line.

These cases provide a clear illustration of various influences of network designs on indicators. In real-world cases much larger differences exist, so that greater variations among network indicators can be found.

4.7.4 Application to Network Analysis

An example of an application of the measures previously presented to an actual metro system is given as follows. In 1980, the Washington Metro network consisted of three lines, Red, Blue and Orange, the latter two with a substantial joint section. Subsequently, two other lines, Green and Yellow, which, similar to Blue and Orange, have a joint trunk section, have been built, so that the network consists of five lines.

Using the first three lines (in their final full length) as the basic network, the question can be asked: What additions to the network will be obtained by the construction of the fourth and fifth lines?

Shown in Figure 4.38 is the initial network of Red, Blue, and Orange Lines, with the additional Green and Yellow Lines indicated by dashes. The elements and indicators (from a and b categories) that change with this addition (given in Table 4.8) show that the number of stations increases by 30% and the network length increases by 40%. Because of the interconnections among lines, however, network complexity increases even more substantially: whereas in the old network there were no circles formed by lines, the new one has three, and the number of OD pairs increases by 70%. Most topology indicators, expressing network complexity (b-2 and b-3), also increase, reflecting a more complex network with more diversified services, while b-4 decreases because of greatly increased number of transferring possibilities with the larger network.

4.7.5 Relationship of Metro Network to the City

An important aspect of metro network evaluation is its relationship to the city: its size and the number of stations in relation to the city size and population and the role of the metro among other transportation modes. Among the numerous measures and indicators, some of which most directly reflect the relationship between a metro network (with the primary emphasis on its geometric form) and the city it serves, several are defined here and listed in Table 4.9.

c-1, density of metro network L_a, is the ratio of the network length to the area of the city. This indicator reflects the extensiveness of a network with respect to the area it serves, primarily center city; for regional networks this indicator is sometimes imprecise because of the difficulty in delineating the served area of the region. The indicator is defined as

Table 4.7 Different network topologies, showing their measures and indicators

No.	Topology	a-1 n_i	a-2 $n_i - 1$	a-3 ℓ_i	a-4 n_m^k, a_m^k, ℓ_m^k	a-5 q	a-6 N	a-7 A	a-8 L	a-9 OD, OD_d, OD_t	b-1 $\overline{S}$	b-2 λ	b-3 β	b-4 δ
1	P_1= 4, 2, 6 (km) 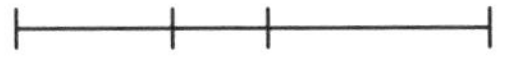	4	3	12	0 0 0	1	4	3	12	6 6 0	4	1.0	0.75	1.00
2	(12) 	13	12	12	0 0 0	1	13	12	12	78 78 0	1	1.0	0.92	1.00
3	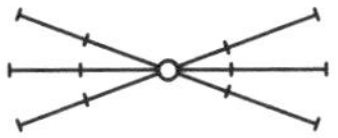	5 5 5	4 4 4	4 4 4	1 0 0	3	13	12	12	78 30 48	1	1.0	0.92	0.38
4	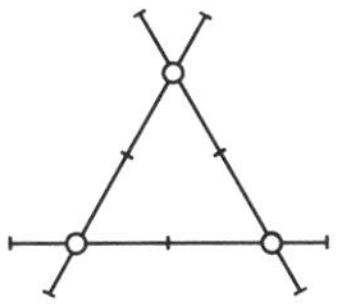	5 5 5	4 4 4	4 4 4	3 0 0	3	12	12	12	66 30 36	1	1.0	1.00	0.45
5	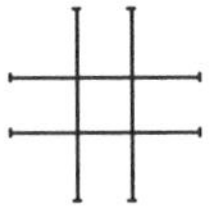	4 4 4 4	3 3 3 3	3 3 3 3	4 0 0	4	12	12	12	66 24 42	1	1.0	1.00	0.36
6	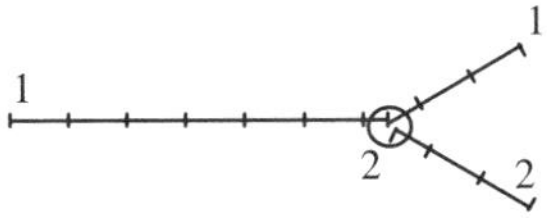	10 4	9 3	9 3	1 0 0	2	13	12	12	78 51 27	1	1.0	0.92	0.65
7	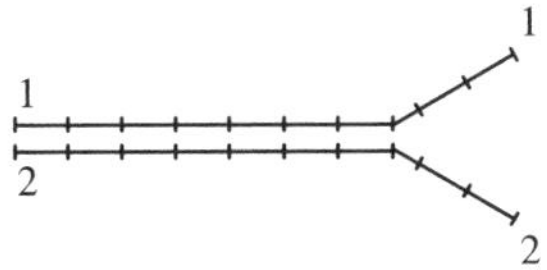	10 10	9 9	9 9	7 6 6	2	13	12	12	78 69 9	1	1.5	0.92	0.88
8	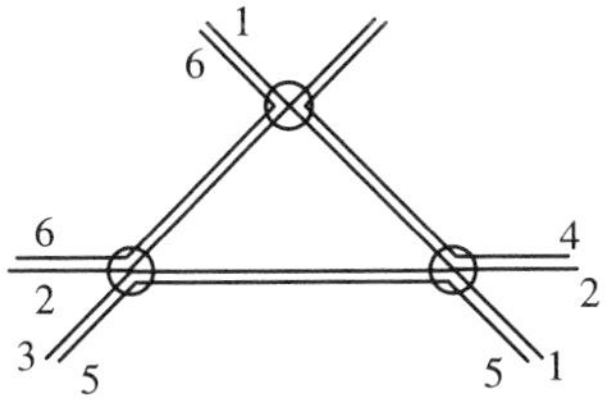	5 5 5 5 5 5	4 4 4 4 4 4	4 4 4 4 4 4	18 12 12	6	12	12	12	66 45 21	1	2.0	1.00	0.68

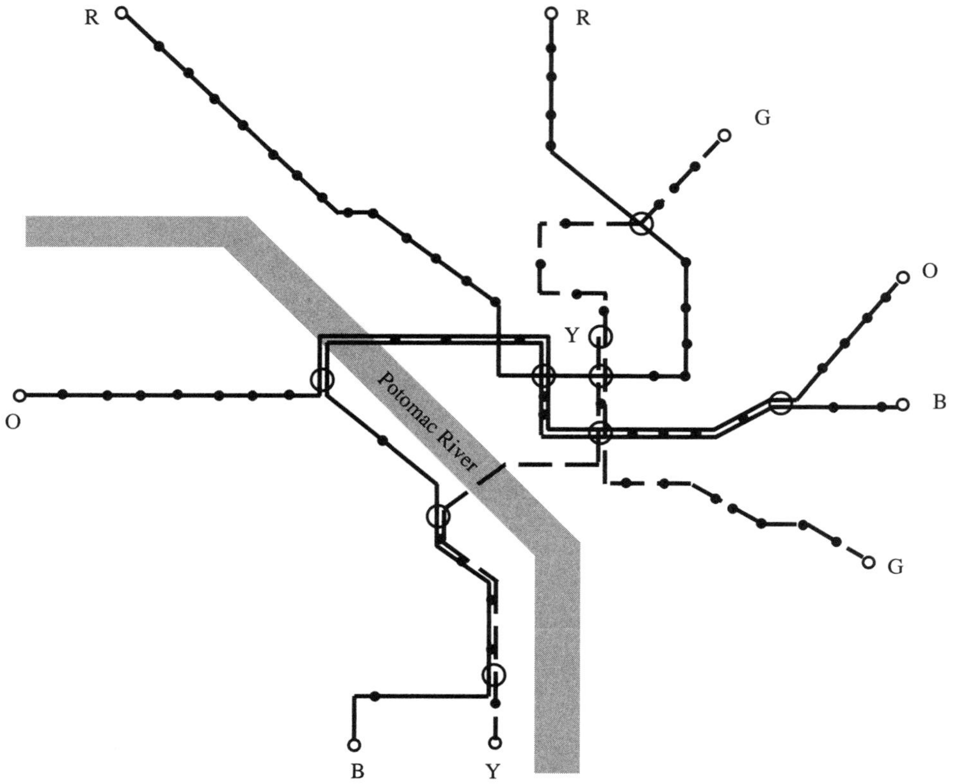

Figure 4.38 *Washington Metro network, highlighting the Green and Yellow Lines. (This 2003 network has a few minor line and station modifications from the 1980 network used in the analysis.)*

Table 4.8 Washington Metro network measures and indicators without and with the Yellow and Green Lines

Item Code	*Item*	*Symbol*	*Without Yellow and Green Lines*	*With Yellow and Green Lines*	*Percent Change*
a-5	Number of lines in network	q	3	5	+67
a-6	Number of stations in network	N	63	82	+30
a-7	Number of station spacings in network	A	62	84	+35
a-8	Length of network (km)	L	115	162	+41
a-9	Number of station-to-station travel paths (origin-destination)	OD	1953	3321	+70
b-1	Average interstation spacing (km)	$\overline{S}$	1.85	1.93	+4
b-2	Line overlapping	λ	1.19	1.24	+4
b-3	Network complexity	β	0.99	1.02	+3
b-4	Directness of service	δ	0.45	0.32	−29

(Based on the network from 1980; Figure 4.34 includes a few modifications through 2003.

Table 4.9 Metro network characteristics indicators

Item Code	*Definition*	*Symbol*	*Expression*
	c. Relationship to the city		
c-1	Density of metro network	L_a	$\frac{L}{S_u}$
c-2	Network extensiveness per population in millions	L_p	$\frac{L}{P}$
c-3	Area coverage	N_a	$\frac{N \cdot S_i}{S_u}$
c-4	Street transit integration ratio	η_t	$\frac{q_s^t}{q_s}$
c-5	Auto access integration ratio	η_a	$\frac{N_p}{N}$
	d. Service measures		
d-1	Operating speed (km/h) weighted by work performed W (veh-km) per day	V_{av}	$\frac{\sum_{i=1}^{q} W_i \cdot V_{oi}}{\sum_{i=1}^{q} W_i}$
d-2	Frequency of service (TU/h) during peaks weighted by stations	f_w	$\frac{\sum_{i=1}^{q} f_i \cdot n_i}{\sum_{i=1}^{q} n_i}$
d-3	Highest design line capacity (sps/h) in network	C	Max $(f_{i\,\max} \cdot n_{TU\,i} \cdot C_{vi})$
d-4	Maximum scheduled line capacity (sps/h)	C_s	Max $(f_i \cdot n_{TU\,i} \cdot C_{vi})$
d-5	Offered work (space-km) on a line per day	W_s	$W \cdot C_v$
	e. Service use indicators		
e-1	Line capacity utilization coefficient	η_c	$\frac{C_s}{C}$
e-2	Riding habit (annual passenger trips, P_{av}/population P)	R	$\frac{P_{av}}{P}$
e-3	Density of travel (annual passenger trips/km of network)	R_L	$\frac{P_{av}}{L}$
e-4	Utilized work (passenger-km) per day (daily trips P_{ad} · average trip length l_p)	W_p	$P_{ad} \cdot l_p$
e-5	Work utilization factor (prs-km/sp-km)	$\bar{\alpha}$	$\frac{W_p}{W_s}$
e-6	Metro daily passenger trips as % of transit daily passenger trips P_t	P_m	$\frac{P_{ad}}{P_t}$

$$L_a = \frac{L}{S_u} \qquad \left| \frac{L_a}{\text{km/km}^2} \middle| \frac{L}{\text{km}} \middle| \frac{S_u}{\text{km}^2} \right| \tag{4.24}$$

where S_u is the area of the city or the served area, as applicable.

c-2, network extensiveness per population L_p, expresses the ratio of network length L to the population of the served area P, expressed in millions:

$$L_p = \frac{L}{P} \qquad \left| \frac{L_p}{\text{km}/10^6\ \text{prs}} \middle| \frac{L}{\text{km}} \middle| \frac{P}{10^6\ \text{prs}} \right| \tag{4.25}$$

Comparing cities with similar populations, the greater value of L_p indicates a more extensive network and generally a more important role of the metro system.

The service that a metro network offers to the urban area and the various forms of access to its stations are more conveniently measured by the three indicators defined as follows; the first one affects access by walking (pedestrian access); the latter two measure the convenience of access to the metro network by street transit and automobile, respectively.

c-3, area coverage N_a, is the percentage of the urban area (S_u) that is within walking distance of metro stations:

$$N_a = \frac{N \cdot S_i}{S_u} \times (100\%)$$

$$\left| \frac{N_a}{\%} \right| \frac{N}{\text{stat}} \left| \frac{S_i}{\text{km}^2/\text{stat}} \right| \frac{S}{\text{km}^2} \right| \tag{4.26}$$

where S_i is the area around the metro station with a radius of 400 m (sometimes a 500-m radius is used as the standard).

Area coverage is the most important measure of the availability of metro services within the entire served area; this indicator is therefore used extensively in the planning of metro lines and networks.

c-4, street transit integration ratio η_t, is the ratio of street transit lines that have transfers to the metro network q_S^t, to all street transit lines q_s:

$$\eta_t = \frac{q_s^t}{q_s} \times (100\%) \tag{4.27}$$

This indicator expresses the relative geometric and functional role of the metro network within the city's total public transport network.

c-5, auto access integration ratio η_a, is the percentage of stations that have park-and-ride (P+R) facilities, N_p:

$$\eta_a = \frac{N_p}{N} \times (100\%) \tag{4.28}$$

4.7.6 Measures of Offered and Utilized Service

Most analyses of metro networks, particularly those that focus on geometric characteristics, involve some consideration of overall service offered by the metro system and its usage. These analyses are usually rather general and are based on selected global measures and indicators such as those given in Table 4.5. For more detailed analyses of transit system performance, other literature can be consulted (Lehner, 1978; Vuchic, 1981; Corazza and Musso, 1981; Tomazinis, 1975).

The fourth group of quantitative network measures and indicators listed in Section 4.7.1 represents quantity and quality of offered service. Several selected items from that group, designated as items d-1 to d-5, are listed and defined in Table 4.9. The most important components of level of service—speed and frequency—are included. They can be measured for individual lines, or for entire network, when they are weighted by vehicle-km on each line for speed, and by the number of stations for service frequency. System performance measures, design and scheduled capacity, and offered work in space-km per day are also included.

The fifth group of network measures, indicators of service usage and utilization, are given as items e-1 to e-6 in Table 4.9. As their descriptions in the table indicate, they include usage of offered services, which influences the economic efficiency of operations and can be strongly linked to the design of metro lines and the topology of its network. Other utilization indicators express intensity of metro system usage in absolute terms and in relation to total transit use in the city. Both groups are also related to the extensiveness and topology of a metro network and the role it plays in the city.

4.7.7 Selection of Evaluation Items for Specific Analysis

Metro network elements and indicators presented in this section can be used for many purposes in the planning, design, and analysis of metro networks. Several typical applications are:

1. Evaluation of an existing network: comparison with other peer cities to estimate its adequacy
2. Change from trunk-feeder into trunk-branch line (or vice versa); evaluation of the impact on services and operations

Table 4.10 Items to be used for different types of metro network analysis

Item	Symbol	1. Evaluation of an Existing Network	2. Change from Trunk-Feeder to Trunk-Branch	3. Adding a New Line to a Network	4. Selecting from Several Alternatives	5. Planning a New Network
a-5	q	x		x	x	x
a-6	N	x		x	x	x
a-7	A	x		x	x	x
a-8	L	x		x	x	x
a-9	OD, OD_d, OD_t	x		x	x	x
b-1	$\overline{S}$	x		x	x	x
b-2	λ	x	x	x	x	x
b-3	β	x		x		x
b-4	δ	x	x	x	x	x
c-1	L_a	x				x
c-2	L_p	x				x
c-3	N_a	x		x	x	x
c-4	η_r	x		x	x	x
c-5	η_a	x		x	x	x
—	Analyze topology	x	x	x	x	x
—	Line geometry	x		x	x	x

3. Addition of a new line to the network; estimation of its impact
4. Selection from among several alternative network extensions
5. Planning of a new metro network

Table 4.10 lists the most useful quantitative and other items for each one of these types of analyses (items a-1 through a-4 are needed only for analyses of individual lines).

In the conclusion of this chapter, it should be noted again that although there are over 100 metro systems in the world today, research and literature on the geometry of their networks is very limited. However, the geometric form of metro (rapid transit) networks and lines can have a major impact on their service for passengers and the efficiency of the system's operation (Lehner, 1969). The complexity of metro systems, the large investments they require, and the permanence of their facilities therefore justify major efforts to perform extremely careful planning and design.

It is very important that metro network designers have thorough knowledge and experience with such systems and their relationships with the city. However, even the most experienced designers should use a number of quantitative measures and analyze their characteristics utilizing many of the materials presented in this chapter: network forms and operating concepts, their quantitative measures of performance, operational efficiency, and relationships to the areas they serve and to the entire city.

EXERCISES

4.1. Define and briefly describe the three major objectives that should be considered in planning transit networks. Which three major parties are affected by the three major objectives?

4.2. List and describe major factors of transit network design that mostly affect passenger attraction.

4.3. A transit network is being designed to serve a 1500-m-wide corridor. There should be two lines parallel to the corridor axis, and it is possible to locate them at any location.

a. What would be the optimal locations of the two lines to minimize access distances perpendicular to the lines, measuring from the left-hand boundary of the corridor?

b. What will be the average and the maximum access distances, assuming that the population distribution in the corridor is uniform?

4.4. Describe the geometric form, service, and operational characteristics of two of the following three types of transit lines:

a. Diametrical line

b. Circle line

c. Trunk radial line with branches.

Find a rail transit network in a city that has all of the described lines and draw a schematic network illustrating each.

4.5. Diametrical and radial lines.

a. Describe and discuss the two major advantages of diametrical lines compared with radial lines.

b. It has been decided that several radial lines will be connected through the city to form diametrical lines. Which factors should be considered in determining which radials will be paired with each other?

4.6. Describe the positive and negative features of circle or ring transit lines as compared to other line types. Be sure to include:

- Three major functions they usually serve
- Their relationship with other, particularly radial lines
- Characteristics of the areas they serve
- Types of trips they attract and the resulting characteristics of their passenger volumes
- Operational and scheduling characteristics, and problems and their solutions

4.7. List and define advantages and disadvantages of trunk lines with feeders as compared to trunk lines with branches.

4.8. An urban sector has a transit network consisting of a trunk line and a number of branches, as shown below. This network is served by buses that come from the branches and continue on the trunk to the CBD.

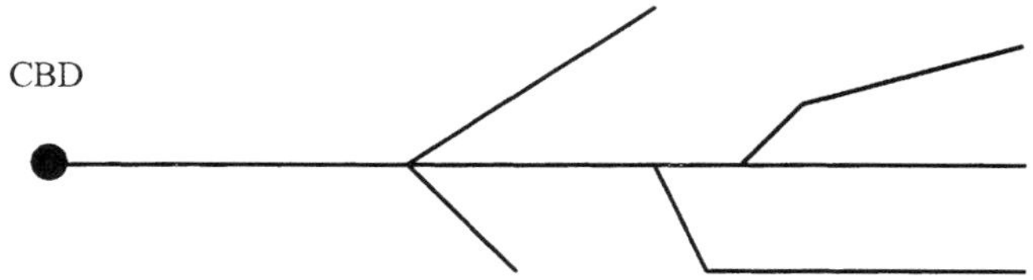

Discuss the advantages and disadvantages of changing the present operation into each one of the following alternatives;

1. The same buses, but branches made to terminate at transfer points and the trunk line operated independently.
2. The same as 2, but it is possible to obtain different types of buses for different lines; for example, articulated buses for the trunk, regular buses and minibuses for different feeders.
3. The same as 2, but introduce a rail system (LRT or RRT) on the trunk.

Which major factors would influence the selection among the three alternatives?

4.9. A city is considering construction of a tunnel that will connect two stub-end commuter rail stations on opposite sides of the CBD, such as was done in Munich, Philadelphia, and Manchester. Explain what changes to the entire commuter rail network this connection will bring (what form the network has now, what it will have in the future) and what specific changes in the operation of the lines and in service to passengers will result from the new connection. Classify the changes into groups for the two major interested parties: the operator and the passengers.

4.10. How many transfer possibilities are there at a joint station with the following numbers of transit lines:

a. Three lines intersecting each other

b. Eight terminating lines

c. Three intersecting and two terminating lines

4.11. A four-track rail line has express and local services. Every fifth station is an express station; in other words, from one express station to the next, local trains have four additional stations to serve. If the additional time used for stopping at a station along the line, T_ℓ, is 75 s, compute the headways for the line that allow express trains to meet with locals at each express station.

Draw such a schedule for a section with two express and intermediate local stations, including full runs of three express trains and all local trains during that time period.

4.12. Timed transfer system (TTS).

a. Define and briefly describe TTS. Which element of scheduling must be coordinated among all lines that serve a given transit center?

b. Compare TTS to a regular network with independent lines in low-density (suburban) areas. Include such aspects as:

- Convenience of travel
- Image of transit service
- Number of vehicles required for the comparable headways
- Sensitivity to delays
- Considerations of headway lengths in planning a TTS

4.13. Three radial bus feeder lines, as shown below, should be scheduled to operate as a single-focus TTS at a rail station that has a service with 20-min headways. The buses should have terminal times of 5 min at the station so that they can arrive 3 min before and leave 2 min after the train arrivals. Terminal times at outer terminals should include 1 min for terminal operations, plus any additional time for adjusting cycle time to a multiple of 20 min.

Bus Line	*Length (One-Way)*	*Operating Speed*
A	6.6 km	21.0 km/h
B	4.2 km	18.0 km/h
C	11.2 km	26.0 km/h

Compute the number of buses required for each bus line to provide the required headways.

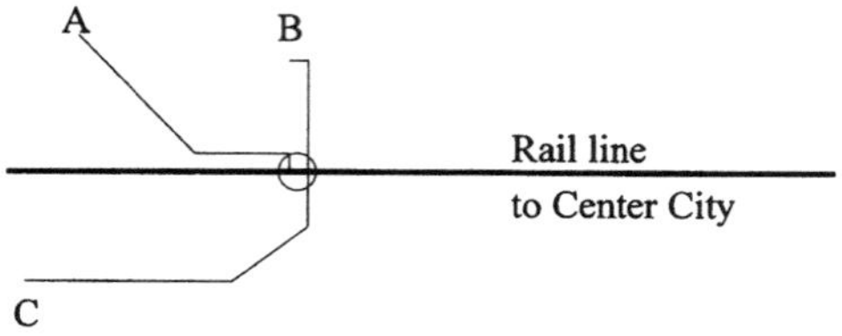

4.14. A timed transfer system (TTS) is to be designed for six bus lines with the following unadjusted cycle times, consisting of operating times and minimum required terminal times, and the longest acceptable headways:

Line	*Minimum Cycle Time*	*Maximum Acceptable Headway*
A	38 min	20 min
B	26 min	15 min
C	48 min	60 min
D	22 min	20 min
E	28 min	30 min
F	54 min	20 min

Determination should be made between a TTS operation with 10/20/40/60-min and 15/30/60-min pulse headways. Remember that each line must have a cycle time that is a multiple of h and greater than or equal to the minimum cycle time.

Assume that the cycle times can be shortened by up to 3 min by speeding up the service or shortening terminal times; or they can be extended to the next integer multiple of the headway. Headways on each line cannot be longer but may be shorter than policy headways if needed for more efficient scheduling. Using the setup from Figure 4.20, draw two bar charts of times up to 60 min, showing the integer multiples of the given headways and cycle times. Use one chart to fit the given values into a 10/20/40/60-min. TTS, the other one for a 15/30/60-min. TTS. Compare the obtained headways using the basic criteria, such as the wasted times, i.e., the times added to fit the headways into the adjusted cycle times, offered headways, and required number of buses (reflecting investment and operating costs). Make the selection between the two and explain it.

4.15. Compare the purely radial rail transit network with the ubiquitous network shown below. Include items affecting users (directness of travel, transferring), operational aspects (line operations, loadings and utilization), and services offered (area coverage).

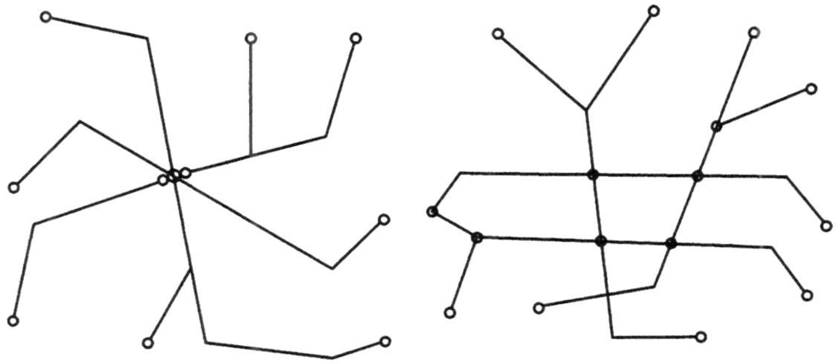

Radial network **Ubiquitous network**

4.16. Compare independent with integrated types of metro networks with respect to:

a. Efficiency of scheduling due to tailoring of services on individual network sections to each specific demand volume

b. Reliability of operations

c. Rerouting of trains in cases of disturbances

d. Economies of rolling stock utilization

e. Passenger convenience in travel through the network

4.17. Sketch, describe, and analyze the geometry of a rail transit network in a city you know well. You should focus on geometric aspects of lines and the network, the relationships of lines, transfer points, area coverage, etc. Do not describe stations, types of cars, and other system elements not directly related to the line and network geometry.

4.18. Select a rail transit system in a city and develop drawings illustrating:

a. Area coverage, assuming that maximum walking distance is 400 m; in other words, draw 400-m radius circles around each station.

b. Different line types for each ROW category if they are not all A.

c. Divide each line in several segments that have different operating speeds. Typically, a line may have a center city section with short station spacings and one or two outlying sections with higher speeds. Compute these speeds and show them on the network diagram by numbers.

d. If graduated fares are charged, show fare contours: the distances from center city to which passengers can travel for a base fare; then another contour for each additional fare increment.

4.19. List and briefly describe the five categories of items that can be used to evaluate geometric characteristics of lines and stations in a metro network and its relationship with the city or its service area.

4.20. Compute the number of station-to-station travel pairs for each one of the following cases. Find the relevant formulae in Section 4.7. Explain your computations.

a. The light rail line in Buffalo has 14 stations. The plan is to add one station in center city and extend the line by two stations to a new suburban terminal. Compute how many new (additional) station-to-station pairs will be created.

b. On the Chicago Rapid Transit and on the Paris Metro, some lightly used existing stations were closed during the 1950s to speed up services on the lines. If a line with 28 stations had two stations closed, how many station-to-station pairs were eliminated?

c. Suppose that a regional rail line, similar to the lines of Metra in Chicago, SEPTA in Philadelphia, and Cityrail in Sydney, has 14 stations, 3 of which are in center city. The line will be extended to get one more station in center city and two on a suburban extension. How many new station-to-station pairs between center city stations and suburban stations will be created? Note: compute only center-to-suburb pairs, not all possible pairs along the line.

4.21. Two transit networks, I and II, are shown below. Compute for each one the following:

a. Number of stations in each network: midline single, transfer, terminal, and total

b. Number of interstation spacings

c. Number of station-to-station trips (OD pairs): without transfers, with one transfer, and total

Network I consists of two lines, Blue (B) and Yellow (Y), which run together on the joint section from B′/Y′ to X , and outer branches from X to B″ and X to Y″. There are 12 stations on the trunk line and 3 and 4 stations on the Blue and Yellow branches (not including joint station X), respectively. Thus, the Blue Line has a total of 15, and the Yellow Line, 16 stations.

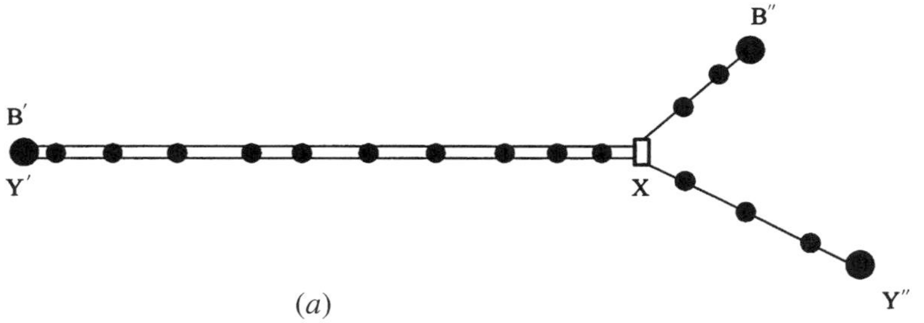

(*a*)

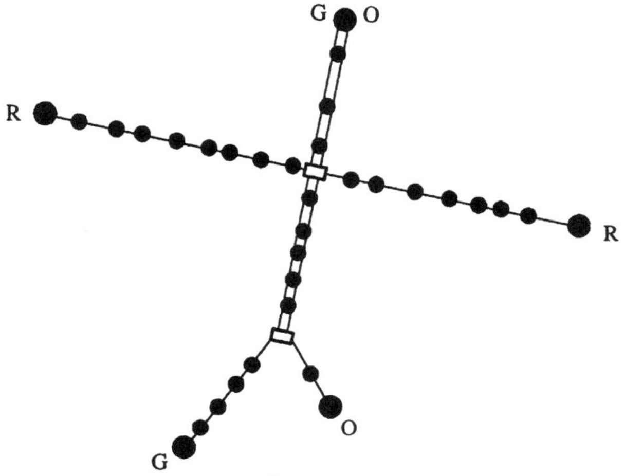

(*b*)

Network II consists of three lines, Red, Green, and Orange, with characteristics as shown on the diagram: the Red Line has 18 stations. On the Green/Orange Line, there are 11 stations on the trunk, 2 stations on the Orange Line, and 5 stations on the Green Line (not including the joint station X).

4.22. Assume a transit network in a city consists of three lines, A, B, and C, intersecting each other at different stations, as shown below. Line A has 17 stations, Line C has 11 stations, and Line B has 15 stations on its trunk. Compute:

a. The number of terminals, transfer stations, and total stations

b. The number of station spacings in the network

c. The total number of different trips (station-to-station pairs) in the network

d. Of the total number of pairs, how many require a transfer

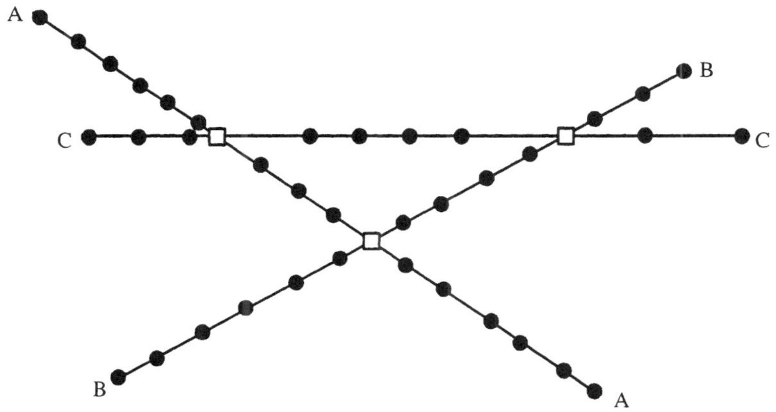

4.23. Define the coefficient of line-overlapping λ in a transit network. What is its minimum value and what does a greater value indicate? Is λ typically different among modes: how do its values usually vary among bus, LRT and metro modes and why?

4.24. (True-False) Rail transit (LRT or RRT) with feeders compared with an all bus service partially on a busway (i.e., bus semirapid transit):

- Offers higher level of service (reliability, frequency, riding comfort) T F
- Is less suitable for downtown operation T F
- Provides travel with fewer transfers T F
- Results in lower labor productivity T F
- Attracts fewer riders T F
- Requires higher investment T F

4.25. (True-False) Timed transfer systems:

- Apply to multicenter areas only; they are inefficient for a single focal network T F
- Apply more to rail feeder stations than to bus only (many-to-many) T F

- Allow fast and convenient transfers among different lines in both directions of travel — T F
- Result in a reduction of the required fleet as compared to unsynchronized transfer — T F
- Improve the image of the transit system and usually increase ridership — T F
- Require higher investment — T F

Explain any of your T or F choices (only if you feel the assumptions of your answer need clarification).

REFERENCES

Bandi, F., et al. 1974. Length of walking distances and distance between stops: their influence on the attractiveness of public transport. *Revue de l'UITP* 3, 248–256.

Berger, R. 1951, *Untergrundbahnen und ihre Einsatzgrenzen.* Berlin: W. Ernst & Son.

Black, A. 1962. A method for determining the optimal division of express and local rail transit service. *HRB Bulletin 347,* Washington, DC, 106–120.

Corazza, G. R., and A. Musso. 1981. Introduction to the structural analysis of metropolitan railway systems. *Revue de l'UITP* 3, 248–256.

Franz, H. D. 1975. *Untersuchungen zur Planung von Verkehrsnetzen unter besonderer Berücksichtigung des öffentlichen Personennahverkehrs.* Bonn: Bundesminister für Verkehr, Abt. Strassenbau.

Garrison, W. L. and D. F. Marble. 1974. Graph theoretic concepts. In *Transportation Geography,* ed. Michael E. Eliot Hurst, 56–80. New York: McGraw-Hill, 56–80.

Girnau, G., A. Müller-Hellmann, and F. Blennemann, eds. 1997. *Sustainable Mobility: People in Motion.* Cologne: Verband Deutscher Verkehrsunternehmen.

Lehner, F. 1969. *Regional Organization of Transport and Urban Development.,* Report 1, 38th UITP Congress in London. Brussels: UITP.

———. 1978. *Der maximale Wirkungsgrad des Personaleinsatzes.* Düsseldorf: Alba Buchverlag.

Leman, C. K., P. L. Shiller, and K. Pauly. 1994. *Rethinking HOV: High-Occupancy Vehicle Facilities and the Public Interest.* Annapolis, MD: Chesapeake Bay Foundation.

Musso, A., and V. R. Vuchic. 1988. Characteristics of metro Networks and methodology for their evaluation. *TRB Record 1162,* pp. 43–54.

Robbins, R. M., and F. Pampel. 1975. *Methods and Theoretical Bases for the Traffic Planning of Rapid Transit Lines.* Report 3b, 41st UITP Congress in Nice. Brussels: UITP.

Rockwell, E. 1959. *Notes on the Principles of Urban Railway Planning* Internal Reports. London: London Transport Executive.

———. 1967. *Urban Railway Planning.* Internal Report. London: London Transport Executive.

Sullivan, B. E. 1976. The timed transfer focal point: a refinement in public transport service design. Revue de l'UITP 25, 43–48.

Tiry, C. 1997. Tokyo Yamanote Line—cityscape mutations. *Japan Railway & Transport Review,* September, 4–11.

Tomazinis, A. R. 1975. *Productivity, Efficiency and Quality in Urban Transportation Systems.* Lexington MA: Lexington Books.

Vuchic, V. R. 1981. *Urban Public Transportation Systems and Technology* Englewood Cliffs, NJ: Prentice-Hall.

Vuchic, V. R., A. Molinero, and R. Clarke. 1983. *Timed Transfer System Planning, Design and Operation.* Report DOT-I-83-28 to UMTA. Washington, DC: DOT.

Vuchic, V. R., S. Kikuchi, N. Krstanoski, and Y. E. Shin. 1995. Negative impacts of busway and bus lane conversions into high-occupancy vehicle facilities. *TR Record 1496,* p. 75–86.

Vuchic, V. R., and A. Musso. 1991. Theory and practice of metro network design. *Public Transport International* 3, 298–325.

FURTHER READING

Bovy, P. 1992. Le reseau ferroviaire Suisse: un modele a suivre? In *Proceedings, the 6th World Conference on Transport Research,* Lyon.

Cervero, R. 2001. Walk-and-ride: factors influencing pedestrian access to transit. *Journal of Public Transportation* 3(4):1–22.

Schneider, J., et al. 1980. *Planning and Designing a Transit Center-Based Transit System: Guidelines and Examples from Case Studies in 22 Cities.* Department of Civil Engineering, University of Washington, Seattle, WA, September.

Vuchic, V. R., and S. Kikuchi. 1986. Planning an integrated regional rail network: the Philadelphia case. *TRB Record 1036,* 51–62.

5

PLANNING OF RAIL TRANSIT STATION LOCATIONS

Stations represent locations at which passengers have access to the transit network. Unlike stops for surface transit, which can be rather easily relocated, rapid transit stations are major, high-investment, permanent structures that often have a strong impact on their surroundings. Moreover, as an element affecting the overall operating speed of trains, both the number and distribution of stations along a line have a major impact on operations: speed, passenger travel time, riding comfort, operating costs, and so on. These operational factors indicate that the location of stations is a very important part of rapid transit system planning.

5.1 OBJECTIVES IN STATION LOCATION PLANNING

Since many transit line characteristics are affected by the number and locations of its stations, their planning is complex and must include quantitative and qualitative factors. It is not possible to develop a purely quantitative methodology for determining optimal station locations. Therefore, a transit designer must be familiar with all relevant factors and determinants of locations. The major objectives in planning station locations are to:

- Serve major centers, activity points, and transfers with other lines
- Achieve minimum passenger travel time
- Provide maximum area coverage
- Attract the maximum number of passengers; this is partly a function of the above two objectives, and partly of competing services
- Achieve minimum system cost (investment plus operating)
- Meet other requirements (e.g., intermodal transfers, planning goals, relationship with land use patterns, economic development, population needs, etc.)

Since the optimal solutions for individual objectives are not always mutually compatible, a compromise must be found by collectively satisfying each objective to some degree; in scientific language, a multiobjective optimization must be performed. Sometimes in practice this complicated task is avoided (incorrectly) by adopting one requirement as absolute and thus practically disregarding others. In some cases (Paris, Hamburg), interstation spacings along entire lines were determined so that a desired area coverage was achieved. In others (San Francisco BART), a design requirement was adopted that the average operating speed between terminals on the line should be at least 72 km/h (45 mph). Such a goal of average speed severely constrains station planning. Because the line alignment is usually fixed, once rolling stock performance is defined, the system's technical characteristics determine the maximum number of stations the line can have while still meeting the average speed goal. That number of stations is then distributed considering local conditions.

These single-objective procedures are overly simplified and may lead to costly errors. For example, average operating speed for the entire line should seldom be the primary objective. Travel speed on the line is an important factor in attracting passengers; however, each passenger is concerned with the speed on the specific section of the line where he/she travels. Therefore, higher speed on heavily used sections is more important than the overall, terminal-to-terminal travel speed. Moreover, other objectives and local conditions must also be considered.

Consequently, the transit planners should consider all the factors influencing station locations and be familiar with the methods for deriving optimal solutions of individual objectives. Based on these analyses, local conditions, and requirements, the designers must determine station locations, usually as a compromise among various objectives. For this purpose, each objective is discussed here separately, leading to a methodology for its application in practice.

The first objective, to serve major activity points, is actually always included in planning of lines and stations: already in selecting alignment for a line, the natural goal is to serve important traffic generators and major activity areas. These include such objects as railway stations, city hall, major shopping areas, transfer points with other rail transit lines, or, in the suburbs, proximity to freeway interchanges where park-and-ride (P+R) facilities can be built.

5.2 PASSENGER TRAVEL TIME

In most cases, minimization of total passenger travel time is the dominant objective since it often represents the largest component of "disutility" to the users of a transit line. Moreover, comparison of travel times is a factor influencing modal split, i.e., the actual choice of one mode compared to another.

There have been several theoretical studies of optimal station spacings on rail transit lines, mostly with respect to minimum travel time for passengers. Hinze (1930), Vuchic (1966), and Wirasinghe (1976) wrote dissertations on this topic. Schneider (1961), Lehner (1969), and Bandi et al. (1974) performed applied research and reviewed this topic. In this chapter, a theoretical analysis of optimal spacings with respect to travel time, area coverage, cost, and other elements is presented, followed by a review of practical considerations in transit line planning and examples of spacings of different modes and types of systems in different cities and urbanized regions.

5.2.1 Two Trade-offs

The travel time of passengers on a transit line PT consists of two main components: access to/from the stations, including waiting for a train PT_a and travel time on the line PT_t:

$$PT = PT_a + PT_t \qquad \left|\frac{PT}{\text{prs-h}}\right| \tag{5.1}$$

The number and locations of stations along a line influence both of these time components. The number of stations on a line with a given distance is often analyzed through the station density g, defined as stations per km, or the inverse of the average station spacing. For example, if the average spacing is 800 m = 0.8 km, station density is $g = 1/0.8 = 1.25$ stat/km.

Considering passenger travel times, station density must be based on the optimum value of the trade-offs between the two travel time elements, access and on-line travel times. More precisely, two different trade-offs are involved.

5.2.1.1 The First Trade-off: Area Coverage versus Operating Speed.

If it is assumed, for clarity, that the number of passengers on the line is fixed (there is no diversion to or from other modes), increasing the density of stations results in a decrease of average access distance and access time to stations PT_a, as shown in Figure 5.1. Travel time on trains PT_t, however, increases because of the growing time losses due to station stops. This increase is linear at first (each station adds a fixed amount of delay, i.e., it increases travel time). When station spacings become so short that trains cannot reach their maximum running speed, the additional delay per station decreases, so that the marginal increase in travel time begins to decline. The total passenger travel time curve PT shows that there is an optimal station density: its value is derived mathematically in the following sections.

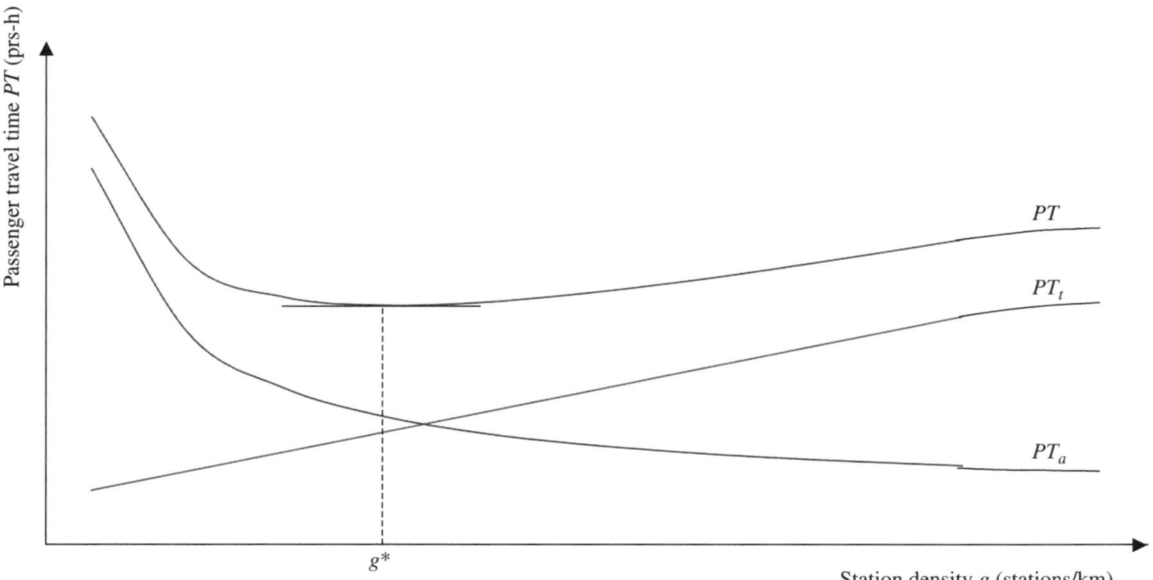

Figure 5.1 *Passenger travel time as a function of station density.*

If passenger distribution along a line is uniform, the optimum station density is determined by this trade-off. However, if passenger distribution is not uniform, this distribution influences the optimal station locations and results in variable spacings.

5.2.1.2 The Second Trade-off: Local versus Through Passengers. Station spacings on any section along a line should be a function of the ratio of the number of passengers with origins and/or destinations along that section (trip-end density) and the number of passengers on the trains passing through the area who would prefer not to stop on that section due to the loss of time for them. The greater that ratio, the more closely should the stations be spaced. Where the volume of through passengers dominates the volume of local passengers, station spacings should be long. In the extreme case, if there is a section without any passengers wanting to board or alight, there should be no stations. An important consequence of this relationship is that, with respect to passenger time, uniform station spacings are seldom optimal with respect to passenger travel time minimization: they should vary with distribution of passenger demand.

5.2.2 The Model

Derivation of mathematical expressions for optimal station spacings will be based on a model of a transit line segment with length L and $n + 1$ stations, shown in Figure 5.2. The distance between some station k and the next station $k + 1$ is S_k. When spacings are equal over the entire length of the line, $S = L/n$.

To analyze station spacings and their area coverage, it is assumed that passengers go to the closest station. The line separating the areas from which passengers go to different adjacent stations is then at middistance between the stations. This line then delineates the market shed of passenger choice, called for this purpose the passenger shed area. The maximum distances passengers walk to the stations represent lateral boundaries of the shed areas.

On the lines where most of the travel is to and from one direction, such as predominantly commuter lines

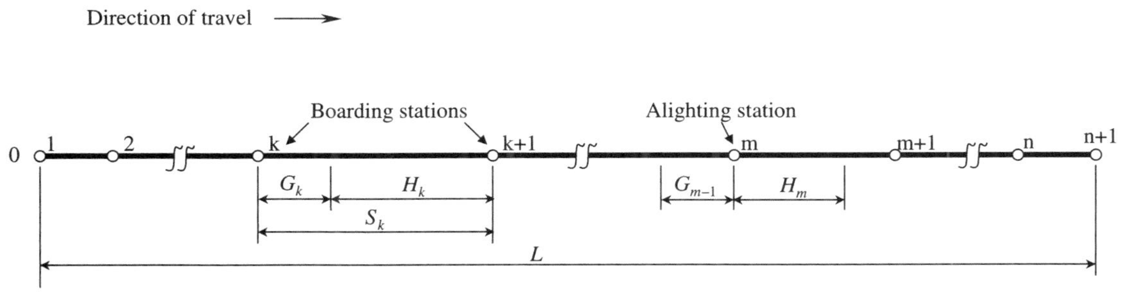

Figure 5.2 *Transit line section model.*

toward city center, the position of the passenger shed line is somewhat different. The assumption is that each passenger selects a station for boarding or for alighting so that his/her *total* travel time is minimized. Thus, the passenger shed, determined by selecting between two adjacent stations, is located asymmetrically between them: it is upstream from the midpoint because access to the station opposite to the travel direction involves travel time on the transit line over that spacing, which people going to the station in the direction of travel do not have to travel. At arrival stations, the situation is reversed, so that the shed line is located downstream from the midpoint.

The two sections of each spacing S_k obtained by the passenger shed line are G_k and H_k. As the figure shows, at boarding, access distance G_k is smaller than section H_k because it requires access opposite to the direction of travel and then also travel on the train over S_k, which passengers going to station $k + 1$ do not have. At the alighting station m, G and H sections are reversed: the shorter section G_{m-1} is shorter because it is upstream from the destination station m, while the section H_m is longer because it is downstream from it.

If it is assumed, for simplicity, that paths for access to stations are via a grid pattern of streets, the distances perpendicular to the line can be disregarded because they are not affected by station locations.

The location of the shed line, or expressions for G and H, are computed by setting access time to the station located in the direction of travel equal to the access time to the station located in the opposite direction, plus travel time on the train between the two stations:

$$\frac{H_k}{V_a} + t_w = \frac{G_k}{V_a} + T_k + t_w \qquad \left|\frac{G,\ H}{\text{km}}\right|\frac{V}{\text{km/h}}\left|\frac{T,\ t}{\text{h}}\right| \tag{5.2}$$

where:

T_k = train travel time on the spacing S_k
V_a = speed of access to stations
t_w = interval between passenger's arrival at station and train departure

From Eq. (5.2) and the relationship

$$S_k = G_k + H_k \qquad \left|\frac{S,\ G,\ H}{\text{km}}\right| \tag{5.3}$$

G_K and H_K can be expressed through known parameters as:

$$G_k = 1/2(S_k - T_k V_a) \quad \text{and} \quad H_k = 1/2(S_k + T_k V_a)$$
$$\left|\frac{G,\ H,\ S}{\text{km}}\right|\frac{V}{\text{km/h}}\left|\frac{T}{\text{h}}\right| \tag{5.4}$$

If the station spacings are of sufficient length that the train can reach its maximum speed ($S \geq S_c$—see Section 1.1.6), using Eq. (1.17), T_k is expressed as:

$$T_k = \frac{S_k}{V} + T_\ell \qquad \left|\frac{S}{\text{km}}\right|\frac{V}{\text{km/h}}\left|\frac{T}{\text{h}}\right| \tag{5.5}$$

where T_ℓ is time lost for stopping at one station (see Eq. (1.18)). Substituting this expression for T_k in Eq. (5.4) for G_k gives:

$$G_k = 1/2\left(S_k - S_k\frac{V_a}{V} - T_\ell V_a\right) = \frac{S_k}{2}\left(1 - \frac{V_a}{V}\right) - \frac{T_\ell V_a}{2} \quad \left|\frac{G,\ S}{\mathrm{km}}\right|\frac{V}{\mathrm{km/h}}\left|\frac{T}{\mathrm{h}}\right| \tag{5.6}$$

A similar expression is obtained for H_k via Eq. (5.5). For further derivations it is now convenient to introduce four new symbols for relationships of parameters that greatly simplify the expressions for G and H and later equations for travel times and optimal station spacings:

$$\rho = \frac{V_a}{V}, \quad \alpha = 1/2(1 - \rho), \quad \beta = 1/2(1 + \rho), \quad \text{and} \quad \gamma = \frac{T_\ell V_a}{2} \quad \left|\frac{\alpha,\ \beta}{-}\right|\frac{\gamma}{\mathrm{km}}\left|\frac{T}{\mathrm{h}}\right|\frac{\rho}{-}\left|\frac{V}{\mathrm{km/h}}\right| \tag{5.7}$$

which, when introduced in Eq. (5.6) and corresponding one for H_k, lead to simpler expressions for the two sections of the station spacing S_k:

$$G_k = \alpha S_k - \gamma \quad \text{and} \quad H_k = \beta S_k + \gamma \quad \left|\frac{G,\ H,\ S,\ \gamma}{\mathrm{km}}\right|\frac{\alpha,\ \beta}{-}\right| \tag{5.8}$$

These expressions for sections G and H show that the position of the line dividing station spacings depends very much on parameters α and β, both of which are determined by the ρ parameter, i.e., the ratio of access speed V_a and maximum train speed V. When this ratio is very small, $V_a << V$, one obtains that $G_k = H_k = S_k/2$, i.e., the shed line is at the midpoint. A typical case for this situation is when passengers walk to stations of a high-speed line, so that V_a is about 5 km/h and V may be 50–80 km/h. For greater access speeds, such as when passengers come by bicycle and by car on streets, H becomes greater and G smaller (for departure from station, the situation is reversed). Within the limits of this theoretical analysis considering minimum travel time only, in the case when access speed reaches travel speed on the line, no passengers would be diverted to it.

The main conclusion here is that for walking access, passenger sheds are close to symmetric about stations. For higher-speed access, such as P+R, the passenger shed line is removed from the midpoint between stations away from the direction of travel for access to the stations, and in the direction of travel for departure from the stations, as shown in Figure 5.2 by stations k and m, respectively.

Using the described line model with its components, optimal station locations for minimum total travel time will be analyzed for a line of length L and P_L passengers with four different distributions along the line axis s. Passenger volume function along the line, $p(s)$, is defined by the distribution of origins, given as boarding function $b(s)$, and the distribution of destinations, or alighting function $a(s)$. The four sets of these distributions are defined here.

a. Uniform and equal boarding and alighting: $b(s) = b = a(s) = a$.

b. Uniform boarding along the line, while all alighting is at one terminus only: $b(s) = b$; $a(s) = 0$, except at the terminal:

$$a(L) = \int_0^L b(s)\,ds = P_L \quad \left|\frac{b,\ s}{\mathrm{km}}\right|\frac{a,\ b}{\mathrm{prs/km}}\left|\frac{P}{\mathrm{prs}}\right| \tag{5.9}$$

thus, passenger volume in trains increases linearly along the line. For travel in the opposite direction, the situation is symmetric: boarding is at one terminus, alighting is uniform, and volume decreases linearly to the outer end.

c. Boarding is variable, defined as $b(s)$; alighting is as in case b. Passenger volume increases as follows:

$$p(s_1) = \int_0^{s_1} b(s)\,ds \quad \left|\frac{P}{\mathrm{prs}}\right|\frac{b,\ a}{\mathrm{prs/km}}\left|\frac{s}{\mathrm{km}}\right| \tag{5.10}$$

d. Boarding and alighting are variable functions, $b(s)$ and $a(s)$, respectively; passenger volume is variable:

$$p(s_1) = \int_0^{s_1} [b(s) - a(s)]\, ds \qquad \left|\frac{P}{\text{prs}}\right|\frac{b,\, a}{\text{prs/km}}\left|\frac{s}{\text{km}}\right| \tag{5.11}$$

5.2.3 Case a: Uniform Passenger Distribution

This is the simplest but also the least realistic case: it may resemble only some segments of real-world transit lines. Since both boardings and alightings are uniform along the line, the ratio of passengers on trains to those along the line is constant, so that optimal spacings S^* must be uniform. To derive S^*, total passenger travel time PT must be expressed as a function of S.

The given line segment with $n + 1$ stations has n spacings (G and H pairs), from which passengers gravitate to individual stations (Figure 5.2). Boarding density being b (prs/km), there are $G \cdot b$ persons from each section G approaching stations with an average access time $G/2V_a$. To obtain the total passenger access time PT_a, corresponding access times from sections H are added, multiplied by two to account for departure times at destination, and by n to include all stations. Note that terminal station 1 has only a G section and terminal station $n + 1$ has only an H section, adding up to one station and making the total of n, instead of $n + 1$, stations.

Waiting time t_w for all passengers boarding along the line, $b \cdot L$, is added:

$$PT_a = \left(G \cdot b \frac{G}{2V_a} + H \cdot b \frac{H}{2V_a}\right) 2n + L \cdot b \cdot t_w$$

$$\left|\frac{PT}{\text{prs-h}}\right|\frac{G,\, H,\, L}{\text{km}}\left|\frac{b}{\text{prs/km}}\right|\frac{V}{\text{km/h}}\left|\frac{t}{\text{h}}\right| \tag{5.12}$$

Substituting n by L/S and rearranging terms yields

$$PT_a = \frac{L \cdot b}{S \cdot V_a}(G^2 + H^2) + L \cdot b \cdot t_w$$

$$\left|\frac{PT}{\text{prs-h}}\right|\frac{G,\, H,\, L,\, S}{\text{km}}\left|\frac{b}{\text{prs/km}}\right|\frac{V}{\text{km/h}}\left|\frac{t}{\text{h}}\right| \tag{5.13}$$

Introducing expressions for G and H from Eqs. (5.8) and recognizing from Eq. (5.7) that

$$\alpha^2 + \beta^2 = (1 + \rho^2)/2 \quad \text{and} \quad \beta - \alpha = \rho$$

Eq. (5.13) can be developed and reorganized as:

$$PT_a = \frac{L \cdot b}{S \cdot V_a}[(\alpha^2 + \beta^2)S^2 + 2\gamma \cdot S(\beta - \alpha) + 2\gamma^2] + L \cdot b \cdot t_w$$

$$= \frac{L \cdot b}{V_a}\left[\frac{1 + \rho^2}{2} S + 2 \cdot \gamma \cdot \rho + \frac{2\gamma^2}{S}\right] + L \cdot b \cdot t_w$$

$$\left|\frac{PT}{\text{prs-h}}\right|\frac{L,\, S,\, \gamma}{\text{km}}\left|\frac{b}{\text{prs/km}}\right|\frac{V}{\text{km/h}}\left|\frac{\rho}{-}\right|\frac{t}{\text{h}}\right| \tag{5.14}$$

The total number of passengers boarding along the given line segment per unit of time is $L \cdot b$. If the average length of travel on the line is L_a, each passenger passes L_a/S interstation spacings; if the spacings are long enough so that V is reached, Eq. (5.5) for travel time applies on each spacing. Thus, the total passenger time on trains is:

$$PT_t = L \cdot b \frac{L_a}{S}\left(\frac{S}{V} + T_\ell\right) = L \cdot b \cdot L_a \left(\frac{1}{V} + \frac{T_\ell}{S}\right)$$

$$\left|\frac{PT}{\text{prs-h}}\right|\frac{L,\, S}{\text{km}}\left|\frac{b}{\text{prs/km}}\right|\frac{V}{\text{km/h}}\left|\frac{T}{\text{h}}\right| \tag{5.15}$$

The two components of passenger time, access time PT_a and on-train time PT_t, are now expressed as functions of S by Eqs. (5.14) and (5.15). To derive the optimal spacing S^* for the minimum total travel time PT, these two times are added up (Eq. 5.1):

$$PT = \frac{Lb}{V_a}\left(\frac{1 + \rho^2}{2} S + 2\gamma\rho + \frac{2\gamma^2}{S}\right) + L \cdot b \cdot t_w + L \cdot b \cdot L_a\left(\frac{1}{v} + \frac{T_\ell}{S}\right) \quad \text{(Units as above)} \tag{5.16}$$

and derivative of PT with respect to S is calculated and set equal to 0:

$$\frac{\partial PT}{\partial S} = \frac{L \cdot b}{V_a}\left(\frac{1 + \rho^2}{2} - \frac{2\gamma^2}{S^2}\right) - \frac{L \cdot b \cdot L_a \cdot T_\ell}{S^2} = 0 \qquad \left|\frac{PT}{\text{prs-h}}\right|\frac{L, S}{\text{km}}\left|\frac{b}{\text{prs/km}}\right|\frac{V}{\text{km/h}}\left|\frac{T}{\text{h}}\right| \tag{5.17}$$

If this equation is multiplied by $S^2V_a/(L \cdot b)$, it becomes

$$S^2\frac{1 + \rho^2}{2} - 2\gamma^2 - L_a \cdot T_\ell \cdot V_a = 0 \qquad \left|\frac{S, \gamma, L_a}{\text{km}}\right|\frac{\rho}{-}\left|\frac{T_\ell}{\text{h}}\right|\frac{V_a}{\text{km/h}}\right| \tag{5.18}$$

Replacing $T_\ell \cdot V_a$ in the third term by 2γ and solving the equation for S, one obtains the optimal station spacing for the minimum total passenger time from origins to destinations, expressed as:

$$S^* = 2\sqrt{\gamma(\gamma + L_a)/(1 + \rho^2)} \qquad \left|\frac{L, S, \gamma}{\text{km}}\right|\frac{\rho}{-}\right| \tag{5.19}$$

If it is assumed that people go to the nearest station (which is realistic for walking to a fast line, i.e., $V_a << V$), expressions are simplified. Since $G = H = S/2$, Eq. (5.14) becomes:

$$PT_a = L \cdot b\left(\frac{S}{V_a} + t_w\right) \qquad \left|\frac{PT}{\text{prs-h}}\right|\frac{L, S}{\text{km}}\left|\frac{b}{\text{prs/km}}\right|\frac{V}{\text{km/h}}\left|\frac{T}{\text{h}}\right| \tag{5.20}$$

while PT_t remains unchanged (Eq. (5.15)). Following the same procedure as above, one obtains the optimal spacing for access to the nearest station:

$$S^* = 2\sqrt{2L_a \cdot T_\ell \cdot V_a} = 2\sqrt{\gamma \cdot L_a} \qquad \left|\frac{L, S, \gamma}{\text{km}}\right|\frac{V}{\text{km/h}}\left|\frac{T}{\text{h}}\right| \tag{5.21}$$

Due to the liberal assumptions of the model, Eqs. (5.19) and (5.21) cannot be used for precise computations of spacings in the planning of transit lines; however, they can be helpful to the planner in finding the general magnitude of spacings for specific cases. Moreover, the equations show some important relationships: spacings should be longer where trip lengths are greater, time loss per one train stopping is greater, and access speed is higher. The correlations are less than linear, however.

5.2.4 Case b: Uniform Cumulative Boarding

In this case, the transit line serves an area with uniformly distributed demand for travel to a single point: the line's CBD terminal or several CBD stations. Although uniform demand is seldom found, this cumulative travel pattern is typical for many radial lines, particularly on regional rail systems.

The basic model is slightly modified. Instead of a line segment, the served corridor with length L is observed, as Figure 5.3 shows. The position of any station k is defined by its distance D_{k-1} from the origin (beginning of the corridor), while the passenger shed line in section S_k lies at a distance E_k from the origin. All passengers travel to station $n + 1$ (CBD) at the end of the corridor. The optimal locations of stations 1 to n will be derived here.

A theoretical solution for this problem, derived by Vuchic and Newell (1968), gives the following relationships among spacings:

$$S_1 = \frac{1}{\beta}S_0 + \frac{\gamma}{\alpha} \qquad \left|\frac{S, \gamma}{\text{km}}\right|\frac{\alpha}{-}\left|\frac{\beta}{-}\right| \tag{5.22}$$

and

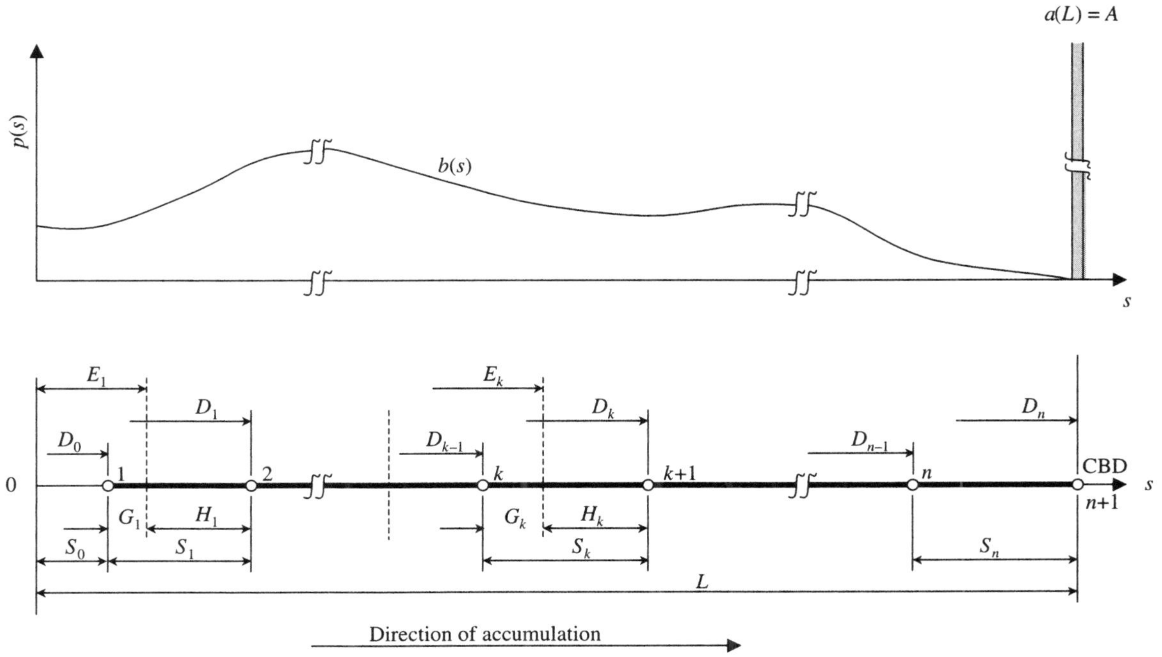

Figure 5.3 *Model of a transit line with cumulative boarding.*

$$S_k = S_{k-1} + \frac{\gamma}{\alpha\beta} \quad \textit{for} \quad k = 2, \ldots, n \qquad \left| \frac{S,\ \gamma}{\text{km}} \middle| \frac{\alpha}{-} \middle| \frac{\beta}{-} \right| \tag{5.23}$$

indicating that the optimal spacings increase in an arithmetic progression: each spacing increases in the direction of cumulation by a constant length. Both equations can be incorporated into one:

$$S_k = \frac{1}{\beta} S_0 + \frac{\gamma}{\alpha \cdot \beta}(k - \alpha), \quad k = 1, \ldots, n \qquad \left| \frac{S,\ \gamma}{\text{km}} \middle| \frac{\alpha,\ \beta,\ k}{-} \right| \tag{5.24}$$

The position of the first station is defined as:

$$S_0 = \frac{2\alpha \cdot \beta \cdot L + \gamma \cdot n(1 - n - 2\beta)}{2\alpha(n + \beta)} \qquad \left| \frac{L,\ S,\ \gamma}{\text{km}} \middle| \frac{\alpha,\ \beta,\ n}{-} \right| \tag{5.25}$$

The optimal number of stations n^* is found by:

$$n^* = \left[\frac{1}{2} - \beta + \sqrt{\frac{2\alpha \cdot \beta}{\gamma} L} \right] \qquad \left| \frac{L,\ \gamma}{\text{km}} \middle| \frac{\alpha,\ \beta,\ n}{-} \right| \tag{5.26}$$

the brackets indicating the nearest integer smaller than or equal to the obtained value.

To summarize, the computations of optimal spacings must start by computing α, β, and γ using Eqs. (5.7). Then n^* is computed by Eq. (5.26), S_0 by (5.25), and all S_k's by Eq. (5.24). It is interesting to note that solutions for this and the previous case are independent of the value of the volume of boarding passengers b, as long as it is constant. This is because the density of stations along a line depends on the ratio of the number of passengers on trains and the density of passenger origins and destinations along the line, and this ratio is the same for uniform distribution regardless what the density of origins and destinations is.

5.2.5 Case c: Variable Cumulative Boarding

This situation is also typical for radial lines, but it is more realistic than case b, since it covers any distribution of boarding: $b = f(x)$. The assumption that $S_k \geq S_c$ for $k \geq 1$ applies both in the preceding case and in this case.

The optimal solution, which can be found either via derivatives of the total travel time or using a dynamic programming approach, is given by relationships of adjacent G and H sections. Each H_k is geometrically related to G_k via parameters α, β, and γ:

$$H_k = \frac{\beta}{\alpha} G_k + \frac{\gamma}{\alpha} \qquad \left| \frac{G,\, H,\, \gamma}{\text{km}} \right| \frac{\alpha,\, \beta}{-} \right| \tag{5.27}$$

while each G_{k+1} is related to the preceding H_k via the total number of passengers boarding in each section:

$$B(G_{k+1}) = \frac{\alpha}{\beta} B(H_k) \quad \text{for} \quad k = 0, \ldots, n-1 \qquad \left| \frac{B}{\text{prs}} \right| \frac{\alpha,\, \beta}{-} \right| \tag{5.28}$$

where functions B are integrals for respective sections:

$$B(H_k) = \int_{E_k}^{D_k} b(s)\, ds \qquad \left| \frac{B}{\text{prs}} \right| \frac{b}{\text{prs/km}} \right| \frac{s}{\text{km}} \right| \tag{5.29}$$

Equation (5.28) shows that the population in any section G_{k+1} should be a constant multiple, α/β, of the population in the section H_k preceding it. Actual computations of these values may be complicated since H_0 ($= S_0$) cannot be determined uniquely. Yet the value of these equations is in the relationships they define. By way of illustration, the optimal spacings for three simple variations in demand distribution, computed using these equations, are plotted in Figure 5.4. The first case is with uniform cumulative boarding: spacings increase arithmetically. Station locations in the other two cases show the influence of the ratio of passengers along the line to those on trains. Note that in the first third of the line station spacings are identical for all three cases, since this ratio of passengers is the same; the value of b where it is constant has no influence on S's. A detailed analysis and derivation of solutions for cases b and c appears in Vuchic (1968).

5.2.6 Case d: Variable Boarding and Alighting

This is the most general and therefore most realistic case. The equations are similar to those for case c, but considerably more complicated because they include boarding as well as alighting functions. Instead of presenting here the complex theoretical equations for the continuous case, a practical method for analysis of lines with variable boarding and alighting is described in the following section.

5.2.7 Application of Theoretical Analysis to Rapid Transit Station Planning

The preceding definitions of travel time components and different types of origin-distribution patterns along a transit line can be used in a practical method for determining station locations.

As already mentioned, the need for a station at any point along a rapid transit line depends on the ratio of two groups of people: those desiring to board or alight from the train want to have a station, while those remaining on the train, i.e., traveling farther, do not want to have any station. The first group, persons wanting to board or alight, is computed by adding up the $b(s)$ and $a(s)$ functions. The second group, passenger on trains traveling through the section, is computed as the passenger volume function $p(s)$ reduced by passengers

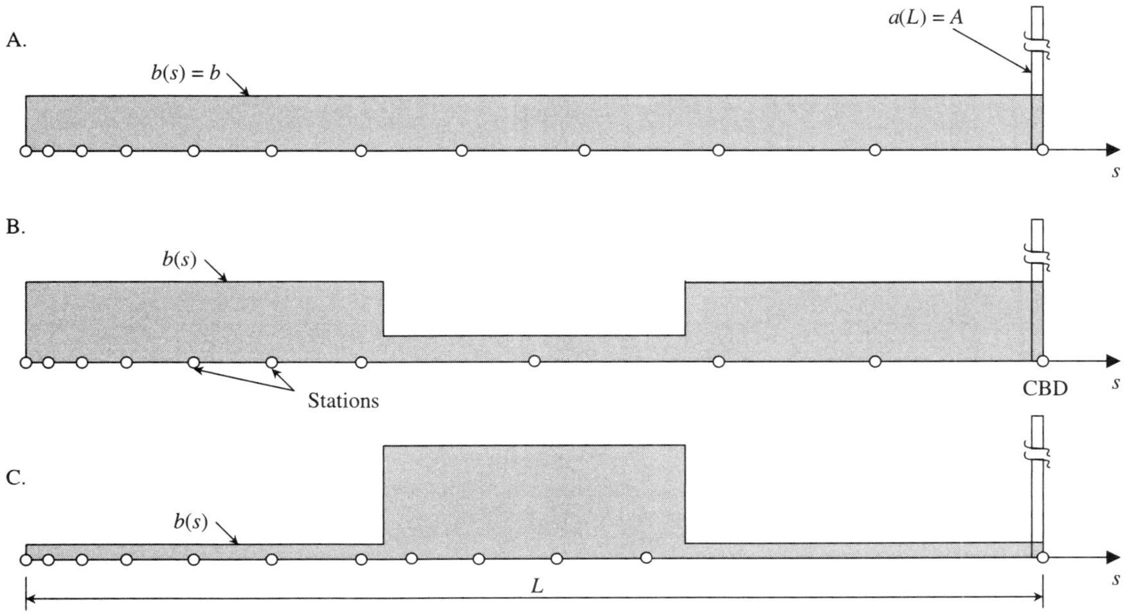

Figure 5.4 *Optimal station locations for minimum travel time with different cumulative boarding distributions.*

alighting function $a(s)$. In the real world, all these functions are not continuous, but discrete numbers at stations.

The planning of station locations should start by tentatively designating a large number of potential station locations. Based on the demand estimate, a diagram is developed with boarding and alighting bars at each station and a step-type plot of passenger flow along the line (see Section 1.1.3 and Figure 1.3). This type of diagram for a set of numbers is shown in Figure 5.5(a).

The values from this diagram are now used to compute and plot a diagram of two groups of passengers which influence the selection of stations as Figure 5.5(b). At each station the sum of boarding and alighting passengers, $b_i + a_i$, is plotted and a line is drawn connecting those values, called the *Stop line*. It represents the number of persons who want to have a stations. The number of persons traveling on trains minus those who want to alight, $p_{i-1} - a_i$, represents the persons who do not want to be delayed by a station. The *Go line* connects those values at stations for visual clarity.

The numbers of the two groups of passengers, highlighted by the Stop line and the Go line, cannot be compared directly, because each station is much more important for the boarding and alighting passengers to use of the line than it is for passengers traveling through who would only lose about one minute by stopping. Thus, the Stop value has a considerably greater relative weight than the Go value at a station, and a higher Go than Stop value does not indicate that that station should be eliminated. Actually, the Go value is considerably higher than the Stop value at most transit stations. Only stations at which Stop is drastically lower than Go should be eliminated.

In the example given in Figure 5.5, locations at which the Stop line is much higher than the Go line are strong stations (e.g., stations F, G, and K in the diagram). Stations at which the Go line is much higher than the Stop line (C, E, I, and M in the example) are weak and should be considered for elimination. After

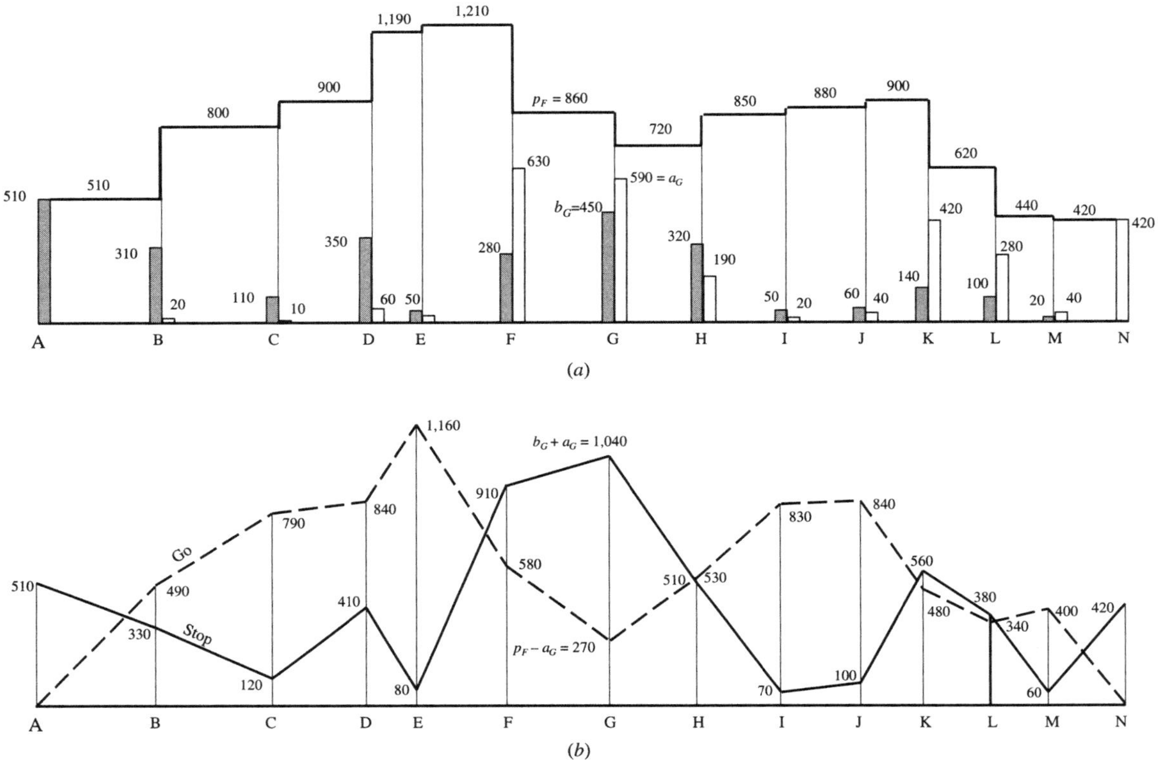

Figure 5.5 *Diagrams for practical planning of rapid transit stations: (a) diagram of boarding* (b_i)*, alighting* (a_i)*, and passenger flow* (p_i)*. In this case,* i = A, B, . . . , N*; (b) Stop line* ($b_i + a_i$) *and Go line* ($p_{i-1} - a_i$).

elimination of each proposed station, its estimated passengers should be reassigned to adjacent stations (unless they are diverted to other modes), thereby strengthening them; their exact locations should then be rechecked. How far the elimination process goes depends on other objectives and considerations in planning.

5.3 OTHER OBJECTIVES IN STATION LOCATION SELECTION

Several other objectives that should be considered are not as complicated as travel time. Their models are rather simple, yet they also require careful attention in planning.

5.3.1 Area Coverage

Maximizing area coverage is an important goal in metro network design because system usage depends on the ability of potential passengers to easily get to stations. Area coverage is a direct result of density of stations, as the following model shows. For simplicity, two very liberal assumptions are made: first, that stations can be placed at any location along a line; and second, that there is ubiquitous access, i.e., people can access any station in straight line. This makes area coverage by a station a circular surface with a radius of maximum walking distance r_a.

If a line segment with length L has only two stations, its area coverage is limited to two circles around them with radii equal to r_a, as shown in Figure 5.6(*a*).

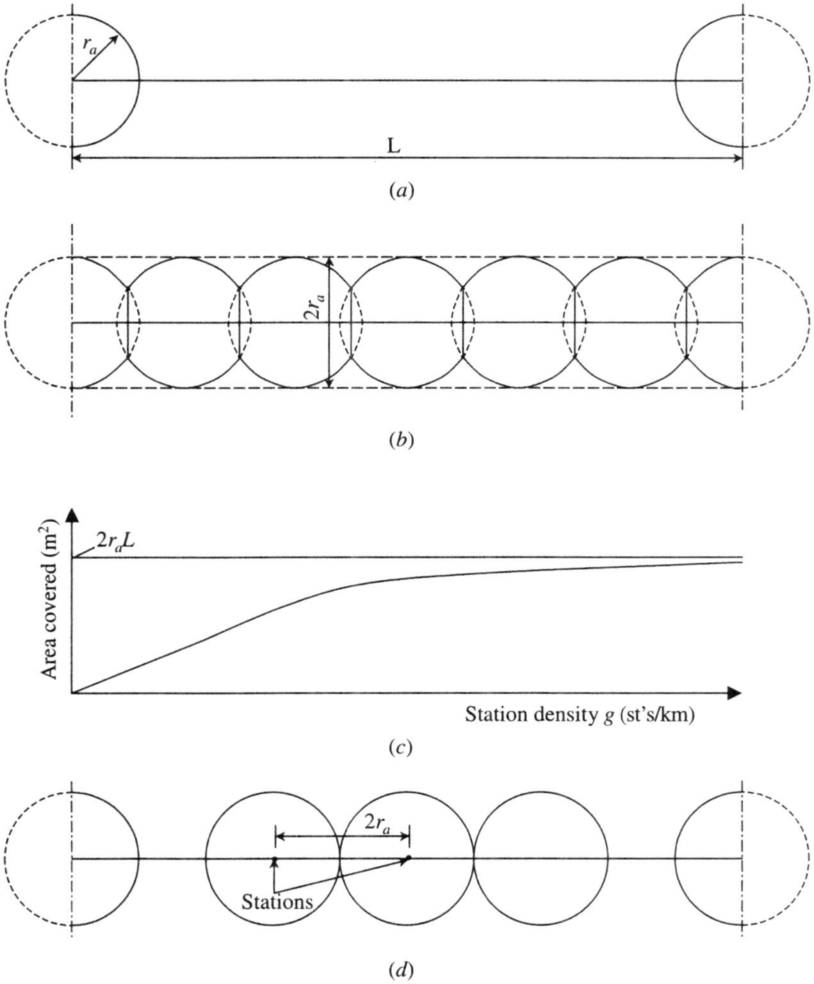

Figure 5.6 *Area coverage as a function of station density.*

Each additional station increases area coverage by another $r_a^2 \cdot \pi$ area, until these areas touch and then begin to overlap (Figure 5.6(*b*)). There the increase in coverage begins to diminish, as the diagrams in Figure 5.6(*b*) and (*c*) show. The theoretical maximum coverage by a line, $2r_a \cdot L$ (Figure 5.6(*c*)), would be achieved by a continuous station (obviously an unrealistic solution for rapid transit service).

Due to the requirement for high operating speed on rapid transit and regional rail lines, station spacings are seldom shorter than two walking distances (Figure 5.6(*d*)) or 800 m, except in high-density CBD areas. When these lines are designed to provide continuous coverage, station locations are often determined by these distances. Thus, most station spacings on metro networks in Paris, Philadelphia, and Hamburg are in the range of 700–1200 m. These short spacings based on passenger access time result in certain sacrifice of operating speed. Such lines are often too slow for trips longer than 10–12 km. Regional rail lines often give continuous coverage of outlying areas, but with somewhat longer spacings (1200–1700 m) to maintain higher speed. For lines that rely heavily on rapid transit or surface transit feeders in the city (St. Petersburg, Paris RER) or on automobile access in outlying areas (Cleveland, Chicago's Skokie Swift, and many regional rail lines), area coverage in terms of walking radius is given less attention.

5.3.2 Attraction of Passengers

The basic element of passenger attraction that is strongly influenced by station density and locations is operating speed on a metro line. This is particularly important for travel that parallels a competing high-speed facility, such as a freeway.

Operating speed for a given set of stations can be computed by physical/mathematical equations or by computer simulation and graphically presented on a time-speed diagram, as shown in Figure 5.7. However, the overall average for an entire line or a network is not sufficient for planning purposes, since operating speeds on different sections vary, mostly as a function of station spacings. Modal split of trips on individual origin-destination (OD) pairs is affected by the operating speed on the line segment serving that specific pair. Therefore, in addition to the average speed for the whole line, operating speeds should be computed for each important segment, as shown on the lower diagrams in Figure 5.7. This analysis allows evaluation of adequacy of speed on each segment depending on physical conditions and competing services.

Naturally, absolute travel time also affects passenger attraction, regardless of competing services. Therefore, design for minimum total travel time also has an influence on passenger attraction.

5.3.3 Cost of Stations

Each station involves certain investment cost (for construction) and operating cost (for station operation and maintenance, for stopping of trains, and for larger rolling stock due to longer cycle time due to stopping). For equal local conditions, the incremental cost per

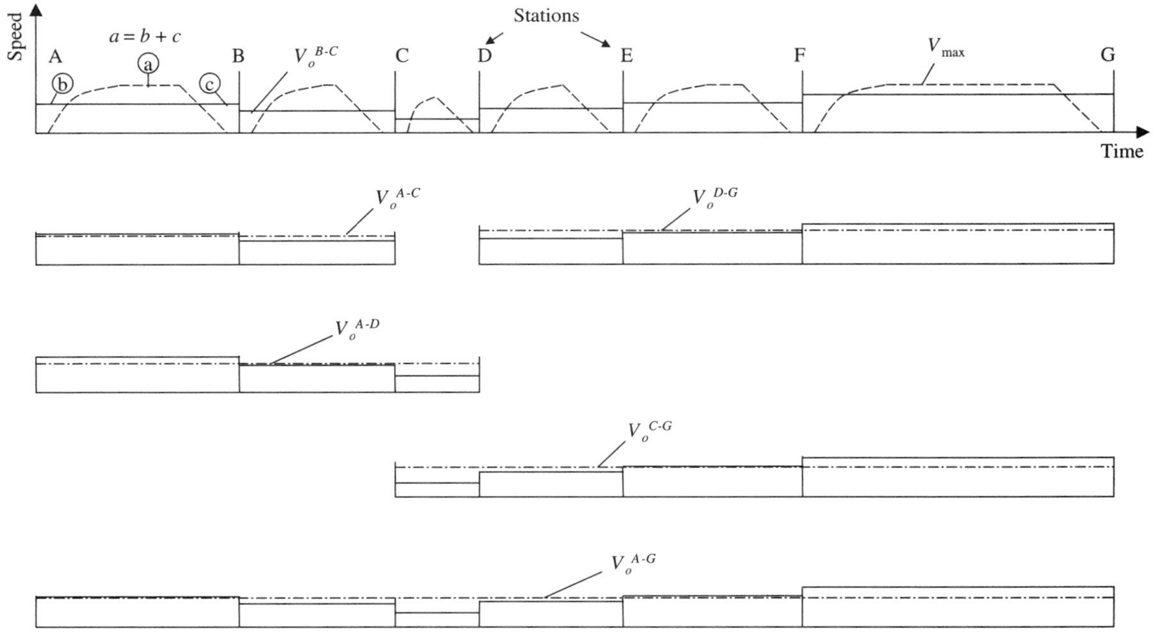

Figure 5.7 *Analysis of operating speed on different sections along a transit line.*

station is constant when stations are far apart, but it decreases slightly when spacings become so short that trains cannot reach maximum speed because the incremental time and energy consumption per stopping are then slightly reduced. Thus, the total cost of stations increases with their density, but at a decreasing rate, as plotted later in Figure 5.8.

5.3.4 Auto-Transit Interface

Since the automobile is the dominant mode of transportation in low-density suburban areas, many trips between these areas and the central city are made most conveniently by a combination of auto and transit modes. To encourage such trips, an intermodal interface should be provided at a number of stations around the central city.

At many older stations at which walking and feeder transit are the main access modes, facilities for park-and-ride (P+R) and kiss-and-ride (K+R) have been added to allow access as well by automobile from dispersed areas. Some recently built lines, however, have auto access as the dominant mode (on the PATCO Line in Philadelphia, 89% of passengers use P+R or K+R). Such heavy use of the automobile for access has a major impact on station planning in three respects:

First, the average distance for access to stations becomes much greater (see Figure 4.1). Some trip origins are 5–10 km from the stations. Long-distance trips from other cities can also be intercepted by well-located P+R facilities. Thus, in planning station locations, the concept of area coverage is broadened by the concept of accessibility to potential station locations from major highways. As a consequence, lines relying heavily on auto access have much longer station spacings in suburban areas than those served by other access modes (see Section 5.7). In central cities, access by walking and surface transit remain dominant modes and spacing should not be much greater than was the case in the pre-automobile era.

Second, stations with heavy auto access represent major traffic generators and therefore require land area and careful design for good traffic circulation.

Third, because of large traffic volumes and corresponding space requirements, P+R facilities are not

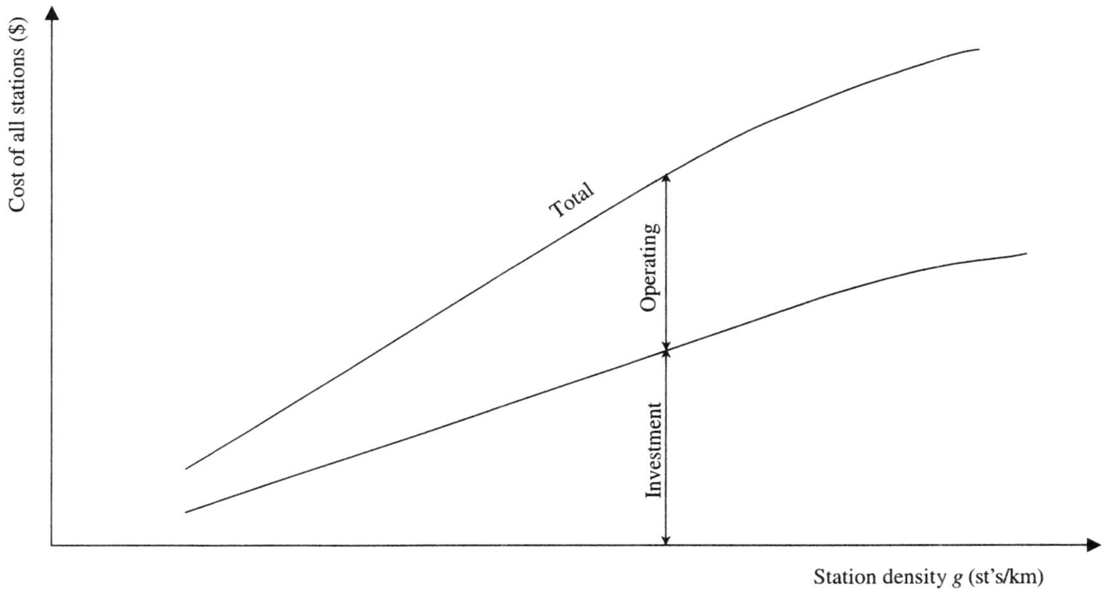

Figure 5.8 *Total cost (investment and operating) of stations as a function of their density along a line with fixed length.*

Photo 5.1 *Suburban rapid transit station with P+R—New Jersey suburb of Philadelphia.*

environmentally compatible with centers of suburban towns, which are often prime locations for rail transit stations. Heavy commuter auto traffic may be disruptive for the residential, shopping, and other human-oriented activities or historic character of these areas. A possible solution to this problem is separation of these two functions among different stations of a line: stations at suburban centers should rely on walking, transit feeders, and K+R but have only moderate-capacity, if any, P+R facilities. At the points where the rail line crosses major highways (typically, circumferential freeways), stations with major P+R facilities are appropriate. The capacity of such facilities may be as high as 2000–4000 parking spaces.

Transit-highway interface facilities should be planned as a system serving all major approaches to the city. Acceptance by auto drivers depends on several conditions. The most important one is that the offered rail transit service is comparable with or superior to driving into the city (including travel time, comfort, fare, auto operating costs, parking fee, etc.). Due to their high-quality service, regional rail and metro systems in many cities have extensive and well-used P+R facilities. In some cases, such as many suburban stations of Washington Metro, San Francisco BART, and Atlanta MARTA, large garages are filled with cars every weekday.

Regional rail systems typically depend on P+R and K+R facilities even more than metros because they predominantly serve remote suburbs and offer services highly competitive to that of private automobile. Virtually all regional rail systems in North American cities depend heavily on automobile access. So do most cities in other industrialized countries. In Japan, bicycles are used massively for access to rail stations. Light rail systems extending to suburbs, such as the Cleveland

Shaker Heights and Boston Riverside Lines, and most of the recently built LRT systems (e.g., Calgary, Portland, Sacramento) and in some cases express buses (New Jersey suburbs of New York, Ottawa, Seattle) have well-planned and well-utilized P+R facilities. An important factor for P+R acceptance is adequate information provided to potential transit users by means of detailed intermodal highway/transit maps and effective roadway signing.

Transit planners must consider the auto-transit interface via P+R facilities as a component of an intermodal system, with its advantages and disadvantages. As discussed, P+R is used extensively in most cities with rail systems, particularly in North America, but it has some serious drawbacks. Automobile parking consumes very large area (20–30 m^2 per space; see Vuchic, 1981), P+R facilities impede developments in the immediate vicinity of rail transit station and attract heavy vehicular traffic. Investment cost per space in surface lots is \$3000–6000 (based on San Francisco BART data). If garages are built, the occupied land area is smaller, but the investment cost is as high as \$20,000 per space. Overall, auto access via P+R is by far the most expensive and disruptive transit feeder mode, much more costly and imposing than feeder paratransit, bus transit, bicycle, and pedestrian modes. Why is it used, then? Because P+R is still much less costly and disruptive than letting auto users drive to center city, congesting highways and imposing the land consumption and costs of parking in center city, with much higher direct and indirect costs than P+R in suburbs.

Some transit systems charge for P+R and apply the revenues for further investment and operation of P+R facilities. However, some systems, such as BART, have a policy of providing "free" P+R. While this policy does encourage use of P+R and contributes to the attraction of transit system users, it is wrong and damaging with respect to its economic impacts and social equity. Free P+R allows use of expensive facilities without any user charge, necessitating use of public subsidies that could be better used to improve transit services by greater frequency, better rolling stock, etc. Considering transit users, the policy of free P+R actually results in passengers who use feeder transit, particularly those who walk to stations, subsidizing auto users, who impose the highest cost; this is obviously a very inequitable situation.

Policies toward providing, financing, and charging for P+R are strongly influenced by the heavy subsidy of automobile travel in urbanized areas, which is particularly the case in U.S. cities. Extremely low direct user charges, cheap gasoline, and frequent subsidies to parking even in central cities force transit agencies to subsidize P+R in suburbs. This situation is more economically efficient and socially equitable in cities that have a reasonable degree of coordination of policies toward different modes, particularly with respect to charges for automobile use. Restriction of parking in center city and charges for driving into center city (such as applied in Singapore, Oslo, and London) are major steps in that direction.

5.3.5 Local Objectives and Requirements

The preceding objectives (minimum travel time, adequate travel speed, station costs, and intermodal integration) should be considered in planning any rapid transit line. In addition to them, however, there are often various local objectives, physical conditions, or other factors that influence station locations. Actually, the most important station locations are usually fixed by the locations of other transportation terminals, major activity centers, pedestrian precincts, office complexes, college campuses, highway intersections, constraints imposed by rivers or hills, etc. Even in outlying areas, stations are often determined by locations of suburban activity centers, availability of land, or a deliberate policy of stimulating development of specific urban form or conforming land use.

Consequently, the locations of most major stations are strongly influenced or determined by various local factors. Generally, passenger time, line speed, and overall cost dominate the determination of intermediate stations in areas of uniform land uses. Hence, analyses are often made for individual segments, rather than the entire lines.

5.4 INTEGRATION OF OBJECTIVES

Since some of the discussed objectives partially overlap (minimum passenger time and maximum operating speed), while others are mutually conflicting (maximum area coverage and minimum cost), the optimal solution cannot satisfy equally all objectives. All of them must be analyzed together with local factors, and a compromise among them must be selected partly by comparing quantitative measures (costs, time) as well as subjective judgments concerning relative importance of such diversified factors as interactions and stimulus of land development, area coverage and enhanced livability in center city. This integration of objectives can be shown conceptually as in Figure 5.9, where the travel times, costs and other forms of disutilities of both passengers and transit operator are plotted as functions of station density to show how they influence the optimal station density.

The diagram shows the following disutilities:

- Passenger time curve from Figure 5.1
- Area coverage: the decreasing area not covered as station density increases (the difference between the theoretical maximum of $2r_a \times L$ and the actual area coverage shown in Figure 5.6*c*)
- Station cost curve from Figure 5.8
- The disutility of decreasing operating speed V_o is represented by its inverse, increasing slowness in min/km as density of stations increases

The curve of total disutility has a convex shape with a minimum that represents the overall optimal density for all these objectives, g^{**}. It is emphasized again that diagram is only conceptual and cannot be interpreted as quantitatively exact; ordinates have different dimensions for different curves and could be summed only if relative weights were given to each one. Although summation is possible, subjective judgment is usually also used in this evaluation of different objectives. Moreover, this diagram yields only average density. For specific locations of individual stations, local factors and the analysis from Section 5.2.7 must be applied.

The basic problem in transit station planning, how to provide high speed without sacrificing area coverage, can be alleviated on some lines by skip-stop operation. When a line operates with high service

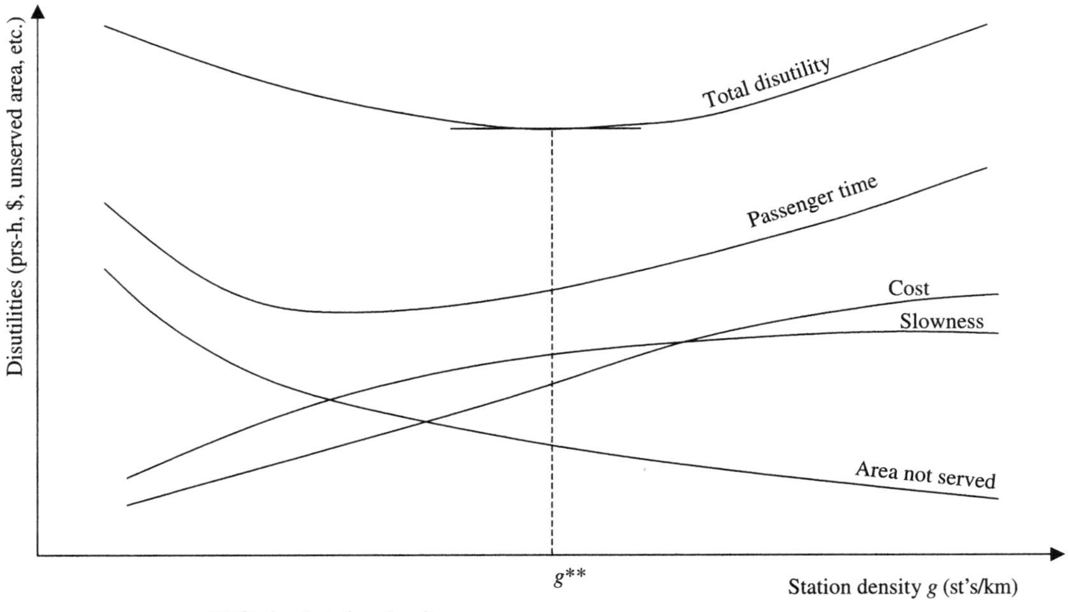

Figure 5.9 *Conceptual diagram of integration of objectives for optimal station density.*

frequency, skip-stop operation allows short station spacings (good area coverage), while train operating speed remains high because each train skips some of the stations. Sacrifice of frequency at less important stations thus permits good coverage and high speed. The methodology for comparing skip-stop with regular operation, which may often be useful in network planning, is presented in Section 2.3.

Application of skip-stop and other accelerated services can be increased where fully automated train operation were introduced. This makes higher station density more desirable with ATO than with manual driving.

5.5 ADDITION OR CLOSING OF A STATION

Major changes in the vicinity of a metro line sometimes create the need for the addition of a new station or closing of an existing one. Since such an action always affects many passengers and other groups and involves substantial investment or saving, it requires a comprehensive evaluation of all its direct and indirect consequences. The systematic evaluation methodology described in Section 3.3 should be used here. It should include positive and negative, quantitative and qualitative, effects on different affected groups.

A basic structure for evaluation of the adding of a new station to an existing metro line is presented in Figure 5.10. Major events caused by the project, given in the first column, include one-time actions (e.g., construction) and permanent changes (operating costs, passenger diversion); therefore, all quantifiable items must be computed on a life-cycle basis.

Consequences of the events are classified by their incidences on the three major groups: passengers (those presently traveling through the section where the station is to be added, as well as passengers generated by the station), operator, and community. Important items that must be included in each evaluation are emphasized by heavy frames. Many of the other items, particularly those framed by dashed lines, are in many cases either difficult to estimate or of negligible importance. Thus, the delay to trains caused by one additional station is negligible; diversion of passengers traveling through that section to other modes is usually close to zero. However, if several stations are to be added, these items become worth examining, since the character of the service may be changed.

As in system evaluation, presented in Chapter 11, the items are evaluated in terms of dollars or other quantitative units, or by qualitative description. The dominant items, usually the increased number of passengers on the positive side with costs of construction, fleet increase, and operations on the negative side, must be supplemented by an estimate of the benefits accrued by the attracted passengers, remaining auto drivers due to lower congestion, improvements in the station's surroundings, etc.

Closing a station has similar consequences, but in the opposite direction: the action causes loss of passengers at the closed station, reduced operating costs, and a very low investment for closing, unless a major building must be destroyed. Thus, the same methodology and items from Figure 5.10 can be used to evaluate a station closing.

5.6 AREA COVERAGE VERSUS OPERATING SPEED

Planning of rapid transit stations always faces a basic dilemma: closer stations (short spacings between them) result in better area coverage and thus easy accessibility for larger numbers of potential passengers; at the same time, however, short spacings cause lower operating speed and possible higher fleet requirements, as well as higher construction and station maintenance costs. Longer spacings result in the opposite situation: high speed and good economy of operations, but with the line passing through areas without serving them, since there are no stations. A portion of potential passengers is lost.

5.6.1 Conventional Approaches to Planning of Stations

With respect to station spacings, metro and regional rail networks can be classified into several groups.

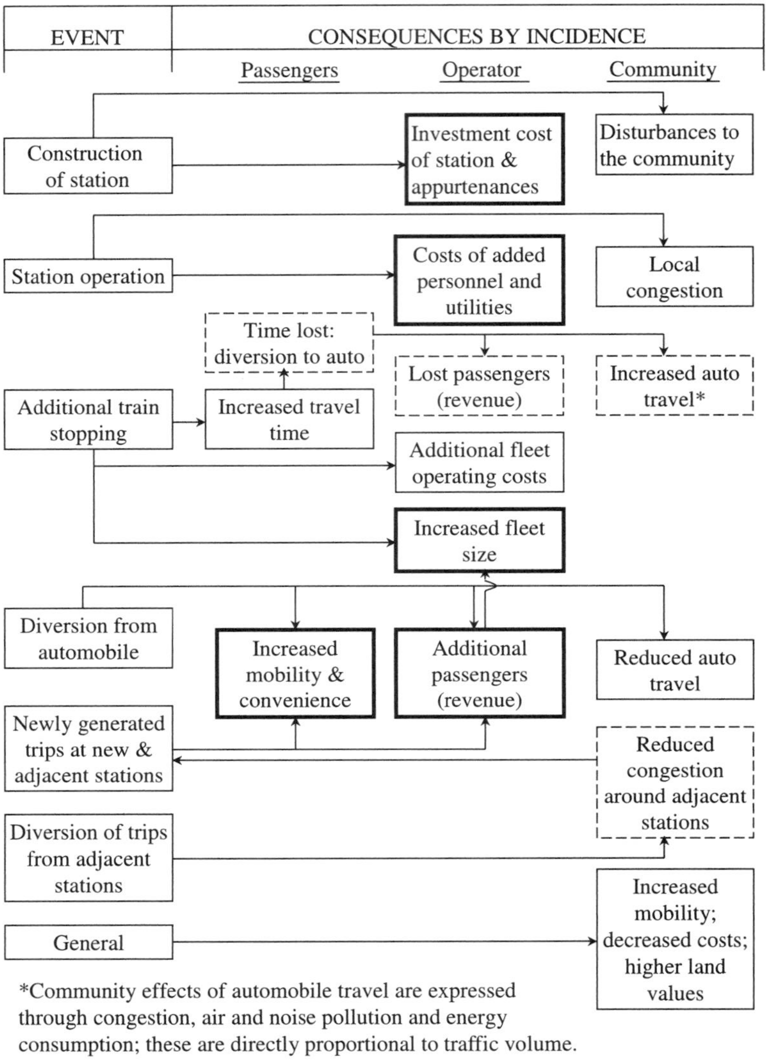

Figure 5.10 *Effects of adding a station to an existing rapid transit line.*

First, metro systems in older cities with high population densities that were built prior to the wide use of the automobile had typical station spacings of only 500–800 m. This was the logical design since the density of travel demand was high and average travel speed was not very important because trip lengths were relatively short.

Second, some recently built metro systems have been designed in a drastically different way due to a number of changed conditions. With rapidly increasing automobile ownership and accelerating growth of low-density suburbs, the freeway mentality of the 1950s and 1960s also dominated planning of metro lines: it was believed that most people would come to stations by auto and travel long distances on transit into central city. Therefore, metro lines should have a speed competitive with that of freeway travel. To achieve such speed, these lines were designed with station spacings averaging more than 1500 m and often as long as 3 km (San Francisco BART has an average spacing of 3907 m).

This latter type of metro line did achieve higher speed and efficient operation, but it was subsequently recognized that even in low-density areas, station spacings exceeding approximately 1.5 km (or 1 mile) bypass many areas that could benefit from better access to rail service and generate many additional passengers. Moreover, construction of relatively few stations dependent on motor vehicle access creates the need for additional automobile-based travel—often with negative environmental impacts on areas surrounding each station, specifically, traffic congestion and related problems. Large parking areas surrounding each station often act as spatial barriers between that station and the surrounding community's shops, office buildings, and other objects.

In an attempt to achieve maximum travel speed for suburban commuters into the city center, these recently built metro lines often go through dense areas immediately around the center city without adequately serving them. Thus, these systems have become more regional than urban: while they adequately serve suburbanites who can easily drive to stations, extremely long station spacings do not satisfy the travel needs of many other potential riders in both suburban areas and the higher-density ring around the central city. Hence, the potential of such lines to foster development (or rehabilitation) of this ring and close-in suburbs through land use impacts is often wasted.

Recognizing the aforementioned problems caused by extremely long station spacings, several metro system managements have partially corrected the situation by adding so-called in-fill stations after the lines were built. Such stations were built on the PATCO Line in New Jersey suburbs of Philadelphia (Woodcrest), on Washington Metro (New York Avenue), and on LRT lines in Sacramento and Dallas.

5.6.2 Use of Skip-Stop Services to Improve Area Coverage

In accelerated services each TU is scheduled to skip some stations along a line (see Section 2.4). Thus, operating speed on the line is higher than the total number of its stations would allow with local (all-stop) operation. Yet access distances to stations are shorter. Consequently, accelerated (particularly skip-stop) services allow building of more stations along a line than for local service and yet retaining of the same operating speed; the only shortcoming is that some of the stations then have longer headways than local service would give them because they are served by every other TU. Comparison of different sets of stations for this type of operation is presented here by a theoretical model.

5.6.2.1 Alternative Sets of Stations. Usually, some station locations are fixed: they may be at major squares, at long-distance transportation terminals, etc. The objective is then to plan the optimal number of stations to be placed on sections of line between the fixed stops. For example, alternative sets of stations can be analyzed on a line segment shown in Figure 5.11, which initially has six stations and a length of six interstation spacings (between the dash lines), shown on the left as the base case). One can then define four different sets, I–IV, of increasing numbers of stations designated for skip-stop operations. The basic set and the four alternative sets are presented in the figure in the order of increasing station density.

The five station sets differ from one another in the number of stations and spacings, station types (A, B, or AB), and consequently in the number of stoppings, travel time, and, in one case, regularity of headways. Sketches and time-distance diagrams for these sets are shown in Figure 5.11. Each one is described here.

The Base Case. Station spacings are L and operation is local. The six spacings comprise the station module, designated AB-AB-AB-AB-AB-AB, which represents the minimum line segment on which four other station modules can fit without any fractional station spacings.

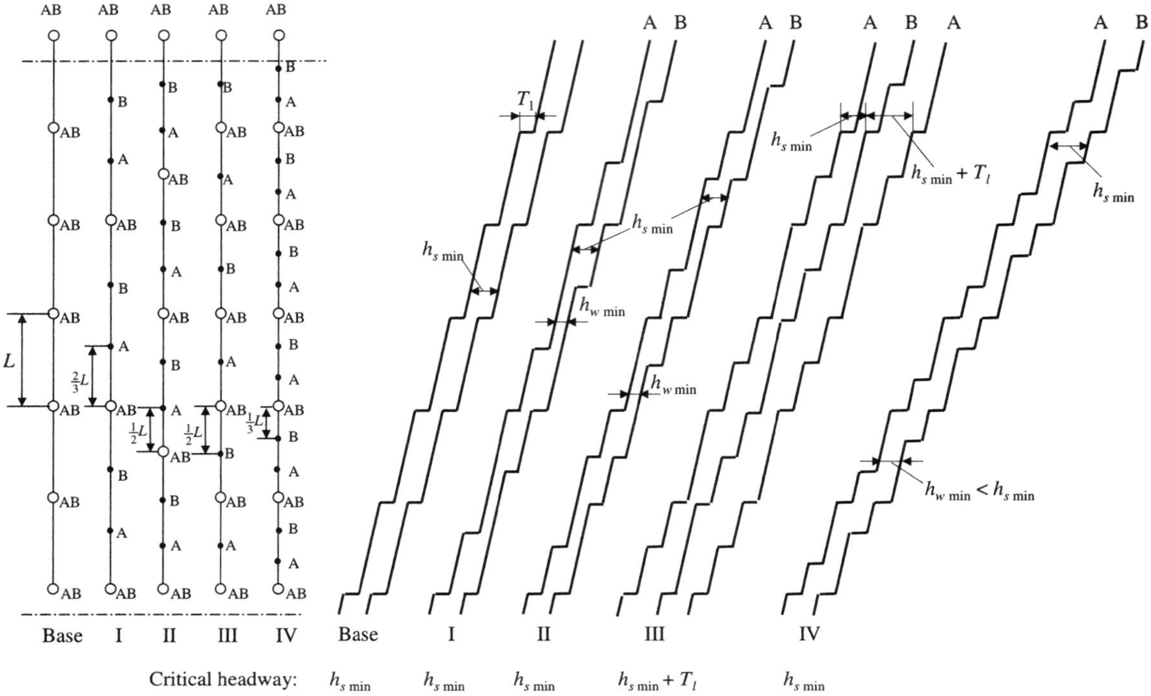

Figure 5.11 *Stopping patterns and time-distance diagrams for different station sets with regular and skip-stop operations.*

Case I. Every other station (in the base case, the second, fourth, and sixth stations from the bottom) is replaced by two skip-stop stations, A and B. Hence, the total number of stations for this case is nine: three of each AB, A, and B stations, with spacings between stations of $2/3\ L$. Since every train stops six times along the observed segment (as designated by the station module AB-A-B), travel time is the same as in the base case. Headways at AB stations are regular because each train stops only once between AB stations: at either an A station or B station.

Case II. In this case, the number (density) of stations is twice as great as in the base case; the spacings are therefore $L/2$. As in case I, every third station is an AB type, and between AB stations is a pair of A and B stations. Thus, the station module on the line remains the same as in case I. However, each train makes eight stoppings along the segment, and headways are regular at all stations. Compared to the base case, travel time increases by approximately $2T_\ell$ (average of A and B fractional stoppings).

Case III. In this case, either an A station or a B station is added between the base case stations (which remain AB type). Average station spacing thus remains $L/2$, as in case II, but every other station is an AB type; stations added are, alternatively, A and B. The station module is AB-A-AB-B.

Each train stops nine times along the segment as defined, resulting in travel time increase by $3T_\ell$. However, headways are not regular at all stations because alternative trains have different travel times on each AB-AB section. Therefore, minimum operating head-

Table 5.1 Alternative sets of stations and their operational characteristics

Case	Number of Stations: AB	A	B	Σ	Average Station Spacing	Number of Stoppings	Average Distance Between Stoppings	Additional Travel Time
Base	6	0	0	6	L	6	L	—
I	3	3	3	9	⅔ L	6	L	—
II	4	4	4	12	½ L	8	¾ L	2 T_ℓ
III	6	3	3	12	½ L	9	⅔ L	3 T_ℓ
IV	6	6	6	18	⅓ L	12	½ L	6 T_ℓ

way on the line is increased, and line capacity is decreased. Assuming regular train departures from an AB_1 station and $h_1 > h_{min}$, headways at the next AB_2 station will alternate as follows:

$$h_2' = h_1 + T_\ell \quad \text{and} \quad h_2'' = h_1 - T_\ell \qquad \left|\frac{h,\ T}{\text{min}}\right| \tag{5.30}$$

The headways at the next AB station will be regular again: $h_3 = h_1$ (see Figure 5.11).

Line capacity is dictated by the longest headway at an AB station, i.e., by $h_{max} = h_2'$. Theoretically, line capacity would have the following value:

$$C_{III} = \frac{60n \cdot C_v}{h_1 + T_\ell} \qquad \left|\frac{C}{\text{sps/h}}\right|\frac{n}{-}\left|\frac{C_v}{\text{sps}}\right|\frac{h,\ T}{\text{min}}\right| \tag{5.31}$$

which is lower than the capacities of other cases which have regular headways:

$$C_I = C_{II} = C_{IV} = \frac{60n \cdot C_v}{h} \qquad \left|\frac{C}{\text{sps/h}}\right|\frac{n}{-}\left|\frac{C_v}{\text{sps}}\right|\frac{h}{\text{min}}\right| \tag{5.32}$$

Case IV. In this case, two skip-stop stations (one pair) are added between every two original AB stations. Hence, the total number of stations on the segment is 18, with 6 AB stations and 12 skip-stop stations. Station density is the greatest in this case, with an average station spacing of 1/3 L. There are 12 stoppings per train, causing travel time to increase by 6 T_ℓ. Headways are regular, for reasons stated previously.

5.6.2.2 Review of Characteristics of Different Station Sets. Reviewing the five cases and proceeding across Figure 5.11 from the base case to case IV, the following parameters change systematically, as shown in Table 5.1: the total number of stations increases, average station spacing decreases (station density and area coverage increase), the number of stoppings per train increases, and consequently, travel time per train also increases. However, travel time remains the same between the base case and case I.

It is important to note that the changes in these parameters are not proportional; for instance, although station density has tripled from the base case to case IV, the number of stoppings per train has only doubled. Consequently, by considering skip-stop operations in the planning phase, transit planners have a greater choice between area coverage and travel time than when they consider local operations only. Depending on local conditions, better area coverage can often be achieved with a less-than-proportional increase in travel time—sometimes none at all.

5.7 STATION SPACINGS ON ACTUAL RAIL TRANSIT NETWORKS

Average station spacings on rail networks in different cities are rather good indicators of the character of these transit systems. In Table 5.2, a number of rail transit systems are listed in four groups. The first group

Table 5.2 Average station spacings and network characteristics of rail transit systems

City	Network Length (km)	Number of Stations	Number of Spacings	Average Spacing (m)
Urban metro systems				
Athens	25.6	44	43	595
Buenos Aires	42.7	63	64	667
Philadelphia (SEPTA)	39.4	53	53	740
Paris (Metro)	211.5	297	284	745
Barcelona (Metro)	88.3	126	111	795
Berlin	146	174	182	802
Boston Rapid Transit	101.5	119	117	867
Munich	92.5	90	98	944
Nagoya	76.5	74	79	968
Montreal	64	65	66	970
Toronto	70	74	71	986
Stockholm	108	100	100	1080
Osaka	113.5	97	103	1102
Oslo	113	94	102	1108
Tokyo (TRTA)	287.3	266	256	1122
Hamburg	101	86	88	1148
Madrid	228	156	196	1163
Chicago	173	151	145	1193
Mexico City	201.7	175	164	1222
Regional rail systems				
Tokyo (TRTA)—suburbs	182.3	165	173	1054
Kiev	54.8	42	39	1405
São Paulo	103.8	2	70	1483
Boston	171.1	119	109	1570
Moscow	269.5	165	157	1717
Cleveland	31	18	17	1750
St. Petersburg	110	60	62	1770
New LRT systems				
Porland Eastside MAX				980
Sacramento (first line)				1130
San Diego				1530
Regional rail systems with P+R				
New York (PATH)	22.2	13	14	1586
Philadelphia (PATCO)	23.3	13	11	2120
San Francisco (BART)	168	43	43	3907

includes urban metro systems that do not extend far into suburban areas; some of these (e.g., Buenos Aires, Paris, and Berlin) provide adequate coverage, reducing the need for many surface lines. Spacings in this group range between 600 and 1200 m, most frequently 800–1000 m.

The second group includes extended metro or rapid transit systems that serve central as well as larger suburban areas, such as Moscow, Cleveland, and St. Petersburg. Station spacings for this group are mostly between 1400 and 1800 m, so that their reliance on surface feeders is greater than for the systems in the first group. Boston and Tokyo are listed in both groups, separately for urban and suburban network portions.

The third group shows several recently built LRT systems in U.S. cities. Serving very dispersed urbanized areas, these systems have longer station spacings than LRT systems in older cities and in European countries. Finally, the fourth group consists of regional rapid transit systems in U.S. cities that rely heavily on access by private automobiles—K+R and P+R modes. Their average spacings in some cases exceed 2000 m, with BART, serving the very large Bay Area Region, having an average spacing of nearly 4 km!

In conclusion, interstation spacings on rail lines depend mostly on the areas they serve, particularly population density in the served areas, and line lengths. Lines serving inner cities tend to have short spacings, which is acceptable when trip lengths are also short. When lines are long, the spacings must be longer to allow higher operating speeds and shorter travel times. Another factor that influences spacings is the submodal split: which modes are used for access to stations. In inner cities walking dominates, and then the spacings are in the range of 600–1000 m. Lines with longer spacings rely on access by bus and other street transit modes. On regional rail lines where spacings exceed 1000 m with individual spacings having lengths of 2000 to over 3000 m, feeder transit and, increasingly, auto access begin to dominate. The designs of stations and area they occupy vary greatly with the type of feeders. Inner city stations are usually in tunnel, and access is by walking, so that they do not occupy large areas. In suburbs, bus feeders and car access, particularly P+R, occupy very large areas.

Photo 5.2 *Rapid transit stations as focal points for urban development—Toronto.*

EXERCISES

5.1. List and discuss the major objectives that should be considered in planning station locations along a rail transit line. Define three major parties interested in the achievement of objectives, and state which objectives affect which of these parties.

5.2. Define the two components of travel time on transit and explain how each one of them is affected by the density of stations along the line. How does the distribution of passengers along the line affect the optimal station density?

5.3. On a diagram of passenger travel time *PT* versus station density *g* along a transit line, plot the general shapes of curves for passenger time for access to stations PT_a, passenger time on trains PT_t, and total passenger time *PT*. Show the optimal density g^* for the drawn curves. Discuss and show on the diagram by dashed lines the changes in the shape of these curves that would result from two of the following three changes in access and operating conditions:

a. Increased train acceleration and deceleration rates

b. Decreased access speed to stations V_a

c. Increased population density along the line

List and define other factors besides travel time that should also be considered in selecting station density and locations.

5.4. Compute passenger shed line positions, or *G* and *H* sections, on the spacings between stations *j*, *k*, and *l*, as shown below. Assume that train operating parameters are: $a = b = 1.1\ \text{m/s}^2$; $V_{max} = 80$ km/h; $t_s = 22$ s; coasting is not used.

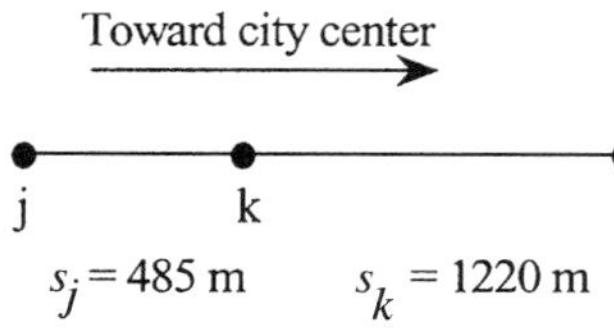

Compute the shed line positions for the following station access speeds:

a. $V_a = 5$ km/h (walking)

b. $V_a = 15$ km/h (bicycle)

c. $V_a = 40$ km/h (car)

5.5. Based on the derivations of optimal station spacing S^* given in the text (Eq. (5.19)):

a. For a line with $T_\ell = 55$ s, $V_a = 5$ km/h, and $V = 80$ km/h, draw a diagram with optimal interstation spacing S^* as a function of average travel distance L_a

b. Draw the same diagram assuming that people go to the closest station (by distance, rather than by minimum travel time).

5.6. Compute the theoretically optimal station spacings S_1 for the minimum travel time, and S_2 for the minimum travel time if passengers go to the closest station, on an LRT line with uniform passenger distribution, assuming five sets of operating and line characteristics:

1. $a = b = 1.0\ \text{m/s}^2$; $t_s = 20$ s; $V_{max} = 70$ km/h; $V_a = 5$ km/h; $L_a = 6$ km

2. Same as 1, but $V_{max} = 60$ km/h

3. Same as 1, except $V_{max} = 60$ km/h and $V_a = 25$ km/h

4. Same as 1, except $V_{max} = 90$ km/h

5. Same as 1, except $V_{max} = 90$ km/h, $V_a = 25$ km/h, and $L_a = 12$ km

Present your finding in a tabular form.

5.7. Several variations in the maximum speed, access speed, and average trip length are considered in planning stations on a rapid transit line. The following passenger volumes are projected for the 12 potential station locations:

Location	*A*	*B*	*C*	*D*	*E*	*F*	*G*	*H*	*I*	*J*	*K*	*L*
Boarding	600	1350	280	920	300	240	2230	240	1220	400	200	0
Alighting	0	20	60	280	40	680	1790	300	1450	140	1990	1230
Loading			2150			2610						
Go		580					820					
Stop					340					540		

To achieve a higher operating speed (ignoring all other factors influencing station spacings), several of these locations are considered for elimination from the planned set of stations.

a. Judging only by these projected passenger volumes, which station should be eliminated first?

b. Would station H or station J be better to eliminate?

Explain how you reached your conclusions.

5.8. The alignment of a new rapid transit line has been fixed. You must determine the number and locations of its stations. List the major factors you would consider. Define each factor, state whether it can be computed or qualitatively explained, and discuss how each would influence the average density and the specific locations of stations and which factors would influence its relative importance in each specific case.

5.9. A radial LRT line is 11.830 km long and has 16 stations. For this line, differently than in other cases, station spacings, rather than stations, are designated by letters A, B, etc. The table below gives lengths of spacings and travel times on them, each including one station dwell time.

Section AG is in outlying suburbs; HL is in close suburbs; MQ is in the city center. Using the formula V (km/h) $= 3.6\ S$ (m)/T (s), compute average speeds on each one of the three sections and plot them on a diagram similar to Figure 5.7. Also draw a diagrams with average speeds for persons traveling on the following sections: AL, HQ and AQ.

Spacings	*A*	*B*	*C*	*D*	*E*	*F*	*G*	*H*
Length (m)	1220	980	1010	890	690	760	820	640
Travel time (s)	75	64	65	60	51	54	57	49
Spacings	*I*	*J*	*K*	*L*	*M*	*N*	*P*	*Q*
Length (m)	740	650	650	810	530	490	420	530
Travel time (s)	54	50	50	57	44	42	39	44

5.10. A large educational institution is planned to be built at a location where a metro line passes but has no station. You have to prepare a feasibility study for a station that is proposed to be built at that location, while the line operation has to be maintained. Prepare a flow chart similar to the one in Figure 5.10 showing all major benefits and costs of that project. Do not include the impacts on auto travel and street congestion, nor the impacts that occur only during station construction (construction jobs, disturbance of line operations, noise and congestion around the site, etc.). Show clearly the items that can be computed quantitatively and show by heavy frames the major permanent benefits and costs.

5.11. A 12-km-long rapid transit line is being planned with stations at 2-km spacings. It is planned that 4-min headways will be operated during peak periods. It has been proposed, however, to consider alternative designs with more stations (better area coverage, greater ridership) and skip-stop operation. Develop the following three alternate plans:

a. Proposed plan with 7 stations, 4-min, headways and local operation

b. Stations at 1-km spacings (13 stations) and skip-stop operation (see case III in Figure 5.11)

c. Stations at 0.667-km spacings (19 stations) and skip-stop operation (see case IV in Figure 5.11).

Describe and present systematically the differences among these three cases in:

- Number of stations and area coverage, assuming passenger sheds with 400-m radii
- Headways at different groups of stations in each case
- Travel times on the entire line

5.12. Find an LRT line in a city familiar to you. Draw its alignment with ROW categories, stations, and their spacings. Describe and analyze the corridor and station areas the line serves. Are all significant points and developments close to LRT stations? Does the line have convenient transfers with intersecting transit lines? What are average speeds on different line sections? Are there some aspects that could be improved by a different station location?

5.13. Select a long metro line in a city you know well. Draw its alignment with stations and their spacings. Describe and analyze its alignment and its relationship with the corridor it serves, particularly with respect to station locations. Include its transfers with other metro lines, feeder street transit lines, and P+R facilities. Is the line predominantly radial with substantial commuting, or multipurpose, serving different types of trips? Are all stations located properly with respect to major aspects discussed in this chapter?

5.14. Choose a regional rail line in a major city. Draw its alignment and station locations, then analyze its character: is it a typical commuter line from the suburbs into city center, diametrical through the city, or a multipurpose urban line such as the regional rail systems in German and Japanese cities? Analyze its station locations and spacings, particularly as they affect area coverage versus operating speed on different line sections. What is its relationship with feeders: transit lines and P+R facilities in the suburbs? What principles were used in determining station locations: good area coverage, service to major focal points, or minimum number of stations to maximize operating speed? Is there any need for possible closing, addition, or relocation of a station?

REFERENCES

Bandi, F., et al. 1974. Length of walking distances and distance between their influences on the attractiveness of public transport. *Revue de l'UITP* 3, 175–181.

Hinze, A. 1930. Rechnerische Ermittlung des für die Fahrgäste günstigsten Haltestellenabstandes bei elektrisch betriebenen Verkehrsmitteln für Stadt- und Vorort-verkehr. Ph.D. dissertation, TH Hannover.

Lehner, F. 1969. Regional Organisation of Transport and Urban Development. Report 1a, 38th UITP Congress in London. Brussels: UITP.

Schneider, M. 1961. Travel time minimization, transit. *CATS Research News* 3(4):10–16.

Vuchic, V. R. 1966. *Interstation Spacings for Line-haul Passenger Transportation.* Berkeley: Institute of Transportation and Traffic Engineering, University of California.

———. 1981. *Urban Public Transportation Systems and Technology* Englewood Cliffs, NJ: Prentice-Hall.

Vuchic, V. R., and G. F. Newell. 1968. Rapid transit interstation spacings for minimum travel time. *Transportation Science* 2(4):303–339.

Wirasinghe, S. C. 1976. A model of a coordinated rail and bus transit system. Berkeley: University of California.

PART II

TRANSIT AGENCY ECONOMICS AND ORGANIZATION

6

TRANSIT AGENCY OPERATIONS, ECONOMICS, AND MARKETING

Transit services must be provided at most times of day and days of the year. Service must be safe, reliable, and attractive. To provide such services, an efficient physical system, competent personnel and good organization are required.

Due to the size and characteristics of each service area, as well as various local conditions and requirements, there are considerable differences in organization, functioning, and operational practices among transit agencies and companies. This chapter presents a review of major aspects of transit agency organization, operations, measures of efficiency, and service quality. The chapter also describes the concept and implementation of full transit system accessibility, and agency's relations with the public through information systems and marketing of services.

The focus here is on large, multimodal agencies. Smaller transit operators, such as bus companies in small cities, generally follow similar principles of operations and measures of evaluation, but at a smaller scale. They also use simpler procedures, involving only one transit mode operating on streets, and thus have little special transit infrastructure.

6.1 ORGANIZATIONAL STRUCTURE OF TRANSIT AGENCIES

When transit developed rapidly from about 1890 into the first decades of the twentieth century, it was predominantly provided by private companies. Only in some cases, particularly metro and other systems requiring large investments, public agencies and government ownership existed from the beginning. This condition was followed by a period of transition from private to public ownership, which became the dominant form of transit organizations in large and medium-sized cities.

Since the 1980s, there has been a trend toward integrated public agencies adopting some forms and practices of private companies for greater operational efficiency. Also, to reduce political pressures and achieve competitive pricing, public agencies contract some sections of transit services to private operators. Yet control by public agencies is retained to ensure that public interest is not subjugated to short-term economic efficiency. Public control also eliminates the need for unrestrained competition, which tends to disintegrate transit network and lower the quality of services (see Chapter 9).

Most transit organizations are public agencies, but some of them may provide all services, enjoying monopoly in the served area, while others may contract out some portions or all services under specified conditions. There are also umbrella organizations that provide integration of services by several transit agencies, which may be public or private (the Transit Federations in Hamburg, Stuttgart, and Munich, the Toronto Transit Commission, and, to some extent, the Los Angeles County Metropolitan Transportation Authority are good examples).

There are many forms and types of ownership of transit organizations. They may belong to the city: municipal transit may be one of the city's departments, with the task of providing mass transit services. Or transit may belong to a semiindependent agency that belongs to a county, state (for example, Delaware and Maryland), or some other governmental unit. Transit organization may also have a form of a special governmental authority or district created by legislation and given certain powers and responsibilities.

In the United States, many variations in the forms and powers of these agencies are used for many public functions: both authorities and districts are used to operate schools, provide parking, water, and sewer services, perform air quality control, operate ports and airports, etc. Their sources of financing also vary greatly, including various forms of user charges (fares in the case of transit), special funds, or allocation from general tax revenues. Usually, but not always, districts are stronger and more independent organizations than authorities because they have direct taxing powers, which authorities do not have.

6.1.1 Board of Directors

Most transit agencies, including independent and dependent public ones as well as private corporations, are governed by a board of directors—a body that legally *is* the agency, or its "owner." Directors of regional transit agencies are typically appointed by the cities and counties that the transit system serves. In the states that provide substantial financial contributions to transit, the state governors may appoint one or more directors. In some regions, directors are elected by the population. In the cities where the transit agency is a part of city government, it may be guided and supervised directly by the city council or by a special body of city officials.

There is no standard size of transit boards. They vary between 5 and 15 persons, the number growing somewhat with the agency's size and complexity. Boards with more than 15 members, however, may show some inefficiencies in their decision-making process.

Studies of transit boards and their efficiencies (TRB, 1981) have discussed the problems as well as good experiences with the selection of directors and the board's performance. Some problems are:

- The representation by county does not reflect the usage, importance, and commitment of different counties to transit. For example, in the Philadelphia region, the city proper, in which 85% of transit users live, has two board directors; the other four counties, with a total of 15% of users, have eight directors.
- The background and commitment of individuals varies from very dedicated public servants interested in and knowledgeable about transit to persons, appointed for political reasons, who have neither knowledge nor interest in transit and never use it.
- As jobs with symbolic or no pay, directorships attract dedicated persons; however, they may lead to influences by special interests.

The following are necessary or desirable qualifications and capabilities of persons to serve on a transit board:

- Strong civic interest (the position usually brings only nominal remuneration), regional orientation (rather than representing interests of a local area only)
- Technical knowledge, or at least good understanding, of transit system operations, such as technical issues, economic aspects, management, and personnel characteristics
- Use of and interest in transit system, positive attitude, activism, and initiative
- Absence of potential conflicts of interest or strong involvement in a political party

To appoint persons who satisfy these requirements and avoid some of the problems listed above, some regions have introduced special procedures for appointment of board members. For example, local governments or county officials consider professional and personal background, personal interests, and desirable requirements of candidates when selecting members for board appointments. Some governments set a requirement that candidates use or have used transit extensively in the past. In some cases, nominating bodies, such as counties, submit candidates to the state governor, who makes the final selection. This creates certain checks and balances between different governmental bodies, making it more likely that the appointed persons are well qualified and eliminating persons who lack understanding of the system they are supposed to guide.

The board has the primary role of setting the policies and making strategic decisions. Its basic goals must include development of a transit system that meets the city's and region's needs for effective and attractive services with high economic efficiency. In the relationship of these two somewhat conflicting goals—high-quality transit service achieving maximum ridership and minimum expenditures—there are typically fundamental differences between public and private transit agencies, as discussed in Chapter 9. Public agencies are expected to have provision of optimal transit service as the basic goal within the available financial resources. Private operators tend to have economic efficiency as the main goal. The purpose of contracting out some services in certain conditions is to achieve the two goals, maximum ridership and high economic efficiency, in a planned, coordinated manner.

6.1.2 Organizational Setup

Another major task of the board is to appoint a general manager (the title may also be executive director, president, or chief executive officer, CEO) who is responsible for day-to-day operations and management. The division of duties between the board and general manager varies, but it is usually efficient for the board to be involved in policy decisions about major investments, fare determinations, labor relations, important tendering, and contracting selections, etc., but to allow the general manager to perform his/her administrative duties with considerable independence. Excessive interference by the board in daily operations creates inefficiencies, confusion, and excessive political influence.

In some cases, operations may be shared by more than one manager, each representing one domain. In Germany, there are often three managers representing engineering, business/accounting, and labor. Inclusion of a representative of labor unions tends to influence management-labor relations to be more cooperative rather than adversarial.

An example of a transit agency organizational chart, based on the San Francisco Bay Area Rapid Transit District as of the year 2000, is shown in Figure 6.1. The highest body is the board of directors, with the general manager (GM) directly under and responsible to it. At the GM's level there are often some positions working directly with him/her. In the case of BART, these are the controller, legal counsel and secretary of the district. The next level consists of departments that vary among agencies in names and domain definitions, but typically include the following functions:

- Operations or transportation, department that schedules and operates all modes or has divisions for each mode, such as bus, metro, and/or regional rail
- Engineering and maintenance, handling maintenance shops, yards, depots, terminals, etc.
- Fixed facilities, including construction and maintenance of infrastructure (mostly rail)
- Accounting and budget
- Information and statistics
- Personnel, training, and labor relations
- Planning and development
- System safety, legal aspects, insurance and claims
- External affairs: public relations and marketing

The number of departments or divisions, headed by directors or assistant general managers, varies considerably, but in large agencies it is about 6–12. Agencies may be organized basically by functions, so that, for example, the fixed facilities department includes rail and bus stations, stops, and yards, while the personnel department includes all supervisors, vehicle operators/drivers, mechanics, and related groups. Or there may be rather independent divisions for different modes (e.g., buses, light rail, and regional rail), each one having its own operations, infrastructure, rolling stock, maintenance, personnel, etc. To some extent, such di-

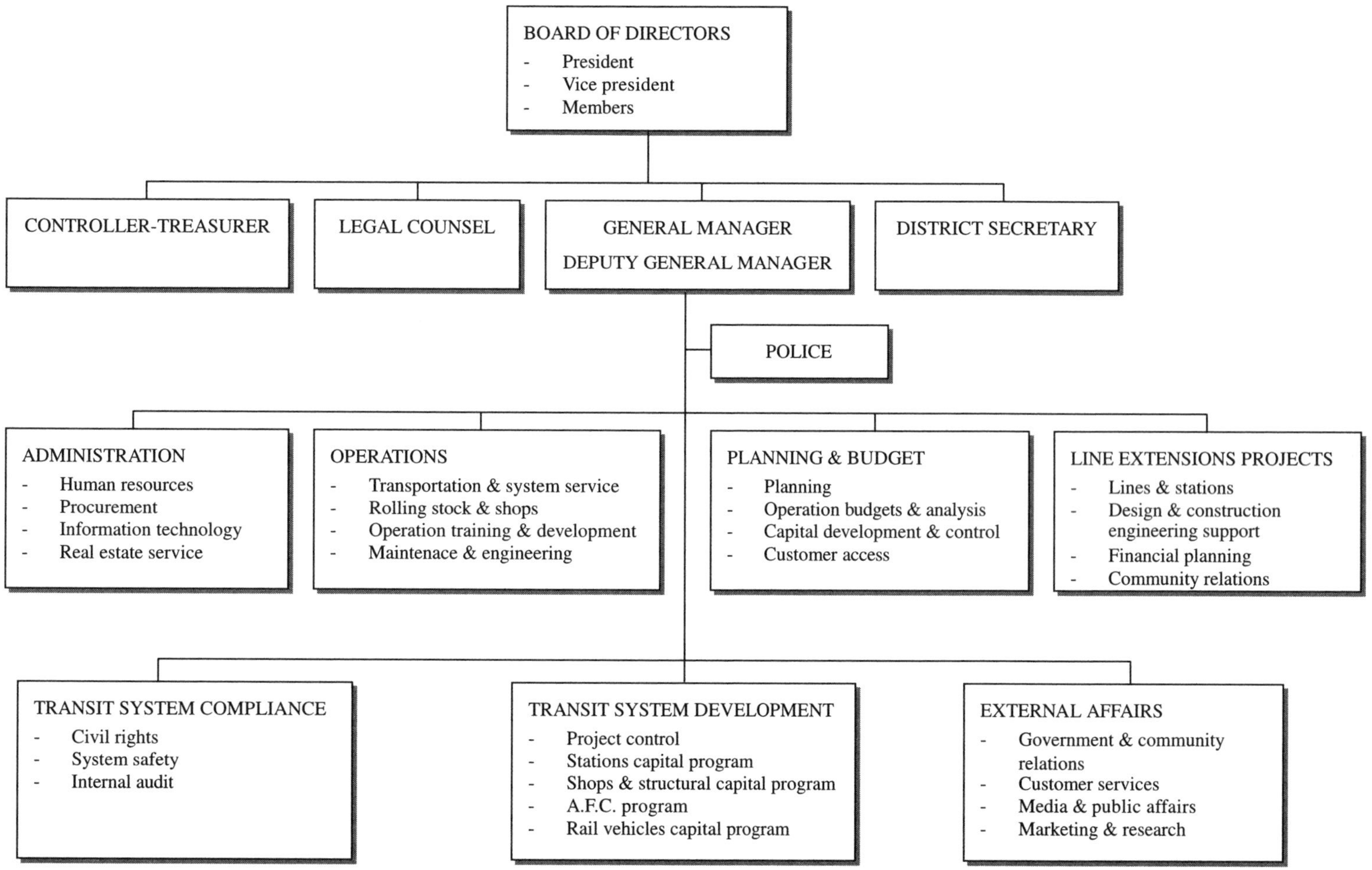

Figure 6.1 *Organization chart of a transit agency (based on the San Francisco BART District).*

visions may be challenged to compete in terms of efficiency, service quality, image, etc.

With respect to networks of lines and services for passengers, all modes in a transit system must be integrated and offer coordinated (rather than competing) services. This is achieved through the departments that cover all modes, such as planning and development, information and statistics, and accounting and budget, through subdepartments that include different modes, such as scheduling, operations control centers, and others. Actual operation of vehicles on lines, assignment of personnel, and supervision are performed by the operations department, which may be subdivided by mode (bus, LRT, metro, etc.).

6.2 MANAGEMENT AND PERSONNEL

As is typical for most companies, transit agencies consist of management and labor groups. Several features of transit systems give certain characteristics to management and labor groups and their mutual relationship. A transit company must provide precisely scheduled services with high reliability; many of its employees—all drivers—must drive under difficult urban conditions with high safety; moreover, a large portion of transit employees are in direct contact with the public. All these conditions require that personnel be carefully selected, well trained, and precisely supervised. To achieve that, transit organization must to some extent operate with paramilitary discipline.

6.2.1 Management Organization

Management can be represented as a pyramid of three levels, the top level being the smallest in number but with the highest-level tasks, then descending to the middle and lowest levels, with larger numbers of persons increasingly involved in service delivery:

- Top-level management defines goals and objectives that transit services should meet, establishes overall operating policies, and gives general direction for relationships with labor groups or unions, including rates of pay.
- Middle management performs strategic planning, guides performance of services, and measures their efficiency.
- Lowel-level management is in charge of daily operations: it performs assignments of duties, monitors attendance, performs surveillance, and enforces discipline.

Transit managements in smaller cities may be rather small and therefore not as formally structured as in large agencies. Qualifications and specialties of management personnel are usually considerably higher in large agencies and particularly in multimodal agencies. Rail systems always require qualified engineers because of the much greater complexity of infrastructure, vehicles, electric power supply, and signal and operations control systems. Further, intermodal operations require persons with experience in complex scheduling and analysis of operations for high reliability and economic efficiency. Therefore, in addition to civil, mechanical, and electrical engineers, multimodal agencies tend to have more qualified computer specialists and accounting and financial experts than agencies operating bus services only.

Transit agencies and the entire transit industry have a special character of operations. They provide public service and a large number of employees deal directly with the public on daily basis. Scheduling of services must consider passenger requirements, intermodal coordination, operating efficiency, and requirements of a large labor body: drivers, controllers, supervisors, and others. Vehicles operate in complex urban conditions. Their control requires coordination with many operating bodies in the city. Economic efficiency is very important. Large infrastructure and rolling stock involve complex engineering tasks. Relations with the public and governmental bodies are important.

Because of these complex and diversified operations of transit agencies, many managerial duties and even apprentice programs in the transit industry require specialized knowledge and, for some functions such as scheduling, operations control, and public relations,

Photo 6.1 *On-board personnel provide public interface—PATCO, Philadelphia Region.*

considerable experience. To achieve needed skills, training programs for different categories of employees are often organized. Programs may involve various levels and categories of management (e.g., organizational, engineering, analytical, accounting), as well as labor: vehicle operators, maintenance mechanics, and others.

6.2.2 Personnel and Labor Unions

Transit is a labor-heavy industry. Operation of buses and trains, maintenance of vehicles and infrastructure, fare collection, and other activities require large workforces. For most transit organizations, public and private, labor compensation is the dominant operating cost element, amounting to well over half of expenses. In some cases, where a transit agency receives considerable public assistance so that it can have low or moderate-level fares, its labor costs may exceed the revenues from passenger fares. Labor productivity, conditions of employment, wage levels, and benefits are major factors in economic conditions of every transit agency.

6.2.2.1 Development and Role of Unions. The transit industry in the United States, Germany, and many other countries is highly unionized. This situation is a result of a series of historic developments that are often disregarded by the present critics who want to eliminate labor unions altogether.

When transit systems developed, mostly as private companies, in the second half of the nineteenth century, workers often had hard working conditions and no work guarantees. Similar to other industries, labor (or trade) unions were later introduced in the transit industry to protect workers from arbitrary management decisions and exploitation through difficult working conditions, low wages, no benefits, and job insecurity.

Labor unions first began to represent workers' direct interests. They worked on developing standards for worker qualifications and training, humane working conditions (specifically shift lengths for drivers), employment benefits, etc. Further, efforts of unions contributed significantly to increased safety at work, protection from health hazards, technical conditions of vehicles, and many other aspects improving performance as well as ensuring safety not only for workers, but for transit services. Finally, labor unions have often been powerful allies of many political leaders who support legislation and financing of transit.

In conclusion, labor unions have changed working conditions and thus quality of transit services. In well-managed transit agencies, drivers and other employees are carefully selected, well-trained workers with higher morale than workers who can lose jobs due to arbitrary decisions of individual managers. When workers are satisfied with their work, that is reflected positively in their attitude toward the public with which many of them are in daily contact. In addition, unions often play a very positive role in developing and improving transit systems and services.

With time, however, transit labor unions developed some practices that have been damaging to the efficiency of transit systems, and even to the safety of transit operations. For example, some labor unions have defended not only employees who were poorly treated, but also those who broke disciplinary rules and even those who caused accidents. In Philadelphia, labor union obtained a court order to the transit agency (SEPTA) to reinstate a train driver who had caused the worst subway accident in the city's history, involving four passenger fatalities. In Boston, San Francisco, and many other cities, labor unions have defended undisciplined and unsafe drivers from dismissals or change of duty.

With respect to operating efficiency, labor unions in many countries, including France, the United Kingdom, Australia, and Brazil, often oppose changes and innovations. They prevent the elimination of jobs that have become unnecessary due to changed technology, new operating practices, or other similar changes. Metro systems in many cities had a conductor opening and closing doors—a function that a train operator could easily perform.

On many regional rail systems, crews of three to five persons are used on six- to eight-car trains because that many persons were needed decades ago to open and close doors, collect or sell tickets, and announce stations. As an alternative that preserves jobs but improves service, the long trains could be broken up into two or three short ones and the large crew split up so that it can operate a greater number of short trains. Thus, the service frequency could sometimes be doubled without any extra cost. Labor unions usually oppose such changes.

Strict division among duties that different categories of workers perform is another cause of inefficiencies that labor unions usually fight to protect.

6.2.2.2 Management-Labor Relations. In the United States, relationships between private companies' managements and labor unions were usually strongly adversarial. Many contract negotiations that included wages, benefits, and work rules led to strikes as a way of solving the disputes. Strikes are often the most damaging means for resolving disagreements and reaching contracts because very often they end up with losses to both sides, as well as to the passengers. The most serious damage is the permanent loss of transit riders, which may be negligible where most passengers rely on transit only, but very large where many transit users can find other means of travel, such as driving cars or carpooling, riding bicycles, or changing travel patterns. Decrease in the number of passengers may lead to decrease in services, thus hurting all other transit users.

With public takeover of transit companies during 1940–1970, it appeared logical to expect that the labor-management relationship would improve. One might have expected the collective bargaining process to change from adversarial to constructive, because both sides had common long-term goal: improvement of transit services to increase ridership. Unfortunately, that did not happen.

Managements in most cases retained an adversarial attitude, as did the labor unions. In some cases, the labor-management relationship actually worsened. With the availability of public funds to support transit services, labor unions increased pressure to obtain a large share of these funds through higher-than-prevailing wage increases, making up for the lower-than-normal increases they had been getting when private industry had been financially ailing (Lieb, 1976).

This situation developed partly because labor unions used political pressure to obtain favorable legislation in many states and at the federal level. This included such laws as allowing transit unions to strike, and the much debated Section 13(c) in the U.S. Urban Mass Transit Act of 1964, redesignated in 1991 as the Federal Transit Act. The latter requires that labor unions approve any application that transit agency submits for federal funds, including operating and capital investments.

The intention of this law is to ensure that "fair and equitable" protective arrangements must be made by the agency to "protect" employees affected by any federal financial assistance. In practice, this ensures that federal assistance does not result in immediate layoffs or worsening working conditions; however, it also gives labor unions the right to oppose and stop projects that increase operational efficiency and labor productivity, which are essential for the agency's overall economic efficiency. As a matter of fact, most major investment projects in transit have as one of the basic goals increase in labor productivity. While such projects as replacement of buses by a rail system with larger vehicles reduce the number of drivers in the short run, the improved quality of service later attracts more passengers and eventually results in additional transit employment.

Consequently, Section 13(c) legislation has been extensively debated and recognized as positive in ensuring fair treatment for transit employees. However, it has also been criticized as a tool to prevent innovations and improvement in operating efficiency of transit systems.

Another case where legislators gave ultimate rights to transit employees at the expense of the public has been legislation in many states giving transit labor unions right to strike. As a consequence, transit strikes occur in many U.S. cities. They have led to their apathetic acceptance as a regular phenomenon, rather than as an irrational collapse of a basic public service that endangers quality of life and safety of citizens. Transit strikes also happen in some West European and other countries, but with far lower frequency and duration than in U.S. cities, where even large cities such as Los Angeles and Philadelphia have experienced strikes of one month or longer. During such strikes, the affected cities get into an emergency situation and incur great losses in productivity, interruption of educational functions, aggravated traffic congestion, reduced tourist activities, and permanent losses of transit riders. Low-income population groups are particularly hard hit.

It is interesting that some transit managers and labor union leaders consider labor strikes to be an unfortunate but normal phenomenon offering the advantage that the threat that it represents increases pressure on both parties to reach a settlement by a certain deadline. Some also argue that when outside arbitrators are used to assist in negotiations where strikes are not allowed may often be biased. This, however, represents the opinions of persons who are deeply involved in labor negotiations but tend to overlook the impact of transit strikes on others than the transit agency and its employees.

The fact is that transit strikes represent a highly irrational phenomenon in contemporary cities, for several reasons. First, they disrupt a fundamental service in a city that is of vital importance for population welfare: it prevents many persons from getting to work, school, or the hospital or purchasing food. It also inflicts long-term damage to the city by inducing some trips to be permanently diverted from transit to car and other modes, involving much higher social and environmental costs. Second, a transit strike is not aimed at a fair sharing of profit. It is against a public agency that does not make profit, but depends on public funds. Third, unlike in private industry, where management's income is affected by work stoppage, transit strike affects mostly a third party, the public, which does not participate in labor contract negotiations and therefore cannot influence them directly. Finally, there are significant negative impacts of strikes, not only during its time, but also before they begin. Several weeks prior to a possible or likely transit strike, most companies and institutions, such as hospitals, universities, and factories, must organize contingency measures to cope with the possible strike, postpone many activities, etc. Economic and social impacts and losses are therefore greater than is often believed.

As a vital social service affecting public welfare, transit should be protected from strikes. They should be illegal in the same way that they are illegal for police and, in many cases, for public school teachers and employees. Disruptions that a strike by traffic police would cause are similar to those created by transit strikes, which prevent people from traveling or force them to use less safe modes.

What means can be used to solve problems in labor-management negotiations? The most desirable may be to establish an agency organizational setup that is conducive more to a cooperative than to an adversarial relationship. An example of such a setup is to include a labor representative in top-level management (often as one of three directors) so that he/she shares the responsibility of making the agency efficient. This has been used in some countries like Germany with very good results and virtually no strikes.

Collective bargaining can be resolved by the use of binding arbitration, the use of a neutral party to develop a contract that parties then must accept. Although this method eliminates strikes, it is not always acceptable to one of the parties, which may find the arbitrators to be not fully neutral with respect to the two sides. To alleviate this problem, in some states several arbitrators are proposed and then each party has a right to eliminate one of the persons it considers not sufficiently independent or neutral. In some cities, the transit agency adopts a key formula for labor wages tied to an indicator based on average wages across the country.

6.2.2.3 Movements Against Transit Unions. In cities and countries where labor unions were politically strong, they used collective bargaining to increase their wages and benefits to the extent that they were considerably higher than labor wages for corresponding jobs on the market. Critics of unions argued that high transit wages meant excessive government spending on support for transit, or that, for given level of governmental support, considerably less service was offered to the public than an organization with nonunionized labor could offer.

This situation was particularly acute in the United Kingdom, where labor costs in bus transit agencies were growing but ridership was decreasing, so that productivity indicators were showing decreasing economic efficiency. However, the critics of this trend ignored the fact that decreasing efficiency was strongly influenced by governmental policies. During the period from World War II to the 1980s, the government gave very little assistance to cities to invest in transit modernization. While most European cities invested heavily in construction of rail systems independent of street congestion (LRT and metro), most cities in the British Isles had transit that consisted of buses operating in mixed traffic. With increasing congestion, productivity of buses on streets had to decrease because of lower speeds and reliability. Incomparably higher investments in roads than in transit encouraged diversion of passengers to cars. With less efficient operations and decreasing numbers of passengers, productivity was decreasing and cost per passenger-km was becoming higher.

In the 1980s, the British Conservative government started its dogmatic deregulation and extreme anti-labor union campaign in many industries. It used the transit crisis to justify a frontal attack on both the established transit agencies and their labor unions. The deregulation of transit, introduced in 1986, broke up agencies that provided integrated intermodal transit systems and allowed entry of transit operators with nonunionized labor. Predictably, this drastic change did result in reduction of transit operating costs and consequently lowered the requirements for government subsidies. Cost reductions were achieved mostly through introduction of much lower workers' wages. However, contrary to the predictions of the reformers that services would be improved and ridership increased, the "British experiment" of deregulated transit systems resulted in lower service quality and loss of ridership, with intensified worsening of traffic congestion in cities and other negative consequences of such a change (see Chapter 9).

With respect to labor, the drastic changes in Great Britain created an opportunity to evaluate the positive and negative features of transit organizations operating with labor unions as compared to those without them. Replacement of transit organizations with service standards and unionized labor by several competing companies without unions results, generally, in the following positive (+) and negative (−) changes:

+ Considerable reduction of operating costs due to lower wage rates, fewer benefits (if any), lower expenses in driver training, etc.
+ Greater flexibility in scheduling and run-cutting, so that more diversified services can be offered
+ Reduction or elimination of quality standards for services.
− Less carefully selected and less trained personnel, resulting in less safe operations, high turnover, and increased training costs
− Drivers and other operating personnel being paid minimum wages, having few or no benefits (health and retirement funds, insurance, etc.), and often having to work under difficult conditions
− Labor unions, as a strong political ally for governmental support of transit, being lost

Some of these negative consequences are very serious. For example, in London, low wages have led to a very high turnover of personnel and inability of transit companies to train applicants for bus driving or mechanic jobs and later retain them. The city government had to give a specific subsidy to increase transit personnel wages.

In summary, transit systems with nonunionized labor tend to be economically more efficient, tighter organizations, but with less qualified and less dedicated personnel. Focused heavily on financial results, the companies have less interest in public service aspects and the overall role that transit should play in the city.

6.2.2.4 Emerging Patterns of Labor Organizations. The changes in transit organizations and their labor relations that started during the 1980s are continuing and developing new forms in the first decades of the twenty-first century. While in the United Kingdom labor unions have been destroyed, no other developed country has followed the same process. In the United States, labor unions are politically very strong, but many subcontractors providing transit services use nonunionized labor.

The most interesting developments are happening in Western European countries, where attempts are made to involve labor unions in eliminating obsolete practices and increasing economic efficiency so that the agency's costs can be competitive with potential private operators. The goal is to maintain all the benefits of labor union presence and practices but eliminate excessive costs and inefficiencies that have developed through union contracts over the years.

The European Union has become very active in setting standards and requirements for transit organizations that require operating efficiencies and allow competitive behavior within as well as between transit operating agencies.

6.3 TRANSIT SYSTEM STATISTICS, PERFORMANCE, AND ECONOMIC MEASURES

For efficient management, monitoring of performance, economic analysis, and planning, a transit agency must have a systematic and accurate collection of data. The collection should include the relevant data about the city and transit service area, the transit system and services offered, the number of passengers, the system utilization and efficiency measures, as well as financial results. A list of the basic data items with their descriptions and uses is given in the next sections. The data items about system performance are followed by important indicators—derived numbers showing various efficiency, consumption, productivity, utilization, and other measures.

6.3.1 Transit Service Area and Its Characteristics

The first set of data describes the area served by the transit system (the term *city* is sometimes used in a broad sense, referring to the city, urbanized region, or metropolitan area).

- *Served area* (km^2) is the area the transit system is required to serve. It may be the area of the city or, now more common, metropolitan area. In many cases, there is a designated service area for the transit system, which may or may not coincide exactly with any other administrative unit.
- *Population of the served area* (persons) refers to residential population in the transit service area. This number is a basic indicator of city/region size, and it leads to classification of the city as small, medium, large, or very large. These categories are not defined by the same numbers in all countries or areas, because the characteristics of a city—its age, type of buildings, activities, and their topographic features—influence the types of activities, living styles, and roles of different modes of transportation. For example, cities designated as medium-sized in North America may be in the range of 250,000–1,500,000 population; in Europe, the designation of medium-sized cities tends to be given to cities with 100,000–800,000. Very large cities, or "megacities," are those exceeding populations of 8 or 10 million persons.
- *Population density* (prs/km^2), or the ratio of population to area, reflects intensity of activities and thus relative roles of different modes of transportation. In general, the higher the density, the greater should be the role of transit. With respect to transit modes, low density is best served by paratransit and buses on streets; as the density increases, the need for and economic feasibility of light rail and metro systems increases.

Population density is an overall measure of intensity of activities, including residential, employment, and all other activities, assuming that they are generally closely correlated. For a more detailed analysis, individual densities may be analyzed using residential density, employment density, and retail activities expressed by the square meters of retail store areas.

For analysis of individual areas that differ greatly in their characteristics, each one of these densities may be analyzed for three different areas: central business district (CBD), central city, and suburban areas (or the remaining urbanized area). Within these areas, densities along individual corridors or in certain zones or districts may be considered for detailed analyses. Transit-dependent population groups are given special attention.

Land use, topography, and climate include such elements as types of land use, major activity corridors, terrain type, rivers, lakes or sea coast, and other components of an area's topography. Narrow corridors and crossings of rivers or bays create the need for high-capacity transit modes. For example, metro systems in the San Francisco Bay Area (BART), Hong Kong, and Rotterdam were logically designed to provide high-capacity links across bodies of water. Wuppertal, Germany, and Caracas are examples of cities that placed rail systems along valleys with concentrations of activities.

Basic climate factors are average annual rainfall and snowfall, average temperature, and seasonal fluctuations and range.

The CBD and other major activity centers are usually focal points of transit networks. Information about the number of employees, office floor area, retail area, manufacturing plants, and major traffic generators such as university campuses, hospital complexes, stadia, etc. are therefore relevant to transit system form and role.

Locations of long-distance transportation terminals (airports, railway and bus stations) are very important because transit should be a major collector-distributor for intercity trips in every city. Neglect of this link between urban and long-distance trips was created when most airports were initially designed without connections to transit, resulting in inconvenience to passengers and congested highways around airports.

Number of on-street and off-street parking spaces in the CBD is important not only for evaluation of modal split and congestion conditions, but also with respect to the negative impacts parking lots and garages may have on the center city's environment and livability.

6.3.2 Transit System and Services

Organization, characteristics, and statistical data that describe a transit system can be grouped in two categories: physical components and services. Physical components or infrastructure encompass lines, networks, and vehicles or rolling stock, and they differ substantially among transit modes. Therefore, all statistical and descriptive information for a transit agency are presented separately for each mode.

6.3.2.1 ***Transit System.*** Major information items and statistical data about transit system are listed here.

- *Transit agency or agencies* that serve the metropolitan area: their names, ownership, modes they operate, and areas they serve
- *Transit modes*

Transit modes are reported separately because they have different physical components, such as ROW categories and vehicles, as well as operating characteristics. One mode may have two times greater average operating speed and carry many more passengers than another mode. The most common modes are, in sequence of ascending performance, buses, trolleybuses, streetcars/tramways, light rail transit, metro (rapid transit), and regional rail. In some cases, a transit agency may also operate paratransit services (minibuses and vans) and special systems, such as AGT, funiculars, monorail, ferryboats, and others.

As defined in Chapter 4, a transit line or route is a fixed path served by transit TUs with a predetermined schedule. A transit network is the total set and length of streets and other transit rights-of-way on which transit lines operate. Consequently, total line length can be equal to, but is usually greater than, network length.

A transit network is described by the following statistical data:

- Number of lines by mode
- Length of lines in km (miles) by mode
- Network length in km (miles) by mode

If some modes use different ROW categories, network lengths for those modes should be given by category also. This is typical for LRT mode, which usually has most lines on street medians or other separated ROW types, but some sections on streets (category C) or in tunnels and aerial structures (category A). Lengths of network portions in each category, in km or in percentages, should be given because that information describes character and operating characteristics of the system. Other feature, influencing operating characteristics and service offered are the stop or station locations and densities, defined as follows.

- *Number of stations on each rail mode:* LRT, metro, and regional rail, which determines area coverage.
- *Average spacing between stops (street transit) or stations (semirapid and rapid transit).* Short spacings are typical for lines serving short trips in high-density areas, and they result in low operating speeds. Long spacings are typical for higher-speed services used for longer trips, such as regional rail or bus lines (see Chapters 2 and 5).
- *Area coverage* is computed as the area within 400 m (quarter mile), or a five-minute walk circles from a bus stop or rail station (see Chapter 4). Since a rail station and its entrances are usually spread out to 100 m or more, area coverage for rail systems is often defined by a 500-m instead of 400-m radius. *Network area coverage* is the total area within these distances from stops/stations, excluding overlapping areas. Again, because of major differences between bus and rail services, total area coverage is usually presented separately for the two modes, rail and bus.
- *Percent of area covered* is the area covered divided by the total transit service areas expressed as percent. *Percent of population served* is the population in the covered area as percent of the population in the transit service area. These two percentages indicate how well are the area and population served.
- *Capacity of park-and-ride (P+R) lots at transit stations* shows how much the transit system relies on private cars for access. Typically, regional rail systems have by far the most extensive P+R facilities. Recently built metro and light rail transit lines also have P+R facilities. Thus, San Francisco BART, Washington Metro, and regional rail systems in the New York area have tens of

thousands of P+R spaces in lots and garages around their outlying stations. Bus lines have P+R only where they offer express services and parking in center city is very constrained, such as in the New Jersey suburbs of New York City and the San Francisco suburbs not served by BART.

- *Capacity of bike-and-ride (B+R) facilities* for parking of bicycles at transit stations. B+R facilities are much more economical and less space-consuming (although less comfortable and more weather-dependent) than P+R. They are used extensively in many developed countries with the widespread use of bicycles, such as Denmark and the Netherlands. However, by far most extensive use of B+R is found in Japanese cities, where very large investments have been systematically made in these facilities in virtually all cities with metro and regional rail networks (Replogle, 1983; Vuchic, 1999).
- *Vehicle fleet or rolling stock* statistics are important for various analyses. The number and types of vehicles reflect the size of the system. Operating fleet size divided by network length (vehicles in operation per km of network or per km of line) shows density of operations. Various measures of system productivity and efficiency are based on the relationship of passengers and passenger-km served to the fleet size.
- *Number of vehicles* should be given by mode and type, including their basic characteristics, such as type of propulsion (diesel, electric, and others), body type (single or articulated, single or double deck, regular or low-floor, etc.), capacity (seats and total), average age, TU size, propulsion, etc.

6.3.2.2 Offered Services. The basic data describing services the transit agency offers to the public are the number of vehicles in service, speeds, and headways.

- *Maximum number of vehicles and TUs operated during peak hours, by mode:* this number expresses the volume of services. Number of vehicles and TUs operated in off-peak service, by mode, reflects the scale of all-day services offered on the network.
- *Operating speeds on lines in km/h, by mode:* this is a major component of quality of service. Generally, operating speeds depend on ROW category, station spacings, and method of operation. Rapid transit, having ROW category A, longer station spacings and fast boardings/alightings, has much greater (often two to three times) speeds than street transit (buses and tramways operating on streets, ROW category C). Speeds on semirapid bus and LRT lines are usually between these two extremes.
- *Cycle speeds on lines V_c (km/h), by mode,* include terminal times and influence operating costs, but not passengers directly.
- *Average network speed V_{av} (km/h)* must be computed as a weighted average by the volume of service provided on different lines. This is computed via vehicle-km or space-km, i.e., by the work performed on each line, using W_i as the number of vehicle-km operated per day or per hour on line I, by the formula:

$$V_{av} = \frac{\sum_i W_i \cdot V_i}{\sum_i W_i} \qquad \left| \frac{V}{\text{km/h}} \middle| \frac{W}{\text{veh-km}} \right| \tag{6.1}$$

- *Headways on lines* represent another important element of service quality. This number is not convenient to give in a statistical summary, but it can be shown graphically. On rail networks this element is often shown clearly by a diagram of service frequency (inverse of headway) plotted on each link of the network.
- *Schedule reliability R (%)* is measured as the percent of TU arrivals with 0-4-minute delays:

$$R = \frac{\text{Number of arrivals 0–4 min late}}{\text{Total arrivals}} \cdot 100\% \tag{6.2}$$

R can be computed by mode and for different periods of time, usually peak- and off-peak. Reliability is one of the basic measures of service quality.

6.3.3 Transit Usage, Work, and Productivity

Every transportation process fundamentally consists of units of service or objects carried a certain distance in a period of time. For a transit system, performance of offered service is measured in such units as speed or vehicle- or space-km, and for utilized service in passengers or passenger-km carried. Table 6.1 presents the basic measures of the transportation process and their units for offered service and performed transport. These units are used in the following statistical measures of transit system performance, presented here in sequence of increasing complexity.

6.3.3.1 Volume of Offered Service. This is expressed by the following measures.

Space-, vehicle-, or TU-hours of revenue service provided (usually per day) expresses service offered to passengers, and it is closely related to the cost of providing the service, because operating costs depend more on labor time than on distance passed by vehicles.

Number of spaces, vehicles, or TUs offered on the lines at a given time, such as peak- or off-peak hours, represents service offered at any one time, and it may be given for peak- and off-peak hours, or on different days.

6.3.3.2 Transit System Usage. This is the most important measure of transit system performance or size, and it can be expressed in several units.

Number of passengers carried per unit of time (hour, day, or year) actually refers to the number of trips during the given time period. Sometimes a distinction is made between fare-paying passengers and total passengers. With the increasing use of prepaid tickets or passes, the definition of nonpaying passengers has become less clear, so that the total number of passengers is now mostly used. In some cases where detailed statistics are available, one person's travel including more than one vehicle is counted as a linked trip, while counting separately each trip on any one vehicle or line is defined as unlinked trips. Logically, the number of unlinked trips must be equal to or greater than the number of linked trips.

If daily passenger volumes are used, separate data are computed for weekdays, Saturdays, and Sundays. Total number of trips should always be used. In some references, daily ridership is divided by two, claiming that it represents the number of persons using transit in a day. This is incorrect and very deceptive, because many persons use transit only once or more than twice. Moreover, thus obtained much smaller number would be incorrect to use when comparing transit with other modes because the numbers of trips by car, bicycle, or airplane are never divided by two.

Annual number of trips or annual passengers represents the sum of trips in a year.

Riding habit, the average number of transit trips per capita per year, is computed as the number of annual passengers (or passenger trips) divided by the served area population.

6.3.3.3 Transportation Work Performed and Transit System Productivity. These can also be computed for offered and for utilized transport.

Offered transportation work is measured by space-, vehicle-, and TU-km. Work performed or "production" of transit system is expressed by passenger-km

Table 6.1 Basic measures of transportation process

Measure	*Offered Service*	*Performed Transport*
Objects	Vehicles, TUs, spaces	Persons (passengers)
Volume, frequency	Vehicles/h, TUs/h	Persons/h
Transportation work performed	Vehicle-km, TU-km	Person-km
System productivity	Space-km/h	Person-km/h
Productive capacity	(Spaces/h) · (km/h)	(Persons/h) · (km/h)

transported. It is computed when the passenger volume is multiplied by average trip length. Both are usually expressed on annual basis.

Transit system productivity is transit work, such as space-km or vehicle-km, computed for any given period of time. This performance measure is usually computed on hourly basis. For example, a bus line carrying 1800 passengers at an operating speed of 14 km/h has a productivity of 25,200 person-km/h. A metro line carrying 12,400 persons at operating speed of 32 km/h has a productivity of 396,800 prs-km/h. The passenger volumes can either be taken from the maximum load section or computed as average volume along the line. Naturally, the former computational method gives higher value than the latter one.

Productive capacity of a line P_c in terms of spaces offered is sometimes measured to compare different lines in a network or on different modes. For offered productive capacity, P_c, line capacity, and operating speed should be used to reflect operating time in service and thus the system's physical ability to produce transportation work:

$$P_c = f \cdot n \cdot C_v \cdot V_o = C \cdot V_o$$

$$\left| \frac{P_c}{\text{veh-km/h}^2} \right| \frac{f}{\text{TU/hr}} \left| \frac{n}{\text{veh/TU}} \right| \frac{C_v}{\text{sps/veh}}$$

$$\left| \frac{V_o}{\text{km/h}} \right| \frac{C}{\text{sps/h}} \right| \tag{6.3}$$

Utilized line productivity in terms of passenger-km per hour carried on a line P_p is computed by multiplying average number of passengers along the line P_{av} and operating speed V_o:

$$P_p = P_{av} \cdot V_o \qquad \left| \frac{P_p}{\text{prs-km/h}} \right| \frac{P_{av}}{\text{prs}} \left| \frac{V_o}{\text{km/h}} \right| \tag{6.4}$$

Vehicle productivity in operation on a line P_v can be computed as offered spaces multiplied by operating speed:

$$P_v = C_v \cdot V_o \qquad \left| \frac{P_v}{\text{sps-km/veh-h}} \right| \frac{C_v}{\text{sps/veh}} \left| \frac{V_o}{\text{km/h}} \right| \tag{6.5}$$

6.3.4 Revenues, Costs, and Operating Ratio

Financial accounting of transit agencies is divided into two major categories: operating costs and capital investments. These two categories consist of different items and separate budgets, and they have different methods of accounting and financing. There is only a limited overlap between them. This section covers statistical items of operating revenues and costs only.

Revenues of transit agencies are usually grouped into the following major categories:

- Income from fares, which consists of direct payments by passengers, so-called farebox revenue, as well as revenues from sales of weekly and monthly passes. Contributions from employers toward purchases of employees' transit passes are also included
- Other income includes revenues from various sources, such as school bus contracts, charter and special event services, advertising, rentals and leases, interest, etc.
- Financial operating assistance, which is obtained from the government (city, state or federal/national) or from other funding sources

Operating costs or *expenses* may be broken down into the following categories:

- Operating personnel wages with fringe benefits. Consisting of wages for drivers, supervisors, and other operating personnel, this is usually the largest item of operating costs
- Fuel and power expenses
- Maintenance and repairs, including mechanics' and other personnel wages, rolling stock, fixed facilities, and miscellaneous equipment
- Fare-collection expenses
- Information, advertising, and marketing

- Vehicle licensing and registration
- Insurance, injuries, and damages, including settlements of liability cases
- General and administrative expenses, consisting of management salaries, space rentals, and other costs of operating the agency.

Operating ratio is the basic indicator of the agency's financial profitability. It is the ratio of revenue from transit fares and other revenues to the costs of operating the service. Operating ratios greater than 1.00 indicate that transit service is operating at a profit. In most cases, however, transit agencies have operating ratios below 1.00 and supplement their direct revenues by the government financial assistance.

6.3.5 Indicators of Efficiency, Utilization, and Consumption

As ratios of two or more measures or items, indicators permit comparison of systems of different sizes or magnitudes, which would not be possible by comparing absolute values of individual measures alone. *Efficiency* (or rates of production) of a transportation system η is defined as the ratio of output to the input consumed in the transportation process:

$$\eta = \frac{\text{Output quantity produced}}{\text{Resource quantity expended}}$$

Thus, efficiency can be computed for various output units, which may be quantities of offered service (seats, seat-km, vehicle-km, etc.) or of utilized service (such as passengers and passenger-km), as related to units of such resources as operating cost, labor (number of employees, driver-hrs), units of time or space, such as passengers carried per hour per meter of ROW width, and many others.

Examples of commonly used efficiency indicators in transit systems offered and utilized service are:

- Vehicle-km per day per operating employee shows labor efficiency in the system.
- Passenger-km carried per vehicle-km expresses density of travel on the network.

Many other measures of labor, vehicle, and investment efficiencies can be used for analyses of operations, scheduling or accounting, such as:

- Space-km per day per operating employee
- Passenger-km per day per vehicle operated
- Passengers carried per day per total number of employees
- Passengers carried per day per dollar of investment in a new transit line

As the term *efficiency* indicates, it is usually desirable to maximize the value of this indicator. Higher efficiency shows greater production for the same expenditure of a resource, or lower consumption of resources for a given system output. This is true, however, under *ceteris paribus* conditions. Increasing efficiency may not be desirable if it results in decrease of quality of service, increased negative side effects, etc.

Effectiveness is also the ratio of output to input, but it relates to the results achieved by a given input. It therefore includes both quantity and quality of the service provided. Typical measures of effectiveness may be:

- Increased mean distance between failures (MDBF) for vehicles achieved by an investment, by improved personnel training, or by reorganization of the maintenance process
- Number of new passengers per day attracted per unit of investment (or an action)

Coefficient of utilization, or utilization ratio, is a kind of efficiency: it is a ratio of output to input, but between the same or very similar units. For example, vehicle utilization (or load factor) is the ratio of passengers to spaces, which is actually ratio of occupied to total (offered) spaces. Several commonly used utilization indicators are:

- Vehicle utilization: vehicles operated in peak hour as per cent of total vehicles owned

- Vehicle capacity utilization: passengers in vehicle divided by vehicle capacity; also known as load factor (see Section 1.1.5)
- Line capacity utilization: passenger-km traveled on a line divided by space-km offered
- Labor utilization: employee hours worked divided by hours paid

As with efficiency ratio, it is usually desirable to maximize the coefficient of utilization, ceteris paribus. However, in many cases there are trade-offs to high utilization ratios, such as higher maintenance costs that high vehicle utilization involves, lower passenger comfort in heavily loaded vehicles, and fewer employee social benefits when labor utilization ratio is high.

Consumption rate represents the ratio of resource input to output, and it may be between different units, such as kWh of energy consumed per vehicle-km operated (product). Thus, consumption rate σ is the inverse of efficiency:

$$\sigma = \frac{\text{Resource quantity expended}}{\text{Output quantity produced}}$$

Typical examples of consumption rates in transit systems operations are the following:

- Energy consumed (in kWh or liters of fuel) per vehicle-km operated
- Number of workers employed in maintenance shop per vehicle serviced in it

In transit systems (as well as in most other processes), the goal is generally to minimize the consumption rate for the given output and product/service quality.

Most analyses of revenues and costs are made utilizing their ratios: revenues or costs to offered service (performed work). These indicators actually represent measures of efficiency or of consumption. The most common indicators related to costs and revenues are:

- Operating cost per space-km or per vehicle-km offered
- Operating cost per vehicle-hr or train-hr
- Operating cost per passenger trip
- Revenue per seat-km or vehicle-km
- Revenue per vehicle-hr or train-hr

These indicators are used in schedule preparations and in analyses of efficiency of operations during different periods of day, for comparisons of services on different lines, by different modes, or among transit systems in comparable cities.

6.4 TRANSIT AGENCY OPERATIONS

This section covers major topics that the transit operator must plan, manage, and analyze to ensure efficient operations that meet passenger requirements. First, the approach to and organization of transit services, are discussed, with emphasis on orientation of the system toward the needs of present and potential passengers. Applications of telecommunications and informatics technology are reviewed. Organization of vehicle maintenance and determination of vehicle life length are then briefly described. Finally, the important aspects of safety, security, and handling of legal liability are included.

6.4.1 Attitudes toward Transit Users

Transit companies, agencies, and their facilities and services play an important role in the shaping, functioning, and economic vitality of cities. Their role in a city's character is very important: livable cities with human orientation are not feasible without good transit services. Quality of service, ease of traveling by transit, and even the image of attractive vehicles, lines, and stations make travel without cars attractive and contribute to the personality of each city. Virtually all visitors to Vienna, Zürich, and Melbourne notice their tramways. In Paris, Moscow, and Washington, the metros are their symbols and parts of residents' lives and visitors' experience, as are London's red doubledecker buses and San Francisco's historic Cable Cars.

In addition to the physical features of transit, its users experience services through the convenience and attractiveness of its use. Among many factors that make transit use attractive are good information, easy transfers, and coordinated schedules among lines and modes, as well as facilities that are in good condition and give a feeling of security. A particularly important aspect is the attitude of and contacts with transit agency management and personnel.

With respect to transit agencies' policies and attitudes toward their present and potential users, a fundamental change has been taking place in recent decades. Although many cities have always had high-quality transit services, there used to be a general feeling that transit provided a basic service that people had to use. Most people were oriented to transit and had no alternative modes for many trips. In recent decades, however, competition from other modes, particularly the private automobile, has greatly increased. At the same time, unplanned growth of many cities has resulted in urban forms less interdependent with transit.

Under the evolving conditions, it has become necessary for transit agencies to change their attitudes toward passengers: instead of assuming that they will automatically use their services, progressive agencies must actively work on retaining the present and attracting new passengers. This coincides with the increased need of cities to encourage transit use and shift travel from cars to transit wherever this is possible. This change has been well designated by Marciani (2002) "from users by necessity to customers by choice."

Transit agencies must have a regular program to examine carefully and evaluate critically their present services. With respect to policies and attitudes toward the public, many agencies have introduced steady improvements and modernizations in recent decades. Every user of transit systems in cities like Paris, Munich, Toronto, Vienna, Zürich, and Portland finds ubiquitous services throughout the urban area, high quality of rides, ample information, and generally good attitudes of transit personnel. In many other cities, however, transit maps are incomplete and signing at transit stops and even on buses and trains is incomplete or confusing. Poor services are compounded by disinterested personnel.

These deficiencies can often be explained by many factors, such as inadequate financing, incompetent management, irresponsible labor union attitude, little support or interest by public authorities, large subsidies of automobile use ("free" parking, municipally paid traffic control, street and intersection design favoring automobile traffic), neglect of pedestrians, etc. These conditions can contribute to resignation from transit authorities. In many cases, however, even with adequate financing and strong public support for transit services, a transit agency is badly organized and inefficient because its management has allowed operations to become gradually adjusted to the convenience of drivers, maintenance workers, and other personnel, rather than passengers. Although explainable, these practices are not excusable. They are costly and damaging.

Management must develop strategic policies and operating plans for introduction and continuous maintenance of transit services that are passenger-oriented. These policies and implementation methods must include all aspects of the transit system that affect services for passengers. Management attitudes and personnel training must create employees who are strongly motivated and feel responsibility for providing high-quality services. This commitment must be present throughout the organization, but particularly with all persons affecting service or being in direct contact with the public, from general manager to station personnel and vehicle operators. For example, train operators must use a positive way of speaking with passengers: "We are providing these special services" or "We will recover the delay shortly," rather than "They don't give us good equipment" or "I have no idea about schedules on that line, they don't tell us about other services." Referring to the agency these persons work for as "they" indicates lack of integration and positive attitude.

System users who notice animosities among persons working for the same company immediately detect an attitude that must be harmful to the provision of an efficient, passenger-friendly service. Agency employees should be aware that poor service is likely to reduce ridership and thus, in time, erode their own employment. As a provider of an important public service, the transit agency must train its employees to have a deep understanding of the broad aspects of the services they offer, rather than behave with a self-centered interest typical of private business, where immediate profit is the only goal. In a transit agency, there must be a synergistic effort aimed at the goal of satisfying present and attracting new passengers while ensuring maximum feasible economic efficiency.

In many transit agencies that have had a long period of operation without many innovations and service reorganizations, a deep inertia develops. With time, organizations have a tendency to develop a pattern of operations that is convenient for personnel, rather than for passengers and long-term operating efficiency. For example, bus and train operators neglect to announce stations. Line designations on buses and trains may be incomplete or not displayed at all. Bus routes and schedules may be adjusted so that the number of vehicles is reduced even where that does not meet service standards. Long regional rail trains needed during peak hours may be kept in operation not only during off-peaks, but during weekends also, resulting in thousands of empty car-km. Maintenance of tracks that could be performed at night or with temporary single-track operations on sections of a line may be performed with closing of a rail transit line for long periods or entire weekends.

This pattern of operations is not easy to change, because in an organization a resistance to change develops that may be designated as "self-defense of incompetence": individuals or units, such as vehicle operators, maintenance shops, and department responsible for public information, resist any improvements and innovations that would require a change in their established ways of performing their jobs. This phenomenon is by no means unique to transit organizations; it reflects some features of human nature in many working positions. The less competent employees are, the more they resist any changes.

In transit agencies, this type of inertia may be very harmful to offered services and the agency's long-term economy. Management must undertake energetic steps to break the pattern of service deterioration, decreasing economic efficiency, and resistance to innovations. In some cases, to introduce changes, management may need support of political leaders, external advisors, citizen advisory groups, and other bodies to get a better perspective on the conditions of service, needed improvements, and obstacles that should be overcome. The obstacles may be inadequate motivation and poor working conditions for the personnel; lax middle management; labor unions that have built an "empire"; lack of support for transit vehicles by poor traffic rule enforcement; and many others.

Specifically, in planning operations and services, management of the transit agency must follow a strategy that considers the role transit should have in the city, adopt increase of ridership as one of the basic goals, and continuously work on introducing innovations in technology and services. Maintenance of the status quo is not sufficient in dynamic urban transportation conditions. The need for shifting more trips in urban areas from cars to transit is great, and opportunities for achieving that goal must be actively pursued.

While competent, dynamic management of transit agency is a basic requirement for its efficient functioning, the effectiveness of its services in attracting passengers and in playing an important role in the area it serves also depend on many other organizations and authorities. Transit services, particularly those operating on rights-of-way B and C, depend greatly on the regime under which transit vehicles operate on streets, cross intersections, stop at stops and stations, etc. On all streets with frequent transit service, it is necessary to have strict control of traffic flow and parking to minimize potential for traffic congestion. Moreover, transit priorities, such as easy entry and exit for buses at curb stops, signal priority for transit vehicles, reserved lanes for rail or bus vehicles, and special signal phases for transit turning movements, should be introduced wherever feasible.

Planning and introduction of these measures requires coordination of many agencies and authorities. In many cities, this function is performed efficiently by a body usually known as a *transit first committee,* which consists of representatives of the transit agency, city's traffic engineering or streets department, police, parking authority, and other relevant organizations. The committee has the goal of analyzing present transit services, defining problems, and developing and implementing measures that will result in improvements of transit services. The committee has a broader view of problems and possible solutions than do individual agencies, and its coordinated efforts are usually more effective actions than transit agency could implement alone.

6.4.2 Organization and Control of Operations

Transit services must be provided with great precision and reliability. To achieve that, the operations department must organize training for the operating personnel. Training should include not only driving technique and technical aspects of vehicles, tracks, signals, power supply, and other system components, but also organizational aspects, such as operating rules, responsibilities, handling of special situations, chain of command, and the role of the operations control center.

For operations there must be a clear chain of command. Operating policies are developed and controlled by operating departments. Staff departments support the operations department's policies but should not set policies. Bus and train drivers obtain schedules for the entire set of runs during their work shift. Line supervisors monitor, control, and intervene when necessary to ensure that services are performed on schedule.

For effective supervision of service and response to any irregular developments, it is very important that line supervisors know positions of all trains and vehicles and that they can communicate with drivers, station attendants, and other personnel in the field. In this respect, operations and control vary greatly among transit modes, particularly between buses and other modes operating on street, or ROW category C, and modes on ROW category A, such as rapid transit, regional rail, and AGT.

Transit line and network control was initially developed for rail systems, particularly metros. Starting with electrically controlled signal blocks for ATP (see Vuchic, 1981, Chapter 5), used from the beginnings of metro systems around 1900, it was possible to have positive electrically transmitted information about the presence of trains in each block, and to follow movement of every train along the line with the precision of block lengths. Around 1950, control systems were introduced that provided automatic track and interlocking control, thus regulating safe train movements through entire networks.

Since the 1960s, the technology of moving (or "continuous") signal blocks and on-line computer control of trains has been introduced on many advanced rail systems. The new Victoria Line of the London Underground and the PATCO Line from Philadelphia into New Jersey suburbs were opened with automated driving under train operator's supervision and initiation of the driving cycle at each station. San Francisco BART, opened in 1972, introduced a higher level of automation: the control center provided on-line supervision and control of all trains in a complex network.

Contemporary rail transit control and command centers are extremely sophisticated and perform multiple functions, such as:

- Monitor all train movements in the network
- Provide real time arrival/schedule information
- Offer continuous communications with train operators, station attendants, and other personnel
- Undertake measures to recover delays and any other irregularities
- Direct operations when there is a service disruption
- Provide communications and leadership in handling emergencies
- In some cases, provide on-line instructions or automated selection of travel regimes for each train
- Control and adjust station dwell times at different time periods
- Provide announcements for the public about services and, particularly, about any irregularities or special situations
- Provide communications with transit police, city's police, fire department, and other relevant organizations

Virtually all rapid transit systems, i.e., those with fully controlled ROW category A, including metro, AGT, and many regional rail systems, now have control and command centers. Most recently built LRT systems also have such centers, although they cannot have full control of LRT train movements on lines with ROW category B and C. In some cases, however, the control center of an LRT system may have an important role in monitoring train movements across major intersections, where they get preferential treatment in cases of delays only. The center's monitoring and intervention is therefore very important.

Control and command centers for street transit modes, such as buses, trolleybuses, and streetcars, were introduced much later than for rapid transit systems because monitoring of transit vehicles in streets was much more difficult. However, this problem has been resolved in recent decades by introduction of various types of AVL systems, described in Section 6.4.3.1.

Advanced control and command centers have sophisticated devices for monitoring all transit lines, even those of street transit modes. For example, the control center of the Zürcher Verkehrsbetriebe in Zürich has computer monitors with graphical schedules for each bus, trolleybus, and tramway line. The schedule is plotted in white lines. Based on the information received continuously from the AVL system, travel of each vehicle is automatically plotted on a time-distance diagram. When a vehicle is more than two minutes ahead of schedule, its path is shown on the diagram in purple. When a vehicle is more than three minutes late, its plot is shown in red. The controller can thus immediately see which vehicles are running on schedule, which are ahead, and which are behind. The latter two require intervention of the control center.

Development of various telematics devices and procedures and their utilization for transit systems (see Section 6.4.3) are usually concentrated in control/command centers. Increasing use of telematics is thus broadening the scope of capabilities and tasks of the centers in transit system operations.

6.4.3 Applications of ITS/Telematics Technology

Urban transportation, and specifically transit, has benefitted greatly in recent decades from advances in electronics, computers, and communications. Many elements of these technologies have found diverse applications in transit systems operations, analyses, and planning.

The field known in the United States as intelligent transportation systems (ITS) and in Europe as telecommunications and informatics (telematics) consists of applications electronics-based systems such as computers, data processing, satellite, and other telecommunications, as well as related technologies, to transportation systems; in this section, the focus is on their applications to transit systems. This area is not precisely defined, because many transit system components that existed long before inventions of computers and electronics are now logically included in ITS. For example, radio communications, automated train protection signals, and control centers for transit operations were in operation years before the broad use of computers, GPS, and other electronics-based devices.

In urban transit, ITS is applied through the use of a variety of technical devices and systems to many different functions. These applications can be classified in four major categories. First and usually most extensive is the application to systems operations. The second function is data collection, which facilitates transit system management, maintenance of facilities and vehicles, and system planning. These two functions directly increase efficiency of a transit agency's activities. A further two functions directly improve services for the public: provision of information for passengers and automated fare collection. Similar to the benefits of the first two applications for the agency, both functions affecting passengers have resulted in major improvements of services for passengers. The four categories of ITS applications are described in the following sections.

6.4.3.1 Operations Monitoring, Supervision, and Control. This is a very broad area of applications, ranging from communications with individual buses or trains to the control of TUs in a network of several lines or traffic control at intersections. Major types of applications and their significance are described here and shown later, in Figure 6.4.

Automatic vehicle location (AVL). A monitoring system is needed to follow and control the movement of transit vehicles on lines. This system is particularly needed for street-running vehicles such as buses, trolleybuses, and tramways. LRT and metro trains usually do not depend on AVL because they can be monitored through signal blocks, various rail contacts, or even

Photo 6.2 *Advanced rail transit control center utilizing extensive ITS technology—San Francisco BART.*

power supply sections, etc. AVLs for buses along their lines have been developed with a number of communication techniques since the 1960s. Initially, scanners were installed on selected light posts along bus lines, which recorded passage of buses carrying encoded transponders and reported each passage to the control center by radio. Another system (see Vuchic, 1981, Section 4.3.4.3) combined this type of location recording with a distance-measuring device on the vehicle, so that each bus continuously records its location.

The control center utilizes a radio-based system that inquires each vehicle by radio and obtains an answer about its position. Each vehicle is scanned about once per minute, except in emergencies, when scanning can be performed every two to three seconds. The vehicles are then displayed on a computerized map as icons. The latest technology utilizes a GPS for scanning buses and recording their positions.

Actually, GPS technology has become so reliable that a new train positioning system has been developed and is being tested on the San Francisco BART (regional metro) system to replace more conventional rail-contact-based signal system for train control. It is expected that the GPS may provide continuous location monitoring so precise that it may result in operation of shorter headways on BART lines than is presently feasible. Thus, various forms of ITS technology provide continuous monitoring of transit TU locations on bus as well as rail systems.

The significance of AVL is considerable. Control of transit operations on lines requires frequent or continuous knowledge of positions of all TU's. The operations supervisor needs this information to detect any problems and to intervene. Radio contacts represent a basic tool for supervisors to communicate with TU operators/drivers, but that link is not continuous nor does it give the supervisor complete information about operations that he/she needs in order to exert control, direct any changes, service recovery, etc. For this reason, AVL is an important technological tool that elevates control and reliability of bus services. Another potential use of AVL is for providing real-time information about arrivals of vehicles/trains at stations or even stops on streets along transit lines.

Computer-aided dispatching (*CAD*) utilizes specially prepared algorithms that allow the dispatcher to determine the optimal sequence of dispatching of TUs under different conditions. This can be used in many different situations, such as when major delays occur, vehicles have to be transferred from one route to another, one or more TUs have dropped out of service, etc. The control center can call or make announcements in individual vehicles, all vehicles on a transit line or in a district of the city, etc.

Closed-circuit television (*CCTV*) can be used for monitoring many station areas for a number of different purposes:

- On busy metro and regional rail station platforms to observe alighting/boarding process either for safety or for train dispatching purposes.
- In some metro systems, train operators use CCTV located on platforms or in the driver's cab to control passenger alighting/boarding, door closing, and train departure. This type of CCTV use enabled metro systems to reduce or eliminate all station personnel and perform their monitoring functions by a few employees observing dozens of TV monitors in the control center. This development started in 1958, when Hamburg introduced one-person trains with assistance in station departures by a station attendant. Gradually, most of these attendants have been replaced by persons in fare-collection booths, using CCTV to observe platforms and signal departure to train operator. Then, in 1969, the PATCO Line in Philadelphia introduced stations without personnel and all monitoring performed in one control center, which uses dozens of CCTV monitor screens observed by two or three employees.
- Boarding areas in large bus terminals may be observed by CCTV to monitor operations at bus berths.
- Fare gates in metro stations are often monitored through CCTV to control fare payments, assist passengers with fare collection, and direct police to apprehend illegal entrants.
- Areas and corridors with heavy passenger flows are observed through CCTV for mass control, congestion control, and safety, particularly around escalators.
- Busy station areas, as well as pedestrian corridors with light pedestrian volumes and limited visibility, are observed via CCTV for security reasons, especially during evening and night periods. Public knowledge that CCTV exists in transit stations can actually be a deterrent to crime, vandalism, theft, and other illegal actions.

With AVL, CAD, CCTV, network control boards showing actual train movements, and other computer and communication devices, modern train control centers have become "brain centers" for rail transit network operations.

Artificial intelligence (*AI*) represents use of special software programs that resemble human thinking. In transit operations, for example, a train operator may report a mechanical or control problem on the vehicle on a line. AI can produce a diagnosis for the problem that occurred and offer a possible action for corrective action. In transit systems, AI is sometimes used by control center supervisors to assist train operators in such cases. Effectiveness of this AI application depends considerably on another ITS element: diagnostic instrumentation of rolling stock. A diagnostic system consists of a complex set of electronic circuits and devices that are designed to check operation of vehicle components and display any irregularities (malfunctioning), so that the operator, or a remote controller, can see what problems exist. Diagnostic equipment thus assists in detecting causes of any problems in the vehicle, so that it facilitates and speeds up the process of determining possible corrective actions.

Street and intersection signal control to give priority to transit vehicles (LRT trains, tramways, trolley-

Photo 6.3 *Operations control using artificial intelligence—São Paulo Metro.*

buses, and buses) can be advanced by the use of ITS devices. These priorities are particularly effective for lines operating in medians or other types of category B ROW. Cities with coordinated intermodal transportation policies have used traffic signal priorities for transit over general traffic to ensure regularity of transit services for several decades. Light rail systems in some European cities have been equipped with signal actuations since the late 1950s, while signal actuations by buses on reserved lanes have been used since the 1970s (Vuchic, 1981, Sections 5.3.3.3 and 4.3.4.3).

More recent ITS devices have facilitated introduction of sophisticated decision making in giving preferential signals. Different types of priorities may be used, such as changing phase lengths, changing sequence of phases, or absolute priority for transit ("signal override"). Some systems allow decisions on when and which type of priority should be used depending on the needs for transit vehicles versus impact on general traffic. Thus, a transit control center can monitor travel of individual LRT trains and select the level of priority that should be granted. When they run on or ahead of schedule, no priority is given, thus avoiding any impact on general traffic. When there is a delay, immediate priority is given to the LRT train by calling for the transit phase ahead of others. Or, though only when the transit vehicle is very late, total override and preemption, similar to provisions for police or fire vehicles, can be used.

Another type of signal priority is when rail or bus vehicles can call for signal phases that allow them to start from stops and go through intersection before other traffic is released from other lanes (advance green). The same type of control is needed where a bus gate is provided for protected left turns by buses (Vuchic, 1981, Section 4.3.4.3). Finally, rail vehicles boarding passengers from the streets may activate a special signal that stops other vehicular traffic from passing through the stop area, allowing passengers to board/alight without conflicts with general traffic. This is called a *signal island.*

Flexible forms of service are transit services in which scheduling or routing of vehicles is sometimes changed to respond to different travel demands or operating conditions. These types of service, mostly used on lightly traveled lines, can be made more efficient with applications of ITS devices. Flexible service scheduling and routing include the following types of operations.

Transfer coordination: Ensuring that a scheduled transfer opportunity among transit lines is achieved requires reliable operation because missing a scheduled transfer is very irritating to passengers, particularly when headways are long. A control system that utilizes AVL and other monitoring and communication devices can reduce the number of missed transfers because it gives dispatcher precise information about positions and movements of all connecting vehicles. The dispatcher can then make the optimal decision whether to delay the departing vehicle to allow the scheduled transfer.

Route deviation: Diverting a vehicle from its route to accommodate a few passengers is always sensitive because it disturbs its regular schedule and delays passengers on the main line section. ITS devices and programs allow computations and predictions of operations that can be used to determine which deviations are possible to make without excessively disrupting the service. This type of service can be useful for providing service for the disabled.

Mixed transit/paratransit, such as dial-a-ride, and pure transit operations can also be operated more efficiently with the use of ITS tools and programs.

Consequently, ITS has potential to improve efficiency and safety of transit operations, as well as make new types of services technically feasible.

6.4.3.2 Data Collection for Management and Planning. The great capability of collecting and processing huge amounts of data and information that ITS offers is used for many operational and planning functions. For example, passenger counting, a costly, labor-intensive, and complicated activity when performed manually, can be easily performed by ITS devices. On rail transit systems with automated fare collection, that system collects complete origin-destination data for all stations with recorded times. Similar but less detailed information can be obtained on LRVs and buses by installing devices that count the number of passengers climbing or descending steps. This information provides a passenger profile (number of passengers on each interstation spacing), but not exact origin-destination of individual passengers.

The availability of passenger counts is a great asset for scheduling of operations, adjustments to introduced schedules in the short run, and medium- and long-range transit planning as well.

Automated data collection is very useful in keeping records of performed services; computing vehicle-km, space-km, and passenger-km; recording operating personnel hours worked and computing wages; analyzing costs of alternative schedules; and other functions.

Operation of yards for storage and shops for vehicle maintenance and overhaul is greatly facilitated by automated data collection and processing. Recording of bus and rail vehicle schedules yields their hours in operation and vehicle-km passed, which are the basis for scheduling their maintenance and repairs by category, as explained in Section 6.4.4.1. This automated data collection is even more useful for metro and regional rail operations because assembling and changing train sizes (so-called train consists) involves a more complicated tracking of performed work by individual vehicles when they may be used in different scheduled runs. Automated records on fleet work also allow easier schedule of different types of maintenance work in the yards and shops.

In addition to statistics of car operations, maintenance shops can utilize diagnostic information to control conditions of vehicles, predict or diagnose failures, and prepare intervention or send a reserve vehicle into the service.

6.4.3.3 Electronic Passenger Information System. ITS technology has virtually revolutionized methods available to provide information about transit services for passengers. Elements of ITS are used in passenger information in planning a trip, in finding a station or stop, during the ride on the vehicle, during transfers, and in finding the destination.

As explained in Section 6.6, transit users in many cities or entire countries can now use the Internet to plan their trips. By entering the origin, destination, and time of a desired trip, one can obtain exact routing, including instructions for walking to the nearest transit stop and from the transit stop to the destination location, any transfers required, schedules on the relevant lines, and the fare. Alternative options can also be easily explored.

Another ITS device for transit users is a personal digital assistant (PDA) or watch using wireless technology to provide real-time information about services. Many passengers in Gothenburg, Helsinki, and an increasing number of cities use these devices to find when the next train or bus will arrive at a station.

During a transit trip, ITS devices can be used for displays of line designation and changeable station announcements on vehicles. The latest, very useful ITS application is for continuous announcements at stations and even major bus stops about expected arrivals of vehicles on each line serving it.

At transfer stations, electronic displays usually provide information about scheduled departures on all routes. Very importantly, however, at stations with coordinated schedules to allow fast transfers, ITS-based monitoring and dispatching can be used to minimize the incidence of missed transfers due to delays of arriving vehicles, so-called transfer protection.

6.4.3.4 Electronic Payment and Fare-Management System. Fare-collection procedure, data collection, accounting of revenues, and related functions have greatly benefitted from the development of telematics. Hardware and software components of ITS/telematics are now widely used for various forms of automated fare collection (AFC) at stations and aboard vehicles.

Fully automated fare collection utilizing magnetized cards with a stored monetary value, which passengers use to enter and leave rail transit systems with stations that have full access control, preceded the recognition and definition of ITS as such. The PATCO Line in Philadelphia opened in 1969, San Francisco BART (1972), and a number of European rail systems introduced these fare-collection systems successfully.

Photo 6.4 *Real-time departure information at stations—Berlin* (source: *G. Girnau, A. Müller-Hellmann, and F. Blennemann, eds.,* Telematics in Public Transport in Germany. *Cologne: Verband Deutscher Verkehrsunternehmen, 2001, reprinted by permission).*

This was a major breakthrough because an automated system replaced a labor-intensive process and provided complete data collection on origin-destination volumes of passengers by any time period.

Prepaid magnetized transit cards with unlimited rides during a given time period, increasingly introduced during the 1970 and 1980s, resulted in increased convenience, faster boarding, and positive data collection about the number of trips not only on fully controlled systems, but also on street-running bus and rail transit vehicles. Note that all fare-collection systems requiring use of magnetized cards provide automatic counts of trips, but only fully controlled systems, requiring use of card for entry and exit, allow application of graduated fares and provide complete origin-destination data. These two features are particularly important on large systems because they cannot use flat fares without major losses of potential revenues.

Electronic fare collection opened up many innovations in types of fares that can be collected, speed and convenience of collection, revenue control and statistics, etc. An excellent example of the benefits achieved through a new method of fare collection was the introduction of the Metrocard in New York City. The card has a stored value that can be replenished, it can be used on rail and bus modes with free transfer, and it can be used to pay for more than one passenger. This innovation alone has resulted in a huge increase of transit ridership on both rail and bus systems. The ridership growth was a result of increased convenience of payment, introduction of free intermodal transfer, certain discounts for some metrocard purchases, and improved services introduced in response to greater ridership.

The latest fare-collection device is the contactless smart card, which does not involve any sliding and thus facilitates and accelerates passage through fare gates (see Section 7.2.3). Smart cards are often equipped for payments of other types of transportation services, such as parking, intercity trains, etc.

The main obstacle to introduction of AFC is the high investment for the new equipment, as well as the required changes in traditional operating procedures, retraining personnel, and teaching the public how to use the new fare-collection hardware and procedures. However, there are also additional benefits of electronic fare collection, such as elimination of the need to carry cash, reduced percentage of unpaid or partially paid fares, reduced need to handle cash by crews or money collectors, as well as easy availability of data on transit trips: origin-destination, transfer, time, etc.

6.4.3.5 Summary. In conclusion, applications of ITS or telematics in transit systems can bring numerous improvements in their operations, efficiency, and ability to increase alternative types of services. Analysis of services, their utilization and costs, testing of possible schedule modifications, impacts of fare changes, etc. can be made much more easily when ITS-type devices are available. Increased service reliability and improved scheduling (or quick service adjustments when needed) can reduce the number of needed reserve vehicles, resulting in reduced operating costs for vehicles and personnel.

On the passenger side, ITS offers numerous innovations in providing information to the public—in marketing and greatly improving transit system image. The resulting increase in the number of passengers brings additional revenues.

Reducing operating costs and increasing revenue improve an agency's operating ratio. These financial results can often show recovery of the investment made in ITS equipment. Such a financial evaluation does not even consider the long-term benefits from an improved image and increased role of transit in the city that are accrued not only by the transit operators, but by the population and the city in general. These benefits are difficult to quantify, but in some situations they may be more significant than the direct financial gains for the agency.

6.4.4 Rail Rolling Stock/Bus Fleet Maintenance and Replacement

High-quality rolling stock is very important for passenger comfort and security, as well as for reliable service and economic efficiency of operations. Mechanical reliability of vehicles is one of their main performance features. It is measured by one or both of the following measures or indicators:

- *Mean distance between failures* (*MDBF*), or the average distance vehicles travel between two breakdowns in service. This number is computed as the total distance of all vehicles in a given fleet during a given time period—usually a week or a month—divided by the number of vehicle breakdowns.
- *Mean time between failures* (*MTBF*), or the average time between two breakdowns in service, obtained by dividing total vehicle-hours operated by the number of failures.

In computing these two statistical items, it is important to use a consistent definition of failure. Mechanical breakdowns of vehicles can be classified in several categories by the severity of their consequences. First, if a mechanical problem occurs that may be tolerated in service (e.g., one of four doors on a vehicle does not operate, and it is kept closed with corresponding information for the public), it should not be counted as a vehicle failure. However, there may be a failure on one car in the train, but train operation is still possible. Such failures have no significant impact on service, and yet the event should be counted as a failure if it is such a mechanical breakdown that the vehicle could not continue to operate by itself.

Similarly, one car in a train breaks down, but it can be closed off and train can continue to run to the terminal station. If the failure is such that train cannot carry passengers any more, the train must be vacated and driven to the first location where it can be taken off the line. Finally, a vehicle or train may break down with loss of power or brakes, so that it blocks the line. Then it must be pushed by the following train, causing a significant delay. All these cases, except the first one, should enter statistics as failure events.

Impacts of different types of breakdowns on service also vary with transit mode. For example, bus breakdowns seldom cause blockages or delays of other buses (except for passenger transloading procedure). When a rail vehicle breaks down, it is likely that the train can continue to run. The cases that a car or train failure requires pushing is very rare, but when it does occur, it can cause a significant delay on the line.

The choice between the two indicators, MDBF and MTBF, depends on the purpose of the desired statistic.

Bus maintenance periods or lives of some vehicle components, brake testing, motor oil change, and tire life are monitored on the basis of vehicle distance traveled, as will be shown below. In some cases, however, vehicle wear and tear, and particularly operating costs, are more closely correlated to the duration of operation, rather than distance traveled. Vehicle cleaning, for example, is scheduled in time intervals. Total operating costs are much more correlated with vehicle-hours rather than vehicle-km traveled, because the main cost component is labor, which is paid by hours worked.

In addition to mechanical reliability, transit vehicles should be efficient to operate and comfortable and attractive for passengers. These requirements are met through systematic vehicle maintenance procedure and regular replacement of vehicles when their useful life expires.

6.4.4.1 Vehicle Maintenance Procedure. The flowchart in Figure 6.2 shows major factors that influence vehicle reliability and positive results of high reliability. Basic influencing factors are the quality of vehicle design, production, and maintenance, as well as quality of maintenance personnel. Good vehicle performance with respect to reliability results in lower investments in vehicles because fewer of them are needed to provide a given service. With a smaller fleet, smaller maintenance facilities and garages are needed. With respect to operations, its cost is decreased and additional passengers are attracted.

Vehicle maintenance is usually organized as a series of four or five different procedures that are performed at intervals defined by distances vehicles have passed. An example of service categories and the intervals at which they are undertaken is given in Table 6.2.

Every vehicle must be cleaned and its brakes inspected at least once per day, with minor cleaning usually performed on vehicles at one of the line terminals, i.e., once per vehicle cycle. Services A, B, and C consist of increasingly more comprehensive sets of inspections, part replacements and repairs. Thus, a vehicle undergoes service A every 12,500 km and every fourth A service is actually a B-type service, including additional elements of maintenance and repair. Further, every other B service is service C, and every fourth service C is general overhaul, when the entire vehicle is dismantled and thoroughly rebuilt, as shown schematically in Figure 6.3.

These service intervals and contents of each work category are in some countries legally prescribed for certain types of vehicles. For rail vehicles technical standards are usually considerably more stringent than for buses. Although such high standards require high expenditures for maintenance, they are one of the reasons for higher safety of rail than bus systems.

When maintenance intervals are not prescribed by governmental standards, they often depend on the maintenance practice a specific transit agency follows. The lowest-quality maintenance is so-called *reactive maintenance,* where parts which failed or are about to fail are replaced or repaired. *Preventive maintenance* is a programmed maintenance procedure based on service intervals, as illustrated in Table 6.2.

With predictive maintenance, when vehicles have diagnostic devices that indicate the condition of different parts, service is performed at prescribed intervals, but instead of all parts being replaced as programmed, only the parts that have deteriorated are replaced. Thus, service standards are maintained but the cost of replacing parts that are not worn out is avoided. This is the most economical, but also the most sophisticated, method, requiring many diagnostic devices and detailed tests.

Transit agencies with large vehicle fleets sometimes analyze the optimal intervals for different maintenance periods. An analytical method for this computation is based on finding the optimum between the cost of maintenance and vehicle performance with respect to its reliability. Conceptually, the trade-off between these two values is the determining criterion. As the length of maintenance interval increases (vehicles are maintained less frequently), the cost of maintenance becomes lower. However, this results in reduced vehicle reliability, which increases costs of operation and costs to passengers through lower reliability of service. Maintenance interval that represents the minimum value of the total cost curve should be adopted as the optimal interval.

6.4.4.2 Length of Vehicle Life. In considering when to replace the existing vehicle fleet, a transit agency should consider the viability or obsolescence

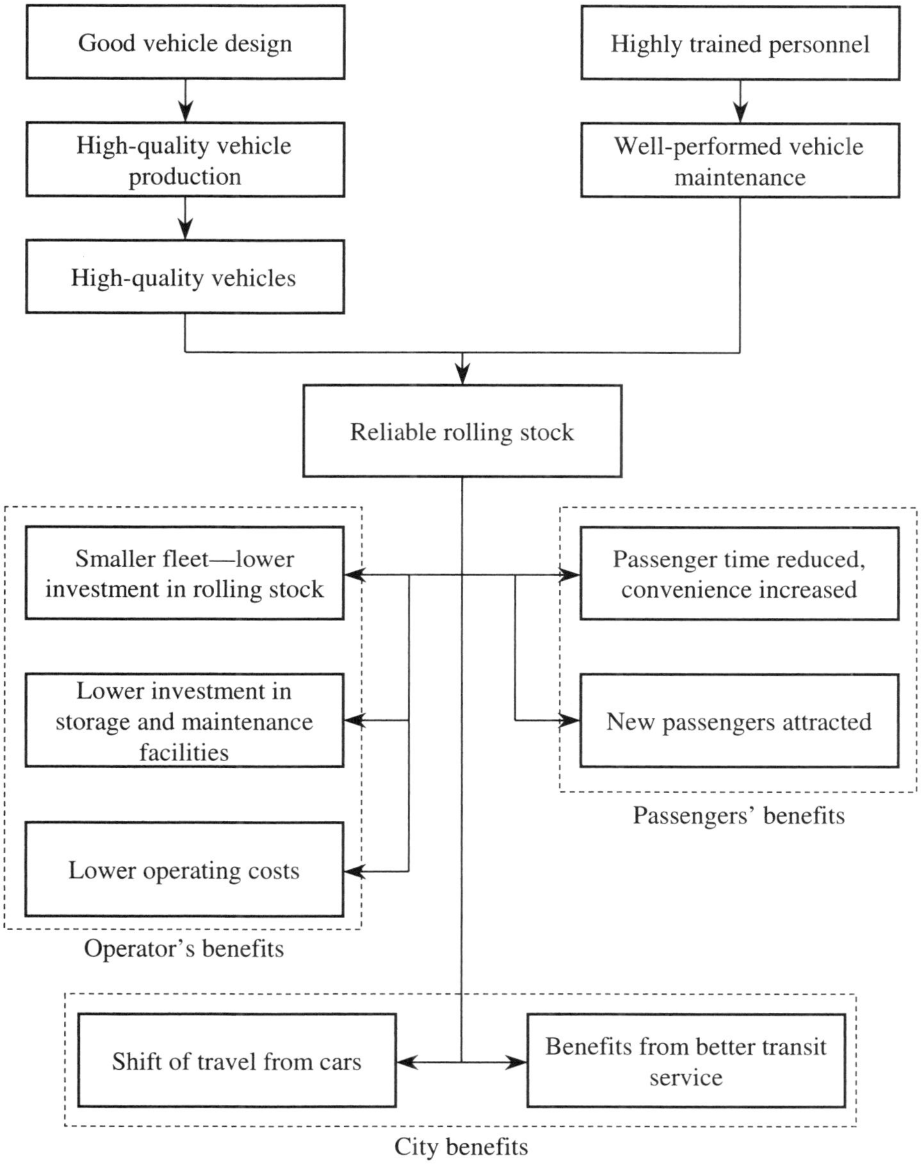

Figure 6.2 *Steps for achieving higher transit reliability and its benefits.*

of its presently owned fleet. In this analysis, three concepts are defined:

- *Economic life:* vehicle life that results in the lowest life-cycle costs, i.e., average costs per year computed for the entire vehicle life.
- *Physical life:* years during which vehicles can be operated with acceptable physical reliability and maintenance procedures.
- *Functional life:* vehicle design, technology, and operation may be superseded by new technology

Table 6.2 Example of maintenance service categories and their intervals

Service	*Interval*
Fueling, cleaning, brake inspection	Daily
Service A	12,500 km
Service B	50,000 km
Service C	100,000 km
Overhaul (D)	400,000 km

or operational requirements. Thus, a vehicle fleet may still have years of economic and physical life but be functionally obsolete. For example, vehicles without air conditioning or without accessibility for disabled may be considered obsolete in some cities.

The shortest of the three lives should govern when vehicles should be replaced. To evaluate all aspects of economic, physical, and functional life and decide how long rolling stock will be operated and when new vehicles should be purchased, a transit agency should consider the following major factors.

- *Annualized investment cost.* Longer life reduces average annual investment for vehicle purchase.
- *Vehicle maintenance facilities.* The purchase of new vehicles may require changes in maintenance facilities, equipment, procedures, and personnel skills. Therefore, retraining of both maintenance and vehicle operating personnel is needed.
- *Cost of maintenance and operation.* These increase as vehicles become older because they require more frequent and heavier repairs, increased energy consumption, and greater probability of malfunctioning.
- *Vehicle resale or salvage value.* If presently used vehicles may have some more years of operational capability, they have a resale value that should be taken into account. Usually this amount is not of significant magnitude in comparison with the price of new vehicles.

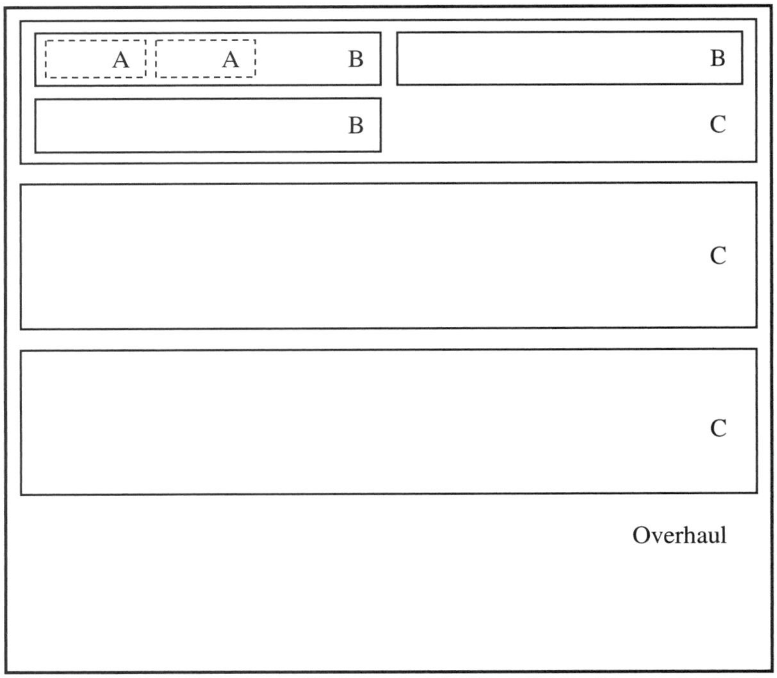

Figure 6.3 *Classification of vehicle maintenance procedure.*

Photo 6.5 Light rail maintenance facility with high- and low-floor cars—Portland, Oregon.

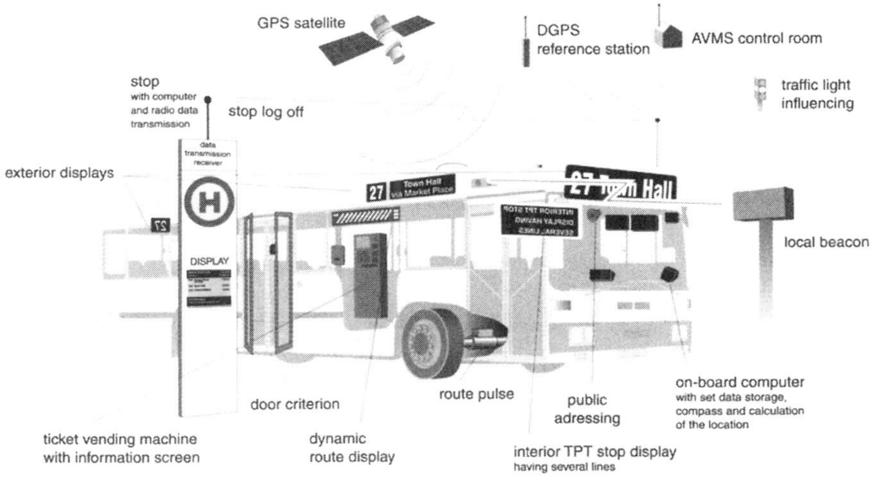

Figure 6.4 ITS technologies on board the vehicle and at the station.

- *Quality of service.* Generally, vehicle interior and ride quality decrease with age of vehicles.
- *Vehicle design and passenger comfort.* New vehicles may have more attractive design and technological features that provide higher service quality. These may be important, such as air conditioning, large windows, low-floor bodies, automated information systems, and more damage-resistant seats. New external vehicle design may give them a more attractive image. In some cases, however, technology changes are not significant, so that the latest designs do not differ significantly from 10–20-year-old vehicles that have been well maintained.
- *New technological features.* New vehicles may offer improved operations and vehicle control. More efficient electric motors, diesel motors with cleaner fuel and better exhaust control, and diagnostic systems to control performance and detect reasons for malfunctioning are such innovations. However, this need not be always the case. Sometimes new vehicles result in higher energy consumption and more complex and even less reliable design of components.

As the preceding two items indicate, all innovative features of prospective new vehicles must be carefully evaluated with respect to their cost-effectiveness. Many are obvious improvements that provide needed increases in comfort, service reliability, energy efficiency, etc. Some innovations may, however, involve very high investment and bring such negative aspects as increased complexity and maintenance cost, decreased reliability due to more sensitive components, accelerated tendency to experience obsolescence, or comfort features that are not necessary for a specific service.

This evaluation is particularly needed for developing countries where a basic, reliable service that conventional vehicles provide at low cost may be more important than the latest advances in high vehicle comfort and complex, sensitive electronic components that must be imported.

The decision about when to purchase new vehicles can also be influenced by the availability of funds. In many cases, government funds are available for transit investments, including rolling stock. This availability of funds does not eliminate the need to carefully examine all the above-discussed factors, but it does place a greater relative emphasis on the factors relating to quality of service, reliability, and introduction of innovations that improve service efficiency and system attractiveness to the users.

Practices with respect to vehicle life and purchases of new vehicles vary considerably among transit agencies. On one extreme, some agencies with large vehicle fleets, such as in London, Paris, and New York, rotate buses through the overhaul process every 5–7 and rail vehicles every 10–20 years. To perform this big industrial operation, these agencies have overhaul shops that are designed and operate like factories, at very high rates of production, such as up to 3–5 buses or 1–2 rail vehicles per weekday. On the other extreme, some agencies try to purchase new buses every 6–8 years to obtain some funds through sale of vehicles that still have considerable usable life, and to avoid all general overhaul work.

Typical vehicle life lengths for different transit modes are given in Table 6.3.

Vehicle life lengths shown as typical are found by most transit operators to be optimal from the point of view of their economic feasibility and quality of operations and service. The lower values shown in the maximum life range are also found in a number of cities. Many Canadian cities operate buses that are 15–20 years old. Some cities in Russia and the United States (Philadelphia, Seattle) operate 20–25-year-old trolleybuses. "Bullet cars," extremely advanced rail cars with streamlined bodies specially designed in 1930s for the Norristown Line in Philadelphia, which

Table 6.3 Typical life lengths for rolling stock of different transit modes

Transit Mode	*Typical Life (years)*	*Maximum Life (years)*
Minibus, van	4–6	7–10
Bus	10–12	15–20
Trolleybus	15–18	20–30
Rail: Tramways, LRT, metro, regional rail	25–30	40–50

operated at speeds exceeding 110 km/h, were in service for about 60 years. Many 40- to 50-year-old cars have operated in metro systems in Paris, New York, Philadelphia, and other cities. Buenos Aires, on one of its metro lines, operated cars built when its metro system was opened in 1913 until the end of the twentieth century. Thus, exceptions exist, but the values shown as typical life lengths can be used as the most reliable for orientation in planning vehicle purchases by most agencies.

6.4.5 Safety, Legal Suits, and Security

Safety is defined as the absence of accidents that may cause injury to persons and/or physical damage. Because transit, as a public service, is expected to be extremely safe, passenger safety is a fundamental element that must be considered and provided for in infrastructure and vehicles design and in system operations. As a result of accidents, transit agencies are faced with personal injury claims. A serious effort is necessary to settle legitimate claims reasonably and defend agencies from excessive or fraudulent ones. Security is the protection of persons and facilities from criminal or malicious attacks or damages.

6.4.5.1 Safety. The requirement to have a very high degree of safety is given great attention in transit systems design and operations. The reasons for this are obvious: passengers, employees, and property must be strongly protected. In addition to ethical reasons and the need for passengers to feel safe in using the system, a transit agency is financially liable for any situation that may be found to be hazardous.

A transit system must be designed for safe operation in several respects. First, vehicle collisions, which represent the highest potential danger to persons and property, must be prevented. Second, the consequences of a collision, however unlikely the event, must be minimized. Third, protection against unforeseen emergencies, such as fire, falling trees, or violent storms, must be provided. Fourth, passenger movements in stations (negotiating stairs, using escalators, or going through fare gates or turnstiles) must be made safe from skidding, tripping, falls, and other mishaps. Finally, contacts between passengers and vehicles, especially during vehicle boarding and alighting, require particular attention.

To minimize the probability of injury or damage, it is necessary to consider safety aspects in planning and design of a transit system's physical components and operating procedures. In addition to this consideration for each component, a systems approach must be used to ensure that the interactions of different components prevent accidents and mishaps, as well as minimize their impacts if they do occur. Major components that must be considered for system safety can be classified as follows:

- Vehicle design and performance
- Travel way design and protection
- TU movement and control
- Line and station organization, operation, and control

Due to the great differences among ROW categories in their characteristics, vehicle technology, and operations, the approach to safety, as well as the degree of safety achieved, vary greatly among modes. Safety aspects of major modes are discussed here, starting from the bus as the simplest, highway-based system operating mostly on regular streets, and ending with metro and AGT, modes on fully separated rights-of-way under signal or fully automated control.

Buses, operated manually by drivers on urban streets, depend greatly on drivers' skills and traffic conditions. The elements to consider in analyzing safety are:

- *Vehicle performance:* the driver's visibility and ability to steer, brake, and accelerate, and the availability of devices that prevent skidding and sliding
- *Bus body design and strength* to minimize damages in case of an accident: the ability to withstand collisions, lateral impacts, or vehicle overturning (extremely rare in urban transit)
- *Fire prevention and resistance:* use of nonflammable materials, emergency exits, etc.
- *Driver training and performance,* including his/her safety record

- *Conditions along the lines:* traffic intensity, control, protection buses may have through special traffic signals, reserved bus lanes, or busways
- *Bus stop design and operations:* stop locations, curbs, passenger boarding/alighting conditions, etc.
- *Communications with control center:* an important element in handling emergencies and providing immediate assistance
- *ITS elements,* such as radar for collision warning and prevention, which may become useful safety elements for buses

Rail transit, because of its different physical and operating characteristics from buses, also differs in its safety characteristics. Being physically guided, rail vehicles do not depend on the driver's steering. For tramways operating in mixed traffic, the precisely defined path has some advantages and some disadvantages with respect to safety. However, LRT vehicles operating on ROW categories B and A can utilize the separation from other traffic much more efficiently than buses: the guidance allows operation of much greater capacity TUs at higher speeds on narrowed paths under protection of automatic safety devices, such as signals and dead-man devices.

The safety devices thus vary among rail transit modes due to their different operating conditions. While all modes have rail guidance and strong vehicles, other elements contributing to safety are increasingly sophisticated, from tramways to LRT, rail rapid transit, and regional rail modes. For example, tramways have no automated signals, LRT systems have signal protection wherever speeds exceed 70 km/h, while metros and regional rail have them on their entire lines.

Because *rapid transit* trains carry large volumes of passengers at high speeds, they must be designed with the highest degree of safety. Major elements that provide such safety are:

- Fully protected ROW (category A), preventing possibility of collision with other vehicles or pedestrians.
- Track signals that ensure fully protected paths for trains through switches and conflict points. They also control spacings between trains and provide automatic stopping of trains if they run through a red signal. This fail-safe mechanism guarantees that even if a train operator makes the serious mistake of driving through a red signal, collision is prevented by automatic application of emergency brakes.
- Dead-man device: a handle or pedal that driver must hold in certain position. If a driver is disabled and releases this device, emergency brakes are activated, bringing the train to a stop.
- Rail vehicle bodies have much greater buffeting strength than highway vehicles to withstand collisions with minimum distortion and strong protection of passengers.
- The communication systems in rapid transit trains, regular and for emergencies, are more elaborate than on surface transit.
- All operations are continuously supervised, including train movements and station activities.

As a result of all these system features—safety devices and fully controlled operating practices—rapid transit generally has such safety that any single physical failure or human error can not cause an accident. Accidents can only happen when two or more very unlikely failures coincide, which is an extremely rare event.

High safety is confirmed by the records of rail rapid transit systems. As an example, the New York City Subway, carrying 4–5 million passengers per day, operated during a 40-year period, 1929–1969, without a single passenger fatality caused by a train accident, such as collision, derailment, or other malfunctions (this excludes accidents at doors and in stations, and suicides).

Fire prevention and rapid intervention, if it occurs, is given great attention because of severity of its possible consequences, particularly in tunnels and underground stations. Fire protection measures encompass a large set of requirements, including that materials in cars and stations be nonflammable. Other fire safety elements include a system of smoke detectors and sprinklers, rapid communications among train operators, supervisors, station attendants, control and communications center, fire and police authorities, etc. Recently built rapid transit systems have ventilation and exhaust equipment in tunnels, which may be activated to blow air into or out of the tunnel, whichever may be the required intervention.

Suicides represent a serious problem on some rapid transit systems. Persons committing suicide usually jump directly in front of the train, so that in most cases the train cannot be stopped. Following this tragic event, the train is not permitted to move until medical and police authorities come and examine the situation, and the death is confirmed or, if the person is still alive, he/she is pulled out from the track. If this intervention lasts long, the delay on the metro line may affect tens of thousands of passengers, creating serious problem in the entire city.

As in most types of suicides, a well-planned act by the involved person is virtually impossible to prevent. Transit agencies in cities where this phenomenon occurs with some frequency therefore organize quick response by a team of specially trained persons, including a transit supervisor, a medical examiner, a rescue team, and a police officer. The delay has in such cases been reduced from 1–2 hours to about 20 minutes.

The only positive prevention of suicides in metro systems is station design with walls along platform edges. This design, implemented on some lines in St. Petersburg and Singapore and on the fully automated Line 14 of the Paris Metro, has some advantages and some disadvantages. The advantages are increased passenger safety, avoidance of costly and disruptive suicide incidents, and easier environmental control, particularly reduced costs of air conditioning in tropical climates, such as in Singapore. The disadvantages are higher investment and operating costs for the walls and specially synchronized doors, as well as for fully automated driving, which is preferred for precise stopping. Some metro systems are considering introduction of walls while maintaining manual driving.

Automated metro and other AGT systems are designed with programmed absolute safety regime. Safety of their TU movements is therefore as high as on conventional metro systems. In stations, the driver's presence and action in emergencies are replaced by automated devices that detect a person or object on the track and automatically trigger emergency braking in the event of detection. Two types of devices have been successfully used for many years: infrared rays above the tracks (used in the Lyon Metro) and contact plates between rails (Vancouver Skytrain).

Statistical data collected by American Public Transportation Association (APTA, 2002) for the 450 largest transit agencies in the United States show that the number of injuries in transit systems operations was in the range of 31,000–34,000 per year for the years 1997–2000, while the number of fatalities amounted to 42–65 per year. Fatality rate on transit buses is estimated at 0.025 deaths per 100 million passenger-km, while for travel in passenger cars that rate is 0.544, or about 22 times higher.

Comparison of safety among transit modes is difficult because of many variations among operational conditions of different modes and types of accidents, as well as limitations in data collection. However, APTA's data collected for the 450 largest transit agencies in the United States show a rather high consistency for the same years. The numbers of passenger trips, passenger-km, number of injuries, and fatalities are computed in percentages among modes and given in Table 6.4. These figures show that about 63% of trips are carried by highway vehicles, consisting of vans and buses in demand-responsive services, regular transit buses, and trolleybuses. About 36% are carried by rail vehicles: streetcars/LRT, metro and regional rail modes. Since trips on rail modes are longer (particularly regional rail), the share of passenger-km is very different: rail modes carry 52%, or more than half of the work performed, while highway vehicles carry only 47% of passenger-kms.

With respect to passenger injuries, 91% of them happen on highway vehicles and only about 9% on rail vehicles. This is explained by the differences between operations in mixed street traffic by buses and mostly separated and automatically controlled rail modes. In fatalities, the distribution is less lopsided: highway vehicles account for about 55%, rail modes for about

Table 6.4 Percent distribution of trips, person-km, number of injuries, and number of fatalities among transit modes in the year 2000

Modes \ *Items*	*Unlinked Trips*	*Person-km*	*Injuries*	*Fatalities*
Demand-responsive vans	1.1	1.8	5.4	22.4
Buses	60.7	44.6	84.9	32.8
Trolleybuses	1.3	0.4	0.7	0.0
LRT/streetcars	3.4	2.8	2.1	8.6
Metro	28.1	29.0	4.3	32.8
Regional rail	4.4	19.7	1.8	1.7
Other	1.0	1.7	0.8	1.76
Totals	100.0	100.0	100.0	100.0

Source: APTA, 2002, Tables 26, 30, 73, and 74.

44% of fatalities. Thus, buses cause many injuries, but they are relatively less severe than on rail vehicles. The higher speed of rail vehicles and boarding/alighting of large passenger masses in stations partly explain these differences in modal safety records.

6.4.5.2 Legal Suits. Transit agencies are required to provide safe travel and are liable for any accidents caused by neglect. The source of most personal injury claims is the operation of transit vehicles on congested streets. An important goal of transit agencies is to minimize the number of accidents and thus reduce the number of claims that divert funds needed to provide services.

In addition to payments for legitimate liability in accidents, however, in some litigious countries like the United States and some of its peers, the legal system is sometimes greatly abused by submissions of excessive and fraudulent claims. In the cases handled by jury trials, there is often a tendency for the jury to sympathize with individuals who sue a transit agency. The agency is sometimes seen by jurors as an owner of large funds; they overlook the fact that expenditures on lawsuits are reflected in either reduced transit services or increased fares. Several examples of just as well as of unfounded legal cases transit agencies face illustrate the problem.

- A child is injured on an escalator that had been repeatedly reported as unsafe, but with no action undertaken to repair it. Due to negligence, the agency is sentenced to pay a large sum in compensation, including a punitive amount.
- A bus with 18 passengers aboard has a collision on a street. The transit agency receives 23 claims for compensation of injuries. Its legal department must find a way to prove which claims are legitimate and that the others are fraudulent. Repeated fraud of this kind has led some agencies to install cameras on each vehicle. If a collision occurs, a picture of the vehicle interior shows who were the passengers present.
- A passenger alights, walks in front of the bus without paying attention to other traffic, and is struck by a car. The transit agency is unjustifiably sued for its contribution to the accident.
- A person with a record of drug abuse falls on subway track and has a leg cut off by an incoming train that was braking for a stop. The plaintiff's "expert" argues in the suit that the transit agency operates trains "at reckless speeds" and claims that trains should slow down before entering every station and travel its entire length at 25 km/h. Initially, the jury awards the plaintiff a large amount of compensation. On appeal, the court rejects the claim of reckless speed and accepts the defense's argument that rapid transit trains are not designed to stop for pedestrians, and that it is impossible to prevent injury to a person intentionally jumping in front of a vehicle, be it in the subway or on the street.

The legal system and practices in some cities stimulate filing of suits, particularly against public agencies, including a significant percentage that are based on so-called soft tissue injuries, in which there is no objective medical evidence to support the claimant's assertion of pain. While some of these injuries are real, many can be easily fabricated. These may be supported by expert witnesses who compromise their technical and ethical standards in expectation of high potential financial awards from suits. Some transit agencies have to cope with "experts" who testify in dozens of such cases. It is thus clear that transit systems must address this problem in a comprehensive and aggressive fashion to prevent unjustified expenditure of limited transit financial resources.

Such a situation was faced by SEPTA, the transit agency in the Philadelphia metropolitan area. During the 1980s, the number of suits was rapidly increasing, so that in 1988, SEPTA received almost 15,000 claims and 5000 personal injury lawsuits and spent $63.5 million to resolve the claims. Amounting to 11% of its operating budget, this was the highest legal expenditure of all transit agencies in the country.

In response, SEPTA undertook an aggressive approach, designated as the Fraudulent Claims Campaign, to solve this problem.[1] The campaign consisted of several parallel efforts:

- Information on all claims was computerized, providing the ability to track claimants, their lawyers, expert witnesses and medical providers, the types of claimed injuries, the frequency and nature of medical treatments, and other factors. This information permitted analysis on a case-by-case basis to determine whether the claim was valid.
- Video surveillance was introduced on vehicles and in stations as an investigatory tool to see whether the claimant was exaggerating the nature of the injury, and whether he/she was on the vehicle at the time of accident at all.
- Community support and cooperation of local law enforcement agencies was developed.
- Court support was obtained in the form of imposition of criminal penalties, including retribution for proven fraudulent claims.
- The Citizens Crime Commission provided assistance through the use of a tip line to report questionable or fraudulent claims.
- Advertising, using billboards, posters, leaflets, newspaper, and radio messages, was used to highlight the problem of fraudulent claims and announce the aggressive campaign to deal with this serious problem. This campaign helped change the perception that the transit system was an easy target for unfounded claims.

This comprehensive approach to claims management had a dramatic impact on the number of claims and amounts spent on their settlements. Total expenditures were reduced by more than 50%, saving SEPTA tens of millions of dollars in subsequent years.

To increase safety and manage claims on a continuing basis, transit agencies, particularly large ones, should consider performing the following activities.

- Inform and remind the public about safe practices and habits.
- Intensify the review of facilities and operating practices with respect to their impacts on passenger safety.
- Improve recordkeeping and collection of information that is applied in the case of accidents. This may include erasable visual recording of vehicle interiors, station platforms, corridors, and passenger waiting areas in terminals.
- Get competent experts and prepare a strong defense against unjustified suits. Avoid settlements to save trial costs, which establish costly and damaging precedents.
- Obtain court or legal protection for the transit agency, similar to limitations or exemption of certain liabilities by public agencies. For example, many departments of transportation have liability limits defined by types of cases, or maximum amounts of their liability.

[1] Information obtained from Mr. James Kilcur, former head of SEPTA's Legal Department.

- Pursue fraudulent claims to obtain maximum penalties for them.

Applications of these measures can result in very significant reduction not only of financial payments, but in the number of filed suits and the public image of the transit agency.

6.4.5.2 ***Security.*** In transit systems operation, security can be defined as absence of or protection from acts that endanger passengers, employees, their property or transit system facilities. Security encompasses protection from a broad range of events, from violent criminal acts to minor acts of theft and fare evasion. For a systematic analysis, major aspects of security can be classified in several categories by their types and objects of potential harm.

Passenger Security. Personal crimes such as assault, robbery, larceny, and harassment are very serious events that must be prevented on transit systems by all available means. Criminal attacks result in harm to the victim as well as in degradation of the transit system's image. In the long run, an image that a system is dangerous results not only in reduced ridership, but in a decreased role of transit, harming the viability of the city.

The problem of passenger security varies greatly among cities and among areas and transit lines within a city. The perception of security may also vary among lines and transit modes, and the security passengers perceive may differ from actual conditions. Factors that influence security on transit systems include the following.

System quality, design, and physical condition have a significant impact on passenger behavior, security, and safety. Physical conditions may create lower perceived than actual security on a system. An old bus with torn seats tends to generate more crime occurrences than a clean bus with large windows and comfortable seats. A subway that has dark stations with trash and graffiti and old noisy cars, is more likely to have lower security than a subway with spacious, well-lit stations and clean, comfortable cars.

The character of the served area and ridership using the transit system are in many cases the main factors influencing security. The ridership that dominates the behavior and relationships among passengers is an important element. Passenger groups may create a safe, pleasant condition or a situation that is conducive to rowdy behavior and incidents, requiring police supervision.

Security elements on transit vehicles and at stops and stations are very important for crime deterrence, prevention, and calls for assistance. These include such devices as CCTV in trains and at stations, pedestrian corridors, and bus terminals, emergency telephones or call boxes, presence of personnel, posted warnings and instructions, as well as voice announcements and communications. In this respect, the use of telematics is also very beneficial: it improves communications, and the use of AVL allows faster response because transit supervisors, city police, or the fire department can locate vehicles asking for intervention very quickly.

The three groups of factors influencing security—system condition, characteristics of passengers, and system supervision—together create an environment that is critical for security, and they must be analyzed and planned together in efforts to provide high security.

Employee Security. Transit vehicle drivers and station booth attendants can be attacked by criminals when they handle cash, particularly in isolated locations and at night. The impact of attacks is very serious for affected employees, costly for the agency, and affect employees' morale.

Reduction or elimination of drivers' handling of cash fares eliminates robbery as a motive for attacks. Protection of employees from criminals has also been increased by enclosed driver cabins and several types of alarm systems connected with the operations control center.

Protection of Revenues. Loss of fare revenue can come in three ways: external theft of revenue by outside criminals, occurring where it is collected or when carried to the agency or bank buildings; internal theft, performed by transit employees in different ways and locations; and evasion of fare payment.

1. External Theft. Fare revenue is collected throughout the transit system—its stations and vehicles—so external theft may occur in various locations and forms. Security is generally improved when fare collection by drivers is replaced by exact fare payments into fare boxes, by use of prepaid passes, and by installation of ticket-issuing and ticket-validating machines. The machines located in stations or in vehicles are usually safe from attacks, but those located along transit lines or at lonely suburban stations are sometimes subject of crime. Various types of alarm systems and telematic devices are increasingly being used to send warnings about attacks and, even more importantly, to discourage burglaries. This is very useful because attacks on automatic fare-collection machines often cause greater losses by damaging the machine than through the theft of cash.

2. Internal Theft. All transit agencies must pay attention to prevent theft of cash by employees. There is always potential for this phenomenon, because in transit agencies many employees handle cash: they receive it from passengers, collect revenues from fareboxes, or ticket-issuing machines, carry it in vehicles, and then deliver it for sorting in the agency offices. Again, the most effective means to reduce opportunities for this theft is eliminating as many employees' contacts with cash as possible. Cash collection from fareboxes into carts, boxes or special vehicles can be designed so that it is fully automated, without the handler directly contacting the cash.

Small-scale pilfering from individual cash transactions is difficult to detect directly, so that it may be necessary to perform testing of personnel by unmarked supervisors or observations through cameras or other types of detectors.

3. *Fare Evasion.* There are always some passengers who try to avoid paying their fare, or try to pay less than the full amount. This phenomenon involves financial loss but inflicts no direct damage to individuals or equipment. It ranges from obvious criminal acts such as jumping over the gates or forcing passage through them, to the use of false passes and overriding fare zones. Prevention of these losses is largely a matter of devising fare-collection techniques and equipment that prevent fraud and checking payments by special controllers, usually on spot-check (random) basis. Police action is used in apprehending and penalizing persons who forcefully enter the fare gates or use other illegal entry.

Damage to transit property can be divided into two categories: theft and vandalism. The motive for theft is personal benefit for the perpetrator. Vandalism is damaging or destruction of property motivated by antisocial behavior without any beneficiary. It can be described as an act performed "with malice toward all," and it is a reflection of social ills and distorted vicious emotions of individuals, often spread by fashions, mostly among youth and delinquents.

Theft in transit systems is usually more commonly done by employees than by passengers. The seriousness of this problem varies considerably among cities and agencies. In developing countries, it may be more frequent because of low living standards and low personal incomes. Protection from it is similar to the protection that industrial corporations carry: fencing off of the properties, control of gates, and well-organized shops and yards with systematic supervision.

Vandalism is a phenomenon found more in developed than in developing countries. It is a social ill of affluence. Although known throughout history, vandalism has frequently focused on transit systems, probably because they are open to the public and difficult to control fully, and because the mobility of vehicles gives exposure to the vandals seeking attention. From the late 1960s up to the late 1980s came a particularly intense wave of graffiti as an oppressive form of vandalism. Social changes and conflicts among generations created many "turned off" youths. Introduction of cheap felt pens and spraypaint made the graffiti phenomenon feasible, and the misnomer of "graffiti artists" for the disturbed vandals gave them a partial cover instead of strong condemnation.

The graffiti plague quickly spread across cities, first in the United States and then in Europe. Transit systems were prime targets in many cities, although buildings, trucks, and railway cars were also attacked. Leading in the development of the problem, U.S. cities were also the leaders in finding the solution and, mostly, overcoming the graffiti vandalism. The case of New York City illustrates this development extremely well.

The underfinanced, decaying New York City Subway system was vandalized extensively during the 1970s. The city's financial crisis and racial and social turmoil created conditions for rebellious vandals. The subway became world-renowned for its extensive graffiti, which covered most of its 6300 cars, the largest rapid transit fleet in the world. This physical condition led to an atmosphere of very low security, and passengers were shifting to any other available mode, from buses and illegal jitneys to taxis, cars, and bicycles.

With secured substantial financing for rehabilitation of transit in the early 1980s, new management decided that physical renovation would not be sufficient: the system had to change its image totally, eliminate traces of vandalism, and recreate an atmosphere of convenience and security. A comprehensive, systematically organized program, proven successful in cleaning up the Philadelphia transit system, was prepared to rehabilitate the system. It consisted of increased supervision and apprehension and penalties for the vandals. But the main component of the program was that graffiti cleaning was not performed any more on single cars and randomly, but one by one train and line. Once a train was cleaned up, all new graffiti was removed within hours. Once a line was cleaned up, it was kept clean: if marked again, cars were not let in service until cleaned.

The program was very successful and confirmed the experience from other cities. Graffiti is seldom produced by vandals on clean surfaces. Once graffiti appears, it attracts more graffiti at an accelerated pace. Partial cleaning on a random basis is therefore ineffective. Without sufficient funds and systematic campaign, the New York City Subway was losing the battle. Initial scribbling developed into elaborate murals, which were obviously well planned and designed and took several hours to paint on unsupervised cars. Random cleaning was considered to be the only possibility because of financial constraints.

Once cleaning was introduced on a systematic basis and followed by instant cleaning of any new occurrences on the clean rapid transit line, the incidence of this phenomenon decreased drastically. The explanation is rather simple. The main reason for the vandals to go to the often considerable effort of painting graffiti was to see the product of that effort going on cruising display throughout the city. With the new cleaning policy, the vandals see that whatever they draw disappears by the next day. This is very discouraging, and eventually the entire phenomenon decreases and disappears. Thus, in the long run, systematic cleaning requires less labor effort and expenses than partial, random cleaning, and the system is permanently clean, rather than only partially clean.

As described by Vuchic and Bata (1989), the cleaning campaign in New York City created by 1989 a rapid transit system that had less graffiti, vandalism, and dirt than many other transit systems, as well as railways, in the world. The lessons from New York can be summarized in the following points:

- Graffiti and other vandalism must not be underestimated. They inflict major expenses, reduce ridership and fare revenue, and hurt the image of the entire city.
- If graffiti appears, a systematic action of cleaning it should be quickly undertaken. Delays in reaction lead to a rapid growth of the problem.
- Cleaning should be done in a systematic and persistent manner. That is less costly and more effective than random cleaning of individual vehicles.
- Graffiti vandalism must be treated as destructive behavior against public property, rather than as the misbehavior of children.
- The negative impact of graffiti, vandalism, and neglect of transit vehicles on the surroundings and the image of the entire city can be very serious. Conversely, cleaning up the system can give a strong stimulus to improvement of other facilities and better city image.

Other vandalism includes scratching of windows and other glass surfaces and defacing or destroying of signs, information devices and maps, lighting fixtures,

Photo 6.6 *Past graffiti problems in NYC have been successfully eliminated, resulting in ridership increases.*

etc. This not only involves costs for repairs or replacement, but also creates problems for users who cannot get information.

In conclusion, security is a very important element of transit systems. Transit is an open, public service that people must consider convenient and safe to use. Any criminal acts, therefore, logically tend to get considerable publicity, which is very damaging not only to the transit agency, but to the entire city, because it may divert some transit users to travel by car, or to avoid coming to the city at all. Transit agencies therefore pay serious attention to the security issues, which vary from criminal acts to theft and vandalism.

Each type of activity damaging security requires a special method for prevention and recovery. Prevention and protection from criminal acts are handled by supervisory units, field supervisors, and police. Some large transit agencies have their own police force, which cooperates with city police. Protection from theft may be more an internal phenomenon, requiring supervision and control mostly within the agency's facilities (buildings, maintenance shops, and yards). Vandalism, on the other hand, is mostly external and is a different type of crime, one without any material beneficiary. Vandalism therefore requires analysis of the motives of the perpetrators, and it must be fought by prevention through education, protection of objects, and immediate reaction when it happens, through repair as well as punishment of the vandals.

Security measures may involve considerable costs. However, feasibility of these activities must be analyzed on a broad system basis. The benefits from high security are not only prevention of direct material losses, but the positive image that a safe and secure transit system has. This results in higher ridership and benefits to the city. Moreover, immediate action in prevention and repair of vandal attacks is usually more cost-effective than its postponement because absence of signs of vandalism is the strongest factor in its prevention.

6.5 TRANSIT SYSTEM FULL ACCESSIBILITY

One of the basic features of urbanized areas and their advantages over rural areas is the provision of mobility—convenient and economical travel of persons throughout the city and its suburbs. As private cars came into wide use, they greatly increased the moblity of their users. However, persons who cannot or do not want to drive cars rely on other modes, such as transit, bicycle, walking, etc. While car users benefitted from the construction of highways and parking facilities, other modes in many countries did not experience similar improvements.

As a reaction to this increasing gap in mobility between car users and others, it became logical to reaffirm or redefine the role of transit as one of the basic services in cities. Actually, the concept matured that transit should be upgraded so that it can serve not only able-bodied persons, but also those with disabilities, persons needing various kinds of assistance, those carrying children or luggage, etc. This evolution of needs and policy considerations led to the development of the concept of *fully accessible transit.*

6.5.1 The Problem of Mobility and Its Solutions

Since the late 1960s, the problem of mobility for handicapped or disabled persons has been raised in many countries. Many physically disabled persons and some of the elderly face obstacles which prevent them from using transit services. As most of these persons cannot drive either, they were seriously mobility-disadvantaged, and their lives were affected by this limitation.

6.5.1.1 Problem Definition. The discussions about the problem of mobility for disabled persons in many countries led to a variety of governmental actions. In the United States, several successive regulations and laws have been adopted. Initially, these discussions referred to the groups designated as ‘‘elderly and handicapped’’ (E&H). This term was later changed because most elderly persons do not require any special assistance in using transit, and passengers needing special accommodations have been designated as ‘‘disabled persons.’’ The mandate and goal to make possible use of transit by the disabled persons are now referred to as ‘‘transit system full accessibility.’’

This definition of full accessibility will be broadened here, however. Long before special considerations for the disabled and elderly were given attention, many transit systems provided assistance to persons with small children, particularly those using strollers/prams, as well as to those with heavy or large pieces of luggage. More recently, great attention has been given to accommodation of bicycles on transit vehicles, and many innovative solutions have been introduced. Consequently, a fully accessible transit system is defined here as a system on which reasonable efforts have been made to accommodate all groups of present and potential users who have special needs. These needs may include different forms of information and guidance, payment of fares, reduced or eliminated steps in approaching, boarding, and alighting from vehicles, seating in vehicles, etc. By far the most complex is provision of access for persons in wheelchairs.

‘‘Reasonable efforts’’ must be interpreted considering economic and physical/operational feasibility as well as impacts of transit system adaptations for accessibility. Being a primarily social measure, transit accessibility cannot be subjected to conventional economic feasibility analysis based on benefit/cost computations. Yet some consideration must be given to the relationship of costs of making a transit system accessible and the number of potential beneficiaries. For example, providing full accessibility on a new metro system and expecting a significant number of users who will be able to use a system with all stations accessible can usually be easily justified. On the other hand, if retrofitting an existing metro system by elevators and other devices that require millions of dollars of investment, while the number of potential users is only 20–30 per day, raises questions whether such an investment would not be more appropriately used for expanding transit services that would benefit tens of thousands of persons, while providing alternate paratransit for the persons in wheelchairs.

A similar evaluation is appropriate for impacts of persons in wheelchairs utilizing buses. Their accommodation on low-floor buses is relatively easy and fast. If, however, use of special lifts on rear doors is required, that process may involve such long delays that bus service would be made very unreliable and unattractive for all other passengers. This could result in passenger losses and eventual decrease of transit service quality, negatively affecting all passengers, including the disabled.

The solutions for different accessibility needs also depend on many physical, operational, and economic factors of individual systems or modes. For example, luggage can be accommodated much more easily on rail vehicles than on buses due to the greater spaciousness of the former. Or a bus line can be made accessible by ordering low-floor buses and a large open area near one of the doors. Opportunities to upgrade vehicle accessibility coincide with the life cycle of rolling stock: bus replacement takes place every 10–12 years, while change of rail rolling stock takes place once every 30–40 years.

Due to the very high cost of retrofitting stations of old metro systems for wheelchair access, making those systems accessible is a more expensive and longer-term process than modification of vehicles, if it is feasible at all. Therefore, regulations require that transit agencies start this process by making some key stations fully accessible and continuing this process gradually as opportunities arise and funding becomes available.

As mentioned above, an exhaustive list of transit user groups who require some assistance or accommodation is actually greater than the elderly and disabled as defined by legislative acts. For planning a transit system with full accessibility, the following groups of users with special needs can be defined:

- Persons with limited mobility or impairments: disabled and some elderly
  - Physically disabled: persons in wheelchairs and those who have difficulty walking or bending their knees
  - Persons with sensory disabilities: visual, hearing and cognitive impairments
- Parents with small children, particularly those using strollers/prams
- Persons with heavy or large pieces of luggage
- Persons with bicycles

The disabled and elderly segment of population is not negligible. Government estimates are that approximately 15% of the U.S. population has physical or sensory disabilities. Surveys of different European countries conducted around 1990 estimated that physically disabled persons comprised between 5 and 15% of the population. The wide range among countries is caused by the differences in the definition of disabled persons. Their handicaps vary by type and their impact on mobility. For example, the hearing-impaired and vision-impaired and persons in wheelchairs face very different obstacles and thus need a variety of assistance methods. The sensory limitations can usually be solved by additional information or special guidance equipment, while the movement of persons in wheelchairs is impaired by any steps, stairways, gaps between platform and vehicle, etc., which require redesign of facilities.

The percentage of persons over 65 years of age was estimated in different countries to be between 12 and 18%, but this number is expected to grow substantially in the early decades of the twenty-first century. Two factors will cause this growth: increase in life expectancy and the aging of the large population segment born in the post-World War II period, the so-called baby boomers. Accommodating the elderly on transit systems is, however, a much simpler problem than accommodating the disabled. Most elderly persons use transit extensively without any problems. Only a fraction of this group has problems, for example, in negotiating steps and in stability aboard transit vehicles. Thus, in most respects the requirements for accommodating the disabled govern the design of modifications, but some of the elderly passengers also benefit from them. A few of the measures, such as better information systems and low-floor vehicles, actually benefit all transit passengers.

6.5.1.2 Policies and Implementation of Solutions. Specific developments of policies for full transit accessibility in the United States are briefly described here. Although the legal formulations, procedures, and standards regarding this problem vary among countries, the general trend toward providing accommodation for the disabled in various forms has taken place in most industrialized countries during the last three decades of the twentieth century.

In the United States, there was a sequence of federal government regulations and specific requirements for accommodating E&H groups was enacted during the 1970s and 1980s. For example, in 1970 the Urban Mass Transportation Administration (UMTA) (now the Federal Transit Administration, FTA) initiated research on the needs of E&H and developed new features in bus design, demonstrated in the Transbus research and development program. This was based on the amendments to the transportation law that stated that transit provision for the disabled was a national policy.

The DOT regulations issued in 1973 mandated that a specific percentage of accessible vehicles be provided in fixed route (regular bus) service. Noncomplying agencies might lose federal support. New regulations in 1986 required that handicapped persons unable to use regular transit service had to be able to use specially services specially designed for this purpose. Fares for the handicapped had to be comparable to regular fares.

The entire sequence of efforts culminated in the approval by Congress of the Americans with Disabilities Act (ADA) in 1990. This new law included requirements for provision of access to disabled persons in all public buildings and on streets and other public areas, as well as for full transit accessibility. It stated that special paratransit services for the disabled had to be complementary with, rather than a replacement of, regular transit services. The complementary service had to extend at least 1200 m (3/4 mile) on either side of the fixed route. Eligibility for paratransit was determined on the basis of a person's ability to use a regular, fixed-route system.

The requirements stipulate, basically, that all new transit systems and services must be fully accessible, while most facilities on existing systems must be made accessible only if they undergo a major reconstruction. The reason for this is that many existing facilities, such as subway and regional rail stations, bus terminals, etc., cannot be made accessible without extremely high investments and disruption of the surrounding areas. However, agreed upon key stations must be accessible to provide partial system access.

The question who should finance modifications of transit systems for full accessibility and costs of operating special services continues to be the subject of discussions. Provision of accessibility for disabled persons is a goal of the entire society, and a very noble and useful one. If the necessary system modifications, or special services needed to achieve that goal, involve expenditures, it is logical that the government allocate the necessary funds. In the United States and most other countries, however, this is not the case. The legal requirement for transit agencies to invest in and continuously provide services for the disabled generally represents an unfunded mandate or unfunded entitlement, a requirement by the government for which no funds are directly allocated.

Requiring transit agencies to finance special facilities and services for the disabled from their general funds creates the illogical situation that the measures to achieve the social goal of accommodating the disabled are financed only by transit users (by sharing of the funds used for the regular transit services) rather than by the entire society. If an agency has limited funds, conflicts may occur over provision of service between factions. When provision of special services requires reduction of regular service used by able-bodied passengers who are transit-dependent, this becomes a socially perverse outcome.

6.5.1.3 Structure and Procedure of Accessibility Planning. The procedure for planning full accessibility can be organized along three dimensions: by groups of users (see Section 6.5.1.1), by transit modes, and by the types of solutions.

Transit modes can be classified for accessibility planning in the following major categories:

- Bus services, including paratransit, buses on streets and bus semirapid transit
- Streetcar and light rail transit
- Rapid transit and regional rail

Other modes, such as AGT and special technologies, generally have the same requirements and solutions for the disabled and other groups as the rail systems.

Types of solutions can be classified in three major categories:

- Physical modifications of present systems, which may be made by retrofitting or by introducing new vehicle or facility designs
- Modifications of operations on the existing services
- Introduction of additional services for the disabled and elderly, mainly paratransit

Procedure for planning full accessibility organized for each specific mode may consist of the following steps:

1. Define transit passengers with special requirements who should be accommodated. ADA requirements must be met, but special accommodations for luggage or bicycles may or may not be planned.
2. Analyze the process passengers go through in using the system: finding information about the service, getting to a stop or station, going through a station, paying fare, etc., and then boarding transit vehicle. Further, the needs of the passenger during the ride, and procedure for alighting from the vehicle and negotiating the way through the station, must again be examined.
3. Define the obstacles or difficulties facing each analyzed group and needed improvements.
4. Develop possible solutions for overcoming the defined problems
5. Select the optimal solutions for each problem, considering physical and economic feasibility, time of implementation, and impacts on all passenger groups for each one.

In the following sections, accessibility of each mode is discussed, reviewing the needs of different groups of users.

6.5.2 Accessibility of Different Transit Modes

Because of different vehicle dimensions and designs, as well as differences in the types of stations and boarding conditions, provision of full accessibility varies in design and operations among transit modes. A brief review of three major groups of modes is presented here.

6.5.2.1 Buses. Conventional transit buses have floor heights from the street of 71–75 cm in Germany and 76–88 cm (30–35 in) in North America. Many buses have high steps: the first one 32–45 cm, others 20–35 cm (EG&G Dynatrend and Crain & Associates, 1995; Spiller and Labell, 1997; Müller-Hellmann, 1999). These are inconvenient for some passengers. Recently designed models have the first steps not higher than 23 cm from the ground and subsequent steps no more than 15–20 cm. Depth of steps is 30–32 cm, subject to the area in the bus which can be taken from the aisle. In some cases, the lowest step may extend outward when the bus stops and be retracted when the doors are closed. Such gradual steps greatly facilitate and speed up the boarding/alighting process.

Another method for facilitating passenger access has been the invention of a kneeling bus. A special mechanism controls the air pressure in the pneumatic suspension of the right front wheel. When the bus stops, the air is released and the corner of the bus with the front door is lowered by at least 7 cm, improving accessibility. The mechanism can be controlled by the

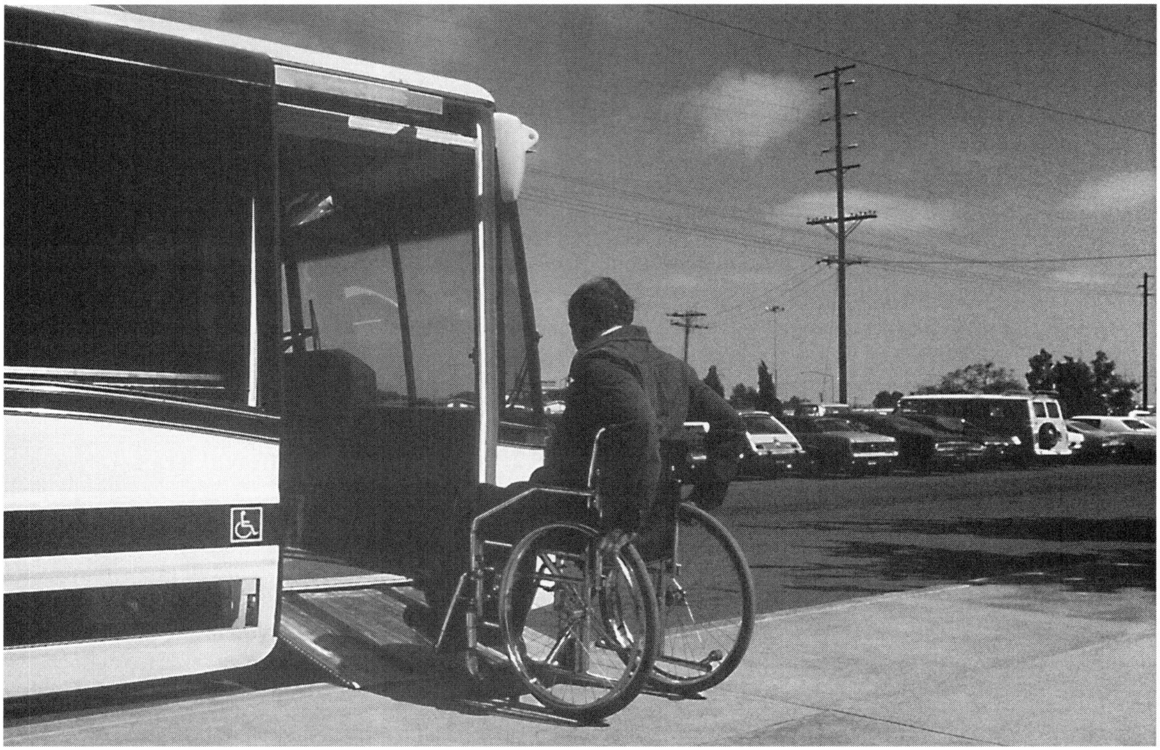

Photo 6.7 *Low-floor technology with wheelchair ramps, more effective than lifts.*

driver, or designed to operate automatically whenever the bus stops. The time for lowering is only four seconds and for raising it is only five seconds. Because this takes place concurrently with the couple of seconds it takes for the doors to open and close, this process does not increase standing time significantly.

For wheelchair accessibility of conventional buses with floor heights of 70–90 cm, special lifts must be provided. They may be built in the front or rear door, depending on the internal bus layout, and there are a variety of mechanisms and methods of operation. Most of them require that the driver activate and assist in its operation. The lifts are complex mechanisms that require frequent operation to be reliable. Taking several minutes to cycle, they may cause several minutes of delay.

A major innovation in accommodating passengers in wheelchairs, as well as assisting all passengers, is the development of the low-floor bus. With a floor elevated only 32–35 cm from the ground (with kneeling, the front door may have a step of only 25 cm), this vertical difference can be overcome without a lift. Usually a plate is used, which is stretched out from the bus or manually lowered so that there is a continuous ramp for rolling the wheelchair into the bus. When the bus stops close to a curb, the elevation difference becomes even smaller. Typically, curbs have a height of 10–14 cm, in European countries up to 20 cm. The maximum slope on ramps or along the aisle in the bus should not exceed 8% for travel without excessive difficulty.

Inside the bus, there are one or two locations for parking wheelchairs and securing them with a clamp, belt, or some other anchoring device from rolling during travel. Their deployment usually requires assistance from another person. Another solution is to have a backward-facing vertical pad backed by stanchions. The wheelchair user can back into it alone. Regardless of this detail, the same location can be used for storing a stroller. Along the aisle, stanchions and handrails provide assistance and guidance for the wheelchaired as well as other passengers.

In addition to physical facilities to accommodate the wheelchaired and other disabled passengers, some bus services can be modified to serve disabled passengers better. Bus lines with low passenger volumes and regular lines during hours of light use can be operated with some route deviations: a bus is diverted from the regular route for a short distance to pick up or drop off passengers who request service. This type of operation may eliminate the need for complementary service, while it also supports the goal of mainstreaming the disabled, i.e., including them in the regular transit system. On the other hand, diverting the bus causes delay to other passengers.

A *service route* concept is a variation on fixed routes that focuses on the needs of the elderly. Service routes meander among popular destinations such as community centers, shopping areas, and medical complexes. First developed in Sweden in the 1980s, this was designed to be an attractive alternative to more expensive paratransit, providing convenient and relatively frequent access, albeit at the cost of increased travel time.

Trips by disabled passengers that cannot be served by fixed-route or route-deviation services must be provided for by some type of paratransit service. Several types of paratransit services can be operated. The most convenient service for users is personal door-to-door with special accessible paratransit vehicles, but that service requires the greatest number of vehicles and is the most labor-intensive, involving very high investment and operating costs. It is therefore used primarily for passengers who cannot make any transfers or use fixed-route transit. Dial-a-ride can also provide door-to-door service in certain areas, but it is more efficient because every ride serves several passengers. Feeder paratransit is usually dial-a-ride many-to-one type of service, which collects persons from an area and brings them to a transfer with a regular bus or rail transit line.

Transit agencies either operate paratransit as a special division or lease it out to specialized operators. In any case, paratransit involves much higher cost per passenger trip than fixed-route regular transit services. The ADA law therefore states that only persons unable to use regular transit lines are entitled to paratransit services, and that paratransit should be provided as supplementary service coordinated with regular transit, rather than duplicating and competing with fixed routes.

Access for persons pushing strollers with little children is facilitated by provision of ramps to platforms and low-floor vehicles. An area for parking the strollers in vehicles is also provided in a similar way as for wheelchairs. However, the boarding of persons with strollers in vehicles with steps can also be performed, often with assistance of other passengers. This practice of voluntary assistance by fellow passengers is quite common and is considered a social ethic in some cities, particularly in European countries. In the United States, the use of strollers is less frequent, but areas reserved for them in vehicles, as well as seats reserved for the elderly, have been increasingly introduced in recent years.

Buses in many cities now have racks above their front bumpers where one or two bicycles can be attached. This method of carrying bicycles has been proven acceptable in suburban areas, and it is being tested in some central cities also.

6.5.2.2 Streetcars/Tramways and Light Rail Systems. The ability to accommodate passengers in wheelchairs on streetcar and LRT systems varies due to different physical characteristics of these modes. Some LRT systems, such as those in Edmonton, Calgary, and Stuttgart, have only high-platform stations. On these systems, access for wheelchairs must be provided from the street to the platform using ramps at stations in the streets and elevators in stations that are grade-separated. Then access into and out of the vehicles is easy unless there is a considerable gap or elevation difference between the platform and car floor levels.

On rail lines with low platforms and conventional, high-floor vehicles, access can be provided by mini-platforms accessible via ramp (Sacramento, California). Alternatively, stationary wayside lifts can be installed at all stations. This type of lift was used in Portland, Oregon, and San José, California, until transit systems in both cities added low-floor sections to their vehicles, making the lifts unnecessary.

Since 1985, low-floor tramways and LRVs have been increasingly introduced in transit systems. They have revolutionized the procedure of boarding and alighting from rail vehicles. Virtually all new tramway and many LRT systems in Europe now use vehicles with low floors in the entire vehicle or a portion of it. Transit systems in North America have somewhat belatedly also introduced this design. Several systems (Portland, Dallas) have rebuilt single-articulated six-axle cars into double-articulated eight-axle cars, adding a middle section with low floor.

The low floor has an elevation of 30–35 cm above the top of rails. Because rail vehicles come precisely close to the platform edge, any station with a curb of 20 cm or raised to match the vehicle floor level provides virtually horizontal entry with a small gap. Inside the vehicles, tramways and LRVs are generally more spacious than buses and provide a smoother ride, so that wheelchairs and strollers can be accommodated more easily and need no anchoring.

6.5.2.3 Metro and Regional Rail Systems. In metro systems, the main problem for passengers in wheelchairs is access from the street to the platforms. In most old systems, this path requires negotiating at least one set of stairs between station levels. Conventional fare-collection gates and exit barriers can also be obstacles to wheelchairs, strollers, and bicycles,

Photo 6.8 *Low-floor LRVs and raised curbs provide easy boarding—Erfurt, Germany (courtesy of Mario Semmler).*

sometimes even luggage pieces. In many recently built metro systems, such as those in Washington, Seoul, and Oslo, stations are designed with elevators between all levels and train platforms. A special wide fare-collection gate allows passage for passengers with special needs.

Entering and exiting from the trains is a problem for the disabled and other passengers with special needs on the metro systems where there is a step from the platform into the cars. The problem exists also where stations have curved platforms, or on regional rail lines where wide freight cars must pass. In such cases, special plates must be provided to bridge the gap. On new systems, or on old systems with recently purchased rolling stock, car floor levels usually match the platform level and the gap between the cars and the platform does not exceed 10–15 cm. Such a gap can be negotiated by wheelchairs and, of course, by bicycles without any problems.

Bicycle access has been allowed in recent years on some metro systems, usually during off-peak hours only. The problem here is not physical access, but the large space bicycles require, as well as imposition on and inconvenience to other passengers.

Accessibility of regional rail systems is rather difficult to describe because these systems have a great variety of station accesses, platforms, types of vehicles with respect to floor level, stairs, and internal layout. Some regional rail systems, such as the German S-Bahn systems, Paris RER, and New York Long Island Rail Road, are very similar to metro systems: they have similar station layouts, high-level platforms, two to four double sliding doors per car, etc. Their accessibility has the same problems as and similar solutions to those of metro systems.

Many regional rail systems have open stations that are easily accessible and have no fare-collection barriers. However, they usually have low-level platforms and mostly high-floor cars. Accessibility for wheelchairs is then usually provided by 2–4-m-long mini-platforms that have a ramp. Because the gap between the car and platform is usually greater than on metro systems, a plate has to be used as a bridge. A crew member therefore must assist in this process.

Regional rail cars are the largest and most spacious transit vehicles and can easily accommodate strollers, luggage in overhead racks, and bicycles. This is important because regional rail lines often serve long distance train stations and airports, so that many passengers have luggage. Also, many bicyclists find it convenient to use regional rail to travel between the city and outlying suburbs or among suburbs. The Caltrain regional rail system in San Francisco and S-Tog lines in Copenhagen have such demand for bicycle carriage that every train has a specially designed car with racks for 12 bicycles. Naturally, such cars have a reduced number of seats.

In Japanese cities, intermodal bicycle-rail transit trips are very popular in many cities, but there bicycles are parked at stations in bike-and-ride facilities rather than carried on the trains. These parking facilities are sometimes mechanized racks with locking devices and weather protection.

6.5.3 Accessibility for Different User Categories

This review of different transit modes shows that the requirements and solutions for accessibility by different groups vary considerably. The ADA and similar laws in other countries prescribe the requirements for accessibility of the disabled and, to some extent, the elderly with limited mobility. It is logical, however, to add to these requirements considerations for other passenger groups with special requirements. They are all therefore briefly reviewed here.

By far the most complex is the accommodation of disabled persons in wheelchairs because they require continuous paths with limited slopes to and through stations and access from the street or platform into vehicles. These accommodations are usually difficult and expensive to provide through retrofitting old stations, but they can be incorporated in the design of new transit systems. Access to street transit vehicles (buses, tramways, and LRT vehicles) has been greatly facilitated by the introduction of low-floor vehicles.

General guidelines for maximum gradients on ramps for wheelchair users are as follows:

- Unassisted, i.e., driven by the user, any length: 8% or approximately 1:12
- Unassisted, short ramps: 16% or approximately 1:6
- Assisted by another person: 25% or 1:4 (based on Spiller and Labell, 1997).

Persons with sensory disabilities have different requirements. The blind and other visually impaired persons are provided with paths through stations designated by special tactile surfaces. These are particularly important to be placed at a safe distance from the edge of high-level train platforms. Large lettering on signs and Braille on various elevator panels and other plates assist visually impaired persons to be more independent. Seeing-eye dogs must be exempted from the prohibition of pets on transit vehicles. At signalized intersections near station entrances or other critical locations, audio guides can be provided for the blind. The deaf and other hearing-impaired persons must be provided with visual displays of announcements that are delivered by loudspeakers in stations and vehicles.

Elderly and other passengers who have difficulty climbing steps or standing in a moving vehicle also benefit from improved designs of steps, low-floor vehicles, and the presence of stanchions in the vehicles. Special design features do not benefit all elderly persons as much as they benefit less mobile persons in wheelchairs; yet their benefits from the improvements are very significant because the number of beneficiaries is great: the elderly in most cities are one of the largest groups of transit system users. Disabled persons in wheelchairs represent the smallest group of users.

Passengers with small children and strollers benefit very much from fewer steps and low-floor vehicles, although they need less assistance than persons in wheelchairs. Designation of one or two seats that passengers should yield to the elderly and mothers with small children usually helps when most seats are occupied; it also draws the attention of passengers to the needs of these persons for courtesy and assistance by others.

Passengers with luggage can be accommodated rather easily on rail vehicles, particularly regional rail, because of the spaciousness of these vehicles. Wider areas around the doors and provision of overhead luggage racks are useful for this purpose. Buses are more constrained in this respect, unless they are designed with an area for parking strollers. Vehicles designed specifically to serve airports and train stations are often equipped with shelves for suitcases.

Bicycles usually cannot be brought into buses, but bicycle racks can be installed in front of the bus, above its front bumper. Racks can be used on many suburban bus routes without problems, but in the central city they are somewhat bulky, and the procedure of placing the bicycles on and taking them off the bus does cause some delay.

As mentioned above, bicycles can be brought into metro and regional rail vehicles, but only when passenger volumes are light. They are not acceptable in crowded conditions. Transit agencies therefore usually place some restrictions on time periods when bicycles can be brought in vehicles or how many bicycles may be brought in a rail car at any time.

6.5.4 Results and Significance of Full Accessibility

Extensive activities aimed at making transit systems accessible for a number of special categories of travelers have taken place during recent decades. Significant progress has been made in defining the problems and finding solutions, including legislation, regulations for implementation, and a variety of physical and operational innovations. The results of this effort, including successes and some problems, are reviewed here.

6.5.4.1 Impacts of Full Accessibility on Regular Transit Passengers. Some measures introduced to accommodate the disabled persons also benefit regular passengers. For example, low-floor vehicles have made access easier not only for elderly and disabled passengers, but also for regular passengers: boarding and alighting has become easier and faster than with traditional high-floor vehicles. However, in some respects accommodation for the disabled persons has negative impacts on transit services. Such cases require various measures to reduce the negative impacts or to introduce certain flexibility in applying the rules and regulations: limits as to where and to what extent such measures can be introduced. Several examples illustrate this problem.

The use of lifts on high-floor bus and rail vehicles usually requires that the driver stop the vehicle, lock its brakes, walk out, activate the lift, assist the passenger, lock the wheelchair in the bus, deactivate the lift, walk back, and start the vehicle. If this procedure takes, conservatively, two minutes, and the passenger's alighting another two minutes, the bus or rail vehicle falls four minutes behind schedule. This is a significant interruption that cannot be made up along the line. The decrease in service speed and reliability is undesirable and costly. This problem makes accommodation of the disabled in wheelchairs through built-in lifts inferior to the use of low-floor vehicles, short raised platforms on LRT lines (San Diego), and other designs.

Deviation of bus routing from its line, which may be operated in different patterns, from a single route deviation to full dial-a-ride, faces the problem of reconciling the convenience for a single passenger with the inconvenience and delay to other passengers. This problem of balance between individual and all other passengers is inherent in these types of flexible routing services not only for the disabled, but for all passengers.

ADA regulations stipulate not only that new transit facilities must be fully accessible, but that whenever any station or other fixed facility is rebuilt or substantially renovated, it must be made fully accessible. On many existing metro and regional rail systems, retrofitting stations for full accessibility is very difficult because it may involve reconstruction, building in of elevators, etc. There have been cases where renovations of stations or corridors linking stations could not be carried out because the funding for the required major retrofitting construction was not available. Thousands of passengers in a number of cities (New York, Philadelphia) have thus been deprived of improvements or had facilities closed and services lost due to rigid applications of ADA rules.

6.5.4.2 Evaluation of Full Accessibility. Policies, investments, and organizational efforts to open up transit systems not only to the general public, but also to persons with various disabilities, introduced in many countries since about 1970, represent a logical step in the progress of contemporary societies. It is a step that is ethically correct because it enhances the concept of equal rights; it is economically sound because it allows disadvantaged groups of people to become more productive; and it is socially progressive because it assists those who are disadvantaged.

The program that started as concern for the elderly and handicapped evolved into elaborate legislation that addresses accessibility of transit as well as many other public and private facilities. The problem of accessibility is most difficult with respect to accommodation of persons in wheelchairs. The number of these persons is rather small in most cities, while provision of accessible stations and vehicles is often very expensive. In the short term, implementation of this program yields a very low benefit/cost ratio, even if all nonmonetary and nonquantifiable benefits are included in the analysis. However, there are several aspects that place a different light on this type of evaluation.

First, the legal requirement to make transit systems fully accessible has triggered major efforts to develop technical and operational innovations, from curb cuts on streets and ramps parallel with stairs to better signing and announcements and better integration of paratransit and regular transit services in suburban areas. Many of these innovations have brought improvements that benefitted all transit passengers. The most drastic example is the development of low-floor rail and bus vehicles, which increases convenience for all transit riders and results in faster transit services. Including these benefits in the analysis of measures for full accessibility makes many of them more feasible, some with benefits easily outweighing the additional costs.

For transit agencies, it is logical to use the ADA and similar government regulations as an impetus to analyze and improve accessibility of all persons with special needs. Thus, sometimes considered, but often overlooked, are the passengers with small children, various categories of elderly persons with differing physical abilities, as well as persons with luggage and those with bicycles.

One caution must always be borne in mind, however. In introducing many transit system modifications for accessibility of special groups of persons, the potential impact of change in regular transit service must be carefully analyzed. As mentioned, some modifications benefit all passengers. However, some others may directly reduce quality of transit services and thus hurt a much greater number of people than they assist. Examples include on-vehicle lifts that cause long delays, route deviation that extends travel time for passengers going through the area, or delays and physical inconvenience to passengers created by bicycles in rail vehicles with moderate or heavy passenger loads. Therefore, implementation of every measure to achieve full accessibility must be carefully planned and the legal regulations reasonably adjusted to the specific conditions.

Finally, a transit system increases its role as a basic public service in the city when it is fully accessible. However, if the legal requirements represent unfunded mandates, they result in the diversion of existing transit funds and thus reduce the funding for the basic transit system. This contradicts the principle that provision of full accessibility as a social goal should be financed by the society (special funds), rather than by transit users only. Efforts should be made to provide special government funds for implementing transit system features that are needed for a special group of users only. If such financing is provided, it will likely lead to further technical innovations and transit system improvements beyond the remarkable progress already achieved.

6.6 INFORMATION SYSTEM FOR PASSENGERS

To attract all present and potential users and perform its important role in urban transportation, a transit system must provide complete, accurate, and easily accessible information about its facilities and services. In recent decades, the changing position of transit in urban transportation has led to increased requirements for more extensive and more sophisticated transit information. Its function has been broadened from specific information for individual services to a comprehensive system connected with marketing of transit services.

Traditionally, many agencies believed that since passengers depend on transit services, they will make any effort required to obtain information and use transit. With the change from "transit users by necessity" to "customers by choice," with passengers now having to be attracted because they have other choices for travel, information systems have become much more elaborate, sophisticated, and user-friendly.

The importance of an effective information system may not be noticeable in the short run. However, experience around the world shows that high-quality information systems yield long-term, permanent benefits in terms of high transit ridership and a system's good public image. Visitors to Paris, the San Francisco Bay Area, Melbourne, or any German city quickly find transit maps showing all transit modes, their stations, schedules, etc. Transit becomes an important part of their experience and impression of the city.

Unfortunately, transit agencies in many cities still follow the obsolete practice of giving minimal information. Or they do not pay enough attention to this aspect of their services, so that information gradually deteriorates to the minimal items needed for regular users only. Thus, in many cities bus stops have no information, transit vehicles operating on streets have designation signs only on the front, entrances to metro stations are difficult to see on streets, and even rail transit lines are not marked on city maps. This situation not only damages the transit system, but also makes the city less attractive. In some cases, inadequate or nonexistent information makes it virtually impossible for visitors and other potential first-time users to use transit.

This section describes a transit information system: its components and planning and testing its effectiveness. At the end, common failures to provide adequate information that can be found in many cities are discussed.

6.6.1 Planning a Transit Information System

Two major goals can be defined for planning a transit information system. The first is to provide the public with information about the system and its services, thus facilitating attractive and convenient transit travel by present and potential passengers. The second is to increase efficiency of operations and utilization of services, which improve transit system efficiency. Increased ridership further results in high mobility in the transit service area, bringing benefits also to the general public and the city.

Persons unfamiliar with transit services (lines, schedules, fare amounts, and method of payment) are often reluctant to use them. An information system must be planned so that people feel that it is easy to use transit. An immediate objective toward achieving this goal is to prepare and effectively distribute information to the present users, as well as to make the system open and inviting to all persons who might potentially use it. Cities with transit systems that are easy to understand and use have a higher level of mobility—an important element of city's livability—than those that rely only on walking (in developing countries), private cars, taxis, and other paratransit modes.

The term *information system* implies that there are several components that complement each other in providing information for the public. These components must be coordinated in their content, method, and places of distribution. Only such an integrated system can provide the information that is necessary for each segment of the public.

In the planning of a transit service information system, three basic questions must be addressed:

1. What types of people are going to use, or may be attracted to use, the system?
2. What information items does each group of users need when it wants to use the services?
3. How can that information best be made available to the public?

These three considerations are discussed in the following sections.

6.6.2 Classification of Present and Potential Users

For the purpose of information needs, four major categories of transit users can be defined. They differ in their familiarity with the transit system and their needs for information:

1. *Regular users on their regular transit lines:* Typically, these are commuters to work, schoolchildren and students, some shoppers, and businesspeople who frequently make the same trips and therefore know the system well.
2. *Regular users making trips on different lines or at different times:* Passengers familiar with the system but traveling to an unfamiliar part of the city comprise this group.
3. *Incidental users:* This category includes residents who are familiar with the city and region but who use transit services only infrequently. They need information about the network, as well as how to use the services, pay fares, etc.
4. *Visitors to the city:* This category represents the persons visiting the city, who may be totally unfamiliar with the city and its transit services.

The four categories are listed in order of increasing need for information. The regular users, transit's steady and therefore best customers, need the least amount of information—only to be kept informed of any changes in schedule, routing, or stop locations. The regular user making an irregular trip needs slightly more information: familiar with the available modes, networks and operations, fare level and method of payment, etc., this passenger may need information about lines, schedules, and stops for the trip to an unfamiliar part of the city. Much the same type of information is needed for the third category, incidental users. They may have to be reminded of transit modes and lines, fare structure and method of payment, and the like. Finally, the visitor to the city requires the most complete information about the city and transit services, since he/she may be totally unfamiliar with the local conditions.

Each one of these groups should be considered in determining the type of information needed, the means by which it can be best distributed, and the locations for its distribution. For basic information about transit system and its operations, the city visitor should be the "design person," dictating the most comprehensive set of information items. This is information about networks showing clearly different modes, because passengers are very aware and have different attraction to modes. Potential passengers are particularly apprehensive about using buses because of their invisible routes and often very complicated networks. An elaborate information system is therefore needed to overcome this problem.

Tramway and light rail lines are quite visible to potential passengers, and they must be clearly marked on transit maps with their stop/station locations. Metro and regional rail systems have large infrastructures and their stations are significant objects in the city, attracting major pedestrian flows. They should therefore be visible from a distance by day and night. Any local special systems, such as cable cars in San Francisco, monorail (Schwebebahn) in Wuppertal, Germany, Skytrain in Vancouver, Canada, or downtown people movers in Miami and Detroit, should always be highlighted because they usually either offer unique services or are a local attraction for visitors. Further, information about intermodal transfers and fares must be provided.

Information for individual trips, particularly schedules, as well as fare amounts and methods of payment, is needed by all four user categories.

6.6.3 Information Items: Their Contents and Forms

To plan and perform transit travel, a passenger needs to have information about the transit network and lines, schedules, transfers between modes, fares, as well as any special services or changes. Major items that provide such information, which are components of an information system, are described here.

Transit system name and symbol or logo can be considered the most important items representing and giving identity to the entire transit system. Simple, convenient names and symbols are particularly powerful in creating a strong image of the agency, as well as of transit service in a city or region. Single system names, such as BART and Muni in San Francisco, Rheinbahn in Düsseldorf, and ATM in Milan, are easy to use in any information. They are also easy to remember for transit travelers and all other citizens.

Simple but characteristic symbols with unique design and color scheme can also be powerful in creating a strong image of transit services. The "T" in Boston, circle with a horizontal crossbar in London, and standard designations of metro and regional rail stations in all German cities—"U" for U-Bahn (metro) and "S" for S-Bahn (regional rail)—are good examples of successful use of such symbols. These names and symbols are used in all information materials, from signs in the city to printed materials and websites. They become well known throughout the areas they serve, as well as among visitors to the city.

Maps with transit lines and network are needed by passengers to plan a trip, during travel, and in any special situations when travel needs to be modified. Transit maps must be clearly presented, logical, and have sufficient details. To achieve that, attention must be given to the following features.

The basic transit map should show all major streets and important landmarks. Superimposed on the street network in darker print should be lines representing surface transit routes: bus, trolleybus, and streetcar. Each of these modes should be shown by a different type and color of line. The lines should have frequent designations of route numbers and stop locations. Turning movements, branching, convergence, and overlapping of lines and routes should be clearly shown. One-way sections should be indicated by arrows. In large cities with many lines converging on a small CBD area, an enlarged central city map should be included. The north arrow and scale should be indicated on each map so that passengers can estimate access and travel distances.

Rail transit lines, such as LRT, metro, and regional rail, often have alignments that do not coincide with avenues or streets. They are shown by heavy lines, usually each one in a different color, with designations or names of lines and stations.

The lines of metro systems and recently built LRT lines usually have specific designations, such as U1, U2, etc. for metro (U-Bahn) and S1, S2, etc. for regional rail (S-Bahn) lines in Berlin and other German cities. In some cities, metro lines have individual names, such as Piccadilly and Victoria in London and Ginza and Marunochi in Tokyo. In recent decades, use of colors for designating metro lines has become very common: Washington has the Blue, Red, Yellow, and other metro lines, while Los Angeles has the Blue, Green, and Gold LRT lines and a Red metro line.

Regional rail networks usually have a single name for the entire system, such as Metra in Chicago, GO Transit, in Toronto, and RER in Paris. SEPTA in Philadelphia uses specific colors for signing and schedules for each regional rail line, although they are designated not by colors, but by mode symbol "R" and individual numbers R1 through R8.

It is desirable that transit maps have, usually on the back side, the basic information about typical headways, times of the first and last services during weekdays, Saturdays, and Sundays, and fare structure and amount. For rail systems, very often a triangular table with station-to-station travel times and fares is presented, thus completing the most important information passengers need for their trips.

In addition to a regular city transit map, rail networks are usually shown on schematic drawings with rail transit lines only, usually not drawn to scale. It is important, however, that the schematic be correct in such details as where the lines intersect and which stations allow transfers. Line designations and station names should be clearly shown. It is very desirable to show on these maps which street transit lines serve each rail station, as well as which stations have park-and-ride (P+R), bike-and-ride (B+R), and full accessibility facilities.

Schematic rail network maps are used extensively in all cities with more than one rail transit line. The Moscow Metropolitan, Tokyo Subway, and Washington Metro schematic maps are used as convenient pocket maps and displayed in all stations, on the respective websites, and on all tourist maps and pamphlets; they are also seen on many T-shirts, mugs, trays, and other popular objects.

Timetables must be published for all lines operating with headways of 10 or more minutes because passengers need them for such services. When several lines overlap on trunk sections, joint headways may be as short as 2–3 minutes, but headways for individual lines may be 7.5, 10, or more minutes, so that published timetables are needed. Transit systems with very reliable services publish timetables even when headways are as short as 5 or 4 minutes, although for such frequent services most passengers do not observe schedules.

Timetables for rail systems have times for all stations. For bus lines with short spacings between stops, times may be given for major stops only. Lower reliability of street transit lines makes great precision of timetables unrealistic.

Timetables are usually given for weekdays, Saturdays, and Sundays. The designation of days should be clearly highlighted. Generally, it is important that timetables be presented clearly and easy to read and that the schedules be made repetitive every hour so that they can be easily memorized for any given station. Printed schedules for individual lines should always show a sketch of the line with its terminals and major stations.

Some transit agencies publish complete books of timetables for all lines. Many residents keep such books handy at home for any trip throughout the city they may have to make. Printed schedules have now been supplemented and largely replaced by Internet information containing all timetables of most large transit agencies.

Fares also concern passengers in two respects, and they must find information about them. First is the type and amount of fare required for a given trip, and second is how fares are collected, controlled, and checked. Fully automated fares and self-service payment systems are quite convenient and fast when passengers are familiar with the ticket-issuing and ticket-canceling machines, fare gates, etc. However, persons entering a transit system for the first time may be more confused than when fares were collected manually by a bus driver or a person in a booth by the faregates, who could provide explanations and assist. For this reason, it should be easy for new passengers to find out which fare they have to pay and how. Ticket-issuing machines require particularly clear and simple explanations.

6.6.4 Locations of Information

Passengers need transit information in four different time periods: before they begin their trips, that is, at home, in an office, or in a hotel; when they need to find a station and board a transit vehicle; when within a vehicle during the travel; and when they leave the transit vehicle to get to their final destination.

6.6.4.1 Trip Planning. While planning a trip, the passenger-to-be should be able to get information from transit maps, via the Internet, or by telephone. Transit maps, at least the schematic ones for rail networks, should therefore be broadly distributed free of charge, so that many residents have them in their homes. The schematic map can usually be found in information literature for the city, and sometimes it is included in city's telephone directory.

Information by telephone is personalized and convenient for users, but it is rather costly to the agency because it requires well-trained personnel and, in large agencies, several positions during most hours of every day if the probability that callers cannot get a response is to be minimized. However, for some passenger needs telephone information service will remain very convenient and important. Therefore, the 511 number is being introduced as a standard transit information source in many U.S. cities.

Although fully automated and impersonal, Internet information for passengers is very powerful because it has extensive data, maps, schedules, and other details that cannot be obtained by other means. It is therefore being increasingly used and is diverting the demand from the telephone information service. It is expected that Internet use will continue to increase, although telephone service will remain indispensable for persons who do not use computers or do not have Internet connections.

In many cities, passengers can not only find on the Internet all facts about the transit system and services, but they can use a program for individual trip planning. A passenger can key in the origin, destination, and time of a desired trip and obtain the entire routing, including walking to the closest stop, departure on that line, transfer locations, and times and departure from the final station to any destination in the transit service area.

Cooperation among agencies can produce joint information for several agencies in a single region or in adjacent cities. In Switzerland, the Netherlands, and some other countries, a national travel information system exists that allows a person at any point to ask for a trip to any other point in the country and obtain a printout for the planned trip, which may include walking access, transit travel, intercity train, and transit travel in the destination city.

6.6.4.2 Information in Transit Vehicles. Vehicles must provide orientation for approaching passengers and information for them during their trip.

The exterior of each transit unit (bus or train) should prominently display the following information:

- Line/route number or designation and destination terminal. This should be displayed not only on the front and on the side where the doors are used, but also on the rear and the opposite side of the used doors. The reason for this is that potential passengers often want to know from a distance which vehicle or train is approaching or standing in a station. These designations should be large, legible, and easily changeable. Electronic ones usually easily satisfy these requirements.
- Transit agency logo, name, information telephone number, and website address

In the vehicle interior, passengers should find, whenever feasible, the following information:

- A displayed schematic map of the line, with all stations, or the entire network in the case of rail systems
- Pocket maps and printed schedules for the given line

Many recently produced buses have GPS-based automatic announcements of the forthcoming stops, transfers to other lines, and other orientation for passengers. On some metro systems, announcements of stations, warnings for door openings and closings, etc. are also programmed. However, many LRT and metro systems continue to use train drivers for making announcements. There are several reasons for this. First, messages are more personalized and many passengers prefer to hear a live person who is obviously aware of momentary conditions and may adapt announcements to the events in boarding, closing doors, weather conditions, or various warnings. Second, rail transit drivers can easily make announcements because their driving is much simpler than that of bus drivers. And third, on metro systems with automatic train operation (ATO), such as San Francisco BART and Washington Metro, making announcements often prevents monotony for the drivers, whose "driving" duty consists only of pressing a single button at each station.

Recorded announcements may be overriden by the announcements made by the driver. In addition, control center announcements may be broadcast in any train or station, along individual lines, or throughout the network. Vehicles may have public address systems not only inside, but also outside in the case of trains, for passengers on the platform.

A major element in providing passenger information is transit agency personnel, particularly drivers on street transit lines and attendants in rail transit stations. These persons provide for many passengers the first and most personal contact with the transit system. If passengers encounter transit employees who cannot or will not provide information and assistance, they have an extremely negative experience of transit system in general. It is therefore very important to train all personnel with respect to information for passengers, as well as courtesy in their public relations role. As a follow-up, transit agencies must maintain a program of testing their personnel with respect to technical knowledge and attitude, using anonymous instructors approaching them as passengers.

Electronic changeable signs are increasingly used inside bus and rail vehicles for station announcements. On rail systems, individual lines or entire networks with all stations are displayed above doors, on side walls, or on the ceiling. Buses usually cannot have such displays because they are used on many different lines, and bus networks are too complicated to display them with details inside vehicles.

6.6.4.3 Information in Transit Facilities and Public Places. Transit stations are locations where passengers first come in contact with the transit system, and they expect to find there any information they need about the network, services, fare payment, etc. This information must therefore be systematically provided.

Transit stops along streets should be marked by a system logo and, desirably, a standardized sign for all transit stops in the city. The minimum information that should exist at each stop is a designation of each route/line that uses that stop. The information should consist of the line number and designations of its two terminals because many potential passengers may not be familiar with transit line numbers. Timetables, or at least headways for each line, should be displayed. On-line electronic display of the arrival of the next vehicle is being increasingly introduced. At stops with large passenger volumes, and particularly at all stops where other bus or rail lines intersect, complete information about alignments and schedules of all lines should be provided.

Transit stations are major facilities at which the transit system interacts with flows of arriving and departing passengers. Passenger transfers with other lines and modes of transportation may also take place. Many urban developments, such as apartments, offices, and stores, are related to and interact with transit stations. As important attraction points, stations must have signs that are easily visible from a distance during the day and night. For example, all metro and regional rail stations in German cities have, respectively, large, colored and illuminated "U" and "S" signs visible from distances of several hundred meters at all times.

In stations, passengers must be provided with full information about transit line or lines serving that station, as well as about the entire transit system: maps with "You are here" markers, schedules, and fare-payment procedures. In addition, stations should have a large-scale map of immediate surroundings (again with a "You are here" marker) so that arriving passengers know which exit to take for their ultimate destinations.

Continuous information about train arrivals is more easily provided than for street transit services because train movements are continuously monitored. This is particularly useful when several lines arrive at the same platform. For example, BART stations in San Francisco have announcement "Richmond train in 2 min," "Fremont train in 6 min," etc.

Photo 6.9 *Clear and conveniently placed passenger information in a subway station—Tokyo.*

Transit terminals, facilities where rail lines terminate or have passenger interchanges with other rail or feeder bus lines, are focal points for information about the entire transit system. Maps must be available on display as well as in distribution that assist passengers to plan their trips, usually involving intramodal or intermodal transfers. Terminals should also have schedules for all lines using that terminal displayed, in printed pamphlets, as schedule books for the entire transit system, or on interactive kiosks available to passengers. Information about coordinated transfers between lines, when it is provided, should be clearly displayed and explained.

Intermodal terminals and the entire functioning of intermodal systems, with emphasis on passenger level of service and information needs, are getting increasing attention. Many transit systems in North America have greatly modernized the design, operation, and passenger processing in intermodal stations and terminals. This has been the case particularly where new rail systems have been introduced, such as Toronto, Washington, and Sacramento, as well as in the cities that introduced bus and rail timed transfer systems or people-mover downtown distributors, such as Edmonton, Portland, and Miami.

Naturally, information about fares should also be complete, including transferring privileges, if any. Complex terminals sometimes require considerable information for passenger movements from the building entrance to each one of the lines coming to the terminal; for transferring movements between lines; and for directions to all terminal facilities and exits. This is required sometimes not only for passenger convenience, but also for directing major flows of passengers in an efficient and safe manner.

Transit systems are increasingly introducing voice-guidance systems for the visually impaired and specialized information devices for disabled persons in wheelchairs.

6.6.4.4 General Information Distribution for the Public. Transit information must also be available to the general public in order to reach potential transit users, those residing in the served area as well as visitors and tourists. To achieve this, the network of rail transit lines and general information about the services and fares are included in tourist information, such as maps, pamphlets, and information about landmarks, museums, stadiums, and other attractions. In addition to the general tourist information, the transit agency must distribute its own, more specific information at strategic locations and by means that reach the broad public.

Public places, such as tourist centers, hotels, recreational and cultural centers, major public office buildings, and newsstands, are examples of locations where transit maps, timetables, and other information

Photo 6.10 *Surface transit with high-quality passenger information—Los Angeles (courtesy of Richard Amodei).*

should be provided for the public in pamphlets, booklets, and displays. Since the wide availability of such information assists and encourages transit use, the cost of this distribution is indirectly recovered from additional fares.

Public media include newspapers, radio, television, the Internet, and other means of public communication. The Internet is by far the most comprehensive medium for distribution of any information passengers may need. However, many current developments and news, such as changes in schedule, bus routings, or stop locations, opening of new lines with their alignments and schedules, innovations in fares and their collection, special services for major events, etc., require broader and immediate distribution than the Internet can provide. For that purpose, special pamphlets and newspaper, radio, and television announcements can be effectively used.

6.6.5 Common Causes of Deficiencies in Information Systems

Availability of and easy access to information about transit system services are major factors in a transit agency's image and its ability to attract passengers. There is, however, a tendency in many agencies to underestimate the importance of information. This leads to insufficient financing of that activity, failure to develop it systematically, or, in many cases, neglect of maintenance and updating of an initially well-developed system, allowing it to deteriorate. Among many factors that may be the reasons for inadequate information systems, four rather common causes for this problem are discussed here. Some of the consequences of these causes are also discussed.

6.6.5.1 Agency's Inward Orientation. Attention that a transit agency gives to its public relations, information it provides, and marketing of its services reflects the orientation of its board and management toward passengers and the general public. An agency that works actively on increasing ridership and improving its role in the city pays considerable attention to passengers' convenience and to the factors that influence passenger attraction. Such an agency tends to have a separate division for public relations and a well-organized transit information system. On the other hand, an agency that is oriented primarily toward retention of its present passengers tends to provide minimal information, usually inadequate for attraction of new users.

Transit agency boards and management with this inward orientation often adopt the attitude that because daily commuters to and from work are the largest category of users, service should be oriented to them. With a commuter as the design person, timetables and other information are largely limited to individual lines. Little attention is given to coordination of schedules among intersecting lines and other factors that facilitate and encourage transferring and travel through the entire transit network.

Furthermore, network maps are difficult to find. Buses and trains carry only a single destination sign, and suburban stations have very few station signs, if any. The explanation for these inadequacies may be, "People using these stations are daily commuters, they are familiar with the line."

While it is true that regular riders represent the majority of passengers and deserve primary attention, it is a serious mistake to neglect irregular and potential transit users. First, their number is often underestimated by agencies that have an inward orientation—those that try to retain existing users rather than also attract new ones. Second, that policy and neglect of information makes use of transit services by other than regular users very difficult and discourages potential passengers from transit travel. Third, most of the incidental or choice riders tend to use transit during off-peak periods, filling the valleys of passenger demand. Therefore, marginal cost of carrying these passengers is lower than the cost of transporting regular, predominantly peak-hour passengers. Their attraction therefore improves the transit agency's operating ratio. Finally, neglect of incidental and new riders results in stagnant and, in the long run, decreasing ridership. The agency is thus destined to a diminishing role in the city and does not have a bright future.

6.6.5.2 Operational Conveniences Override Passenger Needs. Implementation and maintenance of an information system require good planning and coordinated actions, as well as the continuous attention of many employees, including drivers, station attendants, and station and vehicle personnel. If the required activities and discipline are not rigorously controlled and continuously maintained, there is a tendency among this personnel to modify their daily duties and activities to make their jobs simpler and less demanding. For example, announcements of stops along the line are first not made at minor stops, then become increasingly irregular. Some of the destination signs on vehicles are not displayed. Schedules and maps are not available on vehicles, then they may be missing at stations also. Many of these neglected actions and "economy measures" reduce information system scope and effectiveness. To prevent this, a transit agency must never allow various operational changes to be made without checking how they affect information and other conveniences for passengers.

6.6.5.3 Incorrect Selection of the Design Person. The tendency to concentrate on present passengers sometimes leads to errors that reinforce the problem of inadequate information. For example, some agencies organize surveys of the use of information needs by their passengers. Then they claim that elaborate signing of stations is not important because very few transit passengers said that they use them. Such a survey is inherently incorrect because the entire population that may not have found the station and did not use the system is excluded from it.

This example points out the need to define the design person correctly for transit information. Considering the four categories of users as defined in Section 6.6.2, the first category, regular user on regular line, should be used for information system design only on exclusively commuter services, those usually operated during peak hours into and out of center city or another major destination. In all other cases, the second and third categories—persons making trips on lines they are not intimately familiar—with should be the design persons. However, for all major lines and for the use of the transit network, transfers between lines and modes, access to long-distance rail, bus and air terminals, etc., the fourth category—visitor to the city—should be used. In general, the design person in planning any information system should be the one who is least informed. A regular user is therefore virtually never the person for whom the information system should be designed.

6.6.5.4 Inadequate Budget Allocations. The importance of transit information for passenger attraction and the transit system's entire image is not always obvious, because its results can not be easily quantified. Moreover, the benefits may not be immediately visible, although they bring continuous increases in ridership and permanent benefits to the city. For this reason, there is sometimes a tendency to underestimate the value of information and avoid allocation of needed funds. Even more typical are situations when budget cuts have to be made due to financial pressures. There is a tendency in such situations to cut the information system more than other items because it appears to have low importance, particularly in the short run.

Any one of these causes may result in inadequate information and lack of innovations in the items and methods of their delivery to the public. Examples of inadequacies and obsolescence are numerous: missing destination signs on vehicles, transit stops without any information, vandalized maps and signs that are not repaired, etc. In recent decades, many cities have introduced transit line designations that indicate modes (U1, U2, R1, R2) or color designations for rail lines. On the other hand, however, some cities have very inconsistent numbering of streetcar and bus lines, which is confusing for passengers. The reason is often that with changes in lines through several decades, the numbering was changed case by case, resulting in many missing numbers and mixing of letters and numbers, so that there is presently no systematic logic.

6.6.6 Planning, Testing, and Maintenance of the Information System

To ensure maximum cost effectiveness of the program, the transit information system should be planned as a system in which locations of distribution, items that should be distributed, and means of distribution are coordinated, bearing in mind which passenger category is expected to be the primary user of each set of information, i.e., who is the design passenger for needed information. A few examples can illustrate this process.

Minor bus stops in suburban areas are used mostly by regular travelers from adjacent neighborhoods into center city. The information that should be provided must include a sketch of the transit line and schedule of service on that line. For occasional passengers from categories 2 to 4, there should be a telephone number and website for further information. At rail transit stations with similar function, a schematic rail transit map for the region should also be displayed because many noncommuters plan on transfers to other lines in the network.

At a major suburban station where a rail transit line terminates and connects with a number of feeder bus and rail lines, information should again include all items that category 2 and 3 users may need. This includes at least a map of the outlying area with all transit lines and schedules for all lines with locations of their stops. A plan of the terminal with the specific location of each feeder line stop, taxi stand, and, if it exists, a park-and-ride facility within the terminal should also be displayed. In addition to this basic information, it is desirable, if adequate funding is available, to provide on-line information about departures on the rail and all feeder bus and rail lines. When the control system has continuous information about vehicle movements, the public should be provided real-time information about any delays that may occur. It is also desirable to make available to passengers at this location information items for the entire transit system in the region.

In another case, at a railway station that is also a major transit station served by metro and street transit lines, the design passenger should be category 4, visitor to the city. The information should include not only alignments and schedules of lines serving that station, but also extensive information about the entire transit network and its services to major areas and landmarks in the city.

To ensure functional characteristics and effectiveness of an existing or planned information system, it should be periodically tested. Transit stops, stations, and terminals can be grouped by the design passenger categories and tested for whether all the information items that each category requires are provided at respective facilities. In addition, all locations, buildings, and information centers in the city should be checked to see whether they provide information for visitors to the city.

For a major test of an information system, usually focusing on rail transit system and terminals, a test may be designed by inviting a group of three to five out-of-town persons, including transit experts and persons inclined to use transit if they would visit the area, and give them the task of making a tour of the city using various transit services, transferring among lines and modes, paying different fares, etc. The following day, each member of this test group should describe his/her experiences and present a critique, including positive and negative findings.

Other passenger and general public input in the evaluation of the information system may include some of the following methods (Potts, 2002):

- Passenger and public surveys: on-board, at stations, telephone or card-by-mail surveys
- Interviews through focus groups
- Input from employees about their contacts with passengers
- Passenger complaints

Maintenance and innovations are essential elements in an attractive and accurate information system. Poorly maintained, confusing, or vandalized information may be more irritating to users than lack of information. With this in mind, the transit agency should ensure the following steps:

- Design information items that are durable, resistant to vandalism, and easy to maintain and update.
- Protect displayed information from weather and vandalism as much as possible.
- Have a systematic schedule for review and repair of information items.
- Regularly update timetable and other changeable information and introduce innovations in types of signs and methods of distribution, such as updating and expanding the website.
- Periodically check the knowledge and attitudes of agency's personnel in their contacts with passengers, and perform training updates when necessary.

6.7 MARKETING AND PUBLIC RELATIONS

With the evolution of transit from the times when most people had to use it because there were few alternatives for travel, to this time of high car ownership, when transit must attract customers who have different choices, the role of marketing has increased greatly. To compete with other modes of travel and attract new passengers, transit agencies must perform various types of marketing.

Marketing has thus become an important element in transit agencies' orientation toward present and potential passengers. Together with the functions of information and public relations, marketing contributes to the image and role of a transit system in a city.

6.7.1 Definition and Purpose

Marketing of transit is an organized program undertaken by the transit agency in order to achieve the following major goals:

- To promote transit services and inform the public about them
- To ensure that the maximum number of potential riders is attracted to the transit system
- To ensure that transit service is provided where it has maximum overall benefit, considering not only the operator's viewpoint (operating efficiency), but also the passenger and community's viewpoints

Transit information and marketing activities are related and partially overlapping. The basic difference between them is that the information system is oriented to the present transit users, while marketing is aimed primarily at near-term attraction of potential users and creation of a long-term positive, strong image of the transit system, and strengthening of its role in the city.

Development of transit marketing was slow in many cities because transit agencies initially did not understand the competition with the automobile. Moreover, their financial condition deteriorated. When transit agency managers face financial difficulties, concern for short-term financial solvency tends to be considered the dominant criterion in successful transit operation. However, even under difficult financial conditions, there is a great need to attract as many passengers as possible in order to increase revenue and maintain high service quality.

Consequently, good marketing is essential for achievement of effective passenger attraction and high system utilization. Inefficient marketing (or no marketing) has the long-term effect of losing passengers and decreasing an agency's operating ratio.

6.7.2 Marketing Strategies

A marketing strategy coordinates the various marketing activities into a plan that has defined objectives with regard to passenger attraction and provisions of service. The plan should be designed with programs for achieving the objectives. Four types of marketing strategies can be used, individually or in combination.

Undifferentiated marketing is a strategy in which a single product (transportation) is provided and promoted in a way designed to appeal to the public at large, i.e., the entire population of a defined service area (town, city, or metropolitan area). In this strategy, the typical or average consumer is appealed to and served; an attempt is made to attract the most passengers with a single service/promotion package.

For most cases, undifferentiated marketing is the simplest and cheapest marketing strategy to implement. However, its effectiveness as a *sole* strategy for marketing generally depends on a population with fairly uniform characteristics and on an assumed limited role for transit. Undifferentiated marketing is therefore often the base on which to build a more sophisticated and effective marketing program.

Differentiated marketing, in contrast to undifferentiated marketing, appeals to each segment of the population to ride transit based on their differing needs. For example, a campaign to attract more off-peak riders may highlight the benefits of riding transit during those hours for shoppers, housewives, the elderly, and others who would be potential candidates for such a program.

Differentiated marketing is typically more expensive than undifferentiated marketing because of the need for diversity and specialization in the planning and presentation of the program. The higher investment level in differentiated marketing, however, is often offset by increased patronage level and larger revenues than undifferentiated marketing could achieve.

Concentrated marketing is a program that has as its goal the promotion of a single component of transit service or the attraction of selected segments of the ridership market. This type of marketing is most effective for agencies that offer specialized services.

Individualized marketing is a carefully organized program aimed at increasing regular transit usage. The concept was developed during the 1990s by the UITP and the German consulting firm Socialdata (Brög, 1998). The fact is recognized that many people do not use transit because they are not familiar with the transit system and have no experience in using it or motivation to do so. Several groups of potential transit users are first selected and contacted directly, with the intent of motivating them to think about their travel behavior. Then the selected groups are thoroughly informed about transit services, focusing on trips that they may be particularly interested in, and given free monthly transit passes.

This initial experiment was extremely successful: transit use by the selected group nearly doubled. Then the experiences with attitudes obtained in the first project were used to apply the same concept of organizing such an individual approach to informing and motivating potential users to start using transit to several thousand individuals. Significant increase in transit usage brings new revenue, which can then be used to further expansion of the program. Thus, individualized marketing involves a major initial effort and labor-intensive work with individuals, but it results in permanent changes in travel habits of individuals. Persons who had avoided transit because of unfamiliarity with and apprehensiveness about it become occasional or regular transit users. This change benefits both them and the transit agency.

The choice of combination of the four transit marketing strategies depends on four basic factors:

1. *Role of transit:* If transit is to play a dominant role in the community or if there is a community decision to attract as many passengers as possible to transit, it is desirable to invest heavily in a marketing program. Differentiated, concentrated, and individualized marketing are usually effective in these cases. Conversely, when transit is used primarily to supplement auto travel in a city, it is not necessary to have a highly specialized marketing program. An undifferentiated marketing strategy is then acceptable.
2. *Available resources:* The financial resources available often limit the initial marketing effort to undifferentiated or concentrated marketing, depending on community characteristics. Adequate financing, on the other hand, may make individualized marketing feasible and cost-effective.
3. *Community characteristics:* Each economic and social group in a city has different transit needs. In determining the type of marketing program best suited for a city, the analyst must determine how diverse transit needs of the population are. A one-industry city with fairly uniform population characteristics, for example, may find that an undifferentiated transit marketing program is both cheaper and more desirable than a more sophisticated program. Cities that have more variety in social and economic groups often will find a differentiated marketing program essential.
4. *Type of transit service:* Transit services can be differentiated by a number of features, including time of service (peak, off-peak), area coverage (CBD-oriented, dispersed), function (commuting, shopping, recreational), and others. The more diversified a transit service is, the greater the need for a differentiated and individualized marketing programs.

Table 6.5 presents a summary of influences of these major factors on the choice of a marketing strategy. The table should be viewed not as a set of inflexible rules, but rather as an aid when considering alternative strategies. After the best marketing strategy for the defined objectives and conditions is determined, the strategy should be reviewed periodically for its effectiveness, expansion, and possible improvement.

6.7.3 Marketing Activities

Transit marketing can be classified into six types of activities: market research, market segmentation, service adjustments, information distribution, advertising, and public relations. These activities are described next.

6.7.3.1 Market Research. In most market research projects, it is necessary to collect the information nec-

Table 6.5 Influence of different factors upon choice of marketing strategy

Basic Factor \ *Type of Marketing Strategy*	*Undifferentiated*	*Differentiated*	*Concentrated*	*Combination*	*Individualized*
		Role of transit			
Dominant	●	●	●	●	●
Supplemental	●		●		●
		Resources (cost)			
Limited	●		●		
Adequate	●	●	●	●	●
		Community type			
Diverse		●	●	●	●
Uniform	●				●
		Transit services			
Multimodal	●	●	●	●	●
Single mode					●

essary for an effective marketing program. In general, the data are those that define present and projected (or planned) characteristics of the community, its population and transportation patterns, and demand features. Included in this type of information are demographic characteristics (population densities, etc.); physical features of the city (topography, transportation network and land uses, etc.); sociological factors (employment, average income by area); origin-destination data; auto ownership; time variations of demand, particularly peaking characteristics; and so on.

It is also important to remember that for a balanced marketing program, data must be collected on the attitudes and preferences of transit riders already using the system, as well as the potential ridership that may exist in the community. With this information, the analyst can determine the service that may be deficient in the view of the riding public.

The amount and kind of data gathered and the frequency with which it is updated depend to a large extent on the effort needed to collect the data, the stability of the community, and the availability of funds for the project.

There are many ways to gather information for market research. An easy way to start is with information that has already been gathered by others. Government agencies such as the Department of Transportation, city or metropolitan planning offices, and the Bureau of Census or Statistics are good sources. The motor vehicle department can provide information on automobile ownership by area and population group. In addition, it can provide drivers' addresses corresponding to license plate numbers for any survey of automobile driver origins. The Census Bureau has information in varying detail, from overall summaries for regions and cities down to each census tract or transportation analysis zone of a city on such factors as population, age, income, job distribution, and automobile ownership. For large cities, the Bureau also provides aggregate data on trip origin-to-destination information. Regional, state, or provincial offices have data on zoning and land use. Municipal and state departments of streets or transportation have data on street network and traffic volumes under their responsibility.

It is often useful to contact the transit agency's own complaint department, as well as its telephone information service, for relevant information on public needs, attitudes, and values. More specifically, it is possible to find out passengers' most common complaints about deficiencies in service and their suggestions for improvements. However, for much of the

Table 6.6 Characteristics commonly used in market segmentation

Type of Characteristics			
Social / Economic	*Trip*	*Time of Trip*	*Geographic*
Age	Work	Peak	Trip length
Education	Shopping	Off-peak	Trip ends
Sex	Recreational	Weekday	CBD–internal
Family size	Medical	Weekend	CBD–suburb
Occupation	Educational		Suburb–CBD
Income			Suburb–suburb
Auto availability			
Driver / nondriver			

information that the transit agency needs, it is necessary to conduct a market survey. The main techniques for conducting a market survey are questionnaires and interviews, which may be performed on board transit vehicles (short ones), in stations, at homes, or by mail.

Questionnaires should generally be brief and simple, with unambiguous questions. It is generally preferable to provide answer choices that can be checked off by each respondent. This avoids misinterpretation of questions. Questions for which the respondent must compose an answer may reveal more precise or unexpected information, but they require more time and are more costly to read and analyze. If more detailed information is required, an interviewer should assist in filling out the form.

Selection of the type of survey or interview depends on the type and extent of needed information. Surveys taken on board transit vehicles can elicit a high response rate but do not provide information about potential riders. The major problem with mailed questionnaires is that the percentage of returns is usually very low, often resulting in a biased survey. It is possible to ensure a higher response rate by a follow-up campaign of letters stressing the importance of the questionnaire. Another technique that can be used to increase the response rate is to offer inducements (chances for prizes on drawings, free passes on transit, etc.).

Interviews yield considerably higher response than mailed questionnaires, but they require more surveyors' time. Telephone interviews are relatively convenient and inexpensive, but they may have inherent biases that are often not known. Some low-income people, for example, do not have telephones. Other complications with phone interviews are also encountered: how to reach commuters at home at a convenient time of day, multiple phone listings for a single family, unlisted phone numbers, etc.

House personal interviews are very costly and require a large staff, but they usually provide very reliable and detailed information. They are therefore used when there is sufficient need for detailed and highly accurate information. Such interviews are conducted at less frequent intervals than surveys by questionnaires or telephones.

Public hearings organized for some issue, such as a change of bus route, adjustments of fares, etc., may also be used to obtain information on public attitudes or travel habits. Web-based surveys can also be used, but they are limited to certain categories of population only.

6.7.3.2 Market Segmentation. Market segmentation divides the users and potential users of transit into classes or groups, based on social, economic, geographic, or other characteristics that differentiate between the transit needs of individuals. It is after market segmentation that the differentiated marketing strategy or the concentrated marketing strategy is applied.

The degree of market segmentation attainable depends on the kind and amount of data gathered during the market research phase. Table 6.6 shows selected characteristics convenient for market segmentation. A market group may be formed on the basis of one of the listed factors or a combination of any number of them. However, in deciding how to segment the market, the following criteria should be adhered to:

1. Each segment should be distinct in its needs for and attitudes toward transit.
2. Each segment should have a sufficient number of people to justify the cost of advertising to that group individually.

There are many situations in which market segmentation followed by concentrated marketing can be applied. If transit usage on Sundays is low, advertising of possible transit trips to museums, concerts, or sporting events combined with the introduction of special family fares may be very effective. Or more frequent services and low round-trip fares may be offered as a "shopper special" program on Saturdays or in the evening of a weekday when stores are open late, usually Wednesday.

6.7.3.3 Service Adjustments. On the basis of data collected during market research and market segmentation, it may become apparent that certain modifications to service are required. Such modifications include:

- *Transit (usually bus) line modifications:* changing of alignment sections on bus routes, elimination or addition of route segments, introduction of new lines complementing the existing ones, improved transfer locations and facilities.
- *Fare adjustments:* change in fare structure (zonal, sectional, etc.), introduction of special fares (student discounts, tourist one- or three-day fares, passes, etc.).
- *Schedule changes:* adjustments to peak hour service, adjustments to off-peak or weekend service.
- *Operational modifications:* introduction of express, skip-stop, or other accelerated services.
- *Introduction of special services* (shoppers special, special events, etc.). Special services can be a particularly effective means of service innovation. A special service, such as loop bus lines in tourist areas or express metro trains for special events, is usually initiated in response to an expressed or observed need and the potential market is often easily determined. This service is often an excellent means of achieving exposure of potential users to transit services. After introduction, many special services are considered for upgrading into permanent transit lines or changed schedules on existing lines.

Principles to be used in modifications of line alignments, schedules, fare structure, and public information are given in Chapters 1, 2, 4, 6, and 7.

6.7.3.4 Information Distribution. Another essential marketing activity is a program for dissemination of transit information to the public. Such a program, discussed in Section 6.6, can often be adjusted on the basis of marketing studies and combined with some elements of marketing. For example, advertisements of transit service may be combined with on-vehicle information, timetable designs, or information at transit stops and stations.

6.7.3.5 Advertising. The activity most commonly associated with marketing is advertising, which has three related purposes:

1. To draw public attention to the transit service
2. To inform the public of the qualities or advantages of the transit service
3. To create a positive image for transit in the eyes of the public

The main overall goal is naturally to attract more riders to transit.

Public attention can be drawn to the transit service in a number of ways, some of which were discussed in preceding sections. These ways include the distribution of printed media (brochures, pamphlets, newspaper advertisements, posters, billboards, yellow page advertisements, direct mailing, etc.), radio and television commercials, the design of stations and stop designations, and the selection of system name, logo, and color theme for vehicles, signs, stop designations, etc.

In informing the public of the qualities and advantages of the transit service, practical appeals should be emphasized:

- *Money savings:* the annual savings of an actual transit rider over what otherwise would be spent on an automobile, including its dominant, but often overlooked, fixed costs
- *Time savings* (where applicable)
- *Comfort and convenience,* such as ability to read on transit, or avoidance of aggravation in traffic jams and parking problems
- *Increased safety and reliability:* compare accident rates, percent of on-time arrivals by transit, etc.

In some cities, practical appeals are not sufficient to overcome the negative image transit has acquired. Therefore, it is desirable to convince the public that it is not only socially acceptable but even fashionable to ride transit. This can be accomplished if a certain status image can be developed for the transit service. Such an image is developed through advertising:

- By appealing to public spirit by pointing out that using transit reduces congestion, creates a more people-oriented environment, preserves historic values in the city by reducing demand for parking, saves (imported) fuel, helps reduce air pollution, etc.

Photo 6.11 *Award-winning advertisement of commuter rail services in Los Angeles (courtesy of SCRRA).*

- By implying that transit users are smarter than auto drivers
- By using celebrities and prominent community leaders to endorse transit, where in fact they do use it: a mayor or a governor traveling to work by subway or LRT line have been examples that helped transit image and increased these popularity of these leaders

6.7.3.6 Coordination and Budgeting. Coordination of the advertising message with the various media is necessary for a successful advertising program. Each medium is best suited to a different advertising objective or message. Visual media, particularly television, are often best utilized to change or project a desirable image of transit services and the transit agency. Radio and posters can be used effectively for spot promotional advertisements that make the public continually aware of the presence of the transit system. Radio may make a strong real-life impact, but promotional brochures, pamphlets, and schedules can usually be more effective in conveying details of service and its amenities.

The type of medium must also be coordinated with the market segment at which the marketing effort is directed. Each medium is experienced by each segment to varying degrees. Therefore, how and at what time a given market segment will be appealed to will depend on that segment's characteristics. For example, potential riders from a high-income category might best be reached via newspapers, radio stations with news or classical music orientation, or, during evening hours, by either radio or television. Direct contact by mail is one of the few ways to reach all of the potential riders in a market segment, but it may involve considerable cost.

The expenditure devoted to advertising depends on the objectives to be achieved and costs of advertising or other marketing activities. Experience has shown that in most cases an expenditure of approximately 1.5% of gross revenues, although still very low compared with the 4 to 5% spent for this purpose by the beverage, tobacco, chemical, and movie industries, is usually adequate for very successful marketing by transit agencies.

There are various ways by which some advertising expenses can be reduced. One way is for the transit agency to share expenses with another organization. For example, transit service can be promoted by offering its use for going to an advertised sporting event (or series of games). With this approach to advertising, both parties benefit since advertising costs are reduced for each party. The transit agency also benefits from an explicit, well-publicized example of how its service may be utilized.

Another way to keep advertising expenses low is to sell a product that promotes transit usage as a secondary function. Examples include distribution of beer coasters or jigsaw puzzles with transit maps, printing of transit system name and/or logo on all city maps, tourist literature, memorabilia such as T-shirts, etc. In addition to tourist information, free advertising is sometimes available in neighborhood and college newspapers, telephone books, professional circulars and publications, and the like.

6.7.4 Public Relations

The information and marketing activities whose purpose is to project a positive, progressive image of the transit agency are broadly classified as public relations. The basis for good public relations is quality of service provided by the transit agency: a positive public image cannot be established without reasonably good service. Assuming that this basic requirement is satisfied, there are other techniques that can improve the image of the agency and the transit service.

Public relations is based on the transit agency's contact and exposure to three parties: the general public, the press, and governmental agencies and representatives.

The general public's perception of the transit service can be enhanced by the following measures:

- Widely distributed and displayed agency mission statement describing its goal and objectives.
- General manager's or chief operating officer's messages to the riders about innovations, changes in services, or explanations about problems that created service delays.
- Frequent cleaning of vehicles; repair of damage due to vandalism as soon as it occurs. This method of maintenance is often less expensive than it first appears, because a vehicle already damaged by vandalism attracts more vandalism (see Section 6.4.5.2). Thus, in the long run it may cost no more to keep vehicles well-maintained than it would to make infrequent repairs of much greater damage and suffer from continuously poor image.
- Design transit stops, shelters, and stations to be vandal-resistant and require little maintenance.
- Encourage an attitude of helpfulness and courtesy on the part of all employees (bus drivers, station employees, and telephone information personnel) who come in contact with the public. This condition is most easily attained in agencies in which top management reflects a concern for the public welfare and where good employer-employee relations exist.
- Ensure that the telephone information service has sufficient operators that transit users are not frustrated when they call. Getting a continuous busy signal on the telephone information service leaves the impression that the agency is not very interested in assisting its riders.
- Organize a complaint department that actually seeks to eliminate the causes of complaints.
- Provide a lost-and-found service.
- When delays in service occur, provide a prompt explanation as to what the problem is and how soon it will be corrected. Apologies are also in order, especially to passengers aboard a vehicle that breaks down. Printed explanations should be available the next day. They can be used by commuters who may have arrived late to work because of the breakdown. Poor public relations occurs when officials of a transit agency appear indifferent to service breakdowns and the delays suffered by the public.

While the previous suggestions are primarily directed toward improving public relations with the users of the transit system, the transit agency should also maintain a good image in the community at large by becoming involved in civic affairs. Some of the activities in which a transit agency could participate would include:

- Help with charity drives by free on-vehicle advertising space and other assistance
- Support progressive transportation-improvement programs, such as staggered work hours
- Assist in local job training programs
- Cooperate with and assist local agencies in community planning
- Actively participate in special events: holiday parades, civic and sports events, commercial activities, etc.

The press and other media are another important party in public relations. Favorable news stories about a transit agency not only bolster its image, but also reduce the amount of money required for advertising. Thus, maintaining good relations with the press is very important. The following guidelines are suggestions for successful relations with the press.

- Make most board meetings open to the public and media.
- Make sure that board members and top management representatives are available to the press.
- Provide access to all information requested by the press to the maximum possible extent. A policy of openness prevents unfounded speculation and provides for community feedback on anticipated or planned projects.
- Be honest with the press and avoid misleading or ambiguous statements.
- Notify the news media well in advance about important stories.
- Provide press kits with relevant information, brochures, etc. when the agency is issuing an important news release. This improves accuracy of information.
- Keep in mind press deadlines when scheduling news conferences and news releases.
- Do not favor one publication over another in the dissemination of information.
- Do not threaten favoritism due to erroneous or poorly presented publicity unfavorable to the agency. Attempt to have the mistake corrected instead.
- Advise the press of extenuating circumstances when service breakdowns occur.

Government agencies are also very important to good public relations since public transit is a regulated industry and most transit agencies need assistance from different governmental agencies. Actually, civic and political leaders often seek good information from the transit agency in order to implement measures to improve transit legislation and financing. The following guidelines suggest some ways of developing and maintaining these good relationships.

- Maintain an up-to-date list of the key personnel of all relevant agencies, legislative representatives, etc.
- Forward all press releases, new brochures, agency's annual report, etc. to these people.
- On important occasions, such as the introduction of new vehicles, invite some or all of the people on the list.
- Keep key officials informed about developing problems before they become public issues.
- Make political leaders aware of the obligations and constraints under which the transit agency operates and assist them in developing the needed legal and financial solutions.

6.7.5 Conclusions

This section has described the basic components of an effective marketing and public relations program and its various marketing strategies and activities. It is important to remember that the primary prerequisite for good marketing is a good product, i.e., good transit service. With infrequent, unreliable, slow, and uncomfortable service, attracting and retaining passengers through segmentation analysis, advertising, and other techniques is very difficult indeed. Moreover, investment in these activities is not as effective as it will be when good services are offered.

Finally, since attitudes and values of the public are constantly changing, marketing must be organized as a continuing process that must be given constant attention. The marketing program must be continually reviewed and adjusted to adapt to the changing needs and values of the community and its residents.

EXERCISES

6.1. Find information about the board of directors of a major multimodal transit agency. Then describe it, including the following items:

- A chart of the agency's board organization and its relationship with the top management executives
- The number of directors and whom they represent
- How they are selected and who appoints them
- The relationship between the number of directors and the population and the number of transit passengers in the areas they represent

Survey several opinions about the effectiveness of the board and summarize them.

6.2. Select *two* different organizations of multimodal transit agencies. Then:

a. Present their organizational charts, showing the major departments/divisions.

b. Describe the structure of each one: is it predominantly modal (separate divisions for different modes) or functional (most functions unified for all modes)?

c. Compare the two structures: the number, similarities, and differences in the divisions' domains of work.

6.3. Which major problems did workers experience that led to the founding of organized labor unions?

6.4. Define and describe the consequences of the introduction of labor unions in the transit industry:

a. What improvements in working conditions for the employees took place due to unionization?

b. What were the positive consequences with respect to the quality of transit services?

c. What politically positive influence did the unions have on government policies and legislation?

d. Describe some negative consequences in agencies' managerial and operational practices and in the prevention of innovative ideas that were often fought by labor unions.

e. Which political influences of labor unions can be considered as damaging to the efficiency of private and public transit operators? Which ones negatively affect service to the public?

6.5. Describe the methods used to resolve differences between management and labor unions in their contract negotiations, including indexing wage levels, imposing binding arbitration, and allowing labor strikes. How do these negotiating methods affect the concerned parties:

- **a.** The transit agency's financial management, its countering of excessive wage and benefit requirements, and its ability to bring innovations in system operations?
- **b.** The workers' need for job security and reasonable wages and working conditions?
- **c.** The public's need to have high-quality, reliable transit services?

6.6. Discuss positive and negative aspects of legally prohibiting labor strikes in the transit industry.

6.7. Discuss the results of deregulation of the transit industry in Great Britain with respect to employees and labor unions. What were the consequences of eliminating unions on the financial condition of transit companies, working conditions for employees, quality of the labor force, and its attitudes toward the company? How did it affect the quality of transit service as the public experienced it?

6.8. How can statistical data about transit lines, networks, stations, and fleet size be used in comparing transit systems in two or more peer cities? What comparisons can be made through statistical numbers and quantities? Which system aspects can be compared using various indicators more conveniently than through absolute amounts and quantities?

6.9. Define the concepts of efficiency, utilization, and consumption. Give two examples of measures in each one of these categories.

6.10. Define "operating ratio." Why does this indicator reflect public policy toward transit in a given city, state, or country? Specifically, discuss how the operating ratio reflects the efficiency of the agency, the conditions under which its lines operate, and the attitude of government toward the trade-off between the economic efficiency of a transit agency and the goal of maximizing the use of transit through low or moderate fares.

6.11. The productivity of a bus and a LRT line should be analyzed.

- **a.** An 11.8-km-long bus line has a cycle speed of 16.7 km/h and operates with a headway of 5 min. The capacity of the buses, including seated and standing spaces, is 92 spaces, and average occupancy is 36 passengers.
- **b.** An 18.8-km-long LRT line has a cycle speed of 25.1 km/h and it also operates at 5-min headways. Each car has a capacity of 168 spaces, and two-car trains are operated. They have an average occupancy of 68 passengers per car.

Compute for the lines under **a** and **b** the following elements:

- Offered work in space-km per hour
- Utilized work in person-km per hour

Assuming that each line is typical for operating values of its respective mode, compare the two modes with respect to their offered and utilized capacities.

6.12. Compute and compare offered work and utilized work of the following two rail lines:

- **a.** Metro line, 22.4 km long, is operated with a cycle speed of 33.8 km/h, with headways of 4 min and train consists of 6 cars, each with 64 seats and 96 standing spaces. The average utilization coefficient on the line is 56%.
- **b.** Regional rail line, 34.6 km long, operating 8-car trains with 20 min headways, with a cycle speed of 52 km/h. Car capacities are 104 seats and 72 standing spaces. Their average utilization coefficient is 38%.

Compare the offered and utilized work for the two lines.

6.13. A metro network consists of three lines with the following characteristics:

Line	*Length (km)*	*Peak Hour Headway (min)*	*Train Length (cars/TU)*	*Operating Speed (km/h)*	*Average Utilization Factor (% of total)*
Red Line	16.6	6	6	33.1	60.2
Blue Line	12.4	5	4	29.7	66.8
Green Line	8.8	3	8	35.3	77.5

Car capacities are 58 seats and 112 standing spaces.
Considering service during one peak hour and compute the total network productivity and average speed on the entire network.

6.14. The role of transit control centers differs somewhat among modes. Select control center for one of the following two modes:

a. A metro network

b. A network with light rail, trolleybus, and bus services

Obtain information from the literature or by interviews in transit agencies and describe the functions the selected center performs. Then define the role and impacts the center has in system operations. In other words, how does the center improve the efficiency, safety, and quality of service for passengers?

6.15. A major transit agency has planned the construction of a control center for supervising operations of its metro and surface transit lines. The estimated investment for the center is rather high, and the agency's board is expected to examine carefully the project feasibility to determine if the expected benefits will exceed the estimated costs.

You are asked to prepare a systematic estimate of the Center's expected benefits. Prepare a list of benefit items, such as increased service reliability and efficiency, safety, and security. Define which benefits each one of these items will include, then focus on reliability: develop a flowchart of benefits from higher service reliability. For each benefit item define the main beneficiary or affected party: passengers, operator (agency), and the city (community). State which items can be quantified and how.

6.16. Classify major types of applications of ITS/telematics in transit systems operations technology and define the functions of each one.

6.17. A transit agency is considering the purchase and installation of an AVL system for its bus services. The board wants to evaluate the expected benefits in relation to the estimated investment costs. Prepare a list of benefits an AVL system would bring, including operational aspects affecting the agency and service aspects benefitting the public. Include the quantifiable and qualitative aspects and describe how the quantitative results could be computed for a specific system.

6.18. Define and compare the three procedures used for vehicle maintenance: reactive, preventative, and predictive. Compare them with respect to the quality of rolling stock they produce, total cost of maintenance, and under what conditions each one is most likely to be used.

6.19. Define functional vehicle life and compare it with its economic life. List several factors that may influence which one of them will determine when rolling stock should be replaced.

6.20. Describe several measures and methods used to prevent graffiti and other vandalism in transit systems.

6.21. Present and discuss the concept of a fully accessible transit system. How does that differ from the common definition of the provision of accessibility for disabled persons?

6.22. Transit agencies must often finance modifications of their facilities to provide accessibility for the disabled. What is the rationale for this requirement, a so-called unfunded mandate, and how does it impact equity aspects? In other words, define who are the beneficiaries of these investments and who pays for them.

6.23. Select two of the four categories of persons defined by their needs for transit information, such as categories 1 and 3 or 2 and 4. List and explain the items each one of the groups needs, and define those that are different for the two groups.

6.24. Develop a list of information items that should be provided at two types of transit system facilities, such as bus stops, vehicles, or transit terminals. Then make a survey and review how the transit system in your city meets these requirements.

6.25. Define market segmentation and explain how it can be used in transit marketing. Give an example of a market segment with respect to transit use.

6.26. What is undifferentiated marketing and how can it be applied in transit? Give two examples of messages that can be included in this process.

6.27. Define differentiated marketing and compare it to individualized marketing with respect to objectives, techniques, costs, and likely results.

REFERENCES

American Public Transportation Association (APTA). 2002. *Public Transportation Fact Book*. Washington, DC: APTA.

Biggs, D., and D. Levy. 2000. *Transit's Safety and Security Record*. Canadian Urban Transit Association (CUTA). STRP Synthesis S3. Toronto: CUTA.

Brög, W. 1998. Individualized marketing—implications for transportation demand management. *TR Record 1618,* 116–121.

EG&G Dynatrend and Crain & Associates. 1995. *Transit Operations for Individuals with Disabilities*. TCRP Report 9. Washington, DC: Transportation Research Board.

Girnau, G., A. Müller-Hellmann, and F. Blennemann, eds. 2001. *Telematics in Public Transport in Germany.* Cologne: Verband Deutscher Verkehrsunternehmen.

Lieb, R. C. 1976. *Labor in the Transit Industry*. Report to the U.S. Department of Transportation, Washington, DC.

Marciani, M. 2002. The "public" in local public transport: from users by necessity to customers by choice. *Public Transport International* 04, 10–13.

Müller-Hellmann, A. 1999. "The drive towards the non-polluting bus for a better quality of Life in cities. 53rd UITP International Congress, International Union of Public Transport, Brussels, Belgium.

Potts, J. F. 2002. *Customer-Focused Transit.* TCRP Synthesis 45. Washington, DC: Transportation Research Board.

Replogle, M. 1983. *Bicycles and Public Transportation: New Links to Suburban Transit Market*. Washington, DC: The Bicycle Federation.

Spiller, D., and L. Labell. 1997. *Operational Assessment of Paralympics Transit System: Low-Floor Buses, Lift Equipped Buses, and Signage*. FTA Report FTA-DC-08-

9117-97-1. Washington, DC: Federal Transit Administration.

Transportation Research Board (TRB). 1981. *Transit Boards—Composition, Roles and Procedures.* NCHRP Report 75. Washington, DC: TRB.

Vuchic, V. R. 1981. *Urban Public Transportation Systems and Technology.* Prentice-Hall

———. 1999. *Transportation for Livable Cities.* New Brunswick, NJ: CUPR Press, Rutgers University.

Vuchic, V. R., and A. Bata. 1989. US cities lead fight against graffiti. *Railway Gazette International,* January, 39–41.

FURTHER READING

Burkhardt, J. E., A. T. McGavock, C. A. Nelson, and C. G. B. Mitchell. 2002. *Improving Public Transit Options for Older Persons. TCRP Report 82.* Washington, DC: Transportation Research Board.

Intelligent Transportation Society of America. 1998. *Improving Transit with Intelligent Transportation Systems.* Washington, DC: ITS America.

Kittelson & Associates et al. 2003, *A Guidebook for Developing a Transit Performance-Measurement System.* TCRP Report 88. Washington, DC: Transportation Research Board,

Simon & Simon Research and Associates. 2002. *Public Transit Board Governance Guidebook.* TCRP Report 85. Washington, DC: Transportation Research Board.

Tomazinis, A. R. 1975. *Productivity, Efficiency and Quality in Urban Transportation Systems.* Lexington, MA: Lexington Books.

Vuchic, V. R., and T. Potter. 1983. Train crew reduction for increased productivity of rail transit. *TRB Record 980.*

7

TRANSIT FARES

Transit fares are a major factor in attracting passengers; they are also a basic element of transit system operations, affecting the financial condition of the transit agency. The amount of the transit fare, its relationship to the quality of provided service, and the convenience of fare payment greatly influence transit system ridership. Types of fares and their collection also affect the efficiency of operations. The revenue collected from fares influences the method of financing transit operations in an urban area. Finally, in the long run, fares often have a significant impact on the form and development of central cities, their surrounding areas and suburbs. Therefore, planning fares for a given transit system requires careful consideration of numerous interrelated aspects of fares. Their presentation in this chapter is organized in the following five major topics:

- *Objectives, requirements, and constraints* in setting fares
- *Fare collection,* including physical/organizational aspects and types of fares
- *Structure of fares,* representing the amounts as related to distance traveled
- *Special fares* for different categories of users or uses, for different times of day, etc.
- *Level of fares,* or the absolute amounts of fares (regardless of their structure) and their impacts on ridership and revenues

It is important to know the general characteristics of different types of fares and various features of transit services in order to select the most efficient type of fares (their structure, level, and method of collection) for any specific situation in a city.

7.1 OBJECTIVES IN TRANSIT FARE DETERMINATION

In order to plan fares for a given transit system, it is necessary to define what the fares should achieve. The basic objectives for a fare system may be the following:

O-1. To attract the maximum number of passengers

O-2. To generate the maximum revenue for the transit agency

O-3. To achieve specific goals, such as increasing the mobility of the labor force, students, or seniors, improving access among certain areas, promoting the use of a more efficient transit mode, etc.

The common requirements and constraints that must often be considered in setting up fares include the following:

R-1. Elasticity of demand (see Section 7.5.3) may limit the choice of fare level and structure; the quality of service and prices of competing modes play a major role in passengers' willingness to use transit at a given fare level.

R-2. Equity among transit travelers in terms of value of service (trip length, local or express, comfort level, safety and security, etc.) versus the fare paid by different user groups. This may be an important factor for individual users, as well as for groups of population.

R-3. Social and political aspects—the need for service and ability to pay by different population groups—often play an important role.

R-4. Fares should be understandable and convenient for passengers to pay.

R-5. Fare type should allow easy and low-cost collection and control by the agency.

It is obvious that all of these objectives and requirements cannot be satisfied in their extremes since some of them are mutually conflicting. This is most obvious for objectives O-1 and O-2: attraction of the maximum number of passengers in most cases requires moderate or low fares (or, theoretically, free services). However, low fares result in generation of less revenue than higher fares would achieve. Therefore, a compromise must be found between the objectives of maximum ridership and maximum revenue. Similar compromises must be found among the other objectives. Planning transit fares thus requires analyses of many trade-offs among the above listed objectives and satisfaction of requirements and constraints.

Translating these objectives and requirements into real-world situations, trade-offs among major objectives are usually subjects of political decisions. In particular, the objectives of attracting passengers in order to increase the mobility of the population, increase job opportunity or access to schools by students, reduce traffic congestion, and make a city more livable are in most cities considered more important than collecting maximum revenue, or even covering all operating costs. For this reason, in most cities transit obtains public funds for investments in facilities and coverage of a portion of its operating costs (see Chapter 8). Therefore, the objective of revenue maximization in most cities is considered not alone, but in relation to other objectives, and aimed to cover costs in combination with the public contribution to transit funding.

To determine the appropriate fare structure and level within these political decisions, the transit agency that designs the fare system and the public body that approves it and often contributes to the financing must decide, explicitly or implicitly, on the relative importance of each objective.

When fare revenue is adequate for covering all or a required portion of operating expenses, well-managed agencies tend to give due consideration to all objectives. However, when the company or public agency experiences financial strain and there are no other sources of income, the requirement that fares bring the specified level of revenues becomes a question of survival. The transit operator is then forced to focus on that objective alone. If the requirement for maximization of revenue then requires a major fare increase, that may trigger a sequence of events designated as the double vicious circle of fare increases, shown in Figure 7.1.

As the flow diagram of events in the figure shows, if a major fare increase causes a significant diversion of passengers from transit to the automobile, two main consequences occur. First, the increased street traffic congestion may lower transit speeds and increase operating costs (the solid line in the diagram). Second, reduced passenger volume may lead to a reduction in service frequency, making it less attractive and causing

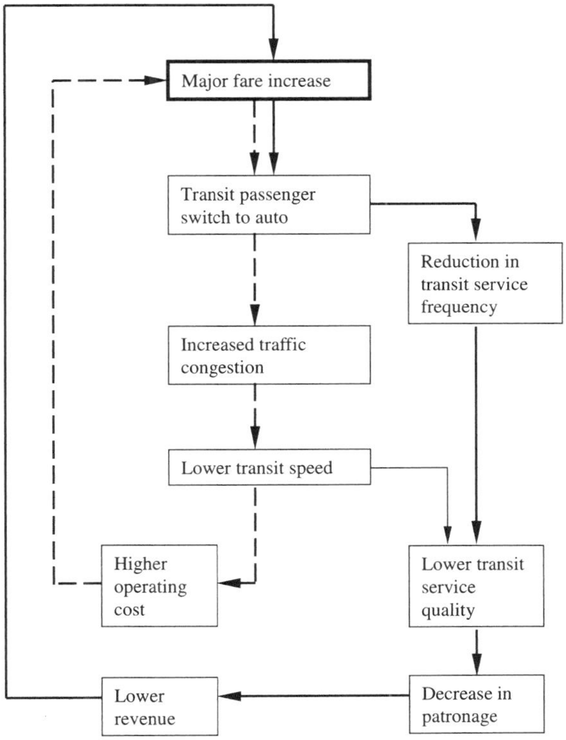

Figure 7.1 *Two vicious circles caused by fare increases:*

further passenger losses (dashed line). Both increased costs and decreased revenues then lead to the need for another fare increase, closing the vicious circle. It is this phenomenon that necessitates assistance by public bodies to maintain transit service at a certain level, rather than allow it to continue to deteriorate and eventually, in smaller cities, to be discontinued.

This problem of transit companies being forced to maximize fare revenues in order to survive was common in many cities, particularly medium-sized and smaller ones, at the time when private automobile came into wide usage and transit was considered to be private business. A similar situation has been created by the deregulation of transit in Great Britain and some other countries during the 1980s. The results have been similar in the two cases: fare revenues were increased and financial conditions improved, but at the expense of significant ridership losses and a decreased role of transit in urban areas.

The other extreme attitude occurs when the revenue objective is completely neglected. If the transit agency considers that it has ensured government subsidies, so that the revenues from fares are not important, it tends to become inefficient and extremely dependent on political support. In the long run, this attitude often leads to deterioration of service quality and/or financial crisis.

Another example of excessive emphasis placed on one objective is reflected in widespread use of flat fare, which is superior to other fares only because it is convenient to the riders and the operator (requirements R-4 and R-5). Although a flat fare does have this advantage, it is in some cases (e.g., in large cities) detrimental to the more important objectives O-1 and O-2, as will be shown later.

In most cities that have rational transportation policies with positive treatment of transit, the objective of attracting riders (O-1) is far more important than maximization of fare revenue. Such policies increase the mobility of population, decrease traffic congestion, and reduce degradation of the urban environment. Yet fare revenue must not be neglected either, and in many cases it is necessary that the fares be as high as possible without causing significant ridership losses.

7.2 FARE COLLECTION

The process of fare collection—the locations and methods of collecting fares and control of payments by passengers—has a major impact on passenger convenience, on vehicle standing (dwell) times at transit stops (which affects operating speed and reliability of service), and on the cost of collecting revenue. Street transit (mostly bus) operations are particularly affected by the delays fare collection may cause at transit stops. When fares are collected in stations (usually rail), fare collection has no impact on boarding, but the cost of collection may be high if it involves personnel or complicated equipment. Self-service systems and the use of smart cards may avoid most of these disadvantages when they can be applied.

Through these impacts the fare-collection method can significantly affect:

- Attraction of passengers
- Quality of service and cost of providing it (fleet size)
- Cost of equipment for collection of money
- Feasibility of using different fare structures

Selection of a fare-collection method is therefore very important. In recent decades, there have been many innovations in the technology and operational concepts of fare collection.

Four basic aspects define a fare-collection system:

1. Time and location of payment in relation to a passenger's trip
2. Method of control of payment
3. Form of payment
4. Time, location, and method of payment control.

These components are interdependent among themselves, and they also interact with the type (structure) of fare, as will be shown below.

7.2.1 Times, Locations, and Methods of Fare Collection and Control

By time and location of payment, fare collection is classified into two broad categories. The first is prior to boarding the vehicle, i.e., off-vehicle. This includes a variety of methods, such as purchasing a transit pass through employer's payroll, at different stores and kiosks throughout the city, in the station at a ticket counter, from a machine, or at the gate or turnstile. The second category is at the time of boarding the vehicle or later on the vehicle, which may be:

- "Pay enter," or at the beginning of the trip
- Paying on-board vehicles (bus or LRT) in a ticket-issuing machine, by canceling a prepaid ticket in a special machine, or by passing a smart card by an encoding register
- "Pay leave," or prior to exiting the vehicle, which is often used together with pay enter on lines with graduated fares.

Payment of fare off-vehicle is more desirable than on-vehicle because it does not interfere with passenger boarding/alighting flows and thus does not reduce vehicle operating speed. This method has always been used on metro systems, but its use has now been broadened to most light rail systems and those street transit lines that utilize self-service fare collection, often in combination with on-vehicle ticket validation.

When on-vehicle fare collection is used, the pay enter method is most efficient on the lines that have a rather even distribution of boarding passengers, so that there is no stop that causes excessive standing delay due to queuing and crowding. The pay leave method may be used on the lines that have a heavy concentration of boarding passengers, while their alighting is rather evenly distributed along the line. Thus, on some radial lines inbound direction may use pay enter only, while outbound may use pay leave collection. The latter method is needed whenever graduated fares are used, unless the self-service method or smart card is used.

Photo 7.1 *Self-service ticketing on board a light rail vehicle—Saarbrücken, Germany.*

Table 7.1 Alternative locations for fare collection and control for different fare structures

Fare Structure	*Location of Payment and Control System*	*Entry of Station / Vehicle*	*On-Board Vehicle*	*Exit from Station / Vehicle*
Flat	Pay enter	√		
	Pay leave			√
	Self-service	(√)	(√)	(√)
Graduated	Pay enter / control exit	√		√
	Pay enter / spot check	√	(√)	
	Self-service	(√)	(√)	(√)

√: spot-check control.

In some cities, the pay enter collection is used on radial lines in the inbound directions until vehicles enter CBD, while the pay exit starts when the vehicles leave CBD. This creates a fare-free zone intended to serve short trips in center city and thus reduce automobile use in that area. The two collections are sometimes also combined on very long suburban lines: passengers pay a base fare as they enter and then pay incremental fares for additional line sections they have traveled when they leave.

Introduction of smart cards has solved the problem of some passengers trying to evade payment when exiting vehicles. Smart cards can allow subtraction from or addition to the stored value of the electronic card. Thus, when entering, the maximum fare on the transit line is subtracted from the card. Then, when leaving a vehicle or station, if the fare is lower than the maximum, the difference is added to the card at the control point. The passenger thus has no incentive to evade, because at exit the value of the ticket is not reduced, but increased. Evasion actually results in loss of money.

Control of payment can be performed in three ways:

1. At the time of payment on the vehicle
2. At the barrier (gate, turnstile) in the station
3. By spot check at any place in the system at random times

The selection of the location and time of fare collection and control depends on the type of fare: a flat fare requires only one payment and control, while a graduated fare requires two controls to ensure that the proper fare amount was paid. As the review of locations of fare collection and control for two different fare types in Table 7.1 shows, flat fares are collected either at the entry (usually on-vehicle for street transit, at gates on metro systems) or, much less frequently, at the exit from vehicle or station. Graduated fares must be collected prior to or at the entry, and they must be checked at the exit, or on a random basis, so-called spot-check control, elsewhere.

It should be noted that it is possible not to have any fixed location for fare collection and control, but to use only a single spot-check control. This is the self-service fare-collection system, discussed in Section 7.2.3.

7.2.2 Forms of Payment

Four general forms of fare payment can be used, individually or in combination, on a transit system:

- Cash fares during the trip
- Prepaid fares
- Self-service fare collection
- Fully automated fare collection and control

While the first two forms, cash and prepaid fares, are only different forms of the same basic collection system, the latter two are quite different from the others in their form and procedure.

7.2.2.1 Cash Fares. Cash is the basic, most direct form of fare payment. It has several important advantages: it is the best understood, it requires the least amount of planning by the passenger, and it is simple to implement and change. However, cash fares also have three significant shortcomings. First, cash handling causes security problems in high-crime areas; second, it is a time-consuming transaction and, when performed on-vehicle, causes major delays; and third, physical collection of cash from fare boxes on vehicle and in stations is a process that requires safety-protected procedures and labor-intensive handling in offices and transfers to banks. Moreover, all cash handling, from drivers to sorting personnel, must be secured against internal theft (discussed in Section 6.4.3).

To solve the first problem, incidence of crime, many cities, particularly in the United States, use exact fares, i.e., direct payments into a farebox, so that the transit vehicle driver supervises, but does not handle any money. The exact fares also have the advantage of faster transaction (no change returns), but they create inconvenience for passengers who must prepare large amounts of change. Moreover, there are sometimes mechanical problems with fareboxes, particularly due to the handling of paper money. This is an especially serious problem in the United States because of the much lower use of coins (25 cents being the largest commonly used coin) than in most other countries, where there are coins up to the values equivalent to U.S. $2 or $3.

Due to the second shortcoming, delays caused by payments, many transit systems have made great efforts to replace cash fares by other forms, mostly by tokens, prepaid fares, and smart cards. In addition to easier handling, prepaid fares have the advantage that they do not require out-of-pocket payment, about which users are particularly sensitive and reluctant.

Payment of cash fares is sometimes discouraged by a significant difference in its amount as compared to prepaid fares. While even a small difference (10–25%) influences passengers to avoid cash fares, the differences are sometimes much greater, even 100% higher. The large difference is very effective, but it very strongly penalizes and discourages travel by incidental transit users. This is undesirable if it leads to a significant loss of off-peak riders or penalizes low-income persons, students, and other categories of incidental riders. Thus, a moderate but not extreme overpricing of on-vehicle cash fares is most desirable.

7.2.2.2 ***Prepaid Fares.*** The trend in recent decades has been toward greater use of prepaid tickets, which can have several forms:

- Tokens, which are prepurchased for cash and used for payment at entry
- Multiple-ride tickets and transfers
- Daily, weekly, monthly, and annual passes and permits valid for a certain number of trips or for unlimited travel on a specific line or on entire network
- Stored value or charge smart magnetic cards

These forms of fares, tickets, or passes may be sold at major stops and public places, stores, hotels, and, in many cities, by employers, often on a regular basis through payroll deduction and at a discount. Presale of fares increases passenger convenience over a strict exact fare requirement and encourages long-term use of the transit system on a regular basis. Transit users are induced to use these types of fares by their convenience, as well as by the discount they get if they make reasonably frequent use of the system. The transit agency gives the discount because it benefits from simplified fare collection as well as the earlier use of the funds collected by presale.

Tokens are coined with the transit agency's symbol and special code. They allow fare payment by a single coin, regardless what the amount of the flat fare is. They have been used in many U.S. cities and sold at a certain discount, but they are being increasingly replaced by electronic tickets, passes, and smart cards.

Multiple-ride tickets can have the form of a sheet of several detachable stubs that are canceled or removed one by one for individual rides; or they can have a form of a single card that is punched or canceled for each ride, or by the number of line sections traveled, or even for several passengers. The canceling is usually done by the passenger in a special machine at the stations or on the vehicle.

Passes are fare cards sold by the transit agency at a fixed price that allows an unlimited number of trips within the given period: most typically, the period is a month or a week, but some cities also have daily and annual passes. In recent years, an increasing number of transit systems have introduced tourist passes valid for unlimited use during a one-, three-, or five-day period. These passes are attractive for city visitors and tourists not only due to the discounts they offer compared to full payment of fares, but even more due to the convenience they offer. Persons who come to the city usually enjoy not having to search how and what amount to pay for each transit ride, particularly if there are several modes and graduated fares are used.

In addition to the passes that allow unlimited travel on a transit network, there may be passes for specific areas or zones, or for individual lines. There are also permits, which are lower-cost passes requiring a small additional cash payment for each ride. These are often used for school and college students, who obtain the passes from their institutions and then are entitled to discounted fares.

Stored value or charge smart cards include magnetized cards or credit cards, which may be valid for a single trip or for multiple trips of the same value, or be debit cards with a stored amount of money. A passenger either feeds the card into the fare gate, swipes it through a slot, or only swings it close to the electronic scanner (contactless cards). The machine subtracts the value of that trip and usually displays the remaining amount of value remaining on the card.

The magnetic cards of different technologies are being introduced by many transit systems, and they are becoming increasingly sophisticated. The smart card used in some cities can be used not only for transit, but also for parking and other similar charges or purchases. This trend will continue, particularly with the introduction of various applications of ITS technologies to transit systems (see Section 6.4.2).

On some transit systems, payment of fare by a token or a multiple-ride prepaid ticket is still done under supervision of the driver (on street transit modes) or by a station employee in metro stations and terminals. Passes and magnetic card tickets either simplify the supervision of payment, or allow replacement of the slow and costly conventional personal supervision of fare payments by a different collection method—self-service, either manual or automated.

Actually, the magnetic cards, particularly contactless ones, are so simple and fast for processing that they allow easy application of graduated fares with full entry/exit control not only on enclosed systems, such as metros with gates at stations, but, even more importantly, on street transit vehicles also. Graduated fares must be checked at exit to ensure that the correct amount has been paid, and with manual control this can only be achieved by spot-check control. With smart cards, the ticket-reading machine at entry subtracts an amount equal to the maximum fare on the line, and at exiting it returns the difference between that amount and the amount actually required for that trip, if it is shorter, to the card. The passenger therefore makes sure that he/she goes through ticket control at exit also, eliminating the tendency to evade that control.

7.2.3 Self-Service Fare Collection (SSFC)

Introduced first on transit systems in European cities during the late 1960s, self-service fare collection (SSFC) is now widely used in Europe. After considerable delay and skepticism (later proven unjustified), transit systems in North American cities began to introduce SSFC, mostly on the new LRT systems, which cannot have manual fare collection because an LRT train has 4–16 doors. Known also as an honor system or proof of payment (POP) system, SSFC is now successfully operated in many cities, including the Los Angeles Metro system and most of the new commuter/regional rail systems. Its applications on buses are very common in Europe, but yet rather limited in North America.

SSFC is a system in which passengers are responsible for making payments or getting their own tickets. Nobody checks passengers for their payments on a regular basis: in most cases they travel without showing their tickets to anybody. However, there is a spot check, or randomly performed control, during the ride or at stations, by one or more agency personnel. Passengers must show to the controller their tickets (proof of payment), and if a passenger does not have one, he/she must pay a penalty. The sample of checking usually varies between 2 and 8% of passengers, and the penalty is determined on the basis of the sample size and tendency for fraud on a given service. Typically, the penalty is 10, sometimes up to 20 times the fare that the passenger had failed to pay.

The comparison below shows that the SSFC has several very significant advantages over the conventional manual collection and control. By eliminating the fare-collection process at entry of street transit vehicles, SSFC allows use of all doors and results in much faster boarding. The reduction of dwell times and of their fluctuations with passenger volumes at stops results in higher operating speed and reliability of service and makes transit service more attractive to passengers.

Careful preparations are needed for introduction of the SSFC system to familiarize the public with this concept of fare payment, particularly when it is introduced in a city for the first time. For example, when the new LRT line was opened in San Diego in 1981, during the first month employees were checking the tickets on board vehicles and explaining to those who had not had fares that later they would pay penalties if they did not pay. After this introductory period, on-board control was replaced by spot-check control of a small sample, and it was highly successful. When passengers get used to SSFC, they find it more convenient than having to pay exact fares, show passes, wait for boarding, etc.

With SSFC, use of various types of passes or other multiple-ride tickets is encouraged by their lower pricing than single fares. The tickets are purchased off-vehicle, and they may be of two types: dated, which are valid for a certain period of time after purchase, typically two hours; and undated, which then must be validated or canceled by the user at the time of boarding. Usually, passengers have a ticket valid for several trips, which may be of different lengths and zone charges.

The passenger cancels the required amount for a particular trip (or some systems allow payment by a single ticket for two or three persons traveling together) in the station lobby or mezzanine or on the vehicle after boarding. The validation machines are small, so they can be easily installed even at several locations in a bus or LRT vehicle. The cancellation imprint shows the time and location of the cancellation, which the controller can later check.

The benefit of SSFC on metro systems is again the simplification and speeding up of the fare-collection

---

Self-service fare collection, compared to conventional manual collection, has the following advantages (+) and disadvantages (−):

- \+ On street transit, all doors can be used for boarding.
- \+ Higher operating speed and service reliability attract new riders and decrease operating costs due to reduced fleet size.
- \+ On rapid transit, stations can be smaller and simpler because no fare gates are needed.
- \+ Reduced manpower requirements result in lower operating costs.
- \+ Any kind of fare can be used because graduated fares do not need double control, resulting in passenger attraction and higher revenues.
- ~ Ticket-selling and canceling equipment must be purchased and maintained, but that cost may be lower than the cost of conventional fare collection.
- − There is an increased possibility of fraud (unpaid or insufficiently paid fares).
- − Control persons/teams are needed; enforcement may sometimes create problems.

---

process, although the process of boarding of trains, unaffected by fare collection, remains the same. In the stations, however, the expensive, space-consuming fare gates are not needed any more. If it is planned for a rail system to have SSFC, its stations can be simplified: in cases where a mezzanine would be provided for fare collection only, such a mezzanine can be eliminated and stations can be much simpler, and thus their construction and operating costs significantly decreased and passenger access distances shortened.

Another, somewhat less obvious and yet extremely important, benefit of SSFC is that it allows use of any type of fare structure. Many types of fares, particularly graduated ones, cannot be used on street transit lines with traditional fare collection because there is no way to check passengers twice to ensure that the proper amount was paid, unless an advanced smart card fare-collection system is introduced. Or the entire operation must be done by using only a single door, resulting in major delays. With SSFC, all doors can be used and any type of fare can be charged. It is the passenger's responsibility to purchase the correct fare and that is controlled by spot-checks only.

It should be mentioned that there was widespread negative prejudice against SSFC in the United States and Canada. It was claimed that control of passengers or imposition of penalties would not be legal, socially acceptable, or physically possible. The highly successful introduction of SSFC on the new LRT systems in Edmonton in 1978 and in San Diego in 1981, followed by a number of other LRT and new commuter/regional rail systems, have now proved that such prejudice was completely unjustified (Padron and Stanger, 1976).

On contemporary LRT systems and many lines served by articulated buses, SSFC is the only feasible way of fare collection at open stops along streets, except when a complete smart-card system is introduced. The reason is that manual fare collection would require large and costly manpower for conductors or result in extremely slow operation, either of which would defeat most of the inherent advantages of the high-capacity, high-performance semirapid transit modes.

The benefits of SSFC increase with passenger volumes on a line or a transit network. They are extremely significant on heavily loaded lines, but quite marginal on lightly traveled suburban lines, where conventional fare collection at entering vehicle does not cause major delays. For that reason, as well as for some errors in the design of circulation areas in buses, introduction of SSFC on all bus lines in Portland, Oregon, was not successful. Basically, the benefits of speeding up the boarding did not outweigh the additional cost and complexity of performing spot-check controls.

Actually, the need to have controllers and potential problems with penalizing the passengers caught without the appropriate ticket are the main negative aspects of the SSFC. Moreover, the tendency for fraud may in some cases be greater than with conventional fare collection, because the apparently free use of vehicles is inviting. Yet these problems have been satisfactorily solved in dozens of cities and countries, from Italy to Norway, as well as in almost all new LRT and regional rail systems in U.S. and Canadian cities, from Edmonton and Buffalo to Los Angeles, Baltimore, and Dallas.

Photo 7.2 *Portland Tri-Met self-service ticketing machine at light rail station* (Courtesy of *Chris Fussell*).

Automated fare collection, compared to self-service rare collection, has the following advantages (+) and disadvantages (−):

+ It has lower temptation to fraud (non-payment or underpayment).
+ It does not require spot-check controls, reducing manpower requirements and costs.
+ It generates much more complete statistical data on ridership.
− It requires more infrastructure: areas for barriers and more complex fare-collection (card reading and encoding) equipment.
− It requires considerably higher investment.
− It is more difficult for disabled passengers.

Moreover, in cities in developing countries with very large transit ridership and high percentage of low-income and unemployed persons, SSFC is probably not feasible. With very low labor costs, fares on large buses and LRT trains are often collected by conductors, or stations have access controlled by fare gates. This is the practice used on an LRT line in Mexico City, as well as on semirapid bus systems in Curitiba, Brazil, and Bogotá, Colombia, and in most other cities in Latin America, Africa, and Asia.

7.2.4 Automated Fare Collection (AFC)

Automated fare collection (AFC) is defined as a system where passengers are automatically charged or checked for payment in an enclosed system: in stations at entry for flat fares, and at entry and exit for graduated fares. Previously performed manually, such as in most metro systems until the 1960s, this automated system became feasible with the development of electronic technology for machine reading of magnetically stored information and automated ticket-vending machines. These machines permitted introduction of unmanned barrier gates/turnstiles. Introduced on PATCO's Lindenwold Line in 1969 and San Francisco BART in 1972, AFC has become the standard fare-collection method for most new and for some modernized older rapid transit systems, such as in Paris and London.

Passengers purchase some kind of magnetic ticket or card with stored value and place it in the gate machine, which, upon virtually instant checking, opens for passage. The same procedure is repeated at station exit, where the machine subtracts the value of the performed trip. Thus, the processing is fast and allows any type and amount of fare, its change with time (peak and off-peak), etc. The ticket-reading machines are sophisticated. It is important that they be very reliable.

AFC represents an ultimate method of fare collection and control. In addition to fast and fool proof processing of any fares for large volumes of passengers, it generates a full record of all trips made on the system, which may be classified by origin-destination, hour, type of fare, etc. Thus, AFC has the same main advantages as SSFC, with some differences (see the box above).

Generally speaking, AFC is a higher-performance, higher-investment type of fare collection than SSFC. Due to its space requirements, AFC can be used only on fully controlled systems, such as rapid transit. It is feasible to use AFC also on street transit vehicles by installing ticket-reading machines at each door (such as reading the entry at the front door of a bus and reading/encoding exit data prior to passenger's exit on the rear door). This system would enable the use of AFC on street transit, but at the price of increased fare-control operations on board vehicles, with all the inherent delays, space requirements and problems if machine malfunctioning occurs.

With further development of smart cards and related electronic equipment, AFC will be increasingly used on transit systems. The conditions that make the AFC particularly effective are grade-separated (metro) systems with enclosed stations, large passenger volumes, and an extensive network with graduated fare structure.

A convenient summary of the great variety of fare-collection characteristics (times and locations of payment, methods of control, and forms of payment) is presented in Table 7.2.

7.2.5 Evaluation of Fare-Collection Systems

The great variety of fare-collection systems allows transit agencies considerable choices in meeting their local conditions and requirements. In evaluating which system is optimal for a particular transit system, the following main criteria should be considered:

- System and user characteristics that affect the fare-collection system (transit mode, passenger volumes, population habits, etc.)
- User convenience
- Minimum vehicle delay for fare collection
- Ease of monitoring payment
- Cost of fare-collection equipment and personnel (SSFC has much lower costs than AFC)
- Security of fare deposits (crime prevention) and safe transfer of cash
- Ability to use different fare structures and types

The large number and diversity of criteria, many of which are influenced by local conditions, explain why there is a great variety of fare-collection systems among cities.

7.3 FARE STRUCTURES

Fare structures are classified on the basis of the relationship between fare amount and distance traveled on a transit line. Using this criterion, there are two main structures: *flat fare* and *graduated fare*. The latter can be further subdivided into *zonal fare* and *sectional fare*. Flat fare is a constant amount independent of trip length. In contrast, zonal and sectional fares change with the distance traveled on the transit network. The

Photo 7.3 *Automated fare collection with smart card capabilities in Seoul* (Courtesy of *Paul Christner*).

Table 7.2 Review of fare collection methods and types

Time of Payment	*Location of Payment*	*Payment Control*
• Prior to travel	• Work place (payroll)	• Driver/conductor
• On-vehicle	• Stores, kiosks	• Spot-check (random)
• Pay enter	• Station ticket offices	• Fare gates, turnstiles
• During travel (validation)	• Ticket issuing machines (TIM)	
• Pay leave	• On vehicle/farebox	

Form of Payment

Cash	*Prepaid*	*Self-Service*	*Fully Automated*
• Exact fare	• Token	• Limited time tickets	• Cash
• Change given	• Multiple-ride tickets	• Prepared tickets with validation	• Swipe cards
	• Transfers	• Unlimited ride passes	• Credit or "smart cards"
	• Stored value and "smart cards"		

three types of fare structures are shown schematically in Figure 7.2.

7.3.1 Flat Fare

As the name implies, a flat fare is constant regardless of the distance a passenger travels on a single vehicle or throughout the transit network. The flat fare is the simplest possible fare; therefore, it is easy for passengers to understand and use. If it is used for individual lines, it is usually collected at fare gates in a station or upon boarding the transit vehicle. Thus, supervision of payment is easy, the work of fare collectors or vehicle operators is simple, and the boarding process is rather fast. When the flat fare is used on a network with enclosed stations, payment is made at fare-control barriers, while exiting from stations is open, without fare or ticket control. When flat fare is used in street transit modes, passengers are issued tickets and these are checked at transfer locations or randomly by controllers.

The basic disadvantage of a flat fare is that it is a fixed amount that does not in any way reflect the quantity of service that the rider receives: a passenger traveling three street blocks pays the same amount as a passenger traveling 8 or 10 km. In cities with limited geographic size, travel distances are relatively uniform, so that the conveniences of the flat fare far outweigh the inequities related to its lack of correlation with trip length. In large cities, however, while the obvious con-

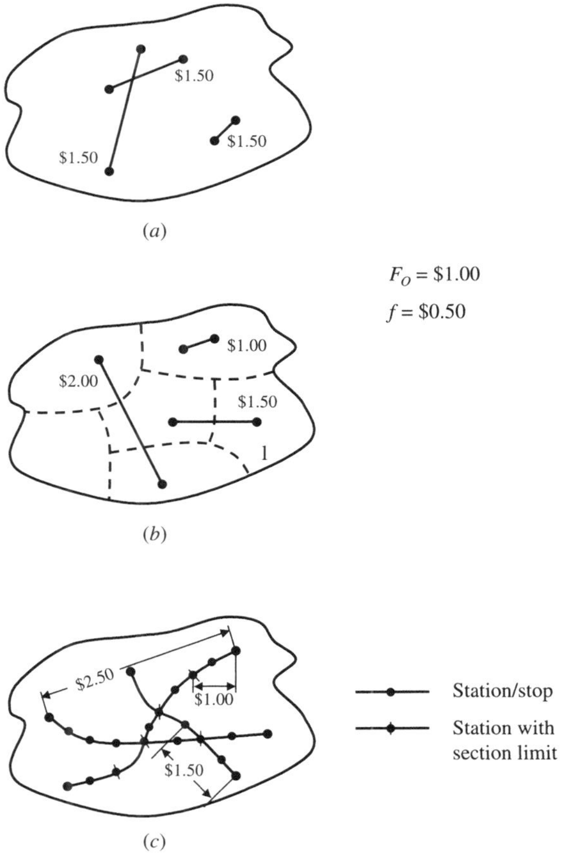

Figure 7.2 *Schematic presentation of alternative fare structures: (a) flat fare; (b) zonal fare; (c) sectional fare.*

venience of flat fare (simple collection) is at least as important as in smaller cities, its disadvantages become very significant. For example, if the level of the flat fare is such that the amount corresponds reasonably to trip lengths from 2 to five km, then such a fare will be excessive for persons traveling several blocks in the center city. At the same time, it will be very low for passengers traveling distances of 8 or 10 km, thus forgoing collection of considerable potential revenue.

7.3.2 Graduated Fares

To avoid the inherent limitation of the flat fare, the inequity of short-distance riders actually subsidizing long-distance riders and inability to collect additional revenue from the latter group, graduated fares can be used. Graduated fares can include many variations, but, as noted above, there are two basic types: zonal fare and sectional fare. They are described in the following two sections.

7.3.2.1 Zonal Fare. The simplest method for charging graduated fares proportionally to the lengths of passenger trips is achieved by dividing the city or urban area into fare zones, sometimes two (central and outlying), sometimes more. Then one fare applies to passenger trips wholly within a single fare zone, a higher fare applies to passengers crossing from one zone into another, an even higher fare applies to passengers whose trips cross two fare zone boundaries, etc. Zonal fare increments should be in convenient amounts (especially under exact fare payment rules), such as $1.50 or $2.00.

The main advantage of a zone fare structure is that it provides a uniform base fare for a given small geographic area (such as within the CBD), as well as a correspondingly higher amount of revenue from longer trips.

A zonal fare system must be carefully planned. As much as practically possible, individual fare zones should be basically self-contained areas delineated by natural boundaries, such as rivers, major parks, arterial streets, etc. It is especially important that there not be a great number of short trips crossing zone boundaries, since these local trips are subject to a much higher charge per unit distance (km) than either intrazonal or long multizonal trips. Sometimes short trips across zone boundaries are exempted from two-zone fares to solve this problem by designation of an area on both sides of a zone boundary as a special zone.

In comparison with a flat fare, a zonal fare has the advantage of charging increasing amounts for longer trip lengths. This makes fares more equitable and acceptable to the users. Consequently, zonal fares usually result in both attraction of higher patronage and collection of greater revenue than flat fares. However, the zonal fare structure has inherently greater complexity in fare collection and control of fare payments. This is a particularly serious problem where fares are collected by fare boxes, such as on buses and other surface transit vehicles. An additional disadvantage may be the disproportionately high fare charges that apply to passengers traveling short distances across zone boundaries.

7.3.2.2 Sectional Fare. Another way of making fares related to travel distance is obtained by dividing transit lines into sections. The fares then increase with the number of sections traveled. Since sections are usually shorter and fare increments smaller than zone diameters, sectional fares are even more closely related to travel distances than zonal fares.

On the other hand, sectional fares are correspondingly more complicated to compute, collect, and control than zonal fares. Consequently, sectional fares often result in longer fare-collection times, require more personnel, and allow greater opportunity for evasion of full fare payment. Sectional fares can therefore be readily used on lines with light-to-moderate passenger volumes with manual fare collection. On heavily traveled lines, it is necessary to have an efficient fare-collection system to apply sectional fares. Two fare-collection methods meet this requirement: self-service and fully automated (see Sections 7.2.3 and 7.2.4).

To recapitulate, both zonal and sectional fares have amounts that vary with trip lengths, but they are computed differently: zonal fares are determined on the basis of geographically defined zones, while sectional fares are based on the distance traveled on a single transit line. Usually, zonal fares have fewer steps, which are larger than sectional fare steps.

7.3.3 Comparison of the Three Basic Fare Structures

The preceding description of the three basic types of fare structures shows that each has its advantages and shortcomings. In general, there are major differences between flat and graduated fares and minor differences between the two types of graduated fares, zonal and sectional. The main differences can be summarized as follows.

Structural definition of the three fare types is illustrated in Figure 7.2. Fare amounts for the three types are plotted as functions of trip distance in Figure 7.3. The diagram in Figure 7.3(a) shows total amounts: the flat fare is constant for any distance, i.e., it is a horizontal line on the diagram. Graduated fare structures

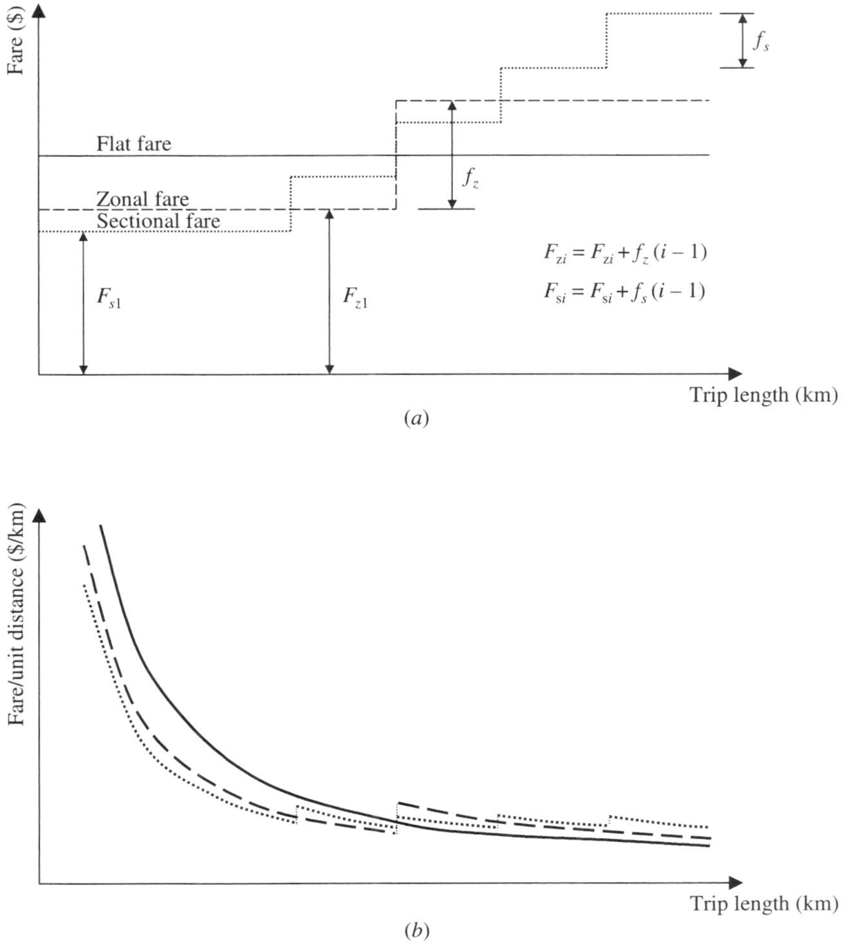

Figure 7.3 *Fare as a function of trip length under different fare structure:* (a) *fare as a function of trip length;* (b) *fare per unit of travel distance as a function of trip length.*

Table 7.3 Comparison of trade-offs among alternative fare-structure types and their optimal applications

Characteristics	*Type of Fares:*	*Flat*	*Zonal*	*Sectional*
Important characteristics	Equity	Poor	Good	Very good
	Passenger attraction	Good	Very good	Very good
	Revenue collected	Variable	Good	Very good
	Simplicity of collection	Excellent	Fair–good	Poor
	Simplicity of control	Excellent	Fair	Poor
	Simplicity for passengers	Excellent	Fair–good	Poor
Desired conditions	Line length	Short (<5 km)	Medium	Long
	Network type	Ubiquitous	Divisible in zones	Long lines
	Travel distance	Short	Variable	Variable

consist of a basic fare F_1 for a short distance and then increase in steps of an amount f for additional zones or sections. Fare for any distance can be expressed as:

$$F_i = F_1 + f\,(i - 1) \qquad (7.1)$$

where i is the number of zones or sections traveled. The two graduated fares, zonal and sectional, have the same form, except that the sectional fare usually has a greater number of shorter increments and smaller fare increments than the zonal fare.

The diagram in Figure 7.3(b) shows unit fares—the amounts of fares per unit of distance traveled in \$/km for the three fare types. The differences among them are obvious: while the flat fare per unit distance is very high for short trips and very low for long trips, the two graduated fares are more even. They also have a decreasing tendency with distance, but at a much lower rate than the flat fare: the decrease in unit fare depends on the length of fare sections and on the size of the increment f, i.e., on the size of the fare steps.

Because of these characteristics, flat fares are advantageous for use in small and medium cities, while in cities with large areas graduated fares prevail: their benefits of greater ridership and revenues outweigh the disadvantages of greater complexity. Exceptions are some large rapid transit systems, like New York City and Paris, where social factors and old traditions of simple and cheap transit that guarantee high mobility for citizens from all parts of the urban area remain a major popular asset in the livelihood of these cities and prevent introduction of graduated fares.

In summary, the flat fare has great advantages of extreme simplicity in understanding by passengers, in collection and control. In fare collection, the flat fare requires only one payment and control point, while graduated fares need two (see Table 7.1).

Being related to distance traveled, graduated fares do not have the problems of overcharging (and thus not attracting) many short riders and undercharging (and forgoing revenue) long-distance riders, which the flat fare has. Graduated fares thus have significant advantages of greater equity, stronger passenger attraction, and generation of more revenue.

For an easy comparison of the three fares, all their significant characteristics are presented in Table 7.3.

The choice between zonal and sectional fares depends greatly on local conditions, particularly the size and form of the urban area and its transit network. Zonal fares are based on areas and airline distances of trips, independently of the transit network form and individual lines. They are therefore simpler than sectional fares, which are based on individual lines, have more increments, and must account for transfers among lines.

7.3.4 Transfer Fares

Fares for trips using several lines may have a variety of forms. A separate fare may be charged for travel on each line, so that each transfer doubles the amount paid. Or transfer tickets may be issued for a certain incremental amount, usually considerably lower than the full fare.

Use of transfer fares varies among cities and is often related to fare structure. In some cities that are rather large but apply flat fares, there are no transfers either between lines or between modes. This partly compensates for the inability to collect more revenue for long than for short trips, but the disadvantage is that passenger trips that involve one transfer have twice-higher fares than those using a single line only. A more common practice in large cities in North America is to have a low or moderate transfer fare for transferring between lines and modes (Philadelphia).

Many metro systems, even very large ones such as in Paris, Moscow, and Mexico City, have a long tradition of flat fares and no transfer charges among all lines. In some cities (Boston), LRT lines are also included. However, some of them have no discounts for transfers to other modes (buses), so that trips using two modes require two fares. New York had this system of fares, but in 1997 it introduced free transfers between its subway network and buses, as well as among bus lines. This change resulted in very large passenger increases on both bus and rail networks.

Some extensive metro and nearly all regional rail systems use sectional fares, which have become easy to handle on the systems with automated fare-collection equipment (London, San Francisco BART, Tokyo). Some of these do not have any discounts for transfers among modes. In general, transfer discounts are somewhat less important on systems with graduated fares because their initial fares are lower than flat fares, so that changing modes does not involve doubling of fares.

Functionally, most advanced fare systems in medium and large cities are, however, those that utilize SSFC or smart cards to charge graduated fares on the basis of origin-destination points of passenger trips regardless of modes. Such fares, used in Hamburg, Munich, and many other European cities, thus minimize the inconvenience of transferring and offer practically a unified transit network with respect to fares, an important feature in competing with automobiles.

7.4 SPECIAL HIGHER AND LOWER FARES

A number of variations on the basic types of fare structure are possible and are used frequently in conjunction with the conventional basic fare systems described above. Such fares are used to achieve one or more of the following objectives:

- Attract additional passengers to increase mobility of population
- Stimulate use of transportation facilities and increase revenue, particularly during hours of low transit system utilization
- Favor a particular group of present or potential transit users, such as tourists or shoppers
- Achieve specified social or economic goals; for example, provide a minimum level of mobility to some population segments, such as students, families, or the elderly
- Change intermodal distribution of trips in favor of transit

The following sections describe the most common special fares and their characteristics.

The rationales for (+) and against (−) charging for transfers among lines are as follows:

+ Transferring imposes a certain cost on the system through the delay it causes.
+ Transfer charges represent a convenient way for getting additional revenue, particularly on the systems without capability to charge graduated fares.
− Transfer payments add to the inconvenience of transferring.
− It is not logical for passengers to pay extra for the trips that are not served by a single transit line.

7.4.1 Fares for High-Quality Services

Transit services that offer higher-quality travel than regular lines, such as express trains or buses, lines with seated passengers only, vehicles with special amenities, etc., are usually operated with higher fares. Their fares reflect the higher value passengers get from using these services, as well as the higher cost of operations (more expensive vehicles, lower utilization when seats are guaranteed, etc.).

Two types of services and fare levels are sometimes used to meet different needs and preferences of passengers. In intercity travel, it is common to have at least two classes: first and second (plus sleeping cars) on railways; first, business, and economy on airlines. Although far less common, two classes with different fares are also used on some transit systems. Regional rail systems in many cities (Cape Town, Munich, Paris) have first and second class. Buses and minibuses often offer different levels of service. In many developing countries, paratransit minibuses offer seats only and faster service at higher fares than transit buses. In other developing countries, buses may offer higher level-of-service than minibuses, which are overcrowded and often with low level of safety.

With increasing quality of transit services in industrialized countries, provision of two classes of service in regular urban transit has become rare. However, there are situations in many cities, particularly in developing countries, where provision of two or more classes of transit services is practically necessary. This diversification is used, for example, where extreme crowding conditions and major differences among population groups exist, so that certain groups of people do not want to use such services. They do everything to acquire a private vehicle and never use transit. But if a high-quality transit service is offered even at a much higher fare, these travelers may be attracted to it, thus reducing auto use and traffic congestion.

Special transit services are sometimes offered with different fares not only reflecting differences in speed of service, level of comfort, or attraction of different passenger groups, but also when they offer very specific services for special target groups of users. Examples of this arrangement are specialized services at different fare levels or packages with distinctly designed buses for a tour of historic areas (Philadelphia), the world-famous San Francisco Cable Cars with about double regular transit fare, or a historic tramway with dinner cruise through the city (Melbourne).

The separation of passenger groups for social, economic, and other reasons is sometimes also provided by regulations, rather than fare difference, due to special circumstances. For example, in Pakistani cities, women have physically fenced off sections in buses due to religious/social customs and standards. In the Mexico City Metro, severe crowding conditions during peak hours have led to reservation of three of the nine cars in each train for women only, to give them physical protection.

Actually, provision of a first class car with a higher fare would be very effective in the Mexico City Metro because it might attract considerable additional ridership. Presently that excellent system has an undeserved reputation for being unsafe, which leads many residents and city visitors to avoid using it. However, social/political pressures usually prevent such social segregation. This is very unfortunate because, as mentioned, when transit services are severely overcrowded or large segments of the population are unemployed and incidence of crime is high, the only way of getting most middle- and high-income population groups to use it would be by provision of a higher-quality service, even at considerably higher fares.

Different fare levels are sometimes charged on different transit modes in the same city. Most commonly, regional rail and express buses have substantially higher fares than urban metro, LRT, and bus services. The main reason for this difference is the fact that the modes with higher fares serve predominantly much longer (mostly commuter) trips, and they offer higher comfort. Moreover, their customers tend to be higher-income suburban residents, rather than all population categories typical for transit lines in central city. Thus, higher fares are justified by the longer trips and higher comfort level offered (purchased product has a higher value) and by the ability of users of these services to pay higher fares and avoid long and costly commuting by private cars.

7.4.2 Peak/Off-Peak and Commuter Fares

With simplicity and convenience of use as one of their basic principles for passenger attraction and efficient operation, transit systems have traditionally had the same fares at all times. Only transit services with long average trip lengths and those heavily oriented to commuters with very sharp peaking of usage, such as regional rail systems, usually have different fares for peak and for off-peak periods. The highest fare on the regional rail system may be as much as two to four times higher than regular transit fare per trip (the ratio of fares per kilometer of travel is usually lower).

The reasons for higher transit fares during peak than off-peak hours are the same as those used on many utilities (electricity) and facilities (hotels) and other transportation services (airlines, railways) with peaked usage:

- Cost of providing peak-period services is higher because they dictate system capacity, fleet size, number of employees, etc.
- Higher fares discourage passengers who have choice of time for travel from traveling during peaks.
- Elasticity of demand is lower during peaks because private automobiles face more severe highway congestion than during off-peak periods.

There are, however, also strong social, economic, and operational reasons for giving reduced fares to commuters:

- The vast majority of commuter trips are to work or school. Being essential for income earning and education, these trips are socially and economically more important than trips for other purposes, such as shopping or recreation.
- Lower commuter fares allow greater choice of work locations and opportunities for all employed persons, particularly for the lower-income groups.
- Work and school trips are less discretionary in time than other trips.
- Being regular, commuter trips allow the transit agency to plan easily and utilize evenly its system capacity (fleet, personnel, etc.).

Consequently, there are conflicting rationales for setting commuter fares: they should be higher because they occur during peaks involving the highest costs to transit agency, but they should be lower because of the above-listed, mostly social, reasons. A reasonable compromise among these requirements is achieved by setting fare levels from the highest to the lowest in descending order for the following traveler groups:

- Infrequent peak-hour passengers
- Regular peak-hour passengers (typically commuters to work)
- Infrequent off-peak passengers
- Regular off-peak passengers

The simplest way to implement these types of fares is achieved by offering commuters discounted weekly or monthly passes and selling multiple-ride discounted off-peak tickets for regular off-peak users.

7.4.3 Child, Family, and Student Fares

On most transit systems, children between 6 and 12 years of age or smaller than a certain height are given child fare, which is usually 50% of regular fare. The reduced fares for children are provided on the basis of the following rationales:

- Raising the young is the responsibility not only of parents, but of the entire society (for the same reason public schools are supported by tax money rather than by tuition paid by the parents).
- The only travel alternatives for children are walking and bicycling (or being chauffeured), so that they are true transit-dependents
- Attracting the young to transit creates a permanent habit for this mode of travel, securing long-term transit users.

Reduced family fares, usually made available for round trips made by several family members, are given for the same social reasons, as well as because travel by three to five family members paying full fares is so high that other modes, such as taxi, become cheaper. Family fare thus attracts trips that would otherwise go to other modes. Given during off-peak periods, family fares are economically attractive to transit agency because they bring fare revenue without involving any measurable marginal operating cost. Another consideration is that the family fares are socially positive because they encourage family travel.

Student fares in the form of discounted tokens, fares, or passes are also commonly provided in many cities for students attending elementary and high schools, as well as colleges in the area. The rationale for these discounted fares is:

- Like raising children, education is a responsibility that is shared by society, rather than by parents only.
- Students, by and large, have no personal earnings.
- Students, making many trips, are even more transit-dependent than children, and attracting them to transit is likely to create future transit riders.

A recent trend, under the conditions of inadequate financing for transit, has been for schools and colleges to purchase tokens or passes from the transit agency at regular rates and then sell them to their students at discounted rates. In this way, the subsidy for students is provided from general (school) taxes, i.e., by the entire society, rather than taken from transit funds and thus actually subsidized by transit users only. Many universities and colleges have recently organized special fare arrangements with transit agencies. The university gives all students transit passes as a part of their enrollment package. This is financed either by the university or by a special surcharge on tuition fees. The pass gives the students either unlimited free travel or deeply reduced transit fares. The transit agency assists the university to purchase the student passes by selling them at a greatly reduced rate, usually 50% of other passes. This is justified by the fact that since students on the average do not use transit as often as other transit riders, the agency will get revenue per actually made trip by the students not much less than from pass holders who are much more frequent transit users. Thus, this type of arrangement for student transit passes may in many situations be beneficial to all three parties. The university makes a certain investment but provides students very high mobility for travel throughout the city and reduces pressures from high traffic volumes and parking demand in the campus area; students may contribute through additional tuition, but they obtain enhanced mobility and convenience; and the transit agency obtains new guaranteed income and more riders, usually far outweighing any marginal costs for additional services it has to provide.

7.4.4 Fares for Senior Citizens and Disabled and Low-Income Persons

These three groups of persons warrant reduced or free fares for social reasons. As a basic element of living standard, most countries attempt to facilitate mobility by transit to all population groups, regardless of their financial and physical capabilities. Assistance in using transit is provided by various forms of discounted fares or no fares at all, usually during off-peak hours. At those times there is sufficient capacity in transit vehicles, so that marginal cost of providing service is very low. On the other hand, the retired and other groups of beneficiaries tend to have flexibility in travel times, so that they do not have to increase peak-hour loads. Some of the above-discussed rationales for transit discounts given to children and families also apply to these groups. Additional considerations and differences in methods of providing reduced fares apply to each one of them, and they will be briefly reviewed here.

In most countries, senior citizens are given special benefits, services, and discounts, ranging from special medical services programs to lower prices for theater tickets. The rationale for this policy is that retired persons generally have reduced and fixed incomes but deserve to enjoy certain basic elements of the living standard that their pensions may not be sufficient to cover. Urban transit, providing mobility as one of the basic services for all categories of population, is particularly important for senior citizens because of their decreasing ability to walk and drive automobiles. Moreover, it is socially desirable to shift their travel to transit wherever feasible, for the well-known benefits resulting from shifting travel from cars to transit.

The reasons for (+) and against (−) higher owl than regular fares are the following:

+ Operation of owl services involves a higher cost per passenger because of low passenger volumes, overtime pay for drivers, and extra security requirements.
+ Many users of owl services have no alternatives for travel, or the only alternative is taxi at a much higher cost.
− Most owl-period trips tend to be travel to work by low-income persons, who should not be charged what the traffic will bear, or the maximum possible fares.

Senior citizens usually enjoy drastically reduced and sometimes free fares on transit services, usually at all times except during peak-hour periods. The funding for this program, i.e., compensation of the expenses to the transit agency, is provided by the city or state governments from general or some special funds, such as sales, employer, tourist or other taxes, lottery revenues, etc.

Discounted or free fares for disabled persons has been one of the numerous measures introduced in many countries in recent years to provide this population group with basic mobility. The basis for these policies is well known and easy to understand: to assist their inclusion into mainstream society. The extensive measures that are legally required in most countries, such as access to stations and transit vehicles (see Section 6.5), are supplemented by lowered or eliminated fares as components of the same policy.

Low-income persons in some countries and cities get transit fares at a reduced rate from their employers as part of their benefits or, if unemployed, from the city or other governments through programs similar to the ones for food stamps and other financial assistance. Sometimes various other public or private organizations provide assistance to get transit fares.

7.4.5 Night, Group, Family, and Other Special Fares

Among other numerous special fares, several deserve some explanation.

Night (popularly known as "owl") fares are charged for travel on owl services, i.e., those during the night hours, such as 11:00 P.M. to 6:00 A.M. They may be regular fares or higher, often double.

Group fares are necessary to attract travel by schoolchildren, excursion and tourist groups, sports teams or fans, etc. If reduced fares are not given to these groups, they may charter a van or bus at a lower price.

Shopper fares apply during off-peak hours for trips in the directions of shopper travel. This type of discount is given in order to give people a greater choice of shopping locations and to attract drivers from their automobiles.

As mentioned, most of these discounted fares are designed to shift discretionary travel from the peaks to off-peaks, when considerable capacity in transit vehicles can be used with negligible marginal operating cost, so that they usually bring revenues higher than marginal cost of providing the service.

Fare-free zones are sometimes given within certain areas, particularly the CBD (see Section 7.2.1). This policy increases mobility and convenience of workers, shoppers, and residents and reduces pressures for parking capacity, which is usually destructive to the environment of intense, diversified activities.

Other special fares may include fares for sport or special events (lower or higher for special, convenient services), promotional fares, including free fares when a new transit line is opened, and others.

7.4.6 Use of Special Fares

Use of different types of fares can bring a number of positive results in the short as well as the long run:

attract new riders, generate more revenue, and achieve certain social goals that a single fare cannot. It must be understood, though, that extreme diversification may create situations where a transit system has 20–30 different fares, which its employees must handle and its users must be familiar with in order to utilize them. Such excessive complexity may be counterproductive, complicating operations and confusing passengers.

The fact that modern methods of fare collection are capable of handling many different types of fares should not lead to the conclusion that such complexity should be implemented. It must not be forgotten that simplicity and clarity of service are among the most important features in attracting transit riders, particularly the potential ones. Therefore, a compromise between the benefits from different fares and simplicity in understanding and handling, must be found for every set of local conditions

7.5 FARE LEVEL

Fare level is the amount of money charged for riding transit, regardless of the type of fare. It affects all objectives for fares listed in Section 7.1. Since it strongly influences transit use and the role of transit in the city, fare level is subject to regulation, and changes to it usually represent major public and political issues.

7.5.1 Influencing Factors

When transit was a private, self-supporting business, fare level depended mostly on the cost of providing services: in principle, fare revenues had to cover costs and provide a reasonable profit. Presently, with the competition of the private automobile, which has a different cost structure and imposes indirect social costs, policy factors often have a greater influence on fares than costs of providing transportation service. Thus, the major factors influencing fare level are the following:

- Governmental urban transportation policies, as reflected in the relative weights given to the objectives in setting fares, as defined in Section 7.1. The policy may vary from total neglect to strong support of transit.
- Methods of financing: basic reliance on fare revenues or on other sources of funds; the types of funding (general city's or state's budget, earmarked taxes, lottery receipts, etc.).
- Type and quality of transit service.
- Costs of providing transit service.
- Quantity and quality of competing transportation services, particularly the automobile, which influences demand elasticity.
- User habits and lifestyle in the served area.

Naturally, relative weights of these factors vary among cities with local political, physical, and social conditions.

7.5.2 The Range and Domains of Fare Level

A typical basic influence of fare level on transit ridership and total revenue from fares is shown schematically, as a conceptual diagram, in Figure 7.4. The diagram shows that with very low fares the maximum potential ridership is attracted and the revenue increases proportionately with the fare (the jump at the zero fare, shown by a dashed line, reflects the fact that free transit generates a certain extra ridership due to the special image and publicity given to such service). At a certain level, increasing fares result in significant decreases in ridership (demand elasticity is discussed in the next section), eventually reaching the low level of basically captive riders making nondiscretionary trips. The revenue increase tapers off with the ridership decrease, reaches a maximum level, and then begins to decrease.

Based on the diagram in Figure 7.4, the following fare levels (values or domains) can be defined:

1. Zero fare (or free transit): transit service is free, maximum ridership is achieved, but there are no fare revenues.
2. Low fares are the fares up to the level at which ridership losses are very small; revenue increases nearly linearly with fare.

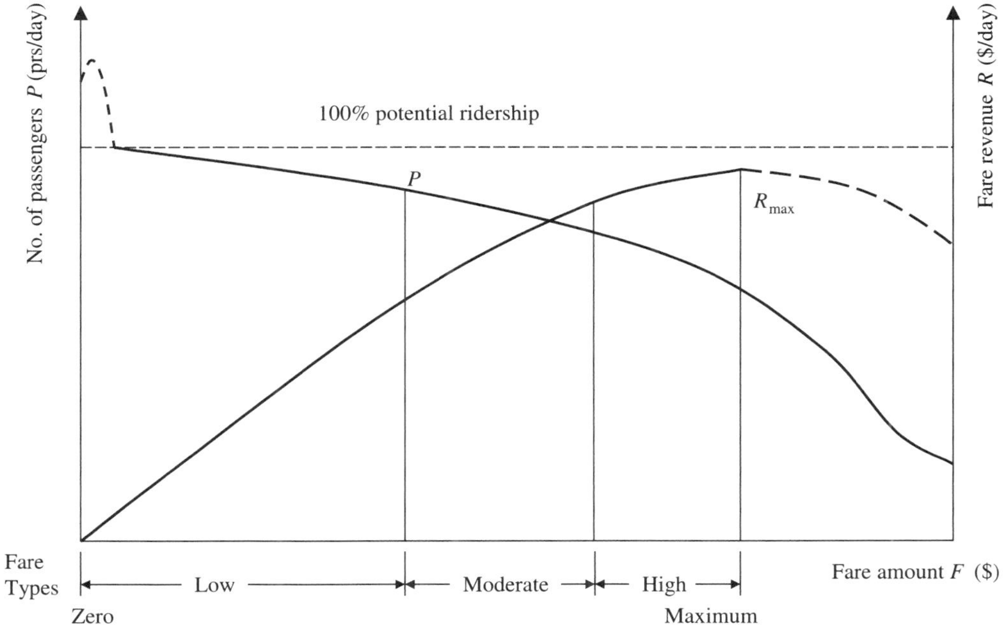

Figure 7.4 *Conceptual diagram of ridership and fare revenue as functions of fare level.*

3. Moderate fares are those that result in moderate ridership losses and still have significant increases of revenues as the fares increase.
4. High fares refer to those in the range where ridership loss is significant while increase in revenue with growing fares is diminishing.
5. Maximum fare is at the point of maximum revenue, where a further increase would cause such ridership losses that the total revenue would actually decrease.

Since it is obviously counterproductive to increase fares above the maximum fare, the range of fares is actually from the zero to the maximum fare, as the diagram in Figure 7.4 shows.

Another concept is the break-even fare, which brings the revenue amount equal to the operating cost of the transit service. As such, this fare cannot be shown on this diagram: it may be anywhere. In cities with high transit usage (e.g., Hong Kong), break-even fare may be in the range of low or moderate fares; in cities with high and subsidized automobile use (most U.S. cities), break-even fare may actually not exist, because even the maximum revenue is lower than operating costs.

Impacts of fare increases at different fare levels are illustrated in Figure 7.5, which shows the two curves, ridership and revenue. Case I is in the domain of low fares; the impact of an increase from fare F_I to F'_I is a small ridership loss ΔP_I and a significant revenue increase ΔR_I. In case II, fare increase is from a low to moderate fare. The ridership loss is somewhat greater and the revenue gain smaller than in case I. Finally, case III is an increase of fare by the same amount from a moderate to a high fare. The ridership loss is very large, while revenue gain is very small.

Naturally, this is a very simplified model of real world situations, assuming the *ceteris paribus* situation with respect to offered service, travel demand, marketing of services, etc. The relationships of values are conceptual rather than quantitatively exact.

7.5.3 Fare Elasticity of Transit Demand

The concept of elasticity of demand with respect to price is used to measure quantitatively the change in

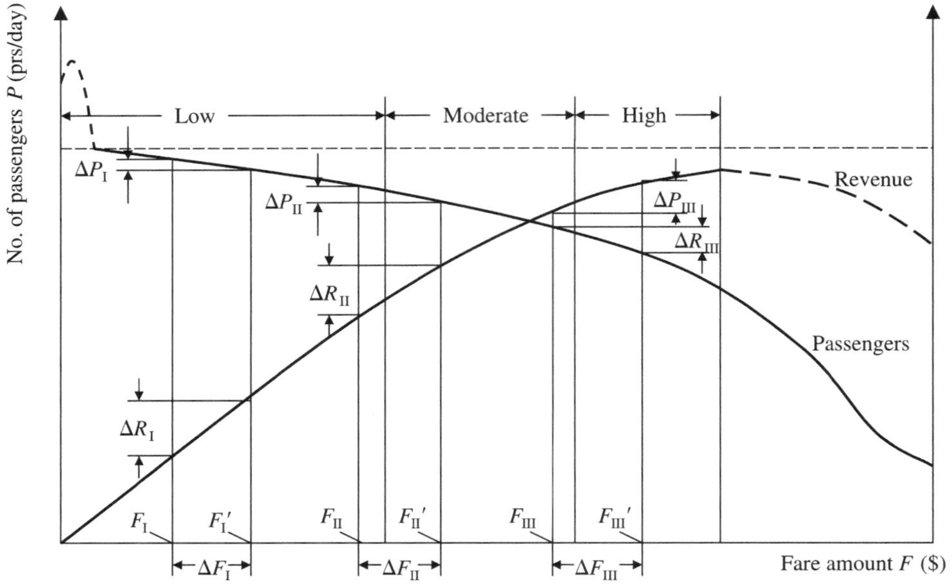

Figure 7.5 *Impact of fare increase on ridership and revenues in different fare level domains (conceptual).*

the number of passengers as a function of the fare level, described conceptually above. Elasticity is defined as the percent change in demand divided by the percent change in the price. This ratio of percentages avoids the problem of dimensions: instead of expressing the change in demand for some units (e.g., pairs of shoes or tickets for a show) due to price increase in the reduced number of those units sold per $1 increase in price, elasticity is dimensionless.

Since an increase in price tends to cause a **decrease** in demand, elasticities usually have a mathematically negative value. The greater the absolute amount of the elasticity value, the more the demand is sensitive to the price. Thus, the values define the following domains of demand:

$E > 1$ (or $E < -1$)	Elastic
$E = 1$ (or $E = -1$)	Unit elastic
$E < 1$ (or $E > -1$)	Inelastic

Applied to transit, elasticity of demand E is measured as the percent change of transit ridership P as a function of the percent change in fare F:

$$E = \frac{\Delta P/P}{\Delta F/F} = \frac{\partial P}{\partial F} \times \frac{F}{P} \tag{7.2}$$

This is defined as *point elasticity* because it expresses sensitivity of ridership to the fare at one point of their relationship. Expressed as the percent of ridership change for 1% of fare change, this concept is sometimes known as *shrinkage ratio* or, in the case of fare reduction, *expansion ratio*. In practice, it is common to use *arc elasticity,* which is measured between two points, i.e., two levels of fare (typically before and after a fare change). For example, suppose that a fare increase from $1.00 to $1.20 (20.0%) has caused a reduction of daily passengers from 10,000 to 9600 (−4.0%), Eq. (7.2) gives the following elasticity:

$$E = \frac{-400/10{,}000}{0.20/1.00} = \frac{-0.04}{0.20} = -0.20$$

Since elasticity represents the percent ridership change for 1% of change in transit fare, arc elasticity in this example shows that 1% of fare increase would cause a ridership decrease of 0.20%.

Two observations about transit demand elasticity to fares are important. First, their values are negative, indicating that the two changes are, as is common with demand/price relationship, in the opposite directions: an increase in fare results in a decrease in ridership (and vice versa). And second, absolute values of elasticities are usually less than 1, meaning that the demand is *inelastic,* i.e., the ridership changes by a smaller percentage than the fare change that caused it.

Many transit systems compute elasticities for each fare change they make through the years. For example, the fare increases on the regional rail system in Philadelphia during the 1980s have shown elasticities between −0.36 and −0.94. The wide range of these elasticities reflects influences of sizes of individual fare increases, quality of offered service and its competition, economic conditions in the area and other influencing factors.

In the simplest case with flat fares, total revenue R_a collected from a given number of passengers P_a with fares being F_a is simply computed as:

$$R_a = P_a \cdot F_a \tag{7.3}$$

Fare elasticities are very useful for estimation of passenger losses or gains due to fare increase or decrease, respectively. There have been many studies of industry-wide average elasticities by travel group categories, such as presented for a hypothetical system in Table 7.4. However, these numbers can be used only in a general manner, for *relative elasticities,* rather than for actual values. For example, the numbers in the table show that the demand is considerably more elastic for short-distance than for long-distance trips and for off-peak than for peak travel; commuter demand is also less elastic than shopper and student demand.

Exact values should not be used from general literature because transit demand varies greatly not only with local conditions, but also with time and changes in attitudes. It depends on such diverse factors as quality of transit service, pricing of parking, and general attitude of the population toward the environment and city in general. Finally, theoretical studies in this area often imply an unrealistic degree of precision: it is questionable how reliably the elasticities of individual groups of users can be determined.

For a reliable estimation of fare elasticity on a given transit system, the agency should measure the impacts of its own fare changes. This can be done by comparisons of before-and-after ridership when fares are changed. Other techniques include aggregated time-series and desegregated cross-sectional logit models. For further analysis of transit demand, see Ferreri (1979) and Kemp (1980)].

Table 7.4 Fare elasticities by ridership groups in a hypothetical system

Fare Structure Element	*Average*	*Commuters*	*Shoppers*	*Elderly*	*Students*
Cash fares	−0.35	−0.18	−0.45	−0.44	−0.25
Prepaid fares					
• Tickets	−0.33	−0.17	−0.43	−0.42	−0.33
• Passes	−0.26	−0.14	N.A.	−0.33	−0.27
Graduated fares					
• Short distance	−0.50	−0.25	−0.62	−0.58	−0.55
• Long distance	−0.23	−0.12	−0.30	−0.25	−0.23
Time period					
• Peak	−0.21	−0.16	−0.25	−0.25	−0.20
• Off-peak	−0.48	−0.24	−0.62	−0.55	−0.52

Source: Based on Ferreri, 1979.

7.5.4 Characteristics of Different Fare Level Domains

There have been numerous discussions, developments, and experiments with fare levels in many countries during recent decades. The experiences can be summarized as follows.

7.5.4.1 Zero Fare. From time to time there have been proposals to eliminate fares entirely in order to maximize ridership. Transit would be financed from general government budget or some dedicated revenue source, such as earmarked sales tax, lottery, etc. There are, however, several reasons against the use of zero fare:

- In most cases, new transit riders are attracted to transit by higher level of service, rather than by lower fares; thus, the most effective use of additional funds to increase ridership would be to improve service, rather than to eliminate fares.
- It is equitable for the transit rider to pay at least some portion of the transit costs, and the general public accepts this concept.
- Zero fare leads to some abuse of the transit-riding privilege and operational problems: unnecessary trip making, possible increase in vandalism, more difficult enforcement of standards of public behavior on vehicles, etc.
- Excessive discretionary travel may cause overcrowding on short sections of lines that may be very costly to serve.
- Fare elimination can result in lower fiscal responsibility by management during periods of ample funding and service deterioration when funding is reduced.

Several demonstration projects of zero fare transit systems were conducted during the 1970s. One was in Denver, Colorado, where fares were eliminated during off-peak hours, i.e., most of the day. This demonstration showed that ridership during the periods with zero fare did increase significantly. Yet the considerations listed above prevailed and the zero fare demonstrations did not lead to permanent introduction of this concept.

Free transit services have been successful, however, in a number of special applications, such as:

- Shuttle services or all transit services within geographically limited, high-activity areas, such as CBDs, university campuses, airports, etc. For example, free transit shuttles operate in the CBDs of Denver (low-floor bus distributors from a large bus terminal); Buffalo, has free service on its LRT section through the CBD; and the entire LRT line opened in 2003 in Tacoma, Washington, operates as a free downtown shuttle service.
- Feeders to major transit lines or terminals, such as suburban access to rail transit and collection-distribution for some air terminals and for large parking areas.
- Entire transit networks for special events. For example, the Washington Metro has been rented by organizers of major events and its services given free to the public on a certain day of national meetings or celebrations.
- Promotional occasions, typically openings of new rail lines.

7.5.4.2 Low and Moderate Fares. Transit statistics show that the highest riding habit in U.S. cities is found in New York, San Francisco, and New Orleans, cities which have traditionally had low or moderate fares. Although the extensive networks and frequent services in these cities partially explain this fact, consistent public policies that enable maintenance of low fares have been a critical contributing factor.

Over the years, several cities, large and small (e.g., Atlanta, Cincinnati, London, and Los Angeles), have introduced temporary or permanent major fare reductions (often as part of local referenda on tax proposals to fund transit improvements). Ridership increases have been quite significant, considerably higher than some theoretical studies of demand elasticities had predicted. For example, a major fare reduction in Atlanta in 1970 (from 40 to 15 cents) generated 25% greater ridership. When services were increased to handle the larger volumes, the better services attracted an additional 8% of new riders.

Most of these fare reductions did not break even financially (i.e., additional fares revenues from increased ridership did not cover revenue reductions due to lower fares and additional operating costs for expanded services, or did not have the same cost recovery as the basic service). Yet most of these changes were evaluated as highly successful because of the improved urban mobility, reduced traffic congestion, and positive impacts on the social and economic vitality of the city.

In some situations, political pressures impose extremely low fares; for example, the fare on the Mexico City Metro was for many years only 1 peso, hardly exceeding the cost of the issued magnetic ticket. Such an extremely low fare has led practically to the same problems as zero fare. When applied to a large transit system, it eventually leads to a financial crisis that forces an inordinate jump in fares (in Mexico City, from 1 to 20, then to 50 pesos, etc.). Such a jump may be more detrimental to ridership than having low-to-moderate fares with gradual increases that keep up with inflation.

7.5.4.3 High and Maximum Fares. Opposite to the policies of promoting transit use are those considering fare revenue maximization to be the dominant objective in providing transit services. Based on the argument that the free market should apply in providing transit as much as possible, these policies provide no or very limited public subsidies to transit, ignoring the fact that the very high (although largely hidden) subsidies to its main competitor, the private automobile, make any consideration of the free market in urban transportation illusory. While application of high fares tends to result in somewhat tighter financial management of transit agencies, it has a distinctly negative effect on transit ridership.

When the maximum fares are reached, the situation becomes directly counterproductive: both passengers and revenues are reduced. Actually, even fares close to the maximum are usually damaging because small increases in revenue cause significant negative nonmonetary impacts: the sizable passenger losses tend to trigger the vicious circles of passenger losses illustrated in Figure 7.1.

7.5.5 Selection of Fare Level

Fare levels depend on a variety of factors, ranging from the cost of providing the service to the policy toward transit in a given metropolitan area. Considering functional aspects of transit system operation, passenger attraction, and financial stability, the experience from recent decades leads to the following general conclusions.

Zero fare is neither practical nor desirable for application on entire transit systems. It can be effectively used for services in smaller, high-activity areas where transit should be the dominant mode and automobiles eliminated as much as possible (CBDs, campuses, fairgrounds), when financing of transit services is provided from other than transit agency funds.

Low fares are very effective in maintaining high transit ridership and thus decreasing the problems of auto traffic congestion. They can represent an important element of a policy fostering a balanced urban transportation system. However, low fares must be backed up by a permanent source of funding for transit other than fares.

Moderate fares represent a reasonable and equitable pricing of transit and provide better financial stability for transit agency than low fares. However, they do deter a certain amount of transit travel and result in more street congestion from auto travel, unless there is a rational pricing policy for the use of automobiles as well. Above all, parking pricing policy may be such that high transit riding is maintained and the need for subsidies is decreased.

High fares and maximum fares are usually applied when governmental bodies consider minimum transit subsidy a more important goal than high transit usage. Such a policy tends to neglect most of the indirect and long-term negative impacts of transit neglect, such as congestion and deterioration of urban life and environment.

Fare increases should generally follow inflation and reflect the pricing of competing modes. Regular small or moderate increases are much more desirable than infrequent but large fare increases.

EXERCISES

7.1. List and briefly discuss the major objectives in setting transit fare level and structure.

- **a.** Do most of them coincide, or do some of them mutually conflict? Give examples.
- **b.** Which objectives or constraints usually have the dominant role in selecting level and structure of fares?

7.2. In setting up transit fare level, a number of goals should be considered. Two of these goals usually dominate, and the trade-off between them is the major subject of policy decisions on fare level. What are these two goals and what is their mutual relationship? Discuss major factors that influence the relevant importance of each of these goals.

7.3. Flat fares and graduated fares.

- **a.** Define both types of fares and describe their basic characteristics.
- **b.** Compare the two types of fares.
- **c.** Define two different sets of conditions that would clearly make each of these fares preferable to the other.

7.4. Opinions about the level of transit fares vary between two extremes:

- Transit should be free.
- Fares should cover all costs of transit operation.

Discuss both of these views and express your stand on this issue.

7.5. How does SSFC affect the selection of the fare structure, specifically flat versus graduated?

7.6. Describe characteristics of magnetic fare cards as compared to cash and token fares. What are their advantages and disadvantages?

7.7. What are the advantages and disadvantages of contactless smart cards as compared to magnetic strip cards?

7.8. A transit agency in a large city (e.g., Detroit, Baltimore) wants to make a comprehensive comparative analysis of flat versus graduated fares (e.g., \$2.00 for any distance versus \$1.50 for the central zone with \$0.50 increments for zones outside the center). List items you would include in the analysis. Specify the data you would need to have and list exactly the advantages/disadvantages of a graduated as compared to a flat fare structure.

7.9. You are in charge of preparing a fare-collection method for a new 20-km-long LRT line, and you are proposing the use of the self-service fare-collection method. Present, in a condensed form:

- **a.** A description of how it works
- **b.** Its advantages and disadvantages in comparison with the conventional fare-collection method
- **c.** Why SSFC is particularly suited to LRT systems
- **d.** Major steps (preparations) needed for its introduction

7.10. (Multiple choice) The main reason flat fares are used in most U. S. cities is that:

- **a.** They maximize revenues.
- **b.** Flat fares are easier to collect and control.

c. They are equitable.

d. Graduated fares would attract fewer passengers than flat fares.

7.11. (Multiple choice) Free transit would:

a. Be more equitable than transit with either flat or graduated fares

b. Help transit agencies, which would receive financing from government funds, to operate as financially healthier organizations than they do now

c. Be the best way to attract auto drivers to transit

d. None of the above

7.12. Transit passes with unlimited use for travel in a time period: weekly, monthly, or annual.

a. What are their advantages and disadvantages for users (passengers) and for the transit operator?

b. Do they have greater importance in transit systems with self-service fare collection than with conventional collection, where every passenger is checked?

c. How do they affect transit system usage?

d. Do the passes have an impact on the competitiveness of transit with car travel in cities?

7.13. Compare unlimited-use transit passes, such as a monthly pass, with debit transit tickets, which are purchased for a certain amount and then have the value of each fare subtracted. Discuss their method and convenience of use and their impact on transit revenue and on the propensity to make transit trips.

7.14. Explain how SSFC application varies among modes. How easy or complicated it isto use it on three major modes (buses on streets, LRT, and metro systems) due to the differences in the fare-control methods (by the driver, at station entrances/exits, and with spot-check control only).

7.15. Discuss the impacts of fare-collection methods on transit system operation and usage:

a. Does the introduction of SSFC influence the ability to use different fare structures? If it does, how?

b. How is the use of different types of vehicles (such as articulated vehicles and LRT trains) in semirapid transit systems affected by the introduction of SSFC?

c. Does the existence of time-passes affect the ability of employers or governmental agencies to stimulate transit use by contributing to employees or residents financially to lower user cost?

7.16. A flat fare requires checking of passengers only once, usually at entry to vehicle or station. This is not adequate for controlling graduated fares. Describe different methods that can be used to control graduated fares.

7.17. Spot-check control of passengers aboard vehicles or in stations.

- What is its basic purpose?
- What are operational difficulties with it?
- Does it have a purpose on systems with flat fares controlled at entry?
- On what types of transit modes is it never needed?

REFERENCES

Ferreri, M. G. 1979. Transit price elasticities revisited, in *Transit Journal,* Summer, 65–71.

Kemp, M. A. 1980. Planning for fare changes: A guide to interpreting and using fare elasticity information for transit planners. Working Paper 142805, Washington DC: Urban Institute.

Padron, M., and R. Stanger. 1976. No barrier fare collection. *TR Record 614.*

FURTHER READING

Deibel, L., S. Stern, S, L. R. Strickland, and J. Sulek. 1979–1980. *Self-Service Fare Collection.* Report No. UMTA-VA-06-0049-79-2–7 and Report No. UMTA-VA-06-0049-80-1. Washington, DC: UMTA, U.S. DOT.

Foote, P. J., and D. G. Stuart. 2000. Impacts of transit fare policy initiatives under an automated fare system. *Transportation Quarterly.* 54(3):51–66.

Hugonnard, J. C. 1988. *Arriving at a Regional Integrated Fare Policy and Structure: A European Experience.* Paris: RATP.

Leopold, H. 1983. Grundsätzliche und aktuelle Fragen der Tarifgestaltung im OPNV. *Zeitschrift für Verkehrswissenschaft* 2.

Lobron, R. 2003. Developing a recommended standard for automated fare collection for transit: scoping study—regional fare management programs. *TCRP Research Results Digest* 57.

Transportation Research Board (TRB). 2002. *A Toolkit for Self-Service, Barrier-Free Fare Collection.* TCRP 80. Washington, DC: Transportation Research Board.

UITP. 2000. *5th International Conference on Automatic Fare Collection.* (CD-ROM.), Brussels: UITP.

8

FINANCING OF TRANSIT

Transit has a vital role and a unique position in urbanized areas. Transit service contributes to a region's living standard by providing personal mobility. On a macro level, public transport stimulates efficient economic activities, promotes equitable social conditions, and creates human-oriented urban environments. Transit is therefore an important subject of public policy and an effective tool for achieving public goals.

To achieve its goals, transit must compete with the private car and other modes that provide different types of transportation services in cities. These modes vary greatly in financing structures, including user costs and payments, as well as in government funding. Transit financing itself is very complex and involves all three levels of government: local (city and region), state or province, and federal or national, as well as private investments and fare revenues.

The motivation for and forms of government funding of transportation in general, as well as specific sources and levels of transit funding, are described in this chapter.

8.1 PRINCIPLES AND TRENDS IN FINANCING TRANSIT SERVICES

In cities, government and public agencies have a major role or full responsibility in providing public facilities, such as streets, pedestrian plazas, parks, and historic areas. Government agencies also finance public services, which include safety and security, public health, clean and attractive environments, and livable and sustainable cities. Governments also participate in financing most transportation systems.

8.1.1 Government Participation in Transportation Financing

There are several reasons for public ownership and participation of governments at different levels in financing many transportation systems. Major factors are presented here.

1. By their physical nature and function, some transportation systems are public facilities. For example, highways, rivers, and coastal areas are public facilities, open to users and the entire population.
2. Most transportation systems require centralized or coordinated control of some system facilities and operations. Thus, even though most highway vehicles and aircraft are privately owned and operated, highways, streets, airways, and air terminals are owned, controlled, and operated by government agencies.
3. All transportation systems—highway, railway, and air and water transport, particularly in carrying passengers—are essential for a nation's economy and living standard. Transportation investments stimulate development and population mobility, so that governments often make financing of transportation part of their economic and social policies. Transportation systems are therefore not purely commercial functions, but a combination of commercial enterprise and social service.
4. Because transportation systems bring benefits to their users as well as to many nonusers and the general public, it is logical and equitable that transportation costs be paid not only by travelers—

the system users—but also by other beneficiaries through taxation and other means.

5. Many of the benefits are qualitative and cannot be easily allocated to different parties.
6. Collecting payments directly from users is in many cases operationally infeasible. For example, the use of streets cannot be charged directly from any users, whether in vehicles, on bicycles, or on foot, to quote an extreme.

The important role of governments in the development of different modes of transportation has been demonstrated throughout history. The Roman Empire built a large network of roads; England, France, and the United States built many canals in the first half of the nineteenth century, railways since their invention in 1825, and air transportation throughout the twentieth century.

Government concern and involvement with urban public transportation started with its beginnings in the nineteenth century. Transit systems were introduced as private companies, and in some cases as municipal agencies. As providers of public services, both privately and publicly owned transit companies were regulated to various degrees by government bodies (see Chapter 9).

Although high ridership in those times provided funds to run profitable operations, there were cases of government subsidies to a transit company with an explicit policy goal: providing greater mobility for population. Thus, already in 1883, the British Parliament allocated an annual subsidy to railways to maintain low fares on suburban services so that workers could afford to live outside the overcrowded center city tenement housing (Weber, 1899).

In cases of major capital investment, such as construction of metro systems, it was common from their beginnings for municipal governments to participate by providing partial or full funding. Many cities considered the provision of streets a government function. Subway tunnels represented another mode, but with the same function as streets: to transport people. Therefore, government should finance construction of metro lines in the same way that it financed streets.

With the increasing competition of private automobiles and other transport modes since the 1930s and 1940s, and the transition from private to public ownership of transit agencies, the role of governments in financing transit has increased significantly.

It is therefore understood and accepted in virtually all countries that the highway, railway, air, and water transport systems, particularly their passenger components, are financed by government funds in various forms and to various degrees in order to achieve or promote society's goals.

In the United States, one still finds widespread misunderstanding of this issue. There is a belief that highways are fully paid by their users through fuel and other taxes (and a few of the highways through tolls), that airlines are independent commercial enterprises, and that transit and the national passenger railroad system are "failing" because they require federal government subsidies. Actually, the fact is that no passenger transportation system, including highways and air transport, covers all of its costs without government funds (see Section 8.1.2).

This distorted picture, which particularly hurts the national railroad passenger system and urban transit, has been created by the policies of the U.S. Congress, which have heavily favored highway and air transportation systems and seriously neglected the passenger railroad and transit modes. While the highway and air transport systems receive billions of dollars for investments and operations every year, Amtrak has never received the financing it needs even for the skeletal services it has been providing. Transit assistance started much later than highway financing, when transit had already seriously deteriorated. It became substantial during the 1970s, but then declined during the 1980s. In the early 2000s, there are pressures to reduce federal financing of transit capital investments to 50%, while highways continue to receive federal shares of 80–90%.

The bias against the national passenger railroad system is particularly remarkable. While the funding of highway construction and its maintenance is referred to as "investment," the same type of federal funding for Amtrak is called "subsidies." The great national need for an efficient passenger railroad system is continuously ignored, while the debates are focused on Amtrak funding for the following year. Thus, the national focus is on the means, rather than on the goal of providing a modern passenger railroad system the country needs. If such political processes had been used for other transport modes, the country would have never developed either the Interstate Highway System or the outstanding national and international air transport system. Both were first defined as national goals, and then the means for their financing were developed, adopted by Congress as laws, and then consistently implemented over many decades.

8.1.2 Funding of Different Transportation Modes in the United States

Government expenditures for transportation are very substantial. Their amounts for the five major modes and breakdowns into federal and state/local, for selected years between 1980 and 1999, are given in Table 8.1. The distribution of total expenditures by mode can be seen in Figure 8.1. Table 8.2 presents a summary of total revenues and expenditures by all three levels of government since 1985. Finally, Figure 8.2 shows the percentages of government expenditures covered by the revenues received. These tables and figures lead to the following observations:

- Federal government expenditures are directed heavily to highways and air transportation, followed by transit and water transportation. Expenditure for the rail mode can be described as infinitesimally small, showing a serious neglect of this mode.
- Expenditures have generally increased during the shown period, but federal expenditures for transit were lower than for other modes, except for rail, where federal contributions decreased drastically. The dominance of highway and air modes intensified during that period.
- State and local expenditures for all modes have increased faster than federal expenditures during the shown period.
- As shown in Figure 8.2, transit has the lowest coverage of the government expenditures among the modes. This is explained by the nature of this mode and the large positive externalities it involves, as discussed in the following section. The diagram shows that none of the modes brings more revenue than governments invest in it, i.e., all modes are subsidized with respect to the three government levels. It should be mentioned that in most other industrialized countries the highway system is not subsidized. Government receipts from vehicle, gasoline, and other highway-related taxes exceed total government expenditures for highway systems (Pucher, 1995).

8.1.3 Principles in Allocating Transportation Funds

Most transportation services have properties that dictate a balance of private and public financing. Transportation modes that provide mostly commercial services are financed mostly by private funds, i.e., by direct payments by the users. Transportation systems providing services that are socially important warrant participation of public financing. This section reviews the balance of different methods and sources of funding for several transportation categories.

Policy principles and financing methods established through long economic development and public policy evolution in many countries have the following general patterns.

8.1.3.1 Passenger versus Freight Transport. The government role in financing is more common for passenger than for freight transportation. The reason for this is that freight transportation is mostly a commercial enterprise, part of a country's production and economy. Passenger transport, while also an economic system, has a strong interaction with social aspects and environmental externalities.

8.1.3.2 Intercity versus Urban Transportation. Long-distance, interurban transportation can be operated much more on the basis of commercial principles than urban transportation. In urbanized areas, car use is subsidized through the free use of streets, often underpriced or free parking, etc., while transit is subsidized through government contributions to operating costs. The role of externalities, such as positive impacts on land use or on mobility of low-income population,

Table 8.1 Government (federal, state, and local) expenditures on transportation by mode (current $ millions)

	1985	*1986*	*1987*	*1988*	*1989*	*1990*	*1991*	*1992*	*1993*	*1994*	*1995*	*1996*	*1997*	*1998*	*1999*
Highway	46,604	50,435	54,032	57,361	59,854	62,563	66,526	68,954	69,991	74,531	79,309	81,550	84,212	89,454	95,494
Transit	16,333	17,586	19,321	16,827	17,594	19,261	20,857	22,322	21,279	25,088	26,162	26,346	26,875	28,108	29,027
Rail	1072	917	817	586	606	541	783	906	819	845	1043	1015	1148	1099	565
Air	7903	8749	9540	10,422	11,240	12,568	13,974	15,916	17,408	17,941	16,960	17,273	18,776	19,593	21,789
Water	5124	5974	5601	5245	5289	5480	5847	6167	6593	7046	6628	6775	6996	7137	7682
Total	77,036	83,661	89,311	90,441	94,583	100,413	107,987	114,265	116,090	125,451	130,102	132,959	138,007	145,391	154,557

Source: Bureau of Transportation Statistics, Government Transportation Statistics 2001.
Note: totals exclude pipeline expenditures and general support.

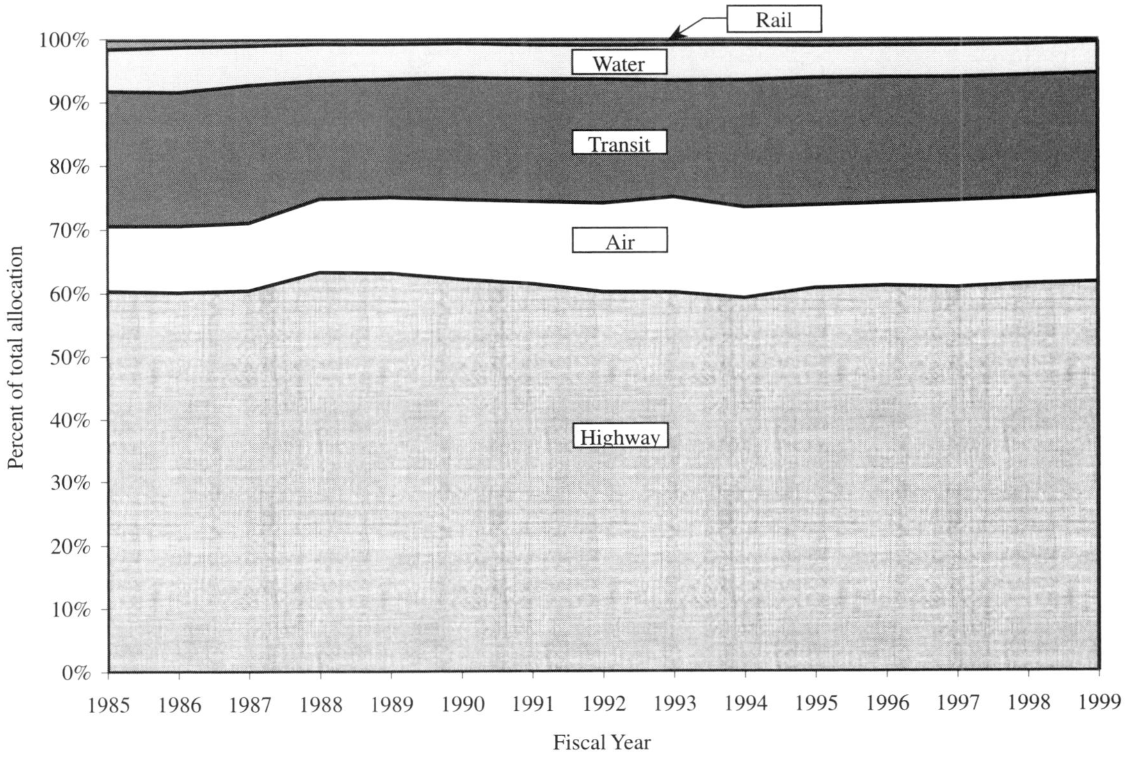

Figure 8.1 *Summary of federal, state, and local government transportation expenditures by mode.*

Table 8.2 Government (federal, state, and local) expenditures on transportation in 1999 (current $ millions)

	Absolute Expenditures	*Percent of Total*
Highway	95,494	61.78
Transit	29,027	18.78
Rail	565	0.37
Air	21,789	14.10
Water	7682	4.97
Total	154,557	100.00

Source: Bureau of Transportation Statistics, Government Transportation Statistics 2001.
Note: totals exclude pipeline expenditures and general support.

and negative impacts of noise and air pollution, influence the decisions about the roles of user charges versus public funds.

Government participation in financing of urban transportation is common with all transport modes in different situations. Many urban streets and freeways require very high investments, far exceeding the revenues that fuel and other user taxes on such facilities can ever generate, so that large government funds are needed. Most transit systems, particularly high-performance ones with complex infrastructure, also depend mostly or entirely on public funding.

8.1.3.3 Operating versus Investment Costs. Revenues from transportation system users are generally applied first to operating/maintenance costs, and then toward capital investments. Government funds are used first for investments, then, when required, also for operations and maintenance. However, structures of contributions among the three government levels have somewhat different patterns.

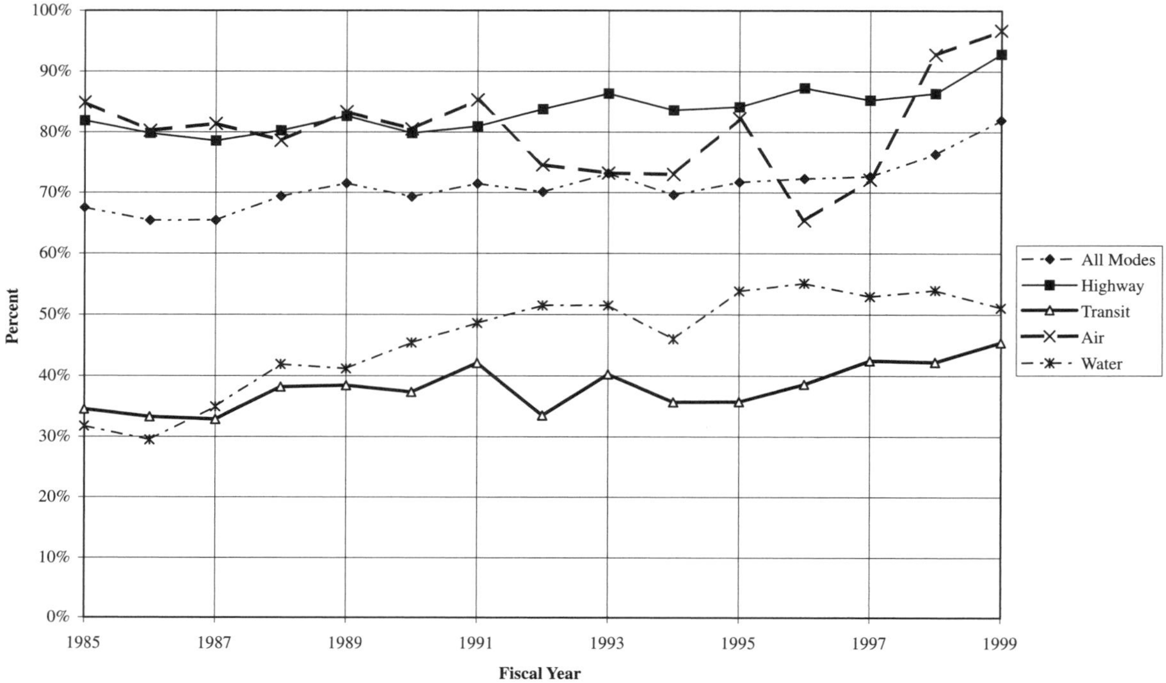

Figure 8.2 *Ratio of government revenues to expenditures by transportation mode.*

8.1.4 Sources of Funds for Public Services

The diverse private and public sources of funds used for financing transit and other public systems and services can be grouped into three general categories. They are defined here, sequenced from direct payments by users to funding from general government budgets.

8.1.4.1 Revenues from Direct Payments by Users. The basic funds for public system financing are direct payments by users in the form of fares, transit passes, rates, tickets, or admission charges. The revenue from these sources pays for operations and investment of many public systems, such as sporting events, electric and water supply systems, taxi, and many other services. In other cases, direct revenues from users cover only operating costs. This is often the case with stadia, museums, and some transit services. Finally, user revenues sometimes cover only a portion of operating costs, as is the case with most transit systems and many civic and cultural events.

8.1.4.2 Special Government Funds. When a tax is placed on users of a system and the collected revenue is used for financing construction or operation of that system, such a charge is called a *user tax*. Examples are the landing fees for aircraft, which are applied to operation or expansion of an airport, and surcharges on passenger fares for the financing of airport access facilities. Somewhat broader based is the tax on gasoline that is used for construction of highways or other transportation facilities.

Governments at all levels (local, state, and federal) sometimes use funds obtained from specially designated sources to finance a given public service. A good example of such earmarked funding is the allocation of a portion of state lottery revenues toward reduced transit fares for senior citizens. In some cases, earmarked taxes are collected from the indirect beneficiaries of the financed system. This is the case when a special tax, often called a benefit-sharing tax, is charged on residents or businesses along a highway or around a metro station because these owners benefit directly or indirectly from the activities such facilities generate. These revenues are then used for investment or operating costs of the respective transportation facility.

8.1.4.3 General Government Budget Funds. Many public systems, facilities, and services are made available to the public without any direct charges (tickets, admission fees, etc.). Their construction and maintenance are financed entirely by the government from its regular budget.

As discussed above, transportation systems consist of different compositions of public and private components and organizational domains. The highway system consists of public infrastructure (streets and highways) and mostly private vehicles. Air transportation is basically the same: public infrastructure (airports, airways, and air traffic control) and mostly private aircraft. Railways in most countries have been owned and operated mainly by national public agencies, but in the United States they have been private, with the exception of Amtrak, the national passenger railroad system. In recent years, railways in many countries have been reorganized in many different ways to improve their economic efficiency, introducing innovative combinations and relationships among private, quasi-public, and public organizations. These changes are still under way.

Consequently, transportation generally falls into the category of systems that are financed by a combination of revenues from system users and government funds.

8.1.5 Relationship of Transit Fare Levels and Public Financing

The amount of investment in transit, the level of transit fares, and the resulting operating subsidies depend on public policies toward transit—the relationship between its commercial enterprise character and social service function. Ultimately, that depends on economic and social policies toward urbanized areas and their population. There are a wide range of opinions about transit pricing and its impact on government financing, as discussed also in Chapters 7 and 9.

One extreme view is that transit fares should cover all costs, i.e., transit should be a purely commercial enterprise. This view is based on the belief that urban transportation should be regulated by free market forces.

Urban transportation is actually an obvious case where free market principles cannot be applied in their simple form, for a number of reasons. First, free market relationships include only monetary aspects, so that they would ignore the public service aspects and all external benefits from transit services. Second, it is not possible to regulate the relationship of competitors with very different capital/operating cost structures by free market principles. That is the case with the relationship between transit and the private automobile. Actually, as Weyrich and Lind (1999) have clearly shown, the use of private automobiles in cities is even less conducive to free market relationships than transit. Subsidies to auto use in many countries are much greater than transit subsidies, albeit less visible.

An even clearer demonstration of the inapplicability of purely commercial and free market relationships in urban transportation is the bicycle. This mode is very desirable because of its many positive features: it is convenient, cheap, and healthy for its users; it requires much less space per passenger than private autos and thus it causes much less congestion; it creates no noise and pollution. Thus, it has much lower negative impacts than the automobile. Public policies therefore usually encourage bicycle use and provide investments in bicycle facilities, although no user charges are imposed on bicyclists. Similarly, infrastructure for pedestrians is always entirely in the domain of public financing. These public policies are so logical that they are never questioned.

Another relevant factor is that ownership of components, methods of financing, and form of user charges for the automobile/street system are quite different from those used for transit. Travelers in most cases decide which mode to use considering marginal costs for a specific trip. In the case of auto trips, the marginal cost amounts to only 10–20% of average costs for that trip. Thus, fixed costs are disregarded. This is even more the case with persons using company cars and not considering the costs at all. To compete with auto travel, transit must charge fares that are comparable with the marginal costs of auto travel, i.e., a fraction of their total travel costs. This makes market-based pricing in urban transportation infeasible. Government measures are needed to create a more even playing field among travel modes.

In conclusion, urban transportation systems have such fundamental differences in organization, cost structures, user payments, and government subsidies that a free market situation among urban transportation modes neither exists nor can be created.

Conditions for self-supporting transit existed at times when there were no significant alternatives to transit travel in cities. Such conditions exist today in many developing countries where travel depends mostly on paratransit systems that require no special infrastructure and have very cheap labor. As a result, both investment and operating costs, as well as level of service, are very low.

The other extreme view is that transit should be offered as a free public utility, i.e., fully paid by government funds. The benefits of such a policy are claimed to be the benefits of high population mobility and avoidance of negative externalities that would be created by travel on alternative modes, primarily the automobile.

This view is in most cases unrealistic because it ignores the commercial aspects of transit systems. Such a setup would treat transit as a free public utility, placing a heavy burden on government budgets. Its implementation would result in inequitable distribution of public funds and lead to inefficient operations.

Several demonstration projects with zero fare transit in the United States during the 1970s showed some advantages (substantial ridership increased during periods of free riding) and some disadvantages (limited funding, some operational problems) of free transit services (see Chapter 7). Nearly free transit has been used in some other countries, however. The cities of the former Soviet Union, as well as Mexico City and a few other cities in developing countries, had such low fares that transit was practically free. Their fare revenues covered a small fraction of operating costs. Economic conditions have in most cases led to raising of fares to more realistic values.

Presently, free transit is used successfully only in special cases, such as in fare-free zones in CBD areas, aimed at facilitating very short trips by transit, decreasing traffic congestion, encouraging tourism, etc. It is also used sometimes for short feeder lines to rail stations and airports, or as semipublic paratansit shuttle lines in or between university campuses, hospital complexes, fairgrounds, and historic areas, usually combined with limited or fully prohibited parking.

Experience has shown that both of these extreme views about transit financing—that its fares should cover all its costs, and that government should cover all costs—have applications only in very special situations for limited transit services. The optimal level of fares is generally determined by a balance between fare revenue and government contribution determined by the local conditions and requirements.

Government subsidies of transit operations were rather limited in developed countries until the 1960s and 1970s. At that time automobile competition intensified and transit faced financial difficulties due to high automobile ownership and extensive indirect subsidies to auto use in cities in the form of construction of major highways and parking garages and direct subsidies to parking and driving (company cars, tax exemptions, etc.).

During the 1960–1980 period in the United States and other developed countries, operating ratios of transit agencies were decreasing steadily due to increasing costs, particularly labor, and the need to maintain moderate fares in order to compete with very low out-of-pocket costs of automobiles. Transit operating subsidies had to increase correspondingly. However, then it became obvious that this trend of increasing subsidies could not continue indefinitely. Stimulated by the requirement in many cities and states that the operating ratio must be maintained at a certain level, typically between 40 and 60%, increasing efforts were made to stabilize financial conditions of transit agencies. Cost reductions were achieved through improved business practices, control of labor wage increases that resulted from excessive demands of labor unions, privatization of certain functions such as vehicle maintenance and station cleaning, contracting out of some transit services, and other measures. This process of searching new ways to improve transit services, increase efficiency of their operations, and find the most efficient ways of applying government assistance has been given great attention in recent decades, and it is continuing to bring innovative solutions.

Construction of high-quality transit systems with extensive infrastructure, such as metros, LRT, AGT, and busway systems, requires major investments, and it is usually not attractive for private financiers because of low returns. Government participation in financing capital investments has therefore been traditionally much greater than assistance in operating costs. Public funds, from the cities, states, or federal government, were used to provide all or large shares of the funding for the construction of the New York, Boston, and Philadelphia subways, for the Paris Metro, and for the U-Bahn systems in German cities. The rationale for this public policy is clear: instead of applying high fares to recover the investments and have low usage of the system, public interest is much better served when transit services are provided at a low or moderate fare, so that higher population mobility, greater intensity of social activity, tourism, etc. are obtained with the benefits of a mobile labor force in the entire region. This social goal is more important than recovery of the sunken public capital.

8.1.6 Roles of Different Government Levels

Degrees and types of government involvement in financing different projects and systems, including transportation, vary among countries and states. Although there are no exact definitions, some typical patterns in the roles of the three government levels can be described.

8.1.6.1 Local and State Governments. The first parties to provide public funds where user revenues cannot pay operating and maintenance costs for transportation systems, including streets, transit, and pedestrian and other facilities, are local and state governments. In the case of transit, these funds supplement the revenues from fares. For capital investments, local and state financing is often dominant, or it is combined with federal financing.

8.1.6.2 Federal Government. The involvement of federal/national governments in financing various public systems and services, including transportation, is usually done for projects and activities where one or more of the following conditions exist (Vuchic, 1987):

1. The project has national significance (rather than benefits only its local area).
2. The project has interstate character: it serves an area including more than one state or province.
3. It requires a major investment in a permanent facility (typically infrastructure) with long-term significance. In this respect, national government represents a high-level entity capable of pooling together national resources so that individual major investments, exceeding capabilities of local areas, can be made one at a time.
4. The project has large positive nonquantifiable externalities.
5. Research and demonstration activities that benefit the country because they can be used by many local areas. This, again, is pooling together of national resources.
6. The project has strong political support at the national level.

These principles are generally followed in federal financing of such objects as major dams, irrigation projects, the Interstate Highway System, and inland waterways. Financing of major transit projects is federally supported because they usually meet several of the above conditions: increased efficiency of urbanized areas is vital to nation's economic strength (1); conditions 3 and 4 exist for all transit projects, and even the physical condition 2, interstate involvement, is met in many large metropolitan areas, including New York, Chicago, Washington, Philadelphia, and St. Louis.

Large cities in many countries have special treatment and financing from national governments. This is the case with Paris, London, the five largest cities in Belgium, and most medium and large cities in Germany. As in the United States, the rationale for these policies is usually based on the conditions listed under 1, 3 and 4.

The role of the U.S. federal government in transit financing has been very significant and diversified. Federal funds are primarily used for capital investments, while the federal contribution to operating costs has been decreased drastically in recent years. This decrease has been replaced only partially by the state and local governments. An area of significant federal role have been research and development, as well as demonstration projects.

8.1.6.3 Interdependence of Government Funding. Different types of transit funding, from fare revenues and local, state, and federal governments, are often interdependent. Every government is concerned about the local interest in the project or service, and it wants proof of its commitment in the form of a financial contribution. An example of this is the requirement that farebox revenues cover at least a specified portion of finances as its local share. Similarly, federal investment in transit projects is in most cases limited to 50–80% of the total project costs. The matching funds for the federal funding in this case are often divided between state and local government in a certain ratio, such as two thirds state and one third local share. These shares are usually based on a certain predetermined formula, but in most cases other factors may result in some adjustments of the formula-derived numbers.

8.1.7 Summary Review of Public Policies in Transit Financing

The preceding description and analysis of the role of transit in cities, public policies, financing principles, and procedures represent very important, often discussed issues. For convenience, some major issues, particularly those related to public financing of transit, are summarized here.

In addition to their important function of providing mobility, transit systems have various positive impacts and externalities that are largely qualitative, not conducive to measuring in monetary units. Transit is therefore operated as a combination of commercial enterprise and social service, and it is financed by user revenues as well as government funds. The question that laypersons sometimes pose, whether a transit system is making or losing money, is as inappropriate as it would be to ask whether a given highway or airport is making or losing money. For a more detailed discussion and analysis of this issue, see Vuchic (1999).

Major reasons and rationale for the use of public funds for transit system investments and operations are summarized here:

- Transit is a public service, providing mobility for the entire population, including all age and income groups, residents and visitors, disabled, and others.
- With its high capacity and reliable service, transit enables development of diverse urban activities, densities, and functions.
- Rail transit systems with special facilities and structures represent basic urban infrastructure, similar to streets, water, sewer, and other utility systems.
- Transit networks, particularly rail, can contribute significantly to desirable urban developments and form, reducing potential for urban sprawl.
- Serving urban core areas of the city and suburban activity centers, transit is oriented to people and thus enhances a city's livability, which a car-dependent city cannot do.
- Transit results in avoidance of large opportunity costs which travel by car would cause in terms of traffic congestion, negative environmental impacts, and occupancy of large areas by parking facilities.
- Transit fare levels are limited by the fact that fares must be competitive with the marginal cost of driving an automobile, rather than its total cost.
- Fare levels that would maximize an agency's revenues would usually cause diversion of a significant portion of trips to the automobile, thus incurring increased social and environmental costs.

As with all public service providers with partial government funding, transit agencies have more complex financing mechanisms and measures of operating efficiency than private, for-profit companies. Recent trends in transit agencies' organizations for increased efficiency are discussed in Section 8.4.

8.2 SOURCES OF LOCAL AND STATE FUNDS FOR TRANSIT

Revenue from fares (farebox revenue), as the direct payment by transit users, represents the basic source of transit operating funds. One of the main goals of every transit agency is to maximize its operating ratio, or the proportion of fare revenues from the entire operating revenues. When the operating ratio is greater than 1, the agency has an operating profit and no other funds are needed. Since the 1960s, however, operating ratios of 30–90% have been common among transit agencies in the United States and most other countries. To meet this need for operating funds, as well as funds for capital investments in transit, governments at all three levels (local, state, and federal) either have to allocate funds from their general budgets or develop new mechanisms for collecting revenues earmarked for transit operating or capital funding. A review of sources for transit funding is given here.

8.2.1 Taxes Dedicated to Transit

Local and state governments have developed a number of different sources of funds for transit financing. Characteristics of transit systems and areas they serve, as well as local political and economic conditions, strongly influence the tax options available for transit financing. This diversity makes generalizations for the entire country difficult. However, it is possible to classify financing mechanisms into several types, grouped into three categories by the types of their sources.

8.2.1.1 Classification of Taxes. By the incidence of taxes and their relationship to transit as a public service, local and state revenues for transit financing can be classified into three categories: benefit-related taxes and charges, broad-based taxes, and other sources. Specific types of taxes can then be grouped as follows:

- Transportation user taxes and charges:
  - Fuel and motor vehicle taxes
  - Parking tax
  - Tolls and area entry fees
- Economic benefit-related taxes and charges:
  - Employer payroll or occupational tax
  - Real estate value increment tax
- Broad-based taxes:
  - Retail sales tax
  - Personal income tax
  - Property (ad valorem) tax
- Other sources (excise taxes, etc.)

Each of these potential sources of funds for transit financing has certain inherent characteristics. In evaluations of possible applications of different financing sources for a given urbanized area, these characteristics must be considered against the criteria, local conditions, and defined goals.

8.2.1.2 Major Evaluation Criteria. Each major criterion for the evaluation of a transit revenue source and its relative importance is briefly reviewed here.

Yield Potential. The amount of revenue a tax has potential to generate is a very important consideration. This depends on the base—how large the subject of taxation is, and what tax rate is feasible to be used. Broad-based taxes have a much higher yield than taxes on a specific item, such as vehicle tire sales.

Economic Effects. Impacts of a tax can be positive or negative. If a tax makes conducting business or residing in an area more expensive, that is a negative impact. On the other hand, taxing of parking in center city may make transit use more attractive and thus have a predominantly positive impact. Similarly, excise tax on cigarettes and alcohol may have indirect benefits on the society. Because these impacts are usually rather small and indirect, relative importance of this criterion is low. It may, however, result in disproportionately strong political opposition, which is considered under another evaluation criterion.

Equity. Impact of a given tax on different groups of citizens, businesses, or travelers is rather important for social and economic reasons, as well as for its influence on political acceptability of that tax. Equity usually focuses mostly on the incidence of a given tax on different population groups: does it affect proportionately the economic status and ability to pay of individual population categories? The most important problem in this respect is that taxes may be regressive, affecting low-income persons more than high-income ones.

Relation to Benefits. The relationship of charges and the degree to which the payers benefit from the financed system, in this case transit, is always a valid consideration because it affects the feeling of equity or justice and thus affects political acceptability. However, in cities where transit is generally recognized as a service that is a basic utility benefitting the entire city, the direct relationship of taxes to benefits is not an important criterion. This is proven by the fact that broad-based taxes, which have the lowest rating in this respect, have been adopted in many cities.

Administrative Aspects. One consideration with any given tax is the complexity of its implementation and administration. In some cases, such as vehicle fuel tax, the mechanism already exists, so that it is easy to add an incremental amount to it. On the other hand, determining the boundaries of a district in which benefit-sharing tax should be imposed may be very complicated. In most cases, this criterion represents basically a technicality and has a low relative weight.

Political Acceptability. This criterion is very important because it is usually critical for adoption of any tax. Most of the preceding criteria, such as yield, equity, and relation to benefits, influence political acceptability of any proposed tax. In addition, numerous local factors, such as types of different organizations, political, business, or neighborhood groups, may have a strong political impact.

A political decision in selecting a tax, or many other matters affecting transportation, tends to have a negative connotation. It often implies that some special interests have overridden an important objective and economic, social, or technical considerations. This critical view is sometimes justified, but by no means in all situations. Responsible political and civic leaders can often have a broader and more realistic view than specialized technical or economic studies can produce. Politicians can also disregard or override views of unrealistic theoreticians, or accept and implement valid and innovative proposals of experienced public servants or consultants.

8.2.2 Characteristics of Different Types of Taxes

The basic types of taxes for transit financing listed in Section 8.2.1.1 are defined here. Their characteristics are outlined and relative weights suggested. It is emphasized that the list of tax types is not exhaustive. Many other taxes are used in different cities and countries. Their characteristics may be valid in most, but not all, localities. The differences in definitions and characteristics may particularly vary among countries.

8.2.2.1 Transportation User Taxes and Charges. Taxes and charges for transit placed on motor vehicles operating in the same area are logical for two basic reasons. First, all major cities should try to achieve an intermodal transportation system that integrates all modes. This requires pooling of resources and coordination of payments in all transportation functions: streets/highways, parking, transit, and other modes. Second, motor vehicle users benefit from good transit services rather directly in several ways. Transit reduces congestion and thus contributes to faster traffic flow (major increases in congestion during transit strikes demonstrate this). Auto users benefit from having transit as a standby service so that they are not stranded when they cannot use their cars. Finally, during emergencies of any kind, such as snowstorms, floods, earthquakes, or even major economic depressions, the role of transit is vital.

Among many types of taxes on motor vehicles used in different cities, three are most common: motor vehicle taxes, parking taxes, and tolls. The first of these is the most common.

Fuel and Motor Vehicle Taxes. Most states and countries already collect taxes on fuel, vehicle parts, and vehicle registration, so that addition of an amount for transit funding is easy to organize and administer. They are not feasible, however, in the states that still have the obsolete nondiversion clause, i.e., the stipulation that all highway-related taxes must be used for highway financing only.

The yield of these taxes can be considerable, although it is not extremely stable, varying with the amount of driving in the region, vehicle fuel efficiency, and fuel price. Economic impacts and equity aspects of these taxes can be analyzed in considerable detail, but that question is rather irrelevant in practice because of the small amounts that tax added for transit financing represents. This is particularly the case where gasoline and other fuel are grossly underpriced, as is the case in the United States. Despite the popular discussions and complaints about gasoline price increases when they occur, gasoline is extremely cheap in the United States. Its increase by several cents cannot produce major economic impacts or social inequities.

Increase in motor vehicle fuel price actually has very positive long-term effects. More expensive fuel leads consumers to purchase more fuel-efficient vehicles and combine trips, thus reducing vehicle-km traveled (VKT), resulting in lower congestion, less urban sprawl, and other environmental benefits, as well as lower national dependency on fuel imports.

Consequently, fuel and motor vehicle taxes dedicated to transit have many practical features and positive aspects, but their feasibility depends in most cases on political acceptability because highway interests oppose any increase in taxes on motor vehicles. It should be borne in mind, however, that from the national policy point of view, energy conservation and reduction of VKT (VMT) are explicit goals. Increasing gasoline and motor vehicle charges is a measure leading toward these goals.

Parking Tax. Parking supply and pricing is one of the most powerful tools for controlling vehicle use in central cities and in shifting modal split toward transit. Cities that have high transit usage and very lively, pedestrian-oriented central areas generally have a policy that limits parking capacity and progressive parking charges, so that short-term parkers are favored over all-day commuters. An additional transit-dedicated tax on commercial parking or on all parking spaces in central city can thus be a logical tool for implementation of a balanced intermodal transportation system.

This measure has been introduced very successfully in Perth and Sydney, Australia, and in a number of European cities. In the United States, the problem is that increasing parking charges weakens the position of the center city relative to suburban areas, where there is extensive free parking and virtually no measures to limit auto use and control suburban sprawl.

An even more fundamental obstacle to implementation of parking taxes in most U.S. cities are weak and uncoordinated transportation policies. The common situation is that parking lots and garages are private property without any control of their rate levels and structures, so that there is no basis for parking taxation that would have positive economic impacts, equity, and political acceptability.

In conclusion, a transit tax imposed on parking is an extremely effective measure where it can be introduced. However, most U.S. cities fail to include parking in a coordinated urban transportation policy, which prevents implementation of intermodal transportation systems.

Tolls and Area Entry Fees. Toll revenues have been applied for transit operating and capital funding for several decades in cities where the same transportation authority, or department of transportation, operates toll facilities, such as bridges and tunnels, as well as the transit system. This organizational setup has led to the use of a portion of tolls for transit, which has many positive features. This type of user tax usually has a substantial yield, it causes some shift of trips from cars to transit, it is easy to implement, and, as a predominantly internal decision, it is usually politically acceptable. Examples of toll revenues used for transit are found in New York City and Philadelphia (port authorities operating rail transit lines), the state of Delaware using toll revenues from an interstate highway for transit in the state, and others.

Another type of tolls used intermodally has been fees for entry driving into central city, collected on a cordon line encircling the city. Singapore was the first city to introduce these fees, designated an area licence scheme (ALS), in 1975. This was a comprehensive traffic regulation measure aimed at encouraging a shift of travel from cars to transit, taxis, and other modes. It actually created a permanent tool for preventing traffic congestion. The ALS program has been extremely successful, but its implementation in other cities has been delayed by lack of foresight and implementation tools, as well as political obstacles. With its very strong features, such as good yield, positive economic effects, and relationship to benefits, this concept eventually began to be introduced during the 1990s in Oslo and several other Scandinavian cities. Following long political debates and technical preparations, London introduced a L5 charge for drivers entering central city before 10 A.M. in February 2003. Other British cities may follow. In all these cases, a portion of fees from electronic road pricing—ERP (formerly ALS) is allocated to transit as a part of this intermodal coordination.

8.2.2.2 Economic Benefit-Related Taxes and Charges. This type of taxes is charged to employers, businesses, or property owners located in areas served by transit, thus accruing benefits from those services.

Employer Payroll Tax. Based on the rationale that businesses benefit from convenient transit services, some states and countries use taxes on employers to generate funds for transit financing. The best example of this tax is found in France, where in all cities with population greater than 100,000, all businesses with more than nine employees pay 1–2% of payroll as a tax earmarked for transit, including capital, operating costs, or reduced monthly tickets. This program, introduced in the mid-1970s, is credited with the revival of transit and great improvements in the economic and environmental livability of cities.

In the United States, businesses have traditionally dedicated large funds to subsidize parking of their employees, and these funds were federally subsidized through tax exemptions, while transit support was limited to $15 per month. In recent years, new federal regulations have greatly increased tax exemption of funds for transit use through employers, thus creating an indirect but very significant stimulus to transit use. Direct employer tax for transit has had limited applications, partly because such tax may decrease the competitiveness of central locations and encourage usually uncontrolled movement into suburban areas.

Real Estate Value Increment Tax. The concept that property owners around new transportation facilities should contribute to their financing because they accrue benefits from the increased accessibility is not new. It was used, for example, in financing the Erie Canal in the beginning of the nineteenth century. In urban areas, this benefit-sharing or benefit-assessment concept has been used in several metropolitan areas to obtain contributions for construction of transit systems. In Washington, DC, the construction of the Metro has stimulated investments in developments around a number of stations, and private contributions have been realized in some cases (KPMG, 1994).

These benefit-sharing or value-added taxes are logical and equitable. In the United States, they have attracted extensive discussion, but their applications have been limited and their yield, although not negligible in amount, has been a rather small share (below 10%) of the funds required for major transit projects, such as metros. In some countries with different economic conditions and legal setups, value increment taxes and private contributions, as well as earnings achieved by the transit agency from purchasing properties and then selling them after a transit system was built, have been extensive. Numerous cases of such applications of joint financing and real estate value-added taxes exist, for example, in Japan, Hong Kong, and Singapore.

8.2.2.3 Broad-Based Taxes. This category of taxes is charged to a very large base of activities or population. Although related to transit benefits only in a broad sense, these taxes are easy to administer and their economic and equity impacts are not strongly negative because their charges are broadly distributed and therefore small. For these reasons, broad-based taxes are by far the most common types of taxes for all state and local revenues for all purposes, as well as for transit support specifically.

Retail Sales Tax. This consumption tax is used by nearly all U.S. states as well as many other countries in somewhat different form (usually included in prices rather than added to the sales bill). The base of this tax is sales of consumer goods and services, as well as purchases by business firms. Its disadvantage is that it is regressive, and some states exempt food, health supplies, clothing, and utilities as the basic necessities to reduce the burden on the low-income population. However, consumption taxes have the inherent advantage that they moderate spending and stimulate savings, a desirable social goal.

Because of the very broad base, sales tax has a very high yield. Transit-dedicated share of sales tax usually amounts to 0.25–1.00% only, but these rates can generate amounts sufficient for sizable investments, operating cost assistance, or, in some cases, direct reduction of fares.

Sales tax has a boundary problem where adjacent states have very different sales tax rates. However, transit tax is often added in metropolitan areas only, and this problem is not significant between this area and surrounding counties in the same state because the transit tax amounts only to a very small increment, at most 1%. Thus, with the exception of a few states that have no sales tax, transit tax has proved to be easy to administer and a very powerful source of finances wherever there has been sufficient political acceptability for its introduction.

Personal Income Tax. Tax on personal income can potentially have a very high yield for transit because it has a very large base. However, this is seldom used because few cities and other local governments use this type of taxation. The cities of Baltimore and Philadelphia have substantial income tax that brings large revenues, but it is considered to represent a significant disadvantage for the city proper in its competition with suburbs. In other countries, some cities have used a special income payroll tax for a limited number of years with a special earmarking for improvement of transit. Administering this tax may be complicated if a metropolitan area has a large number of counties, towns, and other jurisdictions.

Property (ad valorem) Tax. This is a type of ad valorem tax imposed mostly on real estate, sometimes also on business equipment and inventories, and even less frequently on personal property. Property tax represents the main source of tax revenue for most local governments, and the main expenses paid from it are schools. While this tax is proportional to personal wealth, it may represent a burden on persons with fixed income, particularly retirees. Allocation of a portion of property tax for transit is fairly common. Some property taxes to be earmarked for financing major transit projects are submitted to popular referenda, and they have been the main source of local funding for such projects as construction of metro or LRT systems.

8.2.2.4 Other Sources of Funds. Among many other taxes used for transit, various kinds of excise taxes are most common. Excise tax is imposed on different types of sales, goods, or activities, sometimes with the objective of achieving equitable charges by some criteria, or so-called sin taxes on goods or operations that society wants to discourage. Examples of these are tobacco, alcohol, and gambling. Excise taxes are sometimes placed on water, sewer, electricity, gas, and other utilities, where an acceptable amount of surcharge sometimes can bring local government significant revenues. Transit can be one of the recipients of these revenues. A typical case of the application of utility revenues for transit, in the form of excise taxes or direct internal pooling of revenues and expenses, is the joint ownership of electric public utility and transit companies, which was common several decades ago.

8.2.2.5 Review of Different Types of Transit Taxes. The preceding description of taxes shows that each type of tax used for transit has many different aspects that vary greatly among cities and countries. Their appropriateness depends on the type of transit system, economic and political conditions, and attitudes. The task of selecting possible sources of revenues for transit in a given city or metropolitan area is therefore very complex.

To provide a general overview of the described tax types, they are listed and evaluated with respect to the six defined criteria in Table 8.3. Each criterion is given a relative weight from 1 (the lowest) to 4 (the highest), and tax types are evaluated for each aspect, also with 1–4 points. The columns with evaluation points are followed by columns that have products of these points and relative weight of the respective criterion. The total score of each tax type is given in the last row.

Obviously, the presented numerical evaluations are extremely general. Such quantification of qualitative elements is inherently crude and subjective. Because evaluation numbers vary with many factors, this type of table should be developed for each individual case. The numbers presented in Table 8.3 are therefore for orientation only.

8.2.3 Transit Organizations with Taxation Powers

Transit agencies in major metropolitan areas usually cover the central city and extensive suburbs. Because they serve many local jurisdictions, they are established by state legislatures, usually in the form of an authority, district, agency, corporation, or similar title. The names in some cases imply a certain legal status or power. In the cities of the Northeast Corridor, such as Boston, New York, and Philadelphia, as well as in Atlanta, Cleveland, and Honolulu, transit agencies are authorities. They obtain financial support from their cities and states in forms that vary considerably. A few port authorities, namely in New York and Philadelphia, are bistate organizations that own and operate rail transit lines between the respective states and finance them from their budgets. By far the biggest and oldest, which served as a model for others, is the Port Authority of New York and New Jersey, which operates airports, bridges, tunnels (most of them with toll charges), and the Port Authority Trans-Hudson (PATH) rail system.

Many metropolitan areas have transit districts, which usually have legal power of direct taxation. Examples of these are the San Francisco Bay Area Rapid Transit District, Alameda-Contra Costa Transit District, Denver Regional Transportation District, as well as some medium-sized metropolitan areas, such as Albany, New York, and Bridgeport, Conneticut. Their boards have taxing powers to collect public share of their revenues. Tax rates are sometimes limited by specified regulations.

The types of taxes transit districts can impose and their applications vary among states and metropolitan areas. Some districts have powers to collect property

Table 8.3 Evaluation of different types of taxes for transit financing

	Transportation User						*Benefit-Related*				*Broad-Base*						*Other*	
Tax Type / *Criterion and weight*	*Fuel and Motor Vehicle*		*Parking*		*Tolls and Area Entry*		*Employer Payroll*		*Real Estate Value*		*Retail Sales*		*Personal Income*		*Property*		*Excise*	
Yield potential: 4	2	8	1	4	3	12	3	12	1	4	4	16	3	12	3	12	2	8
Economic effects: 2	3	6	3	6	3	6	1	2	2	4	2	4	1	2	2	4	1	2
Equity: 2	3	6	3	6	4	8	1	2	2	4	2	4	2	4	3	6	2	4
Relation to benefits: 2	2	4	2	4	4	8	2	4	3	6	1	2	1	2	2	4	1	2
Administrative aspects: 1	2	2	2	2	3	3	3	3	1	1	3	3	2	2	1	1	3	3
Political acceptablility: 4	2	8	1	4	2	8	1	4	2	8	2	8	2	8	1	4	3	12
Totals		34		26		45		27		27		39		30		31		31

tax, others sales tax, tolls, and other resources. These revenues are earmarked in some areas for capital investments only; in others they may be used for operating expenses, or for both.

Districts that have taxing power are usually more stable organizations than agencies depending on annual allocations of government funds because their revenue relies on the entire region's productive capacity. This guarantee allows districts to sell general obligation (GO) bonds that have lower interest rates than revenue and other types of bonds that have a higher risk factor.

8.2.4 Structure of Local-State-Federal Funding

In the United States, the contributions of these four sources for operating costs for the entire country have in recent years (the figures are for year 2000) been around the following values: fare and other revenues 54%; operating assistance from local governments 22%; state governments 20%; federal government 4%. These percentages vary among states and cities by size and policies toward fare levels.

Funding of capital investments, which include construction of new lines, stations, control centers, and maintenance facilities and purchases of new rolling stock and equipment, is mostly borne by national governments. In the United States, federal contribution was as high as 80% in 1979, but it has been reduced in recent years, while various local sources have increased considerably. These consist mostly of earmarked taxes and sometimes contributions in other forms. For example, the value of a former railroad right-of-way or tunnel is credited as a local share in project financing. Based on all transit capital funds for the country, the shares in the year 2000 were the following: federal 47%, states 11%, local governments 15%, and local special funds 27%. Contributions of the federal and state governments in this category are usually conditioned by provision of certain local share, commonly although incorrectly designated as matching funds. For example, a new LRT line was paid in some states 80% by the federal government, 16.67% by the state, and 3.33 % by the local government.

8.3 FEDERAL FUNDING OF TRANSIT

In many countries around the world, national or federal governments have traditionally been involved with planning and financing transit systems in their capital cities. From Moscow, Athens, and Stockholm to Brussels, Paris, and Mexico City, national governments have been strongly involved in planning, financing, and even implementing their transit systems, primarily metro and regional rail, but also LRT and bus systems. In all these cases, the government feels that the capital city is of great national significance and efficient and attractive transit is one of its vital components.

In many countries, national governments have an active role in developing and financing transit systems in all cities and metropolitan areas. This is the case in the United States and Germany, where federal governments not only finance transit systems in all cities, but also carry out very important activities in research and development and technological and operational innovations. Governments in France, Norway, the United Kingdom, and other countries also have various forms of national programs for transit financing and advancement.

8.3.1 Development of the Federal Role in Transit

The federal role in highway financing, which was initiated in 1917, became extremely powerful and legally anchored in the economy, transportation, and urban development of the United States with two pieces of federal legislation in 1956. The Federal Aid Highway Act initiated construction of the national highway system, and its companion Highway Revenue Act ensured highway financing, mostly through gasoline and other highway-related taxes. This was a very effective method of financing, but it was limited to the construction and maintenance of major highway categories. Urban streets, pedestrian and bicycle facilities, and the entire transit industry had no support: actually, they suffered from the strong pressures of increasing auto dominance.

Federal highway legislation thus closed a vicious circle of urban transportation: the more highways were built, the more traffic volumes increased and generated funds for building more highways. At the same time, all other modes of travel in cities deteriorated, further increasing the demand for travel by auto. Despite the higher mobility by car, urban areas suffered because all other modes deteriorated and more cities were overwhelmed by cars, becoming less livable.

The sharpening deterioration of urban areas, with transportation as one of its causes, led to the introduction of the federal involvement in transit systems during the mid- and late 1960s. This represented the turning point in transit development in the United States. The federal agency in charge of transit affairs, initially the Urban Mass Transportation Administration (UMTA), then its successor, the Federal Transit Administration (FTA), has pursued a broad set of activities, which may be grouped in three functional categories.

8.3.1.1 Financing of Transit. The FTA establishes criteria and distributes federal funds for many different categories of transit capital assistance, from capital investments in new rail lines and purchase of new buses to the construction of new maintenance shops or control centers. This program has resulted in construction of new LRT, metro, and regional rail systems in more than 20 U.S. cities, purchases of a large portion of rail and bus rolling stock, rehabilitation and modernization of old rapid transit lines, bus depots, and rail maintenance shops, construction of modern control centers, etc.

8.3.1.2 New System Concepts, Research and Development, and Operating Procedures. This group of activities is very broad and has also played an extremely important role in modernizing transit systems. It was initiated in the late 1960s, when obsolete transit systems had no ability to develop innovations or even transfer them from other countries. The UMTA, and later the FTA, were instrumental in bringing to North America the concepts of light rail transit, articulated and low-floor rail and bus vehicles, self-service fare collection, applications of ITS technology, new marketing methods, and many others.

8.3.1.3 Transit Improvements for Urban Livability. Many federally sponsored programs and projects have been designed to promote positive interactions between transit and urban economic activities, social conditions, and livable environment. This broader aspect of transit has been increasingly supported since the enactment of ISTEA in 1991, the federal transportation act that emphasized intermodal aspects of transportation and required a comprehensive approach to urban transportation planning. Through these programs, the FTA now finances not only transit systems, but also related developments, such as pedestrian areas in CBDs, transit-oriented developments in suburbs, land use developments around intermodal terminals, etc.

8.3.2 Trends in Federal Funding

From its initiation in the late 1960s, federal funding for transit was increased gradually to meet the need that had resulted from the neglect of transit during several decades. In 1974, federal assistance for operating costs was also introduced. The growth of federal transit assistance was, however, stopped in 1980 and its trend was reversed during the 1980s. As the diagram in Figure 8.3 shows, federal funding for operating costs has stagnated in current dollars since 1980, which means that it decreased in constant dollars. The shares of state and local governments, and particularly revenues from fares, increased since the 1980s. As a result of these trends, the percent distribution of operating costs obtained from fares and other revenues has increased to over 50%, while the federal share dropped to under 10% (Figure 8.4).

The absolute amounts and percent distribution of capital funds from the four sources, shown in Figures 8.5 and 8.6, respectively, show a different pattern. The shares of local and state governments have decreased somewhat, but the locally generated funding has increased substantially. In this case, these funds are not from fare revenues, but from various local taxations and other contributions. Federal funding has decreased from over 60% in 1988, and yet it remains at about

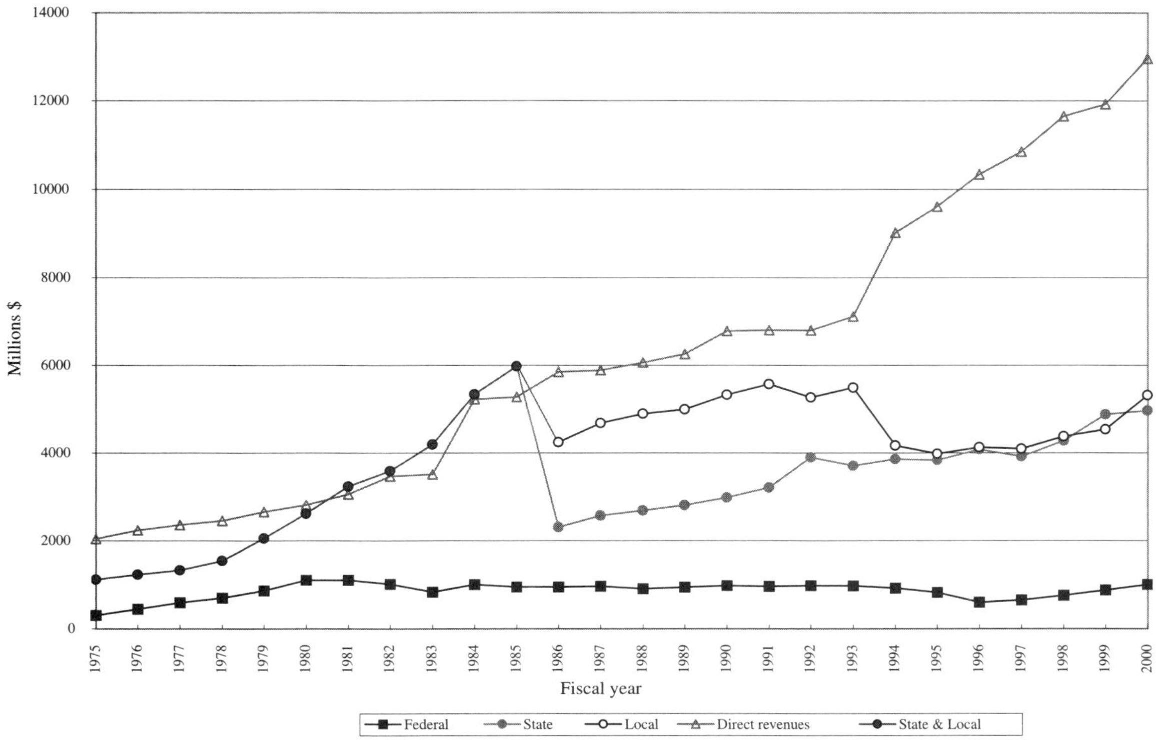

Figure 8.3 *Sources of transit operating funds, 1975–2000.*

50% the largest single contributor, following the funding policy principles discussed earlier in this chapter.

Exact percentages of the four categories of fund sources for operating and capital costs in 1990 and 2000 are shown as pie charts in Figure 8.7. The diagram shows the dominance of the farebox and other revenues in financing operating costs, and the major role of federal share in the capital funding.

The discussed trends in federal financing reflect federal policies. During the 1980s, federal transportation policies returned to a strong highway bias similar to that from the 1950s. Under the slogan "Users should pay a greater share of costs," there was strong pressure to increase transit fares. The same slogan was not applied to highways, however, and proposals to increase fuel taxes were criticized as a hardship on highway travelers. Federal funding for transit decreased from $4.2B in 1980 to $3.6B in 1988 (both amounts in constant 1987 dollars), while highway funding continued to increase, going from $12.3B in 1980 to $13.8B in 1988.

The adoption of ISTEA as a law in 1991, and its successor TEA-21 in 1997, reversed the downward trend of transit financing. Federal transit appropriations increased from $4.6B in 1995 to $6.7B projected for 2002.

8.3.3 Criteria for Capital Fund Allocation

However significant federal funding for transit in the United States has been, its allocations for major capital investment projects have not been sufficient to meet the demand for such funds. The FTA has therefore

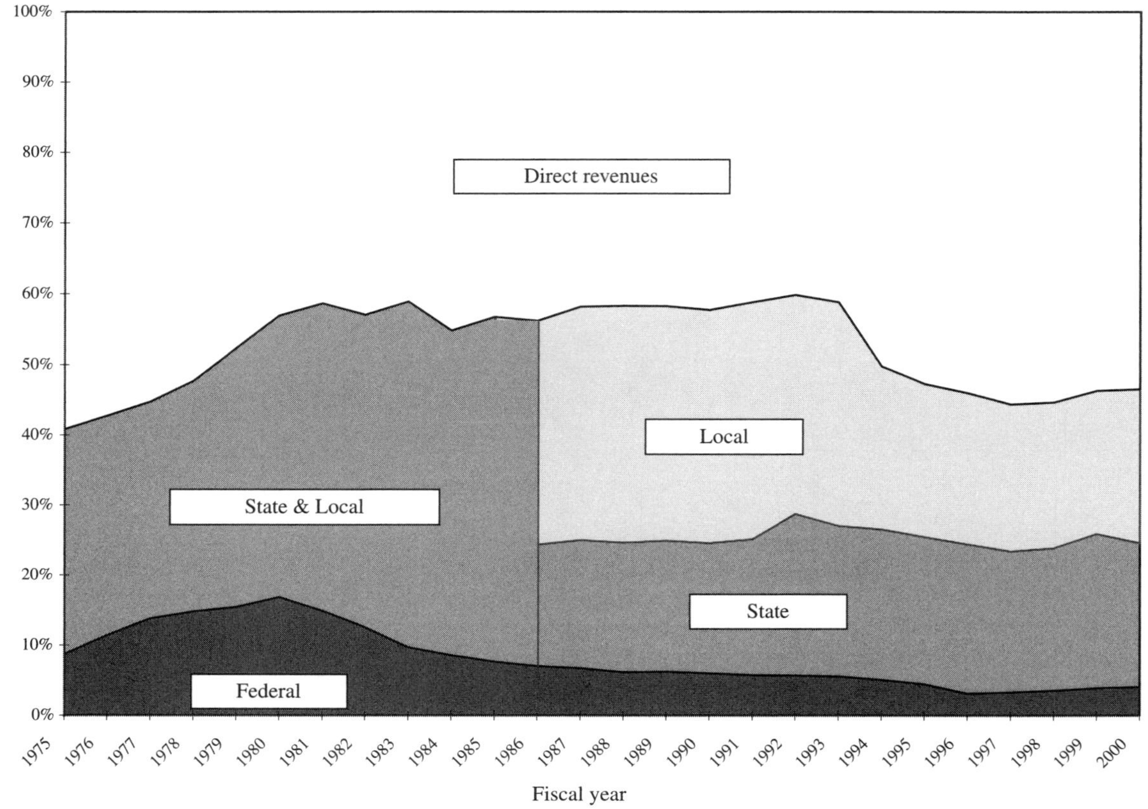

Figure 8.4 *Percent distribution of transit operating funds among sources, 1975–2000.*

used various types of criteria to evaluate and rank the submitted requests from different cities. There were attempts to use a number of indicators based on investment and operating costs, projected ridership, cost effectiveness, etc., so that the evaluations would be fully objective. This proved to be unsatisfactory for several reasons. First, since many of the indicators are based on projections of future conditions, such as future cost, ridership, revenues, diversion, and impacts on other transit services and on highway traffic, such estimates vary with the projection techniques used, and the reliability of their findings is often debatable. Thus, such projections are actually subjective in an indirect way. And second, conditions, needs, and goals established for major transit projects vary greatly among cities, so that they cannot be based on any quantitative measures.

Three major criteria are presently used:

1. *Project cost-effectiveness:* projected investment cost per projected new transit rider.
2. *Justification of the locally preferred alternative.* It is still required that all reasonable alternative modes be considered for the planned project, but the selection of the proposed alternative can include nonquantitative aspect of local preferences in selecting mode and other basic design characteristics.
3. *Local financial contribution.* The FTA wants to stimulate transit agencies and local government authorities to demonstrate their interest in and commitment to the selected project. The main proof of their commitment is the amount of funding that is locally prepared. Thus, a project

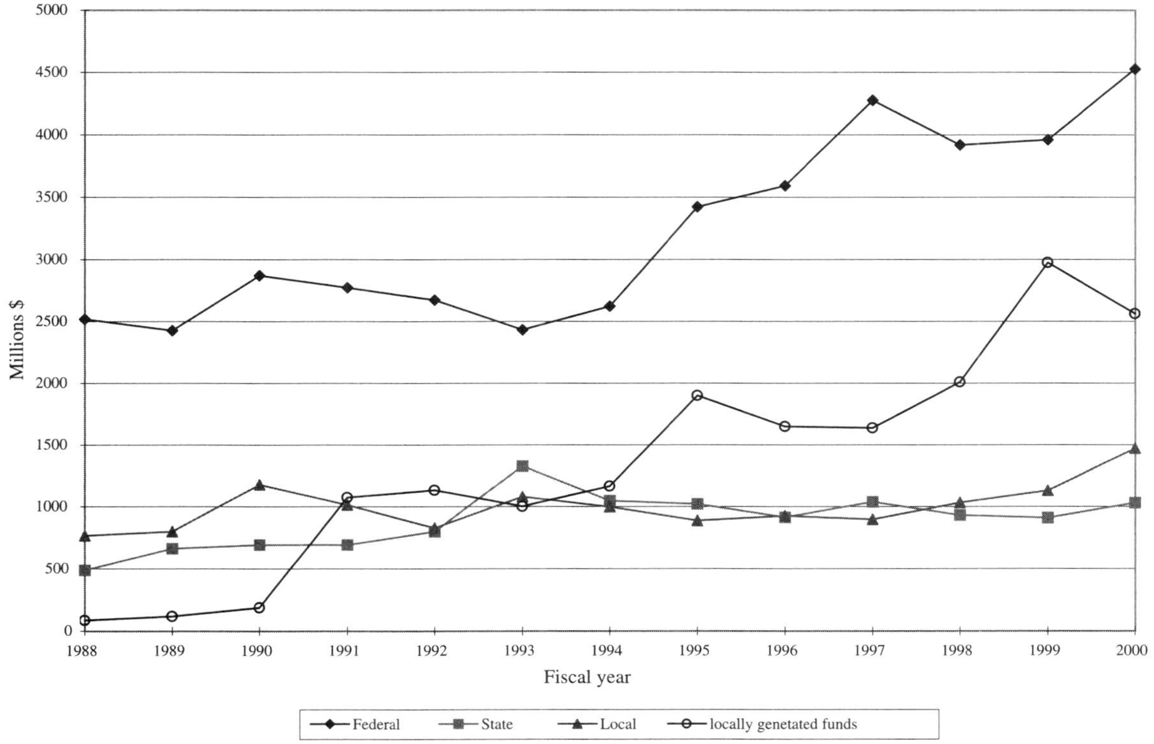

Figure 8.5 *Sources of transit capital funds, 1988–2000.*

that has 40% local share is ranked lower than a project that has 60% financing from the agency, local and state funds, and any other local sources of funds.

In most other countries, transit investments by national governments or by international investors, such as the World Bank and the European Union, are also subject to careful examination and criteria related to economic feasibility of the projects, rate of return, their noneconomic aspects, and externalities, including short-term social as well as long-term intermodal and land use impacts.

8.4 IMPROVING THE EFFICIENCY OF TRANSIT AGENCIES

The reasons for public financial assistance to urban transit are logical and powerful. However, the availability of public funding must not be allowed to diminish the need to operate transit systems efficiently following principles of commercial enterprise to the maximum possible while meeting the required social service aspects.

8.4.1 Requirements for Efficiency

To satisfy operating efficiency requirements and at the same time provide high-quality service and achieve a city's transportation goals, efforts should be made to satisfy the following principles:

- Transit systems must maximize their operating efficiencies regardless of the availability of public funds.
- Labor wage rates should be competitive with current rates in comparable jobs, considering em-

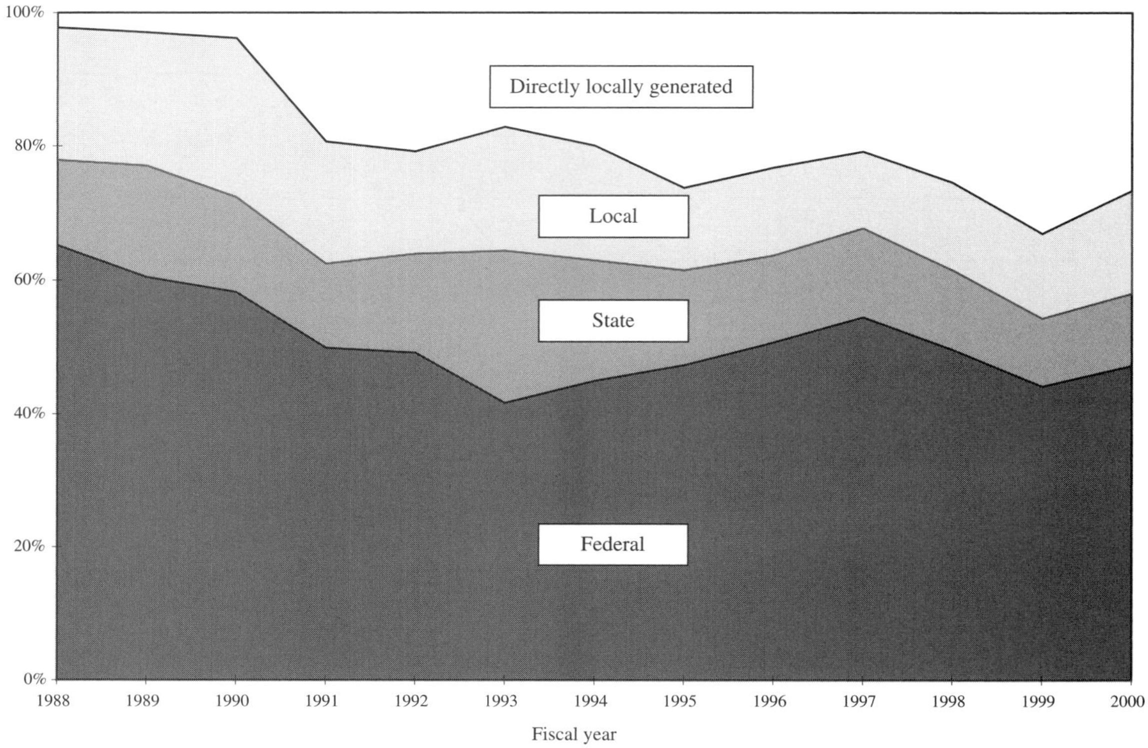

Figure 8.6 *Percent distribution of transit capital funds among sources, 1988–2000.*

ployment benefits, job security, and other benefits as well as responsibilities of public duties transit employees perform.

- In determining fare levels, the trade-off between the amount of revenue and the number of passengers should be determined on the basis of the city's transportation policies, considering elasticity of demand.
- Designs of high-quality transit systems should be checked by value engineering and other reviews to ensure efficient designs and prevent unjustified "goldplating" of systems. However, this does not imply that transit systems must always be built to the lowest standards. As parts of the city and its environment, transit systems must be attractive and contribute to the livability of areas they serve. Moreover, aesthetically pleasing transit stations and vehicles result in higher comfort, safety, and attraction of passengers.
- To facilitate operations and provide high service quality, transit system design must be made with careful consideration of maintenance costs, prevention of vandalism and crime, and easy repairs.

Many aspects of policies and financing practices presented here have been known and applied for many decades. However, a number of new ideas and procedures of financing transit have been developed in recent decades, and the process of innovation is likely to continue in many cities and countries.

8.4.2 International Efforts to Increase Transit Efficiency

Government assistance to transit in most industrialized countries was increasing during the 1970s. However, around 1980, it became obvious that the trend of increasing government assistance and decreasing oper-

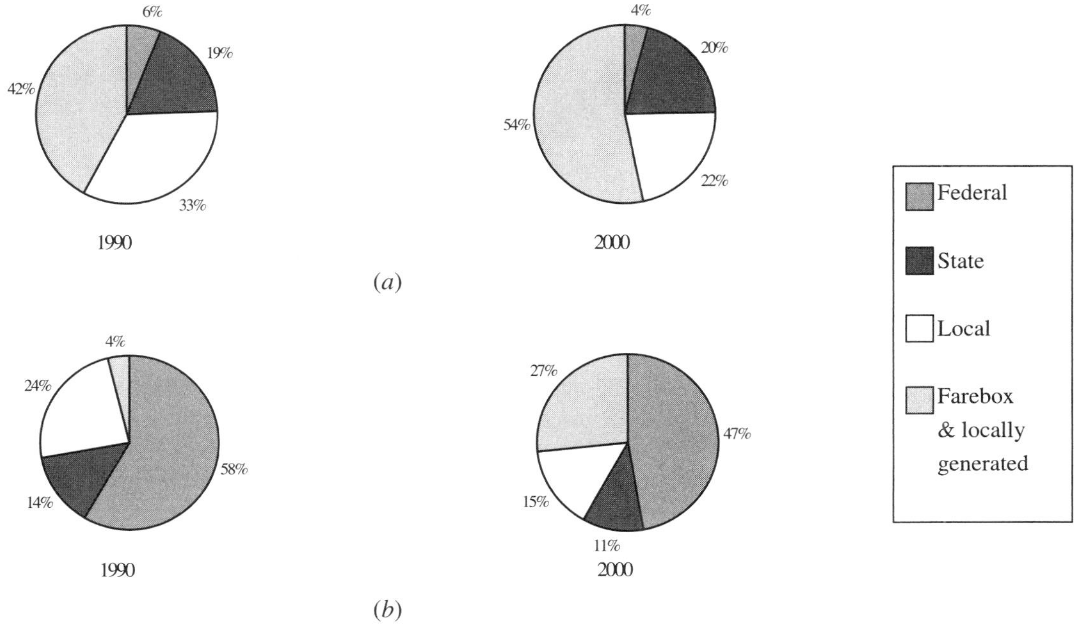

Figure 8.7 *Distribution of sources of transit operating and capital funds, 1990 and 2000: (a) operating funds; (b) capital funds.*

ating ratios could not continue indefinitely: the financing procedures had to be stabilized. The steps to achieve that goal have included pressure on transit agencies to improve their operating efficiency and increase operating ratios; reorganization of agencies by changing their ownership and legal status, such as partial privatization and leasing of services; modification of regulatory practices; and, in some extreme cases, even deregulation.

While many transit systems in the United States have made organizational changes, internal as well as through reorganization and partial leasing or privatization, the most innovative developments have taken place in Europe. There, parallel with diverse improvement efforts in individual countries, the European Union has introduced new standards and requirements for transit system organization, competition in purchasing rolling stock, and tendering of certain types of services and other aspects, which have been the subject of extensive discussions, reviews, and revisions. The entire situation is likely to undergo further developments for a number of years.

The privatization and deregulation of transit in British cities (see Chapter 9) has been one extreme in this array of transit reorganizations. These were fundamental changes in the legal and organizational setup of transit companies, and they resulted in significant operating cost reductions, at least at the time of transition of ownership. However, the price of that was disintegration of transit networks and substantial loss of passengers, with many negative consequences (Fawkner, 1995).

While organizational break-up of transit systems was common for bus companies, similar attempts were made with other modes in a few instances, including cities in other countries. For example, Melbourne even divided its Tramway system into two companies, with mixed results. The greatest political battle about fundamental organization of a transit system in recent times, however, has been fought about the London Underground system: the national government has been forcing break-up of the Underground network into three companies. The mayor of London has resisted this, arguing strongly that short-term efficiency must not be pursued at the cost of system disintegration that results in deteriorated operations, poor service quality, and loss of passengers.

Other European countries have rejected the British model of deregulation because it pursues financial efficiency as the only goal, often causing deterioration of services. Other countries have been making efforts to increase operating efficiency of transit while maintaining or raising service quality for passengers. This approach recognizes transit service as the *raison d'être* of transit systems, while their financial efficiency is considered an important, but not the dominant, requirement. Extensive discussions and new rules have been under development in the European Union about the introduction of some elements of competition while retaining various forms of municipal ownership, public regulation and, above all, integrated transit systems.

Specific improvement measures that have been considered and introduced in many European countries can be grouped in several major categories:

- Internal organizational changes to improve efficiency of transit agencies
- Contracting out of portions or entire transit organizations to professional transportation organizations
- Improved internal operations through technological and organizational innovations (changed labor practices, automation, ITS, etc.)
- Introduction of transit vehicle priorities in street traffic, resulting in improved speed and reliability of services
- Investments in modern transit system infrastructure that attract more riders and decreases operating costs.
- Competitive international bidding for transit rolling stock purchases

Germany is an excellent example of applications of successful measures to improve financial condition of transit systems. Major efforts have been made to retain integrated public agencies and transit federations but increase their efficiency through a variety of organizational changes. These include competitive setups of different divisions of transit agency, privatization of some functions such as fare collection or station cleaning, control of labor union pressures for wage increases that exceed predominant labor wage rates, and resolution of labor unions' opposition to technical innovations such as fare-collection automation. Underlying the very successful results of these changes in German cities is, however, the superb infrastructure of transit systems, which allows high labor productivity and low operating costs while providing attractive services that result in high ridership. This condition has been achieved through a consistent policy of steady investment in transit systems and coordinated urban transport implementation since the late 1960s.

Many cities in France, New Zealand, Sweden, and other countries have reorganized transit systems by contracting them as entire companies or their individual divisions to locally or internationally based operating companies, such as Connex, Stagecoach, and many others. The new organizational setups have generally resulted in significant cost reductions while maintaining integrated, high-quality transit services (Ministère de l'Aménagement, 1995; Domstad, 1996).

As a result of these improvements, many transit agencies in Europe have increased their operating ratios from 40% to 70% or even as high as 90%. It is emphasized, however, that as in Germany, the high quality of transit services has been a major element in increasing transit efficiency and attracting new ridership. In most cities in Switzerland, Belgium, the Netherlands, the Scandinavian countries, and elsewhere, transit systems have rail networks largely independent of street traffic (LRT, metros, and regional rail), attractive stations, long trains with one-person crews, automated station operations, extensive applications of ITS technology, and full integration of rail and bus services. These features offer high quality of service to passengers at much lower operating cost than the systems with obsolete infrastructure and rolling stock, labor-intensive operations, etc.

In most countries, the role of federal/national government remains very important in financing capital investments, usually complemented by certain shares of state/provincial and local governments. Public/private partnerships and a variety of new types of organizations of transit project implementation, such as design-build-operate-manage (DBOM) and similar arrangements, have been increasingly used.

EXERCISES

8.1. List and discuss at least three reasons for the use of public funds to finance transportation systems, and describe historic cases in which those reasons were applied in policy decisions.

8.2. Compare and comment on the financing of the national passenger railroad system (Amtrak) relative to financing of other modes and financing of passenger railways in other countries.

8.3. In most urban areas, transit is financed from transit fares and public funds. What is the rationale for providing public financing for transit systems?

8.4. How do public policies influence the ratio of transit funds collected from fares to the amount provided by governments?

8.5. In most cities, if sufficiently high fares were charged, transit could be financed from fare revenues only. Why do most cities not apply such pricing, but instead have policies to provide financial support for transit services?

8.6. Discuss the goal of collecting maximum fare revenue and the goal of serving the maximum number of passengers. Specifically analyze:

a. Which goal is more appropriate for transit and why

b. What negative results are produced if maximum revenue is used as the goal

c. What indirect benefits to the city are achieved with the maximum passengers goal

8.7. Describe the concept of the character of transit and its role in the city that is held by the proponents of two extreme policies in transit financing:

a. Transit should be self-supporting and fares should be high enough to achieve that.

b. Transit should be fully financed by public funds and provide free services to the public.

8.8. Why do free market principles not apply in the competition between cars and transit?

8.9. Is the share of public assistance typically greater for operating funds or for capital investments in transit? Explain why.

8.10. Discuss the application of user taxes for transportation financing and describe their positive and negative features compared to the use of general taxes.

8.11. Discuss the appropriateness of using benefit-related taxes for transit financing.

8.12. What are the three major categories of broad-based taxes commonly used for transit, and what are their characteristics?

8.13. Find an actual (or hypothetical) city in which construction of an LRT line is planned and several different types of taxes are considered for its financing. Describe which conditions you would take into account in selecting the criteria for evaluating different types of taxes for this project.

8.14. List and comment on the arguments for and against the use of fuel and motor vehicle taxes for transit financing in a given city.

8.15. Central city entry fees for automobiles have been used in Singapore since 1975 and have recently been introduced in several other cities. Describe these fees as a source for transit financing and as a tool of an intermodal transportation policy.

8.16. Which type of bonds, revenue or general obligation, has a higher interest rate and why? Which power must transit agencies have to be able to sell bonds with the lower rate?

8.17. What are the policy principles for the federal/national government to participate in financing urban transit?

8.18. What are the reasons for the requirement that transit fund allocation by a government often be conditioned by the provision of certain contributions (matching funds) by lower-level governments or from farebox revenues?

8.19. List and describe major categories of changes in transit systems introduced in recent years to improve their operating ratios.

8.20. Which types of internal changes have many transit agencies made to improve their financial results?

8.21. How do investments in system capital improvements influence operating efficiency and operating costs of a transit system? Give two examples of investments that reduce operating costs. Can some investments increase operating costs? Explain.

REFERENCES

Domstad, R. 1996. *Tendering for Bus Services in Gothenburg*. Gothenburg City Traffic Authority, Sweden.

Fawkner, J. 1995. Bus deregulation in Britain: profit or loss? *Public Transport International* 6:18–23.

KPMG Peat Marwick. 1994. *The Impact of Metrorail on the Commonwealth of Virginia.* Northern Virginia Transportation Commission.

Ministère de l'Aménagement du Territoire, de l'Équipement et des Transports. 1995. *Urban Public Transport in France, Institutional Organization*. Paris: Land Transport Administration.

Pucher, J. 1995. Urban passenger transport in the United States and Europe: a comparative analysis of public policies. *Transport Reviews* 15(2):89–107 and 15(3):261–277.

Vuchic, V. R. 1987, Urban transit: a public asset of national significance. *Urban Resources* 4(1):3–6, 63.

———. 1999. *Transportation for Livable Cities,* New Brunswick, NJ: CUPR Press, Rutgers University.

Weber, A. F. 1899. *The Growth of Cities in the Nineteenth Century*. New York: Macmillan, Repr. Ithaca, NY: Cornell University Press, 1965.

Weyrich, P. M., and W. S. Lind. 1999. *Does Transit Work? A Conservative Reappraisal*. Washington, DC: Free Congress Foundation.

FURTHER READING

American Public Transportation Association (APTA). 2002. *Annual Report for 2002*. Washington, DC: APTA.

Hartman, R. J., E. M. Kurz, and A. B. Winn. 1994. The Role of Performance-Based Measures in Allocating Funding for Transit Operations. TCRP Synthesis 6. Washington, DC: TRB.

Stanford Research Institute (SRI). 1956. *Organizational and Financial Aspects of the Proposed SFBA Rapid Transit System*. Menlo Park, CA: SRI.

UITP. 2003. "PT financing: exploring all the options." *Public Transport International,* special issue 06.

9

TRANSIT OWNERSHIP, REGULATION, AND ORGANIZATION

The primary reason for the existence of transit in cities is to provide passenger transportation services. As a basic public utility, transit has a major role in the functioning of the city: its economic and social activities. This important role places on transit certain requirements with respect to its quantity and quality of service (network extensiveness, service frequency, capacity, speed, safety, etc.) as well as to the price (fare) that can be charged. On the other hand, there is a requirement that transit be provided in the most efficient and economical way, so that its revenues from passenger fares cover as much of the costs of transit as is set by political decision in each specific city or region. This may be, for example, a certain portion of its operating costs, usually balancing the objectives of maximizing transit usage and minimizing the public expenditures for transit operations.

These two aspects of transit systems—the service they provide and the funding they receive—have a major influence on the ownership and organization of the company or agency that operates transit, on government regulations, and on various financial aspects of transit systems. Public and private ownership, degree and form of government regulation, and related issues have been the subjects of extensive discussions among transit professionals and government officials, and many changes have been introduced in recent years.

To understand these issues, this chapter starts with a brief overview of historic developments of transit ownership and its transition to public agencies in most cities. In the following section, forms of publicly owned transit agencies are defined, followed in Section 9.3 by a review of methods for integrating services to provide the public with a single, coordinated transit network. Section 9.4 gives a systematic classification and definitions of government regulations of transit. The controversial debate about deregulation and a critical review of its results comprises Section 9.5.

The last section, 9.6, presents a description of recent trends in making regulations more sophisticated to allow greater diversity in agency ownership and management while retaining protection of public interest in providing the required services. It also gives an overview of the numerous new forms of transit agency management arrangements, public-private partnerships, tendering, contracting, and other arrangements that are bringing diversity and improvements of transit services in many cities.

9.1 PRIVATE TRANSIT COMPANIES AND PUBLIC TAKEOVER

Most of the transit services in the nineteenth century (horse-drawn omnibuses and, from the late 1880s, electric streetcar/tramways) were initially provided by private companies. The demand for travel in cities was great, and transit was the only means people could use for trips on distances greater than could be walked. Under such conditions, it was possible for private companies to invest in infrastructure and vehicles, operate transit services, charge reasonable fares, and make a profit. This was even the case with some of the rapid transit/metro systems. In most cities and countries, however, the high investments and the need for long-range planning for metros and areas they were designed to serve involved government participation or leadership from the early beginnings of this mode.

9.1.1 Transit Origins and the Need for Government Control

The significant inventions of mechanized transit (primarily electric transit from the 1880s and into the early 1900s) coincided with the fast growth of cities and resulted in rapid development of transit. The great demand for transit services and its rapid growth made transit systems an attractive business venture. Most transit companies in the United States were therefore private, sometimes founded in connection with land development and growth of suburbs, which new transit lines were designed to serve and help promote. In many other countries, public (usually municipal) ownership existed from the introduction of electric transit services.

During this period of transit growth, the transit industry was very dynamic and innovative and played a major role in the development of cities. In general, when transit could make a profit, the initiative, drive for expansion, and efficiency of private enterprise operations were used effectively for provision of transit services. However, conditions varied greatly among cities and with governmental policies in different countries.

Initially, some cities allowed free entry into the transit business without any government regulation. These cities experienced several types of problems, which, in time, required either various kinds of regulation, or public takeover. The problems were as follows.

First, with the dominant concern of private companies for economic efficiency and profit, services were often offered mostly on lines with heavy ridership. Services that were socially important but financially nonremunerative, such as lines in low-income areas, low-density areas and services during late evening hours, were often given inadequate attention.

Second, a similar phenomenon existed among companies. Where entry to transit markets was not limited, many companies crowded the busiest lines and arterial streets, while none wanted to serve lightly used suburban routes, offer evening and weekend services, etc. This situation led to the classical cutthroat competition phenomenon: a large company would lower its fares to undercut the competitors in its service area and force them to cease their operations. Once the competitors were eliminated and the large company became the only operator, it would use its monopoly position to maximize its profit at the expense of its riders: fares would be increased and services reduced.

This cutthroat competition led to bankruptcies of weaker competitors, instability of service, and, eventually, uncontrolled monopolies in some areas or entire cities.

Third, uncontrolled competition resulted in many duplicative, uncoordinated services that neither represented an efficient way of providing service nor offered to the riders convenient, connected networks. To the contrary, passengers were often forced to transfer several times, wait for connections, pay several fares, and have difficulties finding information about different lines and systems (Cheape, 1980; Cudahy, 1990; Yago, 1984).

City governments gradually realized that for a number of reasons, even during growth conditions, transit could not be an entirely private industry in which free market practices operated in their rudimentary form. Several factors give transit specific characteristics and distinguish it from purely private businesses.

The first distinguishing characteristic is transit's need for high investments. This is particularly the case with the construction of metro systems, but it is true for other modes as well. The required investments are very high and the role of metros is important for entire cities. Therefore, they have to be provided, or at least partially supported, by the government. Arrangements have varied: some facilities (lines, stations, rolling stock, etc.) were constructed by private developers; in other cases, government guaranteed bonds for their construction; but in most cases, even where a private company operated the system, construction of metro tunnels and other major transit structures was fully financed by the government.

Second, similar to other public services and utilities, such as water supply, sewer system, electricity, etc., transit is not conducive to the classical forms of market competition. While various forms of taxi and bus services can compete effectively against each other in some areas, higher-quality transit services, such as LRT or metro systems, require significant infrastructure and involve substantial investments; they are natural monopolies and therefore are not conducive to competition among different operators providing services on the same lines.

This point became obvious when streetcar, and later bus, companies were competing and eventually had to be taken over by public agencies and integrated. Metro systems were built as partly competitive systems only in very large cities, when there was no auto competition, labor was cheap, etc. In New York, there were at one time three rapid transit companies, two private and one public. They were merged in 1942 to improve financial conditions but continued to operate as separate companies for several years. In 1948, the fare was integrated and increased from 5 to 10 cents. Operational integration was even slower: it was introduced in several stages over the following couple of decades, leading to better services and greater operating efficiency.

Multiple transit services paralleling each other are found in very large cities where concentration of travel is so high that the demand justifies construction and operation of parallel rail lines in many major corridors. However, even in those cases the parallel services often differ in the type and quality of service, station spacings, etc. Typical are parallel local and express rapid transit lines in New York and Chicago; metro and regional rail lines in Paris, London, and New York; and the Tokaido section in Tokyo where four different rail lines, two intercity and two regional, today carry a total of up to 170,000 prs/h in the peak direction.

The third transit characteristic that distinguishes it from market-based systems is that it is an essential public service for the welfare of the population and the economic viability of the city. As such, it is necessary to ensure that transit services are safe, reliable, and stable. Sudden discontinuance of service due to a company's bankruptcy cannot be allowed because of the problems such an interruption would cause. For this reason, transit had to be regulated by the government in various ways from its early beginnings.

Finally, transit has many positive and negative externalities, such as lower air pollution than other modes of travel and positive impacts on land use development, which are of direct interest to the city government but are not reflected in system pricing.

Initially, government regulation usually concerned various aspects of safety and company's financial/managerial capability. Companies meeting these requirements obtained franchises to operate transit services. In some cases, franchises that cities gave included financial benefits to the city, such as obligation of the transit company to maintain entire streets. Aspects of service coordination among different companies were not considered, however. Thus, a number of companies providing transit services would be competing with each other, even though fares were usually controlled.

9.1.2 The Problems with Private Companies

It is sometimes said that transit flourished under private enterprise: it was operated efficiently and it provided excellent services. Historic accounts show, however, that this was not always the case (Yago, 1984). The quality of transit systems varied greatly, and their condition changed in time.

Many transit companies were civic-minded, had efficient organization, and worked on continuous improvements and expansion of services. For example, transit and railroad companies in the United States. built extremely attractive stations in many cities (New York, Philadelphia, Los Angeles) and provided superb passenger services. Some others, however, were primarily interested in maximum short-term profits and sacrificed service quality to their own interests.

The problems with private transit companies and pressures for public takeover were growing in many cities, but there was also strong opposition to this change by the transit company owners and their lobbies. Thus, in Chicago there were three referenda for public takeover of transit systems as early as 1902, 1906, and 1907. Although each of them was approved, the takeover took place only in 1942 (Cheape, 1980; Yago, 1984).

Several developments sharpened the conflict between serving the public interest and pursuing maximum profit, a major issue in transit systems planning and operations. First, fare levels, regulated by the government, became a major political issue because they particularly affected the welfare of large working and low-income classes. This situation led to prevention of fare increases that operators needed to remain financially solvent. In other words, the problems increased when private companies were forced to provide public services that could not be reimbursed from fares and the government did not compensate for such services. When this happened, companies were forced to reduce maintenance of vehicles and infrastructure and decrease quality of services, thus losing riders and further aggravating the problem.

Particularly well known is the national financial crisis of the electric railways (urban transit companies) at the end of World War I, which led to a federal investigation in 1920 (U.S. FERC, 1980). The investigative committee, appointed by President Wilson, confirmed the important role of transit for American cities and suggested a number of very progressive policies; however, during the 1920s and 1930s, these recommendations were largely ignored.

With the subsequent introduction of the private automobile into wide use, public policies in U.S. cities failed to protect transit from the difficulties created by intensified street congestion and reduced transit usage. Moreover, public policies were increasingly favoring the automobile not only at the local but at the state and, particularly forcefully, the national level. Federal assistance to highways was increasing, culminating with the 1956 Federal Interstate Highway Act, which financed nationwide network of freeways and established permanent financing of highways, including many urban freeways and arterials. No financial assistance was provided by federal government for urban transit until the mid-1960s.

Transit also suffered from a major conspiracy by General Motors, Firestone, and Standard Oil companies, which formed an organization, the National City Lines, to purchase transit companies and replace streetcars with buses. A well-documented description of the National City Lines conspiracy, proven by courts, is given by Snell (1974). In most cities, entire streetcar networks, even rail lines with rights-of-way categories B and A, were abandoned and replaced by buses on ROW category C. Mixed with traffic, with lower comfort and much weaker image than streetcars, buses suffered a steady loss of passengers even during these years when they were greatly expanded to the former streetcar networks, from 1946 to the early 1960s.

The deteriorating financial condition of transit companies, degraded services through the closure of many streetcar lines on separate rights-of-way, and intensifying competition of private automobiles led privately owned transit companies to focus more on survival than on the public interest and innovations. Actually, in some cases they began to work against the riding public in order to lose riders and thus obtain government permission to cease unprofitable operations. This was a common phenomenon among railroad companies that wanted to discontinue commuter lines. Put very simply, private transit companies without public assistance could not compete with the automobiles operating on publicly provided and subsidized highways.

9.1.3 Public Takeover of Transit Companies

The growing use of private automobiles and the increased dependence of transit on public policies, which in many countries did not realize the need to balance the relationship between private and public transportation, led to the creation of a vicious circle: increasing auto use → decreased transit ridership and revenue → decreasing services → further diversion of trips from transit to auto. This situation led to the takeover of many private companies into an integrated public transit agency. This change should, in principle, have many advantages and some disadvantages, as is summarized in the box.

Compared to many private operators, a single public transit agency has the following advantages (A) and disadvantages (D):

A-1. A public agency unifies different lines and modes and thus provides an integrated transit network with coordinated services, fares, and information. If this is implemented, the public can use a true intermodal transit system.

A-2. The main goal of the public agency should be providing services the city and its residents need, rather than focusing only on optimal financial results of operations;

A-3. Considerable economies of scale can be achieved by consolidation of many different lines, vehicle fleets, company managements, etc. into a single agency.

A-4. Network integration allows profits from heavily used lines to be used to support lightly used lines that are essential for area coverage, social, or other reasons;

A-5. Governmental public policies can be better coordinated and subsidies controlled with a public agency than with many private companies.

D-1. Financial aspects of operations are more difficult to define and their neglect may lead to serious inefficiencies and deficits.

D-2. Labor unions may utilize their monopolistic position to force excessively high wages and inefficient labor practices. In some states, legislatures revoked the prohibition of strikes against public transit agencies, contributing to this problem.

D-3. The stimulus of competition, which existed mainly in bus operations, is eliminated. If agency managements are not continuously innovative and alert, agencies may become obsolete in operations and insensitive to the needs of the riding public, thus largely defeating advantage A-2, provision of good service as the dominant goal.

D-4. Large regional agencies sometimes orient their attention to regional systems at the expense of local services, which are essential for vital for urban as well as suburban neighborhoods.

Given growing public pressure for better transit services and the increasing complexity of urban transportation, the advantages of the two changes, transit system consolidation and transition to public ownership, greatly outweighed the disadvantages. The transition in some cities, however, did not bring about all the listed benefits unless there were continuing public pressures over the coming years. As mentioned, actual integration of the three rapid transit companies in New York occurred years after their formal organizational consolidation. Acquisition of bus companies and their integration through easier transfers and joint fares actually took several decades to implement. Moreover, many publicly owned transit agencies have not been as public-oriented as had been expected when they were founded.

In some cities, consolidations of many small companies into one large private company had already occurred under private ownership; in others, the two events, consolidation and change to public ownership, occurred simultaneously.

Some cities, mainly in Europe, had municipal ownership of transit companies since the introduction of electric streetcars during the 1890s. Most others had public takeovers between the two World Wars. In Hamburg, many streetcar companies and a rapid transit company were merged in 1918. In London, the London Transport Board was founded in 1933 with the task of "operat[ing] all modes of public transport in the London Transport Area except for taxis and railways."

In the United States, public takeovers occurred roughly 20 years later. The earliest public ownership was in San Francisco (1912), but the Municipal Railway Company had to compete with private companies until 1947. In New York, the three rapid transit companies (BMT, IND, and IRT) were unified in 1942, but the citywide public transit agency, the New York City Transit Authority (NYCTA), was founded in 1948. In Philadelphia, the public agency, SEPTA, began operations in 1967, when it took over the largest private company, PTC. SEPTA took over several other transit operators later, completing regional unification when it acquired the regional rail system from the railroads in 1983.

9.2 OWNERSHIP AND FORMS OF TRANSIT SYSTEMS

Similar to other developed countries, the United States followed the trend of integration and public takeover of transit agencies in all large and medium-sized cities. By 1970, there were 159 publicly owned transit agencies. Although this represented only 15% of the total number of transit operators in the country, they carried 77% of unlinked transit passenger trips in the country. This disproportion is due to the fact that in most large and medium-sized cities transit agencies were public, while most small-town operators were still private.

By 1990, the public takeover had progressed further. At that time, the American Transit Association (now the American Public Transportation Association, APTA) reported that there were 5086 transit systems in the United States; of these, 1580, or 31%, were owned or subsidized by municipalities, counties, regional authorities, states, or other governmental agencies. This number included a category that has also been increasing: transit systems operated or managed by private firms under contract to a governmental agency. These 31% of operators carried 94% of unlinked passenger trips (APTA, 1993). More recent statistics on ownership are not available because their form and categories of statistical information have been changed.

Thus, while the majority of transit systems by number were still private, they were all small, carrying only a fraction of transit passengers. Public ownership included virtually all large and medium-sized transit systems. The dominance of public agencies was similar in other developed countries, with a few exceptions. In Great Britain, deregulation has resulted in dissolution of integrated multimodal agencies (see Section 9.5.2.1). In Japan and Hong Kong, there were different types of transit organizations and special conditions that made operational integration of public and private rail companies possible and successful (see Section 9.2.1), although cities in these countries generally lack intermodal integration of rail and bus systems.

Since the 1980s, however, there have been numerous changes and innovations in transit ownership, regulations, and organizational forms. A significant shift toward privatization in transit systems has taken place. Previously, many public agencies contracted out to private companies various functions and operations, such as vehicle maintenance, repairs, cleaning of stations, fare collection, and marketing, but retained all service-providing functions. More recently, many fundamental changes in transit system regulations, organizations, contracting out of services to private operators, or bringing of specialized companies to manage transit agencies have occurred in many countries.

Ownership and forms of transit organizations depend on many factors, including economic system and legal setup in the country and metropolitan area, traditions in operating private and public enterprises, and physical conditions and political boundaries. Another factor that strongly influences types of transit organizations is the source and method of transit financing. A review of various forms of organizations and types of ownership of urban transit systems, from traditionally private companies to city and regional public transit agencies, is presented here. The new developments in ownership, such as private-public partnerships, changes in regulations, and new forms of transit agencies, are discussed in Section 9.6, following a systematic review of transit service regulations in

Section 9.4 and of unregulated transit services in Section 9.5.

9.2.1 Private Transit Companies

Private transit companies have operated successfully under several sets of conditions, broadly grouped and described here.

Transit companies in small U.S. cities are usually more efficiently operated under private ownership than they could be by special public agencies at small scale. Because of the strong orientation of U.S. cities and towns to the private automobile, transit companies generally cannot cover their investment and operating expenses. Therefore, they usually have some form of direct or indirect financial assistance from the government, such as compensation for special services, including transportation of schoolchildren, charter services, or other arrangements.

Many urban and regional rail and bus transit companies in European countries also operate under private ownership. They either have public assistance or serve under special geographic conditions (mountainous terrain, narrow corridors, across bodies of water, etc.) where they have high ridership. In some cases, private railways operate freight, which indirectly supports passenger services. The best-known examples of this transit category are Swiss private railways, as well as regional rail and bus transit services in Germany, Australia, and some other countries.

Many extensive rail and bus transit systems in Hong Kong and Japan operate successfully under private ownership. Some of the conditions that distinguish these systems from their peers in most other countries and make such operation possible are:

- Population densities are extremely high in Hong Kong and Japanese cities such as Tokyo and Osaka.
- Use of automobiles in these cities is not only limited by space, but it is more expensive than in U.S. and most European cities. Car travel is less subsidized by direct and indirect measures, such as tax exemptions for many trip categories, company car ownership, cheap or free parking, etc.
- Land uses, including major activity centers, are planned with rail transit lines and located around their stations.
- In Japan, many regional rail companies own housing complexes, department stores, shopping centers, amusement parks and other commercial developments whose income is used for partial support of transit operations.
- Although privately owned, many transit companies have various arrangements for cooperation, financial support or guarantees by the government in such aspects as infrastructure investments (e.g., construction of metro systems), social fares, etc.

Recently privatized transit organizations are discussed in Section 9.6.

9.2.2 City Area Public Transit Agencies

Some transit agencies are based in the city and generally serve networks that are contained within the city boundaries. They can have several forms.

A *quasi-private company* is an arrangement to operate transit in the form of a private shareholders company with the majority of shares (51% or more) owned by the city. Used in some German and other European cities, this form combines the advantages of efficient private operation and autonomy from direct political interests with the benefits from city's protransit policies.

A *utility division,* usually the electric company of the city, was in charge of transit system in some cities when transit consisted of electrically powered modes. This arrangement was convenient for financing because electric utility could obtain surplus for its power sale and apply it to the financing of transit. This setup allowed cities to adopt a policy of low-fare transit, which greatly influenced very high riding habits of the population. For example, the city of New Orleans, using this type of organization, for a long time maintained one of the lowest transit fares and achieved one of the highest transit habits in the United States. This form of organization is not common anymore.

A *municipal agency* is owned by the city and operated as a department of the city's government. The advantages of this form, exemplified by the San Francisco Municipal Railway organization and by many *Stadtwerke* transit organizations in Germany, is that the city is directly interested in and responsible for transit, and its policies tend to be more protransit than in the cities with independent transit agencies. Financing is direct and coordination with other municipal functions tends to be easier. On the negative side, transit has to compete for funds with many other pressing needs of the city, such as schools, housing, and police. Another problem is that transit is often subject to intensive local political interests, as well as to the strong influence of labor unions, which have a very powerful position in many city governments.

City-based transit agencies have the advantage of local control over local transit services and good coordination with other municipal functions. Their shortcomings are that they have less financial independence and stability. Moreover, as the cities grow into regions, transit must also grow into a regional system and therefore exceed jurisdiction of the central city (or any other municipality).

9.2.3 Regional Transit Agencies

In most developed countries, cities have grown into spatially large metropolitan areas, but governmental structure has not followed that growth. City governments' jurisdiction remains within the original city boundaries, while surrounding areas have a multiplicity of units: cities, townships, counties, etc. Thus, while these urban regions have common regional functions, problems, and services, such as taxation, air pollution, and transportation, respectively, these issues must be handled by a great number of governments. In most large metropolitan areas, there are dozens and even hundreds of governmental units.

In some metropolitan areas, such as Toronto (Metropolitan Toronto) and Miami-Dade County, there are regional governments with executive roles. In most other areas, there are metropolitan governmental agencies or metropolitan planning organizations (MPOs), which guide local governments in functions, issues, and policies concerning the region. They usually have limited powers to coordinate and approve plans and distribute funds, but no executive powers. In addition to such government or agencies, many individual functions, such as transit, water utility, and air pollution control, are handled by special-purpose government organizations that cover areas concerned with that function, including different government jurisdictions. These governmental organizations, such as school districts, clean air districts, and bistate port authorities, may have the form of an authority, district, or bistate compact, as discussed below.

With respect to transit, metropolitan growth made it necessary for the former urban transit systems to be expanded or supplemented to serve entire regions. Public agencies serving such regions can have several different forms with respect to the governmental owner and type of organization. The most common forms are described here.

A *county-owned transit agency* is founded and owned by one or several counties, and it serves the areas of those counties. When such an agency serves areas in more than one state (this is the case, for example, with many public agencies in New York, Philadelphia, St. Louis, and Washington), the agencies are called bistate or tristate compacts. Characteristics of organization and financing of county transit agencies are similar to those of the municipal ones, but in relation to the county instead of city government.

A *regional transit authority* is an organization created by a state legislature for a special purpose (in this case transit). Its service area is defined geographically (usually as one or several counties), and its task is to provide (build and/or operate) transit systems. Authorities usually operate entire transit systems, such as New York City Transit (NYCT), the Washington Metropolitan Area Transportation Authority (WMATA), the Southeastern Pennsylvania Transportation Authority (SEPTA) in the Philadelphia region (which also includes the Regional Rail Division), and many others. Or a public authority may be an umbrella agency for several transit operators, coordinating their planning, operations, and, in some cases, such as transit federations in Germany, financing. In the New York region, the Metropolitan Transportation Authority (MTA) is an umbrella agency with jurisdiction over NYCT, the Long Island Rail Road, the Metro-North Railroad, Long Island Bus, and MTA Bridges and Tunnels. In the Chicago area, the Regional Transportation Authority (RTA) coordinates the Chicago Transit Authority, which operates rapid transit and buses, the regional rail agency Metra, and the suburban bus agency PACE. The transit federations in Germany (see Section 9.3.2) have the same role but with heavier involvement in planning and financial transactions of their members than the umbrella agencies in U.S. metropolitan regions.

An authority is usually created by an act of the state legislature. Its governing board is appointed by regional or state governments. This type of organization is relatively independent with respect to the network it serves, service standards, fares, and financing. However, most authorities are not financially independent. They obtain funding from local or state governments, but they do not have a taxing power, so that they can issue only revenue bonds, but not general obligation (GO) bonds.

In some cases transit authorities are parts of larger, well-financed authorities. For example, the Port Authority Trans-Hudson (PATH), an agency operating a high-capacity rapid transit system between New Jersey and New York, is a part of the New York-New Jersey Port Authority. This authority owns and operates a number of airports, bridges and port facilities in the New York-New Jersey-Connecticut Tri-State Region. The Port Authority Transit Company (PATCO), operating the Lindenwold Line between Philadelphia and its New Jersey suburbs, is a part of the Delaware River Port Authority, a bistate agency that owns and operates the Delaware River Port and several bridges.

A *regional transit district* is similar to an authority: it is a special-purpose organization with a service area exceeding single governmental jurisdiction (city or county), created by an act of a state legislature. The essential differences from an authority are that a district usually must be approved by a voter referendum by 50, 60, or 66.7% of the votes (this requirement varies among states and cases), and it has taxing power. That power gives the district legal right to issue not only revenue, but also GO bonds. Being backed up by the tax base of the district's residents, GO bonds are very secure and are therefore sold at lower interest rates than revenue bonds. With the stronger financial base and ability to issue low-interest bonds that most districts have, they are more powerful organizations than authorities.

An excellent example of transit ownership and organization, closely related to financing, is found in the San Francisco Bay Area. When it adopted the law that a rapid transit system should be built in the Bay Area, the California legislature also prepared a law about its organizational form and method of financing. Based on a study of many possible organizational forms (SRI, 1956), the legislature decided to create a Bay Area Rapid Transit District (BARTD), which would serve three counties (San Francisco, Alameda, and Contra Costa) and have taxing power to raise funds for the construction of the rapid transit system (BART).

This plan was submitted to the boards of supervisors of the three counties and adopted in 1961. The $792 million bond issue to finance the project was then brought to a popular referendum in 1962 and was approved by 61.2% (a majority of 60% was required). BARTD then issued GO bonds to obtain capital for the construction of the system. Additional funds were obtained from the tolls generated by the Bay Area bridges except the Golden Gate Bridge for the construction of BART's Transbay Tube from San Francisco to the East Bay, paralleling the Bay Bridge. Revenue bonds were also issued to cover the costs of the rolling stock (no federal funds were available at that time).

The BARTD organization successfully built the system and has been governing its operation since its opening in 1972. The most important subsequent extension has been to the San Francisco airport, opened in 2003. That extension required special organizational arrangements because it lies outside the three counties that initially comprised the District. San Mateo County, where the line extends, made a special contract and payment to join the BART District and contributed funds for partial payment of that construction.

Most characteristics of authorities and districts, such as their legal status and method of their approval by legislature, appointment of their board members, and even taxation powers, vary somewhat among states. The given features are most typical, but there are cases in which authorities also have taxing power, that districts need no voter approval, etc. Some of these specific characteristics depend on the laws and constitutions of different states.

A *state-owned transit agency* is a logical type of transit organization for relatively small states with many towns and urbanized areas, such as Delaware, Maryland, New Jersey, and Rhode Island. Although Maryland and New Jersey are not geographically small, they are so developed that full integration of their transit services into one agency (with several divisions, contracted services, and other subdivisions) was the logical organizational setup.

9.3 INTEGRATION OF TRANSIT SERVICES PROVIDED BY DIFFERENT OPERATORS

Despite the strong trend toward consolidation and integration of multiple private transit companies, as well as some public agencies, into single systems between the 1920s and the 1960s, transit services in many cities, particularly large ones, remained partially fragmented. For example, regional rail systems in most European cities remained under ownership of their national railways. In New York, the three rapid transit systems were united with a public takeover in 1942, but bus systems remained separated from rapid transit with respect to fares well into the 1990s. Integration of the three major regional rail systems (LIRR, Metro-North, and NJT) is still not complete. These railroads share tracks and stations, but there are no through services and their fares and information are only partially integrated. In Tokyo, there are two major subway agencies (one public, one private), which have separate fares. Their services are well integrated with the Japanese Railways and several private railway companies, but there is little coordination between rail and bus systems. Actually, the separation of rail and bus services is typical for most cities in Japan, Korea, South Africa, Mexico, and many other countries. In general, the greatest progress in integrating services has been achieved in the Western European countries, the United States, and Canada, while in many other countries the process of integration is either limited or nonexistent.

This lack of full integration among transit systems has serious consequences. While passengers are most directly affected, the operators also incur inefficient operations and losses; ultimately, the entire urbanized regions suffer indirectly but significantly from such conditions.

There are many different obstacles to achievement of full integration of transit systems. They can be classified into several categories: historic (traditional type of operation, incompatible rolling stock, etc.), political/legal (different geographic jurisdictions, legal restrictions on areas and types of services), and organizational (different labor unions, city transit and railroad company as operators, etc.).

When full service integration cannot be achieved, several solutions can be applied to achieve partial service integration or coordination. These partial solutions may be permanent, or they may be used as steps toward full integration when the legal and other obstacles can be resolved.

9.3.1 Limited Functional Integration

Two rather common forms of offering to the public services that are unified to various degrees are tariff associations and transit communities.

Tariff associations consist of contracts on joint fares or tariffs and the distribution of the jointly collected revenues. These associations are limited to situations where the partners do not compete with each other but usually have end-to-end line connections.

Transit communities bind the partners not only to a common tariff, but to coordinate routes and schedules; if appropriate, they pool their rolling stock. This arrangement is used by freight railroads in the United States and most other countries, as well as by subway and regional rail services in Tokyo and several other Japanese cities.

Full integration of all aspects of services for the public provided by different agencies is known as *transit federation* or, by its original name in Germany, *Verkehrsverbund*.

9.3.2 Full Functional Integration: Transit Federation

A formal federated agency is established with certain powers related to planning, tariffs, revenue distribution, etc. Thus, the transit federation is an organization that provides fully integrated services although its partners remain separate agencies.

The basic philosophy that led to the invention of the transit federation was developed in Hamburg in the mid-1960s (Pampel, 1969; Homburger and Vuchic, 1972; Interplan Corporation, 1973; Pucher and Kurth, 1996). The motivation to create a federation was recognition of the fact that while the automobile offered direct travel among all points in the city, transit not only required access to stops/stations and waiting, but often also involved indirect travel, double fare payments, and poor coordination among lines. These organizational obstacles existed due to historic ownership conditions, which actually did not interest the user. In order to compete better with the automobile, transit must provide a single information system, unified and logical fares, and minimum friction in transferring among lines and modes (seamless intermodal travel). For that purpose, the first federation, Hamburger Verkehrsverbund (HVV), encompassing the main transit agency operating metro, tramway, bus, and ferryboat lines (HHA), regional rail (S-Bahn operated by the German Federal Railways), and several public and (then) private bus and ferry companies, was founded in 1965.

The transit federation, HVV, is legally required to provide for the region an effective and economical transit system independent of administrative boundaries within the region and ownerships of facilities. HVV is an umbrella organization that has the following major functions:

- Planning of transit networks, lines, and transfer points
- Preparation of coordinated schedules and fleet assignments
- Preparation and introduction of a joint tariff
- Collection and redistribution of the collected revenues on the basis of a contractually adopted formula
- Public relations, including schedule publication
- Preparation of applications for financial assistance from different government levels

The partners provide services and collect and transfer fares to HVV. The formula for the redistribution is based on the provided services multiplied by unit costs of bus, rail car, or train kilometers, and a historic index that expresses the operating ratios of the partners prior to joining the Federation. Thus, the share of total revenues due partner a, R_a, is expressed as:

$$R_a = (R_t - C_o) \frac{I_a \sum_{x=1}^{m} A_{ax} c_x}{\sum_{a=1}^{n} I_a \sum_{x=1}^{m} A_{ax} c_x} \qquad (9.1)$$

where:

R_t = Total revenue of all operators given to HVV
C_o = Operating expense of HVV
I_a = Historic index of partner a, i.e., his operating ratio prior to joining HVV
A_{ax} = Quantity of service item x, such as line-km, vehicle-, train-, and seat-km, provided by partner a;
c_x = Unit costs or relative weights of respective A_x quantities
m = Number of service items used in cost computations
n = Number of partners in HVV

The removal of barriers among modes and services provided by different agencies brought by the introduction of HVV in Hamburg allowed more direct and faster travel. A unified fare system and better information and image all led to excellent results: improved operating efficiency and attraction of new riders. The success of Hamburg led to the founding of transit federations in many cities, including Munich (MVV) in 1972, Frankfurt (FVV) in 1974, Stuttgart (SVV) in 1978, Vienna in 1985, Zürich in 1990, and many other urban regions in Germany, Austria, Switzerland, and other countries. One of the most comprehensive federations is in the Ruhr Region in Germany (VRR). Founded in 1980 after several years of planning and negotiations, it covers the Ruhr Region, including large cities like Düsseldorf, Duisburg, and Essen, includes transit operators from a total of 22 cities, and serves a population of 7.4 million persons. Each federation varies somewhat in its legal, financial, and organizational arrangements, but the basic concept, provision of fully integrated transit services for customers by different agencies, is the same in all cases. Naturally, in many regions, such as Hamburg and Rhein-Ruhr, the initial organizational setup and financial arrangements, particularly the formulas for revenue redistribution, unit costs of different modes, and fare structures, have been renegotiated and revised from time to time to adjust to the changing travel patterns, economic changes, new federal financing programs, organizational changes with partial privatization, etc.

The transit federation has been considered the most effective form of organization for providing integrated services for the public where complete merger of different operators is not possible. Other forms of integration of services provided by different operators vary with types of services, agencies involved, particular legal opportunities and limitations, etc. For example, the three first RER lines in Paris, A, B, and C, are fully integrated in services, fares, information, etc., although one of them is owned by the transit agency (RATP), another is owned by French Railways (SNCF), and the third is under their joint jurisdiction. Moreover, these three RER lines are also integrated with RATP's extensive metro system with respect to transfers (joint stations), fares, and information.

9.4 REGULATION OF TRANSIT SERVICES

As a public service affecting the welfare of not only transit users, but the entire metropolitan area, various aspects of transit organization and operation must be regulated by government agencies or other regulatory bodies. The aspects and degrees of regulation vary among countries, cities, transit modes, and system characteristics, as well as various local conditions.

9.4.1 Purposes of Regulation and Its Categories by Subject

The purposes of transit regulation are numerous and often mutually overlapping, but they can be classified into four major categories, listed in the box with their main objectives.

Due to the diverse nature of regulations, they are performed by several different agencies, which may be at the local, state, or federal/national level. Each of these four categories will be discussed, with emphasis on economic regulation, which has been the subject of considerable debates and experimentation since about 1980.

9.4.1.1 Safety Regulation. Ensuring a high standard of safety is the basic function of governmental control of transit, as well as of highways, air transport, infrastructure, and many other systems and services.

Safety standards apply to various items concerning:

- Structural elements and design of transit infrastructure, particularly for the exclusive rail transit facilities, such as bridges, tunnels, tracks, switches, and signals
- Vehicle design and performance, including structural strength, braking capabilities, fire resistance, etc.
- Frequency and scope of vehicle maintenance
- Operational practices, such as crew hours and duties, vehicle control, and signals
- Training of drivers and of other personnel

Safety standards are usually the highest for fully controlled metro and regional rail systems as compared to buses and paratransit vehicles. Their control tends to be stricter and easier to maintain in large than in small agencies or companies.

There is sometimes justified criticism of excessively strict or obsolete safety regulations and standards, which make operations slower and less efficient. Even more seriously, some types of operations cannot be performed because of excessive regulations. An example of obsolete regulatory measures is often found

By their subject and purpose, regulations can be classified in four categories:

1. Safety and security:
   - To ensure safety of the traveling public and other potentially affected parties
2. Economic (franchise, operations, fares):
   - To ensure reliability and permanence of services and prevent individual and societal losses that temporary or permanent service cessation would cause
   - To protect the public from being overcharged (excessive fares)
   - To prevent losses to the operators due to duplication of services, unfair competition practices, cutthroat competition, etc.
3. Technical and social aspects:
   - To achieve or test certain technical/operational innovations
   - To provide certain level of mobility and special services for some population groups, geographic areas, types of trips, etc., which are considered socially necessary or desirable; for example, accommodation of the disabled, discounts for schoolchildren and the elderly
4. Impacts:
   - To minimize negative and maximize positive impacts on surroundings, environment, urban character, and quality of life

in rail transit safety rules. In some cases, safety rules and standards developed for long-distance trains are applied to regional rail and rapid transit systems, where they are not necessary because of different operating and physical conditions. For example, the PATH system between New York City and Newark, New Jersey, is subject to railroad standards and rules, although it is a typical rapid transit system. These rules impose very high required buffeting strength of vehicles, causing excessive weight and increasing operating cost. Unlike long-distance railroads, however, PATH has fully protected rights-of-way, no grade crossings, short signal blocks, emergency brakes and fail-safe signals that allow operation of very short headways, etc.

Another example of regulations in the United States that need modernization is the case of the tram-train system: LRT vehicles run in city streets and then proceed to railway tracks where long-distance high-speed trains also operate. This tram-train system has been made feasible in several European countries through technical innovations making LRT vehicles compatible with railway track, power supply, and signaling systems, and through modifications of railway safety standards and rules: instead of excessive attention to minimization of damages in the case of collision, the control system has been perfected to eliminate the possibility of collisions. This allowed joint operation of high-speed trains and LRT transit units on the same tracks. In the United States, railroad standards still do not allow this type of operation.

Consequently, there is sometimes a need to review and modernize safety standards, many of which are based on railroad operations 100 or more years ago, and eliminate those that unnecessarily impede or prevent joint operations of transit systems (LRT, metro, or regional rail) not meeting railroad standards that should not apply to operations in urban areas in transit regime operations.

There is no question or controversy about the need for government regulation of transit safety. But there are many cases in which traditional railroad standards need revisions to reflect contemporary technology (particularly signal and other controls of vehicle movements), particularly in transit systems' uses of railroad tracks.

9.4.1.2 Economic Regulation. This category of regulations is needed to ensure that the public obtains a desired level of service and that uncontrolled competition, often ruinous for the competing operators, is prevented. It consists of controls over:

- *Franchise,* or the legal right and obligation for a company to operate services
- *Fares* that are charged for each service
- *Quantity and quality of service*

Franchise, or permission for an agency or company to provide service on a given alignment, gives the right and obligation to the operator. It can be of two basic types with respect to the scope of control: minimal control, limited to safety and reliability of service, or also considering system aspects—coordination and quality of service in the transit network.

In the case of minimal regulation, the requirement for obtaining franchise is limited to proof of economic and managerial ability to provide service with certain prescribed safety and other standards. Transit is treated as a free market, with entry open to any qualified operator. This type of minimal regulation (or even no regulation) may result in dynamic competition and interest of potential transit operators, but it can also lead to cutthroat competition, low service quality, and bankruptcies of some competitors. It exists in many developing countries.

A more advanced type of regulation considers systems aspects. It either limits, controls, or prevents competition in order to achieve an integrated, fully coordinated system. This type of regulated monopoly can provide an optimal system if the transit agency is adequately funded and competently managed. If these conditions are not met and services deteriorate, it may be necessary to review and modify methods of funding, regulations, and management practices.

There are variations between these two types of franchise regulations, combining different degrees of entry control and system coordination. There may be regulated operation of two services that partially compete on sections of lines. In developing countries, a public agency often competes with partially controlled or even uncontrolled competing private buses, minibuses, and a variety of jitneys.

This combination of a single public agency with uncontrolled entry for private competitors was abandoned as counterproductive in most developed countries several decades ago (although it was revived with deregulation in Great Britain, which is discussed later). The reason is that in such cases the public agency is obliged to perform services on an extensive network of lines during all daily hours. However, private buses and jitneys concentrate on most lucrative lines and during peak times, while they leave less lucrative services to the transit agency. This practice, popularly known as skimming the cream, naturally results in clear advantages for the private operators and places the public agency in a greatly inferior position. When an agency requires public subsidy, it is often unjustly criticized for its "inefficiency." This combination of one regulated operator competing with unregulated ones usually results in provision of two types of services with different qualities or classes and unstable economic conditions of all operators.

Regulation of fares also varies greatly in its extent and form. The rudimentary situation, without any control of fares, results in overcharging of passengers, inequitable fares, and the well-known phenomenon of fare wars among transit operators.

Fare wars have happened in unregulated transit services, as well as in other transportation systems. They occurred frequently among competing passenger and freight long-distance railroads in the nineteenth century, and they were one of the main reasons for introduction of government regulation (the Interstate Commerce Commission was founded for that purpose in 1887). Fare wars have become a common phenomenon and cause of bankruptcies among U.S. airlines since their deregulation in 1978 (Eastern, Pan-Am, and dozens of others). It has also led to mergers and takeovers and caused bankruptcies of many foreign airlines (Swissair, Sabena).

In an unregulated transportation market, with free entry of many competing companies, the situation often leads to cutthroat competition (see Section 9.1.1) with lowering of fares at the expense of service quality.

When the financially stronger company eliminates competitors, it obtains a monopoly, which allows raising of fares to the level of "what the traffic will bear," without service improvements. This phenomenon, widely used by passenger airlines, would have socially detrimental impacts if used in urban transit. This is observed in many developing countries, where unregulated competitors, particularly jitneys and other forms of paratransit, undercut major bus and rail companies and provide services with low reliability and safety.

Thus, fare wars and cutthroat competition are destructive for several reasons. They create inequities and hurt mostly low-income people because they are based on the principle of charging maximum fares that any one category of passengers will bear. They usually have little relation to the cost of providing transportation and thus lead to economic distortions and eventually the bankruptcies of some, as well as losses to all competitors.

One type of regulation is control of minimum fares (floor on fares). This prevents cutthroat competition, but it does not protect passengers from being overcharged. If there is a single operator on a line, he can charge any amount; if there are two, they may do the same through collusion if the control prohibiting that practice is not adequate.

Another type of limited regulation may prescribe the maximum fares (ceiling) to prevent excessive charges for the public; this control, however, does not prevent fare wars. This fare level always eliminates many riders, creates inequities, and hurts mostly the low-income groups.

Finally, the most common type of regulation is full control of fares. This allows rational decisions about fare level and structure and prevents fare wars. A problem may occur if political pressures are allowed to have a strong influence on the regulatory body. Political influences often keep fares excessively low, preventing a transit agency from obtaining the funds needed for normal operations, expansion, and modernization.

Quantity and quality of service are often regulated or defined by technical standards to ensure that transit offers the level of service deemed required for the city as a whole, for individual lines, areas, and neighborhoods, and for different population groups. The standards in this category include such items as service frequency, reliability, speed, comfort (e.g., percent of passengers seated), ventilation and air conditioning, vehicle cleanliness, etc.

9.4.1.3 Regulation of Technical and Social Aspects. As a regulated public service, transit is often used by government to develop, test, or demonstrate new equipment, operating standards, and various innovations. Examples of these regulations are prescribed maximum noise and air pollution levels; fire safety standards through selection of fire-resistant materials, requirements for fire protection, and evacuation procedures; testing of new types of cleaner bus engines; operation of new designs of electric motors and their controls; continuous signal block systems; fully automatic train operation; and many others.

Extensive regulatory requirements are placed on transit with respect to various social aspects of services, such as special fares for schoolchildren, discounts or free travel for senior citizens, and special equipment as well as accommodations or special services for the disabled (see also Section 6. 5).

In many cases, the government provides special funds and mandates demonstration and introduction of certain technical innovation. This facilitates improvements of transit systems and often provides spillover benefits—innovations for other transportation systems too.

A major problem for transit systems, however, is when changes, innovations, or special services are mandated by the government without allocation of special funds that would cover the additional costs of meeting the mandate. Examples of such unfunded mandates are in many cases construction of facilities to accommodate disabled persons, purchase of buses with stricter air pollution control than required from other vehicles, and compliance of some rail transit systems with railroad standards, discussed above.

In many cases, these unfunded mandates represent an illogical societal policy. Generally, most such mandates concern service provision, technical innovations, or social improvements that are very progressive and socially desirable. As such, these measures should be provided and financed by the society as a whole. However, failure to provide any special funds for implementation of the required provisions on transit systems actually transfers this problem from the society to transit users only: funds for them directly reduce the quantity and quality of basic services a transit agency can provide. This situation is not only politically and ethically incorrect, but it aggravates the problem of inadequate funding of transit in the United States, particularly in comparison to the ample and seldom critically evaluated financing of highways.

9.4.1.4 Regulation of Impacts. With the growing complexity of modern economies and societies, there is an increasing need for careful examination of the numerous impacts transportation systems have on their surroundings, on urban development, and on the quality of life in the short- and long-run. These impacts are discussed in Section 11.3.

In most developed countries, there are now numerous laws that regulate the impact of transit systems on such aspects as air pollution, noise, nature, and displacement of residents. In the United States it is a legal requirement that for each project involving federal funds an environmental impact study (EIS), a detailed, comprehensive study of its impacts, be performed. It is important to bear in mind, however, that these impacts should include not only negative, but also positive impacts. In some cases, opportunity costs and benefits of the project must also be included. For example, if a rapid transit line is not constructed, what will be the impacts of further growth of automobile use on the city? Or how will construction of a high-performance transit system improve the environment and human-oriented character of the entire urban area in comparison to the alternative—heavy dependence on the automobile only?

9.4.2 The Role and Importance of Transit Regulation

Since the beginnings of urban transit, it has been clear that these services must be subject to certain government regulations, for the reasons defined in Section 9.1. As time passed, cities grew and expanded into multigovernmental regions. Economy, society, and environment have become more complicated, and the need for regulations has grown to encompass not only safety and franchise, but also environmental, social, and economic impacts.

Regulations were generally expanded in response to the need for ensuring safety, reliability, reasonable fares, and other features of a contemporary urban transit service. On the other hand, as the complexity of regulations increased, some regulatory processes became very complicated and slow. Thus, regulations became an instrument for securing high-quality transit systems in response to the needs of modern urban areas; on the other side, in some cases regulations began to slow down operations, make them more expensive, prevent increases in labor productivity, impede planning and construction of new facilities, etc. Consequently, the problem in complying with regulations has become how to provide the necessary functions and desired effects of transit without incurring excessively complex procedures, costs, and delays.

There is a general consensus that the government must regulate safety aspects of transit services, although the degree of such regulations varies among countries, transit modes, and transit organizations. For example, safety standards and their enforcement tend to be higher in developed than in developing countries. Among transit modes, due to their size, technology characteristics, and ROW control, regional rail and rapid transit systems are subject to the most detailed and strictly enforced safety standards; the LRT and bus modes follow in descending order. Paratransit modes are far less regulated and vary from strict controls and high quality of taxi services in some cities, such as London, to very little control in others or operation of unregulated "illegal jitneys" with low standards of service quality and even basic safety in many developing and some developed countries.

Technical and social regulatory measures have become more sophisticated in recent decades. They include testing or use of prototype vehicles, fuels, other equipment, ITS devices or programs (see Section 6.4.3), operations control, communications, and public information systems. New equipment or technologies are sometimes targeted for different user groups, such as information systems for regular users, auto drivers who may be attracted to transit, and tourists visiting the city.

Some social regulatory measures have improved accountability of revenues and governmental funds. For example, low-fare or free transit for senior citizens during off-peak hours is now common in cities of most countries. These fare reductions are often financed from special earmarked funds or taxes. Reduced fares for students are also very common in many cities, but there has been a trend for school authorities or government to reimburse transit agencies for these services. Thus, instead of forcing agencies to absorb these costs and practically pass them into general subsidy, reimbursement allows transit agencies to have easier accounting and control of costs. Tax exemptions on certain types of fares have been introduced, partially compensating for a variety of costs in automobile use (gasoline, parking, tolls, etc.) that are exempted from taxing.

Impact controls, particularly environmental protection, have also been increasingly applied and supported by the public, although there are considerable differences in their scope and methods among cities of different countries. It is likely, however, that the environmental impact regulations will continue to be very important and will require the special attention of transit operators.

Among all categories, economic regulations are subject to the greatest controversy, which is largely based on the degree of society's development and on prevailing ideology of different governments. Many free market proponents are opposed to economic/financial regulations, while public interest advocates support them strongly. Two situations without economic regulations—unregulated transit in developing countries and deregulated transit in a few developed countries—are discussed in the following two sections.

9.5 UNREGULATED TRANSIT SERVICES

Transit services that are subject to limited regulation or no regulation at all by governments are reviewed in this section. They are divided into two groups: those in cities in developing countries that have never had established, well-organized transit systems, and those in cities in industrialized countries that have had well-organized transit systems but in recent years have applied deregulation and experienced fundamental changes in the organization and type of their transit services.

9.5.1 Unregulated Paratransit in Developing Countries

In many cities of developing countries in Africa, Asia, and Latin America, a large, sometimes dominant share of transit services is provided by a great number of competing private companies, organized groups, and individuals. They utilize buses, minibuses, vans, passenger cars, and other road vehicles in a variety of transit and paratransit service types, such as bus and minibus lines, jitney services, and group taxis. In most cases, unregulated transit operators compete with publicly owned transit systems; in some, rather exceptional cases, the two are coordinated so that their services complement each other.

These diverse paratransit services, particularly unregulated ones, are used extensively in developing countries where the following conditions exist:

- Very high demand for travel by captive riders (low auto ownership)
- Lack of satisfactory, or even basic, transit services
- Low fares more important than high service quality in terms of safety, comfort, reliability, etc.
- Low-cost labor
- Lack of a rational transportation policy that would coordinate modes and increase efficiency of the entire system, the condition that favors short-term, immediate pay-off solutions over long-term, more efficient solutions
- Lack of organizational ability to introduce priorities for transit vehicles on streets (bus lanes, traffic signal priorities, etc.)
- Lack of funds for investment in higher-performance modes, such as BRT, LRT, or other rail transit modes

Photo 9.1 *Unregulated minibus services—Mexico City.*

9.5.1.1 Services to Users. Competitive private paratransit services vary greatly among different countries and cities with respect to their ownership, organization, quality of service, safety conditions, etc. Yet most of them have the following common positive features that make their role in many cities important:

- Paratransit provides very frequent service wherever high demand exists: along major travel arterials, between and within central cities and major activity centers.
- Stopping only on call, they travel faster than transit buses which serve many stops.
- Most paratransit vehicles have seating for all passengers.
- Some paratransit services distribute passengers from major lines to different destinations on their outlying sections.

The limitations and deficiencies of paratransit services, particularly unregulated ones, are usually the following:

- Paratransit services on routes, in areas, and at times when travel demand is low are infrequent and unreliable
- Safety standards are low and liability insurance for passengers is usually nonexistent.
- In some cases, overcrowding causes poor riding conditions.
- Since competing services do not represent an integrated system, passengers have great inconvenience in getting information about routes, transferring, paying different fares, etc., especially because there are no schedules, maps, or other written information. Their use by other than regular transit users is very difficult.

- If fares are not regulated, operators may overcharge passengers.

9.5.1.2 Transit System Aspects. Observing paratransit as a system, one can define its following positive characteristics: (+) and deficiencies (−):

+ Private operators or organizations in competitive market situations can provide services in response to travel demand in a short period of time.
+ Private paratransit (jitneys and taxis) are self-sustaining, needing no government subsidies.
+ They provide employment for drivers and mechanics, albeit often at substandard wages or owner-income levels.
− In cities with poor traffic regulation and enforcement, jitneys greatly contribute to chaotic street conditions (Bangkok, Caracas, Lagos, Manila).
− In large cities, paratransit is much less efficient than regular transit because its vehicles occupy more space and produce more pollution per passenger-km than regular transit.
− Paratransit does not represent a system that can be coordinated and upgraded by provision of transit lanes, preferential signals, and higher-quality vehicles.
− Paratransit operators create a powerful lobby that opposes any upgrading of services by introduction of higher-quality regular bus or rail services.
− When operating conditions become unfavorable for paratransit, a large number of operators may go out of business in a short period of time, leaving the city without services.
− When paratransit is unregulated, planning of transit is extremely difficult because there are no data on the number of operators, their operations, passenger volumes and origin-destination patterns, etc. Without information and data about existing conditions, it is not possible to perform any planning, and it is difficult to implement any changes.

9.5.1.3 Policy Aspects. Opinions about competitive paratransit services are rather sharply polarized. One school of thought, particularly represented by market-oriented economists, gives paratransit, including unregulated services, a very positive evaluation. Their main argument is that jitneys and taxis usually require no public investment or subsidy and very little intervention by governments, which are often not very efficient in policy-making, particularly in many cities in developing countries. They also point out that paratransit provides fast and frequent service and that its supply responds to demand in a short period of time.

Transportation planners and urban experts analyzing quality of life and social conditions in these cities tend to have a different view. These professionals approach transit system evaluation through the service it offers and role it has in the economy and the social and environmental character of the city. They agree that paratransit, including unregulated services, is vital in many cities where the public sector cannot satisfy the demand. However, they point out that jitneys and similar services are in most cases unreliable and offer very low comfort and safety. Paratransit does not represent a system that is easily accessible for all potential users, it is difficult to control, and it is virtually impossible to coordinate or upgrade. In summary, paratransit is a collection of independent services, rather than a system that can be planned and controlled.

Many large cities in developing countries, like Bangkok, Lagos, and Lima, are chronically congested and have poor transit but extensive, largely unregulated paratransit services. How satisfactory are these services for the cities' populations? Generally, the cities' residents use such services extensively (if their fares are low), despite their usually low reliability and service quality, because they have no other choice. Persons not familiar with local services have great difficulties finding which services are available and how to use them. Car owners tend to avoid paratransit because of its poor quality and frequent overcrowding.

Government officials and transportation experts in these cities realize that improvement of regulated transit is important to improve mobility of population, attract some car owners, and decrease street congestion. Many cities in developing countries have made major efforts to reduce the role of paratransit, particularly unregulated services, by introducing high-performance transit. Metro systems have been constructed in cities like Mexico City, Cairo, São Paulo, and Santiago; light rail lines have been constructed in Manila, Mexico City, Tunis, and other cities. Bus services with high performance have been introduced through investments in infrastructure and major organizational changes in São Paulo, Curitiba, Bogotá, Quito (trolleybuses), and other cities. All these systems serve very large volumes of passengers and provide much more efficient transportation than paratransit.

Limited funding is often a great problem in developing countries, and some international financial institutions oppose infrastructure investments and exert pressure to stimulate paratransit services, which are by their nature difficult to regulate, particularly in developing countries. Encouragement of unregulated services results in lower government expenditures for transit, but also in lowering the quality and image of paratransit. Very significantly, extensive use of unregulated transit results in increased separation of travelers into two classes: predominantly low-income paratransit users and middle- and high-income auto users. Traffic conditions in such cities continue to be disorganized, and congestion chronic, without any preferential treatment for transit or construction of higher-performance transit systems.

Introduction of unregulated paratransit services in cities that had good transit systems, such as Santiago, Mexico City, and Cape Town, has resulted in diversion of many transit trips to unregulated services. While government subsidies have been reduced, the quality of services has degraded and become disorganized and unreliable. Authorities are trying to correct these conditions by a more fundamental organization of all services. For example, in Santiago a major reorganization of bus services has been prepared by the city's government. A major effort has been made to overcome minibus owners' lobbies and introduce a coordinated system. An extensive, very chaotic network of overlapping and competing independent bus lines is being replaced by a coordinated network of trunks and feeders with superior service, greater efficiency, and less street congestion.

In perspective, competitive private transit and paratransit in developing countries offer valuable services, and in many cities they provide vital mobility for the population. They do, however, suffer from the same problems that had existed in the cities in the western industrialized countries prior to the introduction of transit regulation and integration of services: most of the competitive private services are not coordinated with regular transit services and tend to collect the cream of the crop, leaving less remunerative lines, evening services, etc. to be served by public transit agencies.

Unregulated services are not reliable, transferring among them is inconvenient, and they are not easily usable by first-time travelers. Drivers of unregulated buses and jitneys are often inadequately trained and paid substandard wages. Most importantly, the overall urban transportation system cannot be coordinated and upgraded until regulation is introduced. Private competitive transit is not easily conducive to intermodal integration.

9.5.2 Deregulation Experiments in Developed Countries

The function of transit regulation, which is related to its ownership (private or public), has been debated in some developed countries since the 1970s. Because this problem is closely related to political ideology, there has been a broad spectrum of views and considerable controversy on the subject.

There is a general consensus on one side that control of safety in public systems such as transit is clearly the responsibility of the government. Therefore, regulation of safety aspects is not questioned by any group. There is also a consensus on the other side that there have been cases where regulation has been excessive and caused inflexibility in operations, prevented innovations, and caused inefficiencies.

There are many cases of this type of regulation. Examples are long legal procedures for changing alignments of individual bus lines required in many countries, rigid interpretation of some requirements for accessibility by disabled persons, classification of some rail transit systems as railroads, legal requirements that labor unions approve all investment projects, etc.

Except for these agreements about the basic need for safety regulation and about its harmful excesses, there is an intense controversy about most aspects encompassed by economic regulation of transit services. As defined in Section 9.4.1.2, this area includes aspects concerning franchise, entry control, ownership of transit systems, fare levels, schedule and service standards, etc. There are two groups of opinion about these aspects: proponents of deregulation and proponents of integrated transit systems (i.e., opponents of deregulation). Their respective arguments and proposed solutions are summarized here, referring mostly to the debate and events in Great Britain, where it has been most intensive and where deregulation has been implemented. In the following sections, the developments in Great Britain are described and arguments for and against deregulation are summarized. Then actual results of deregulation in that country are presented and analyzed.

9.5.2.1 Deregulation in Great Britain. The strongest proponents of transit deregulation have been political leaders and a group of economists in Great Britain. With the electoral victory of the Conservative party in that country in 1979, the government began to introduce what was promoted as a free market in transit. This marked the beginning of the most intensive debate on and largest experiment in transit deregulation in a developed country, which has been watched by other countries.

As a part of the Conservative government's campaign to privatize various sectors of the economy, reduce the role of government, and reduce the power of or completely eliminate labor unions, the British Department of Transport (DOT) issued in 1984 a White Paper entitled *Buses* (U.K. DOT, 1984). The report presented a critical analysis of the condition and trends in bus transit in Britain, as well as a government proposal for its drastic reorganization—privatization and deregulation. The report started an extensive debate and controversy and led to enactment of deregulation and its implementation, with results that have been analyzed and debated by transit operators, civic and political leaders, and academic researchers ever since.

The approach of the report and subsequent arguments by proponents of deregulation have been limited mostly to the questions of bus systems and their operations. Other, broader relevant issues, such as the need for intermodal integration, the role of transit in cities, transportation policy, and general problems of urbanized areas, are hardly mentioned. Transit is treated as a commercial enterprise that should consider public and social interests only in some cases.

Actually, even transit is not treated as a system: *Buses,* as its title indicates, presents bus transit as an entity independent from other transit, such as urban rail and regional rail, then operated by British Railways; competition of private transport (automobiles) was not considered to be an important factor, nor was the problem of negative impacts of excessive automobile use on cities discussed!

Buses correctly pointed out some serious problems in bus transit in Britain (which also existed in some other countries to various extents):

- Operating costs were increasing faster than inflation.
- Services were not always responsive to public needs because they were overregulated and transit operators had a monopolistic position.
- Some transit agencies were not innovative because of their protected monopolistic position.
- Minibus services, which could be cheaper and thus offer higher frequency, were not used because of high wages and lack of initiative.
- Labor unions blackmailed agencies into excessive wages, often capturing a large share of any increases in public subsidies.
- These conditions resulted in a decreasing share, or even in a decreasing absolute number, of transit passengers, further eroding economic viability of transit systems.

The reason for all these problems, claimed the report, was the organizational setup of transit services: government intervention through regulation and subsidies created a monopoly and lack of competition, labor unions wielded too much power in their demands for wage raises and benefits, there were various demands for public services, efforts to coordinate modes were too costly, etc.

While most transit and urban transport experts agreed that these problems existed, the vast majority of them argued and testified in Parliament that the important role of transit in cities is far better represented by public agencies than by private companies dominated only by profit interests. There were major differences in both the explanation of the causes of transit problems and the proposals for their solutions put forth by the two groups.

Actually, there is a fundamental philosophical difference between these two views. The proponents of the free market claim that transit should be a commercial enterprise; it must give some considerations to the public interest, but that is not its main *raison d'être*. If it operates with profit, the public will also benefit and the ridership will actually increase.

The proponents of regulation and intermodal integration consider transit an essential element of urban economy and quality of life. They contend that transit should be operated as efficiently as possible, but that the entire system exists to serve passengers and the city, rather than to maximize profit. Orientation toward profit does not always coincide with the public interest, and if it is pursued, transit use will decrease significantly. Moreover, strict financial accounting for transit is neither logical nor feasible without a much stricter financial accounting for other modes of urban transport, particularly private automobiles and trucks.

Thus, there is a basic conceptual difference between the two groups on what the primary and secondary purposes of transit are—commercial enterprise or public service. The specific arguments of each group are summarized next.

9.5.2.2 Arguments for Deregulation. Bus deregulation in Great Britain, introduced by the Transport Act of 1985, was a part of the British government's drive to privatize and deregulate as many segments of the nation's industry and services as possible. The philosophical/political basis for this was the dogma that the governmental organizations are inefficient and social services are excessively wasteful; private businesses are, according to the dogma, always efficient (otherwise they go bankrupt) and their economic health eventually benefits the public more than government-provided social services.

The British government's plan for introduction of deregulation of bus services and its arguments for its benefits can be summarized as follows (U.K. DOT, 1984; Pickup et al., 1991):

- The government would ensure that conditions for free market competition existed by allowing free entry and free establishment of fares by service providers. Introduction of the free market in bus transit would bring many benefits of competition and eliminate the inefficiencies of regulated operations.
- All bus routes would be open for bidding, and all for which there were qualified bidders (those approved by the government as "responsible operators") would be given to them. The routes for which there were no bidders (those not expected to be remunerative) would be operated by the government-supported public transit agency.
- The government would ensure that all bus operators had equal access to bus stations and other fixed facilities. It would also supervise operations to prevent "cowboy driving," competing bus operators trying to overtake each other in order to get the most passengers.
- To create equal conditions among competing bus operators, bus services would have to be completely separated from rail or any other transit modes.
- Such free entry by any bus company would result in competition, particularly on major routes.

Transit services provided by these principles would have the following specific results:

- Due to competition, operators would provide services more responsive to public needs.
- Entry of nonunionized companies would result in considerable reduction of operating costs (a reduction of about 30% was predicted), and com-

petition would force these cost reductions on all operators.

- Lower operating costs would allow a great increase in transit service frequency and diversity, thus attracting new riders to transit.
- The profit motive would stimulate technical and operational innovations (such as much greater use of minibuses).
- Elimination of cross-subsidies among bus routes, i.e., carrying financial accounting for the entire network, rather than separately for each line, as well as lower operating costs and competition, would result in lower fares.
- More frequent and diversified services and lower fares would result in increased transit ridership, which would increase revenues and allow further increase in services.

9.5.2.3 Arguments Against Deregulation. The opponents of deregulation argued that in order to create a free market, at least the following three conditions must exist:

1. The purchaser of service must pay at least the marginal cost of the service he/she utilizes.
2. The ratio of operating costs to capital costs should be approximately equal for all competitors.
3. Externalities (spillover costs and benefits) are not significant.

In the proposed case of transit deregulation, none of these conditions exists. First, unregulated pricing of services can often lead to fare wars and great underpricing, which violates the first condition. Such a phenomenon has been clearly demonstrated with deregulated airlines in the United States and has been ruinous for the industry, causing bankruptcies of many newly founded airlines, as well as of some of the old, well-established lines, such as Pan-Am, Eastern, and Braniff.

Second, although most bus systems do not have large infrastructure facilities, new contenders for transit services must make a considerable investment in purchasing buses. And third, all transit, including buses, has in most cases very significant social and environmental externalities that are not reflected in their fares in any way.

With many competing bus companies, including private and public ones, and their separation from rail lines, integrated transit systems with coordinated services and fares would disappear. This would destroy a great asset of transit systems created after several decades of efforts for integration that had attracted considerable additional ridership.

The opponents of deregulation further argued that uncontrolled competition would have quite different results from those that its proponents claim. Their major points were as follows:

- Due to cost-cutting measures to maximize profits, private operators would have lower operating and safety standards and less trained personnel than the public agencies.
- Elimination of fare controls would result in fares that maximized revenues and profits, rather than those that attracted maximum ridership. Thus, fares would increase rather than decrease, particularly on routes served by one company only.
- Increase of services on main routes and their deterioration on lightly traveled routes would increase segregation among urban areas and negatively affect the lowest-income groups.
- Without cross-subsidies among routes, services on lightly traveled routes would require even higher government subsidies than they already received. Thus, the cost reductions expected to be achieved on the main lines would be lower when the entire bus network was considered.
- The benefits from introducing more frequent services (mostly minibuses with lower operating costs) would be more than offset by the elimination of integration of services through coordinated transfers, joint fares, comprehensive information, and other assets of modern area-wide transit systems.
- Breaking up of transit systems into sets of independent bus lines would diminish the image of transit, reduce the role of transit in urban trans-

portation planning, and lead to even greater dominance of the private automobile, with all the problems it creates in urban areas.

- Absence of labor unions would result in lower wages, fewer employment benefits, and no job protection, negatively affecting the level and quality of employment in the transit industry.

In overall assessment, the defenders of regulation and integrated transit systems pointed out that the free market in transit failed many years ago, when there was no automobile competition. Now, without elimination of the main causes for that failure and with the added competition of subsidized automobile use, reintroduction of the free market would result in an even greater deterioration of transit services (Pickup et al., 1991; Cartledge, 2002).

The deregulation opponents' claims that public ownership provides far better transit systems than unregulated private market were supported by real-world examples. All contemporary high-quality transit systems, such as those in Brussels, Stockholm, and Munich, are publicly owned; the cities with free market or unregulated transit are exemplified by Manila, Bangkok, and Lima. The superiority of the former ones is quite obvious to every person visiting these cities.

9.5.2.4 Testing the Arguments: Results of Deregulation. Many studies of the impacts of deregulation in Great Britain have been made. After several years of changes and adjustments, surveys of a number of conurbations indicate that the following major results have occurred (Fawkner, 1995; Pickup et al., 1991).

- The number of new transit companies grew rapidly at the time of deregulation, but very few large companies became true competitors. Gradually, the competition changed from actual, on-the-road competing buses to "boardroom competition" of a few large bus operators.
- There has been a considerable increase in the use of minibuses, and their services have become more frequent on major lines, benefitting the public. However, minibus drivers have lower training than regular bus drivers (Banister and Mackett, 1990).
- Services on many lightly traveled routes have been discontinued, hurting transit riders.
- Operating costs (and subsidies) decreased by about 20–25%. Following that drop, it appears that they resumed a similar trend of growth as had been experienced prior to deregulation.
- Bus fleets have become older because investments in new buses have decreased.
- Innovations, except for introduction of minibuses, have not occurred; actually, some demonstrations of new concepts, such as a section of O-Bahn operation in Birmingham, have been discontinued.
- Fares have increased substantially, particularly immediately following deregulation. This negatively affected the ridership and created problems for low-income population groups.
- Ridership losses have occurred in nearly all conurbations. This was caused by elimination of many bus routes, fare increases, and much more public confusion due to disintegrated services with poor information. These factors greatly outweighed the benefits of more frequent minibus services on major routes.
- Bus-km traveled increased mostly on parallel competing routes, rather than on new routes and new types of service. The number of persons employed in the bus industry decreased, and their wages fell compared to the wages in other industries.
- Costs of operating a bus-km decreased, but density of use, or average passengers per bus-km, also decreased. Thus, buses became cheaper to operate, but their utilization decreased.
- It is estimated that costs to non-bus users fell by about 20% (due to decreased subsidies), but costs to bus riders increased (due to rapidly rising fares) by at least 40%. During the same period, data show major increases in car ownership, particularly in the areas where fare increases were the greatest. Analyzing indirect impacts of deregulation, Pickup points out that the resulting hardship

Table 9.1 Results of transit bus service deregulation in British cities versus privatization through contracting in London

Expected Results	*Effects of Deregulation*	*Effects of Privatization / Contracting*
Encourage cost reductions	Yes: 25% reduction or more	Yes: 25% reduction
Reduce fares	No: Fares up 30%	No: Fares up 6%
Provide services that correspond better to customers' needs	No: Worsened regularity and poor information have offset the benefit of a substantial increase in bus-km	Yes: Large improvement in service quality
Arrest decline in bus travel and reduce reliance on private cars	Mixed: Bus ridership down 30%, although in a few areas (usually where one operator has monopoly) there have been large increases	Yes: Ridership maintained
Encourage service and technical innovations	Little: Increased use of minibuses the only important service innovation	Mixed: Some service innovations by more enterprising operators

Source: Based on Fawkner, 1995.

for low-income bus riders and increased costs of congestion and environmental damage should be included when evaluating decreases in transit subsidies (Pickup et al., 1991).

Looking at the entire system of transit bus services in Great Britain, it is obvious that deregulation has resulted in reduced coordination with and increased separation from rail systems. Information and exact data about bus systems are difficult to obtain because the numerous unregulated operators do not report them. Thus, bus services cannot be considered as an integrated system.

Probably an even more significant problem is that deregulation has practically removed buses from urban transportation planning and thus weakened multimodal planning and systems approach to transportation in general. Following the enactment of the Bus Deregulation Act in 1986, the British Parliament approved a major investment for highways. The critics claimed that the deregulation was just one of the disguised measures of the highway lobby to decrease further the already highly inadequate role of transit in British cities.

It is significant that the British government realized that if transit deregulation were to be applied to London, in which bus transit plays a critical role, its negative results would greatly outweigh its benefits and serious problems would occur. Therefore, the sweeping deregulation of buses applied to British cities, removing government controls over fares, schedules, and intermodal coordination, was not applied to London. Instead, bus services in London were partially privatized through tendering of sectors of services, with the publicly owned London Transport (now Transport for London) allowed to bid against private companies.

The resulting contracted services were thus partially privatized, but not deregulated. A comparison of the results achieved in British cities by full deregulation with the results of partial privatization through contracting is presented in Table 9.1. The table clearly shows that deregulation has decreased costs but has clearly been damaging to passengers and the role of transit. The results of privatization through contracting have also brought cost savings, but their impacts on passengers have been generally positive.

In conclusion, if the main goal of deregulation was to reduce taxes and government costs for transit in the short run, it has been successful. However, with respect to its impacts on transit ridership, on the welfare of the low-income transit-dependent population, as well as on the role of transit in the cities in the short and in the long run, bus deregulation was clearly counterproductive and increased transportation problems in conurbation areas of Great Britain (Pickup et al., 1991; Fawkner, 1995; Vuchic, 1999).

It is interesting that in the early 1990s, when the deregulated situation stabilized somewhat, there were strong signs that a number of diverse groups, such as engineering societies, automobile clubs, and political parties, generally argued for the need to improve transit services, reduce negative impacts of automobile use, introduce road pricing, etc. These measures to balance urban transport modes, however, were more difficult to introduce with disintegrated transit services and little control over them than they would have been with regulated and coordinated transit systems.

9.5.3 Transit Deregulation in Other Countries

The concept of transit deregulation has been discussed in many countries, but in most cases this term has been used to mean reexamination of regulations and their revision in the countries and cities where they were excessive and unjustifiably burdensome. Some changes, adjustments, and reductions in regulatory requirements and procedures were found to be useful. Regulatory reforms in countries like Australia, France, Germany, and Sweden have increased efficiency of operators, broadened organizational options, and stimulated various elements of competition, resulting in service improvements (see the following section).

There are, however, fundamental differences between transit deregulation in Great Britain and the processes sometimes referred to as deregulation in most of its peer countries. First, the sweeping deregulation in Great Britain included not only elimination of economic and fares regulations, but also major loosening of franchise approvals and, due to the nature of fully deregulated services, lowering of standards and weaker supervision of safety regulation implementation. Second, the main goal of deregulation in Great Britain was to reduce government expenditures for transit. This goal overrode the public interest in having coordinated, high-quality transit system services, as was expressed in the title of a paper by one of the strong deregulation promoters: "The Market Alternative to Integration" (Hibbs, 1986). Finally, complete deregulation weakened transit planning and its role in intermodal urban transportation policies.

Most other countries introducing revisions to their regulatory standards and procedures had increased economic efficiency as a major goal, but secondary to the provision of an integrated transit system that is aimed at maximization of ridership. Therefore, their scopes of deregulation have been smaller and subject to retention or improvement of integration. The presence of coordinated transit services allows treatment of transit as a system in intermodal transportation planning and policy formulations.

9.6 EVOLVING FORMS OF TRANSIT OWNERSHIP, ORGANIZATION, AND MANAGEMENT

The transit industry has been undergoing significant organizational changes in recent years. The traditional organization of public transit—a governing board overseeing a single public agency that provides all transit services in an urbanized area—has been challenged because of sharpening financial problems and increasing public dissatisfaction with offered services. Discussing this problem, a recent study of transit organizations sponsored by the Transportation Research Board (Stanley et al., 2003) defined this problem as focusing on three basic issues:

- *Why* is fundamental change in traditional transit organizations necessary?
- *What* is the nature of changes likely to emerge?
- *How* can established transit organizations be transformed to acquire a new organizational paradigm?

These questions about the organizational aspects of transit are discussed in the following three sections. They cover, respectively, the factors that make changes necessary, various aspects of changes, and specific changes that have been taking place in recent years.

9.6.1 Organizational Problems and Their Causes

A review of conditions in transit operations typical for many cities can show *why* there is a need for changes in organization of transit, as well as other urban transportation services. During recent decades, transit agencies in many cities have faced increasing deficits. Under financial pressures, agencies have adopted defensive policies and developed the mentality of preserving the existing services, rather than undertaking progressive measures to attract more passengers. Thus, while the need for subsidies was increasing, passengers felt that the agency did not give them sufficient attention and did not introduce innovations in services. This has particularly been the case in cities where public funding in high-performance transit modes, such as rail and upgraded bus services, was traditionally inadequate, so that transit consisted mostly of bus services on streets, which were not competitive with auto travel.

The increasing problems transit organizations faced were of two types. The first set of problems was that unsatisfactory performance of transit agencies themselves led to increasing inefficiencies in operations and financial difficulties. The second type of problems was caused by a lack of clear transportation policies. On the one side, many Western European and some other countries, such as Australia, Canada, and Japan, have made considerable advances in creating attractive transit systems. On the other side, some countries have continued to favor the auto-highway system over transit by policies and government financing decisions. The progress with the important steps to introduce intermodal coordination has also varied greatly because of many political and organizational obstacles.

Most of the recent organizational changes focused on transit agencies and services, but in many cities increased attention has also been given to comprehensive urban transportation policies and to the development of new organizations for their more effective implementation.

The trend of decreasing efficiency and unsatisfactory transit services triggered extensive discussions about organizational, regulatory, and ownership aspects of transit. A critical review of two opposing views, one critical and the other supportive of traditional public transit agencies, is useful to eliminate influences of criticism based on different ideologies and defenses based on self-interest. The focus is placed on a realistic evaluation of the problems and possible solutions to transit systems' organizational issues.

The critical views, often referring to transit organizations as "public monopolies," often include ideological overtones in a sweeping statement that public ownership causes poor management and inefficient operations. However, these critics correctly point out the problems resulting from the disadvantages of having a single public transit agency, listed in Section 9.1.3 as items D-1 through D-4. Financial aspects are more difficult to control when they depend on political decisions, particularly if urban transportation policies are not clearly defined and pursued. Labor unions, counting on public subsidies, often exert pressure on management to obtain excessive wage increases, using strikes as a powerful tool. A further claim, valid for some cities, is that lack of competition results in neglect of customers and deteriorating quality of transit services.

As already discussed, some proponents of this line of thought argue that deregulation and privatization of transit would solve these problems because private companies are inherently efficient, labor unions can be broken up, and profit-driven operators would bring innovations and be very sensitive to public needs, particularly if competition is introduced. As shown in Section 9.5, application of transit deregulation in Great Britain and elsewhere has shown to be counterproductive in solving transit problems, so that other solutions are being developed and applied.

Many transit operators reject these critical views as exaggerated and unrealistic. They correctly claim that integrated public transit agencies have generally achieved what was expected from them when they took over a number of private transit companies several decades earlier, listed in Section 9.1.3 as A-1 through A-5. This is proven by the fact that the most advanced integrated intermodal systems, such as those serving Paris, Munich, and Toronto, were developed by public agencies. However, many transit operators agree that the problems of financial efficiency and insufficient responsiveness to the customers' needs are serious and must be addressed. The leaders of the most progressive public transit agencies emphasize the fact that innovations in organizations and changes in their attitudes toward passengers are needed.

Intense activities in reorganization and changing the attitudes of transit agencies are now under way in many countries, as well as at the international level. Extensive articles, reports, and policy statements of the International Association for Public Transport (UITP) strongly emphasize the need to improve attention to passenger needs greatly, as well as to improve efficiency of operations. Contractual relationships between authorities and operators, the key issue in introducing new forms of transit organizations, are one of the subjects discussed extensively.

Reviewing the views and arguments proposed by the critics and the defenders of traditional public transit agencies, a number of facts about the required features of transit services and their present deficiencies that should be resolved can be summarized.

First, the main purpose for existence of transit agencies is to provide high-quality services to the public. This requires that an integrated and coordinated system be offered to the present and potential transit users.

Second, many transit agencies operate under organizational and financial conditions that in many cases create operational problems, such as:

- Legal setups and appointment procedures for members of managing boards sometimes create agency governance that is technically incompetent or insufficiently interested in providing high-quality transit service.
- Management is more oriented to preservation of the status quo than to improvements and innovations.
- Labor unions abuse their power to oppose technical and operational innovations and demand higher wage increases than justified by productivity increases.
- With time, convenience of work by operating and maintenance personnel tends to override the requirements of customer-oriented services.

Finally, political support and financing procedures for transit represent an important foundation for provision of high-quality transit in an efficient way. Even the best-organized transit systems cannot efficiently meet customers' needs for transit if they do not have adequate organizational and financial support within the region's intermodal transportation system.

Placing the discussed problems in a broader perspective, the quoted TRB study (Stanley et al., 2003) defined the problems that are forcing the search for innovations in transit organization as the following six points (abbreviated here).

1. The quality of life and economic vitality of urban areas are increasingly threatened by sprawling development, rising traffic congestion, and increasing cost and decreasing quality of public services.
2. Socioeconomic trends tend to diminish the relevance of traditional transportation services.
3. The enabling environment, or process of providing services in surface transportation, is replete with fragmented responsibilities.
4. The organizational culture and dynamics of many transit organizations have historically posed a barrier to change.
5. The quality of customer experience has not had a dominant focus in transit service delivery.
6. Transit organizations are inherently slow in the deployment of state-of-the-art technology.

Understanding these problems and their causes is important for searching new solutions.

9.6.2 Reevaluation of Transit Agency Structures

To analyze *what* changes are needed, a brief analysis of transit agency functioning is useful. Activities of transit organizations can be presented by a simple model defining three levels or tiers of activities, as shown in Figure 9.1. The highest, strategic level should focus on the basic goal of the transit agency: service to passengers or customers. Agency's policies should be formulated at that level. The second, tactical level consists of designing networks, lines and their scheduling, fare collection, etc. Actual provision of service (conducting operations) represents activities at the third level.

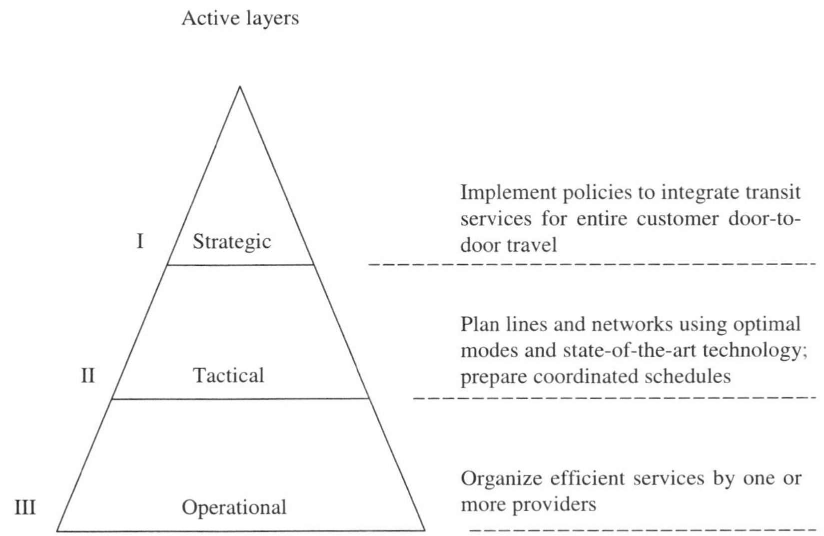

Figure 9.1 *Conceptual model of transit service provision.*

In traditional transit agencies, all three levels of activities are performed by the same organization. While the levels are interdependent and need to be coordinated, it is desirable to define them and maintain a certain independence among responsibilities for them in the agency. Experience has shown, for example, that direct involvement of board members in technical and operational issues can be disruptive. Also, separation of the three levels may allow a transit agency, or an umbrella organization that performs tactical planning, to call for tenders or contracts, so that operations may be performed either by the transit agency or by other public or private operators. Major activities on each level are shown in Figure 9.1.

In review, the changes considered necessary in transit organizations encompass a number of aspects, from the broad consideration of urban transportation policies toward different modes to the quality of offered services and use of new technology. Major items requiring changes in many cities can be classified in seven groups or dimensions, which are briefly outlined here.

9.6.2.1 Broadening the Mission. The present basic mission of transit agencies, provision of transit services, should be broadened to encompass responsibility for managing mobility in cities. It has become obvious that better management of multimodal systems requires coordinated management by one intermodal agency. In some cities, a reorganized transit agency may be the logical body to perform that duty. In several cities, including San Francisco and London, new agencies have been formed to handle transit, parking, and some other transportation functions. Oslo has a separate agency collecting road charges for entering the city, which are then distributed to street and transit financing. In Singapore, the Land Transport Authority performs policy and coordination functions over roads (including Electronic Road Pricing), transit, and taxi companies.

9.6.2.2 Focus on the Customer. Instead of it being expected that passengers who need transit will use it, transit services must be organized and operated on the basis of careful analysis of the needs of present and potential transit users. The difference between the two approaches, one from the passenger side and the other from the operator's point of view, is illustrated in Figure 9.2. Passenger C in the figure may use one or several different lines or modes, shown in the figure as a bus, LRT, and metro line, and he/she judges system service quality by the experience from all trip components (heavy horizontal frame).

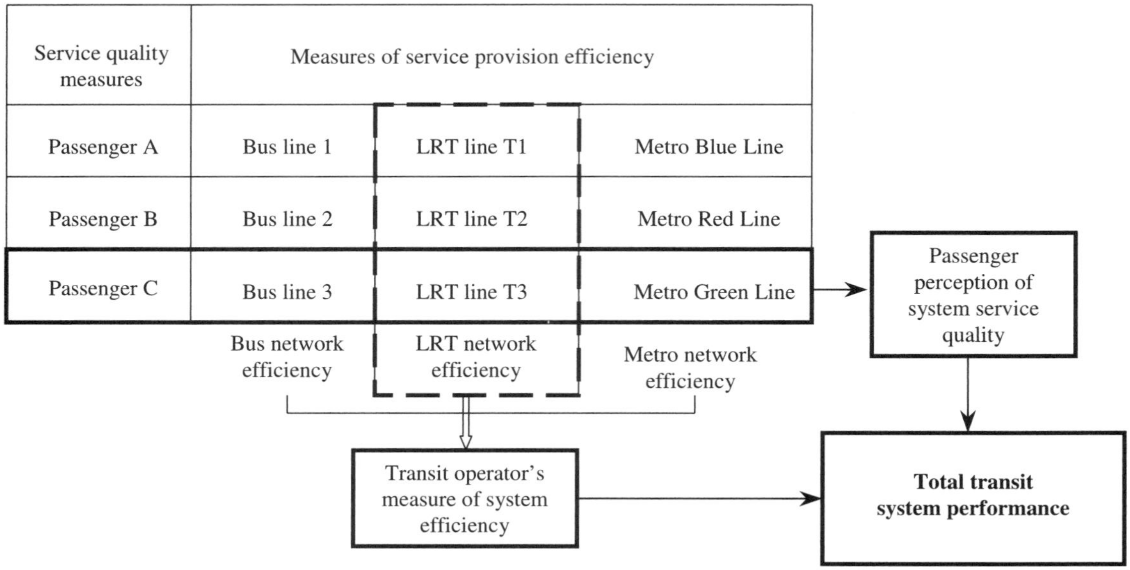

Figure 9.2 *Concept of system efficiency, service quality, and transit system performance.*

The transit provider (one or more companies) operates and analyzes the efficiency of individual services or modes (vertical dashed heavy frame). Evaluation of the overall transit system performance should include both of these. As the diagram shows, the total transit system performance correlates the two, service quality and efficiency, involving the users and provider(s).

The point here is that transit operators should evaluate the service they offer not only through their efficiency measures of individual services (shown vertically), but also through the points of view of passengers (shown horizontally) who use different services comprising individual trips.

9.6.2.3 Competition and Other Incentives. Different forms of competition usually effectively prevent degradation of transit agencies into self-protective bodies opposing changes and innovations. Competition can be organized among departments of a single agency, or through a competitive bidding/tendering process. Competition in transit should be for specifically determined lines, schedules, service quality, and fares, so that it is performed for the lines, rather than in the streets.

9.6.2.4 Cooperation and Integration. Cooperation among transportation agencies operating different modes and functions, such as transit, traffic, parking, pedestrians, etc., is essential to achieve a broader mission of managing mobility. For effective cooperation, services and sometimes facilities and assets can be shared or integrated.

9.6.2.5 Use of State-of-the-Art Technology. Transit systems technology has been very innovative in recent decades. New tunneling techniques, low-floor bus and LRT vehicles, and automatic train operation are a few examples. The most elaborate innovations presently come in the area of ITS or telematics (Girnau et al., 2001). ITS applications in transit include on-line operations control, smart-card and other contemporary fare-collection systems, passenger information about services via telephone, by Internet, and at transit stops and stations, security systems, vehicle diagnostics, and many others (see Section 6.4.3).

Most of these technical innovations require investment and an effort to change practices and train personnel. However, the pay-off in greater efficiency and better services for passengers often justifies efforts for their introduction in a short period of time. Transit organizations must allow and encourage initiatives for introduction of innovations.

9.6.2.6 Changes in Organizational Structure. Reevaluation of internal organizational structure of many transit agencies is very useful. Revisions may be needed in areas including legal composition, duties, and methods of appointing members of the board; management-labor relations; and stimulation of interest in providing services attractive to passengers.

9.6.2.7 Monitoring of Service and Operating Efficiency. To ensure that transit system performance meets established goals, standards, or contract provisions, transit agencies or their supervisory bodies should maintain a regular program of monitoring regular transit service quality and operating efficiency (see Figure 9.2).

9.6.3 Privatization and Other Forms of Transit Reorganization

Recognizing the need to upgrade traditional public transit agencies to organizational forms that increase their operating efficiency and improve services they offer, the question is faced of *how* that upgrading can be done.

Reviewing the transition from private companies to public agencies several decades ago (see Section 9.2), one can see both positive and negative results of that change over the years. The benefits from the creation of integrated intermodal transit networks for the public have been great. Transit systems in New York, Paris, Berlin, and most of their peer cities are far better organized than were the disjointed services offered by many private operators. In establishing fare levels, public transit agencies generally attempt to find a reasonable compromise between meeting social needs of the population and maximizing revenues.

With time, however, the negative aspects of single public agencies became increasingly serious. As discussed above, their operating efficiencies as well as their concern in meeting the customers' needs have often decreased. This sharpening crisis led some economists and politicians to argue for deregulation of transit, without explaining how such a reversal to the unregulated competition would avoid the problems that led to the public takeover.

With obviously unsatisfactory results of transit deregulation demonstrated in several countries, many cities around the world have concentrated their efforts on finding the answer to the following basic organizational question:

How can the provision of transit services be made economically and organizationally more efficient while continuing to provide integrated, reasonably priced services to the public with increased attention to consumer needs?

Competitive contracting to private companies has been done by many transit agencies for different functions, such as maintenance of infrastructure, fare collection, and vehicle overhaul. These arrangements, resulting in improved operations and reduced costs, are continuing and increasing in many countries.

In recent years, public-private partnership has entered a new phase: contracting out operations of bus transit lines. This concept has had very positive results and has been increasingly applied in many cities in the United States, France, Germany, the Scandinavian countries, and many others. In Great Britain, London has used competitive contracting under coordination by the central transit agency, Transport for London (as distinguished from the deregulation of transit systems in other British cities).

9.6.3.1 The Gothenburg Example. A variety of organizational and procedural forms of contracting and privatizing bus transit services are used in different cities. As an excellent example of this process, the case of the city of Gothenburg, Sweden, is described here.

As early as 1978, the Swedish Transport Act mandated that "every county has to form a public transport authority that will be in charge of organizing local and regional transit services." The deficits had to be covered by the city and the county, 50% each. The authority, designated as PTE, could be a publicly owned shareholding company or a city/county department. It could either buy services from operators or operate its own buses, directly or indirectly. PTE had to coordinate the network of lines and determine fares for all operators.

This act demanded preparation for separating the strategic and tactical from the operating level of transit service provision, as illustrated in Figure 9.1. GS, the city's transit agency, was given an annual bonus of SEK 20M (US $3M) to prepare for the competitive bidding. Ten years later, the Transport Act of 1988 went further in defining organization of transit services, and it required their implementation. The main instructions of the 1988 Act were:

- Counties' PTEs must be in charge of concessions for transit operations.
- PTE can but does not have to go out for tendering.
- PTE defines all specifications for tendered services: network of lines, sizes of vehicles, operating hours, service headways, fares, drivers' uniforms, designations and logos, etc.
- Tendering must be open to state-owned, city-owned, or private operators.
- Service quality and contract cost should be the main criteria in selecting operators.
- Politically elected boards are responsible to their constituencies to choose the best operators.

The first round of contracting consisted of five different sets of bus lines in the western part of Gothenburg. Four operators competed: GS (the city's transit agency), one private and one state-owned company, and a consortium of four small operators. GS won 55% and the private competitor 45% of the tendered lines, and they started operations in 1993.

The results of this restructuring of services were very positive. GS lost some lines, but proved to be competitive with private operators. The new arrangement of services was estimated to have reduced operating costs from the conditions in 1989, prior to the change, by about 45%. These savings were achieved through several changes. GS was restructured as a shareholding company, with all shares owned by the city. Internally, competitive divisions were formed by mode: light rail, bus, and ferryboats (this change had been rejected by the unions before). Operations were streamlined and a portion of the staff was reassigned to other functions in the city government. The light rail and ferryboat divisions continued to be owned and operated by GS.

A second set of tenders implemented in 1997 improved coordination of local and regional transit lines and increased the required standards for bus emissions.

9.6.3.2 Transit Organization in French Cities. The national government in France has assisted cities to finance and organize transit systems at the local level by several laws. In 1973, it introduced the *Versement de transport* (transportation financing) tax, which required that in cities with population over 300,000 all companies with more than nine employees had to pay a payroll tax of approximately 2%, which was earmarked for transit development. This reliable funding source resulted in a strong revitalization and expansion of transit systems across the country.

With local financing ensured, several subsequent laws stimulated organizational decentralization of transit to the regions and promoted the formation of regional or local transit authorities that provide budgeting, set policies, fares, and service standards but delegated operations to separate public or private companies. The 1982 Law for Organization of Local Transport (*Loi d'orientation des transports interieurs,* LOTI) regulated the current organizational structure of urban transportation in France. It introduced competitive contracting procedure for selection of operators of urban transit systems mandatory.

Despite LOTI, however, many French cities and private operators did not support a competitive procedure for operation of urban transit systems. This led to companies operating in groups of cities determined by collusion within the industry, rather than through a competitive process. Cities would issue request for proposals and receive a single bid, often from the current operator. There were lawsuits about collusion and bribes in several cities. In 1993, the Sapin Law (named after its promulgator Michel Sapin, Minister of Economy and Finance at that time) was introduced, which focused on improvement of the delegation of public service, including transit. This law specifically stipulates that local authority is responsible for providing transit service by competitive contracting of the operation to another entity.

The Sapin Law is intended to ensure fair competition and prevent corruption. It is specifically aimed at achieving two goals: balancing market efficiencies with public needs and government regulations, and achieving the transparency and accountability of contracts.

As a result of this organizational setup of transit agencies, most French cities now have transportation authorities that contract out transit system operations, in most cases as entire, integrated companies rather than through several bus and rail transit operators. Three large operators, Connex, Keolis, and TRANSDEV, operate transit systems in over 60% of cities. Other cities have a variety of operators, sometimes with participation of management teams or transit agencies from other cities and other countries, or private companies operating sets of bus lines. Rail systems are in all cases operated by a single company. The contracts are given out for certain duration, usually five years, with reviews of performance and renewal subject to satisfaction of the contract with respect to quality of services offered and economic efficiency.

9.6.3.3 Examples from Other Cities. As these examples show, privatization and contracting of transit services have been introduced in a variety of forms. Some selected features are described here, but a comprehensive systematic review is impossible because of the great variety of solutions and continuing introductions of new laws and paradigms.

Tendering of bus services is increasingly used not only to introduce competitive operations in traditional public agencies, but also to improve unregulated services. This organizational effort is actually from the other side: as opposed to replacing monopolistic public agencies, it is aimed at improving the chaotic services and inefficient operations of multiple unregulated companies. Many cities in Latin America are working on such reorganization. For example, Santiago, Chile, has had chaotic operations of hundreds of bus companies competing among themselves and with the metro on many redundant routes without schedules. A major effort has been undertaken to upgrade these services to a coordinated intermodal system. Following extensive negotiations with private operators and the public agency, a central authority has designed a network of trunk and feeder bus routes coordinated with the metro system. Some of the new services will be opened out for tendering.

Contracting is sometimes done for entire urban transit systems. As described above for the case of France, the city or transit authority hires a professional transit-managing company to operate the entire system under conditions specified by the contract. Companies like Arriva, Stagecoach, Connex, TRANSDEV, Keolis (a subsidiary of French Railways, SNCF), and many others operate transit systems in a number of cities in France, Great Britain, Australia, New Zealand, and other countries.

Some transportation agencies have branched out into management of transit in other cities. For example, the National Rail Passenger Corporation (Amtrak) has operated regional rail services in several cities. The Hamburg Transit Agency (HHA), in cooperation with other neighboring agencies, manages operations of transit systems in several smaller German cities.

Tendering of rail services has been much more limited and less successful than bus tendering. Most transit operators oppose breaking up rail systems into different agencies. They have a valid argument: rail systems are by their nature integrated systems, not conducive to multiple operators and competing services. Experiences from several cities where rail systems were broken up to achieve competition confirm this argument.

The extensive Melbourne tramway network was divided between two competing companies. The result was considered to be a failure and return to a single agency is being considered. In Singapore, the Mass Rapid Transit (MRT) network has been divided between two companies, which also operate buses. The reason for splitting the rail system into two companies was claimed to be to create competition, but this is questioned because the two companies serve different areas, so that they actually do not compete. The question whether the London Underground system should be divided into three networks has developed into a bitter dispute between the national government and the mayor of London. Although leading world transit experts support the mayor in strong opposition to the division of the network, the national government has prevailed, again supported only by the ideology that competition is the only method to achieve efficiency.

There are attempts to separate ownership of railway infrastructure from the operating companies, which may have influence on organization of rail transit systems also. This setup is still subject to many discussions. While such reorganization of British Railways has been a failure, the Swedish and German railways have cautiously introduced changes that enable some opening of tracks to different operating companies.

It is obvious that there is a considerable difference between managing bus and rail transit services. The distinctly lower success of tendering rail than bus services appears to be caused by the inherent differences between the two modes. Important factors are the differences in the physical system characteristics (special ROW and more elaborate system technology), cost structure (different investment/operating cost ratios), management and operating methods (much more centralized for rail), as well as fundamentally different public images of the two modes.

Political and even ideological factors influence transit organization in many countries. For example, the European Union has adopted a number of regulations aimed at stimulating competition in purchasing vehicles and providing services. While some of these regulations have invigorated the transit industry and its suppliers in certain respects, some of the regulations were directly opposed by many large transit agencies because they would have seriously damaged their superb integrated services (Stanley et al., 2003). The discussions are expected to continue for many years.

9.6.3.4 Review of the Competitive Privatization Procedures. The main common features of the procedures used in the present efforts to reorganize transit systems are summarized here.

In many cities, specialized authorities have been founded to perform mobility management, i.e., to analyze population mobility needs and coordinate transit with other functions, such as traffic, parking, and bicycles. In many cases, specialized authorities have been founded to specifically organize all transit services. These authorities make strategic decisions and organize service providers, separating the two functions.

Transit policy formulation and organization of transit services may be performed by a city or state agency or an umbrella organization, or, transit agency may also perform service tendering if it does not participate in the competition.

Service contracting is typically based on precisely specified tasks. The central transit authority (or transit agency) prepares networks of lines, schedules for each one, fares, and a number of details, including quality and cleanliness of stations and vehicles, attitudes toward customers, information, marketing, etc.

Two sets of specifications are particularly important. The first, system performance, includes condition of stations and vehicles, speed, and reliability of service. The second consists of all elements that affect passenger experience. These two sets of provided services must be monitored by surveys performed by the transit authority or by an independent survey company. It is common for the contract to foresee penalties for the operating company on the basis of survey results.

In the contracting of services, where attitude toward passengers can have considerable influence on attraction (or loss) of passengers, contract payments may be related to the number of served passengers to maintain operator's incentives to attract customers.

Contracts are usually given for three- to five-year periods. A clause may be included that the contract may be canceled if the operator does not perform satisfactorily. The operator is obliged to perform service and in the case of discontinuance must give a specified notice, usually several months ahead.

In summary, contract with an operator defines all requirements for performance and quality of service. Financial arrangements may vary between operator's payment to serve busy lines and the authority's payment for operating lightly traveled lines that require subsidies. The duration of the contract and conditions of termination must be included. Sometimes the penalties for unsatisfactory performance and rewards for increased ridership may also be included. Review and rebidding of the contracts are performed at prescribed periods, such as semiannually or annually.

In most large cities, leasing is done for certain portions of the transit bus network. Since the authority or transit agency must monitor, coordinate, and plan the entire network, leasing services out to many different contractors may create such a complex supervisory and coordinating task that its costs and organizational intricacies outweigh the benefits of reduced operating costs. This aspect often presents the upper limit on the extent of leasing out services. In some small cities, large portions, or even entire transit services, may be operated under a single contract.

EXERCISES

9.1. What were the public benefits from the mergers of many independent operators into a single integrated transit agency? What were the benefits to the operating companies (what economies of scale could be achieved)?
Note: discuss the benefits of integration, regardless of the ownership (private or public).

9.2. Which problems in public transit services made government regulation necessary from their early introduction? Considering the conditions and requirements for public transit services when they were introduced, define the objectives for their regulation.

9.3. What are the basic differences between privately and publicly owned transit agencies with respect to:

- The relationship of public service (attraction of passengers) and financial results
- Primary goals and objectives in providing the services
- Methods of operation and internal organization

9.4. List and describe:

a. Three of the major benefits from the public takeover of private transit companies that took place in large cities several decades ago.

b. Two of the disadvantages that are more likely to occur in a large public transit agency than in private transit companies.

9.5. List and describe various forms of city-regional transit agencies and their ownership.

9.6. What are the reasons for forming regional transit agencies?

9.7. Define transit districts: their purposes, forms, and domains of operation, i.e., areas they serve.

9.8. What are the advantages of public agencies that have taxing power in their service areas, i.e., how does that power assist them in financing investments in their transit system?

9.9. Define a transit federation: its purpose, the operating agencies it may encompass, and its basic method of operation. Where is founding of a federation most appropriate, and where is it unnecessary?

9.10. Define the general purposes of transit regulation by governmental bodies and list their main categories.

9.11. List and describe each of the four major categories of government regulations in transit.

9.12. Which category of transit regulation is most essential and what major items does it include? List and define major groups of this regulation.

9.13. What are the reasons for economic regulation of transit from the passengers' (users') point of view? What are the benefits to transit operators of that regulation?

9.14. Describe the goals and give examples of technical and social aspects of transit that are sometimes regulated.

9.15. What problems exist when there is no economic regulation of transit services with respect to:

- The behavior of operating companies
- The quality and pricing of the services the public receives

9.16. What can be negative aspects of regulation in the following areas:

- Safety
- Economic (fares)
- Social aspects
- External impacts

9.17. What is the principal difference between transit deregulation in Great Britain and transit privatization measures in the United States and other developed countries? Which approach is superior and why?

9.18. In the cities of many developing countries, many paratransit and some transit operators are unregulated. Describe the advantageous and unsatisfactory features of these services.

9.19. What are the main reasons that city governments want to regulate presently unregulated transit services in developing countries? Who opposes regulation and why?

9.20. Define different forms (or degrees) of transit deregulation of formerly regulated transit services, such as those in Mexico City and Manchester. Using the form applied in the United Kingdom, briefly discuss the following:

a. According to proponents of transit deregulation in that country, what benefits would be realized by deregulation?

b. After deregulation was introduced in 1986, which of these benefits were actually achieved?

c. What problems and losses did the transit deregulation create?

9.21. What major problems have led to the partial privatization of traditional public transit agencies?

9.22. List and discuss the purposes of privatizing or contracting out one or more bus transit lines through a bidding or tendering process. What are the advantages of such arrangements? Why does this process usually result in lower operating costs?

9.23. What may be some disadvantages of contracting out transit services, and what arrangements can be used to prevent or resolve them? Include the problems that may appear between operators, as well as those affecting the passengers.

9.24. What has been a major positive change of attitude among transit operators in many cities toward passengers, and what have been the results of such a change?

9.25. What may be the negative and what the positive consequences of contracting out transit services for the public transit agency?

9.26. Describe the importance of the following aspects of privatized services:

- Competition and economic efficiency
- Attention to passenger attraction

9.27. Why are passenger surveys about transit services important in contracting private operators?

9.28. Observe and evaluate the attitude of the transit agency in your city toward existing and potential passengers. If some aspects are not satisfactory, suggest improvements.

9.29. What organizational arrangements have been introduced in some rail transit systems to increase efficiency of their operations, and what have been the results?

9.30. What is the purpose, and what have been the results, of establishing agencies that manage mobility, i.e., organize transit as well as other transportation functions in an urbanized area?

9.31. (Multiple choice) A transit federation is:

a. A special transit district covering more than one state

b. A transit authority including operators of different modes

c. An integrated organization of private transit companies

d. An integrated transit organization that may include both public and private operators, regardless of mode

REFERENCES

American Public Transportation Association (APTA). 1993. *1993 Transit Fact Book.* Washington, DC: APTA.

Banister, D. J., and R. L. Mackett. 1990. The minibus: theory and experience and their implications. *Transport Reviews* 10(3):189–214.

Cartledge, J. 2002. "Privatizing public transport—the British experience. *Public Transport International* 1, 18–23.

Cheape, C. W. 1980. *Moving the Masses*. Cambridge: Harvard University Press.

Cudahy, B. 1990. *Cash, Tokens, and Transfers: A History of Urban Mass Transit in North America*. New York: Fordham University Press.

Fawkner, J. 1995. Bus deregulation in Britain: profit or loss? *Public Transport International* 44, 18–23.

Hibbs, J. 1986. The market alternative to integration. *City Transport,* May/June, 23–24.

Homburger, W. S., and V. R. Vuchic. 1972, Transit federation—a solution for service integration Revue de l'UITP 2.

Interplan Corporation. 1973. *Integration of Transit Systems.* Report to UMTA. Washington, DC: U.S. Department of Transportation.

Pampel, F. 1969. *The Hamburg Transport Community, an Example of Co-ordination and Integration in Public Transport.* Report 1b. 38th Congress of UITP in London. Brussels: UITP.

Pickup, L., S. Meadowcroft, P. Goodwin, B. Tyson, and F. Kenny. 1991. *Bus Deregulation in the Metropolitan Areas.* Aldersot: Avebury/Gower.

Pucher, J., and Stefan Kurth. 1996, Verkehrsverbund: the success of regional public transport in Germany, Austria and Switzerland. *Transport Policy* 2(4):279–291.

Snell, B. C. 1974. American ground transport: a proposal for restructuring the automobile, truck, bus, and rail industries. *Hearings before the U.S. Senate Committee on the Judiciary, Subcommittee on Antitrust and Monopoly of the Committee on the Judiciary,* 93d Cong., 2d sess., Appendix to Part A, A1–A103.

Stanford Research Institute (SRI). 1956. *Organizational and Financial Aspects of the Proposed SFBA Rapid Transit System*. Menlo Park, CA: SRI.

Stanley, R. G., Matthew A. Coogan, and Michael P. Bolton. 2003. *Emerging New Paradigms—A Guide to Fundamental Change in Local Public Transport Organizations.* TCRP Report 97. Washington, DC: TRB.

United Kingdom Department of Transport. 1984. *Buses*. London: HMSO.

United States Federal Electric Railways Commission (FERC). 1980. *Proceedings and Final Report*. Reprint. Washington, DC: U.S. GPO.

Vuchic, V. R. 1999. *Transportation for Livable Cities*. New Brunswick, NJ: CUPR Press, Rutgers University.

Yago, G. 1984. *The Decline of Transit*. Cambridge: Cambridge University Press.

FURTHER READING

Bayliss, D. 2002. Review—urban public transport competition. *Public Transport International* 3, 4–9.

Girnau, G., A. Müller-Hellmann, and F. Blennemann. 2001. *Telematics in Public Transport in Germany* (in German and English) Cologne: Verband Deutscher Verkehrsunternehmen.

Mees, P. 1999. *A Very Public Solution: Transport in the Dispersed City*. Melbourne: Melbourne University Press.

Rinqqvist, S. 2002. Scandinavian solutions: a model of cooperation. *Public Transport International* 42, 31–33.

UITP. 2002. *Relationships between TOCs (Train Operating Companies) and IMCs (Infrastructure Management Companies)* Report by the Regional and Suburban Railway Committee; International Association of Public Transport, Brussels, Belgium.

Vuchic, V. R. 1986. Deregulation: a return to the primitive. *City Transport,* May–July, 18–20.

YTV Transport Department. 2001. *Competitive Tendering of Bus Services in the Helsinki Metropolitan Area 1994–2001*. Technical report by the Helsinki Metropolitan Area Transport Department, Helsinki, Finland.

PART III

TRANSIT SYSTEMS PLANNING AND MODE SELECTION

10

TRANSIT SYSTEMS PLANNING

This chapter presents the organization, basic procedure, and methodology of transit planning. A general process of long-range urban transportation planning is described, with a focus on transit system planning, including ridership estimation and development and selection of plans.

10.1 PURPOSE AND ORGANIZATION OF TRANSIT PLANNING

In this section, the importance of comprehensive planning is discussed. Then the legal requirements, government's role in transportation planning, and the evolution of this process in recent decades are reviewed.

10.1.1 Short- and Long-Range Transit Planning

Transportation and transit planning is usually classified into two categories by planning period. They differ in their nature, types of objects, and characteristics.

1. Short-range planning encompasses projects and measures that can be implemented in 3–5, sometimes 8 years and usually do not involve major investments and infrastructure construction projects.
2. Long-range planning consists of planning of major infrastructure objects, lines, networks, and similar facilities. Long-range plans may involve large investments, construction, or development of new vehicle types. The planning horizon may be 10–25 years.

In some cases, medium-range planning refers to projects involving moderate investments and an implementation period of 5–10 years.

Short-range transit planning may include service schedules, purchase of new vehicles, modification of existing or opening of new lines and networks, some infrastructure changes, as well as organizational aspects, such as fare types and methods of collection and intermodal integration measures. Thus, short-range planning is dependent on present conditions and near-future trends since the changes that are made are easily modified or even reversed. It is not directly focused on long-range developments, except that there should be a check that short-range plans are compatible with long-range ones.

Long-range planning is aimed at meeting the city's transit needs in the year of the adopted planning horizon, such as 10–25 years in the future. By its nature, this planning deals with projects involving capital investments, physical and organizational modifications, or new components of the transit system. Since such developments are permanent, they have many impacts and interactions with other activities. Their long-term effects are usually modeled and carefully evaluated. Financing of long-term transit projects must be evaluated along with the economic, social, and environmental benefits and costs to the community as a whole. Examples of long-term transit projects are constructing of a new rail line, network, or major transit terminals, reorganizing the network, building a control center for operations, installing a new signal system, and introducing various ITS technologies.

A long-range plan is usually reviewed in certain time intervals, such as every 5 years, and revised or modified if some changes have occurred.

(*a*)

Photo 10.1 *Long-range planning results at Eglinton Station on Yonge Line in Toronto: (a) 1953 (Courtesy of Toronto Transit Commission); (b) 1983 (Courtesy of Toronto Transit Commission).*

This chapter focuses mainly on long-range planning.

10.1.2 Need for Comprehensive Planning

Planning is a basic function in many areas of society. Public systems and facilities, such as cities and transportation systems, generally require more complex planning than private systems and organizations because of their greater complexity and their diverse clients—the general public. Moreover, the goals and objectives of public systems are more diversified than private ones: they involve not only financial results, which dominate private systems, but also impacts on the public and on the functioning of the entire city or metropolitan area.

Intermodal transit systems consist of networks of different modes, have complex infrastructures, and occupy considerable space. To function efficiently, the networks, infrastructure, and vehicles must be planned as coordinated systems. Moreover, since transportation strongly interacts with virtually all activities in urban areas (housing, industry, commerce, services, and others), transportation planning must be coordinated with land use plans, urban form, and the area's character and lifestyle. Therefore, all these aspects must be planned and designed through comprehensive, long-range planning procedures.

Specifically, transit systems that have permanent facilities (ways, terminals) require long-range, comprehensive planning. Without such planning, transit systems in large and medium-sized cities can seldom meet the transportation needs of their population. One

(*b*)
Photo 10.1 (*continued*)

reason for this is that the needed coordination among urban form, land use, and transit system can only be achieved through long-range planning. Another reason is that networks of fixed facilities usually require investments and considerable time to be planned and built.

Transit system efficiency also requires careful planning. Operation of individual lines, or even of the whole transit system, separately from other modes of transportation in the city may create deficiencies and operating problems. Planning must therefore incorporate the whole transit system with other modes as well as its interactions with land uses. Due to the increasing complexity of cities and strong interdependencies with other activities, transit planning must be related to all relevant city planning, economic, and social aspects of cities. To encompass the whole transportation system, planning must include all geographic areas that function together, rather than be limited by the administrative boundaries. Consequently, contemporary long-range urban transportation planning must be comprehensive in several senses of the word, as is further elaborated below.

10.1.3 Legal and Administrative Aspects

Organization of transportation planning—its legal basis and procedures—vary from country to country. The product of planning may also vary with time and condition of transportation in cities. Planning is sometimes aimed at the production of master plans with physical definition of systems and networks: the National System of Interstate Highways, adopted by the U.S. Congress in 1956, or master plans of rapid transit networks for Bay Area Rapid Transit, the Munich U-Bahn, and the Washington Metro, are good examples. In other cases, a set of policies and financing methods is adopted as the basis for planning and construction of urban transportation systems. For example, several federal laws in the United States have defined not only the financing, but also planning procedures for transportation planning. Similarly, the Federal Act for Financing of Urban Transportation, adopted by the German Parliament in 1967, has been the legal and financing basis for development of transit and other transportation systems in urban areas.

10.1.3.1 Development of Federal Planning Legislation in the United States. The initial legislation with respect to metropolitan area transportation planning, which established its structure and procedures, was the Federal-Aid Highway Act of 1962. This act stated that after July 1, 1965, federal aid would be allocated for highway projects in urban areas of over 50,000 population only if they were based on "a *continuing comprehensive* transportation planning process carried on *cooperatively* by the states and local communities" (emphasis added). Specific guidelines for conduct of these studies were established by the Bureau of Public Roads and modified in subsequent laws of 1981, 1991 (ISTEA), and 1998 (TEA-21). These laws imposed more extensive requirements but attempted to give more planning powers to the MPOs for coordination at the regional level.

The introduction of the "3C" (continuing, cooperative, and comprehensive) process was important in the development of an integrated transportation system. "Continuing" specifies that the planning process be ongoing, frequently reevaluating and updating the transit plan to reflect changes in the urban area. "Cooperative" is defined as the need for coordination not only between the various levels of government (local, state, and federal), but among individual agencies at the same level. It also implies cooperation between disciplines during the planning process and mandates that planning teams be interdisciplinary.

The definition of "comprehensive" is more precise than that of the previous two terms. It is described as a set of 10 elements for which inventories and analyses are required:

- Economic factors affecting development
- Population
- Land use
- Travel patterns
- Existing transportation facilities, including general traffic and transit
- Terminal and transfer facilities
- Traffic control features
- Zoning ordinances, subdivision regulations, building codes, etc.
- Financial resources
- Numerous qualitative elements, including social and community value factors, such as preservation of open space, environmental amenities, and aesthetics

The intent of this requirement and the listed items to be included can be translated into the following specific practical requirements. Comprehensive planning approach and organization should:

- Include the whole system (network) rather than individual lines or routes only
- Incorporate all modes of transportation, i.e., public and private, passenger and freight, etc.
- Include not only technical and economic (cost) factors, but also related elements, such as social and environmental factors and urban form
- Cover the entire metropolitan area, which represents a functional system

To include related factors and cover the entire metropolitan area, the planning must be performed by a multidisciplinary team and involve the cooperation of different administrative jurisdictions and units.

The planning of transportation, and transit in particular, is intertwined with the planning of many non-transportation-related elements. Thus, an area's transportation plan is an input to other planning processes, as the other area plans are inputs to transportation system planning.

There are two important effects of this legal requirement. Since the plan is developed for an entire urban area, setting the scale at the regional or metropolitan level led to treatment of transportation as a system. The act called for cooperation between state planning agencies and local communities. Because of the lack of agencies capable of performing the required comprehensive planning, this act led to the creation of special agencies in all urbanized areas, which were dedicated to performing transportation and other planning for the entire metropolitan area or region.

Before 1962, urban transportation planning actually produced plans and provided financing for highways only. This was a result of the two 1956 Highway Acts, which produced the plan for the 66,000-km (41,000-m) Interstate Highway System and provided federal funding for 90% of its investment. By inclusion of all modes of transportation, the 1962 Act made the first step to broaden the exclusive highway emphasis toward intermodal planning, and financial aid. It also required the founding of metropolitan planning organizations (MPOs) in all metropolitan areas with urbanized areas of more than 50,000 persons.

A major step in transportation development was the establishment of the Federal Department of Transportation in 1967. That was followed by the Urban Mass Transportation Assistance Act of 1970, which introduced for the first time long-term federal financial commitments to transit projects. The Act also declared that population categories with special needs, including the elderly and handicapped, have the same right to public transportation services as the rest of the population.

Another important federal action was the adoption of the Joint Highway-Transit Planning Regulations in 1975. The urban transportation planning process was required to produce a long-range transportation plan, reviewed annually to confirm its validity. The transportation plan contained a long-range element and a shorter-range transportation systems management element. The short-range element was aimed at improving the operation of existing transportation system without new facilities. This part of the regulations was important because it differentiated short-range decisions from long-range planning decisions. It also had to be a plan including all modes, forcing planning agencies to examine the operation of all the modes and their interfaces.

A multiyear transportation improvement program (TIP) was required to be developed, consistent with the plans, goals, and objectives established in the transportation planning process. The TIP's planning horizon was five years, and it included all projects in that horizon. It became a link between the planning and programming of urban transportation projects. This report brought together all the transportation projects into one document that could be reviewed by decision makers. It also contained an annual element that provided a basis for deciding the amount of federal funding for the following year.

The Surface Transportation Assistance Act of 1982 gave states and local governments a greater latitude in the use of both highway and mass transportation funds. It also led to regulations that greatly reduced the federal requirements with respect to the planning process and concentrated on what the planning goals should be.

The Intermodal Surface Transportation Efficiency Act (ISTEA) of 1991 represented a major step forward in transportation, particularly benefitting urban areas. For the first time, this law strongly stated the need for intermodal transportation systems, emphasizing their importance for efficiency, quality of service for travelers, and positive impacts on the quality of life in urbanized areas. The same general spirit was retained in the subsequent Transportation Efficiency Act for the 21st Century (TEA-21) and is expected to be continued in the 2004 transportation legislation.

ISTEA stipulated that every state must establish a statewide planning process in six areas, including public transportation and intermodal transportation facilities, to provide "seamless travel" among modes. MPOs are required to incorporate their TIPs for short-term improvements and their long-range plans, which should consider land-use policies, intermodal connectivity, enhanced transit services, and management systems.

ISTEA greatly increased and facilitated the use of federal funds for projects aimed toward enhancing the environment, such as air quality-improvement measures, highway beautification, and wetlands protection. An innovative and highly successful program has been endorsement and financing for creation of bike lanes and bikeways, which have been used very effectively to stimulate and change the status of bicycle travel in cities across the country.

Intermodal facilities, such as park-and-ride, bicycle storage at stations, pedestrian facilities, and other measures for transit improvement and increased livability of urban areas, have also had major positive impacts. Many programs financed by ISTEA funds have been undertaken by the Federal Transit Administration not only to improve transit, but to enhance the livability of communities and plan and implement transit-oriented developments (TODs) and other innovative concepts in urban transportation and city planning.

ISTEA also allocated funds for high-speed ground transportation, including a demonstration project for a magnetic levitation system, although it failed to establish a permanent adequate financing of Amtrak, the National Rail Passenger Transportation Corporation. That financing still remains subject to separate legislation and to unstable policies due to extreme political pressures and lobbies promoting competing highway and air transportation modes.

10.1.3.2 Metropolitan Planning Organizations. MPOs, established by the Federal Aid Highway Act of 1962, perform long-range transportation planning in all urbanized areas, defined as standard metropolitan statistical areas (SMSAs) with populations exceeding 50,000 persons. Planning for small cities is sometimes centralized in the states' departments of transportation (or highways) to ensure greater expertise and economy of work than the individual towns could obtain locally.

Typically, on MPO is headed by a policy committee consisting of representatives of the city, surrounding counties, and state (or states in SMSAs on state boundaries) administrations and other related agencies, usually appointed by an established procedure. The public interest is represented either on this committee or through a special citizens advisory committee, consisting of various representatives of organizations, leading citizens interested in transportation, etc. The appointments to this body must be sensitive to local political conditions and should ideally represent a cross-section that will form the grassroots support for any decision that the committee must make.

Transportation planning studies are performed by a staff headed by a director who follows directions of the policy committee. He/she is usually assisted by a technical advisory committee consisting of transportation experts from different agencies and departments in the metropolitan area. The staff usually has a sizable team of transportation planning and related professionals.

10.1.4 The Initial Set of Comprehensive Transportation Plans

Beginning in the early 1950s to about 1970, most metropolitan areas in the United States had large transportation planning agencies that produced comprehensive transportation plans for their regions. This generation of plans, known as area transportation studies, included such plans as CATS (Chicago), PATS (Pittsburgh), TCATS (the Twin Cities), BATS (San Francisco Bay Area), DVRPC (Philadelphia), and many others. The planning process has continued since that time, but in recent decades there has been less emphasis on the production of comprehensive plans and more on planning individual corridors, systems, and facilities.

10.1.4.1 Progress and Significance. The work on the development of these initial metropolitan area transportation plans was very significant for several reasons. First, stimulated by the pressure to plan the Interstate Highway System and other freeways, the need to plan the entire transportation network, including streets, arterials, freeways, and transit systems, was recognized. Second, the new capabilities of computers to handle large volumes of data and operate complex models permitted a new type of planning: instead of architecturally oriented and largely qualitatively evaluated plans, there were new possibilities of performing planning and evaluation based to a large extent on data and computations. Travel volumes, traffic conditions, and transportation costs could be computed to much greater accuracy than before. And third, because of the new computational capabilities, it became possible to develop many alternative plans and then select the optimal one.

With these new opportunities, urban transportation planning entered a new era, attempting to respond to the much more complex requirements of the time. Transportation systems consisting of many different modes required coordination, as well as analyses of their interactions with land uses. The rapidly increasing mobility of the population, particularly with growing auto ownership, began to exert greater impacts, both positive and negative, on the development, character, and livability of cities and urbanized regions.

As a consequence of the abundant funding for the interstate highways and lack of any federal support for transit, this initial planning was not fully intermodal, however. Extensive criticism of this highway domination led to the 1962 Act, which formally mandated the 3C planning described above. Greater comprehensiveness, which had been attempted but not at all achieved in the initial planning, now became mandatory. Yet even this legal requirement did not sufficiently correct the weaknesses that had been introduced through the initial set of metropolitan area transportation planning efforts.

10.1.4.2 Conceptual and Technical Weaknesses. The basic planning philosophy and procedure from urban transportation planning processes in the early period have become standard planning practices in subsequent decades. However, many of their aspects were found to have fundamental problems in their conceptual approach and in the procedures used. They have been subject to many revisions and corrections. Discussions and changes in planning philosophy and techniques are still continuing. Major deficiencies in that planning are important to point out.

Trend Extrapolation Was Used Instead of Planning. The procedure used in most of the comprehensive metropolitan area transportation studies consisted of the following major steps. Extensive inventories and data were collected from the preceding years. The obtained historic trends in population size, auto ownership, transit usage, etc. were then extended into the future. They were used to define the scenario in the target year for which planning was then performed. Generally, linear extrapolation of the past trend was used, unless there was an obvious reason that the past trend would not continue at the same rate of change. This set of projected conditions was considered to be the demand that the future transportation network should be designed to satisfy.

This procedure has several fundamental conceptual flaws. By its purpose and function, transportation planning represents a set of activities aimed at producing a desired transportation system or facility. Planning should consist of a review of the present conditions and the existing transportation system, as well as an extrapolation of past and present trends into the future. A very important step is then to examine whether the extrapolated trends lead toward the adopted goals. If they do, the plan should provide for the accommodation of trends. However, if the trends do not lead toward the goals, which is usually the case, the plan should consist of policies, actions, and construction of objects that would not encourage, but modify or reverse the present trends so that the adopted goals will be achieved.

Consequently, trend extrapolation is a useful analysis in one phase of planning, but making plans on the basis of extrapolated trends is not planning. It is a set of activities that encourages present trends regardless of the direction in which they lead. Such "planning" can therefore lead in undesirable or infeasible directions. Actually, in many cases plans and recommended policies may focus on measures aimed at changing past trends in order to achieve certain goals.

Travel Demand Was Considered as Fixed, Independent of the Provided Transportation System. Extrapolated travel demand was considered as a function of land uses and activities, but not dependent on travel conditions. Actually, the construction of new transportation systems and facilities has a significant impact on the volume of travel and its distribution among modes. Planning must take this into account as feedback that incorporates the impact new facilities have on travel demand. Similarly, if services deteriorate, travel demand tends to decrease. This phenomenon may occur if traffic volumes reach highway capacity and travel time increases, or if the quantity or quality of transit service is reduced.

The Desired Roles of Different Modes Were Not Clearly Defined. The dominant planning philosophy in the 1950–1970 period was that urban travel would be auto-based, with transit supplementing it during peak periods and serving the nondrivers. Transportation planning therefore focused on highways, particularly emphasizing construction of new freeways. Only in a few very large cities did transportation plans include consideration of rail rapid transit. In others, it was assumed that bus services could be provided on existing streets and highways. Light rail transit was not included in any plans. This approach resulted in an intensification of highway dominance, with transit plans for systems that were not competitive with auto travel. Thus, instead of attempting to produce plans for balanced transportation systems, the comprehensive plans usually led to increasing auto dependency.

Other Modes, Such as Paratransit, Bicycles, and Pedestrians, Were Generally Disregarded. This was a serious oversight because pedestrians alone sometimes account for 20–25% of trips (and even more in other countries), particularly in central urban areas.

The City-Transportation Relationship Was Not Fully Understood. The impacts of transportation on urban form, such as land use distribution, formation of subcenters and corridors, establishment of pedestrian zones, or decentralization and urban sprawl, were not given adequate attention nor even fully foreseen. For example, the negative impacts of the construction of high-capacity parking garages in central cities are still not understood in most U.S. cities. Awareness of environmental issues intensified in the late 1960s and was later translated into many laws concerning environmental impacts, clean air requirements, and others.

10.1.4.3 Modeling Focused on Demand, Neglecting System Design. As mentioned above, the sophisticated analysis and modeling made possible by the use of computers created new, powerful planning techniques. However, this capability also had some negative consequences.

First, the availability of computer-based models and quantitative tools created a belief that transportation networks could be planned mechanically, using design principles or algorithms. This led to a drastic reduction in innovation and creativity in transportation systems design. Actually, quantitative models can be used to evaluate plans, but for their design, professional knowledge, experience, and judgment are indispensable.

Second, estimation of travel demand attracted a great amount of intellectual attention. Travel demand forecasting became a major area of modeling, analysis, studies, and theoretical papers. The models are behavioral, rather than normative, so that their use strengthened the procedure based on trend extrapolation, rather than on planning toward rational goals.

Extensive efforts put into theories and procedures for travel demand forecasting are in many cases difficult to justify because extreme accuracy in this process is neither possible to achieve, due to the nature of behavioral analysis of future human behavior, nor critical in many cases for transit planning, as will be discussed in Chapter 12. Actually, the accuracy achieved in travel demand forecasts in transportation planning studies is difficult to estimate because very few studies have compared actual travel volumes on the planned facilities in the planning horizon year with the forecasted ones.

Third, while the demand side of transit planning has been given great attention, most aspects of the supply side have been largely ignored in long-range planning. The existing policies and practices have usually been considered as given and fixed. Possible improvements in policies toward transit development and innovations in transit system hardware, such as vehicle designs, information systems, and fare-collection equipment, and impacts of pricing policies in transit and other transportation modes, such as parking, receive relatively little coverage. As a result, the United States has lagged behind its peer countries in the attitudes, policies, and practices of transit systems planning and development.

To some extent, the same bias toward demand while neglecting transit system development also exists in international planning literature. Many high-quality publications covering urban transport models, such as BTE (1998), include excellent coverage of demand forecasting models and their relationship with land use planning, but do not include any models of transportation systems design and operations, either highway or transit. Many aspects of transit line and network forms, number and location of stations, intermodal relationships between transit and highway networks, and operational issues of transit systems, receive much less attention in academic and professional literature.

Most studies on planning and design of transit networks are approached empirically, with little theoretical basis and few innovative concepts. Moreover, among the limited number of publications on formulating transportation plans, many are so theoretically oriented that they cannot be used in actual planning for any city. There are very few publications with materials that transit system designers can directly use in planning transit networks and facilities. This area requires greater attention and bridging of the present gap between actual plans for specific areas and applied research for analysis and development of transit system design and evaluation procedures.

10.1.5 Evolution and Trends in Urban Transportation Planning

Urban transportation and transit planning passed through many phases during the second half of the twentieth century. Revisions of the initial planning efforts resulted in increasingly sophisticated and comprehensive planning processes. Social, economic, and political conditions led to many changes in procedures, legal requirements, and policies toward urbanized areas. Finally, the constant dichotomy between the interests of individuals and the efficiency of overall systems (individual equilibrium versus social optimum) and government protection of the public interest from various lobbies affecting transportation have stimulated innovations in transportation policies. For a detailed description of these developments in the United States, see Smerk (1991), Weiner (1999), and Papacostas and Prevedouros (2001). For analyses of urban transportation policies and their impacts on cities, see Newman and Kenworthy (1998), Pucher (1995), Vuchic (1999), and Katz et al. (2003). Some of the significant changes in transportation planning that have taken place in the United States and other countries in recent decades follow.

A broader concept of comprehensive urban transportation planning, started by the 3C requirement in the 1962 Highway Aid Act, was strengthened in subsequent legislation, with ISTEA being the most significant step in that development sequence.

Transportation planning was influenced considerably by the laws regarding environmental protection, clean air, facilitation of transit use by the elderly (including the development of low-floor vehicles), guaranteeing of accommodation of the disabled, etc.

The energy crises during the 1970s caused a shift of emphasis to the planning of individual projects that can be implemented in a shorter term than long-range comprehensive plans.

Very significant technological and planning innovations have resulted in the development of new medium-capacity transit modes, such as LRT, AGT, and BRT. Introduction of these modes in many cities terminated the period of polarized transit systems, which consisted only of buses on streets and rail rapid transit without modes with intermediate PCP characteristics (see Section 12.1). These developments created a much broader family of modes.

Since the 1980s, transit planning has been broadened to include various forms and processes of private sector participation. For example, implementation became easier with the use of design-build-operate-maintain (DBOM) and similar arrangements.

Many countries experiencing a rapid increase in auto ownership in recent decades, such as Russia and other Eastern European countries, are repeating to a considerable extent the highway-dominated planning at the expense of transit that prevailed in the United States during the 1950s and 1960s. While their highway domination is much less intensive than it was in the United States, their problems may be even greater because those countries do not have the space and economic conditions to accommodate extensive construction of highways and parking facilities in cities. China's present suppression of bicycle and bus use for the benefit of accommodating private automobiles is a drastic example of misguided planning policies.

In the developed countries with most advanced transportation planning, attention has been focused in recent decades on the reduction of the conflict between vehicular traffic and urban environment. Pedestrian areas, traffic calming, and new design concepts have resulted in an increasing role of transit in the cities that are placing great emphasis on their livability, consisting of improved quality of life, economic efficiency and environmental sustainability.

ISTEA in 1991 and subsequent developments in the United States have strongly emphasized the need for greater use of different modes and their linking into coordinated systems, which requires comprehensive intermodal planning.

10.1.5.1 The Role of Transit Agencies in Long-Range Planning. While it was initially intended that MPOs primarily perform long-range planning for all modes of transportation and thus leave to the transit agency only short- to medium-range planning, in most cases this arrangement did not work out satisfactorily. MPOs tend to have a broad orientation toward comprehensive regional planning, which involves land use plans, economic analyses, social considerations, and environmental aspects, with less extensive expertise in the field of transit systems. Transit planning, which requires extensive data, deep understanding of intermodal transit operations, and experience in transit system design and operations, is most effectively performed by the transit agency (if it has a planning department) or a special transit planning agency, often using consultants under its direct supervision. Transit systems designed without the participation of a transit agency are more likely to have problems in operation than when they are designed by persons with operational experience.

For these reasons, it is necessary that all transit agencies that have or are planning to have permanent way lines (basically, rail systems) have their own planning departments. Those departments develop transit plans and give them to MPOs for testing and coordination with plans for other modes. The transit agency then proceeds with detailed planning and design (determination of lines, station locations, development of specifications for right-of-way alignment, structures, rolling stock, operations, etc.) and then carries out plan implementation.

Experience has shown that planning of major new rail transit systems has been done by specially formed bodies within transit agencies, or independent bodies closely related to the transit agency and MPO. Good working cooperation between these organizations is essential. The transit agency must supply data on its network, operations, and ridership. It should also assist the planning agency in developing and evaluating the plans. This cooperation usually takes place through direct contact between the two staffs, or through the members of the two boards or technical advisory committees, when these exist. The MPO usually participates in evaluating the plans, analyzing intermodal aspects and intergovernmental activities required for obtaining the financing and leading to implementation of plans.

Transit planning should not be limited only to large transit agencies that have special departments for that purpose. To be effective and efficient, each agency must perform planning even if it does not include major capital investments but concentrates on short-to-medium range improvements.

10.1.5.2 Implementation of Transportation Plans. Planning agencies usually produce a master plan for a target year 15–25 years in the future. Its implementation is usually done separately for different modes because their funding comes from different sources. For example, highways and streets receive funds from certain combinations of local, state, and federal funds, depending on the category of facilities. Bicycle facilities often have a special funding source.

Transit plans also obtain funding for each major project as it is approved by different government levels. In the case of major transit projects, such as the Washington Metro or the LRT in Denver, funds are available from a special government allocation or local tax funds that have been approved by voters. Then the

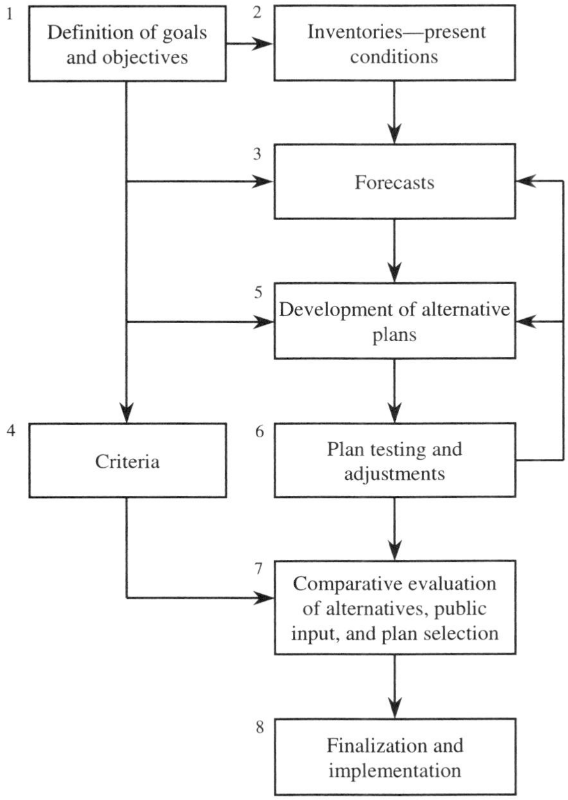

Figure 10.1 *Sequence of basic steps in comprehensive urban transportation planning.* (Source: *based on Creighton, 1970.*)

plan for a line or an entire network is implemented following a prepared schedule over 5, 10, or, in the case of the Washington Metro, 25 years.

Most metropolitan areas in the United States incorporate a large number of governmental units: towns, cities, counties, and sometimes states. In many states, all of these bodies have partial or nearly full independence in transportation planning. This highly decentralized decision making presents serious problems in both planning and plan implementation. Under these conditions, many MPOs adopt a general regional plan for a future year that represents an advisory guideline rather than a fixed, coordinated plan that will be implemented. Then the MPO collects all locally developed transportation projects and reviews them with respect to their own feasibility and their relationship to the master plan. This evaluation results in ranking of the projects for implementation.

Consequently, metropolitan areas that lack regional, coordinated decision making have the serious problem that they can perform only locally generated, partial long-range planning. Such planning often makes achievement of major, coordinated transportation networks very difficult. Therefore, in the states and metropolitan areas having these conditions, achieving efficient planning requires not only improved technical planning procedures but, more importantly, a governmental reorganization to make decision-making and plan-implementation processes that produce regional plan not advisory, but agreed upon and obligatory for all administrative units in the region.

10.2 PLANNING PROCEDURE AND METHODOLOGY

Urban transportation planning encompasses a number of different planning processes depending on the objects and scope of planning, basic comprehensive planning or plan updating, time range, and other factors. The procedures and sequences of steps for several basic types of planning are presented here.

10.2.1 The Basic Steps in a Transportation Planning Process

Transportation planning can be presented in its generic form as a sequence of steps, as shown by the flowchart in Figure 10.1.

1. Definition of goals and objectives for the transportation system in the future
2. Collection of inventories: data about the existing city and its transportation system
3. Forecasts of changes and conditions in the selected target year for plans
4. A set of criteria for plan evaluation derived from the goals and objectives
5. Development of several plans, usually referred to as alternative plans, for the projected future conditions, meeting the defined goals

6. Technical elaboration and testing of alternative plans, considering their impact on projected demand and urban development
7. Comparative evaluation of alternative plans using the set of goal-based criteria and public hearings, resulting in selection of the preferred plan
8. Finalization of the selected plan and preparation for its implementation

This is a conceptual organization of a planning process that includes the entire planning from data collection to the production of the final plan and its implementation. Each of these steps further consists of data handling, modeling, technical analysis, etc.

The planning process is shown with more technical details in Figure 10.2. Starting from the definition of planning goals and objectives, the process can be grouped into four major phases. Phase I, focusing on inventories and data, usually consists of collecting comprehensive information about the present transportation system and urban area. This includes infrastructure elements, such as land uses and transportation system; general data on population, economic, and social conditions; and travel volume, characteristics, and modes of travel, as well as transportation system services.

Phase II is the development of analytical, computer-based models for forecasting travel. These models are needed for estimation of present and future travel on the existing or on newly designed networks of different modes. The models also allow evaluation of the present conditions on the networks: traffic volume/capacity ratios, operating speeds, and reliability, showing how adequate the conditions are and where deficiencies exist. The models are also used to predict future travel volumes on the existing and then on alternative newly designed networks. The new designs are tested through travel forecasts obtained by these models.

The most important and creative part of transportation planning is phase III, the development of alternative plans, based on the present conditions, projected changes, and objectives for the future metropolitan area. Short-term improvement planning also belongs in this planning stage.

The final planning step, phase IV, is the evaluation and selection of the preferred plan, using the criteria derived from the adopted goals and objectives for the future transportation system.

10.2.2 Transit System Planning Procedure

While comprehensive urban transportation planning includes all modes, it has traditionally been heavily oriented toward the highway/automobile system. Some modes, such as bicycles and walking, are not included in most models. Transit is seldom given sufficient attention. Therefore, long-range transit planning is usually organized as a special project by the transit agency, MPO, or a specially founded agency, usually with strong involvement by consultants. Moreover, in several stages of this planning, citizen participation is included for suggestions, reviews, comments, and approval of the planning inputs and outputs. In the United States, this is a legal requirement for the official planning process.

The roles that transit and other modes play in a city are closely related not only to its physical form, but to the type and character of urban activities and living. Conceptually, the sequence of planning a transportation system for a city and its region should logically start by defining the type of the city/region. This vision should be translated into the goals for the transportation system, specifically its intermodal composition, because the roles of different modes have a major impact on the city's physical characteristics, activities, and environment. For example, a city with concentrated office buildings, large manufacturing plants, and high-density residential areas requires an efficient, high-capacity transit system. On the other hand, an auto-oriented region tends to rely on lower-density residential areas and commercial and business activities dispersed along highways. In another case, an urban area with moderate climate, flat terrain, and many educational institutions may be appropriate for design with an extensive system of bikeways. Naturally, in each case these major roles of individual modes are supplemented by networks of other modes with relative significance depending on local conditions and adopted goals.

Intermodal urban transportation planning can be seen as a process consisting of four levels of analysis

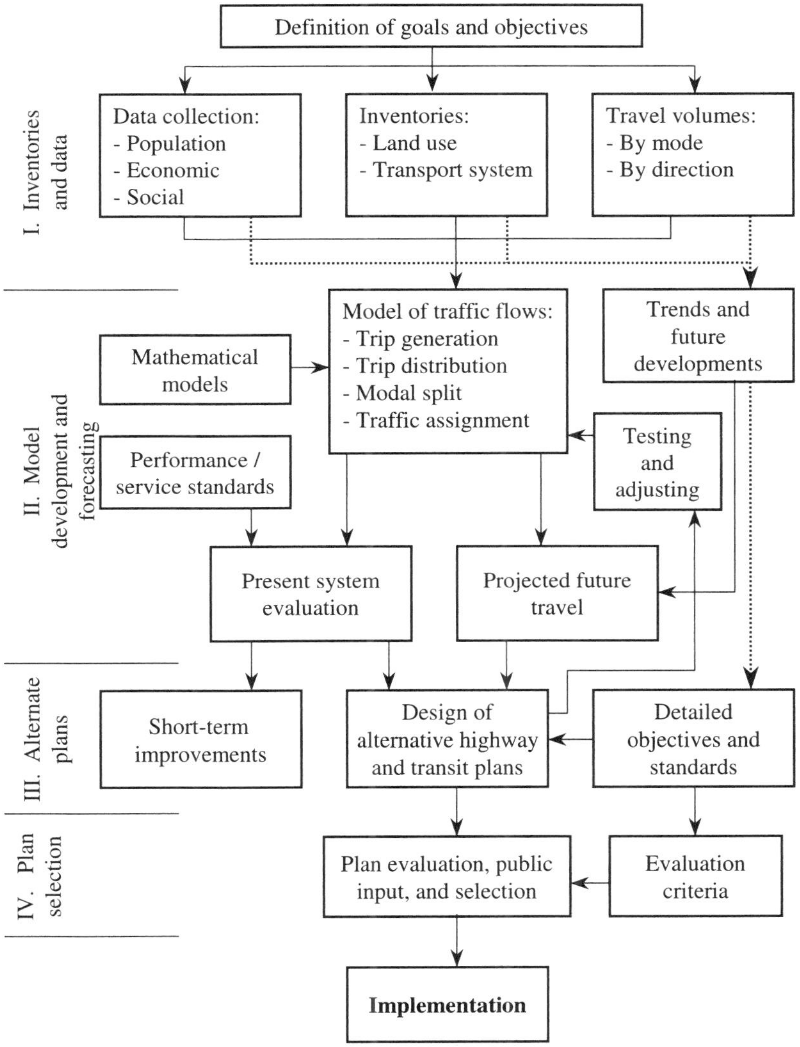

Figure 10.2 *Technical procedure of long-range urban transportation planning.*

and decision making, as shown in Figure 10.3. Starting with the vision of the desired urbanized area in the future at level I, goals based on that vision should be translated into the plan for relative roles of different modes and their relationships at level II. The decision at this level determines the role transit and other modes should have, i.e., what their interrelationships should be. Level III then consists of the design of each mode individually. A transit system is formulated here, determining the modes and their networks. Finally, individual transit lines and facilities are planned and designed at level IV of the planning procedure.

A more detailed sequence of transit planning, including technical steps, is illustrated in Figure 10.4. As in the general planning procedure shown in Figure 10.2, the first step (box 1 in the diagram) is the collection of data and information about the present transit system: modes and their networks, service, and usage,

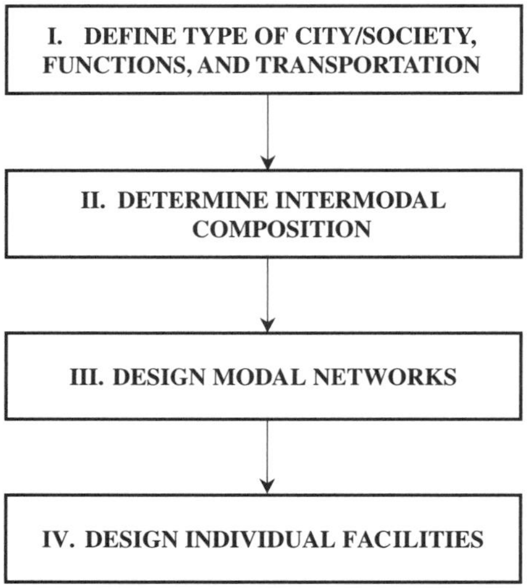

Figure 10.3 *Four-level planning sequence from urban transport system to individual projects.*

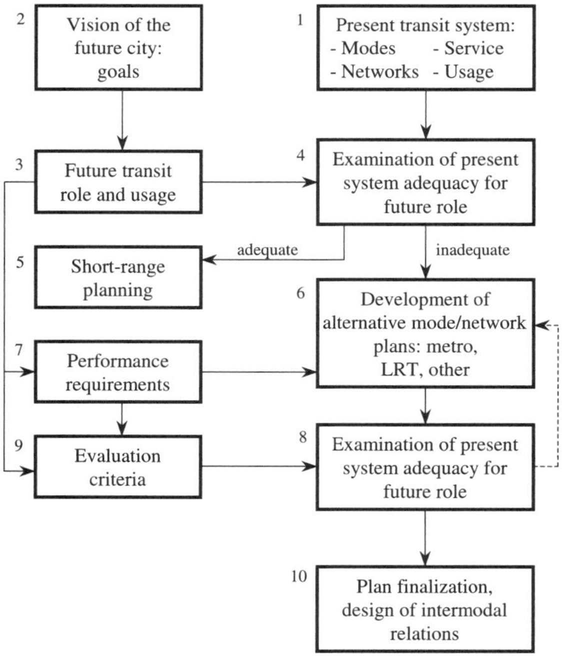

Figure 10.4 *Transit planning process.*

i.e., passenger volumes by directions and time periods. Parallel with this effort, a vision for the future city/ urbanized area is developed (box 2), defining the future role of transit (levels I and II in the sequence described above). Based on this vision, a travel-forecasting model is used to estimate future transit's role and usage in step 3. This projected volume is used in step 4 to examine the adequacy of the present transit system (its modes and networks) to serve the future demand and play the desired role in the city's functioning. If the system is generally adequate, the procedure goes to step 5, terminating in short-range planning. This is usually the case when only a bus network is needed.

If the existing system will not be able to meet the future demand and, most importantly, cannot provide a high-quality transit service competitive with auto travel, long-range transit plan development is the next step, shown as box 6. In this step, one or more plans are developed that can satisfy the projected travel volumes and meet the performance requirements to achieve the desired role of transit in the city (box 7).

The development of alternative transit plans represents a very sophisticated process that requires special expertise in transit systems. It involves two aspects: selection of mode and design of network (see chapters 12 and 4, respectively). These two features are mutually interrelated: even in similar roles for transit services, different modes, such as metro, LRT, and BST, would have somewhat different optimal networks. This process usually produces plans with different modes and networks. They are then evaluated in step 8 with respect to the criteria based on the future system role (box 9). In the final step, 10 in the flowchart, the selected plan is finalized and its intermodal relations are defined.

As this review of transportation planning procedures shows, the approach and flow of the planning process vary with the conditions, scope, and goals of transportation planning. Transit planning is more focused than comprehensive urban transportation planning, and it requires special expertise in the plan development, selection of modes, and design of networks. It is emphasized that the presented sequences are not absolute, and the work plan, shown by the flowcharts, should be adjusted to special conditions for any specific planning process. However, the concepts described here are useful as a basis for most of the transit planning tasks.

Setting of goals and data collection are discussed next. Other phases of planning, such as development of models, demand forecasting, and alternative plans, are presented in subsequent sections.

10.2.3 Setting Goals, Policies, Objectives, and Standards

The foundation of any well-designed plan is a clear definition of the goals and objectives the planned system is required to meet in the target planning year. The policies and standards should be the guides to how the planning agency intends to meet the adopted goals and objectives. While the goals and policies are likely to be politically driven and often imposed on the planning agency, the objectives and standards are basically technical, and they are more likely to be at the discretion of the planning agency. Standards are technical values, usually numerical, that the planning agency is likely to have the most control over.

Goals are broad statements addressing the desirable conditions with respect to mobility, travel opportunities, and transit service performance characteristics and outcomes upon a plan's implementation. For example, the general goals referring to the role of transportation in urbanized areas may be:

- Adequate mobility should be available to all categories of the area's population to enhance their opportunities for employment, recreation, commercial, and other activities.
- Transportation in the city should permit diverse activities that require various densities.
- Transportation within the central business district (CBD) should stimulate its functioning and growth; over time, it should secure improved environmental and economic conditions.
- The CBD and other major activity centers should be accessible without chronic traffic congestion.
- Transportation should serve suburban residential areas without imposing excessive negative impacts, such as air pollution, noise, and safety hazards.
- Low-income areas should be provided with adequate transportation to the areas of potential employment.
- In medium-sized and large cities, intermodal transportation systems should be provided to achieve the required services efficiently.

To meet these goals with respect to mobility, more specific objectives for the transportation system can be formulated, such as:

- Networks of each type of highway (e.g., freeways, arterial streets) should be continuous and coordinated with other networks as well as with terminal facilities.
- The efficiency of the street network should be increased through specialized uses of facilities, such as individual streets designed predominantly for fast automobile traffic, others primarily for local access and parking; some streets reserved exclusively for transit, or pedestrian malls without any vehicles.
- Pedestrian travel should be encouraged in the CBD and other subcenters, as well as in residential neighborhoods, through the provision of safe and convenient walkways and other facilities.

The goals with respect to the role of public transit in different categories of cities include the following:

- In developing countries, transit represents the basic mode of travel. It should offer reasonable quantity and quality of service in all towns and cities.
- In countries with high auto ownership, the primary role of transit in small cities is to provide a social service: to transport mostly non-automobile owners.
- In medium-sized cities, transit not only provides social services, but it replaces some automobile travel, thus decreasing congestion and excessive parking demand.
- In large cities, transit must provide high-quality transportation services competitive with automobiles throughout the urban area. Without this type of service, the functioning and growth of large cities would be severely constrained.

Based on these goals, some specific objectives for medium and large urbanized areas may be defined as:

- Transit should offer service competitive with or superior to that of private automobile in the central urban area and along major corridors.
- To achieve competitive service, transit should be provided with maximum possible right-of-way separation so that it is independent of traffic congestion.
- Where feasible, parking should be provided at rail transit stations and in the immediate vicinity of the central area, while it should be reduced within the CBD.

The adopted goals and objectives can be used as the basis for the derivation of technical system specifications and service standards, considering existing conditions, such as the size of the city or transit service area, population number, density and characteristics, trip purposes, as well as uses of other modes (auto ownership, taxis and other paratransit, bicycles, and walking habits). These specifications are determined on two levels.

System specifications define the basic level and quality of service transit system should offer, and thus the role it will have in relation to the city and other transportation systems. For high-performance modes, such as rail and other guided modes and BST systems, the specifications include:

- Right-of-way category—the basic feature determining the transit mode and performance (see Chapter 12)
- Infrastructure geometric elements, such as minimum radius and maximum gradient of tracks/guideways, vehicle profile, TU length, type and dimensions of stations, etc.
- Maximum and operating speeds
- Area coverage (area within a 5- and 10-minute walking distances of stations) and population served by the high-performance mode
- Investment and projected operating costs of the considered alternative systems or modes

Each transit mode, such as metro, LRT, and regular bus, has a different set of specifications, i.e., each one represents a distinct performance-cost package (PCP).

Service standards are a set of definitions, expressed mostly as quantitative values, that transit system should meet in its operation. These standards are also developed separately for each transit mode. Examples of standards are:

- The average and longest acceptable headways for each category of lines and times of day (peak, midday, evening, owl service);
- Number and locations of stations/stops with transfers among different lines and modes, and degree of coordination among lines.
- Average and maximum load factor, often related to the number of seats, for each period of day;
- Reliability of service for off-peak and peak periods.
- Average travel speeds. This can be measured for specific trips, for example by the lines showing how far from the center a passenger can travel in a given amount of time. These access time contours, or isochrones, can be mapped and can help to determine transit network shape. An example of access time contours is given for Washington in Figure 10.5.

The preparation of system specifications represents a major decision in the long-range transportation planning process. The goals and objectives for the future transit system are translated into plans through such specifications and selection of transit mode. Service standards are more a short-term decision because, unlike the system specifications, they can vary for any given system depending on passenger volume, available funds, etc. Thus, determination of service standards is mostly an operating decision, but it is considered in long-range planning because it determines the expected level of transit service and thus the modal split and projected passenger volumes.

The detail to which these goals, standards, and objectives are specified varies among planning and transit operating agencies. Precisely drafted plans and documents eliminate confusion at later stages.

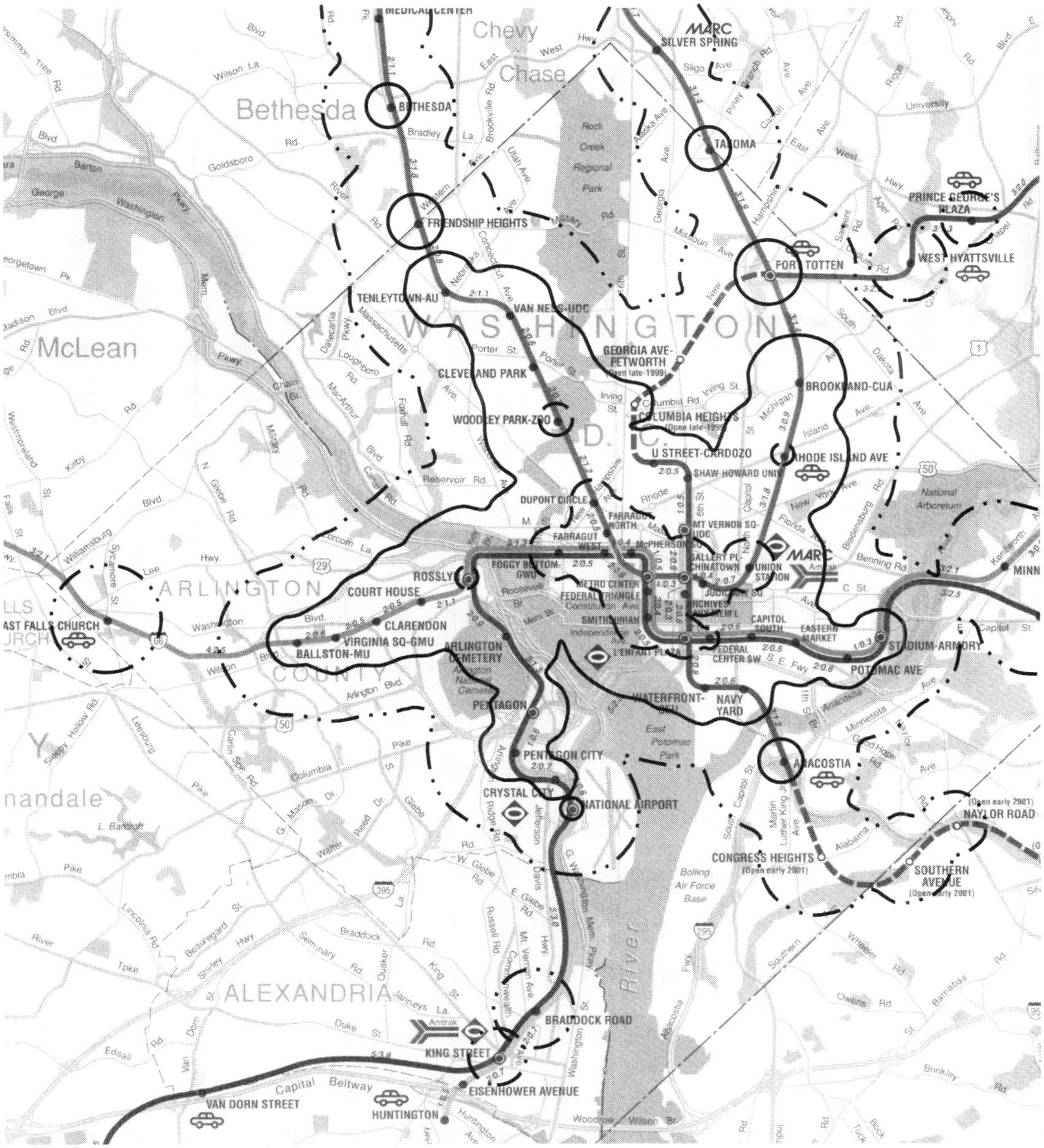

Walking speed = 80 m/min

Time spent in getting out of the metro station is 1min

Waiting time is 2 min

Transfer time is 5 min

— — — 10 min contour

——— 20 min contour

— · · — 30 min contour

Figure 10.5 *Travel time contours from the Metro Center Station on the Washington Metro network.*

10.2.4 Data Collection and Evaluation of the Present System

Information and data collection about the existing conditions are needed for planning a transportation system, as well as for making a model to be used for testing various plans. Data collection in comprehensive studies is always a large task, requiring numerous personnel. It is often the most costly part of the whole planning study. However, the detail to which the data are collected varies among studies, so that planning of a specific transit project may not require collection of data for the entire region and the entire transportation system.

Information and data can be grouped into several categories, each one having several items.

Population, economic, and social data include a breakdown by sex, age, household size, auto ownership, etc. Economic factors relating to transportation are, for example, employment data, household incomes, and forecasts for development of the area or its individual sectors.

Land use data consist of the type and intensity of individual activities (number of employees for office blocks, size of retail area for stores, school enrollment, etc.). In addition to these inventories, forecasts and projections of all these items for the future are also collected, as far as they have been made.

In most cases no field surveys are needed for the collection of these data. They are obtained from agencies, governmental units, or chambers of commerce within the respective jurisdictions.

Transportation system infrastructure, including networks, stations, terminals, parking facilities, pedestrian areas, truck, and bicycle facilities, is collected. Again, all planned changes and expansions are also collected to facilitate projection of the system as it is expected to be in the foreseeable future.

In transit planning, special emphasis is placed on transit system facilities and their operations, while the street/highway system is included only where it is directly relevant to transit: when it is used by transit lines or for access to stations and stops.

The facilities inventory data are supplemented by operational factors, such as traffic engineering features (regulations, control devices, etc.). Finally, travel patterns are surveyed to obtain origin-destination (O-D) tables, modes used, trip purposes, trip lengths, etc.

Transit system infrastructure and operations data include the following specific items:

- Transit network of lines by mode and its area coverage
- Individual line descriptions, including designation, length, number of stations, operating time, hours of service and headways by period of day, operating and cycle speeds, and reliability of service
- Transit terminals and intramodal and intermodal stations
- Transit yards, maintenance shops, vehicle storage areas, and other facilities
- Number of vehicles by type, size, and age
- Fare types and amounts

Transit system usage consists of data about the present transit ridership. The data should consist of passenger volumes on each line, including:

- Directional volume profile along each line, including the MLS (see Chapter 1)
- Hourly, daily, and seasonal variations
- Boarding, alighting, and transferring passenger volumes, especially at major stops/stations
- Passenger characteristics and travel patterns

Most of the data about the transit system infrastructure, lines, vehicles, etc. are available from the transit agency. Operating conditions along transit lines and some of the detailed passenger volumes are usually obtained through field surveys, described in Chapter 1.

The collected transportation data, which are usually quite voluminous, are entered in a computer database so that they can be easily manipulated to develop a travel forecasting model, as well as to derive various relationships for transit system analysis and planning. Among the numerous relationships that are of interest are, for example, the number of daily trips generated by different types of land uses, the time distribution of trips by purpose, and transit riding related to automobile ownership.

Some automated fare-collection systems have computerized data summaries that provide detailed data on passenger travel and volumes. The rail systems with automated fare gates for station entry and exit usually provide all information on passenger volumes, including even specific OD data for any time period. The costs of data collection and entry into the computer are avoided when the transit agency uses such a system. Some of the data needed for planning exist in census data already in machine-readable form, but other data have to be collected by the transit agency or the planning organization through ridership counts and other surveys.

The remaining planning steps, shown as boxes 3–8 in Figure 10.1, are more complicated. They are described in the following sections.

10.3 TRAVEL AND TRANSIT DEMAND FORECASTING

Forecasting future travel volumes, also referred to as travel demand prediction, is a basic element of the road and transit planning process. Decisions about upgrading an existing transit line, extending a network of lines, or building a new network require certain knowledge about the future passenger volume on such facilities. The tools for predicting future travel demand are procedures based on logic and knowledge about travelers' behavior, which are based on theoretical relationships adjusted to each specific situation. For small-scale demand predictions, such as trip estimation for a neighborhood, shopping center, small hospital, or college campus, direct (manual) computations can be used. For planning of a large corridor, a major activity center, or an entire urban area, computer-based models must be used.

Travel-forecasting models can be used in the course of plan development, in testing plans with respect to their individual features, such as changes in line alignments, number of stations, mode features, land use developments, etc. They are also used in evaluating alternative plans and in finalizing the selected plan for its implementation.

Extensive intellectual effort, studies, and scientific papers have been dedicated to travel demand forecasting. Yet the results of most demand forecasting studies have limited reliability, and very few studies have examined the accuracy of projections after a transportation network had been built (Boyce et al., 1970). To understand the nature of travel estimation fully by any mode, one must bear in mind the following facts.

First, estimation of future travel demand can only be as accurate as prediction of any future developments: population growth (or decrease), economic developments, social activities, trends in needs, habits and attitudes of the population, national and international developments, etc.

Second, the standard planning procedure is based on models representing the present conditions and relationships. This is valid only if the assumption that these relationships will not change in the future remains valid.

Third, travel volume is not only determined by the mentioned external influencing factors, but also affected by the type of transportation system provided, so that a feedback mechanism may be needed in the course of transportation planning.

Fourth, as a consequence of the above, a high-performance transit system should not be planned only as an outcome of various independent developments, because the new system itself usually has an ability not only to generate new trips, but also to influence land use and other developments. Transit should therefore be considered as an interdependent element of urban planning. In other words, transit planning should not be based merely on a set of independent trends; rather, it should be a component of a plan of urban development with normative elements. A well-planned transit system has a tendency to anchor itself to its alignment and stations by its positive impacts on their development. This element of transit impact on activities and land use can often change the demand forecasting process from empirically based trend projections to modeling with normative elements of planning.

10.3.1 Introduction to Four-Step and Other Demand-Estimation Models

The traditional method for transit demand estimation is a process consisting of four steps, which estimate the number of trips generated in each area, their distribution, the transport mode they use, and the exact routes they follow (Meyer and Miller, 2001). Because this method is rather complicated and requires extensive data, many variations of its steps, as well as different procedures, such as disaggregate models, have been developed. Yet the four-step method and its individual steps have generally remained the most common procedure. Knowledge of this methodology and its major models, particularly the modal split estimation procedure, is therefore very important for transit planning, and it will be briefly explained in the following sections.

The four-step model and most other travel demand-estimation models are initiated by defining the study area: an entire metropolitan area, one county, a corridor, or an area that a rail transit line or network will serve. Planning models are then developed that can simulate and forecast activities and travel in the planning area. To develop such models, the area is divided into zones, defined as contiguous areas, desirably with homogeneous population groups or activities, such as commercial, industrial, office, or residential with one type of housing. Often the units defined as population census tracts are used, so that some population and activity data information are already available. Zonal boundaries should, desirably, follow major roads, water bodies, or other geographic dividers.

For each zone, data on transportation facilities relevant to the planning are collected: street and highway network, transit lines, stations, transportation terminals, and other facilities for all modes. Then a central point, representing a center of gravity of the activities and referred to as a *centroid,* is selected. The data about the population and its characteristics, number of trips and their purposes, times, etc. are also collected for each mode. Traffic and passenger volumes, street conditions, and transit service characteristics are usually counted and surveyed. If population surveys are undertaken, it is common to use samples of 1–4% of residents. Collection of these data is usually a major effort, involving an extensive organization, carefully prepared methodology, and a large team of surveying persons. Cost and time for this phase of planning are considerable.

Several phases of transportation planning are based on theoretical models representing relationships between various factors and trip production. These models are developed on the basis of the present conditions: their parameters are computed from the conditions counted or measured at the time of planning. This process is called *model calibration.* Then such calibrated models are used to project travel for the future planning target year, using the assumption that the relationship of people behavior and activities to the production and characteristics of travel will be the same in the future as they are now. If there are indications that these relationships will change, it may be necessary to adjust the travel forecasting models accordingly.

Traffic volume forecasting may have a different sequence of steps, depending on the assumptions that the planning agency makes at the outset of the planning process. The most common sequence is presented clearly by the classical flowchart for this process shown in Figure 10.6 (Fertal et al., 1966). The four steps in this process, shown in hexagonal boxes in the figure, are the major steps that each study must include. They are defined here and described in the following sections.

Trip generation is the estimation of the number of trips originating from or destined to each zone of study area (or another land parcel) in a designated period, such as a given weekday. This estimate for different classes of households can be made utilizing various factors, including, for example, size of household, age of its members, occupational status, etc. The number and type of residences (e.g., apartments, single-family homes), employment, retail area, or sales volume in the zone is also used to estimate trip generation.

Trip distribution is the process of connecting the origins and destinations obtained from trip-generation estimates. The result is a trip table whose rows and columns are the zones in the study area. Many different types of models have been used for estimation of trip

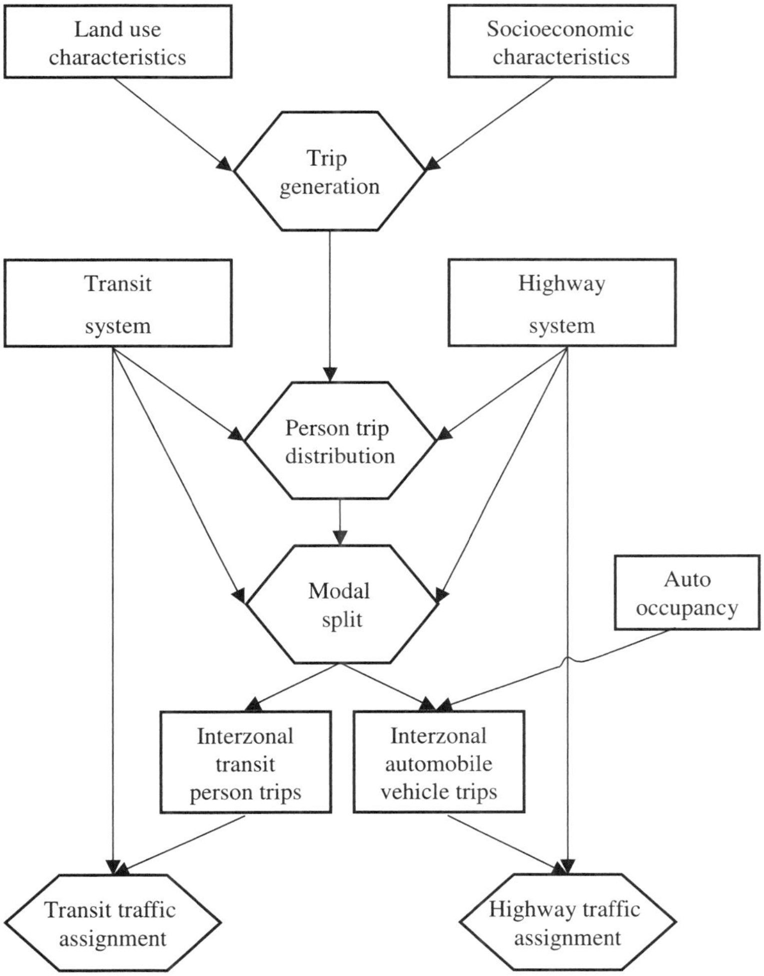

Figure 10.6 *Sequence of steps in travel volume forecasting.*

distribution. The gravity model in several different forms, and sometimes growth factor models, are some commonly used ones. They are introduced below and described extensively in the literature (in particular, Meyer and Miller, 2001; Ortuzar & Willumsen, 2001; and Papacostas and Prevedouros, 2001), as well as in most metropolitan area transportation study reports.

Modal split is the planning phase for estimating the distribution of trips to different modes. Most commonly, the distribution focuses on transit and automobile modes, but in many cities other modes, such as walking and bicycles, must be included because they may amount to 10–30% of trips.

Trip Assignment is the process for predicting the specific routing of trips on the basis of their origins, destinations, and travel conditions on the network of each mode. These four steps in the travel-estimation modeling are described in the following sections.

Mathematical models are used as a basis for all major steps in traffic flow forecasting, shown in Figure 10.6. All these models are based on an assumed mathematical law of behavior of urban travelers.

10.3.2 Trip Generation

The goal in trip generation is to estimate for the present and for the future planning target year the number of trips generated and attracted by each zone in a given time period, not considering their destinations. This includes both trips generated internally within a zone and trips generated between zones (Dickey et al., 1983; Papacostas and Prevedouros, 2001; Meyer and Miller, 2001; Ortuzar and Willumsen, 2001). The basic form of the models used in the trip-generation phase can be simply written as (BTE, 1998):

$$T_i = f(S_i) \quad \text{and} \quad T_j = f(S_j) \qquad (10.1)$$

where T_i and T_j represent, respectively, the numbers of trips generated by zone i and trips attracted to zone j, and S_i and S_j are the socioeconomic characteristics of zones i and j.

The socioeconomic and land use characteristics used to estimate trip generation vary with types of zones and activities in them, available data, and possibilities, of collecting local information. Generally, based on Stopher and Meyburg (1975) and BTE (1998), socioeconomic variables may include household size, household income, car ownership, number of drivers, and type of residence structure. Land use variables include types of activities, such as offices, industry, retail outfits, medical centers, educational and public buildings, and recreational and open space.

These characteristics expected in the planning target year are employed to estimate the number of trips. The basic assumption is that the relationships between influencing factors and trip production in individual zones will be stable over time and that future land uses can be predicted with reasonable accuracy. Moreover, it is implied that the demand for transport is inelastic with respect to the total amount of activities and the quality of the transportation system. When these assumptions are not accurate because major construction of buildings, upgrading of the transportation system, or similar changes are planned and the influences of those changes on trip generation may be different than the basic models predict, it may be necessary to adjust the models.

Several types of techniques developed for computations of trip generation are briefly described here.

Multiple linear regression models are most commonly used for trip-generation forecasting. There are two types of this model: the aggregate zonal version and household-level regression models. The general form of these models is:

$$Y_i = a_0 + a_1 X_{1i} + \cdots + a_n X_{ni} + E \qquad (10.2)$$

where:

Y_i = Total number of trips generated by the inhabitants of zone i
X_1, X_{ni}, etc. = Explanatory variables for zone i, such as its population, auto ownership, etc.
a_0, a_1, a_n = Regression coefficients (estimated by least squares techniques)
E = Error term, assumed to be independently and normally distributed with an expected value of 0

The planner inserts the historical data for the region in the X and Y values (to estimate the coefficients, one must also know the Y values) and determines the coefficients a. Using the coefficients determined from the historical data, future trip generation is predicted by substituting the predicted values of the X variables into the formula.

There are a number of problems with this model. The aggregate values may not be sufficiently precise. Also, maintaining the statistical validity of regression is difficult, and if the stringent assumptions are not met, accuracy of the results is affected. Sometimes, even if the model is statistically valid, the planner may obtain a result that is not logical or physically possible.

The household version of the trip-generation model is similar to the zone-based model. It simply substitutes household characteristics, mentioned above, in the X_i's for the aggregate values in the zonal version.

The household category model selects important variables and stratifies them arbitrarily. From these categories, the average values of the trip rate for each category are computed. These averages are assumed to be constant over time and location. Then projections are made based upon average trip rates. An example of the values for the trip rate variable is given in Table 10.1.

Although these socioeconomic characteristics may better approximate trip-maker behavior, they suffer from the same problems as the zonal regression model. This model also requires a lot of detailed information for each household, making the acquisition of a statistically significant amount of data expensive and time-consuming.

Both linear regression models, zonal and socioeconomic, are not transferable spatially, and they may be affected by a change in cultural values and attitudes of the population.

Cross-classification analysis is aimed at construction of a multidimensional matrix with each dimension representing an independent variable, stratified into a number of discrete classes and categories. Examples of variables for trip production are automobile ownership, household size, and income, and for trip attraction employment type and employment density are often used (Kanafani, 1983).

This method makes it easy to understand the impact of different factors on and their relationship to trip generation. Also, it avoids the linear assumption of the regression models. Its drawbacks are suppression of variances among households in a specific zone, sensitivity of the results to grouping, and lack of a statistical measure to assess reliability of the results (Stopher and Meyburg, 1975; BTE, 1998).

Direct estimation of transit trips can be used in areas with a considerable degree of transit use. Since most of the travel-forecasting models described here have been developed for metropolitan regions with high auto ownership, they are heavily oriented to auto trips with less attention to transit and often complete neglect of walking and other modes. In urban areas with high-quality transit systems where most households generate some transit trips, planning of transit systems can use models that directly compute transit trips as a function of household characteristics, employment, closeness to transit stations, etc. For example, surveys of areas served by metro lines can produce trip-generation rates that can be applied to the areas a planned metro line will serve.

Short-term transit planning for bus lines in inner cities can also rely on statistical data on present transit use, preferably using service quality characteristics such as speed, frequency, reliability, and system image as variables.

This procedure can be much more practical and less expensive than comprehensive area transportation studies and yet yield results with accuracy that is sufficient for planning many types of transit services in large cities with high-quality transit, such as Boston, New York, and Philadelphia, many cities of other industrialized countries, and, even more efficiently, in most developed countries, where bypassing the modal split analysis is not a problem because of dominance of transit, bicycle, and walking trips over auto trips.

Table 10.1 An example of daily trip rates for different family sizes and auto ownership rates

	Daily Trip Rates for Families with		
Family Size	*0 Autos*	*1 Auto*	*2+ Autos*
1–2	1.15	2.43	4.21
3–4	1.23	3.18	5.55
5–6	3.43	6.43	6.98
>6	4.88	7.33	8.99

10.3.3 Trip Distribution

Following the estimation of the number of trips produced by each zone or attracted to it, the next step is to predict how are the generated trips from each zone i distributed to each zone j as its attracted trip. In other words, this phase links the produced trips with attracted ones. Generally, trip distribution is considered to be a function of three major groups of factors: the type and extent of transportation facilities connecting zones; the land use patterns; and socioeconomic characteristics of the population.

Models that have been used to forecast trip distribution can be classified into three types:

- The *growth factor model,* which applies a constant growth rate to existing traffic flows. It assumes that travel patterns will not change in the future.

- The *gravity model,* based on the presumption that the number of trips between each pair of zones is proportional to the activities (or trip generation) of those zones but inversely proportional to the distance and other resistances among the trips to potential destinations.
- The *intervening opportunity model,* which bases traffic flows between zones on the relative opportunities in each area for shopping, entertainment, employment, etc.

The growth factor model simply applies a constant growth rate to the current traffic flows. This growth factor is determined empirically from past data. The growth model, however, has two problems. First, assumes that either there will not be any changes to the highway and transit systems or that those changes will not have any impact on the number of trips individual systems or facilities attract. More importantly, the model assumes that the rate at which traffic has grown in the past will be the rate at which it will grow at in the future, i.e., it is based on extrapolation of past trends.

The gravity model, with many variations in its exact form, is by far the most commonly used model for forecasting trip distribution. The law, when it was first used for analysis of intercity railway trips by Lill in the nineteenth century (Lill, 1891), had the exact form of Newton's law of gravity:

$$T_{ij} = K \frac{M_i * M_j}{d_{ij}^2} \tag{10.3}$$

where:

T_{ij} = Number of trips between zones i and j per unit time
K = A constant reflecting local conditions which must be empirically determined
M_i, M_j = Populations of zones i and j
d_{ij} = Distance between zones i and j

With extensive applications of this basic model to trip distribution in urban areas during the 1950s and 1960s, it was noticed that travel behavior of people does not follow this law as precisely as physical bodies relate to gravity. To achieve a better fit, two adjustments were made:

- The form of the equation was modified. For example, each of the *M*'s is either changed in definition (e.g., to include employment in the zone) or is multiplied by a constant factor.
- One or more coefficients in the model are computed from the data obtained for the specific city. This calibration step actually fits the model to a given area at a given time.

After various modifications, the gravity model used most commonly in planning studies has the following basic form:

$$T_{ij} = T_i \frac{A_j F_{ij} K_{ij}}{\sum_{j=1}^{n} A_j F_{ij} K_{ij}} \tag{10.4}$$

where:

T_i = Total number of trips generated in zone i
A_j = A measure of the attractiveness of zone j (e.g., available employment)
F_{ij} = Friction factor, or impedance to travel between zones i and j, usually computed as inverse function of the costs of travel consisting of travel time, distance, direct cost, etc.
K_{ij} = Specific zone-to-zone relationship factor, empirically computed

Thus, the equation expresses the number of trips from zone i to zone j as proportionate to the relative number of attractions in zone j to the sum of attractions of all zones. A good review of different distribution models is given in Papacostas and Prevedouros (2001).

The main limitation of this model is that after modification of the initial equation and calibration of its coefficients, the equation represents a mathematical model that has the main flaws of the constant growth model: it is basically empirical, fitted to a specific situation and time. Its use for predicting travel behavior at a different time, i.e., 10–20 years in the future, is therefore only as reliable as the assumption that the basic behavior of people in the study area will remain the same as it is at present.

The *intervening opportunities model* is an attempt to correct the deficiencies of the two previous models. It states that the number of trips between zone i and zone j, T_{ij}, is:

$$T_{ij} = \frac{P_i \, (e^{-LVj} - e^{-LV(j+1)})}{1 - e^{-LVn}} \qquad (10.5)$$

where:

L = Probability of accepting any particular destination/opportunity (e.g., shop, accept employment, etc.)
V_j, $V_{(j+1)}$ = Number of opportunities passed up to the zones j and $j+1$, respectively
V_n = The total number of opportunities
P_i = The population in zone i

It is important to note that as the total number of opportunities V_n goes to infinity, the number of interchanges becomes a function of the relative number of opportunities in the respective zones and the population of the zone under study.

This method of estimation is more complicated and prone to computational errors, but if used correctly, it is more powerful and useful to the planner than the models based on the present conditions only. If the planner is concerned about values 10–20 years in the future, he/she can adjust the population (available from census projections) and the number of opportunities in each zone (including developments under construction or planned to be built in the planning period). Although the number of current opportunities may be available, the number being planned may be more difficult for the planner to obtain. Thus, if the data are available, the planner can adjust traffic flows for future changes in the structural environment outside the transportation system, and can even project changes in development caused by changes in the transportation system.

10.3.4 Modal Split

Estimation of the distribution of travel among different modes is one of the most important steps in multimodal transportation planning. It produces the basic information on the demand for each mode for given transportation networks. Most commonly, modal split refers to the distribution between transit and automobile travel, but it may sometimes encompass more modes than these basic two. Estimates of the distribution among different modes, such as walking, bus, bicycle, kiss-and-ride (K+R), and park-and-ride (P+R), for access to a major mode, usually a rail transit station, are referred to as *submodal split*.

10.3.4.1 Factors Determining Modal Split. Since each urban traveler decides which mode to use on the basis of various factors concerning him/her and his trip, the modal split estimate is based on the evaluation of various factors that can be defined and estimated in some quantitative manner. Which factors are used in a given study depends on many local factors, such as the characteristics and roles each mode plays, their relationship, as well as the availability of data. Thus, the factors used vary widely among studies performed for different cities. They can be classified into four major categories.

The first category of influencing factors is tripmaker characteristics. For example, plausible correlations may often be established between such factors as a person's auto ownership, age and income, and travel habits with respect to modal choice. The second category concerns the characteristics of each trip. The third is characteristics of the available alternative modes, while the fourth category encompasses certain indicators related to zones of trip origin and destination. The most commonly used factors in the four categories are the following:

- Trip-maker characteristics: auto ownership, family size and composition, age and income
- Trip characteristics: purpose, length, and orientation (direction)
- Transportation system characteristics: travel time (ratio or difference among modes); travel cost (including parking restrictions) and accessibility ratio
- Zone characteristics: residential density, employment density, and distance from CBD

In practice, each study can incorporate only a few of these factors since the derivation of relationships and complexity of models increases with the number of factors. Moreover, many of the factors are mutually correlated so that using several of them would introduce either unnecessary complexity or a bias. For example, automobile ownership is usually closely related to income and residential density, so that when automobile ownership is used, the latter two factors are indirectly also considered. On the other hand, some factors may have a different correlation to modal split under different conditions. For example, most surveys of U.S. cities show that higher income causes lower transit riding, since automobile ownership increases. However, this is generally true only for low-quality transit service, primarily buses. High-quality transit sometimes has the inverse correlation. For example, regional rail or metro lines serving a suburban area often attract riders with higher income than automobile drivers traveling in the same corridor.

The planner must therefore be extremely careful which factors are used and what generalizations are made from individual surveys. The planner must also be aware that indirect correlations are less reliable and precise than direct ones. Thus, some studies used time of day as a factor. Actually, time of day influences modal split mostly through different relationship of service factors (travel time, service quality) and trip purpose composition. Therefore, the latter factors should preferably be used directly in an analysis.

10.3.4.2 Modal Split Models. Several mathematical methods have been used for mode choice models and travel forecasting. They can be grouped into two basic categories: aggregate and disaggregate. Most of the aggregate versions are of the trip-interchange type, which use multiple regression, cross-classification, or diversion curves to estimate the division of trips between modes. By contrast, the disaggregate type models, also known as individual mode-choice models, use probabilistic methods to estimate behavior of individuals. They are based on utility theory. All these models include some of the factors listed in the four categories of characteristics of trip maker, trip, transportation system, and zone.

Among the aggregate models, the trip-interchange models in various forms are most commonly used. Trip-end models were used in a number of cities, but they have been largely abandoned because of their inherent weaknesses and flawed underlying logic.

Trip-interchange modal split models analyze travel conditions between individual zone pairs, obtained in the distribution step, and determine the portion of trips attracted by each mode, transit, and highway. The relationship of modal split with other steps in the travel forecasting procedure was shown in Figure 10.6.

The simplest type of trip-interchange model uses the diversion curve method, which utilizes a diagram of trip distribution between transit and automobile travel as a function of the travel characteristics on the two modes. The independent variable is usually the travel time ratio or the travel cost difference. Some studies have used an accessibility index, which incorporates both travel time and cost or some other relevant factors. One of the commonly used definitions of the accessibility index for zone i, a_i, is:

$$a_i = \sum_{j=1}^{n} A_j(F_{ij}) \qquad (10.6)$$

where:

A_j = The number of trip attractions in zone j

F_{ij} = A travel time friction factor associated with travel from zone i to zone j over the considered transportation system

n = Number of zones

This index is computed for trips on both modes, transit and automobile. Then their ratio is obtained, trips with the same ratios are grouped together, and the composite percentage of transit usage is computed for each group. When this percentage is plotted against the accessibility ratio, the diversion curve is obtained. A conceptual form of such a diversion curve is shown in Figure 10.7. The curve shows the percentage of trips divided between modes A and B, such as auto and transit, as a function of the ratio of service quality, usually measured by travel times, costs, or some other elements. It must be borne in mind that this diversion curve depends only on the used parameter, such as travel time ratio or difference, disregarding such factors as quality of service, individual preferences, etc. The curve does not go from 0 to 100 because some travelers use one of the modes regardless of travel times or cost. These travelers, including nondrivers using transit and persons carrying special equipment using cars, are designated as transit captives and automobile captives, respectively.

One drawback of a single diversion curve is that it is very crude and does not incorporate any effects of the trip and trip-maker characteristics. This can be remedied by developing different diversion curves for different trip purposes, car ownership, the economic status of the trip maker, or other characteristics affecting travel. In some transportation planning studies, a large number of diversion curves for different categories of trips were developed. However, while a greater number of curves introduces more factors influencing modal split, it also decreases the sample size for each category and thereby the reliability of the results. The planner must be careful not to spend excessive effort on highly elaborate computations and diagrams if the basic assumptions and input are not very reliable. For example, if the estimated travel cost has the price of gasoline as a substantial element and that price cannot be reliably predicted far into the future, the development of elaborate curves may not be justified. Another shortcoming of the diversion curve method is that it cannot deal with an entire transit and highway network at one time.

Yet despite these limitations and deficiencies, trip-interchange models, of which diversion curve methods are the best-known representatives, are based on real-

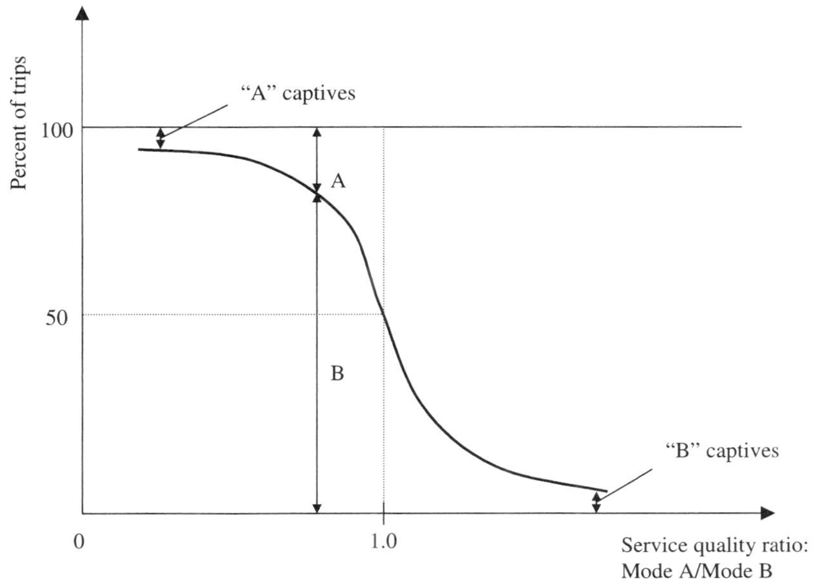

Figure 10.7 *Modal distribution curve for two alternative modes.*

istic relationships and are intuitively clear. Although their computational techniques and factors vary widely, most of the trip-interchange models originate from the same basic assumptions about relationships in mode choice behavior.

Trip-end modal split models, when used, were performed before the trip-distribution step. They estimate the modal split of trip-ends in each zone on the basis of user characteristics only, such as income, residential density, and automobile ownership. In some cases, a general level of transit service was included as an accessibility index. The underlying assumption of this type of model, that the number of transit trips is made independently from the form and extensiveness of transit network, its quality of service, etc., is highly unrealistic. The main reason for a rather extensive use of this model during the early comprehensive transportation studies in the 1960s was its greater simplicity and lower cost because it requires fewer data.

An underlying factor in the use of trip-end modal split models was the serious neglect of transit during the 1950s and 1960s, when most of the initial comprehensive transportation planning studies were performed. This attitude is described by Ortuzar and Willumsen (2001) as follows:

"In the past, particularly in the USA, personal characteristics were thought to be the most important determinants of mode choice and therefore modal split models were applied immediately after trip generation. . . . As at that level there was no indication to where trips might go, the characteristics of the journey and mode were omitted from these models." The authors correctly conclude that "This type of model is, to a large extent, defeatist in the sense of being insensitive to policy decisions; it appears that there is nothing the decision maker can do to influence the choice of mode. Improving public transport, restricting parking or road use charges would have no effect on modal split according to trip-end models."

Since this model could not predict any benefit from improvements of transit services, its results could not be used to design a transit system plan. Moreover, this insensitivity to transportation system characteristics makes trip-end models poorly suited for testing alternative plans and transit system planning in general. Consequently, in all cities that used it, the trip-end modal split model intensified the bias against transit.

These inherent weaknesses in the logic, assumptions, and capabilities of trip-end models were later recognized, and they have been abandoned in favor of trip-interchange and disaggregate modal split models.

10.3.4.3 Evaluation of Modal Split Forecasts. Extensive analyses of modal choice and the development of techniques for its forecast have been done in recent decades. As a result, there is now a wealth of data on the behavior of urban travelers and the relationships among different factors and the usage of the two basic modes, transit and automobile. Numerous models have been developed to mimic *present* behavior and attitudes of people towards *existing* modes. The insights into these problems represent a valuable contribution for analysis and planning of urban transportation systems.

Despite this progress, however, the state-of-the-art in this area needs further improvement. Major deficiencies exist in the underlying orientation in modeling and forecasting methodology, as well as in the models themselves. As the quote in the previous section points out, modal split models were developed for studies heavily oriented toward highways and automobile traffic while giving little attention to transit. Metropolitan area transportation studies have in most cases failed to define precisely the objectives that planned transit systems should fulfill and what service quality they should offer. The models developed under such conditions, based and calibrated on the situations where transit had been seriously neglected, were descriptive rather than creative. They were attempts at explaining quantitatively the modal choice at that time. As an observer commented at that time, the models were "no more than a reflection of today's transportation-consumer reaction to today's transportation system for today's purposes." If the models are calibrated on the conditions where transit service is inferior to the standards that it should ideally meet, they tend to be inadequate for planning improved or new systems. The demand projection based on them was automatically lower and justified less investment in transit than a balanced intermodal transportation plan would require.

The modal split models also have technical deficiencies. Based on quantitative factors, such as travel time, fare, service headway, etc., they are not sensitive to such qualitative factors as reliability of service, information, amenities at stations and in vehicles, marketing, and image. Many of these factors have been shown to have a major influence on modal split.

Another frequent problem has been that modal split analysis is often limited to two modes, auto and transit. In many cities, particularly large ones with strong centers, other modes often play a significant role. For example, in many Western European and Asian cities, walking trips may represent as many as 25–35% of all trips.

In spite of these shortcomings, modal split models and demand-estimation procedures represent a valuable tool in planning if they are developed and applied correctly. When forecasting models are used for certain quantitative changes of services on an existing system, they are adequate. However, when they are applied to forecasting patronage on a qualitatively different system (e.g., express service, new mode, modern information system), it is important that a predictive model be used that includes attitudinal variables based on studies of passenger behavior, values, etc. If such models are not available, the planner should complement formal models with personal experience, knowledge about the system, and judgment.

This leads to some general conclusions about transit forecasting and planning methodology. Forecasting demand by models of present attitudes toward existing modes may represent a self-justifying prophecy rather than planning. It may not be adequate to produce a plan for a transit system adequate for the future. To achieve the selected goals and objectives, planning must have normative elements, rather than be based only on models that extrapolate the present conditions and attitudes. Thorough experience and knowledge of transit systems characteristics is, and will remain, the *sine qua non* of effective transit planning.

10.3.5 Trip Assignment

Following modal split estimation, the forecasted amount of travel by each mode among zones is known, but the exact routings of trips must be determined. There are usually several alternative paths that a single trip could take. The purpose of traffic assignment is to allocate all trips to specific paths, thus deriving traffic volume forecasts for each section in street and transit networks.

The trip-assignment models are calibrated on the present network and then used to estimate travel volumes in the future network. Thus, they provide a tool for testing the adequacy of alternative transportation networks to serve traffic volumes generated by different land use plans and zoning ordinances at present and in the future. An important role for these models is to check whether traffic volumes reach or exceed capacity of individual links and to compute the equilibrium state of flows under different network modifications and travel conditions.

To perform traffic assignment, the planner decides which criteria best represent the actual decision-making behavior of the trip maker and what data are available to him/her. Usually, the most important criteria for route selection are, in sequence:

- Shortest travel time
- Shortest travel distance
- Minimum trip cost

There are three basic methods for traffic-assignment procedure. The all-or-nothing method assigns all traffic to one route based on the selected criterion (e.g., minimum cost). The assignment curves method assigns a portion of the traffic to each route based on the comparative values of the criteria between the best and the next-best routes. The capacity restraint method assigns traffic based on the travel times and capacities of available routes and assigns a portion of the traffic to alternative routes as the primary route nears capacity. The two capacity restraint models that are used most frequently are the stochastic model and the user-equilibrium model.

In the all-or-nothing method, the best route gets the entire load. Although this method is the simplest, it is also the least accurate and worst representation of the decision-making process of the trip makers of all the mentioned models. As traffic is assigned to individual links, some of them may become loaded so much that their volume exceeds its capacity and level of service decreases. Thus, the initially assumed travel time on such facility cannot be achieved, making the results of the assignment unrealistic and physically infeasible.

The assignment curves method assigns traffic between the best and the next-best routes. Based on how the two routes compare quantitatively in regards to the assignment criteria, the planner determines the percentage of trip makers who will opt for one route or another. The number of trips assigned for given alternate routes is based on past data and the planner's own experience. This method has been shown to have good results for urban and quasiurban situations, but it has been found inadequate for recreational trips in rural areas.

The capacity restraint models take into account the fact that travel times on each link change with assigned traffic volumes. The assignment is therefore made gradually, and assumed travel times on each link are adjusted, i.e., lengthened as the assigned volumes increase. The closer the assigned traffic comes to capacity on a given route, the more additional volume must be diverted to the routes that now may have shorter times than on the heavily loaded route or link.

This process is iterative and requires a number of traffic assignments. In practice, four assignments are made and the average volume on each link is taken. The federal highway administration recommends the following formula for the determination of the congestion-modified travel time:

$$T = T_0 \, [1 + 0.15 \, (V/C)^4] \tag{10.7}$$

where:

T = Travel time modified for congestion
T_0 = Travel time assuming no congestion
V = Assigned volume
C = Capacity of each link

The formula must be applied for each link and related to the flows determined for the other links in the system. Although the procedure appears complicated, this model approximates closely the behavior of trip makers and therefore is the most realistic trip-assignment method.

Capacity restraint models differ in their assumptions of how trip makers choose paths between their origins and destinations. Stochastic models assume that trip makers choose paths that they *perceive* to have the minimum travel time. However, the travel time of a single path may be perceived differently by each trip maker due to variations in information they have and nonmeasurable factors (e.g., weather). Therefore, trip makers may choose different paths between the same origin and destination, even though each trip maker believes he/she is using the shortest path. Stochastic models determine the probability that a trip maker will choose a given path, with the underlying assumption that perceived travel times are randomly distributed. Based on these probabilities, trips are assigned to each path in the network.

User equilibrium models assume that trip makers always choose the shortest travel time path between an origin and a destination (individual equilibrium, or Wardrop's first principle of traffic distribution in networks: Wardrop, 1952). It is implicitly assumed that all trip makers perceive travel time in the same way and that there is no random variation in perceptions. However, this does not imply that all trip makers will choose the same path between each origin and destination. Travel time on each path changes with the volume of traffic (as shown in the equation above), so that as the shortest path reaches capacity, its travel time may exceed that of alternative paths. A user or individual equilibrium exists when no trip maker can improve his/her travel time by choosing an alternative path. This implies that in equilibrium, all paths between an origin and destination will have the same travel time (although their speeds may vary because of different levels of service), otherwise a trip maker could choose an alternative path that would take less time.

10.3.6 Alternative Methods to Four-Step Model

The four-step modeling procedure for travel forecasting, developed in the 1950s and 1960s, has been widely adopted in practice because of its simplicity. There are, however, several fundamental problems with the four-step approach as it is currently implemented.

First, aggregate models are unable to predict the behavior of individual travelers. By using macro-level analysis on regional zones, aggregating assumptions are made about an individual traveler's value of time, modal preferences, household forms, etc. While these simplifying assumptions are somewhat mitigated by the introduction of several trip purposes or classes (home-based work, home-based school, etc.), significant predictive capabilities are lost through aggregation and average values.

To overcome these shortcomings, alternative model formulations based on individual behavioral properties have been developed. These *disaggregate* or *discrete choice models* compute for individual travelers the relative utility of completing a trip. These utility functions, which typically include a linear weighting of travel time, out-of-pocket costs, and modal preference, can be computed for not only mode choice, but also for trip distribution and trip assignment. A traveler's behavior may be deterministic—he/she maximizes his utility in selecting a destination, a mode, or a transit route or highway path. Alternatively, the model formulation may be stochastic, so that the probability of choosing a destination, mode, or route may be a function of that trip's relative utility. Computing the relative utility of one trip parameter, such as modal choice, is done through logit models. The utility of several trip parameters (i.e., destination and mode choice) is computed simultaneously using nested logit models, where the total utility is based on the utility of destination and the conditional utility of the mode, given that destination choice.

These discrete choice models represent individual behavior far more accurately than does the aggregation commonly practiced in the four-step procedure. The mathematics and implementation of the discrete choice model are a major field of transportation operations research studies.

A second major shortcoming of the four-step procedure is the treatment of travel cost in predicting auto driver behavior. As currently implemented, the four-step procedure predicts trip distribution using a cost that primarily involves only distance. Similarly, many modal split computations are made using empirical evidence or socioeconomic data. Even when travel costs are used to compute trip distribution and modal split, the travel costs assumed are not in an equilibrium (i.e., loaded network) condition. Because in the traditional four-step model trip assignment is completed last, the equilibrium cost of travel by different modes cannot be known during the trip-distribution and modal split steps.

This weakness in the model tends to underestimate the impacts of congested highway conditions on automobile travel costs. As a result, the cost and potential benefits of alternative transportation modes are underestimated.

To overcome this problem, practitioners have begun to use feedback in the modeling process, where the equilibrium travel costs are fed back into the distribution and modal split calculations. The model is then solved iteratively until a network equilibrium condition is approached.

In practice, the concept of feedback has had mixed results. Because of the time needed to solve the assignment problem to sufficient accuracy using common algorithms, there has been some resistance to solving the entire four-step procedure multiple times. In general, insufficient guidance and resources have been expended on defining and implementing the feedback procedure.

An alternative to feedback procedure exists and is now being implemented in many locations. A field of so-called combined models, originally formulated by Beckmann, Winsten, and McGuire in the 1950s, has recently received considerable attention from travel forecasters (Beckmann et al., 1956). The models use a nonlinear, constrained optimization approach to solve the trip distribution, modal split, and trip-assignment problems simultaneously, typically using nested logit formulations. This approach ensures that the cost structure is consistent in each decision the traveler makes. The model can be solved for system optimal or user equilibrium conditions with either deterministic or stochastic (dispersion) travel behavior assumed.

A third weakness of conventional models (both the four-step and combined models) involves the definition of trips. Research suggests, for example, that most commutes include one or more stops between home and the workplace. These intermediary destinations are chosen dependent upon their proximity to the commuting path. The mode used for these intermediary trips is also dictated by the commuting mode. The cost structure for making intermediate stops, especially by transit, also can vary widely. None of these trip-chaining phenomena, which research suggests are becoming far more prominent, are adequately treated by contemporary models.

A more robust, activity-based approach has been suggested. In this instance, multiple combined trips would be treated as a single activity. The destinations, modes, and paths of each segment of the trip would be interdependent. One approach to implementing this dependency is to define trip classes that incorporate trip chaining. For example, a trip may be defined as beginning at workplace, stopping to shop, and then continuing home: "home-based work-shopping trip."

As with all complex modeling work, efforts to improve and validate travel forecasting models have been limited by the availability of data. The household travel survey and the U.S. census information provide only a very small sample, typically yielding insufficient data to calibrate models of great complexity.

10.4 TRANSIT PLAN DEVELOPMENT, EVALUATION, AND SELECTION

Similar to a street, highway, or any other surface transportation system, a transit network serves two basic functions: it provides service for access to individual areas where it passes (area coverage), and it carries through the served areas those passengers traveling on the network who have origins and destinations outside of it. The former function requires primarily stops or stations in the served areas, with high travel speed as a second objective. The latter function is better satisfied with high travel speed and as few stations as possible. Thus, the density of the transit network, the spacings of its stations, and line performance (capacity, speed, reliability, service quality, etc.) on its lines depend on the ratio of these two functions.

In small cities and low-density suburban areas, the service function is important because the volume of through traffic is relatively small. Therefore, street transit, consisting of buses, trolleybuses, and streetcars, is the most efficient and often adequate to meet the needs for transit travel. When passenger volume is greater in the individual corridors typical for medium-sized and large cities, it is necessary, and economically feasible, to provide not only street transit, but also semirapid transit systems on partially separated ROW (category B), which have intermediate capacity and higher performance than street transit. Light rail transit, improved buses, and, in some cases, AGT modes are candidates for planning such transit lines.

Finally, in large cities a transit system must provide not only the area service function, but also high-capacity and high-performance transport to meet the need for travel of large passenger volumes on longer distances and successfully compete with automobile and other modes. This task can be achieved only by rapid transit modes that are entirely separated from general traffic (ROW category A), such as metro and regional rail.

Street transit networks do not require long-range planning because they have very limited infrastructure. Long-range transit planning consists of plans for semirapid and rapid transit networks, which have extensive infrastructure, such as separate rights-of-way, tunnels, stations, and related facilities. They require substantial investment and have long life. Their permanence requires very careful planning because mistakes in the design result in inefficiencies that are difficult to correct.

10.4.1 Planning Procedure

The transit system planning procedure can be defined by the general steps shown in Figure 10.8. Starting from the definition of goals and objectives for the transit system and its network, the planner considers local conditions, which include a broad number of factors, from the topography and existing street and other networks to land use characteristics, the city's character, public policies, and resources allocated (or potentially available) for transit system construction. Another major input is the passenger volume forecast with origins-

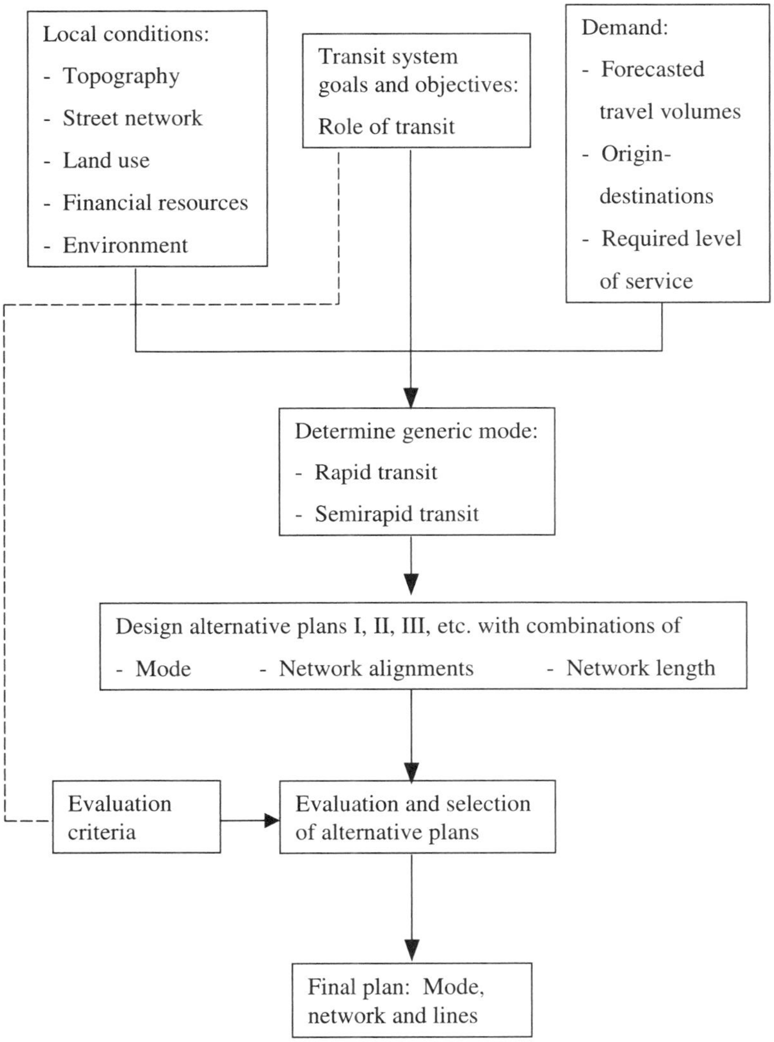

Figure 10.8 *Procedure for transit plan development.*

destinations along each line. That volume is the key element in defining the required level of service, strongly influencing the choice of transit mode.

10.4.1.1 Design of Alternative Plans. With this basic material, planning and design of the transit system can start. Since there is no exact methodology for development of an optimal transit network, in most cases planning consists of development and evaluation of several alternative plans. The purpose of developing alternative plans is to examine a range of possible solutions and compare them in order to find the most advantageous one. The alternative designs or plans may differ among themselves with respect to several elements or characteristics. Typically, the alternative plans differ in the following elements.

Generic Modes: Semirapid and Rapid Transit. This is the basic decision because the two generic modes differ in many respects. Rapid transit has a considerably higher performance-cost package (PCP) than LRT, but it also has larger facilities and different alignment elements, so that the two modes have different feasible alignments and stations. For a given level of investment the metro will offer a system with higher capacity, speed, and reliability, but on a significantly smaller network. Or the alternatives have similar networks by alignment but vary in investment levels and performance characteristics. This is discussed further in Chapter 12.

Transit Modes with the Same ROW Category That Vary in Technology. The alternative systems are rather different in the case of rapid transit from those in semirapid transit category.

In the case of rapid transit, the basic alternatives are several types of guided systems with different technologies:

1. Rail rapid transit (conventional)
2. Rubber-tired rapid transit, such as is used on some lines of the Paris Metro, in Lyon, Montreal, Mexico City, and a few other cities designed by French companies, and in Sapporo, Japan
3. Monorails with different technologies (suspended, supported, and others), used on single elevated lines in several cities, such as Wuppertal in Germany, the Tokyo-Haneda Line, and Seattle, Washington

This is usually not a difficult choice because conventional rail has been selected in a vast majority of cases. Further decisions in system design then concern such technical components as specific alignments (aerial, tunnel, cut, at-grade, or other, but always fully protected), station designs, rolling stock characteristics, method of operation and control, etc.

For semirapid transit, or modes often referred to as medium-capacity systems, the choice of modes is broader and more complicated. These modes include systems mostly with ROW category B, but sometimes also A and C on line sections. Even some guided modes with exclusively ROW category A but with smaller vehicles and TUs than metros, thus representing "minimetros," are included in this category. Their capacity range is generally between 3000 and 20,000 sps/h. Major modes in this category are the following:

1. Light rail transit (LRT), including light rail rapid transit (LRRT). The former uses ROW categories B, A, and C, while the latter has ROW category A only, with correspondingly higher PCP.
2. Bus (semi)rapid transit (BRT), also with a variety of technical characteristics.
3. Automated guided transit (AGT), with rubber-tired or rail wheels operated automatically on ROW category A guideways. There is considerable diversity of technologies in this category.

These modes have PCPs greater than any street transit mode, but considerably smaller than PCP of rapid transit systems. Although in the same range, PCPs of these medium-capacity modes vary considerably among themselves. Investment cost per kilometer is the highest for AGT because of its exclusive ROW, followed by LRT and BRT. However, their different types of ROW and technology give them very different physical feasibility: BRT can be built at a lower cost and with less disturbance than LRT in existing streets, but it cannot be placed in tunnels. AGT, with its large concrete guideways and very high station structures, is much more bulky than the other two modes, so that it cannot fit easily in high-density center city streets. For more detailed characteristics of these modes, see also Chapter 12.

Different Networks for the Same Mode. These, varying in alignments, geometric forms, lengths of lines, or types of operation, are always useful to consider. The alignments and extensiveness of lines and network are usually designed to follow major forecasted transit travel volumes, considering lines and area coverage by the existing rapid or semirapid transit network. An example of an existing metro area coverage map that can be used in designing new lines or extensions is shown in the example of the central area of Washington, DC, in Figure 10.9. It clearly shows, for example, that there are considerable areas to the southeast, northeast, and

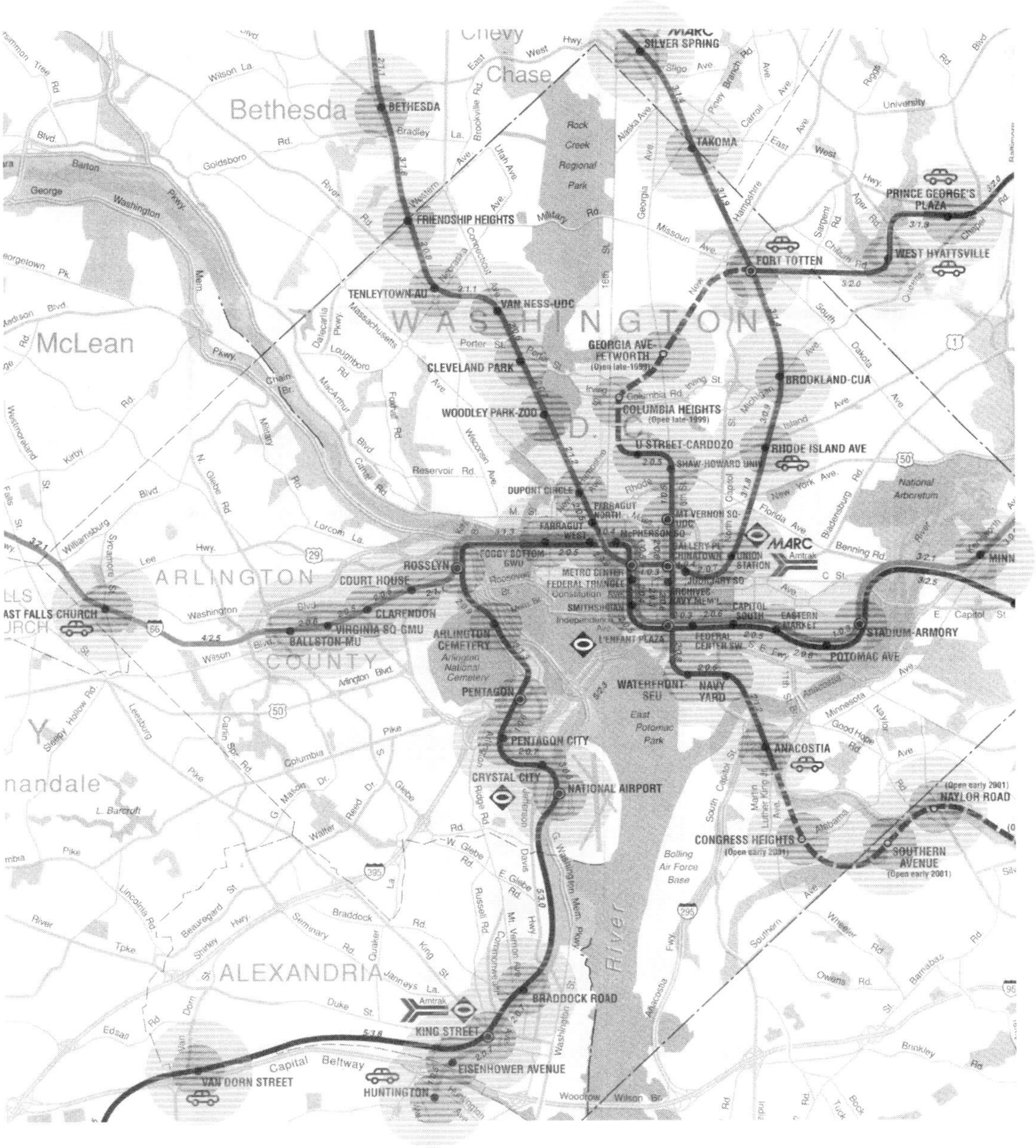

Radius of station area coverage:
Primary—400 m
Secondary—800 m

Figure 10.9 *Metro area coverage in the central area of Washington, DC.*

northwest of the center that are not served by the Metro. The feasibility of serving these areas would be analyzed by an analysis of land uses and corridors in these areas.

Characteristics of different line alignments and network extensiveness, geometric forms, and operational patterns are described in Chapter 4. It must be borne in mind that network form can have a great influence on transit system passenger attraction, operating efficiency, and the shaping of city's land use patterns.

Various Levels of Investment in the Planned Transit Network. These lead to alternative plans that differ in all these elements—modes, network extensiveness, and forms. The budget for the network construction may be constrained to a certain amount, or it may be defined by other factors than the estimated cost. Following the development and adoption of a plan, the required funds may be secured from various sources, such as contributions from different government levels, special taxation, or private sources.

The procedure for mode comparison and selection is given in Chapter 11, while more detailed descriptions of different modes and many comparisons of their characteristics are presented in Chapter 12. In actual planning, procedures for system design, comparative evaluation, and selection are so complicated that they seldom follow an exact sequence of distinct steps. The procedure and sequence of steps may vary considerably among different cases. Moreover, the steps are not always either distinct or independent from each other. There is usually a strong interdependence among them (selection of generic transit modes, specific mode technologies, design of network form and its extensiveness), so that there are often interactions, feedback, and adjustments among them.

10.4.1.2 Evaluation and Selection of Alternative Plans. As Figure 10.8 shows, in most transit planning processes, several alternative plans are developed, which must then be compared mutually and the preferred one selected for implementation. Plan evaluation and comparative analysis are made on the basis of a set of criteria derived from the transit system goals and objectives.

Plan evaluation can be performed by a number of different methodologies, utilizing three types of measures:

1. Monetary, expressing investment and operation costs in dollars, Euros, or other currencies
2. Other quantitative measures, such as number of served passengers, passenger-kilometers or passenger-hours, forecasted reduction of auto trips and parking spaces required in areas served by the planned transit line, etc.
3. Qualitative evaluations, which include many sets of items, from the quality of transit service and increased mobility for certain population groups to the expected air pollution reduction due to shifting travel from highway vehicles to electric transit, and increased image of transit, and its impacts on land use development and the city's livability

Economic evaluation of plans, i.e., estimation of investment and life costs and benefits, such as benefit-cost analysis (computing the ratio or difference between benefits and costs—see Chapter 11) or computation of the project's rate of return, is usually done for alternative transit plans. They produce a basic measure of the investment and its effectiveness. Economic evaluation is given the dominant, sometimes the sole, significance, particularly by some economists. It is popular because it gives a single, easily understood number for each alternative. However, for many transit projects, economic evaluation of plans is either inadequate or misleading.

Providing public service in complex urban areas, transit systems have long life and exert numerous and permanent impacts that are difficult to measure in monetary units. Evaluation of such projects based on estimation of life-cycle monetary values has several fundamental difficulties. First, estimation of public benefits and costs, such as value of mobility of persons, increased job opportunities, and increased social equity due to high-quality transit services, is by its nature mostly subjective, so that its quantification produces an unfounded impression of accuracy. Second, such externalities as reduced traffic congestion and air pollution due to the shift of travel from automobiles to electric transit, or increased livability of cities, can be very significant and yet extremely difficult to quantify. Finally, discounting future benefits and costs to their present values is very sensitive to the assumed discount rates, which are never easy to predict even for the short-term future.

Among these difficulties in monetary evaluations of project benefits and costs, the impacts and various externalities are more complex to estimate than direct costs. As a consequence, there is an inherent bias toward estimating these costs while giving less attention to or even omitting many benefits. This results in unfavorable evaluation of transit projects, particularly ones that offer higher levels of service and have longer life, such as rail transit. This bias is seen in the fact that the economic evaluation is strongly emphasized and used as a popular, although simplistic, measure of alternative projects by the opponents of transit improvements. It is quite common for lower-cost projects to be claimed to be superior to higher-cost ones without mentioning the differences in their performance and positive and permanent impacts on the city and its character being mentioned.

How should this problem of transit project evaluation be handled in the real world? In evaluating alternative systems that differ mostly in technical characteristics, but have very similar system performance and serve the same function, economic evaluation can be done rather precisely because the role of unquantifiable factors is limited. With major transit projects involving different transit modes, network forms, and services, however, the problem cannot be solved by comparing individual numbers for each alternative system.

Long-range transportation planners and civic leaders interested in broader aspects of transit system improvements and their permanent impacts consider economic aspects of alternative projects together with their broader values, and the decision is often made by civic and political leaders or, in some countries, such as the United States and Switzerland, by popular referenda. This complex process of decisions about construction of high-performance transit systems (semirapid and rapid transit) is illustrated in Section 10.4.2 by brief descriptions of several cases in different cities.

10.4.1.3 Procedure for Preliminary System Design. When the general line or network layout and mode have been selected, planning proceeds to preliminary engineering. This phase consists of a review and detailed determination of the alignments and projected passenger volumes by line sections and design of system components. An example of this procedure applied to planning of a new LRT system is shown as a flowchart in Figure 10.10. Three sets of LRT system components are defined as vehicles, operations, and infrastructure. The flowchart lists the major items that are included in each of the three sets of components and their mutual dependencies. Naturally, this process is very elaborate and contains many feedback steps, rather than a simple one-way flow of the design process. However, the chart can serve as a general guideline for organization of the preliminary design process.

10.4.2 Examples of Long-Range Transit Planning

The transit-planning procedure and sequence of discussed steps will be illustrated here by several examples of actual planning performed in different cities in recent decades. The cases described here have attracted attention in professional circles either for their comprehensive and creative approaches or for a variety of factors that influenced the planning process, ranging from technical/functional to financing issues, political and special interest pressures, and considerations given to the impacts on the surroundings and the entire city that the planned transit system was intended to achieve. Some cases of mistakes, failures and corrections of potential failures are also included.

10.4.2.1 San Francisco Bay Area Rapid Transit (BART). During the 1950s, when the entire country was focused on the construction of the Interstate Highway System and most cities were trying just to salvage their declining transit systems, the San Francisco Bay Area was a pioneer in planning a balanced intermodal

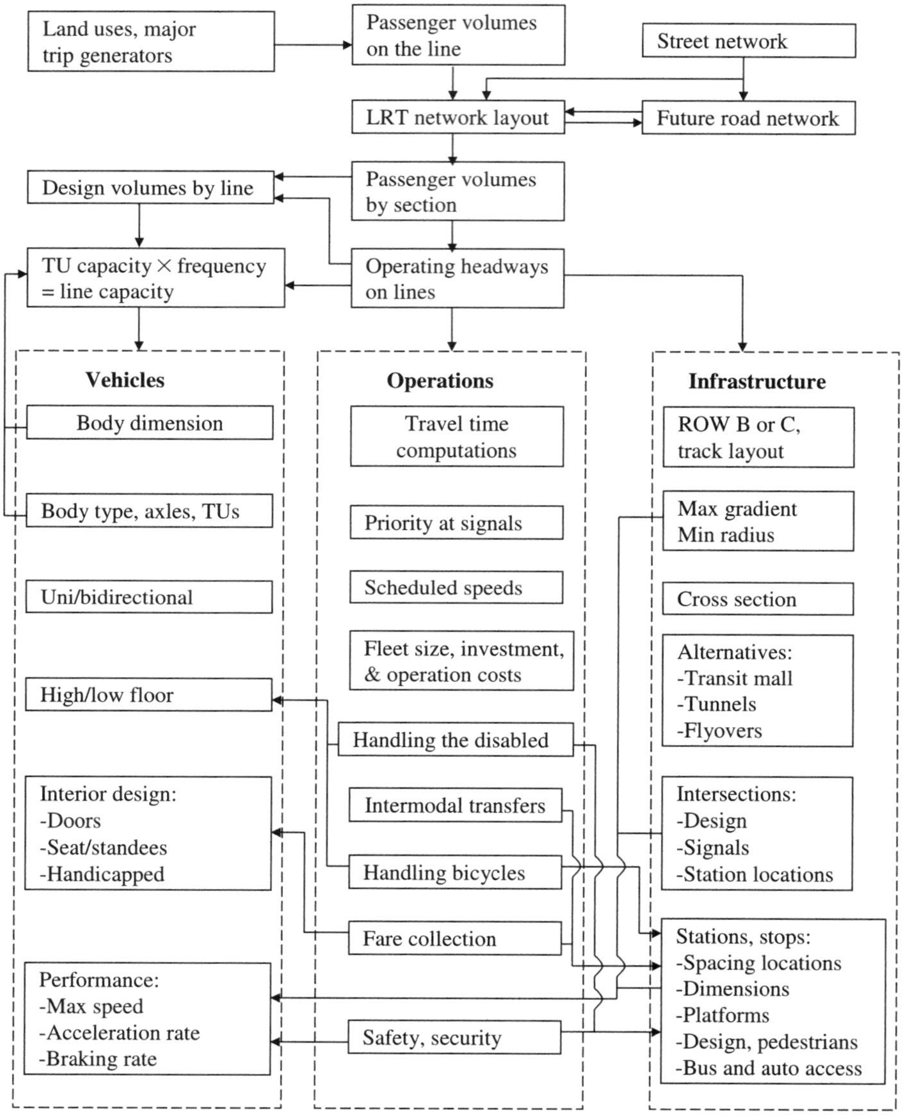

Figure 10.10 *Procedure for preliminary engineering design of a light rail transit system.*

transit system. Planning of a regional rapid transit system, started decades earlier, progressed so far that the California legislature adopted a law requiring planning of a "high-capacity, high-performance regional rapid transit system which will be independent of the highway system."

This political decision thus actually defined the rapid transit generic mode, so that system planners could start their design from the next step—selection of the specific technology of rapid transit systems. They compared four different modes: standard rail rapid transit; rubber-tired rapid transit, which had been developed at that time in Paris; supported monorail (Alweg type); and suspended monorail, whose prototype had been developed and tested in Texas.

The selection of rail rapid transit over the other alternatives was easy because the adopted maximum speed of 130 km/h eliminated rubber-tired systems. Moreover, in tunnel operation, monorails would require much greater profile and therefore involve higher construction cost. Also, rubber-tired rapid transit would have problems with excessive energy consumption and heat production. Monorails were found impractical for operation of a network with branches that involved complex networks and switching, which is very difficult with monorails.

While the network form did not have many alternatives because of many topographic constraints (crossing the San Francisco Bay in a tunnel, alignments of main streets in San Francisco, Oakland, and Berkeley, etc.), network extensiveness was eventually decided on a political basis when two suburban counties decided not to participate because they could not afford the cost. The initial BART network thus remained within three core counties: San Francisco, Contra Costa, and Alameda. San Mateo County made an agreement with the BART district some 40 years later to pay a substantial amount to offset the initial system development costs that the three counties had initially paid alone and then participate in the construction and operating for the extension of BART line to San Francisco Airport, which is located in San Mateo County.

The cost of the developed rapid transit network was then estimated and the plan submitted to the population for approval by a referendum. The vote was positive and construction of the system was funded through property taxation in the three counties and from diversion of the Bay Bridge tolls. These highway tolls were used for the construction of the 8-km-long Transbay Tube, paralleling the Bay Bridge. This diversion of funds was logically supported by the fact that two tracks of the previous rail system on the Bay Bridge were converted into highway lanes and the BART tube would unload traffic volumes from the Bay Bridge.

Over four decades after this transit plan was implemented, it is recognized as one of the most farsighted planning processes in the country. While several later BART impact studies tended to focus mostly on direct, measurable quantitative impacts of this system, it is generally recognized that the most significant impacts of BART are reflected in the outstanding livability of the city of San Francisco and its region.

10.4.2.2 Washington, DC. Having abandoned its last streetcar lines in 1962, the city of Washington and its region had only with buses on streets, serving a large network with very low performance and poor image. In particular, as the nation's capital, the city had a large number of visitors for whom the bus services were extremely difficult to understand and inadequate to provide mobility. The need for a high-performance transit system that would correspond to the needs and image of the capital city was widely recognized, and preparations for planning of a metro system intensified during the 1960s.

In most U.S. cities, there is strong opposition to transit progress by lobbies promoting auto use, highway construction, or promotion of other transit modes, sometimes even unproven technological inventions. The lobbying particularly focuses its attacks on rail transit, the highest-quality transit mode (see Vuchic, 1999, Chapter 5). This was particularly the case with Washington because a modern metro system in that city would have a particularly strong national visibility. The critics of rapid transit planning in Washington brought up, among others, the following three arguments, typical of transit opponents' arguments in many cities.

First, it was claimed that a system using buses would be cheaper. Because buses on surface alignments could not provide sufficient capacity, an elaborate technical plan for an underground bus network, including large stations with multiple bus bays, was developed. A technical review easily rejected this plan because bus operation is not only inefficient in narrow tunnels, but infeasible because of bus exhaust fumes.

Second, as the plan for rapid transit (later named Metro) was to be approved, the owner of the private bus company serving the city came up with a printed booklet showing monorails and contending that this "mode of the future" had not been considered. This claim was also dismissed on technical grounds.

Third, the criticism of the large investment on the Metro system was repeatedly brought up. It was argued that the cost would be lower if all expected trips on the Metro were served by taxis. This simplistic single-dimensional analysis of a complex public project did not consider the much inferior service taxis would offer, the inability of streets to accommodate such traffic volumes, or what kind of city Washington would become with that type of transportation. In a debate about excessive cost of the Metro, the following answer was given (Vuchic, 1991): it is correct that the Metro could have been built at a considerably lower cost; but it is also correct that the Capitol building could have been built at a much lower cost if it had a flat roof!

The plan for a 165-km-long Metro network built over 25 years proved to be very farsighted. The Metro has provided excellent service and upgraded the character and image of the entire region. Nobody seriously challenges the selection of transit mode, network alignment, or its length. The main debates continuing at present are about which of the proposed extensions should be built first and how it should be financed.

10.4.2.3 Frankfurt, Germany. Transit planning for Frankfurt, Germany in 1960 produced several alternatives based on different transit modes: metro, LRT, and Alweg-type monorail. The main lines of the network were common to all three plans, with some differences due to the characteristics of each mode. The three alternative plans were then compared on the basis of investment and operating costs, quality of service they offered, and impacts on the city. The LRT system proved to be the preferred mode. Although it had somewhat lower performance than metro, its costs were substantially lower and its network fitted better in the city's physical characteristics and environment. Monorail was inferior to metro because it was operationally impractical and had more negative impacts due to its large aerial structures. Today, Frankfurt (city population 640,000), has an excellent, extensive LRT system built by gradual upgrading of the former streetcar lines by their relocation to several types of ROW category B and some short tunnels in the city center. In addition, an extensive regional rail system (S-Bahn) serves its broader region, with a total population of 1.9 million.

10.4.2.4 Large Cities with Metro Systems. Transit planning in very large, rapidly growing cities that required extensive networks with high-performance services, such as Hong Kong, Madrid, São Paulo, Seoul, and Tokyo, considered only metro, i.e., the rapid transit generic mode with rail technology. It was clear at the outset, without many analyses, that none of the semirapid transit modes could provided the very high required capacity. Alignments of lines and form of the network were planned on the basis of analysis of alternatives in each case, however.

10.4.2.5 Mexico City. With a population exceeding 10 million during the 1960s (and reaching 18 million in 2004), Mexico City logically selected the metro as its basic high-performance mode. However, it uses rubber-tired metro technology, which limited car size. Similar to the cars in the Paris Metro, they are only 15 m long and 2.50 m wide. The system is very popular, with its beautiful station designs and fast trains, but its capacity, even with nine-car trains, is insufficient for the large passenger volumes on several lines. Thus, the mode choice was correct, but the selection of specific technology is still being debated.

10.4.2.6 Strasbourg, France. Similar to several other medium-sized cities in France, Strasbourg went through very extensive technical planning, city planning analyses, and political discussions, including very heated and long debates, in its process of planning an upgraded transit system. As a compact city with a city population of 260,000 and regional population of 460,000, Strasbourg considered semirapid transit, medium-capacity modes. It was clear that buses could not provide adequate capacity and level of service in mixed traffic in urban streets. Moreover, the physical characteristics of buses, including the noise and air pollution they produce, would negatively impact the unique pedestrian-oriented environment of that historic city. The major debate then developed between LRT and AGT (VAL system). An initially adopted LRT/ tramway system was rejected in favor of the VAL system when the proponents and manufacturers of this mode exerted pressure on the city's administration. Several years of technical, planning, and political debates followed.

Proponents of the VAL system argued correctly that this system would provide faster and more frequent service. The tramway system proponents brought up another set of important arguments for their plan. A tramway system would have a two to three times lower construction cost per kilometer of line. Thus, for a given investment, a much longer network of tramway than for VAL could be built. Moreover, the tramway would increase the city's livability and enhance its historic character, while VAL would be in tunnels or on large aerial structures. Light rail can be extended more easily and can have more convenient transfers with buses and railways at the main station.

A mayoral candidate in an election made adoption and construction of the tramway system one of her goals, and that issue was apparently a significant factor in her electoral victory. The tramway was then adopted again. Following its construction, which included reorganization of traffic in the entire central city with pedestrianized streets and squares, as well as visually attractive, quiet low-floor cars with stations directly on streets, the planners point out an interesting aspect of the winning alternative. They claim that if VAL in tunnels had been adopted, the streets, squares, and entire central city would have been auto-dominated, with suppressed pedestrians. With tramway, pedestrianization was made feasible and it greatly enhanced the city's livability.

10.4.2.7 Ottawa, Canada. As Canada's capital city, Ottawa has a large number of office employees and therefore higher transit ridership than other cities comparable in population size. With a regional population of about 1 million in the 1980s (in 2004 it has grown to 1.1 million), it was clear that a medium-capacity transit system was the most appropriate generic class of modes. Transit planning therefore considered LRT and a systematically upgraded bus network, corresponding to the present definition of BRT.

Analysis showed that while LRT offered a higher quality of service and upgrading of transit image, it required a higher investment and more complicated construction than buses. On the other hand, local conditions were conducive to a very pragmatic design and implementation of an advanced bus system. Wide streets in center city allowed provision of strictly enforced exclusive bus lanes. Some sections of radial freeways that were not congested were also utilized. These facilities were supplemented by newly constructed sections of elaborate busways with grade-separated stations. With a substantial investment in these facilities, systematic gradual expansion of the network and strict operational supervision, the city obtained a system with high-quality service.

The extensive network of bus lanes, ramps, roadways, and streets in Ottawa represents today by far the most extensive and efficient BRT system in North America. Its further expansion and upgrading is likely to proceed through further improvements of the bus network as well as introduction of rail lines on some regional rail corridors.

10.4.2.8 Munich. In the early 1970s, Munich embarked on construction of a rapid transit (U-Bahn) network. This led to a discussion about the changing roles of different transit modes in the city. The promoters of the rapid transit mode in the transit agency and the city insisted that the logical removal of streetcars from the corridors served by the new U-Bahn lines should lead to total abandonment of streetcars. The city would be served by a two-mode system: U-Bahn and buses. This followed similar transitions made in West Berlin and Hamburg, based on the thinking that transit in large cities should have only two modes. (Interestingly, following German reunification, Berlin has upgraded streetcars in the former East Berlin into a modern LRT system.)

This proposal caused major opposition and led to a debate that lasted over a decade. Following many arguments and a very thorough technical-economic study, the city decided that the two-mode philosophy

***Photo 10.2** Transit planning in Strasbourg resulted in high quality LRT on center city streets.*

would result in a loss of many lines on which the rail system was superior to buses. An important role for an intermediate mode was confirmed, and the streetcar system has since been upgraded into a modern LRT system that is coordinated with the U-Bahn and buses. Today, the Munich Tram has some of the most modern low-floor cars. It is very attractive and popular.

10.4.2.9 Singapore. In the mid-1970s, Singapore introduced a comprehensive plan for urban intermodal transportation coordination and improvement. It included street traffic congestion abatement through road pricing, control of parking by regulation and parking rates, and strong development of transit.

The main transit project was construction of a rapid transit network. The plan for the rapid transit system was prepared by the consulting firm Wilbur Smith Associates utilizing very sophisticated trip-estimation models and careful design of line alignments, stations, train sizes and capacities, scheduled speeds, schedules and line capacities, etc. Just prior to its implementation, two economists from Harvard University submitted another study and a claim that the rail rapid transit was "wasteful" because a system of buses on busways could provide better service for a much lower investment. They supported their claim by an analysis of bus system operations utilizing a computer-based model.

This challenge to the largest investment the city had ever made drew the attention of some civic and political leaders, who called for a project reevaluation. Following extensive discussions that involved the highest government officials and a TV debate between the economics professors and transit experts, it was clearly shown that the computer-designed bus system model and its cost estimates were founded on unrealistic assumptions and produced incorrect conclusions. The proposed bus system design was a concept infeasible for implementation that could not produce levels of service even remotely comparable to those of rapid transit.

Presently, the city-state of Singapore, with a population of 3.2 million, is efficiently served by one of the most advanced rail rapid transit systems in the world. Its transit experts and citizens talk with amusement about the computer-designed bus proposal that only caused a delay: how could anybody imagine that a bus system through center city could have offered any similar service to its present rapid transit, and what would this attractive city have looked like with hundreds of buses daily congesting its streets?

Nearly three decades after its pioneering intermodal approach to urban transportation improvements using several concepts that no city had used before, Singapore continues to be in the forefront among its peers with respect to its urban transportation policies and their implementation, as the following items illustrate.

- The road pricing introduced in 1975, known as area licensing scheme (ALS), was upgraded in 1998 by the electronic road pricing (ERP) system, which allows easy activation, deactivation, and rate change. Traffic surveys measuring traffic volumes and speeds are performed every three months, and the ERP rates are adjusted to maintain desired levels of traffic volumes and traffic speeds: when speeds are high, the rates are lowered to allow more cars into the central zone; if the speeds are low, rates are increased to decrease the volume and increase level of service (Menon and Chin, 2004). The road pricing concept has been introduced by several other cities, such as Oslo and Trondheim, Norway, and London, more than 20 years after Singapore.
- Rail rapid transit lines are being built and extended, carrying in 2004 about 1.1 million passengers per day.
- Extensive bus services serve the entire city with frequent and reliable service.
- A number of bus-rail transfer stations, known as interchanges, have been built. Some design aspects affecting passenger convenience are not yet satisfactory and their improvement is being discussed.
- The only medium-performance mode offering better service than buses at lower investment than rapid transit is several AGT lines serving as rapid transit feeders in high-density residential areas. BRT and LRT modes, which could be competitive with respect to their PCP, have not yet been given sufficient attention.
- On all transit services (rapid transit, AGT, and buses), the most advanced smart card contactless system has been introduced. It has great advantages with respect to passenger convenience, speed of entry and exit in stations and on buses, application of sophisticated graduated fares, and statistical information the system collects.
- The most remarkable progress achieved in Singapore during the last three decades has been consistent development and implementation of coordinated intermodal transportation policies. This has been possible to pursue with such continuity mainly due to the organizational setup of transportation authorities and competence of transportation professionals. Organizational changes have taken place from time to time to respond to the changing conditions and needs of this city-state. Presently, one semigovernmental organization, the Land Transport Authority (LTA), develops transportation policies, performs transportation planning, and determines the roles of transit and other operating agencies. Implementation of policies, operations, and control of transportation systems are performed by LTA (for highway and traffic), two large transit agencies operating buses and rapid transit, and taxi companies.

In conclusion, Singapore came to the forefront of urban transportation planning and management through its intermodal policies and several operational and technological innovations in the mid-1970s. Since that time, this city has consistently pursued systems approach, intermodal coordination, and numerous technical innovations in many areas of transportation. Thus, it remains one of the world leaders in this area.

10.4.2.10 Portland, Oregon. Among all U.S. cities, Portland, Oregon has been one of the best examples of a farsighted interdisciplinary approach to urban planning that incorporates transportation as its major element. It adopted an intermodal approach to transportation during the 1970s and made systematic progress toward adopted goals of creating a livable city.

Intermodal transportation policies in Portland have consisted of auto use disincentives and transit use incentives. The former have included elimination of several highways from the plans, notably the Mt. Hood Freeway, traded in for federal light rail funding under the Interstate Transfer Program, and, more recently, a controversial Westside Bypass. Traffic-calming measures and parking controls have been introduced in some center city streets and in neighborhoods throughout the city. Transit use incentives consisted of a number of short- and long-term improvements, including the following.

Major improvement of existing bus services has been carried out. The bus network was divided into specially designated zones with distinct names and symbols. Lines were organized into trunks and feeders with synchronized transfers at specially designed transit centers. Two one-way streets in the CBD were reconstructed into exclusive bus facilities with specially designed multiple stops and leapfrogging bus movements along them to provide very high stopping capacity. These streets carry up to 185 buses per hour.

A new LRT line was built from the CBD eastward to the town of Gresham, then extended to the westward through a long tunnel to the town of Hillsboro, creating a major diametrical trunk line through the region. This was followed by the construction of a line to Portland International Airport, using private investor contributions for its financing. Another line from the CBD to the boundary of the state of Washington on the north, named the Interstate Line, was opened in 2004.

In outlying areas, these lines have ROW category B. Where they converge in the city streets, their tracks are protected from other traffic so that they still enjoy similar independence from other traffic. The tracks also go through and are part of a pedestrian-dominated square in the downtown area, and they have short sections of street running, ROW category C.

A new streetcar line has been built to provide easy travel and support walking in a growing downtown Pearl District. That line has become a major promotional symbol for the development of businesses and residential complexes in the areas it serves. A short extension is under construction, with further extensions and lines planned.

The transportation system is coordinated and mutually supported with land use planning that controls developments and limits growth in surrounding areas to prevent excessive sprawl.

As a result of these coordinated planning measures and consistent public policies toward land use control and transportation planning, Portland today has a reputation as one of the most livable cities in North America.

10.5 REVIEW OF THE PROCESS AND TRENDS IN TRANSIT PLANNING

This chapter has described the process and methodology for planning medium- and high-capacity transit systems. Since these systems have special infrastructure, this procedure is usually classified as medium- and long-range planning. This material and its relationship with technical materials presented in other chapters are briefly reviewed here.

10.5.1 Planning Process and Its Components

The first section in this chapter defines basic concepts in transportation planning and describes its organizational and legal aspects. It points out that in the United States disproportionate effort has been given to demand-forecasting methodology, while actual transit system planning and design have obtained much less attention. In some peer countries, particularly those that have more positive controls over urban planning and its implementation in general, transit is given more attention and plays a stronger role in cities.

In the preceding description of the planning procedure and methodology, the tendency to base planning on extrapolation of past trends is given considerable attention. This approach is criticized in several respects. Conceptually, the purpose of planning is to guide developments toward adopted goals. Extrapolation of past trends downplays and in some cases ignores the goals, so that trend extrapolation actually does not represent planning. From the practical point of view, mechanical trend extrapolation may lead in undesirable directions (as has been the case in some aspects of urban transportation planning), can lead to physically or economically infeasible conditions, and tends to discourage innovative solutions. Clear definition of goals and objectives is therefore emphasized as essential for a rational, constructive planning process.

Section 10.3 presents a review of demand-estimation techniques. As mentioned above, this complex phase of planning has received great attention in academic and professional circles. Yet, due to complex relationships of many factors affecting personal travel and uncertainties in predicting future human behavior and economic and social conditions, the presently used methodologies are debated and criticized inside planning circles. The reliability of demand forecasting results is often challenged by planners and civic leaders.

With respect to transit, however, one can argue that extremely high accuracy of demand forecasting (which cannot be achieved anyway) is in most cases actually not as important as is often believed, for the following reasons.

First, the basic modal choice decision depends on rather broad ranges of critical passenger volumes on a transit line or network, rather than on a precise figure. Generally, high-performance medium-capacity transit systems are needed when volumes reach 3000–5000 prs/h, which regular buses in mixed traffic cannot easily handle. These systems can serve up to 18,000–20,000 prs/h (under special conditions, such as wide rights-of-way and special organizational measures, even higher volumes can be achieved). But already at 10,000–15,000 prs/h rapid transit systems may be the optimal choice if other factors contribute to the advantages of that mode. Thus, there are overlaps among the three generic mode classes (street transit, medium-capacity, and rapid transit), which mostly correspond to systems having ROW categories C, B, and A, respectively.

Second, the passenger demand volume is not the only, and sometimes even not the dominant, factor in selecting a transit mode. Many other factors, such as transit system image and its impacts on city's livability, can have a strong influence on mode selection. Further, high-performance systems with strong image have a significantly stronger passenger-attraction ability than buses and streetcars in mixed traffic. This impact of service quality and system image on ridership, known as the rail factor, can attract as much as 30% higher ridership (see Section 12.2.4.1). Few demand-forecasting models can handle this factor, so there should be a feedback step in the demand-forecasting procedure that adjusts the demand at the time transit mode is chosen.

Third, high-performance transit, particularly rail systems, is planned in many cities as part of positive, normative planning of land use development, efforts to increase livability and sustainability of the city, etc. In such cases, again very high accuracy of demand projections is not of critical importance. For example, metro (T-Bana) in Stockholm was designed and built together with several towns around its stations in suburban areas. The recently opened extension of the Skytrain ALRT system in Vancouver has rather low ridership, but that is not criticized because the line is considered to be an investment to stimulate growth of suburban areas it serves.

The transit plan development process is described in Section 10.4 in a general overview referring to technical details used in planning that are presented in greater detail elsewhere in this book. Technical characteristics of lines and networks are given in Chapter 4. Operations and scheduling are covered in Chapters 1 and 2. Chapter 11 presents characteristics of services and methodology for comparison of modes, while Chapter 12 contains descriptions and comparisons of specific transit modes, again focusing on high-performance ones, which are subject of medium- and long-range transit planning.

Plan development and selection methodology are followed by a set of brief descriptions of selected cases of transit planning and mode selections in different cities, including the successes and problems of plan implementation. The cases are intended to illustrate two critical steps in transit planning: the methodology of alternatives analysis and mode selection, and various influences and actual process of plan implementation.

The cases show that the four major decisions (selection of generic mode, selection of a specific mode, establishment of general line alignments and network form, and determination of the amount of required investment funds) are made in a variety of sequences. For example, decisions about the mode selection or about a line or network alignment may be made prior to the development of a detailed transit system plan. However, the network for each mode (metro, LRT, bus, AGT, or monorail) must be optimized even if the main corridors of lines are considered fixed. Thus, alternative plans generally represent systems that differ in varying degrees in network form, size, investment cost, type of operation, etc.

The described cases have been selected to show rational, well-conceived and organized planning, as well as situations where many difficulties were encountered and confusion and conflicts caused delays or prevented plan implementation. Obstacles common in many cities include a variety of interested groups. Antitransit lobbies, particularly strong in U.S. cities, use various tools, from factual misinformation to political pressure, to prevent investment in high-performance transit systems. Promoters of individual modes, who have either financial or emotional motives, sometimes produce positive impacts toward implementation of individual modes, or work against alternative modes and create confusion. Finally, lack of information and expertise about the different modes that are available limit planning in many cities to the two extremes, buses on streets and rail rapid transit, without any other modes filling the large gap between these two modes with very different PCP characteristics.

To overcome difficulties and obstacles in transit planning and implementation, it is important to improve knowledge and expertise about the entire family of transit modes, the variety of services they offer, and their optimal roles in different sizes and types of urban areas.

10.5.2 Trends in Transit Planning

Urban transportation planning, as well as transit planning, evolves with the society, its economic and social development, and environmental concerns. Changing values of the society have a direct impact on the scope of and trends in transportation planning. As discussed in the first two sections of this chapter and elsewhere in this book, transit planning has evolved from serious neglect during the freeway-building era in the 1950s and 1960s to a revival during the 1970s. Activities stimulated by federal government programs and local initiatives have resulted in many changes in attitudes toward transit. Many innovations have been created in the United States and followed by other countries, while our transit developments have also benefitted greatly from innovations in other countries, particularly those that have had more advanced developments in transit (Vuchic, 1999; Pucher, 1995; Cervero, 1998).

In an excellent overview of trends in urban transportation planning during the last 40 years, Meyer (2000) lists a number of important changes and innovations that have taken place in the United States. The following have special relevance to transit planning (adapted from Meyer and expanded).

From: Emphasis on methods and data in producing plans and investment programming

To: Improved information on a wide-ranging set of impacts on a variety of investment, operational, pricing, land use, and lifestyle decisions

From: Focus on highways and their levels of service

To: Multimodal system integration, operation, and broad set of performance measures

From: Basing transportation systems on vehicle movements

To: Considering trip makers as the basic unit and object in urban transportation

From: Vehicle and system technology viewed as given

To: Innovative technologies used to influence system operation and travel behavior

From: Consideration of land use patterns as a given, independent of the solutions set

To: Use of growth-management tools in connection with corresponding transportation policies as a major strategy

From: User benefits and costs

To: Equitable distribution of benefits and costs within a community or city

From: Transit considered as a mode serving captive riders, except in a few large cities

To: High-performance transit as a powerful tool in creating balanced transportation and livable cities (Vuchic, 1999).

Some of these changes and trends refer primarily to the United States. However, there has been an increasing exchange of contacts and experiences among countries. Generally, the problem of the collision between cities and cars has occurred in Europe and other developed countries later than in the United States, and they were in position to use many experiences and solutions from North America. There have been, however, significant differences in the attitudes toward cities, planning, and particularly the importance given to transit between the United States and many industrialized countries. In this respect, and particularly in the much more positive policies toward balanced transportation, the United States has been learning a lot from its peers. With respect to environmental concerns and sustainability, both the United States and its peer industrialized countries have made significant progress and benefitted from mutual exchanges of experiences. In the future, further progress in transit planning and cooperation in approaches to problem-solving and innovations can be expected.

EXERCISES

10.1. Define short-range and long-range planning. How is the planning of different modes divided between the two and why?

10.2. What does the 3C requirement for urban transportation mean? What is the rationale for this requirement, and how does it affect the planning process?

10.3. Why was the adoption of ISTEA considered a very significant development and a milestone in the planning and financing of urban transportation in the United States?

10.4. Describe the role of metropolitan planning organizations in the process of transit planning and its implementation.

10.5. Discuss the positive achievements and shortcomings of the urban transportation planning studies performed in the United States during the 1950s and 1960s.

10.6. Many urban transportation planning studies have been based on extrapolation of past trends. Discuss the basic conceptual basis of that approach. When can it be used well? When is it inadequate or even damaging and why?

10.7. What is the problem in establishing goals and objectives for a transportation plan with respect to urban form and character when the planning procedure is based on trend extrapolation?

10.8. Present a flowchart and discuss the sequence of the basic steps in the urban transportation planning process. Explain the reasons for any feedback steps in that process.

10.9. Define goals, policies, and standards and discuss their significance in the process of urban transportation planning. Explain why the definition of the vision of the future city is particularly important for long-range transit planning.

10.10. What are models in transportation planning? How are they calibrated? What are they used for? Give two examples.

10.11. Describe the four-step demand-forecasting model. Define each step and discuss the logic of their sequence.

10.12. Describe the original gravity model and one of its later forms used in transportation planning.

10.13. List and describe groups of factors that can be used in modal split estimation.

10.14. Briefly describe trip-interchange and trip-end modal split models; then compare them.

10.15. How are diversion curves used in demand estimation by trip-interchange models?

10.16. The trip-end modal split model was used by a number of major metropolitan area transportation studies, but it was later mostly abandoned. What was its advantage, and what were its serious weaknesses?

10.17. For modal split analysis, compare the service that an automobile generally offers as compared to transit for:

- Peak hour travel to center city
- School trips
- Recreational travel with families on weekends

10.18. What is the difference between all-or-nothing and capacity restraint traffic-assignment models?

10.19. What is the user equilibrium condition in traffic-assignment models?

10.20. Discuss some weaknesses of the four-step demand forecasting model.

10.21. What is the fundamental difference between aggregate and disaggregate traffic forecasting models?

10.22. List the major steps in transit system plan development and selection; then draw a flowchart of the planning process, using your own brief descriptions of each step.

10.23. The differences among transit alternative plans can be grouped into four major categories. List and describe them. Find two cases of long-range transit planning and explain how the considered alternatives differed among themselves.

10.24. What basic alternative modes can be considered in transit planning in these two generic categories?

a. Rapid transit

b. Medium-capacity or semirapid transit systems

10.25. Why is the selection among transit modes generally simpler in very large than in medium-sized cities? Hint: review the described planning cases.

10.26. Based on the described transit planning cases and your own knowledge of other cases, list and describe several difficulties that transit project plans can encounter in their implementation.

10.27. Discuss the changes in transit planning that have occurred over the last 40 years with respect to:

a. Focus on different transportation modes: highway, transit, paratransit, pedestrians, and others

b. Basic units computed in traffic flows on highway and other modes

c. Considerations of land use in transportation planning

d. User and community/social benefits and costs

e. Assumed role of transit in urban transportation

10.28. (Multiple-choice) The percentage of transit trips as a share of all vehicular trips is higher for:

a. Work than shopping trips

b. Off-peak than peak

c. Circumferential than radial

d. None of the above

10.29. (Multiple-choice) In a city with well-coordinated intermodal transportation, the automobile generally offers better service than transit for:

a. Peak hour travel in any direction

b. Recreational travel

c. Trips inside high-density areas

d. Trips along heavily traveled corridors

e. None of the above

10.30. (True-False) Characteristics of auto-transit modal split:

Auto attracts a greater share of peak than off-peak trips	T	F
Transit attracts a greater share of business than recreation trips	T	F
Auto attracts a greater share of circumferential than radial trips	T	F
Auto always serves school trips more efficiently than transit	T	F

10.31. (True-False) Characteristics of trip attraction by transit:

Most transit trips are for business purposes	T	F
A second car in the family reduces transit riding more than the first car	T	F
In urban areas school trips can be served efficiently by transit	T	F
Transit attracts more family recreational trips than auto	T	F
Transit attracts the greatest share of radial work trips	T	F
Transit riding is always inversely related to income	T	F

REFERENCES

Beckmann, M., C. B. McGuire, and C. B. Winsten. 1956. *Studies in the Economics of Transportation*. New Haven: Yale University Press.

Boyce, D. E., N. D. Day, and C. MacDonald. 1970. *Metropolitan Plan Making*. Philadelphia: Regional Science Research Institute.

Bureau of Transport Economics (BTE). 1998. *Urban Transport Models: A Review*. Working Paper 39. Canberra: BTE.

Cervero, R. 1998. *The Transit Metropolis*. Washington, DC: Island Press.

Creighton, R. L. 1970. *Urban Transportation Planning*. Urbana: University of Illinois Press.

Dickey, J. W. 1983. *Metropolitan Transportation Planning,* 2nd ed. Washington, DC: Hemisphere.

Fertal, M. J., E. Weiner, A. J. Balek, and A. Sevin. 1966. *Modal Split: Documentation of Nine Methods for Estimating Transit Usage*. Washington, DC: U.S. GPO.

Kanafani, A. 1983. *Transportation Demand Analysis*. New York: McGraw-Hill.

Katz, B., R. Puentes, and S. Bernstein. 2003. *TEA-21 Authorization: Getting Transportation Right for Metropolitan America*. Washington, DC: The Brookings Institution.

Lill, E. 1891. *Das Reisegesetz und seine Anwendung auf den Eisenbahnverkehr*. Vienna: Spielhagen & Schurich.

Menon, A. P. G. and K.-K. Chin. 2004. ERP in Singapore—what's been learnt from five years of operation? *Traffic Engineering & Control,* February, 62–65.

Meyer, M. D. 2000. Refocusing transportation planning for the 21st Century. In *Conference Proceedings #20*. Washington, DC: TRB.

Meyer, M. D. and E. J. Miller. 2001. *Urban Transportation Planning,* 2nd ed. New York: McGraw-Hill.

Newman, P. and J. Kenworthy. 1998. *Sustainability and Cities: Overcoming Automobile Dependence.* Washington, DC: Island Press.

Ortuzar, J. D. and L. G. Willumsen. 2001. *Modeling Transport.* New York: John Wiley & Sons.

Papacostas, C. S., and P. D. Prevedouros. 2001. *Transportation Engineering and Planning*. Upper Saddle River, NJ: Prentice-Hall.

Pucher, J. 1995. Urban passenger transport in the United States and Europe: a comparative analysis of public policies. *Transport Reviews* 15(2):89–107 and 15(3):261–283.

Smerk, G. M. 1991. *The Federal Role in Urban Mass Transportation,* Bloomington: Indiana University Press.

Stopher, P., and A. Meyburg. 1975. *Urban Transportation Modeling and Planning*. Lexington, MA: Lexington Books.

Vuchic, V. R. 1991. "Recognizing the value of rail transit." *TR News,* September–October, 13–19.

———. 1999. *Transportation for Livable Cities*. New Brunswick, NJ: CUPR Press, Rutgers University.

Wardrop, J. G. Some theoretical aspects of road traffic research. *Proceedings, Institution of Civil Engineers, Part II* 1:325–378.

Weiner, E. 1999. *Urban Transportation Planning in the United States*. New York: Praeger.

FURTHER READING

Beimborn, E. A. 1995. *A Transportation Modeling Primer*. Milwaukee: Center for Urban Transportation, University of Wisconsin.

Ben-Akiva, M., and S. Lehrman. 1985, *Discrete Choice Analysis: Theory and Application to Travel Demand*. Cambridge: MIT Press.

Gray, G. E., and L. A. Hoel, eds. 1992. *Public Transportation,* 2nd ed. Englewood Cliffs, NJ: Prentice-Hall.

Hutchinson, B. G. 1974. *Principles of Urban Transportation Systems Planning*. New York: Scripta Books.

Voorhees, A. M. 1955. A general theory of traffic movement. In *Proceedings, Institute of Traffic Engineers*. New Haven: ITE.

Vuchic, V. R. 1987. Urban transit: a public asset of national significance. *Urban Resources* 4(1):3–6, 63.

World Bank. 1996. *Sustainable Transport*. Washington, DC: World Bank.

11

ANALYSIS, EVALUATION, AND SELECTION OF TRANSIT MODES

Transit mode is a generic concept describing transit systems which have a specified common set of physical and technical components and operational features (Vuchic, 1981, Chapter 2). The transit system in a city is a specific transit network of lines, vehicles, and operational elements, which may be of one or several modes. Thus, one refers to the Chicago or London transit system. These two systems are different, each serving its city with a different network configuration and combinations of physical and operational characteristics. However, both cities utilize common transit modes, such as buses, rapid transit, and regional rail, as well as a paratransit mode, taxis.

Selection of a transit mode is the central decision in planning new or expanding existing transit systems. This decision is very important. Selection of a mode not only determines the technological, operational, and network characteristics of the planned system, but through these elements it also influences the role transit will have in the development of a city's physical form, its economic activities, and social and environmental conditions. Characteristics of modes are diverse, there are several interested parties, and there are a variety of impacts that must be considered in their planning. Both the absolute and comparative analyses of modes are thus quite complex.

Some elements analyzed in mode comparisons can be quantified and thus compared exactly. However, there are also many qualitative elements, so that their evaluation must include experience and value judgments. The procedure, therefore, cannot be formulated as a quantitative model, nor can the results of the comparative analysis be expressed by a single numerical value. The desire to simplify this process by using a mechanized procedure and basing the comparison on a single criterion (usually direct cost) has sometimes prevailed, but it has often produced erroneous results, contrary to real-world conditions and sound judgment.

After a brief review of the methodology for public project evaluation, this chapter focuses on analysis and comparison of transit modes, including conceptual procedures and definitions of elements that must be considered. Evaluation and selection of actual semirapid and rapid transit modes, illustrated by examples from different cities, are given in Chapter 12.

11.1 EVALUATION AND SELECTION OF PUBLIC PROJECTS

In the private sector of the economy, investment decisions are relatively simple, at least in principle. Although such goals as providing jobs, improving living standards, and promoting new technology may be considered, profit maximization is usually the dominant, sometimes the exclusive, goal. The main criterion for an investment is thus based on projected costs and revenues.

In the public sector, evaluation of projects, particularly those involving investments in infrastructure with long-term benefits, is considerably more complicated. While the maximum financial effectiveness of the investment is an important goal, many other goals must be considered and often given more importance than the monetary one. The goals of public project planning usually include such elements as providing essential public services (mobility), improving economic vitality and environmental quality in the city, achieving in the long run a desirable urban form and quality of life, etc.

Benefit-cost analyses were used for some transportation projects many decades ago. Particularly well known was the methodology recommended for estimation of highway user benefits and costs, developed by the American Association of State Highway Officials in 1952 and updated in 1960. It was published in a book popularly known as the Red Book (AASHO, 1960). This methodology converted all items into monetary values for computation of the project worth. Kuhn (1962) published an extensive critique of such analyses and showed the deficiencies of comparisons using monetarized values only. He emphasized the importance of including not only quantitative factors and illustrated the difficulties in quantifying the qualitative elements, such as pain and suffering from accidents or indirect impacts of projects on the surroundings and nonusers. Kuhn then recommended a comprehensive methodology illustrated by a framework for a comparison of a freeway and a rapid transit line.

Consequently, evaluation of public projects must include not only direct, short-range costs and benefits, but also indirect and long-range ones. The evaluation must consider different affected groups, often defined as system users and nonusers, or the general public. A brief theoretical framework for such evaluations is presented here. Detailed discussions of this topic have been done, for example, by Thomas and Schofer (1990).

As discussed in Section 6.4.4.2, for each facility or infrastructure object, optimal life lengths can be computed on the basis of its economic, physical, and functional lives. These lives are different, and the shortest of the three is usually critical and determines when an infrastructure object, facility, or vehicles should be replaced or rehabilitated.

In project evaluation, total life of the planned facility, i.e., the shortest of the three (functional, economic, and physical) must be studied. Throughout that life there is a stream of gains or benefits on one side and costs on the other. As mentioned above, the benefits and costs must include quantifiable as well as nonquantifiable elements in quantitative terms. If the benefits from the planned project in year t are B_t, then the present value of those benefits, BPV, is the sum of the discounted stream of benefits through the project life:

$$BPV = B_0 + \frac{B_1}{1 + r} + \cdots + \frac{B_t}{(1 + r)^t} + \cdots + \frac{B_L}{(1 + r)^L} + \frac{S_b}{(1 + r)^L} = \sum_{t=0}^{L} \frac{B_t}{(1 + r)^t} + \frac{S_b}{(1 + r)^L} \qquad (11.1)$$

Correspondingly, the present value of the project costs, CPV, is:

$$CPV = \sum_{t=0}^{L} \frac{C_t}{(1 + r)^t} + \frac{S_C}{(1 + r)^L} \qquad (11.2)$$

where:

r = Discount rate
L = Years of project life
S_b and S_c = Salvage benefits and costs, respectively

The discount rate accounts for the time value of money. Each year in the future money is worth $[1/(1 + r)]$ times its value in the previous year. This rate is different depending upon the user and the purpose. If the only purpose is to pay back the money at the same rate at which it was borrowed, r is the loan interest rate.

Generally, however, there is a minimum allowable rate of return (MARR). A private firm will set this rate based on its opportunity cost, that is, the rate it would earn investing elsewhere. A public agency is usually given an MARR as public policy. It is important to note that a public transit agency almost always has a lower MARR than a private firm. Hence, publicly funded project proposals are feasible much more often than privately funded ones.

The present worth or net present value, NPV, of the project is:

$$NPV = BPV - CPV = \sum_{t=0}^{L} \frac{B_t - C_t}{(1 + r)^t} + \frac{S_b - S_c}{(1 + r)^L} \tag{11.3}$$

When NPV is computed or estimated in quantitative terms, the first question in project evaluation should be whether NPV > 0. If the answer is negative, the project should not be built. If the answer is positive, the computed NPV should be compared with corresponding values obtained for alternative projects competing for the same investment. Assuming that all evaluation criteria have been included in this project worth-assessment procedure, the project with the highest value of NPV should be adopted for implementation.

An alternative evaluation is to define NPV of the project as the benefit-cost ratio (BCR), rather than the difference between the projected streams of benefits and costs:

$$BCR = \frac{BPV}{CPV} = \sum_{t=0}^{L} \frac{B_t/C_t}{(1 + r)^t} + \frac{S_b - S_c}{(1 + r)^L} \tag{11.4}$$

Between these two methods of computing project worth, the former one, using the difference BPV − CPV, is preferred to the latter, based on the ratio BPV/CPV. The main reason for this is that some items in evaluating a project can be considered either as benefits or as reduced costs. Using the difference method, the result is the same regardless of this allocation. However, if the ratio method is used, adding a certain amount to the numerator (added benefit) gives a drastically different project worth than subtracting the same amount (reduced cost) from the denominator—for example, if a project has a benefit/cost ratio of 6/4 = 1.5 and a value of 2 should be included as an additional benefit or as a cost reduction. The former would change the ratio to 2.0, the latter to 3.0!

Another common procedure for capital investments analysis is the internal rate of return method, which consists of setting NPV of a project (Eq. (11.3)) equal to zero and solving for the discount rate r. If the computed rate is less than r, the project is not viable; if it is at least equal to r, it is acceptable. This method is not always reliable, because it is possible that there are multiple roots of the equation, giving more than one discount rate. Also, the value of r is not a reliable indicator of relative worth. For example, project A could have a greater r than project B, but project B could still have a larger NPV!

Several important notes of caution about these methods of project evaluation—net present value and internal rate of return—are in order here. First, the methods require quantification of all elements; this involves subjective judgments and allows introduction of biases, which are then disguised under numerical values and appear to be exact and objective. Second, discounting future benefits and costs of a project is extremely sensitive to the assumed interest rate, which is quite uncertain. And third, using a single number for evaluation of a project is very attractive to decision makers, which may lead to overlooking of the serious limitations of this evaluation method.

Another serious weakness of all these economic analysis methods is that they have an inherent bias against the benefits and value added of the projects in the distant future, and thus its value to future generations. Even low discount rates reduce their future value considerably, sometimes making it negligible. This is not a major problem if the competing projects have similar lives and they are not longer than 10–15 years. However, transportation systems infrastructure is used for 50–100 years. In particular, exclusive rights-of-way for transportation ways can be considered to have a permanent worth. For example, subway tunnels in several cities are now more than 100 years old, and their use is projected beyond any foreseeable future. In several countries, roads from ancient Roman times have been in continuous use. The value of projects with long life for future generations is actually ignored by the current discount techniques. A similar problem on the side of costs exists with respect to evaluation of projects that involve the consumption of nonrenewable resources: their value in the future is diminished by the discounting evaluation techniques.

This problem has been very damaging to many cities in developing countries, where international financing banks use the internal rate of return method in comparing transit projects with different useful lives, such as purchase of buses versus construction of bus or rail infrastructure. Financing of the long-term projects has often been rejected because of lower computed rates. The fact that a city will have continuing benefits and often be shaped differently if it has a high-performance transit system physically independent of street congestion has been ignored by many economic analysts.

Consequently, the economic project-evaluation methods presented here can be used only for projects with similar characteristics and life lengths for which the benefits and costs can be reasonably monetized. For projects of different natures, these methods can be used for conceptual clarification, rather than as an operational procedure for transit system evaluation. Various project feasibility studies should always be checked with respect to the definition of feasibility. All projects that include many qualitative factors and values, such as environmental improvement projects, construction of transit systems, as well as many other public facilities, must be evaluated on a much broader (and more complex) basis than quantitative economic analysis. Particularly for investments in projects with long life or virtual permanence, due consideration in a nonmathematical sense must be given to long-term benefits and costs to the community and the ecology.

It is sometimes possible to compartmentalize a project in such way that a part of it can be evaluated by different private companies and each one can provide the public agency with its best project proposal. If given to a firm in its entirety for a long period of time, such as a design-build-operate-maintain (DBOM) contract, each company has an incentive to identify the approach that has the highest NPV. The remainder of the project is then evaluated in a multidisciplinary approach that attempts to account for all benefits and costs, both monetary and nonmonetary.

11.2 TRANSIT MODE EVALUATIONS

In principle, evaluation of transit modes and systems should utilize the same basic method as the described evaluation of all public projects: all benefits (gains) and costs of each transit project or mode must be estimated to find its worth, which must then be compared with the corresponding worths of alternative modes or systems in the planning process (see Section 10.4). However, transit systems have a particularly complex set of characteristics and impacts, many of which cannot be given monetary or any other quantitative values, so that they cannot be summed up in the evaluation process.

To include systematically all benefits and costs, their incidences or affected parties must be determined. For this purpose it is helpful to approach evaluation as an analysis of a demand and supply situation. The demand is the need for travel and for an area to be served, characterized by location of the transit facility and all physical and other relevant conditions. The supply consists of transit modes, which may have a variety of characteristics. Since the conditions in the served area and transit system characteristics interact, evaluation of system performance and selection of the best mode must be based on a detailed analysis of mode performance for the given conditions.

11.2.1 Classification and Review of Mode-Comparison Studies

The studies involving comparison of transit modes vary considerably in their approach and purpose, as well as in the validity of their findings. They will be classified here by their objectives, intents, and methodologies. Then each group will be described and evaluated, particularly with respect to its procedural correctness and applicability to real-world situations.

Four basic types of mode evaluations and comparisons can be defined. They are shown schematically in Figure 11.1: different areas for which a transit system is planned are shown as A_1, A_2, . . . A_n, while M_1, M_2, . . . M_n represent different modes. Lines combining different sets of areas and modes indicate different types of comparisons. The following comparative analyses are defined.

I. Theoretical comparisons of typical modes (M_1, M_2) in one hypothetical area (A_h)

II. Analyses of general characteristics of different modes

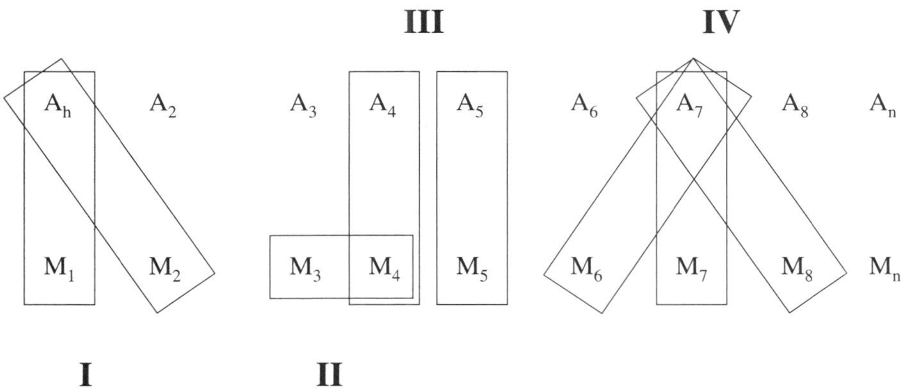

Figure 11.1 *Types of comparisons of transit modes (M) for different conditions sets or areas (A).*

III. Comparisons of different modes serving similar areas

IV. Comparison of different modes planned for a given area

The methodologies, purposes, and realistic validities of these studies vary considerably. The first two types attempt to define optimal domains of different modes in general: the former utilizing theoretical models, the latter actual modal characteristics and real-world data. The second two types focus on actual systems operating or planned for certain areas or corridors.

11.2.1.1 Comparisons of Different Typical Modes in a Hypothetical Area. When properly performed, with realistic assumptions, this comparison can give useful insights into relative advantages and disadvantages of different modes under a set of model corridor conditions, as well as some indication of the optimal domains of different modes. Such comparisons are, however, very difficult to perform correctly because definition of hypothetical conditions is complex and conducive to many biases.

Several theoretical economic studies of mode comparisons have been performed for hypothetical urban corridors, utilizing average costs from different cities, or with some assumed values. Started by Meyer et al. (1965), this type of cost-based study was repeated with the same conceptual and methodological errors by several groups of economists in subsequent years, and even now arguments favoring lower-cost modes while overlooking quality of service and other characteristics of compared modes are often heard.

The assumptions and models used in most economic comparisons of modes are so unrealistic that their findings are, in most cases, at sharp variance with the studies of transit in actual cities. Their simple model and seemingly clear results give these studies unjustified credibility in the popular press. It is therefore necessary to explain the major deficiencies inherent in this methodology.

These economic comparisons of modes are intended to find optimal domains for individual modes, defined by the number of passengers they carry during the peak hour. Actually, choice of mode must be based on a number of factors, such as local conditions, alternative means of travel, service quality throughout the day, and short- and long-term impacts of transportation systems on the served area. Optimal domains of modes in terms of passenger volumes, therefore, are not delineated by a fixed number for all conditions. Therefore, the objective of these studies, to find a domain of optimal use of a mode in terms of peak-hour one-way passenger volume, is incorrect.

The sole criterion used for determining the optimal mode, the minimum cost per passenger-trip, is valid only in the rare cases when modes with identical performance are compared. However, in most cases, each mode has a different performance-cost package (PCP). For example, if mode M_1 has a lower cost but also lower performance than mode M_2 under given conditions, it would be incorrect to conclude that mode M_1 is better because it is cheaper. If the difference in performance of mode M_2 is worth its additional cost, mode M_2 should be preferred.

Since many important mode characteristics cannot be converted into monetary units, the problem of including these characteristics into mode comparison cannot be disregarded. For example, the mentioned economic studies that compared rapid transit and auto-highway systems as alternatives must include many other factors. The authors, however, often ignore the externalities of these modes, such as limited space in CBDs to accommodate both freeways and parking facilities for up to 30,000 automobiles arriving per hour from a single corridor. This makes such hypothetical analyses highly unrealistic.

The computational analysis and the diagrams used in these economic comparisons also have conceptual flaws. The basic diagram used presents average cost per trip as a function of passenger volume for different modes, as shown in Figure 11.2. However, since each mode has a different performance, it therefore attracts, under any given set of conditions, a different number of passengers. Rail rapid transit and LRT lines can attract greater numbers of passengers than does a typical bus route. All three modes are so different from the automobile in type of service and potential users that plotting them on the same diagram is meaningless; it amounts to comparing apples and oranges.

More specifically, the diagram implies that it presents costs of different modes for any given passenger volume. Thus, in Figure 11.2, it appears that one can compare cost per trip C for a bus on streets with cost B for a bus on a busway and cost A for rail rapid transit. The fact is that there cannot be a served area in the real world in which these three modes would allow cost comparison for any given passenger volume. If a street bus line would attract P passengers in a given area, then a bus on busway would attract a volume $P + \Delta P_1$, and rail rapid transit would attract a volume of $P + \Delta P_2$. Unit cost C, therefore, should not be compared with costs B and A, but with B' and A'. Moreover, the criterion is not whether, for example, C is lower or higher than B', but whether the cost difference $B' - C$ is worth the attraction of ΔP_1 passen-

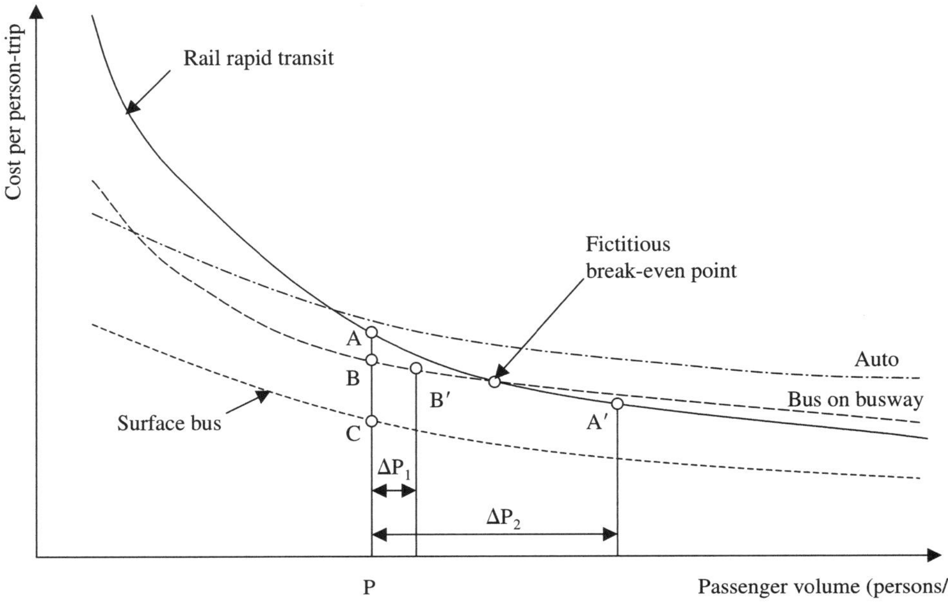

Figure 11.2 *Comparison of modes based on their costs, disregarding differences in level of service and passenger attraction.*

gers. Incidentally, this cost difference may sometimes be negative, which makes the higher-quality mode clearly superior even for volumes well below the break-even point. Actually, the break-even point is a fictitious concept here because the plotted curves are on different LOS surfaces, i.e., they cannot be plotted on the same plane because they do not intersect.

Another problem is that this type of diagram is highly unreliable when it is applied to hypothetical typical conditions because of the extreme sensitivity of the curves to the assumptions made by the analyst. It is easy to change relative positions of curves for different modes by manipulating assumptions made for the model. This characteristic allows use of this methodology to develop biased arguments in favor of or against any mode.

Many of these hypothetical economic studies confuse technologies with modes (see Chapter 12). Since the costs of modes are mostly dependent on their type of ROW, the curves in Figure 11.2 refer more to types of ROW than to bus and rail technologies. For example, a bus on an exclusive busway may involve a higher investment and operating cost than light rail transit on ROW category B, such as a street median.

Because of these deficiencies, the results of the economic studies have little realistic value. Since such important factors as reliability and frequency of service, comfort, and safety enter the economic models only as cost items, while their major influence on passenger attraction is disregarded, these studies greatly distort the relationships among modes in favor of low PCP.

The results of these theoretical analyses, implied to be correct, have often been used by various interest groups to argue against improvements of public transportation infrastructure, particularly of rail transit. However, their validity has been proven incorrect not only with respect to methodology, but also by real-world events. Comprehensive studies for many cities in the United States and other countries, discussed in subsequent sections, have clearly shown that real-world conditions are far more complex than the economic hypothetical studies assume: there are a variety of modes, steered and guided, that represent viable alternatives for different cities. Several cities have found that upgraded bus systems (BRT) represent the best solution for many of their corridors (for example, Curitiba, Ottawa, Pittsburgh); others, or even the same cities (Pittsburgh), have found that they also have corridors that are best suited for rail transit.

Another fact is that there are a variety of rail modes, rather than one standard rapid transit. Thus, examples of successful rapid transit systems built in recent decades include those in Toronto, San Francisco and Washington; light rail has been used in San Diego, Portland, Sacramento, and St. Louis; and Vancouver and London (Docklands) have developed an innovative mode, automated light rail transit, a rail system with driverless trains. This development is continuing in many cities in North America and Europe, as well as increasingly, in developing countries.

In recent years, hypothetical mode comparisons have become less popular even in academic circles. With increasing investment in high-quality transit systems, the need for practical planning has increased and practical methodology for comparison of modes, broadening it from the cost basis to inclusion of other factors, has become more prevalent.

11.2.1.2 Comparison of General Characteristics of Different Modes. Comparison of mode characteristics in general, without reference to specific conditions, can give useful insights into generic differences among modes. Some of the differences can be quantitatively defined (e.g., capacity, speed); others that are heavily area-dependent (e.g., type of operation, capital and operating costs) can be compared only in general terms. In most cases, however, a range of values, rather than a single value, should be used for this comparison since no mode has one set of fixed characteristics for all applications. This comparison should define the trade-offs that indicate ranges of conditions making one mode superior to another. Complete inherent dominance of one mode over another does not exist, except in comparisons of real-world modes with hypothetical, unsound technologies or concepts, such as, for example, the personal rapid transit (PRT) concept.

In an early comparison of this type, Lehner (1975) made a comprehensive comparative analysis of all major features of LRT with rapid transit, utilizing actual data from many operating systems. Vuchic (1981) presented a systematic definition of modes and their main components: ROW categories C, B, and A, technology (particularly guided versus steered), and types of operation. Impact of each component on modal performance and service characteristics was presented. This definition was then used to describe different modes and discuss trade-offs among them and optimal domains of different modes in the family of transit modes. Further and updated comparisons of modes and their characteristics are presented in Chapter 12.

Transit capacity and quality of service for many modes are presented in great detail in TCRP Report 100 (Kittelson & Associates, 2003).

11.2.1.3 Comparison of Different Modes Serving Similar Areas. Comparison of two different modes, such as LRT versus regional rail, or buses on busways versus metro or LRT, operating in two corridors with similar conditions can also be used for comparison of characteristics of two modes in general, not necessarily related to a specific planning project. This kind of mode comparison can be much more realistic than the first method (hypothetical modeling) for two reasons. First, real-world systems are designed optimally for the local conditions, considering characteristics of each mode. This is not always possible when different modes are planned for one given area. And second, it permits comparison of real-world systems for which performance, cost, and attracted passenger volumes are demonstrated and known, rather than estimated for hypothetical situations. The numerous assumptions needed for abstract modeling are avoided.

In this type of comparison, it is particularly important to separate the differences among systems caused by different operating concepts from those attributable to different transit mode characteristics. For example, the differences between a commuter bus line on an HOV roadway and an LRT line in a similar corridor may be caused more by the commuter-type service of buses versus regular transit LRT service than by the differences in the technology of transit modes, bus and rail.

There have been few comprehensive comparative studies of actual systems involving different modes operating in similar areas. One such study, comparing rail and bus systems serving similar areas with different technology and in networks and types of operation, is summarized in Section 11.4.4.

11.2.1.4 Comparison of Different Modes Considered for a Given Area. With diversification of modes in recent decades, planning of major transit systems must include this type of comparison as an important phase of work. Instead of traditional planning based on a decision to build a transit system utilizing a predetermined mode, contemporary procedure usually requires a systematic analysis of several candidate modes for a given corridor or network.

In some cities, the transportation problem is so serious and complex that it is clear that a high-capacity, high-speed, reliable system is needed that will be independent of external influences. These requirements can be met only by modes with fully controlled ROW. Candidate modes, thus limited to those with A category ROW, are reasonably simple to compare since only guided technologies should be considered. An early example of this type of comparison was done during the 1950s for San Francisco BART. The requirement that the new regional transit system be independent from and competitive with highway traffic was mandated by the California legislature. The choice was therefore limited to four guided technology systems, and the choice of rail was logical (see Section 10.4.2). In other cases, if satisfactory street conditions and moderate projected transit demand clearly indicate that exclusive ROW is not absolutely necessary nor can be economically justified, LRT or BRT may be serious candidates that should be considered.

As the choice of modes became greater, comparative evaluations began to include both different technologies (e.g., rail, bus, proposed modes) and different ROW categories. The methodology for selection became more sophisticated for several reasons. First, due to their technical differences, each mode had to be fitted separately into the planned corridor; an optimally designed bus system usually has a somewhat different alignment than an LRT or metro system. Second, the differences among the candidate systems have become in some cases greater and more complex. Finally and most importantly, there are considerable differences among modes in their abilities to attract passengers by such characteristics as ROW categories B and A, rail technology, electric traction, and distinct stations (see Section 12.2.4.1).

Among the early studies of this type was Leibbrand's study of LRT, rapid transit, and monorail for Frankfurt, Germany, in the early 1960s (summarized in Leibbrand, 1967). Another very comprehensive comparative analysis of different technologies and types of operations was performed for Newcastle-upon-Tyne, England (Voorhees & Associates, 1972). Similar studies of mode comparison and selection were subsequently performed in a number of U.S. and foreign cities, such as San Francisco, Atlanta, Pittsburgh, Singapore, Strasbourg, and Ottawa (descriptions of several cases are given in Section 10.4.2). The tendency in most of these studies was still to base mode selection mostly on monetary costs, but the trend has been toward including an increasing number of nonmonetary factors.

During the 1970s, a comprehensive comparison of modes, known as alternatives analysis study, became federally required in the United States for all transit systems requiring federal participation in financing. These studies have become increasingly comprehensive and detailed. Examples of such studies in recent years are those performed for Baltimore, Los Angeles, Denver, Portland, Sacramento, Honolulu, Dallas, Houston, and Phoenix. These studies have produced a number of excellent conceptual definitions and methodologies for comparisons of modal characteristics.

11.2.2 The Basic Principles and Methodology for Transit Mode Evaluation

Several basic principles should be followed in most studies analyzing and comparing transit modes:

1. Each evaluated system/mode should be considered with comparable assumptions. For example, a comparison of a new bus system with a 20-year-old rail system and its original vehicles (or vice versa) does not give a valid comparison of the two modes.
2. If different technologies or modes are compared, each one should be optimized to the local conditions; for example, an AGT, LRT, and metro systems would have somewhat different operations, and possibly alignments, even in the same corridor due to their different characteristics.
3. Differences among candidate systems should be defined by categories such as those due to different technologies (guided versus steered, electric versus diesel), different types of operation (automated versus driver-driven), type of service (regular versus commuter), etc. Then each set of differences should be evaluated separately.
4. Attention must be given to the entire system: feeders in suburbs and distribution in central areas should be considered in mode evaluation.
5. The evaluation should include monetary and other quantitative, as well as qualitative items. In many cases, a transit system alternative that involves considerably higher costs is superior to the lower-cost alternatives due to the importance of nonmonetary objectives, such as performance, service quality, or strong image.
6. A transit system's ability to attract passengers and to influence land use development and quality of urban life should be included because they are often major goals.
7. The greater the differences among compared modes (e.g., if modes with different ROW categories are compared), the more comprehensive must be the evaluation procedure.

Each city, area, or corridor to be served by a new transit mode has a unique set of characteristics. For selection of the optimal transit mode, it is necessary to define all the site-specific conditions, requirements, and constraints, designated as a conditions set. This set may be considered as the demand side of the mode-selection process. On the supply side, the candidate modes must be specified with respect to their physical and operational characteristics, service quality, costs, and impacts, and one of them must be selected for the specified application. Thus, the selection procedure consists of three major phases: definition of the conditions set (demand), preliminary design of candidate modes (supply), and comparative evaluation and selection of the optimal mode. These three phases are described in the following sections.

11.3 DEFINITION OF THE CONDITIONS SET

There are a great number of diverse factors that should be considered in transit mode selection. These factors, which may have different relative importance among cities and served areas, can be classified in the following categories:

- Planning goals and objectives
- General route alignment and delineation of the service area: its physical characteristics, such as topography, climate, street network, major objects, etc.
- Economic, social, and environmental factors: cost aspects such as availability of capital and operating costs, as well as the trade-off between capital-intensive and labor-intensive modes.
- Volume and elasticity of potential transit travel demand, which depend on land use and population characteristics, as well as on availability and quality of competing transportation services

Based on each set of local conditions, requirements and standards for the planned transit system can be established. For example, a certain city's framework, defined by its physical characteristics and goals, will have an influence on the form of the transit network, number and locations of stations, and service quality required. Hilly terrain requires higher-performance modes or vehicles. Areas with medium- or high-income population may require higher-quality transit services but also allow charging of higher fares than transit service in low-income areas.

To allow a systematic evaluation of all modal characteristics, the local conditions must be disaggregated into a set of transit system requirements and standards. The number and definitions of requirements to be analyzed may vary considerably among different locations and types of transit systems. Traditionally, transit operators used to consider that to attract maximum number of passengers, offered service should be frequent, punctual, comfortable, and fast and have low fares. While these five requirements remain important for analysis of transit services, a much more comprehensive approach must be undertaken to ensure that the transit system is competitive with alternate modes, is efficient in operation, and has positive impacts on the area it serves. Therefore, a considerably greater number of requirements must be analyzed.

On the other hand, it is also important to keep the number of requirements limited, for two reasons. First, each requirement should be clearly defined and distinguished from others to minimize duplication. And second, many characteristics of transit modes cannot be precisely measured, so that their analysis in excessive detail has a diminishing value. Attempts to analyze transit modes through as many as 100 requirements have shown that there was a great overlap among them and that many of them could not be adequately defined and measured. Thus, the complexity of such analysis may not be justified by any detectable improvement in study results.

Due to the complexity of transit systems, their evaluation requires that the analyst have knowledge and experience with transit systems. That knowledge cannot be fully substituted for by a rigid, mechanistic methodology.

The requirements for the transit system are classified into three categories by three major affected groups or interested parties, which have somewhat different sets of requirements:

The first, passengers, are the most important party. They want to have good service at a reasonable price.

The second party, the transit system operator, must provide the service and meet a certain service quality level to attract passengers and at the same time maximize efficiency of operations, i.e., minimize costs for a given service quality.

The requirements of these two groups are interrelated but by no means identical. For example, the operator must make sure that each line provides adequate capacity, mechanical reliability, and labor utilization;

passengers are not concerned about these aspects as long as they enjoy service that is fast, reliable, comfortable, etc.

The third interested party is the community or city. This group represents city government and the entire population of the served area, including passengers. It is the most diversified group since it encompasses people living close to the transit lines, who feel their positive and negative side effects directly, as well as all citizens affected by the level of mobility transit provides, its impacts on land use patterns, etc. Finally, this group includes governments, most directly local, but also state and federal, which are, or should be, interested in promoting an economically and socially viable urban environment, quality of life, energy conservation, etc.

The set of specific requirements to be used in transit system evaluation, classified into three categories by the interested parties, is given in Figure 11.3. These requirements must be carefully defined to distinguish them from each other and make their evaluation possible. It is pointed out again that the same set of requirements may not be applicable to all situations nor to all times: community requirements may vary greatly. For example, prior to the energy crises of the 1970s, energy efficiency was not considered an important factor for mode evaluation in most U.S. cities. Since that time, this factor has usually been included in the criteria. Its importance also varies among countries, depending on their energy availability and price. Similarly, vehicle exhaust and the major advantage of absence of exhaust by electrically propelled transit vehicles in urban environment have become increasingly important, particularly since many cities introduced transit into pedestrian zones in central cities.

Despite these variations in the definitions and relative weights given to different requirements, the basic framework of this type of analysis has general validity for virtually all modes of transportation: with some modifications, one can analyze the suitability of walking, minibus, or helicopter for a three-block trip or for a long commuting trip across a metropolitan area. Naturally, the more precisely the conditions set is defined, the more detailed the analysis can be.

In the following section, each requirement listed in Figure 11.3 is defined and its values and significance are discussed.

11.3.1 Passenger Requirements

Passengers must be considered as the most important party since they are the customers for whom the system is provided. If their requirements are not met and transit service does not attract enough passengers, both the operator and the community suffer losses.

11.3.1.1 Availability. This requirement has two aspects:

1. *Locational availability:* Transit stops or stations must be reasonably close to both passenger's trip origin and destination.
2. *Temporal availability:* This is the period during which service is offered. Most transit lines should operate 18–20 hours per day. In major cities, a skeleton network, consisting of major lines, often operates 24 hours per day, while commuter transit is usually limited to 2–4 hours during the morning and afternoon peaks.

Availability is the basic requirement for every transit user: if it is not satisfied, people physically cannot use a transit system. For good availability, users must have both close stations and service throughout the day.

11.3.1.2 Frequency/Headway. High frequency (i.e., short headways) is a very important element for passengers because people dislike long waiting. For short, intraurban trips, passengers will often tolerate low comfort if the service is very frequent, so that they do not have to wait or orient their activities around transit schedules. For short urban trips, headways of up to 6 minutes are always fully adequate. Services with headways of 6–12 minutes are still satisfactory for urban trips up to 5–10 km. For longer trips, such as regional commuting trips for which passengers use

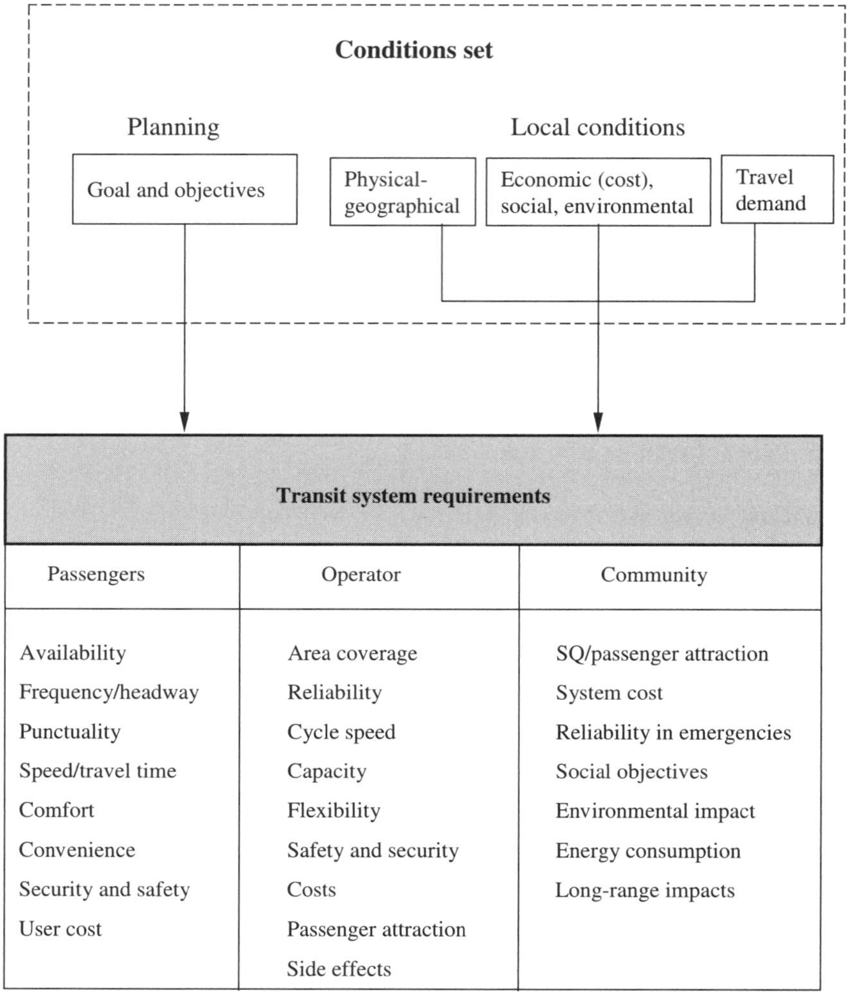

Figure 11.3 *Conversion of goals and local conditions set into transit system requirements.*

schedules, headways of up to 20 and even 30 minutes are fully acceptable. Headways over 30 minutes become rather unattractive, particularly for people who have an automobile as an alternative. Commuter transit satisfies the work trips of commuters, but its infrequent and irregular service cannot attract virtually any noncommuting trips.

To be attracted to transit, potential passengers should ideally have both reasonably close stations and frequent service. When demand is not very high, the operator must decide on the trade-off between the two. At one extreme, there exists a dense network with low frequency: such a system is not available for long intervals of time. At the other extreme, there is frequent service to few points; users far from a terminal do not have the service unless they use feeders. The optimal solution varies with local conditions, but generally passengers prefer higher frequency over shorter access: they accept considerable walking distance to arrive at a line with frequent service (see Section 4.2.1).

11.3.1.3 Punctuality. Also referred to as on-time performance or, in a broader sense, reliability of service, punctuality is measured as percent of vehicle arrivals within a certain interval after the scheduled time. Most transit systems define this interval for measuring on-time performance as a delay of 0–4 minutes for transit lines operating with headways of 5–30 minutes. For short headways, such as 1–5 minutes, this interval must obviously be shorter, such as 2–3 minutes, while hourly services and intercity public transportation may define this period as 5–15 minutes. On lines with short headways, however, passengers are less inconvenienced by service delays, except that vehicle overloading may occur.

Various surveys have shown that passengers rank punctuality very high on the list of attributes they require from transit service. Low punctuality is reflected not only in inconvenience it causes to passengers, but also in increased travel time. A good illustration of this is a person who has to arrive at an airport at a given time and has two options, a limousine and metro, both having the same travel time under normal conditions. If street congestion is likely to occur, the person will have to leave considerably earlier by limousine than by rapid transit because of its higher probability of encountering a delay.

Since traffic interference is by far the most frequent cause of delays, on-time performance is directly related to ROW category. Metro and regional rail systems usually operate with punctuality rates in the range of 97–99%. For semirapid transit, the range of 90–96% is more typical, while street transit varies widely, sometimes being as low as 75–80%. The latter two mode categories usually have a considerable difference between punctuality rates during peak and off-peak periods, while metro and regional rail are seldom affected by peaks.

11.3.1.4 Speed (Travel Time). Travel speed on transit lines, measured in km/h, is in many cases the most important single characteristic influencing modal split. However, each individual passenger considers the total door-to-door travel time, which consists of the times for access to transit stop, waiting, travel on vehicle, transfer, and departure. Thus, in addition to transit operating speed, several other service characteristics, such as network coverage and frequency of service, also influence this service element. Relative weights of these time intervals vary since passengers perceive them differently. Some passenger surveys indicate that a factor of 1.5–2.5 should be used for waiting and transfer times to obtain perceived times corresponding to on-vehicle travel times. The relative weight of walking time depends heavily on the attractiveness of the area, including aesthetics, security, topography, weather, etc.

11.3.1.5 Comfort. Broadly defined, comfort is an absence of mental and physical strain and the presence of pleasant experiences. For transit passengers, this factor incorporates a broad range of qualitative elements: from such physical factors as air conditioning to psychological ones such as image of a given service. The major elements of passenger comfort during an entire passenger trip are as follows.

Walking to and from transit stations is usually, but not always, a negative factor of comfort: people dislike long walking, particularly through an unpleasant environment or inclement weather. However, many people enjoy walking moderate distances through an attractive environment, such as parks or shopping streets.

Attractive stations in terms of appearance, weather protection, safety, etc.

Boarding/alighting comfort depends on time and effort needed to board or alight from the transit vehicle. Number and height of steps and width of doors are major influencing factors.

Availability of a seat during the ride is often the most important single component of transit passenger comfort. Standing is particularly objectionable for long trips and on highway and rubber-tired guided vehicles because of the greater sway they have than rail vehicles.

Dimensions of seats, together with their design, softness, etc., influence comfort. Typical area per seat in buses is 0.25–0.35 m^2, on light rail and rapid transit vehicles 0.30–0.40 m^2, while on regional rail it may be as high as 0.45 m^2.

Comfort and ability of standing passengers to move are determined not by design but by the policy on passenger density introduced as the load factor in setting schedules on transit lines. The degree of crowding (and discomfort) and ability of passengers to circulate in vehicles is measured by the average number of persons per square meter of floor area, or its inverse value, area per standing passenger. These values with the corresponding conditions of passenger comfort or crowding are given in Table 1.2 (Chapter 1).

Tests with actual loading of transit vehicles have shown that values below 0.20 m^2/prs are in many conditions physically impossible to achieve. Actually, the value of 0.15 m^2/prs (1.6 ft^2/prs) is sometimes used for *computation of structural strength* in designing vehicles. The most commonly used standard for computation of maximum vehicle capacity is 0.20 m^2/prs in heavily loaded transit systems, usually in very large cities. In most currently designed systems in industrialized countries the value of 0.25 m^2/prs is used.

Vehicle interior includes several elements of comfort, such as aesthetic appearance, cleanliness, and illumination. The minimum illuminance should be 215 lumen/square meter or lux (20 foot-candles). In reading areas, the illuminance should be between 320 and 430 lux (30–40 foot-candles). The exact ratio of these units is 1 lux = 10.76 foot-candles.

Temperature should most desirably be in the range of 16–25°C (61–77°F), but the human comfort zone is also a function of humidity: higher humidity lowers the range of tolerable heat. The recommended range of temperature-humidity is specified in publications of professional societies such as ASHRAE (American Society of Heating, Refrigeration and Air-conditioning Engineers). Passenger discomfort increases with deviation from this range and heating or air-conditioning become desirable, but their feasibility must be determined on the basis of duration of deviations from the comfort zone, passenger trip lengths, cost of energy and other relevant factors.

Ventilation is necessary by itself, as well as a means of assisting temperature control. Commonly used standards for ventilation, expressed by the volume of air circulated per minute per person, are given in Table 11.1.

Table 11.1 Level of comfort and volume of air circulation

Condition	*Ventilation Related to Load (m^3/min/prs)*
Desirable—very comfortable	>0.85
Acceptable level	0.70–0.85
Minimum tolerable	0.45

Riding comfort of vehicles is measured by their vertical, transverse, and longitudinal vibrations. It depends on transit system technology, type and condition of vehicle suspension and supporting way, route alignment, etc. It is desirable that vehicles not exceed acceleration of 1.5 m/s^2 (0.15g) in any direction. The maximum jerk (change of acceleration) rate should not exceed 0.7 m/s^3.

Since riding comfort varies considerably among different vehicles of the same mode depending on their design, age, and condition, comfort of different modes must be measured on vehicles of comparable age and condition.

Interior noise in vehicles and at stations can be annoying to passengers if it exceeds a certain level (for definition of noise and its characteristics and measurements, see Section 11.3.2). The commonly adopted noise level standards for passengers are listed in Table 11.2.

Modern transit vehicles are extremely quiet, particularly the electrically powered ones (rail vehicles and trolleybuses). For example, the noise measured at ear level of seated passengers in contemporary rail rapid transit vehicles running on tangent track with welded rails and wooden ties is between 60 and 65 dBA at a speed of 60 km/h, and it rises to 64–70 dBA range at

Table 11.2 Noise levels and their impacts

Condition	*Noise (dBA)*
Desirable—very quiet system	<75
Acceptable	75–85
Maximum tolerable	95

Note: for the definitions of noise measurement units, see the discussion of side effects in the next section.

100 km/h; thus, it is below the noise level considered acceptable.

Relaxation during the ride, instead of mental strain often connected with urban auto driving, is another element of comfort.

Psychological factors may also play a significant role. They include choice of company—either privacy or desirable social contacts, depending on the situation; a negative feeling while traveling on a dilapidated vehicle; or a feeling that transit is an important part of exciting urban living when using an attractive system, such the metros in Washington and Montreal, LRTs in Stuttgart and Strasbourg, or buses in Curitiba and Portland. Aesthetic elements of stations, vehicles, and surroundings, cleanliness, image of the transit system, its role in urban life, and many other elements represent psychological factors.

11.3.1.6 Convenience. This aspect of service quality is closely related to comfort since a lack of convenience often results in discomfort. For example, luggage space in a transit vehicle is convenient for passengers; if it is not provided, their riding comfort is reduced. The following are the main components of convenience.

- *Choice of alternative travel options* gives passengers an opportunity to select the mode or routing of travel that best suits their needs.
- *Information* about the system that is easy to obtain, understandable, and accurate.
- *Regularity of schedule,* allowing passengers to memorize departure times easily.
- *Use of travel time* for reading, writing, dictating, or just resting is an important element that practically decreases the net time loss for travel.
- *Availability of parking* at outlying transit stations is often the critical factor for modal choice of some potential users. For persons living in an auto-dependent part of the metropolitan area, P+R is the necessary condition to use transit when going to central city.
- *Direct travel or easy transfers* between routes or modes.
- *Ability to accommodate disabled persons,* which includes easy, safe, and fast boarding and alighting and accommodation of wheelchairs in vehicles. This aspect is legally required in many countries.

It is obvious that convenience includes many important elements. Since they are mostly qualitative, they must be evaluated descriptively.

11.3.1.7 Security and Safety. From the passenger's point of view, security and safety are very significant aspects of transit service. If they are not perceived to be very high, many people may be discouraged from using transit. Traditionally high security on transit systems deteriorated on some systems in the United States and some other countries during the 1970–1980 period, but it has been generally improved in recent years. Older, inadequately maintained systems tend to suffer from the problem of crime more than new systems designed for easy surveillance, especially if the new systems appear modern, high-quality, and heavily used, such as the Montreal Metro, Singapore Mass Rapid Transit, and Munich U-Bahn.

Safety refers to absence of accidents, which is ensured by safe operation of vehicles, adequate design of stations and stops (no slippery surfaces, adequate geometry and visibility of steps, protection of passenger islands, etc.), easy boarding of vehicles, good departure control, etc. On most transit systems in developed countries safety is considerably higher than on other modes. For other aspects of safety and security see Section 6.4.5.

11.3.1.8 User Cost. The cost of transit travel for passengers is expressed in dollars per trip or per year, depending on the type of analysis performed. It consists usually of only the fare. However, when a passenger has other out-of-pocket costs, such as parking at station, they must be included. If a commuter owns an automobile specifically for commuting to a station, total cost of owning and operating it represents part of the cost of travel.

Transit passengers accept fares if they are closely related to the value of service they use. However, travelers are much more sensitive to out-of-pocket than to indirect costs. Most automobile owners count only parking and gasoline expenses as costs of individual trips, while overlooking insurance, depreciation, maintenance, and other long-run costs. For this reason, transit fares must be competitive with marginal, rather than total, cost of automobile travel.

A composite of all the above-defined service elements affecting passengers, including quantitative ones (speed, cost, etc.) and others referred to as quality of service (convenience, image), comprises the level of transit service. An interesting method for defining different levels of transit service, which is very difficult due to the diversity of service elements and the ways they affect users, has been developed by Botzow (1974). By far the most comprehensive analysis of pedestrian characteristics and pedestrian levels of service was done by Fruin (1971).

11.3.2 Transit Operator's Requirements

While the system must be designed and operated so that it meets passengers' requirements, its operation must be economical and technically efficient. To achieve that, a transit operator has a set of requirements that differ somewhat from those passengers have, although some items overlap but have different forms. Examples of such overlapping items are punctuality and reliability, travel speed and cycle speed, and safety. Operator's requirements are defined here.

11.3.2.1 Area Coverage. Primary transit coverage is usually defined as the area within a 5-minute access distance (see Sections 4.1.1 and 10.4.1) from transit stations and stops. Secondary coverage includes areas between 5- and 10-minute walking distance. Only regular transit lines should be considered for this coverage. If there are a number of commuter transit lines not operated during off-peak hours, two different coverages should be analyzed separately: regular transit coverage and peak-hour coverage.

A more precise measure of transit availability through area coverage is obtained by computing population of the served area as percentage of the total urban area population.

11.3.2.2 Reliability. The strict definition of reliability is one minus probability of failure. However, failure in transit systems is complex and difficult to define. It can be of several types and must be measured by its consequences: persons affected and time lost, material losses, or, the most frequent measure, time needed for recovery of regular, on-schedule service.

While passengers experience transit system reliability through punctuality of service, the operator analyzes this characteristic of his system through several of the following features.

Interference from other traffic is in most cases the dominant cause of low reliability of transit service. Use of ROW categories B and A is the most important factor for operational reliability on a transit line.

Furthermore, reliability is related to technology, particularly with respect to inclement weather. Rail systems have a considerable advantage in this respect over rubber-tired vehicles, both steered (buses) and guided (e.g., rubber-tired rapid transit), since their adhesion is much less affected by rain, snow, and ice and snow removal is more easily done on rail tracks than on streets.

Breakdowns of transit operation are defined as irregular and unpredicted delays in operation that are much longer than regular variations in travel time caused by traffic interference, passenger boarding, etc. Frequency of breakdowns is measured by their number per unit time (usually day or week) and depends mostly on three factors:

- Technological and operational complexity of the system: greater complexity always increases frequency of breakdowns, *ceteris paribus*
- Quality of design and manufacturing of the system and its components
- Operating and maintenance practices and competence of personnel

Technical reliability of vehicles is usually measured by the mean time (or mean distance) between failures

(MTBF or MDBF). See Section 6.3. Frequency of breakdowns on the system depends largely on this factor, but it also includes other factors such as operational breakdowns in stations, failures in the control system, inadequate field supervision, and communication impedances.

Each breakdown represents a problem in transit operation, and it is important that the system be capable of minimizing its consequences. In particular, there must be a possibility of restoring operation by removing the vehicle that breaks down or rerouting the service around the area of delay. In this respect, diesel-powered buses have an advantage over electric rail systems: breakdown of a bus does not affect other buses; breakdown of a rail vehicle or electric power failure delays other vehicles or the entire line, respectively.

Reliability is a very important feature of transit systems. If it is unsatisfactory, an operator faces considerably higher operating costs, as well as losses of ridership.

11.3.2.3 Cycle Speed and Line Capacity. While the operator must provide adequate operating speed to attract passengers, his/her costs are directly dependent on cycle speed of the system. For definitions of speeds and their importance for the operator, see Section 1.1.

To meet passenger demand without delay and discomfort, the operator must provide adequate capacity on transit lines. A definition of capacity and a description of its relationship to speed, safety, and comfort of transit systems are given in Section 2.1.

11.3.2.4 Flexibility. Defined as ease or inexpensiveness of adapting a system to change in conditions or operations, flexibility of transportation systems can refer to a great number of their characteristics. Thus, a transportation system may be flexible with respect to space (degrees of freedom of movement, way flexibility, number of terminals, extensiveness of network), time (possibility of changing the system through time), or operations (flexibility of routing, schedule, capacity, and type of control). Because of this diversity of meanings, the concept of flexibility should always be further specified: a statement that "transit mode X is flexible" is poorly defined and often misused.

In most cases, higher flexibility of a mode is related to lower efficiency in some sense. Thus, dual-mode (diesel/electric) bus vehicles have greater flexibility in using rights-of-way with and without electric wires, but they are much more complex, heavier, and more expensive than either buses or trolleybuses; or a system designed for possible modification at a later date usually involves a higher total investment. Consequently, flexibility is almost always connected with trade-offs in other characteristics, which should be included into evaluation together with the feature of flexibility.

It is very common to hear that flexibility of transportation systems is implied to be a very positive feature. This is incorrect, not only because the trade-offs of flexibility must be considered, but also because flexibility may be a negative feature. For example, to have positive impacts on land uses and economic development, a transit system must be permanent and fixed, rather than flexible, i.e., temporary. Regular and reliable schedules of transit service are more desirable than flexible (irregular) ones, except when demand is very low or service is tailored to special groups of users.

The fact that flexibility as a concept may be positive or negative is clearly shown by a review of concepts that are contrary to flexibility in many applications. They are, for example, permanence, reliability, regularity, and predictability. All of these concepts are in most respects positive system characteristics. As will be discussed in Section 12.2.1.1, flexible transit systems are inferior to permanent, reliable ones in many respects and many applications. Thus, most aspects of traditional bus systems that must be changed to upgrade those systems into BRT are exclusive busways, regular headways, and all-day service, rather than flexible use of lanes, changeable schedules, and operations during some daily periods only. Clearly, many bus services can be upgraded by reduction in their flexibility and increase of their permanence and service reliability.

In conclusion, if flexibility of transportation systems is analyzed, it should be defined precisely (e.g., flexibility in expanding system capacity) and considered together with its trade-offs, such as higher operating costs, lower safety, etc. A detailed discussion of this concept is given by Vuchic (1971).

11.3.2.5 ***Safety and Security.*** Safety of a transit system is measured by statistical records of the number of accidents, fatalities, injured persons, or material damage per 10,000 or 100,000 vehicle-km, per 1 million passenger-km, or on a similar basis (see Section 6.4.5). The transit operator's major concern with respect to safety is to ensure that operating accidents do not happen. Prevention of vehicular accidents is the most important requirement, and well-managed systems never compromise on it.

Safety of street transit modes (buses, trolleybuses, and streetcars or trams) operating in mixed traffic depends largely on the conditions in the streets where they operate, although various traffic regulatory measures and good training of drivers also play a very important role. Operations on ROW category B have considerably higher operating safety than running on streets, but by far the highest safety is achieved on ROW category A by metro and other electric guided transit systems with full signal control and fail-safe devices. These systems actually have the highest safety record of all passenger transportation systems. Many metro systems have operated for several decades without a single train accident.

Passenger accidents may also occur during boarding or alighting or on board because of high steps, door failures, floor condition, or vehicle jerking. Safety must therefore be carefully considered in vehicle design, maintenance, and operation (e.g., supervision of doors during departures from stations).

Security, measured by the number of crime incidents per unit of time or volume of travel, is sometimes a serious problem in many cities, and it has led transit operators to undertake many crime-prevention and control measures. Consideration of security requirements during the design and construction of a transit system, particularly stations, vehicles, and P+R lots, can greatly increase a system's security and reduce the cost of crime control on it.

Inadequate security can result in serious losses of passengers. Around 1980, many transit systems, including New York, London, and, to a lesser degree, Paris, suffered from a wave of vandalism and crime, causing major passenger losses. When special efforts were undertaken to curb these phenomena, ridership returned. Thus, since the late 1990s, New York transit has carried record numbers of passengers due to increased security compounded by improved service reliability and introduction of new fare cards allowing intermodal transfers.

11.3.2.6 ***Costs.*** Total cost of providing transit service consists of capital and operating costs. In recent years, capital costs for transit and other urban transport modes have been increasingly supplied by public sources (various levels of government) while the operator has been directly affected by operating costs only. In evaluating different modes, both capital and operating costs, expressed as unit costs ($/sp-km or $/prs-km), are usually considered.

As discussed in preceding sections, transit system cost, particularly investment, is by no means the only evaluation criterion. Yet cost is always a major, and sometimes the single most important, criterion of transit system evaluation. Regardless of where the funds are coming from, the transit operator must be interested in securing optimal use of funds through maximum efficiency of operations.

11.3.2.7 ***Passenger Attraction.*** The number of people who use a transit system is a function not only of potential demand, but also of the degree to which it satisfies the passenger requirements defined in Section 11.3.1. In addition to that, however, the operator may stimulate patronage by making his/her system known and easy to use. This includes such aspects as marketing, public relations, and the physical image of the system. Visual distinction of transit lines is very important. Similar to the stronger attraction freeways have in comparison with city streets (not only by higher level of service, but by their distinct image and easy orientation), transit modes on separate ROW have a much stronger image and attraction of passengers than transit services on streets. Rails and even overhead wires for trolleybuses also give such distinction. Image of bus services is particularly deficient in this respect, except where there are exclusive bus lanes.

Passenger attraction of planned transit systems is difficult to predict quantitatively. However, experienced transit planners and operators are aware of this factor and consider it in comparing modes.

11.3.2.8 Side Effects. This is a broad category of items. It includes all physical and aesthetic impacts of the system affecting both system users and nonusers, including persons in the immediate vicinity of the line as well as the entire city. Despite their importance, side effects are often given inadequate consideration. The problems resulting from this neglect led to the adoption of legislation requiring that for each public project an environmental impact statement (EIS), a comprehensive analysis of all projected side effects, including environmental, physical, social, etc., be provided prior to project's evaluation. Major factors in this category that the operator must consider are briefly described here.

Physical effects of transit ROW and stations may include upgrading and renovation of the corridor or, on the negative side, division of neighborhoods, deterioration of properties because of obstructed view, poorly maintained structures, etc.

Aesthetic effects may be negative (old structures and vehicles) or positive if an attractive system is successfully incorporated in urban design. The aesthetic experience of passengers should not be overlooked either. Thus, elevated rail transit structures have often been criticized because of inherited preconceptions of unattractive structures built many decades ago. However, if modern design is applied with much lighter, open aerial prestressed concrete structures, these facilities may be more attractive for passengers than travel through tunnels.

Noise, or undesirable sound, may have a negative impact on persons and property adjacent to a transit line. Noise represents pressure waves propagated through the air, caused by surfaces vibrating at frequencies within the range the human ear can register. It is measured by its amplitude, frequency composition, and duration. Since perception of sound is not linear but logarithmic, sound pressure levels are measured in different frequencies of the audible spectrum by a logarithmic unit designated as a decibel (dB). A pressure of 2×10^{-4} microbars or $2 \cdot 10^{-5}$ N/m^2 (10 microbars = 1 N/m^2 = 1 Pa), being approximately the weakest sound human ear can perceive at a frequency of 1000 Hz, is usually adopted as the reference sound pressure: $P_0 = 0$ dB. Other sounds are then measured as logarithms of the square of their ratios to P_0. Thus, sound pressure level (SPL) for a pressure P would be computed as:

$$SPL[\text{dB}] = 10 \log(P/P_0)^2$$

$$\text{dB } re\ P_0 = 2 \times 10^{-5}\ N/m^2 \qquad (11.5)$$

The sound created by two vehicles, each one producing pressure P, is:

$$SPL[\text{dB}] = 10 \log 2(P/P_0)^2$$

$$= 10 \log 2 + 20 \log (P/P_0) \qquad (11.6)$$

For example, two vehicles producing 80 dB each together produce a noise of 83 dB (*not* 160 dB), which is an increase barely perceptible by most humans. It is evident that the loudest vehicle dominates, and each additional vehicle makes only a small contribution to the overall sound level.

The SPL is often measured by so-called A-weighting, i.e., the relating of sound to the note A (440 Hz), designated as dBA. The measuring instruments use this to weight the amplitude of sound as a function of frequency to approximate the response of the human ear. This measurement is used mostly for nondirectional sources outdoors. (For description of noise measurement techniques see Peterson and Gross, 1972.)

In comparing noise impacts of different transit modes, the duration and intensity of the noise must be taken into account. If locations of the compared modes are different, the number of persons exposed to each one must be considered. Furthermore, a thorough analysis, such as required for EIS analyses, would account for the pattern of noise intrusion over the course of the day and night, including any reductions in automobile traffic due to traffic reduction when a new transit line is built. This can be done by time averaging over the course of the day, calculating an equivalent level L_{eq}. For existing facilities, instrumentation can be placed in the field that will collect and process the data automatically. For planned facilities, the noise can be estimated analytically using the current situation as a reference and introducing the projected changes. There are several variations on L_{eq} in use, depending upon local conditions and regulatory requirements.

Buses are often major offenders as noise producers on urban streets, particularly when they have frequent stops. Applying maximum engine rotation for acceleration, they generate a high level of noise while in the close proximity of pedestrians. Most of the bus noise is produced by the fan, engine, transmission, exhaust, and, at higher speeds, tires. Diesel engine noise levels depend mostly on the engine speed and on how the exhaust is muffled. The noise produced by buses with reduced fan noise and good exhaust control is quite acceptable. However, since these controls reduce performance and increase fuel consumption of buses, many agencies tend to use the efficient version of buses, with noise levels sometimes as high as 90 dBA. Yet the trend has been toward production of quieter buses, particularly in the developed countries.

The trolleybus, being electrically powered, has virtually no motor noise. When it is well maintained, it produces only tire and very slight trolley pole noise. The trolleybus is therefore always much quieter than a diesel bus, and it is usually the quietest of all transit vehicles.

Another factor influencing noise production of all highway vehicles is the condition of pavements: well-maintained asphalt surfaces are usually the best, while worn-out concrete surfaces and particularly potholes in roadways are the worst with respect to noise and vibration production.

Rail vehicles, in their early stages, used to be known by the high level of noise they produced. In recent decades, however, this problem has been practically eliminated through advanced technologies used in vehicle design and in the now-standard use of welded rails and switches with flexible points without joints, insulation in rail supports, special tunnel wall surfaces, etc. On tracks with welded rails and insulated base, rail vehicles are extremely quiet. Typical modern LRT vehicles are so quiet that their humming while operating at low speeds in transit/pedestrian malls (Calgary, Canada; Mannheim, Germany; Sacramento, California) is not objectionable for the pedestrians. At higher speeds, up to 80 km/h, both LRT and rapid transit systems produce noise well below 80 dBA.

The diagram in Figure 11.4 was computed for the San Francisco BART, a representative of modern high-speed regional rapid transit systems. It shows that the hourly equivalent noise level for 6-car trains operating every 20 minutes in each direction (a total of 6 trains passing per hour) varies from 53–59 dBA for the speeds of 65–130 km/h. The equivalent noise level increases to the range from 64 to 69 dBA for 2-minute headways (60 trains passing/hour).

Table 11.3 presents some other very interesting results of the study of noise on the BART system. Using the six-car BART train as the baseline case with its L_{eq} value, several other cases were measured: different train lengths, from 2 to 10, running on aerial structure, train going through track switches on the ground and on aerial structure, and train running through a mild curve. The table shows the differences in noise produced in these cases. It is interesting that track position on aerial structure, track curve, and switch has as great an influence as the increase of train length from 2 to 10 cars: each of them increases noise by 5 dBA.

Sound barriers reduce rapid transit noise appreciably, although they sometimes may increase the sound levels at a greater distance from the tracks.

It is sometimes believed that rubber-tired guided systems are much quieter than conventional rail systems. This is a misconception originating from old rail technologies. Measurements of modern systems have shown that noise levels of the two technologies, rail and guided rubber-tired, are very similar: the original rubber-tired lines of the Paris Metro, built from the late 1950s into 1960s, were considered to be very quiet, but noise levels on the Berlin and Hamburg U-Bahn systems were found to produce even lower noise levels (Operations Research, 1964). Subsequently built systems, such as the Montreal Metro with rubber tires and concrete running surfaces, Mexico City with rubber tires on steel surfaces and Munich with rails, have been improved even further. However, the relationship in noise levels between rubber-tired and rail vehicles have not changed significantly.

Vibrations may have a negative impact on both persons and property adjacent to a transit line. In extreme cases, vibrations may cause noticeable shaking of

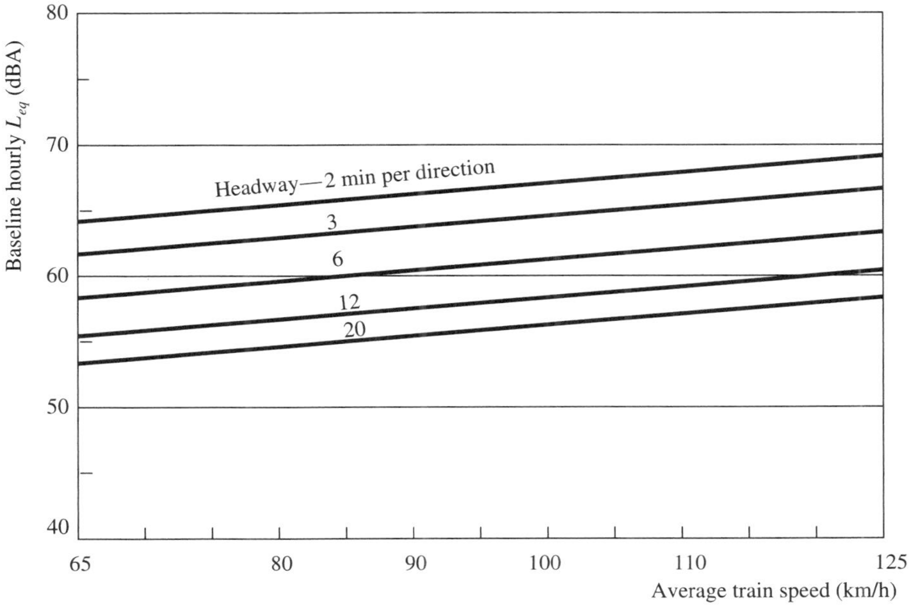

Figure 11.4 *Hourly equivalent noise level as a function of train speed and headway.*

Table 11.3 Impacts of train lengths and track elements on noise level: BART system

Train Length and Track Characteristics	*dBA Changes*
Two-car train	−3
Four-car train	−1
Eight-car train	+1
Ten-car train	+2
Aerial structure	+5
Switch on berm or grade	+3
Switch on aerial	+5
Curve ($R < 1500$ m)	+5

Baseline case is 6-car train running on wooden ties, ballast track, straight level alignment.
Source: Gruen Associates and DeLeuw, Cather & Co., 1979.

buildings. Beyond a certain amplitude, vibrations contribute to eventual structural damage of buildings through fatigue.

The vibration caused by vehicles running on streets depends very much on the condition of the road surface or rail tracks. For example, buses running along streets with many potholes can create major shocks and vibrations in the surrounding buildings. Similarly, rail tracks without adequate insulation can create vibrations. Modern tunnel and aerial structures for rail systems transmit very low vibrations to the surroundings. Thus, vibrations are minimized not only by modern vehicle design, but also through careful design and good maintenance of rights-of-way, streets, rail tracks, and other types of guideways.

Air pollution is even more difficult to measure than noise because different emitted gases have different impacts on humans and on the environment, buildings, historic monuments, etc. Moreover, their dispersal (or concentration) depends on climatic and geographic conditions. On the other hand, pollution produced per person-km can be obtained by dividing vehicle pollution by its occupancy; this cannot be done so easily for noise because of its logarithmic measurement.

Comparison of modes with respect to air pollution must be based on the type and quantity of pollutants produced by vehicle per passenger-km during its travel in the city, which depend on the following factors:

- Transit mode: propulsion system and type of vehicle

- Regime of vehicle travel (speed, frequency of stopping)
- Vehicle occupancy

With respect to *propulsion,* urban transportation modes can be classified into three general types: *electric,* used by trolleybuses and all rail modes except some regional rail systems; *diesel,* propelling buses, trucks and some regional rail systems; and *gasoline,* used by private automobiles. Electric propulsion is Opollution-free in the city (power-generation plants are usually far from concentrations of people). The latter two propulsion types utilize internal combustion engines (ICEs), which produce several kinds of pollutants. Their exhausts are delivered on streets, at stations and terminals, and in parking facilities, where they directly affect people.

A number of factors influence compositions of exhaust that highway vehicles produce: type of engine and fuel, vehicle age and maintenance, pollution-control devices, etc. In many countries, there are very strict laws stipulating the quality of fuel (e.g., lead-free gasoline, ethanol fuel and propane for buses). Moreover, considerable changes in ICE characteristics have been introduced in recent years, and a number of new propulsion systems and types of fuel have been under development and testing. It is therefore difficult to quote either average values or even ranges of emission components that would be useful for computations of pollution in a specific situation. An estimate of average pollutants production on work trips by different highway modes, given in Table 11.4, should be used more to see the relative pollutants production than their absolute amounts.

The table shows that buses (diesel ICE) produce more nitrogen oxides of various forms (NO_x) than automobiles (gasoline ICE). However, automobiles produce more carbon monoxide (CO), known to be among the most harmful pollutants, than buses. Finally, automobiles are also much higher in the production of hydrocarbons (HC) and sulfur dioxide (SO_2).

Generally, the average values, such as those in Table 11.4, are higher for buses in off-peak hours (their ratio to the single-occupancy vehicle (SOV) values is somewhat lower) because bus occupancies are much lower at those times. On the marginal basis, however, every switch of a passenger from SOV to bus results in a reduction of all pollutants produced by the auto because the bus pollutants do not increase; during the peaks, shift of passengers to buses may require addition of buses, slightly lowering the marginal reduction of pollutants.

Regime of travel has a strong effect on both the quantity and composition of emitted pollutants: slow, interrupted travel on congested streets results in much higher production of pollutants than moving at constant higher speeds. Congested CBD streets therefore usually have the highest degree and the most harmful types of pollution, which affect the greatest number of people. Separation of transit vehicles from other traffic in the CBD, resulting in an increase of their speed, is therefore also desirable for reduction of air pollution.

Vehicle occupancy is a major factor that makes transit modes far superior to auto travel with respect to air pollution. Occupancies vary considerably. Buses in medium-sized cities, for example, may typically perform in the range of 10–40 prs-km/veh-km during off-peak periods, while this occupancy range increases to

Table 11.4 Air pollutants emitted by different modes on commuting trips (grams/100 prs-km)

Mode	*Nitrogen Oxides*	*Carbon Monoxide*	*Hydrocarbons*
Transit bus[a]	95	189	12
Vanpool[a]	24	150	22
Carpool[a]	43	311	43
Single-occupant auto	128	934	130

[a]National averages for commuting trips.
Source: RPA, 1991.

25–60 during the peaks. Typical peak period occupancies for private automobiles in U.S. cities are 1.1–1.2 prs/car; in off-peak periods they are slightly higher. The average vehicle occupancy of buses in urban travel is thus on the order of 7–30 times higher than auto occupancy. The specific values depend on the direction of travel, time of day, bus scheduling, etc.

Consequently, the only way to provide transportation service in an area without creating vehicular air pollution is through the employment of electric vehicles. This was done several decades ago when Boston oriented its transit in the CBD mostly to electric vehicles in tunnels (rapid transit and light rail); the same goal resulted in creation of pedestrian malls with light rail lines in many European and North American cities (e.g., Amsterdam, Buffalo, Kassel, Mannheim, Portland, Sacramento, Zürich). However, every policy leading to a significant shift from auto travel to buses also results in a major reduction of air pollution per person-km of travel.

There have been arguments that a comprehensive analysis of total pollution caused by transportation should encompass pollution caused by power plants for electric energy generation (if they are oil- or coal-powered) and oil refineries for producing diesel fuel and gasoline. In a broader sense, pollution caused by the strip mining of coal, oil spills, or spills from land transportation of fuels should also be considered. These studies are, however, extremely complex and cannot be performed with a satisfactory degree of reliability.

Even the most comprehensive comparison of air pollution generated by ICE as compared to electric propulsion is, however, made irrelevant when one considers actual impact of that pollution on people. The pollution generated by buses is far more damaging since it is emitted directly into the air that great numbers of people (pedestrians and vehicle occupants) breathe. Electric power plants and oil refineries are generally located outside of populated areas, and their exhausts can have good pollution control, thus affecting a much smaller number of people to a far lesser degree. Diesel vehicles are therefore distinctly inferior to electric vehicles with respect to air pollution and its impacts. This difference is also obvious from the fact that use of tunnels for passenger transportation in cities is possible for electric vehicles. Diesel buses cannot use tunnels because of their exhaust (as well as inefficiency of steered vehicles on narrow paths).

11.3.3 Community Requirements

The most important requirements of this diversified interested party overlap largely with the requirements of the other two parties, the passengers and the operator. Above all, the three parties are united in requiring that the transit system offer suitable service. However, with respect to minimizing negative side effects, community requirements are usually more stringent than those that the agency imposes on its own system. Finally, goals such as long-range impacts of the transit system are held only by the community; neither passengers nor the operator is directly interested in them. The major requirements of the community are defined here.

11.3.3.1 Service Quality/Passenger Attraction. The basic community requirement is usually that the transit service must be available to all residents and be of such quality that it attracts many passengers. Achieving a maximum number of transit passengers, representing high mobility and reduced traffic loads on highway systems, should always be an important goal of the community. This requirement therefore coincides with the level of service requirements of passengers and of the operator, although in an overall sense, rather than in detailed elements such as speed, reliability, comfort, capacity, etc.

11.3.3.2 System Cost. Each transit system can be considered as a performance-cost package (PCP). Evaluation of each mode must be made in terms of this ratio of its performance and positive impacts to its costs. The dominant items among costs are the monetary ones, consisting of investment and operating costs, but ROW acquisition problems and potential negative impacts on the surroundings must also be included.

Most of the major investments in transit are provided by governments, i.e., from public sources. The public share of operating costs is also substantial in most countries. Thus, it is obvious that the community is directly interested in system costs.

The most important cost item for system evaluation is total (lifetime) unit cost, i.e., depreciation of capital cost plus operating cost per unit product (person-trip or person-km), also considering all differences in service quality of alternative modes. In addition to total cost, investment cost must also be analyzed by itself. The basic questions that the community (i.e., government) must decide are, first, whether the required investment capital is available, and second, which alternative investment—among transit projects, or sometimes including other possible investments—is most desirable, considering how each one satisfies its respective requirements. A conceptual framework for this analysis was given in Section 11.1.

If two compared modes or alternatives in transit system planning provide similar service quality, the community prefers the one with lower total unit cost. There are, however, usually two trade-offs to consider in such comparisons. First, there is a trade-off between investment and operating costs: a system involving higher investment should require lower operating cost, assuming comparable service is provided. The decision between different combinations of these costs depends, among other factors, on availability of capital funds. And second, alternatives usually have different PCPs: a mode that requires higher cost offers higher performance than another mode. The decision is then based on evaluation of whether the additional number of passengers attracted by a higher mode performance and its other positive impacts are worth the additional cost.

11.3.3.3 Reliability in Emergencies. In addition to its general interest in having a reliable transit service, the community must specifically ensure that its total transportation system is capable of providing a basic service in cases of emergencies. Several types of emergencies must be considered.

Reliability in inclement weather is a significant feature of transit systems, particularly in cities that have considerable rainfall, snow, and ice conditions. If in such cities transit consists of surface buses only, a snowstorm affecting highways often paralyzes the entire city. Total economic and social loss from two or three days of lost production in the entire city every year may alone be sufficiently great to justify a major investment into a transit system with exclusive rights-of-way, which represents the basic prerequisite for greater reliability.

Other types of emergencies are generally less frequent and predictable, but they may be much more severe. These include natural catastrophes such as earthquakes, floods, major fires, hurricanes, major economic crises, energy shortages, and civil disorders. Since 2001, international terrorism has been added to this list as a particularly serious danger since transportation objects have become likely targets for attacks. Since these events seldom happen, they are often disregarded in urban transportation planning and operation. However, this is usually an extremely costly oversight when such an emergency occurs.

Most emergencies impose a major role on a transit system that is capable of providing adequate basic transportation (i.e., the required speed, capacity, and reliability) throughout the urban area. Measures for keeping a basic system of highway routes operational are also important. In general, a transportation system consisting of two independent modes can always better respond to emergencies than a unimodal system. This is particularly the case when the private automobile is the only system because the highway/auto system requires most resources, it is vulnerable to shortages, and it is private, so that it is available only to vehicle owners, rather than to public at large.

Since presently most U.S. cities have an unbalanced reliance on the private automobile and their transit systems have limited excess capacity, a serious emergency could practically immobilize urban areas. The 1973/74 energy crisis gave only a minor indication of this possibility.

Labor strikes may also represent a negative element in long-term reliability of transit services. Labor-intensive transit modes, such as buses, are more vulnerable to this problem than rail and AGT modes.

11.3.3.4 Social Objectives. The basic social objective, which translates into a requirement for transit systems, is that all major population groups, particularly those who do not own or drive cars, have an acceptable means of transportation. Other objectives may include reasonably priced commuting, increasing mobility in economically depressed sections of the city, providing access to recreational and cultural facilities, and providing services for masses attending special events.

11.3.3.5 Environmental and Energy Aspects. Environmental aspects, referring primarily to aesthetics, noise, and air pollution, were discussed in the preceding section. They usually affect people in the vicinity of transit lines as much as they affect passengers. Therefore, desirable aesthetic features and minimum negative impact of noise are requirements of all three groups: passengers, operator, and community. Absence of air pollution primarily concerns the community, and a requirement to minimize it falls on the transit operator.

Energy efficiency, resulting in direct savings to the operator, also concerns the community through the total national goal for energy conservation. Public policies and federal laws in this respect are often bypassed and contradicted in actual developments. For example, very low price of gasoline stimulates driving in single-occupancy vehicles instead of carpooling, and SUVs are not subjected to federal standards on fuel efficiency. Yet it is likely that the trend is toward increasing importance of energy efficiency in the future.

11.3.3.6 Long-Range Impacts. Although difficult to define precisely, the impacts of a transit system on an urban area's economy and land use may in some cases be the dominant goal in long-range transit planning. Even a cursory analysis of form and character of different cities shows that in the long run, a good transit system strongly influences land use distribution and fosters the development of a healthy CBD, various metropolitan subcenters, and attractive suburban residential areas. It permits many different types of development by providing adequate mobility among them.

Permanent facilities of any mode (rail transit stations, bus terminals, or freeway interchanges) always make intensive development likely to take place in their vicinity. However, since a number of other factors also influence land use distribution and shaping of urban form, there is no guarantee that the most desirable development will automatically happen as soon as a transit system is built. In some cases, the desired subcenters will not be built or their physical form will not be well coordinated with a transit station; in others, overbuilding may occur at stations in residential areas where it is not desired.

The only way of ensuring that the great potential for positive long-range impacts of a transit system is achieved is to plan and control land use development together with the transit system and thus ensure full mutual coordination of the two.

11.3.4 Selection of Requirements

Many of the requirements for transit system performance included in the preceding sections are qualitative and difficult to define precisely. Some partially overlap. The definitions are therefore neither all-inclusive nor equally applicable for analyses of all transit modes. Their precise definitions and relative importance vary somewhat with the type of modes and their prospective applications. For example, punctuality and comfort have much less significance when different modes are evaluated for CBD shopper routes than when mode selection is made for a regional transit network; capacity is not very important for suburban services, while it is critical for heavily traveled radial lines in central city.

Additional differences are caused by the character of the planned transit system: is it an individual line, several lines, or a major urban network? Also, the character of the city changes the importance of requirements. For example, side effects, such as aesthetics, noise, and air pollution, have greater importance in a city attractive for tourists, such as San Francisco, Blackpool, or Salzburg, than in industrial cities. Or a fixed limit may be imposed on investment funds, giving this factor an infinite relative weight relative to other factors.

The requirements listed in Figure 11.2 should therefore be considered a general guideline or a checklist of items, rather than an absolute or exact set applicable to all transit system evaluations. For each specific analysis, the requirements should be examined and some or all of them used with their appropriate relative weights.

11.4 FORMULATION, COMPARISON, AND SELECTION OF CANDIDATE MODES

Traditionally, transit modes were defined by their popular image: buses, streetcars or trams, metro, etc. These definitions were based mostly on the technological characteristics of modes: guided (rail) and steered (bus) vehicles, diesel-powered and electric, etc.

Numerous technical, infrastructure, and operational innovations of transit systems in recent decades have resulted in a creation of a diversified family of transit modes (see Vuchic, 1981, Chapter 2), which vary among themselves in many characteristics. This diversity in many features requires that modes be defined analytically, based on their three generic characteristics: right-of-way (ROW) category R, technology T, and type of service/operation S. Each transit mode consists of a different group of these characteristics; thus, each mode can be defined as an RTS set.

These basic elements of transit modes and their influence on characteristics of modes are discussed further in Chapter 12, which focuses on the descriptions and comparison of actual transit systems and modes. This section explains the conceptual procedure for formulation of alternative transit systems and their selection.

11.4.1 Formulation of Alternative Modes or Systems

As explained in Section 11.3, in transit planning the conditions and requirements for the planned system, referred to as the conditions set, must be defined first. Once they are evaluated and translated into a set of transit system requirements, general characteristics of the planned system, consisting of performance (capacity, speed, reliability, and others) and level of service (operating speed, comfort, image) are known and are used to select the candidate transit systems for the specific planned project.

This procedure for development of candidate modes or alternative plans is shown schematically in Figure 11.5. In the first steps, the ROW category, technology, and operation/service, or RTS, elements are selected on the basis of system requirements. They are mutually interdependent. The basic one is ROW category, which influences the choice of technology and operations.

Two examples will illustrate this step in planning. If a capacity of 7500 sps/h, operating speed of 20–25 km/h, and rather high service reliability are required, the candidate systems should be medium-performance modes with ROW category B or A. The logical candidates may be some variations of LRT, BRT, and AGT modes. In another case, line capacity of 18,000 sps/h, operating speed of 35 km/h, and stations in several major activity centers along a major corridor are required. These requirements clearly show that a system with fully separated ROW is needed, which then determines that a rail system is the logical technology choice. Several designs of rail rapid transit, with variations in station locations, train lengths, car types, degrees of automation, etc., should be developed as alternative systems.

These two examples show that the choice is sometimes among alternative plans that use different transit modes, while in some cases alternatives represent variations in elements of the same mode. This procedure therefore refers to the selection of modes or systems, respectively.

For each of the alternative modes/systems, a preliminary design must be made for the planned application. Due to the differences among modes, each alternative's specific design may vary with respect to the number and locations of stations, type of operations, and even to some degree exact alignment in the planned served corridor. This preliminary design results in the preparation of several candidate modes, or alternative systems, which should be compared for the final selection.

11.4.2 Mode Evaluation and Selection

The next phase of the mode-selection process is to evaluate each candidate mode with respect to each requirement from the prepared conditions set. Type and

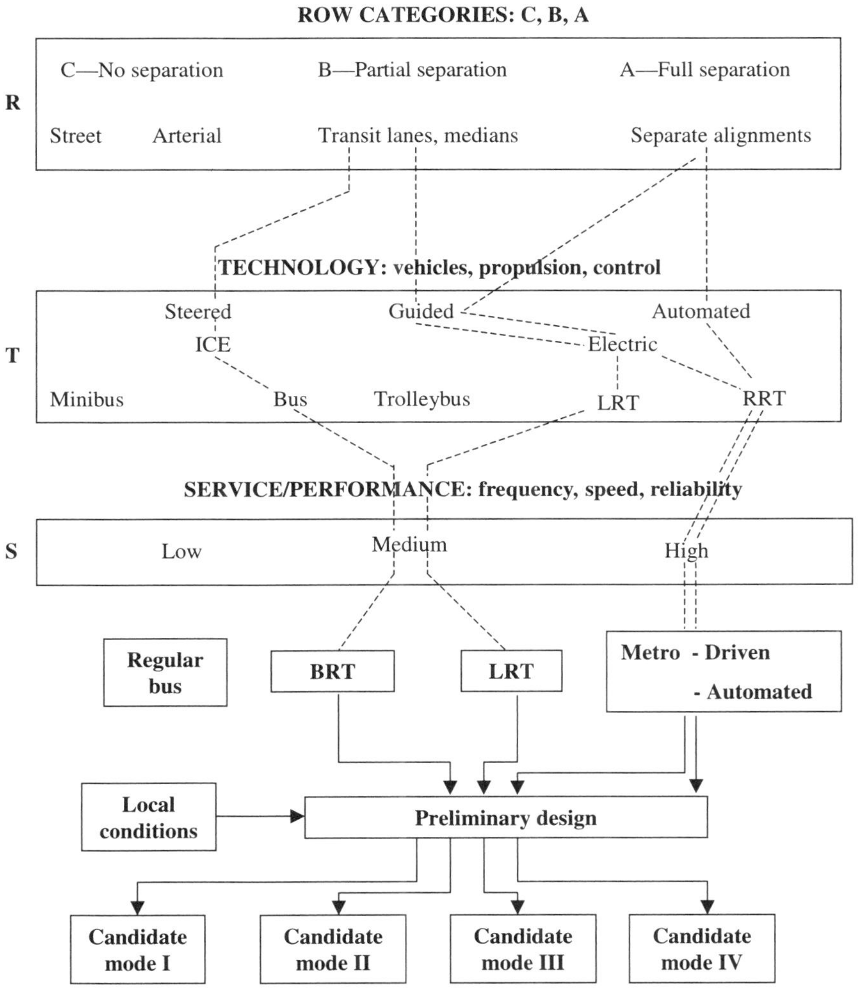

Figure 11.5 *Conceptual presentation of modes defined as RTS.*

depth of evaluation must be selected that are reasonable and practically possible with the available data and objectivity of evaluation of the quantitative factors. Each element evaluation should be expressed and evaluated in one of three basic ways:

1. Monetary units (dollars, euros, yen, etc.)
2. Other quantitative units, such as service frequency, projected numbers of passengers, or person-hours
3. Qualitative (descriptive) terms

All evaluations expressed in quantitative measures can be summed up in respective units, while qualitative evaluations must be considered separately.

This evaluation is followed by the weighing of individual items, either by quantitative units or by descriptive terms. Relative weights are even more value-based than the evaluations of individual elements, particularly because relative weights may vary not only in time, but also among locations and types of transit systems. For example, operating speed is more important for regional than for urban transit lines,

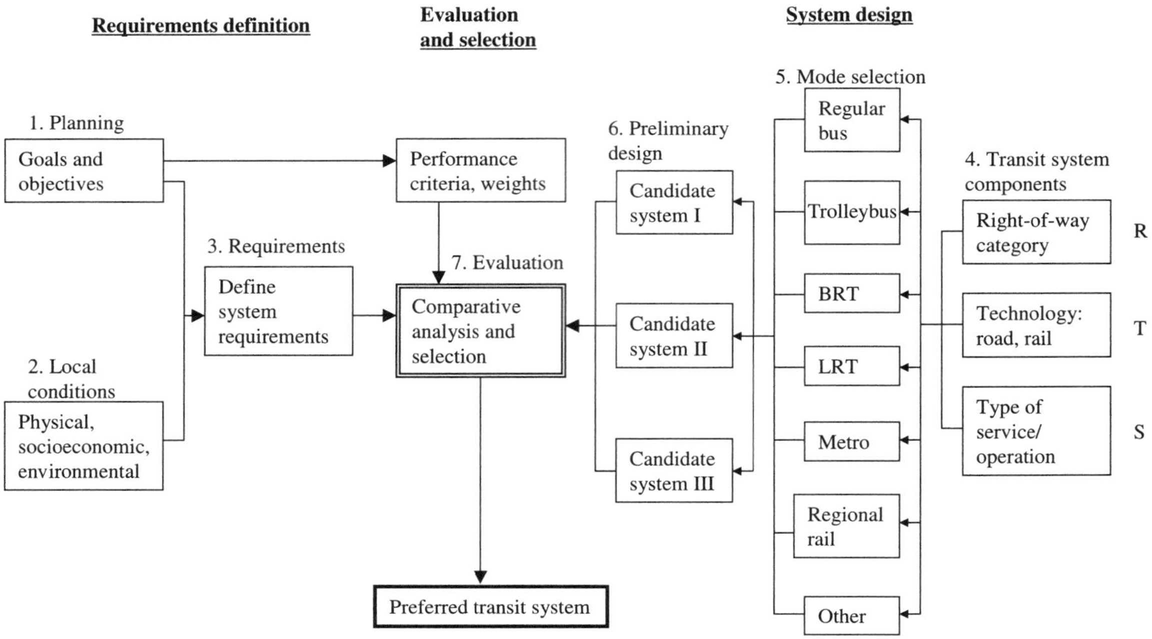

Figure 11.6 *Transit mode/system selection procedure.*

while frequency of service is more important for the shorter transit than for long regional lines. Local transit services attract more passengers by conveniences such as good information, easy transfer, etc., than by high riding comfort.

Various point systems can be used to derive an overall evaluation of each mode, expressed by a single quantitative value. This method requires quantification of all parameters, as well as their relative weights. It is thus more sophisticated than evaluation based on monetary costs only. Yet that is seldom an acceptable method of evaluation because it still tends to neglect or underestimate some qualitative elements, particularly the long-range impacts. In some relatively simple cases of project evaluation, such evaluation may be satisfactory, but with most transit system evaluations this is not the case. The reader, analyst, or decision maker can usually get a better picture of compared modes through a complete list of evaluated items than from a single number based on internalized subjective values whose validity cannot be examined.

11.4.3 Summary of Mode-Selection Procedure

The methodological procedure for comparative analysis and selection of transit modes presented here consists of three major phases: definition of the conditions and requirements for the service (demand), selection of candidate modes or systems (supply), and evaluation and selection of the preferred mode. The entire process can be defined as a sequence of seven steps, which are illustrated in the flowchart in Figure 11.6 and outlined briefly in the box.

Naturally, the procedure for the selection of the preferred mode or system is usually more complex and less regular than this conceptual sequence indicates. There are many reasons for many variations in actual planning and their deviations from this idealized pro-

Sequence of seven steps in transit mode/system planning, design, and selection:

1. Based on urban transportation policies, develop goals and objectives for the transit system.
2. Define conditions for the area to be served and their relative importance.
3. Utilizing the results from the preceding steps, define the requirements and standards for the planned system.
4. Based on the system requirements, select one or several RTS sets for candidate modes.
5. Select several alternative modes/systems that correspond to the defined RTS sets.
6. Develop preliminary design for each alternative system to optimize it for the local conditions.
7. Based on the system requirements, criteria, and their relative weights, perform comparative analysis and selection of the preferred transit system.

cedure. For example, project goals, local conditions, and requirements often interact: financial conditions may influence the goals. ROW category, system technology, and type of operations are also interdependent. The process of evaluation must therefore in many cases be modified and include feedback analyses or iterative procedures, returning to preceding steps and modifying some conditions or decisions to reach a satisfactory relationship between the local conditions and the selected preferred mode.

11.4.4 An Example of the Mode-Comparison Procedure: Lindenwold Rail Line and Shirley Busway

In 1969, two new transit systems were opened in Philadelphia and Washington, DC, that had very similar alignments and functions but different modes and different network/operational concepts. The Lindenwold Line, serving New Jersey suburbs for access across the Delaware River to Philadelphia, was a regional rail rapid transit line. The Shirley Busway consisted of an extensive network of bus lines serving Virginia suburbs, which converged on a busway in the median of the Shirley Highway, crossing the Potomac River to access Washington.

Vuchic and Stanger (1973) made a detailed comparison of these two lines: types of services they offered, passenger attraction, and economic efficiency. As Table 11.5 shows, both served areas are low-density, auto-oriented suburbs with similar population sizes. The transit lines are drastically different, however. The Lindenwold Line was only 22.8 km long and had 13 stations, while the buses had dozens of lines and branches and hundreds of stops in the suburbs and in Washington CBD, but none along the Shirley Busway (Table 11.6). The rail line was operated 24 hours a day with short headways (10 minutes or less) except during the night, while buses offered frequent service during 6 peak hours, minimum skeleton service for another 12 hours, and no service during 6 night hours.

The authors visited the sites and made extensive analyses of characteristics of the two lines: speed, headways, comfort, image, operating cost, and others. Utilizing a set of requirements affecting passengers, operating agencies, and areas they were serving, they found that the Lindenwold Line had access at very few stations (only nine in the suburbs) and 89% of its riders used cars for access (K+R and P+R). The Shirley bus service covered a very large area and had hundreds of bus stops, but operated at long and irregular headways, had lower image, and depend nearly exclusively (84%) on walking access.

The Lindenwold Line demonstrated very superior results to those of the Shirley buses: its ridership was substantially (24%) higher, its fares were slightly lower, and its operating ratio was slightly higher.

Table 11.5 Basic characteristics of the compared transit systems

Characteristics	*Lindenwold Rail Line*	*Shirley Busway*
Location	From southern New Jersey northwest into central Philadelphia	From northern Virginia northeast into central Washington
Types of lines	Single radial line with nine suburban stations via bridge to CBD with four stations	Dozens of lines and branches converge on a busway then via bridge to CBD; hundreds of stops
Type of service:		
Line-haul	Rail on ROW A	Buses on streets and on busway
Access to suburban stations modes (%)		
Feeder transit	9	0
P&R	37	8
K&R	43	8
Walk	11	84
Transferring between modes (%)	89	16
Downtown distribution	4 stations, discount on transfer to rapid or street transit	Stops at each one of 15 blocks, possible transfer to city buses
Primary service area (km^2)	337	246
1970 population	442,000	356,000
1970 population density (estimated prs/km^2)	1312	1447
Average cars per household	1.3	1.5
Employment centers served (1975 jobs)	Central Philadelphia (380,000) Camden, NJ (45,000)	Central Washington (400,000) Crystal City (40,000) Pentagon (20,000)
Opening date	February 1969	September 1969

A summary of major system characteristics and their comparison, showing the mode that was more advantageous for each characteristic, is given in Table 11.7. The comparative analysis and its conclusions were that the Lindenwold Line was more successful (it was superior to buses in 12 out of 16 individual characteristics). This is a result of two sets of characteristics. First, rapid transit as a mode has a stronger attraction due to its service quality, including frequency of service, speed, and reliability. And second, the single rail line with strong image and frequent service throughout the day also represents a superior type of service to the extensive network of bus lines without a strong image and reliability, oriented to commuters with minimal services at other times.

The Shirley Busway (later degraded to an HOV facility) thus suffered from some of the misconceptions about organization of transit services discussed in Section 12.2.1.1: many lines with weak image and few possibilities for transferring, thus focusing on many-to-one peak hour commuter services. The Lindenwold Line, on the contrary, offered high-quality service concentrated at relatively few stations, but providing many more intermodal transfers, serving all-day, many-to-many trips in the served area and center city.

The present efforts to develop the BRT mode have paid considerable attention to changing bus services from the commuter type similar to the Shirley bus system to services more resembling the ones offered by RRT and LRT lines, such as the Lindenwold Line in this case.

Table 11.6 Operational characteristics and system use

Characteristics	*Lindenwold Rail Line*	*Shirley Busway*
Weekday hours of operation	19 + 5[a]	6 + 12[a]
Daily trips		
Weekdays	288	992
Saturdays	230	200
Sundays	156	83
Average travel speeds (km/h)		
Line-haul	76	56
Suburban collection	Auto: 40	Bus: 21
Downtown distribution	39	19
Overall (typical)	48	32
Average speed on competing highway (km/h)	20–50	15–35
Number of vehicles	Rail: 75	Bus: 340
Peak hour seated and standing passengers per TU	456 + 192	48 + 5
Average peak hour floor area/passenger (m^2/prs)	0.59	0.51
Number of passengers/weekday	41,500	33,570
Fare ($)	0.35–0.75	0.40–0.80
Revenue/passenger ($)	0.57	0.70
Operating cost/passenger ($)	0.51	0.68

[a]Hours with very low frequency.

Table 11.7 Summary of comparative analysis

Requirement	*Lindenwold Rail Line*	*Shirley Busway*	*Higher-Rated System*
	Passenger		
Availability	Good	Limited	LRL
Speed-travel time	Good	Very good	SB
Reliability	Very good	Moderate	LRL
User costs	Good	Very good	SB
Comfort	Good	Moderate	LRL
Convenience	Good	Fair	LRL
Safety/security	Very good	Good	LRL
	Operator		
Area coverage	Good	Very good	SB
Frequency	Very good	Very low	LRL
Speed	Very good	Fair	LRL
Cost: investment	Very high	Moderate	SB
Cost: operating	Good	Fair	LRL
Capacity	Good	Limited	LRL
Side effects	Good	Fair	LRL
Passenger attraction	Very good	Limited	LRL
	Community		
System impact	Very good	Good	LRL

EXERCISES

11.1. In evaluating investments in public facilities such as a new transit line or a highway:

- What are the major goals to consider?
- Which major interested parties are included in the analysis?
- How does this procedure differ from investment decisions in the private sector?
- In what respect is the analysis about investments more complex in the public than in the private sector?

11.2. Explain the meaning of net present value of a public project. How is it computed and what is its purpose?

11.3. Several economists have used the diagram shown in Figure 11.2 to find the optimal domains of different modes of urban transportation. Is this method correct? Should it be used for guidance in planning different modes of urban transportation? If not, what are its deficiencies?

11.4. What are the main weaknesses of transit mode comparison studies based on hypothetical urban transportation models? Which factors make the results of these mode comparisons unreliable and subject to analysts personal views?

11.5. Define the concept of performance-cost package (PCP) for transit modes. Why is this concept important in comparing transit modes?

11.6. List and explain the elements of the conditions set in transit planning. Define the three main interested parties whose requirements must be considered in developing and evaluating transit plans. Which of the three parties is usually the most important and why?

11.7. Select one of the three interested parties, then list and define its requirements for transit system characteristics. Explain how each of the requirements can be measured, quantitatively or qualitatively.

11.8. Take three pairs of related requirements from the passengers' and operator's point of view, respectively:

1. Service availability and area coverage
2. Punctuality and reliability
3. Travel speed and cycle speed

Define the differences between the requirements. Explain how each of the two parties defines each concept and how each measures the parameters.

11.9. How is the characteristic of flexibility defined? List its several meanings and explain which are positive and which are negative. Which system features are opposite to flexibility?

11.10. Discuss various transit system characteristics that influence passenger comfort. Which characteristics of a passenger's trip (length, purpose, time, etc.) influence the weight given to individual comfort components?

11.11. What are the consequences of unreliable service on passengers? Does it affect their travel time? Can it be so important that it influences a passenger's modal choice?

11.12. An investment into a control center for a rapid transit system is considered. In evaluating this project, the question is raised what benefits it would bring. Major items include increased operating efficiency, security, safety, and service reliability.

You are asked to assess the significance of service reliability, measured as percent of trains operating with, say, 0–3-min delays. Thus, 96% reliability means that only 4% of trains are more than 3 min late or ahead of scheduled time. Assuming that the system's reliability will increase noticeably due to the control center's construction, develop a flowchart representing the impact of increased reliability, classifying them by the affected party: passengers, operator, and city (or community). State which items could be quantified and how.

11.13. Find in your city, or in another city that you know well, two points that generate a considerable number of trips and are served by two to three different transit lines, preferably different modes. Compare travel between these two points by the alternative modes, using a selected set of passenger requirements. Define each requirement and analyze it in detail using quantitative elements wherever it is applicable. Develop a metric of alternatives and characteristics. Then conclude which alternative line or mode is preferable and why. Note that the choice may be different depending on trip conditions: time of travel (day or night), weather conditions, or trip objective.

11.14. Find a major transit project that is being planned in a city. Define and analyze which project aspects and results the community (through its government representatives and public participation) should consider in evaluating the project and decide how strongly the project should be supported.

11.15. What transit system elements does the RTS set consist of? What is the sequence of RTS elements choice and how do these lead to the definitions of alternative candidate systems?

11.16. What are the advantages of evaluating a mode through a single quantitative (usually monetary) value? What are the limitations of this method? Why can serious mistakes be made in applying this method under certain conditions? Give an example of using such an evaluation for a transit project.

11.17. Discuss the differences between the described Lindenwold Rail Line and Shirley Busway with respect to the services offered. Which differences are a result of the modes (rail and bus)? Which differences are caused by line or network designs, and which are caused by different operating methods?

REFERENCES

American Association of State Highway Officials (AASHO) 1960. *Road User Benefit Analysis for Highway Improvements*. Washington, DC: AASHO (now AASHTO).

Botzow, H. 1974. Level-of-service concept for evaluating public transport. *TRB Record 519*. 73–84.

Fruin, J. 1971. *Pedestrian Planning and Design*. New York: MAUDEP.

Gruen Associates and DeLeuw, Cather & Co. 1979. *Environmental Impacts of BART*. Report to Metropolitan Transportation Commission, Berkeley, CA.

Kittelson & Associates. 2003. *Transit Capacity and Quality of Service Manual*. TCRP 100. Washington, DC: Transportation Research Board.

Kuhn, T. E. 1962. *Public Enterprise Economics and Transport Problems*. Berkeley: University of California Press.

Lehner, F. 1975. *Light Rail and Rapid Transit*. Light Rail Transit Special Report 161, 37–49. Washington DC: Transportation Research Board.

Leibbrand, K. 1967. "An analysis of different forms of rapid transit. Summary by W. S. Homburger. In *Urban Mass Transit Planning*. 197–203. Berkeley: Institute of Transportation and Traffic Engineering.

Meyer, J., J. Kain, and M. Wohl. 1965. *The Urban Transportation Problem*. Cambridge: Harvard University Press.

Operations Research, Inc. 1964. *Comparison of Noise and Vibration Levels in Rapid Transit Vehicle Systems*. Report to the National Capital Transportation Agency, Washington, DC.

Peterson, A. P., and E. E. Gross. 1972. *Handbook of Noise Measurement*. Concord, MA: General Radio Co.

Regional Plan Association (RPA). 1991. *The Renaissance of Rail Transit in America*. New York: RPA.

Thomas, E. N., and J. L. Schofer. 1990. *Strategies of the Evaluation of Alternative Transportation Plans. NCHRP Report 96*. Washington, DC: Transportation Research Board.

Voorhees, Alan M., & Associates. 1972. *Tyneside Rapid Transit Analysis: Technical Report*. London.

Vuchic, V. R. 1971. Concept of flexibility in transportation systems analysis. *High Speed Ground Transportation Journal*. 5(1):53–61.

———. 1981. *Urban Public Transportation Systems and Technology*. Englewood Cliffs, NJ: Prentice-Hall.

Vuchic, V. R., and R. M. Stanger. 1973. Lindenwold Rail Line and Shirley Busway: a comparison. *HR Record 459*, 13–28.

FURTHER READINGS

Eco Northwest, Ltd. and Parsons Brinkerhoff Quade & Douglas, Inc. 2002. *Estimating the Benefits and Costs of Public Transportation Projects: A Guide for Practitioners*. TCRP Report 85. Washington, DC: Transportation Research Board.

Parkinson, T., and I. Fisher. 1996. *Rail Transit Capacity. TCRP Report 13*. Washington, DC: Transportation Research Board.

Powell, T. 2001. *The Principles of Transport Economics*. London: PTRC.

Vuchic, V. R., and O. A. Olanipekun. 1990. Lindenwold Line and New Jersey transit buses: a comparison. *TR Record 1266*, 123–138.

12

PLANNING AND SELECTION OF MEDIUM- AND HIGH-PERFORMANCE TRANSIT MODES

Following the theoretical basis and methodology for comparative analysis and selection of transit modes in Chapter 11, this chapter presents a systematic review and comparison of medium- and high-performance transit modes, which closely coincide with semirapid and rapid transit, respectively. Having partially and fully separated rights-of-way, B or A, these two mode categories require considerable investments in infrastructure and result in construction of permanent transit systems. These systems provide important transit services and represent a strong element influencing a city's and region's functioning and growth pattern. Semirapid and rapid transit networks always represent the basic skeleton of their cities' transit systems. Therefore, their design and mode selection represent the most important and creative phase of transit planning.

Construction of these transit systems often represents the highest-investment public system a city or region has ever made. Its planning always attracts considerable attention. It is often subject to pressure by different interested groups or lobbies and sometimes reflects different political philosophies. Strong emotional views for or against different transit modes may further complicate this phase of planning. The purpose of this chapter is to provide an objective description of all major medium- and high-performance transit modes and their mutual comparisons, including their relative advantages and disadvantages.

12.1 BASIC ELEMENTS OF TRANSIT MODES

Planning and selection of medium- and high-performance transit systems starts by a functional definition of the planned transit system, mainly consisting of:

- Level of service (LOS) required for passengers
- Performance-cost package (PCP)
- The role transit should play in the city's functioning and development

Once these functional characteristics are defined, one or several transit modes have to be selected. For this selection, characteristics of modes need to be analyzed. For that purpose, a systematic review of the basic transit system elements that define transit modes is given in this section. These elements and their technical options are listed in Table 12.1. Following the three groups of components described in Chapter 11 as the RTS set, ROW is item 1, technology is items 2–5, and service is item 6.

The following sections discuss each one of these modal elements in sequence, including their descriptions and comparisons among modes.

12.1.1 Right-of-Way Categories and Their Significance

In most cases, the ROW is the most fundamental transit system element because it strongly influences each

Table 12.1 Basic elements of transit modes and their technical options

Element	*Options*	*Descriptions and Comments*
1. Right-of-way	C—street with mixed traffic B—partially separated A—fully separated	Increasing investment and mode performance
2. Vehicle guidance	Steered—highway vehicles Guided—mostly rail, some rubber-tired	Some guided modes use guiding tires or other types of guidance
3. Guided vehicle support	Steel wheels on rails Rubber tires on roadways or running beams	Rail-guided technology allows use of any ROW A, B, or C. Guided rubber-tired vehicles are limited to ROW A, except guided bus when steered
4. Propulsion	Internal combustion engine with diesel fuel, CNG or LNG Electric Linear induction motor (LIM)	Diesel mostly on highways, electric on rails, but with overlaps, such as trolleybus and diesel commuter rail systems
5. TU driving and control	Driver-driven Driver with ATO ATO, no driver	Automated requires ROW A, guided and electric propulsion
6. Service—type of network and operation	Many overlapping lines Rail trunk lines with feeders Rail networks supplemented by bus lines	With increasing passenger volumes, trunk lines play a greater role

mode's PCP. Moreover, the ROW type interacts with the other two elements, system technology and service/operation.

Transit modes operating on ROW category C represent *street transit,* the basic type of transit, exemplified by buses, trolleybuses, and streetcars or tramways. These modes require low to moderate investment. In addition to vehicles, the only items requiring investment on bus lines are equipment at transit stops and, in some cases, signs/signals provided specifically for transit vehicle operations and their priorities over other traffic. Trolleybuses also require investments in overhead wires, while streetcars also need more substantial investment in tracks.

Street transit performance depends greatly on traffic conditions along its lines: if there is congestion, service is unreliable. Operating speed is always lower than the speed of other traffic because of the time transit vehicles lose at passenger stops.

As a result of such performance and the lack of permanent distinguishing features of these services, street transit is generally not competitive with respect to speed and reliability with auto travel on the same streets. The balance between the two modes can be maintained by providing transit priorities at intersections and other critical locations, as well as by parking charges in the served areas.

Modes having ROW category B, including LRT and buses on physically separated busways, represent *semirapid transit.* Since most or all of the ROW for these modes must be constructed, they involve considerably higher investment costs than street transit. However, the performance of semirapid transit is also much better than that of street transit due to the nearly total absence of interference by other traffic. Particularly important for passenger attraction is the fact that semirapid transit has a distinctive, separate, and permanent ROW and therefore a much stronger image than street transit. The differences between the two ROW categories are summarized in the box.

Most modes with ROW category A represent *rapid transit:* independent systems with high-capacity trains operated at short headways. This group, also known as metro (and in individual cities as subway, subway-

Photo 12.1 ROW C: mixed traffic bus operation in Buenos Aires.

Transit systems with ROW category B compared to those with category C have the following advantages (+) and disadvantages (−):

+ Considerably higher performance: speed, reliability, capacity, riding comfort, safety, etc.
+ Ability to operate trains of two to four vehicles (rail modes)
+ Stronger identity and image, which, combined with higher performance, results in higher passenger attraction
+ Lower operating costs per passenger
+ When rail technology is used, these modes are electrified, providing high vehicle performance and no air pollution, so that they can be operated in tunnels
− Space required for ROW
− Involve higher investment
− May require special signal or other control and priority measures

Photo 12.2 *ROW B: light rail on separated median with grass track—Hannover, Germany* (source: *G. A. Girnau, A. Müller-Hellmann, and F. Blennemann, eds.,* Light Rail in Germany. *Cologne: Verband Deutscher Verkehrsunternehmen, 2000. Reprinted by permission*).

elevated, MRT (mass rapid transit), U-Bahn, etc.), in most cases has rail technology. In some cities with relatively small vehicles (typically 16 m long, 2.50 m wide) and speeds not exceeding 70 km/h, such as Lyon, Montreal, and Mexico City, rubber-tired rapid transit (RTRT) is used. In a few cities, individual lines are operated by different types of monorails. The oldest monorail line (1901) is the Schwebebahn in Wuppertal, Germany, utilizing suspended rail vehicles. Others include the Tokyo–Haneda Line, the Kitakyushu monorail in Japan, and the monorail presently under design in Seattle, all of which use the Alweg-type rubber-tired vehicles riding on a concrete beam.

Systems with ROW category A have generally similar differences from ROW B as B has over C, but expressed to a greater extent (see box on next page).

Transit planning varies greatly among the three basic mode categories. Street transit systems, not involving much infrastructure, are generally planned in the short term, and they can be relatively easily extended or modified. Semirapid and rapid transit systems, however, involve infrastructure planning and construction, which require substantial financing and take time. Those systems are not flexible, but fixed and permanent and have major positive impacts on the city. Their efficient design is therefore very important. Mistakes, such as an alignment that results in inefficient operations or has undesirable impacts on the surroundings, may have permanent negative consequences.

Inherent differences among the basic characteristics of most modes grouped in the three ROW categories are expressed by simple descriptive terms in Table 12.2.

Some transit modes, such as most rubber-tired AGT systems, have ROW category A, but they have medium capacity (3000–10,000 sps/h, exceptionally higher) because they are operated with small vehicles and short trains. Examples are the AGT systems in Kobe and Osaka, Japan, Lille and Toulouse, France, and Taipei, Taiwan.

Several rail systems operated with LRT rolling stock also have ROW category A. Defined as light rail rapid transit (LRRT), these systems can provide capacities of up to 20,000 sps/h. Examples of these medium-performance rail systems with ROW category A include the Norristown Line in Philadelphia, with a maximum running speed of 120 km/h, and the more recently built lines in Vancouver (the Skytrain),

Photo 12.3 *ROW A: Fully separated, exclusive right-of-way—Singapore Metro.*

Transit systems with ROW category A compared to those with category B have the following advantages (+) and disadvantages (−):

+ Free from any obstacles, systems on ROW category A have the highest performance (speed, reliability, capacity) of all modes.
+ All rapid transit systems utilize electrically propelled guided technologies with very high dynamic performance.
+ The absence of any interaction with other traffic and the use of automatic train protection results in the highest safety among all transit modes.
+ Exclusive ROW allows operation of long trains with multiple doors at platform level, resulting in rapid passenger exchange and short dwell times.
− They require by far the highest investment costs, due to exclusive facilities along entire line length.
− Rapid transit line alignment is more geometrically rigid, requiring in central cities mostly tunnels or aerial structures and longer access to stations.
− Stations are grade-separated and require longer access than LRT on ROW B, which can penetrate pedestrian areas.

Table 12.2 Characteristics of modes with different ROW categories

Characteristics \ *ROW Category*	*C*	*B*	*A*
System performance: capacity, speed, reliability, safety	Moderate	High	Very high
Investment cost	Low	High	Very high
Level of service	Moderate	High	Very high
Image/identification	Moderate	Good	Very strong
Passenger attraction	Moderate	High	Very high
Potential impact on urban form	Weak	Strong	Very strong
Full automation possibility	None	None	Full

London-Docklands, Kuala Lumpur, and Copenhagen, which represent automated LRT (ALRT) (see Section 12.3.1.2).

An important note on designations of transit modes is in order here. Technical terms for transit modes utilizing ROW categories C, B, and A are, respectively, street transit, semirapid transit, and rapid transit. Popular designations, often intended to enhance popular image, confuse these technical definitions. Thus, several cities in the United States have "rapid transit districts" that operate regular buses only. Similarly, several cities in Germany use the term "U-Bahn" (rapid transit) for LRT (Stadtbahn) systems to increase their image. Dallas Area Rapid Transit (DART) operates an LRT system and a network of regular bus lines, but not rapid transit.

The entire current BRT promotion refers to buses operated in semirapid transit mode. By the strict definition, no bus system is rapid transit because none has ROW A on its entire length. Technically, they are bus semirapid transit (BST).

Thus, popular names of transit systems sometimes deviate from their technical definitions. Technical terms are also sometimes adapted for personal designations of some transit systems. For example, the term "metro," standing for "metropolitan railway" is also the personal name of metro systems in Paris, Moscow, and Washington, as is "Max" for the LRT in Portland, Oregon, and "Metrolink" for the St. Louis LRT, as well as for the Los Angeles commuter rail.

12.1.2 Transit System Technology

The four major components that influence system performance and interact with ROW category and system operations, listed as items 2–5 in Table 12.1, are reviewed in the following sections, emphasizing their operational characteristics.

12.1.2.1 Vehicle Guidance. With respect to guidance, transit systems are classified into two very distinct categories: vehicles steered by driver and vehicles/TUs physically led by the guideway. The steered technology systems include mainly paratransit vehicles and buses. Being road-based systems, they can operate on nearly all streets and need investment only for vehicles and little special infrastructure for their physical movement. Bus services can be quickly introduced, modified, or suspended. These factors reduce the need for long-range planning of driver-steered transit systems.

While the flexibility that driver-steered vehicles have is advantageous on urban streets, their mixing with other traffic and the absence of a fixed path or guideway allow these modes to give excellent area coverage and provide extensive services throughout cities. On the other hand, lack of fixed facilities is also a strong disadvantage with respect to their performance, image, and passenger attraction. Many passengers are apprehensive about using buses operating on a complicated, interwoven network. Exclusive bus lanes, trolleybus overhead wires, and especially the tracks of streetcars and other rail modes give LRT and RT lines a strong image that allows passengers better orientation in travel. Transit line image becomes particularly strong with independent metro lines. For example, many visitors to Mexico City, Moscow, Paris, or Washington use only their metro networks, transferring to buses only where the metro network does

Guided transit systems compared with bus (driver-steered) systems have the following advantages (+) and disadvantages (−):

+ Ability to use larger vehicles and operate trains, thus greatly increasing line capacity, providing lower operating costs per space, and offering economies of scale with increasing passenger volumes.
+ Electric traction can be used, bringing major benefits (see below).
+ Have higher overall performance: speed, capacity, and reliability.
+ Automated signalization provides fail-safe operation, preventing accidents due to driver's error.
+ Require narrower ROW.
+ With exact ROW delineation, no air pollution and noise, better suited to operation in pedestrian areas.
+ Operation in tunnels is feasible.
+ Provide much higher riding comfort.
+ When combined with ROW A, guidance makes fully automated driving possible.
− Require separate ROW (B or A), involving much higher investment cost (except for tramway/streetcar mode).
− Require higher investment due to track/guideway and station construction and electrification.
− Network is much more limited, requiring more transfers.

not extend. This attraction is not only caused by the higher level of service rail systems offer, but also by their strong image and easy orientation.

Another limitation of steered technology systems is that they are labor-intensive and have no economies of scale. They can be operated as single vehicles only (exceptionally, there are cities that operate trolleybuses with trailers or in double traction (Lausanne, St. Petersburg, Kiev), but in most countries transporting passengers in trailers is prohibited for safety reasons). Therefore, for each additional vehicle an additional driver is required.

Although road-based transit vehicles do not need special infrastructure for their movement, for their operation in busy corridors it is desirable to provide for them such infrastructure components as designated bus lanes, special signals with priorities, bus stop bays, and signs. When passenger volume is so great that special busways are required, replacement of buses by rail technology systems becomes a realistic alternative that should be examined.

Guided technology (mainly rail) systems, compared to steered highway vehicles, require a considerably higher investment for the guideway construction and usually electrification, but they have major physical, operational, and economic advantages.

In conclusion, construction of guided (rail) system to replace a steered (bus) system requires investment, but it leads to a considerably higher-quality transit system that attracts a significantly greater number of passengers (see box above).

Consequently, buses and minibuses are superior to guided systems for light to medium passenger volume transit lines. Their lower investment cost outweighs all advantages of guided systems. For high present or potential passenger volumes, however, guided technology systems represent a superior solution with respect to passenger attraction, economic efficiency, and impacts on the city.

12.1.2.2 Guided Vehicle Support. While all highway vehicles ride on rubber tires, guided systems have several possible types of support and guidance:

1. Steel wheels on steel rails, which is by far the most dominant

2. Rubber tires for support, usually supplemented by the same number of additional horizontal tires for guidance against side or center guiding rails (monorails and some AGT systems)
3. Magnetic suspension, used only on a few transit systems and later discontinued

A summary comparison of the two used vehicle support technologies is given in the box.

For the reasons shown in this comparison, rail technology is by far the dominant mode for high-capacity metro systems, while AGT modes, mostly with small to medium vehicles (capacities 30–100 spaces, including standees) and short trains have advantages for shuttle lines in airports, exhibition grounds, and similar activity centers and on short transit lines, such as center city distributors. Rail and rubber-tired systems overlap on some medium-performance regular transit lines.

12.1.2.3 Propulsion. Most transit systems are powered by different types of internal combustion engines, usually diesel, or by electric motors. The introduction of compressed natural gas (CNG) or liquid natural gas (LNG) engines on transit buses helps lower their emissions of air pollutants. Similar to the differences among ROW categories and steered versus guided systems, diesel traction requires lower investment but has lower performance than electric traction, as the comparison in the box shows.

In general, electric traction is superior in TU performance, in the absence of negative impacts, particularly in high-density areas in cities, and in making operation of transit lines feasible in tunnels and attractive in pedestrian zones. Again, the lower-investment diesel-powered vehicles are optimal for lightly traveled transit lines. As passenger volumes increase, the investment in electrification becomes economically justified and higher-performance electric traction can be introduced, usually together with the introduction of rail systems with ROW category B or A.

As a part of numerous campaigns in many U.S. cities against high-performance transit, particularly rail, there have been claims that electric traction does not have advantages over diesel traction even when considering air pollution caused by electric rail versus diesel bus modes. For example, a consulting firm claimed that replacement of an LRT line operating in a tunnel under a major city's CBD by articulated buses on city streets with over 300 trips per day would not increase air pollution. Another computer model-based study supposedly found that electric transit usually produces more air pollution than diesels because most

For guided transit modes, rail systems compared with rubber-tired ones have the following advantages (+) and disadvantages (−):

+ Rail systems can utilize ROW categories A, B, or C; rubber-tired ones can use only ROW A.
+ Rail vehicles have two to three times fewer wheels with lower resistance, resulting in lower total resistance, energy consumption, and heat production.
+ Rail tracks have very simple switching and crossing, while rubber-tired guideways require large space and grade separations in yards. Rail systems are therefore much more conducive to efficient line branching, network, and yard layouts. Therefore monorails and AGT systems operate mostly on single lines.
+ Rail systems can have larger vehicles and operate at higher speeds than rubber-tired systems.
+ Rail vehicles have considerably higher riding comfort.
− Rubber-tired systems can negotiate sharper curve radii and higher gradients than rail; this allows them to have more flexible line alignments.
− Rail vehicles produce more noise and vibration in sharp curves.

Electric compared with diesel traction of transit systems has the following advantages (+) and disadvantages (−):

+ Produces higher acceleration and higher operating speed.
+ Allows multiple-unit propulsion and operation of trains.
+ Regenerative braking recovers energy, reducing its consumption.
+ Produces no noise and vibration, as most diesel engines do.
+ Produces no air pollution in the city, while diesel buses produce it in areas of concentration of people.
+ Absence of pollution and good dynamic performance makes use of tunnels possible.
+ Electricity can be produced from any primary energy source, thus reducing oil dependency.
− Requires additional investment for electrification of lines, construction of substations, etc.
− Power failure stops all vehicles on a section of line.
− Service is limited to electrified lines, requiring more transfers to feeders.

electricity is produced in old coal-burning generating power plants (Vincent and Walsh, 2003).

These conclusions have been reached on the basis of the following basic and many smaller fallacies in study approaches and model formulations.

1. They claim that air pollution should be considered on a "total system basis," i.e., not only vehicle tailpipe emissions, but including the pollution produced in energy production, i.e., power plants for electricity generation. However, air pollution generated by oil drilling and refineries that produce diesel oil are not even mentioned. Thus, the gross pollution for electric traction is compared with the net pollution by diesel traction.
2. The impacts of produced air pollution depend very much on the exposure of persons to the source of pollution delivery. Buses emit pollution in areas of the highest concentration of pedestrians and residents, producing impact on breathing orders of magnitude greater than pollution at electric power plants in remote rural areas.
3. Computer simulation models used to examine impacts of, for example, replacing an electric rail transit line with diesel buses have been made for entire metropolitan region, spread to such geographic area and all air pollution sources that any change of one transit line is of negligible magnitude. By increasing the base of the simulation model, it is possible to make any change of one component negligible. Such models of the urban universe are so insensitive that they cannot detect impacts of huge garages, highly polluted sections of individual streets, etc.
4. The much higher noise pollution from diesel vehicles as compared to electric vehicles is not mentioned in any of these studies.

Any one of the described fallacies is sufficient to invalidate the authors' claim that electric propulsion has no advantages over diesel propulsion. Scientific facts and experiences of cities worldwide over many decades clearly show that electric propulsion has very significant advantages over diesel and other internal combustion engines with respect to performance and direct and indirect negative impacts on transit users and on the city. For that reason, electric propulsion is used whenever the higher investment cost of its introduction as compared to diesel can be justified by high passenger volumes, better performance, and higher benefits.

12.1.2.4 Transit Unit Control. All transit modes operating on ROW categories C and B are manually driven with advisory signals only. Some guided modes use automatic train operation (ATO), which is usually introduced to optimize travel regime with respect to propulsion, braking, and travel time. Fully automated transit systems, operating without any crew members, are also getting increasing attention. Initially, AGT systems were used extensively for local transportation within airports, university campuses, and hospital complexes. Since the 1980s, however, several regular transit lines have been opened with fully automatic operation. In addition to the AGT rubber-tired and rail transit systems (VAL in Lille and Toulouse, Skytrain in Vancouver, Docklands in London, and several others), fully automated regular rapid transit lines have been opened in Lyon, Paris, and Singapore, while Berlin and Nuremberg plan to convert, for the first time, existing manually driven lines to automatic operation. Actually, Nuremberg plans to have mixed operation of manually driven and fully automated trains on sections of one line.

Clearly, the greatest advantage of fully automated train operation comes from the fact that with it, the cost of operating trains is proportional to the number of cars; therefore, long trains, which are more economical when drivers are used, can be replaced by a greater number of short trains without any cost increase. Thus, frequent operation of short trains is feasible even with a much lower passenger volume than on regular transit lines. This feature lowers the boundary of feasible application of high-performance guided transit. The benefits from full automation are thus particularly important on medium-capacity systems. On metro systems with long trains, the presence of an operator is usually desirable for supervising passengers, handling service interruptions, and other reasons. Moreover, the cost of an operator represents a much lower percentage of operating costs than on medium-capacity systems. Operation without train crews is therefore not as beneficial on high-capacity systems as it is on medium-capacity guided and typical AGT systems. These advantages and disadvantages of automation, summarized in the box on next page, are not related to system technology: the same factors are equally valid for rail as well as rubber-tired guided modes.

The second potential benefit of fully automated operation is the ability to change train sizes according to passenger volume on a line. A line controller can remotely couple or uncouple trains, adjusting the offered capacity to meet the demand so that high average load factor can be achieved. This results in a significant reduction in operating costs. Paradoxically, while this feature is used on the Vancouver Skytrain system and a few others, most fully automated trains have been designed with fixed train consists (lengths), forgoing this potential operating saving. For example, most AGT systems in Japanese cities operate with fixed trains, such as 6 cars on the Kobe Portliner and 10 cars on the Yokohama system.

12.1.3 Types of Network and Service

Bus services on streets are often operated as a large number of overlapping routes. As discussed in Section 4.2.4, such networks are designed to fit passenger desire lines with minimum transfers, but such route patterns result in irregular and often long headways. When passenger volumes are large and separate ROW is provided, it becomes advantageous to consolidate transit routes into fewer lines with higher-quality service: shorter and more regular headways, larger, more comfortable vehicles, more reliable operation, and, particularly, the introduction of higher-performance modes, such as rail transit.

Three different patterns of transit lines/networks and types of operation in one corridor or segment of a city are shown in Figure 12.1. The first pattern, 12.1(*a*), consists of a large set of bus lines converging on an arterial or freeway toward center city. This network, typical for commuter transit, serves many-to-one radial trips efficiently, but it does not allow for any convenient transfers nor trips among different points along the same corridor. The second pattern, 12.1(*b*), consists of a trunk line, usually upgraded to rail technology, with high performance (regular headways, high capacity and reliability) and transfer stations with numerous bus feeders. The trunk line may, however, also be a BRT line, while some feeders may also be LRT or even AGT modes.

Fully automated train operation compared with driver-driven systems has the following advantages (+) and disadvantages (−):

+ Very frequent operation of short trains is feasible even during off-peak periods.
+ Quick adjustment of TU schedules to changing conditions is possible.
+ Driving regime can be optimized for all conditions.
− Fully automated operation requires ROW A.
− Investment cost is considerably higher.
− Presence of a crew member is often desirable for security, passenger information, etc.
− Mechanical and control systems are more complex and require more expensive maintenance.

The highest type of network and service is shown as case 12.1(*c*). It shows two radial corridors with rail lines that are connected by a circumferential rail line and served by several timed transfer stations (for details of timed transfer systems, see Section 4.5). Some of the feeder lines also serve as connections between the two corridors, so that this set of lines represents an interconnected network covering the entire area and offering travel among most points using different lines with convenient simultaneous transfers.

Consequently, with improvements to transit system performance, provision of separate ROW is usually introduced with consolidation of transit services from extensive overlapping routes to fewer, high-performance lines.

12.1.4 Interdependence of ROW and System Technology

A matrix of the three ROW categories and three groups of technology by vehicle guidance (highway steered, partially and rubber-tired guided, and rail) is presented in Table 12.3. The table shows that the ROW category and system technology with respect to vehicle guidance are interdependent. On ROW category C, the steered modes (buses) are mostly used because their flexibility of movement and low investment cost are usually more important than the advantages of vehicle guidance. Streetcars/tramways also operate on ROW category C, but to utilize their advantages of guidance effectively, they usually need good traffic engineering and regulation coupled with enforcement.

On the other hand, systems operating entirely on fully separated ROW are always guided. Strictly speaking, there are no bus rapid transit lines (i.e., bus systems operating entirely on ROW A) because for collection and distribution they are always extended onto surface streets. This is logical, although it eliminates fully controlled operation of buses. The reasons for the use of guided modes on exclusive rights-of-way are obvious: once TUs operate on exactly specified paths, flexibility of movement is not needed, while the advantages of guided technology, such as narrower paths, operation of trains, electric traction, and fail-safe signal control, are very important. Since ROW category A is in most cases provided for a smaller network length of heavily used lines, the benefits of these guided systems features can outweigh the higher investment cost for their additional infrastructure construction and equipment.

The basic relationship between investment costs of the three ROW categories and mode performance is shown graphically in Figure 12.2. The box for each ROW category shows generally the area where the PCP for the systems in that category belongs. ROW category C, street transit, requires distinctly lower investment than medium- and high-performance modes using ROW categories B and A, respectively. The diagram shows the problem that occurs when a city uses only buses on streets and rapid transit: the former has relatively low performance, while the latter requires

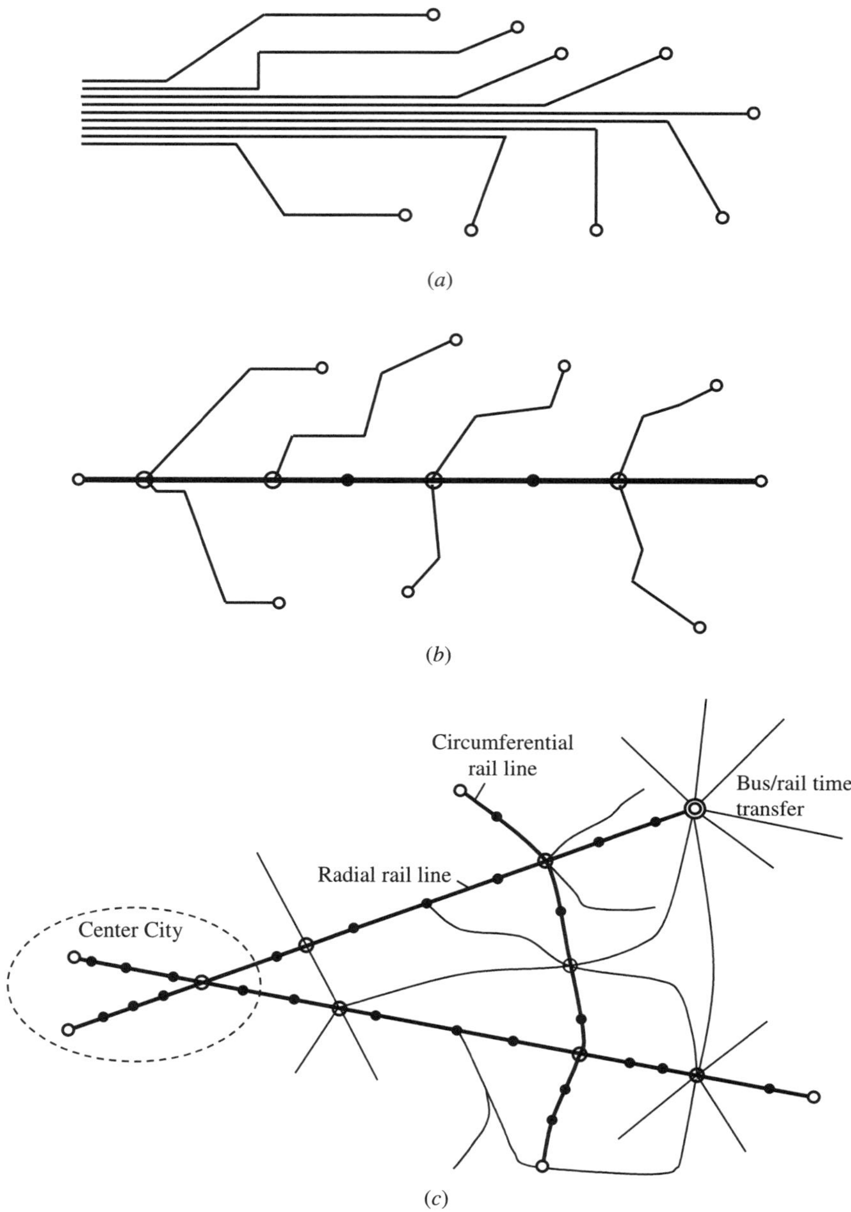

Figure 12.1 *Different types of transit lines and networks:* (a) *trunk line with branches, direct service to center city, poor connections between suburban area points—transfers not accommodated;* (b) *trunk line with frequent, regular service, usually rail; feeders with transfers to the trunk, but not among feeders;* (c) *rail radial and circumferential lines with coordinated bus TTS at major stations: an intermodal network.*

Table 12.3 Classification of transit modes by ROW category and guidance technology[a]

ROW Category \ Guidance/Technology	*Highway: Driver-Steered*	*Partially Guided, Rubber-tired Guided*	*Rail-Guided*
C	*Paratransit* *Regular bus* *Express bus on streets*	*Trolleybus*	*Streetcar/tramway*
B	*Bus (semi-) rapid transit (BRT)*	Guided bus (O-Bahn)	*Light rail transit (LRT)*
A	Bus on busway on entire line[b]	*Rubber-tired rapid transit (RTRT)* Automated guided transit (AGT) *Personal rapid transit (PRT)*[c]	*Light rail rapid transit (LRRT)* Monorail *Rail rapid transit (RRT)* *Commuter rail (CR)* *Regional rail (RGR)*[d]

[a] Modes extensively used are in italic type.
[b] Mode not used in any urban area.
[c] Nonoperational mode (theoretical concept).
[d] Many commuter and regional rail lines have grade crossings, but with unconditional priority, so that they have no effect on operating speed and reliability.

large investment. A large area of medium-performance modes at medium investment cost levels is provided mainly by LRT, BRT, and AGT modes with their relative PCP positions as shown.

This diagram, like most tables, figures, and comparative statements about modes given in this chapter, is by its nature general, rather than numerically precise. Most numbers are given in ranges that encompass many transit systems. In most cases there may be examples that are outside the quoted values, but these are usually exceptions and their values usually apply only under certain specific conditions. The descriptive statements about mode characteristics and their mutual relationships are those drawn from exact physical facts and actual experiences from many transit systems.

12.1.5 Review of Technological and Operational Features

Following the classifications of modes by their ROW and technology, an overview of the technical components of major transit modes, including their guidance, propulsion, control, vehicle, and line capacities, is presented in Table 12.4.

Most of these elements, grouped in RTS groups, are presented graphically in Figure 12.3. The diagram shows the sequence of upgrading the system components. Each component has two or three steps, from lower to higher performance. Interdependence of some components is shown by arrows in the right-hand column. For example, operation of long trains requires ROW A, guided technology, and, with few exceptions, uses electric propulsion.

This diagram shows several basic transit modes as lines connecting a series of components. Regular buses on ROW category C (RB) can be considered the basic system, which utilizes steered rubber-tired small-capacity ($<$100 spaces) vehicles. A variation of bus technology is BRT, which utilizes category B ROW and often has medium-capacity (articulated) vehicles ($>$100 spaces).

Most modes with ROW categories A and B, on the other hand, have guided vehicles, steel rails, and electric traction. They operate large vehicles and medium to long trains.

Metro or rail rapid transit has all the highest elements except that, with a few exceptions, it does not have fully automatic operation (crewless trains). AGT modes are an interesting combination of these com-

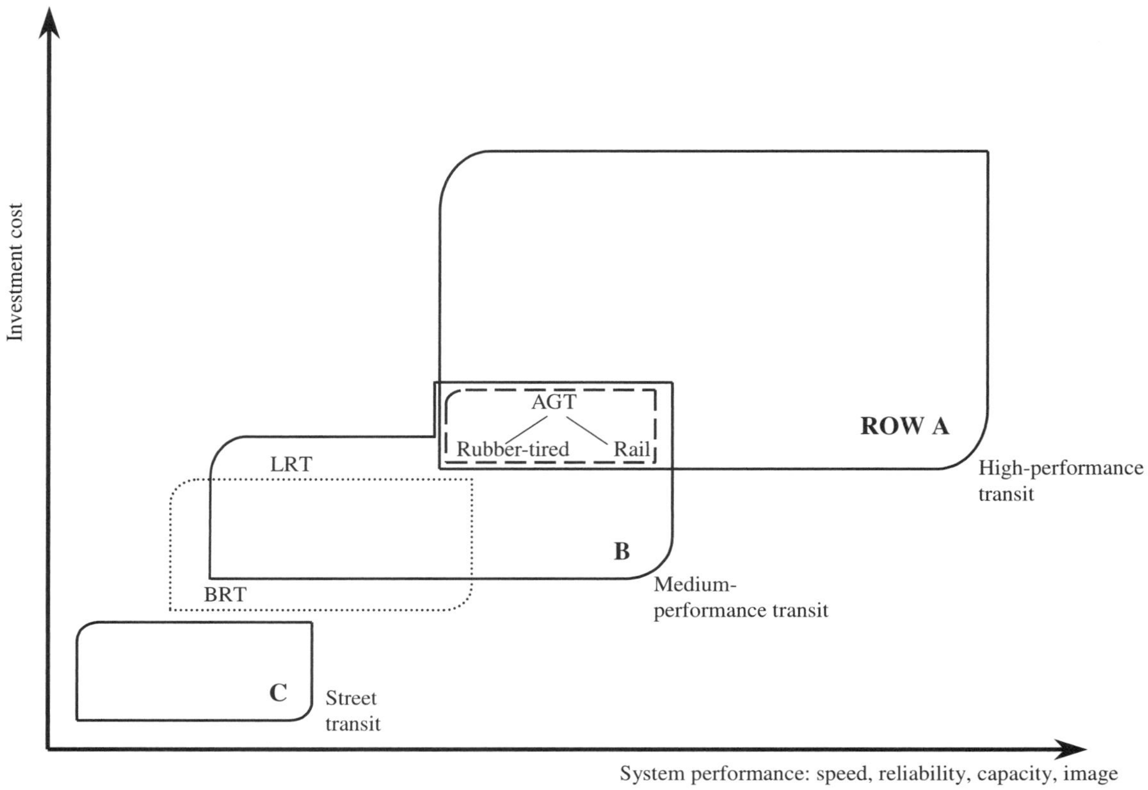

Figure 12.2 *Performance-cost packages (PCPs) of different generic classes of transit modes.*

Table 12.4 Characteristics of different transit modes

Characteristic / *Mode Category*	*ROW Category*	*Mode*	*Support and Guidance*	*Propulsion*	*TU Control*	*Cars per TU*	*TU Capacity (spaces)*	*Line Capacity (sps/h)*
Street transit	C	Bus	Road/ steered	ICE	Driver/ visual	1	80–125	3000–6000
	C	Tram	Rail/ guided	Electric	Driver/ signal	1–3	100–300	10,000–20,000
Semirapid transit—medium performance	B	BRT	Road/ steered	ICE	Driver/ visual	1	80–180	6000–24,000
	B	LRT	Rail/ guided	Electric	Driver/ signal	1–4	100–720	10,000–24,000
	A	AGT	Guided	Electric	ATO	1–6 (10)	50–480	6000–16,000
Rapid transit—high performance	A	LRRT	Rail	Electric	Signal	1–4	100–600	10,000–28,000
	A	Metro	Rail	Electric	Signal/ ATO	4–10	720–2500	40,000–70,000
	A	Regional rail	Rail	Electric/ diesel	Signal	1–10 (14)	150–1800	25,000–40,000

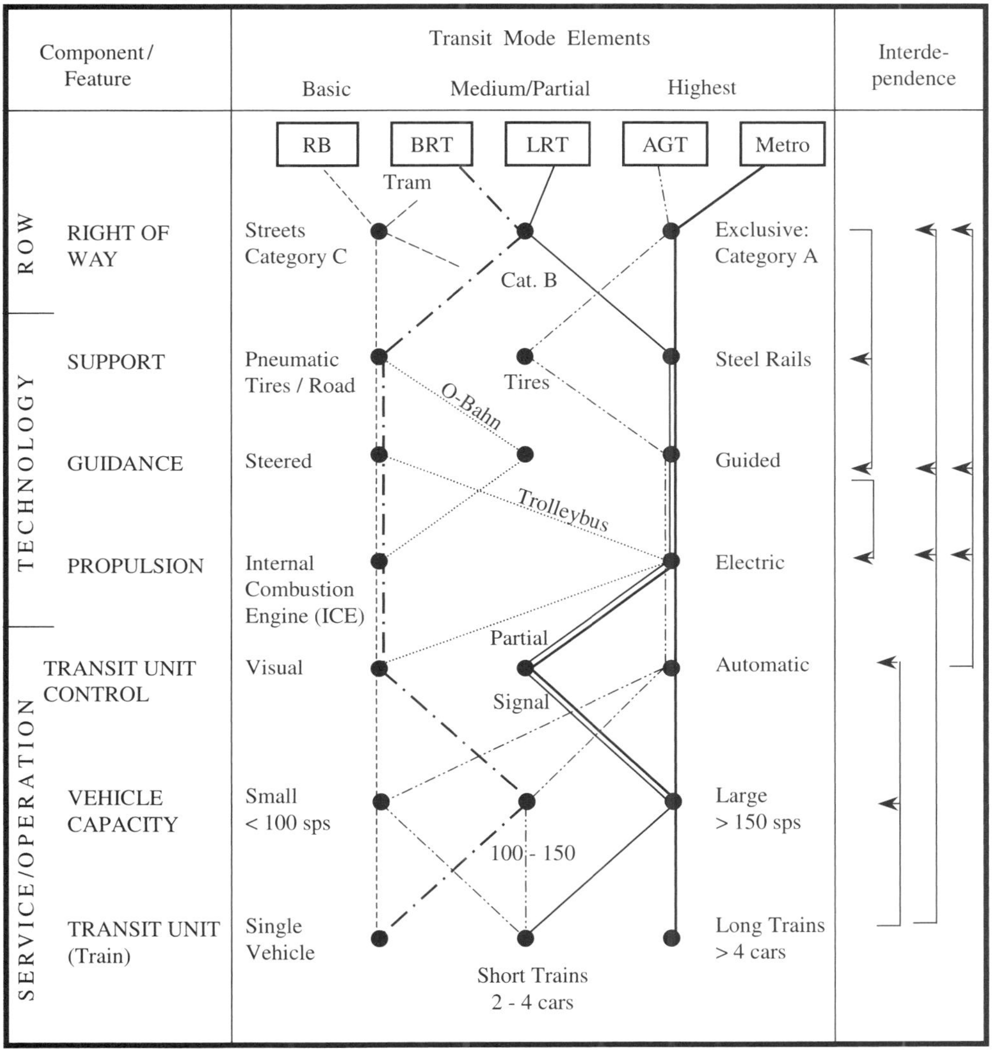

Figure 12.3 *Graphical presentation of the physical and technological features of different transit modes.*

ponents. They have ROW A, with rubber-tired electric vehicles that are guided and automatically driven but with small capacity and operated in short trains. These features are determined for reasons discussed above: serving medium passenger volumes with medium-capacity TUs makes automation necessary for their economic feasibility.

This diagram also shows the differences in components among the shown modes. For example, streetcar/tramway and LRT differ in ROW category (C versus B); their basic technology (guidance, rails, electric, large vehicles) is the same, but LRT has partly automatic control, which it needs due to its higher technical speed.

It is emphasized that the modes with basic components (ROW C, steered, small vehicles without signal control) and low PCPs should not be considered inferior to those with higher PCPs. They are only better suited for transit services that have lower passenger volumes and require lower speeds or less comfort, etc., and where the higher PCP modes are not physically and economically feasible.

A great variety of transit technologies and modes can be considered in planning high-performance transit. However, in many cities the planning process has been complicated, sometimes confused, and greatly lengthened by considering an excessive array of systems. Often, proposals and political pressures for construction of specific modes, including proprietary ones, create this problem. Actually, in practical planning the selection usually comes down to three or four modes, each with some technical or alignment alternatives. Therefore, brief descriptions of virtually all operational transit systems are given in the next sections, grouped into two generic categories: medium-performance modes, mainly semirapid transit (Section 12.2), and high-performance modes, or rapid transit (Section 12.3). In each category, the most common systems are compared to each other with respect to their main characteristics.

12.2 MEDIUM-PERFORMANCE TRANSIT MODES

As cities grow, they need to develop transit modes that have better performance and service than buses in mixed traffic can provide. Separate ROW is the critical element for improved performance, but other system components, such as steered or guided technology, stops, vehicles, and operations, must be considered as components of a coordinated system. Since fully separated ROW usually involves very high investment, a logical step for the first level of transit system improvement is building medium-performance modes, using partially separated ROW.

The most common role of medium-performance modes is the basic transit system in medium-sized cities, where network extensiveness is more important than the very high capacity and performance that rapid transit offers. For example, LRT is the main carrier in many cities with populations between 200,000 and 2 million in dozens of cities in Germany, the Netherlands, the United States, Canada, and other countries. BRT systems serve as the main transit lines, for example, in Curitiba, Brazil, Bogotá, Colombia, and Ottawa, Canada, while Lille and Toulouse, France, and Vancouver, Canada, use the AGT mode for their main lines. Many large cities with rapid transit also deploy medium-capacity modes for complementary networks. Examples of these roles are LRT systems in Boston, Budapest, Moscow, Prague, San Francisco, and many other cities.

Planning of medium- as well as high-performance modes usually starts with the definition of the maximum improvements of the existing system considered to be the base case alternative. This set of improvements, known as transportation system management (TSM), consists of traffic engineering measures, design and control of physical facilities, transit system operation, parking regulations, and pricing policies.

Consisting of diverse activities, TSM is complex organizationally, but it is usually cost-effective. If the organizational problems can be solved, TSM measures can result in significant improvements of traffic flow, transit services, and transit usage. However, in many cities there is a tendency to introduce various traffic engineering and transit priorities measures, but then fail to maintain and enforce them adequately. This results in decreasing efficiency in street and highway traffic flow, which may also affect transit operations: its speed and service reliability gradually deteriorate and operating costs increase. In such cases, TSM cannot be considered an effective and permanent solution to improve transit services.

Consequently, if TSM can be maintained and enforced, it can be used for maintaining efficient traffic conditions and transit services, but it cannot fundamentally change the basic character and roles of automobiles and transit. TSM is therefore usually considered the best condition of the transportation system without major investments, but it does not represent a real alternative to provision of medium- or high-performance transit.

Medium-performance transit generally consists of three modes: BRT, LRT, and AGT. These are described in the following three sections.

12.2.1 Bus Rapid Transit (BRT)

Since the mid-1990s, intensive promotion and publicity have been given to the bus rapid transit mode, particularly by the Federal Transit Administration of the U.S. Department of Transportation, often presenting it as a new highly promising concept. This promotion is a logical development because most of the existing bus services need improvements. The vast majority of bus services in U.S. cities consist of buses running in mixed traffic on many overlapping lines with different headways. In many cities, the obsolete practice of having bus stops on every street corner is still used. Bus travel speeds are often low and reliability depends on traffic conditions. With complex line alignments and confusing networks, buses often cannot compete with car travel, so they serve mostly captive riders.

In many other countries, regular bus services also suffer from excessive complexity and poor image. In many developing countries, particularly those with unregulated services, the problems are even more serious. Buses provide service on many poorly marked lines and have no schedules and no coordination with rail systems. Maps of these lines usually do not exist. Their service quality and safety are often low, and government regulations either do not exist or are not enforced.

Efforts to improve bus systems and the concept of BRT are not new. There have been numerous efforts to upgrade bus systems during recent decades, and it is useful to describe such efforts and discuss their effectiveness.

12.2.1.1 Correcting the Misconceptions about Transit from the 1950–1970 Period. During the era of rapid increase in auto ownership and the most intensive construction of urban highways in the United States, 1950–1970, transit was neglected and concepts about different modes and their types of operation were often misunderstood. During that period, some misconceptions were introduced in transit systems planning and network design, and they created serious obstacles to efficient operations and passenger attraction. These misconception particularly affected bus services because at that time bus networks were being greatly expanded, replacing streetcar and trolleybus lines. Introducing bus lines, which, unlike fixed guideway modes, could be easily changed, led to the concept that they should be dispersed to many routes. The fact that such networks usually degrade transit services and diminish their image was overlooked. It is therefore useful to review critically some fundamental misconceptions that were dominant and caused erroneous decisions. Four major misconceptions are discussed here.

Misconception I: Transfers Are Not Tolerable: If Passengers Must Transfer, Choice Riders Will Not Use Transit. This belief has resulted in attempts to provide more door-to-door services by running large numbers of complicated bus routes with low service frequency, irregular headways, poor image, and low passenger attraction. This pattern still prevails in many U.S. cities, and it is carried to even greater extremes in many developing countries with little or no transit regulation and planning, as mentioned in the preceding section.

Fact I: Intermodal, High-Quality Transit Networks Involve Considerable Transferring. With good station design, frequent service, and reliable operations, passengers easily accept transfers. The best transit systems (New York, Toronto, Munich, Paris) involve extensive transferring among lines and modes. Millions of passengers make transfers daily to perform the most efficient travel through multimodal networks.

Misconception II: Transit Can Be Provided Either by Buses on Streets or by Fully Separated Rapid Transit. This planning misconception created bipolarized systems that have services represented in the diagram in Figure 12.2 with only ROW C and A modes. It left many cities unsuited for rapid transit without any options for transit upgrading.

Fact II: Many Lines and Networks Need Better Service than Buses on Streets Can Offer, but for Much Lower Investment than Metros Require. This need has resulted in the development of LRT, BRT, and AGT in recent decades; these modes have filled the gap between street transit and rapid transit.

Misconception III: Bus Lines on Urban Streets Must Have Stops at Every Corner, while Rapid Transit Competes with Car Travel and Must Rely on Kiss-and-Ride and Park-and-Ride Access in the Suburbs; Therefore, It Needs Few Stations. This is an extreme polarization of modes by type of operation. The problem of excessively short bus stop spacings still exists in many U.S. cities. Such bus lines sacrifice speed for easy access: stopping at every corner creates slow, creeping bus service. Rapid transit, on the contrary, sacrifices access for high operating speed among few stations. With long station spacings, some metros designed in the 1960s do not serve many areas along their lines and discourage access by walking, bicycle, and transit feeders. The former loses choice riders because it is very slow and unreliable; the latter loses potential riders because it bypasses them, resulting in poor area coverage.

Fact III: This Bipolarization in Access/Operating Speed Ratios Needs Correction: Fewer Stops for Buses, More Stations for Rapid Transit.

Misconception IV: Flexible Transit Systems Are Needed. The vague term *flexibility* has been falsely proclaimed to be always a major goal in transit planning (see Section 11.3.2). It has been used not only to criticize transit systems, particularly rail, as inflexible, but also to imply that transit services that are changeable in alignment and schedules are superior to fixed, permanent, reliable ones.

Fact IV: Concepts Opposite to Flexibility Are Permanence, Reliability, Durability, Efficiency, Simplicity. These Are Desirable Features for Most Transit Services (Vuchic, 1971). Thus, while some transit services, such as commuter lines, can be diversified in scheduling and dispersed in routing, most transit services aimed at attracting incidental users and the general public must have fixed, permanent lines, fixed schedules, and known fares. They must be simple to understand and use. Moreover, the more fixed they are, the stronger are their impacts on land use, as well as on quality of life in the city.

12.2.1.2 Evolution of the BRT Concept. During the 1970s, the need for upgrading bus systems operations and services was already obvious, and several reports described possible improvements and innovations (Wilbur Smith and Associates, 1975; RATP, 1977; VÖV & VDA, 1979; Vuchic, 1981; Vuchic and Kikuchi, 1994). Numerous improvements to bus system operations were made, particularly with respect to the most important component: provision of ROW category B, from reserved lanes on streets to exclusive busways.

Many events that have occurred in bus system improvements and changes during recent decades deserve careful attention because they indicate which innovations can be efficiently introduced and which face opposition in their implementation, are impractical, or cannot be maintained. Ten major developments are listed and analyzed here.

1. *Reserved bus lanes on streets and in bus/pedestrian malls* were introduced during the 1970s in many cities, such as Los Angeles, Mexico City (Ejes Viales with buses and trolleybuses), Philadelphia (Chestnut Street Mall), Paris (extensive network of bus lanes with excellent results), and São Paulo (Comonor with buses and trolleybuses carrying over 20,000 prs/h). The general experience has been that these facilities are successful when large volumes of buses are operated, efficient police enforcement continually protects bus lanes, and there is strong political support for them. These conditions existed, for example, in Paris and São Paulo. Poor enforcement led to degradation of bus lane operations in Mexico City and abandonment of the Chestnut Street Mall in Philadelphia. The HOV lanes used by buses on the Santa Monica Freeway, evaluated positively by detailed professional engineering studies, were eliminated in the 1970s by a court decision, i.e., by a person who was a layman with respect to transportation system operations.

2. *Priority treatment at intersections* has been introduced with excellent results in many European countries, such as Switzerland and the Netherlands. In the United States, introduction of priority measures for buses as well as many LRT systems have faced considerable difficulties. Although the technology for signal and other priorities has existed for decades, even

today in Boston, Baltimore, and Los Angeles, buses with 80 persons and LRT trains with 300–400 persons are treated at intersections with the same rights as cars with an average occupancy of 1.3 persons. Bus priorities on the exclusive busway in Miami were discontinued after a few accidents, supposedly for safety reasons.

3. *Exclusive busways as separate facilities or in freeway medians* were build in Ottawa, Pittsburgh, Washington, DC (Shirley), Los Angeles (El Monte), Lima, and a number of other cities. Particularly successful have been busway facilities specially designed as ROW category B (with intersections at grade) providing bus priorities in São Paeblo and Curitiba, Brazil, an O-Bahn line in Adelaide, Australia, and in Bogotá, Colombia (Transmillenio). The busways in Ottawa, Curitiba, and Bogotá have attracted worldwide attention with their excellent comprehensive planning, efficient operation, and very positive image. However, most busways in U.S. cities, including Washington and Los Angeles, were degraded when the HOV facilities concept, discussed next, was introduced.

4. *The concept of HOV lanes and roadways* was introduced in the United States during the late 1970s. High-occupancy vehicle (HOV) facilities are lanes on streets or freeways that only vehicles carrying more than a prescribed number of passengers (4, 3, or only 2) are allowed to use. This concept, introduced in many U.S. cities since the 1970s, gives relative priority to more productive vehicles. This is a correct policy for improving urban transportation when regular lanes for general traffic are converted to HOV lanes. When new lanes for HOV are constructed, however, they benefit the HOVs, but general traffic also benefits from the elimination of HOVs. The reduced congestion then results in generation of new traffic and an increase in

Photo 12.4 *Busway with controlled intersection—Miami.*

Photo 12.5 *High-capacity bus/trolleybus lanes—São Paulo.*

vehicle-km traveled (VKTs), which is contrary to the goals and requirements stated in U.S. laws (the Intermodal Surface Transportation Efficiency Act [ISTEA], Clean Air Amendment [CAA], and their successors).

Another major problem with the HOV concept is that it has led to degradation of exclusive bus lanes and busways. The auto/highway interest groups have claimed that when buses do not follow each other closely on busways (which is nearly always the case, and it is the condition that ensures high speed, safety, and reliability of service), other vehicles can be let into these lanes to utilize the space. This claim, known as "empty lane syndrome," is based on a false belief that filling lanes with vehicles and introducing general drivers in addition to professional bus drivers does not affect the safety and reliability of traffic flow. Actually, allowing other vehicle categories into busways drastically changes the features of bus services.

When the claims that HOV facilities are more productive than busways were accepted by the Federal Highway Administration, nearly all busways in U.S. cities were converted into HOV facilities. The consequences of this change are summarized in the box on next page.

This degradation of busways into HOV facilities was paralleled by similar pressure for bus lanes on streets to be opened to other vehicles, such as HOVs, taxis, trucks, and bicycles. In recent years in some areas of the U.S. even HOV lanes on freeways are under attack by single-occupancy vehicle (SOV) motorists who are tempted when they see free-flowing lanes next to the congested lanes in which they are traveling.

5. *Land use impacts and interactions with surrounding areas* have varied. Excellent coordination between transportation and land use planning in Curitiba (Rabinovitch and Leitman, 1996) and Ottawa enabled BRT systems to have positive impacts on land devel-

Conversion of a busway into an HOV facility results in the following positive (+) and negative (−) changes (the affected vehicle classes are given in parentheses):

+ Increased capacity for vehicles, resulting in improved LOS, i.e., speed and reliability (vans, HOV autos)
+ Increased LOS on regular lanes due to elimination of HOV vehicles (SOVs, trucks)
+ Greater transporting capacity of the entire facility in terms of vehicles and passengers, although usually with decreased average vehicle occupancy (vans, autos)
− Increased vehicular volume, introduction of mixed traffic, police intervention to pull out illegal users from HOV lanes, etc. cause reduced LOS in the former bus lanes (buses).
− Bus passengers are often stolen from bus stops by single drivers who want to qualify for HOV lanes by increased occupancy ("casual carpooling"). Ridership and revenue losses (buses) result.
− The strong image and identity of the exclusive bus facility and competitive transit service are lost (buses).

In review, all the benefits from changing busways into HOV facilities are accrued by vehicles other than buses (vans, all classes of automobiles), i.e., bus competitors; buses experience only a deterioration of service and the loss of passengers and revenue.

opment around major stations and along the served corridors. In São Paulo, on the other hand, corridors along the highest-capacity bus/trolleybus lines have deteriorated economically and environmentally due to intensive noise, pollution, and physical separation of the two sides of the avenue.

6. *Bus vehicle design* has had very significant advances, including new vehicle types such as "push" articulated buses, double-articulated buses, and low-floor buses. In addition, many buses now offer increased comfort, large windows, improved appearance, and cleaner engines as compared to their predecessors from the 1970s (Hondius, 1995). Dual-mode buses, such as diesel/electric articulated buses in Seattle, Washington, have had limited success due to their great complexity and high cost. However, increased comfort and special demands result in reduction of seats in a standard 12-m-long bus from 47–50 to only 30, increasing investment and operating costs.

7. *O-Bahn or guided bus,* developed in the early 1980s (Vuchic, 1985), has had limited success with respect to implementation. The Adelaide system has remained the only major facility with guided buses. The interest in the concept of guided buses, however, continues to exist, and future applications are possible. An important advantage of this mode is that it physically prevents introduction of other vehicles into the bus facility. Optically or electromagnetically guided buses, which are also technically feasible, do not have any visual distinction, which is actually a disadvantage for ROW separation enforcement and system image.

8. *Organizational changes of bus transit* have had direct impacts on bus system operations and service quality. Deregulation of bus services in Great Britain and some developing countries has resulted in breaking up bus systems and making their technical and organizational upgrading difficult or infeasible. For example, a very effective busway in Lima, Peru, and a guided busway section in Birmingham, England, were discontinued when deregulation was introduced. Privatization and other forms of competition in bus systems have, on the other hand, resulted in improved services and innovations in operations in many cities.

9. *Applications of intelligent transportation systems (ITS) technology for upgrading bus services* have already brought significant results, and there is consid-

Photo 12.6 *Guided bus—O-Bahn in Adelaide (courtesy of Alex Wardrop).*

erable potential for their wider use in BRT operations, safety, fare collection, and enhanced passenger information.

10. *Introduction of BRT as a system concept* has resulted in the most comprehensive approach to bus services as a coordinated system, and it is the culmination of many bus improvements developed during recent decades (see the following section).

These 10 bus system developments and innovations since the 1970s are presented in the box with the following designations of their results: + Success; ~ Partial success, some failures; − Failures and setbacks.

This review of the historic developments of the BRT concept and elements since the 1970s shows that many efforts to upgrade buses have been made. Some represent significant progress, while others met difficulties and were only partially successful or were discontinued. Major experiences from these developments and lessons for the future can be summed up as follows.

1. Bus services are upgraded when extensive but infrequent flexible routes are replaced by fixed lines with reserved lanes, fewer but more distinct stations, and frequent service.
2. The main obstacles to upgrading bus services have been organizational and political rather than technological.
3. Provision of separate bus lanes in roadways must be followed by their continuous protection from pressures by lobbies to share their facilities with other vehicle categories.
4. Most steps of bus upgrading, such as provision of ROW category B, stations, trunk lines with transfers from feeders, and a stronger image, make bus lines more similar to LRT. However,

Bus system developments and improvements since the 1970s:

+ Bus lanes introduced—performance depends on their utilization, enforcement, and political support.
~ Bus/transit priorities along streets: successful in some countries, limited in the United States.
~ Construction of exclusive busways: successful unless converted into HOV facilites.
− Conversion of bus lanes and busways into HOV facilities degrades bus system performance and destroys its image.
~ Coordination with land use planning and impacts on surroundings: cases of successes and failures
+ Bus vehicle development: many innovations in vehicle body design and technology, such as low-floor buses and motors with reduced air pollution and noise.
~ O-Bahn or guided bus: applications very limited, but the potential for specific cases exists.
~ Organizational changes: deregulation has caused direct setbacks; competitive organization, such as tendering for coordinated services, brought improvements in services and efficiencies.
+ Introduction of intelligent transportation systems technology has brought many service and operational improvements.
+ Development of the bus rapid transit concept represents the most comprehensive approach so far to bus services as a coordinated system.

buses always suffer from a weaker image and lower performance than rail systems, and they are often subjected to pressures to share their reserved lanes with HOVs and other modes. LRT with well-designed ROW B is physically protected from such pressures and therefore has permanently separated tracks and ways.

5. Successful BRT systems are found in cities that have very strong planning, good traffic engineering, and a clear policy of prioritizing modes on the basis on the number of persons they carry. Ottawa and Curitiba illustrate the importance of these conditions.
6. The successful BRT applications in developing countries are also a consequence of the fact that labor intensity of this mode is not a serious problem because of low wages and social policies aimed at increasing employment.

12.2.1.3 Definition of BRT. With strong promotion of this mode, there is a tendency to use the term *BRT* for any bus line that has partial separation of lanes, new buses, or distinctive line designation. This proliferation of the term actually downgrades and hurts the BRT concept. It is therefore important to provide a technical definition of BRT. Because of many different system elements that define a transit mode, such a definition cannot be very precise; however, it must include certain elements that are *sine qua non* for the mode listed in the box above, and others that must exist on a system to various degrees to consider it a BRT.

In addition to these elements, BRT must also have some or most of the following features:

- Stations and stops that allow simultaneous stopping of two or more buses, their overtaking or platooning to provide operations with short headways (less than two minutes), and a line capacity greater than 3000–5000 spaces per hour
- Engines with low levels of air pollution and noise
- Convenient transferring to other bus and rail lines
- Prepaid or contactless fare collection, which minimizes delays during boarding and alighting

12.2.1.4 BRT Evaluation and Its Applications. Physical compatibility of buses with other highway vehicles is an advantage of bus transit because it

Bus rapid transit must have the following elements:

- Predominantly ROW B, which is not shared by other modes, such as HOVs, and only limited line sections with ROW C
- Distinct stops or stations with good passenger protection, information, and fare-collection equipment spaced on the average at least 300–400 m apart in central city with greater distances in suburban and outlying areas
- Distinctive lines with frequent, reliable service and regular headways during all daily hours
- Buses with attractive design, number and size of doors, low floor, or high platform that allow fast passenger exchange at stops and stations
- Preferential treatment at intersections
- Use of ITS techniques for monitoring vehicle locations and movements, passenger information, and fare collection

allows buses to use all streets and highways. At the same time, that compatibility is a major disadvantage when it is necessary to create a BRT line, which depends on extensive use of ROW category B.

Because since the 1970s bus lanes and busways in most U.S. cities have been converted into HOV facilities, it can be concluded that BRT systems can be introduced and maintained only in cities where at least one of the following conditions exists.

First, there must be a strong commitment to favor transit over private transport, bus facilities must be considered permanent, and police enforcement of bus priorities must be effective. This condition exists, for example, in South American cities such as Curitiba, São Paulo, and Bogotá. Second, busways must have physical characteristics that prevent their use by other vehicles, such as the trolleybus tunnel in Seattle and the use of guided buses such as the O-Bahn in Adelaide, Australia. Both of these conditions, permanent enforcement and physical separation, exist on large segments of BRT facilities in Ottawa, Pittsburgh, Brisbane, and cities in several Western European countries.

The most extensive and highest-capacity busways have been successfully developed in several Brazilian cities, notably São Paulo, Manaos, and Curitiba, as well as in Bogotá, Colombia, and in several Asian countries, such as Turkey and Indonesia. Several conditions have led to that development. First, in many rapidly growing cities in developing countries, extremely high passenger volumes have to be transported, but the countries lack investment funds and the time required to build rail systems. Busways on trunk lines are then clearly justified by the volumes. The high frequency of buses easily justifies keeping all other vehicles out, and popular support for better transit and government control are usually stronger than opposition by proautomobile groups. An additional important factor is that low labor wages do not make operating costs of large number of buses as much inferior to rail as is the case in industrialized countries. In addition, the higher service quality that rail modes offer is not as important as the need to provide transporting capacity faster than would be possible by constructing rapid transit systems.

In the United States, the current strong promotion of the BRT mode promises good results in applications of this mode in medium-sized cities that need considerably higher-quality transit than regular buses offer, but where a higher investment in LRT and the complexity of introducing a different technology is not necessary. A good example of such a BRT application is Eugene, Oregon. In several cities in this category and in some medium-density corridors in large cities, such as Wilshire Boulevard in Los Angeles, various versions of BRT have brought very cost-effective service improvements and passenger increases.

In several cities, strong promotion of BRT has led to some illogical developments. BRT is promoted as a lower-cost alternative than LRT. However, to make BRT services more similar to LRT, the design of BRT is made increasingly more elaborate. As a result, some busways with generous dimensions, aesthetic improvements, and large stations, such as in Pittsburgh, have required investment not much lower than average construction costs of LRT lines. Bus vehicles with elaborate body design, luxurious interiors with comfortable seats, and special low-polluting engines designed for Boston and Eugene, Oregon, have been estimated to cost over $2 million, which exceeds the cost of LRT vehicles, particularly when the much longer life and larger capacity of the rail vehicles are taken into account. In such cases, when the investment cost of BRT approaches the investment cost of LRT, it is logical to make a moderately higher investment and build LRT obtaining its significant advantages over BRT, such as electric traction, less obtrusive ROW compatible with pedestrian areas, superior ride quality, and greater passenger attraction.

The large potential for introducing many individual elements of BRT into regular bus systems is often overlooked. A bus line or network can be upgraded by increasing its stop spacings, making its stops more attractive, and improving its information system even if it does not qualify as a BRT mode. In many cities, investments in such improvements may be more cost-effective than the much higher investments for attempts to promote BRT over rail systems.

12.2.2 Light Rail Transit (LRT)

In recent decades, LRT has become the dominant mode of medium-performance transit, playing a large role between services offered by regular buses and those of rapid transit or metros. It is a complex and very diversified transit mode, so that its definition here will be followed by a description of its development during recent decades, innovations in its elements, and growth from streetcar mode to its present form, more similar to metro than to streetcar and bus systems.

12.2.2.1 Definition and Description. Light rail transit is an electric rail transit system consisting of high-capacity, spacious vehicles operating as one- to four-car trains mostly on ROW category B, but sometimes with ROW A and C categories on the same line. Its line capacities are in the range of 5000–24,000 sps/h. Quiet vehicles without exhaust can operate in tunnels and in pedestrian areas, having a distinct image as an element of urban livability with good passenger attraction.

Due to the great diversity of physical and operational features, the mentioned LRT characteristics are reviewed here in some detail with their typical ranges.

ROW category B is the most common, but LRT often has sections of ROW A (tunnels, aerial structures) or ROW C (street sections, pedestrian malls). There are lines that incorporate all three ROW categories, so that travel regime and speeds vary greatly along their alignments.

Typical vehicles are articulated, bidirectional, and 18–45 m long. Electrically powered, the vehicles operate in one- to four-car TUs, with total capacities of 50–240 seats and 120–720 total spaces.

Typical passenger volumes range from several hundred persons per hour in off-peaks to line capacities in the range of 5000–24,000 sps/h. With technical maximum speeds of 70–100 km/h, their operating speeds vary from 20–35 km/h depending on ROW category, local operating conditions, and station spacings.

Driving is manual, but control varies from visual to full signalization. At the high end of the LRT mode, fully automated driving is used on light rail rapid transit (LRRT) lines that have fully separated ROW. Recently, full ATO has been introduced on the LRT lines of the San Francisco Muni, which operate on ROW categories C, B, and A on different sections. When LRT cars enter the Market Street tunnel, the driver switches on the ATO. Emerging from the tunnel, the driver switches back to manual operation.

Investment cost for LRT systems at grade, along existing rights-of-way with few new infrastructure objects, such as the systems in Baltimore, Denver, San Diego, Salt Lake City, and San José, has been between \$8 M/km and \$28 M/km, averaging about \$15 M/km. LRT lines that required extensive structures, such as in Dallas (including some tunneling), Los Angeles, and Portland West Side (a long tunnel), were from \$21 to 38 M/km, averaging about \$30 M/km.

Vehicle costs are in the range of \$2–3 M, depending on the size, design, and complexity of the rolling stock.

The diversity of ROW categories, vehicles, and operations makes LRT adaptable to use in different corridors, from pedestrian/transit malls to high-speed suburban sections, from short transit lines to long regional lines, and from neighborhood services with simple stops to major lines with elaborate transfer terminals, etc.

12.2.2.2 Evolution of LRT Since 1960. The LRT concept is still often misunderstood and confused with its neighboring modes, streetcar and metro. To clarify this, a brief review of its development is useful.

Streetcars or tramways, the basic transit mode in most cities until the time of World War II, began to suffer from street congestion when automobile traffic intensified. In a naive analysis of street traffic conditions, most cities in the United States, Great Britain, France, and many other countries decided that streetcars (rather than private automobiles!) caused congestion and that this problem could be solved by replacing streetcars with buses, because buses are flexible and can better fit in mixed, congested traffic. Streetcars were proclaimed to be an obsolete mode, while buses were presented as the transit mode for the automobile era.

On the other hand, several central European countries, such as Austria, Belgium, the Netherlands, Germany, and Switzerland, took a different approach to urban transportation. Already at that time (approximately 1955–1970), cities in these countries realized the importance of creating transit that was competitive with the automobile for economic and urban livability reasons. This realization led to different policies with respect to transit modes. Most cities in these countries insisted that separation of transit from other traffic should be the goal, instead of "flexible mixing of transit vehicles with other traffic." This goal had a direct impact on mode selection. Instead of abandoning tramways/streetcars, these cities began to upgrade them by developing modern, high-capacity articulated vehicles, constructing separate rights-of-way, special signals, intermodal terminals for transfers with buses and metros, etc. For example, Hannover and Bremen, Germany, Göteborg, Sweden, and many other cities built ROW B for many lines, while Cologne, Frankfurt, Brussels, Essen, and other cities also built tunnels (ROW A) in central cities.

This extensive effort to upgrade traditional streetcars resulted in the creation of a mode that had much higher performance and better image. Although having the same basic technology as streetcars (electric rail transit), this mode is in some cases more similar in its speed, reliability, comfort, and safety to rapid transit than to conventional streetcars. It was therefore given a new name: light rail transit (LRT) (in German, *Stadtbahn,* and in French, *metro leger*).

Considerable confusion is created when some people refer to all nonmetro rail transit systems, including typical streetcar systems such as most lines in Toronto, Melbourne, and Zürich, as "LRT." This confusion comes because in some cities many streetcar lines were gradually upgraded into LRT, so that there is no sharp boundary between the two modes. Another source of confusion is that the boundary between LRT and metros is not sharp, either: there are systems that by their characteristics are between these two modes. Good examples of this are Metro Line 2 in Rotterdam, Skokie Swift in Chicago, Circumvesuviana in Naples, and several others.

Although there is today virtually a continuous range of rail transit systems, from historic trolleys and conventional streetcars to metro and regional rail systems, there are distinct differences between typical representatives of all major modes discussed here. The Toronto and Zürich streetcars are drastically different from typical LRT systems such as those in Calgary, Hannover, and St. Louis, which again differ by their infrastructure and service from typical metros, such as the Paris Metro, New York Subway, and Tokyo Subway.

When the U.S., British, and French cities not only replaced all streetcars, but in many cases abandoned excellent transit category B rights-of-way (e.g., Woodward Avenue in Detroit, a large network in Los Angeles), it was practically a capitulation of transit to auto traffic. These cities continued to suffer from decreasing transit ridership and increasing urban decay, to which traffic congestion was a contributor. In the mid-1970s, it was realized that separation of transit vehicles from general traffic is the basic element for improvement of its service; therefore, the concept of special bus lanes was adopted in several cities. The significant advantages of LRT over buses in terms of capacity, speed, and reliability were found particularly important for major, heavily traveled corridors. Most significantly, it was realized that rail technology and fixed guideway greatly facilitate provision of separate ROW. Rail tracks and separate ROW give LRT a much stronger distinction and image, resulting in greater passenger attraction than buses can achieve. Finally, its infrastructure gives rail systems permanence, influencing intensification of land use developments around stations where other conditions support it.

Thus, the streetcar mode went in many cities through a cycle of an ugly duckling becoming a swan: degraded and considered obsolete during the confusion created by the onslaught of the private automobile, it has reappeared in a new form after its inherent assets of high service quality, passenger attraction, and permanence have been rediscovered. Since the full development and recognition of this mode during the 1970s, new LRT lines and networks have been built in about 20 North American cities, in many cities in France, Great Britain, Spain, Ireland, Israel, and Australia, and in many developing countries, including Turkey, Egypt, Tunisia, the Philippines, and Mexico.

12.2.2.3 Review of Technical and Operational Innovations. In the course of LRT development during recent decades, new concepts in alignments and infrastructure, new vehicle designs, and operational practices have been introduced. These developments of LRT, its description and implementation in European countries, followed by its construction in many North American cities, are described in extensive technical literature. A particularly good source is the Proceedings of nine national conferences on LRT organized by the Transportation Research Board and Federal Transit Administration, six similar conferences organized by the International Association of Public Transport (UITP), Girnau (2000), and Vuchic (2001).

A summary review of the most significant developments that resulted in the creation of LRT is presented in the box.

Interestingly, the trend has clearly been from streetcar-type systems for in-city services to higher-performance LRT systems for longer, higher-performance lines. However, in recent years some interesting trends have developed. Following the invention of low-floor cars in the mid-1980s, tunnels in central cities have been forgone in some cities by placing LRT lines in transit malls (Calgary, Buffalo, Sacramento, Portland), thus utilizing a LRT system as a focal element for CBD activities. LRT has actually been successfully used as a symbol of city's rejuvenation, vitality, and enlightened planning.

Even more recently, during the 1990s, there also occurred a revival of streetcar-type services. Thus, the new rail systems in several French cities have most elements of LRT, but they operate on streets and in center city areas with functions falling between LRT and traditional, although technically greatly modernized, tramways. A similar development has occurred in some U.S. cities. The best known is the extremely popular new streetcar line F in San Francisco, crossing the entire CBD on Market Street and extending to Fisherman's Wharf. In Portland, Oregon, the very successful high-speed LRT lines have recently been supplemented by a typical streetcar line. Referred to as "Streetcar," this line has become a symbol of redevelopment and fashionable living in center city. Another such development has occurred in New Orleans, which in 2004 reintroduced streetcars with modern technology but classical vehicle design on its famous Canal Street.

12.2.2.4 LRT Productivity. The above-listed developments of LRT have had a major impact on operating characteristics and productivity of this transit mode. These developments are presented by a diagram and a

Major steps in the development of LRT into the dominant medium-capacity mode:

- Consolidation of extensive streetcar networks into fewer, higher-quality LRT lines
- Systematic replacement of ROW C by ROW B and A
- Integration of tunnel, surface, and aerial alignments on the same lines
- Intermodal integration with buses and metros
- Use of high-quality tracks and switches, drastically reducing noise production and vibration
- Articulated vehicles offering high comfort operating in one- to four-car trains
- Change from unidirectional streetcar-type vehicles to bidirectional metro-type rolling stock
- Self-service fare collection, allowing rapid boarding/alighting and use of any fare types
- Low-floor (0.30–0.35 m above top of rails) vehicles, facilitating boarding/alighting of all passengers and meeting the need of the disabled
- Operation of LRT in central cities initially in short tunnels superseded in some cities by direct running in pedestrian malls and zones
- Integration (track sharing) of LRT with regional rail lines for service to suburbs

Photo 12.7 *Light rail operation in Frankfurt: ROW C.*

Photo 12.8 *Light rail operation in Frankfurt: ROW B.*

set of numerical values in Figure 12.4. The traditional streetcars consisted of one- to three-car trains (case I in the figure), the latter having a driver and three conductors serving a train capacity of 300 spaces, resulting in a productivity of 75 spaces per crew member. Pressures to decrease operating costs and improve service quality led to increasing use of four-axle cars with a single trailer (II), and then technical development of articulated vehicles resulted in introduction of 30-m-long double articulated cars served first by two crew members (III). Then, with the introduction of self-service fare collection, the crew was decreased to only one person (IV), doubling the productivity of the crew. As the demand for higher comfort grew, some cities, such as San Diego and Sacramento began to operate three- and four-car trains with capacity of up to 760 spaces (V). This resulted in crew productivity of 760 spaces, about 10 times greater than the old streetcars had up to 1960. Introduction of low-floor vehicles led to further diversification of vehicle designs and TU consists, shown as cases VI–VIII, the last being a single 45-m-long car with 12 axles and five articulated body sections.

If the increase in operating speeds is considered, which approximately doubled (from 12–14 to 24–30 km/h) from conventional streetcars to LRT during this time period, crew productivity in terms of space-km offered per hour increased by an overall factor of 20. This sequence of developments has been a result of upgrading ROW from C to B category, preferential treatment of LRT vehicles at intersections, longer trains with up to 16 double-channel doors per side on which passengers alight and board without any delays for fare payment, etc. Thus, the innovations in infrastructure, vehicle technology, and operations have resulted in the emergence of the modern LRT mode,

Photo 12.9 *Same line on Row A—tunnel station with high platform.*

which has been built in recent decades in many cities as the most effective medium-performance system.

12.2.2.5 Types of LRT Systems and Their Applications. As the preceding sections show, LRT systems have a number of different forms because of their ability to be adjusted to functional system requirements, local conditions, and relationships to other modes in the city. A broad range of LRT systems, including their neighbors on the lower-performance side (streetcars) and the higher-performance side (LRRT and ALRT), are listed in the box as a set of nine different types with examples of cities in which they are used.

12.2.3 Automated Guided Transit (AGT)

Fully automated transit systems have been proven feasible since the 1970s. While full automation brings some benefits in better programmed travel regime, optimized travel with respect to energy consumption, and other operational aspects, these benefits are not vital and would seldom be significant enough to justify the higher investment and greater complexity of automatic than driven TUs and line operation. The most significant benefits of full automation are generally two.

First, elimination of train crews allows operation of many short trains at short headways to be performed at the same cost as operation of long trains at long headways. For example, if a metro line with six-car TUs is operated with 12-minute headways, full automation would allow that the TUs be shortened to two cars and service with 4-minute headways be provided without any cost increase.

Second, fully automated operation allows easy coupling and uncoupling of TUs to fit changes in demand

Years (approx.)	Cities	Length of TU (m)	Capacity (spaces)	Crew size	Capacity/ crew mem.
1910–60	I Hamburg, Vienna	3×10=30	240	4	60
1920–75	II Belgrade, Cleveland	2×14=28	220	3	73
1960–75	III Cologne, Düsseldorf	28	220	2	110
1965–	IV Cologne, Düsseldorf	28	220	1	220
1990–	V Sacramento	4×23=92	680	1	680
1996–	VI Cologne, Portland	2×28=56	330	1	330
1997–	VII Saabrücken	2×37=74	486	1	486
2003–	VIII Dresden	45	260	1	260

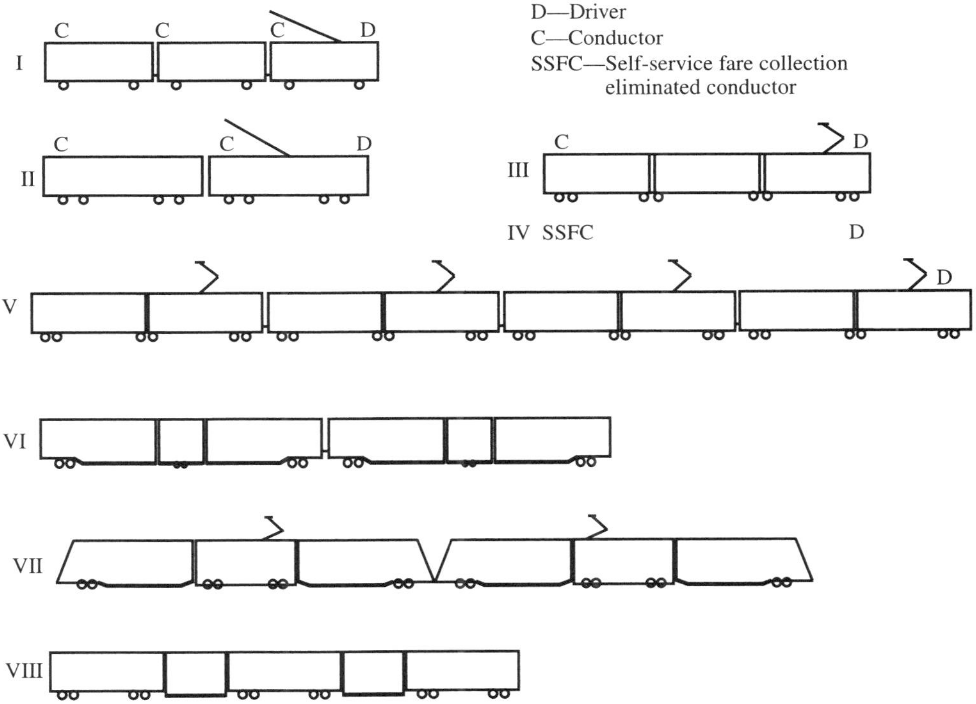

Figure 12.4 *Increasing performance and labor productivity of tramways/LRT through vehicle and operational innovations from 1950 to present.*

on a short notice simply by pushing command buttons in the control center, rather than sending crews to maneuver trains in yards. This results in better service and significant reduction of empty car-km operated.

The fact that short trains can be operated at a cost proportional to the number of vehicles per TU makes provision of high-quality, frequent service feasible not only on heavily used metro lines, but also on medium-

Photo 12.10 *Modern 70% low-floor LRV—Dresden, Germany* (*Vukan R. Vuchic*).

Main categories of light rail transit systems and its neighbors, streetcars and minimetro modes, with ROW category A only:

1. Conventional tramways: Moscow, St. Petersburg, Toronto
2. Upgraded conventional tramways: Amsterdam, Melbourne, Oslo, Vienna, Zürich
3. New tramway systems: Grenoble, New Orleans, Portland Loop, Valencia, Spain
4. LRT networks developed from tramways: Berlin, Boston, Cologne, San Francisco, Stuttgart
5. New LRT systems: Birmingham, Calgary, Dallas, Denver, Nantes, San Diego
6. LRT systems in suburbs of megacities: Hong Kong (Tuen Mun), London (Croydon), New York (Hudson-Bergen), Paris (Bobigny)
7. LRT/regional rail integrated systems: Bern, Karlsruhe, Manchester, Saarbrücken
8. Light rail rapid transit (LRRT): Essen-Mülheim, Manila, Göteborg (Line 8), Philadelphia (Norristown)
9. Automated light rail transit (ALRT): Copenhagen, Kuala Lumpur, London (Docklands), Vancouver (Skytrain)

capacity lines if these are automated. For this reason, most efforts to apply fully automated guided transit have focused primarily on medium-performance systems.

A great number of manufacturers developed automated transit systems, named automated guided transit (AGT) or automated people movers (APM). Initial efforts to use small cabins—vehicles with 20, 10, or even personal rapid transit (PRT) with three to four spaces for only individual parties, was shown to be operationally and economically infeasible as soon as a few systems were built (PRT was never even tested because of its obvious flaws). Most of the AGT systems have therefore been designed with vehicle capacities of 80 or more spaces operated as single vehicles or short trains. With this size, most of the proprietary systems were designed with rubber-tired guided technology.

Several AGT systems have also been developed with conventional rail technology. Their vehicles and trains generally have greater capacity than rubber-tired vehicles.

12.2.3.1 Rubber-Tired AGT. A number of AGT systems with different arrangements of rubber tires and running and guiding surfaces have been developed in recent decades. Most of them have been used in airports, amusement parks, and fairgrounds and as shuttles on university campuses and hospital complexes, etc. However, since the 1980s, several rubber-tired AGT systems became prominent representatives of this mode: Bombardier (former Westinghouse), Siemens (former MATRA) VAL, and Airtrans (including its version adopted as standard for several Japanese cities). Recently, Mitsubishi has also entered the field. Following their successful application at about 25 airports, these systems have also been built as regular transit lines in several cities.

The Bombardier system, formerly known as the Skybus, or Westinghouse C-100, has been the most common system for airport AGT applications. It has been built at about 20 airports, most notably in Atlanta's Hartsfield, where the airport was designed with AGT lines as an integral part of its functional layout. The same system also serves as a shuttle in London Gatwick, Miami, and Jacksonville airports. Most recently, several European airports, such as Frankfurt and Rome, have opened their shuttles with Bombardier AGT systems.

The same AGT system, built by Westinghouse, also operates as a regular transit line—the Downtown People Mover in Miami—and as a feeder to a rapid transit line in a suburban residential area of Singapore. This system's vehicles, running on four dual tires, can have capacity of up to 100 spaces. They operate in one- to three-car TUs (Atlanta). Longer TUs are technically possible but are not used anywhere. With minimum headways of two minutes, this system, operating three-car trains, can offer a theoretical capacity of 9000 sps/h.

The Siemens VAL system, developed by MATRA in France, was designed for a medium-performance transit line in the city of Lille. To minimize tunneling costs, cars were designed with very narrow bodies (2.06 m), so that capacity of its 12.5-m-long car with wide longitudinal seats is only 90 sps, including standees. The basic TU is a married pair of two cars. To compensate for small TUs, the control system has been designed to operate very short headways. The line in Lille has been operating with one-minute headways, so that its theoretical capacity is 10,800 sps/h.

For its later applications (Chicago O'Hare Airport and a major line in Taipei), MATRA developed 2.56-m-wide cars with correspondingly greater vehicle capacity. The VAL system has also been built at Orly Airport in Paris and in the cities Rennes and Toulouse, France.

The Airtrans system, originally designed by LTV/Vought Corporation, was the first AGT operating on a complex network: it was opened in 1974 as an integral part of Dallas-Fort Worth Airport, serving several loops that connect many terminals. Automated trains consisted of one or two cars, each with a capacity of 40 spaces, maximum speed of only 28 km/h, and very low riding comfort. The system also transported freight transport among airport terminals.

The Airtrans system found wider applications when its technology was adopted by the Japanese government as a standard AGT system for its cities. Airtrans is now used on regular transit lines in several Japanese cities (Osaka, Yokohama). It is interesting that these lines use fixed train consists (four or six cars) at all times, thus forgoing one of the main potential advantages of the AGT mode: tailoring train lengths to fit variable demand at different times of day or for special occasions.

Several medium-capacity automated rail systems have also been used for regular transit lines. They have considerably greater capacity than the rubber-tired AGT systems and are therefore discussed in Section 12.3 on high-performance modes.

Table 12.5 lists significant developments of automated guided transit systems, showing that the AGT concept actually spans a broad range of transit systems. The small to medium-sized rubber-tired vehicles, which are usually identified with AGT, were initially introduced on short shuttle lines. During the 1980s, this concept was extended to include rail systems with medium-sized vehicles and higher capacity than rubber-tired AGT systems, and full-size regular metro systems operated with fully automated control. Thus, the AGT concept actually spans from medium-performance to high-performance modes, described in Section 12.3.

12.2.3.2 AGT Evaluation and Its Further Applications. Rubber-tired AGT systems with low to medium capacity will continue to dominate shuttle lines in major traffic generators, particularly airports. With their ability to negotiate sharp curves and steep gradients with very little noise, they offer better performance on such lines than rail modes.

In transit applications, AGT systems are likely to be applied to rather short lines as feeders to rail transit and intercity terminals. Their shortcomings on regular transit lines as compared to rail systems are several. First, their vehicles are smaller and less comfortable than rail vehicles. The ride quality of guided rubber-tired vehicles is inherently less stable than that of rail systems. Second, their investment cost is higher than the cost of LRT by a factor of two to three. Finally, to

Table 12.5 Historical development of fully automated guided transit systems

City, Line	Year	Train Length: Cars	Meters	Spaces	Crew Size	Operation; Line Capacity	Type of System
Dallas-Fort Worth, Airtrans	1974	2	13	80	0	ATO, ATS; low capacity	Interterminal network in airport
Morgantown	1975	1	5	21	0	ATO, ATS; low capacity	Shuttle line on university campus
Atlanta Airport, Westinghouse	1980	3	36	420	0	ATO, ATS; medium	Shuttle in airport building complex
Lille, VAL	1983	2	28	172	0	ATO, ATS; low	Regular rubber-tired AGT
Vancouver, Skytrain ALRT	1986	4	51	440	0	ATO, ATS; medium	Regular rail AGT, roving attendant
London, Docklands LRT	1988	2[a]	56	254	(1)	ATO, ATS; medium[a]	Regular rail AGT, attendant on each train
Lyon Metro, Line D	1993	3	50	450	0	ATO, ATS; medium	Automated rubber-tired metro
Paris Metro, Line 14	1998	5	75	750	0	ATO, ATS; high	Automated rubber-tired metro
Singapore, Northeast Line	2002	6	120	1200	0	ATO, ATS; high	Automated rail metro

[a] Being lengthened to four-car trains in 2004 to double line capacity.

Photo 12.11 *AGT as airport shuttle in Frankfurt.*

avoid the very high cost of tunneling, AGT systems are placed on aerial structures, which create problems where passing along buildings in cities terms limits deployment of AGTs in urban centers and residential areas.

As has been the case since the 1980s, it is likely that rubber-tired AGT systems will be built in some medium-sized cities, but their broad application does not appear likely. Fully automated rail systems are likely to be built in more cities, as will be discussed in Section 12.3 on high-performance modes.

12.2.4 Comparisons of Medium-Performance Modes

As a result of numerous inventions in recent decades, the set of transit systems that comprise the family of transit modes now represents practically a continuous range of systems, from the bus, trolleybus, and streetcar modes of the street transit category to medium-performance transit modes including BRT, guided bus, and a series of LRT variations. There is a clear distinction of high-performance modes by their exclusive use of ROW category A, although the AGT modes span across this division: with some rapid transit features, most AGT modes still belong to medium-performance systems because of their considerably smaller capacity and speed.

In spite of this near-continuum in performance among modes, comparisons of medium-performance modes are not simple because specific technological modes differ in many basic physical and operational features. Several comparisons of neighboring modes, focusing on medium-performance ones, are given here.

Photo 12.12 *AGT as Downtown People Mover in Miami.*

Individual characteristics are defined by numerical values, usually in ranges, by descriptions, and by qualitative evaluations.

12.2.4.1 Light Rail Compared to Buses. Comparisons of all major medium-performance modes are presented here. The first comparison, given in Table 12.6, includes three modes: regular bus (the typical bus system used in most cities around the world), BRT, and LRT. Their characteristics are grouped in three categories: system components, lines and operational components, and system characteristics.

This table clearly shows the elements in which BRT differs from RB: ROW category, somewhat larger maximum vehicle capacity, different line designs and stop spacings, higher investment, and higher passenger attraction. LRT has even more significant characteristics that are superior to those of BRT: it is an electric rail system with much larger TUs and fewer lines with higher service quality. It requires even higher investment than BRT but has a much stronger image, passenger attraction and positive impacts on the city. The three modes are rather different as PCPs, increasing from RB to BRT and then to LRT. Therefore, RB dominates applications on lines that require basic service. Because higher investment is made to obtain higher performance modes, BRT and LRT represent two distinct steps (see the box on next page), with considerable differences among all three of them.

Among these differences between the two modes, the higher passenger attraction by LRT, a so-called rail factor, is of considerable importance. It is useful to give some examples of this phenomenon from actual developments in cities when LRT systems were introduced. A comparison between Columbus, Ohio, and Sacramento, California, is appropriate because the two cities have many similarities but have had very different transit developments.

As the figures in Table 12.7 show, Columbus and Sacramento have very similar populations in their urbanized areas and in transit service areas. Both are very auto-oriented, with extensive low-density suburbs. Both are state capitals with sizable concentrations of office jobs and economic bases. Both cities had all-bus transit systems in 1987, but transit ridership in Columbus was 25% higher than in Sacramento.

Transit developments in the two cities were quite different in subsequent years. Sacramento built one long U-shaped diametrical LRT line (another line was opened in 2003) and reorganized the entire transit network to take full advantage of the new high-performance rail line. The results were quite dramatic. While transit riding in Columbus increased by only 7%, Sacramento experienced an increase of 101%. The riding habits in the two cities increased by the same amounts, 7 and 101%, respectively.

The statistics in the table show that Sacramento experienced a drastically different, much more successful development of transit during the 1987–2000 year period. Three major events can be credited for that success. First, LRT was built, introducing much higher-quality service than regular and express bus ser-

Table 12.6 Comparison of regular bus, bus rapid transit, and light rail transit characteristics

Characteristic \ *Mode*	*Regular Bus (RB)*	*Bus Rapid Transit (BRT)*	*Light Rail Transit (LRT)*
	System Components		
ROW	C	B (C, A)	B (C, A)
Support	Road	Road	Rail
Guidance	Steered	Steered	Guided
Propulsion	ICE	ICE	Electric
TU control	Driver/visual	Driver/visual	Driver/signal
Vehicle capacity (spaces)	80–120	80–180	100–250
Max TU size	Single vehicle	Single vehicle	1–4-car trains
Max TU capacity	120	180	$4 \times 180 = 720$
	Lines/Operational Elements		
Lines	Many	Few	Few
Headways on each line	Long/medium/short	Long/medium/short	Medium/short
Stop spacings (m)	80–250	200–400	250–600
Transfers	Few	Some/many	Many
	System Characteristics		
Investment costs/km	Low	High	Very high
Operating cost/space	Medium	Medium	Low
System image	Variable	Good	Excellent
Passenger attraction	Limited	Good	Strong
Impacts on land use and city livability	Least	Moderate	Strongest

LRT, compared to BRT, has the following advantages (+) and disadvantages (−):

+ Due to rail tracks instead of roadway lanes, separate ROW (B or A) for LRT is easier to design and requires no protection by police enforcement as do bus lanes and busways.
+ Due to electric traction, LRT has better vehicle performance than BRT.
+ LRT produces no exhaust along the line and much lower noise than BRT (except that in a few cities on high-speed LRT lines trains are required to use horns at grade crossings).
+ LRT is more acceptable in pedestrian streets and areas than buses.
+ LRT can use tunnels, BRT cannot;
+ LRT vehicles are more spacious and comfortable and have better riding quality than buses.
+ LRT has a stronger image, it is more popular, and attracts more riders.
+ LRT has a stronger positive impact on urban development than BRT.
− Investment costs for LRT are higher than for BRT.
− As a different technology, LRT requires more extensive construction of infrastructure, new equipment, and longer implementation.
− For the first LRT line in a city, introduction of new technology requires construction of special yards and maintenance facilities.
− LRT services are limited to track network and involve more transfers than buses.

Table 12.7 Comparison of two cities of similar size and impact of LRT introduction

Item	*Columbus, Ohio*	*Sacramento, California*
Urbanized area population (1990)	945,237	1,097,005
Service area population	961,437	937,146
Annual passengers (millions of unlinked trips)		
1987 Bus + Rail = Total	17.5 + __ = 17.5	14.0 + __ = 14.0
2000 Bus + Rail = Total	18.7 + __ = 18.7	19.5 + 8.6 = 28.1
Percent increase 1987–2000	7%	101%
Riding habit (annual trips per capita)		
1987	18.2	15.0
2000	19.5	30.2

Source: Adapted from Federal Transit Administration (FTA), *Transit Profiles,* 2000 Transit Database Report Year. Washington, DC: FTA.

vices had offered. Second, the bus network was reorganized to feed LRT as the trunk line in two major corridors. And third, a timed transfer system (TTS; see Section 4.5) was introduced, greatly improving intermodal and intramodal connectivity of the entire transit network. Thus, the introduction of the LRT was the catalyst for improved transit service quality, more efficient network structure, and stronger image of the system. The impacts of these improvements caused by introduction of LRT on passenger attraction are obvious from Table 12.7.

Portland, Oregon, experienced similar growth of transit passengers when it introduced an LRT system. Transit ridership in 1987, the year the LRT opened, was 42.0 million passengers on buses and 5.9 million on LRT. In 2003, there were 62.7 million passengers on buses and 26.1 million on LRT. Total annual ridership increased from 47.9 to 88.9 million, or 86%.

Most cities that introduced the LRT mode in recent decades recorded similar increases in the number of passengers. In most cases, a new LRT diverts bus passengers and attracts a substantial number of new transit users. High service quality and strong image of LRT, as well as improvements of all transit services caused by LRT introduction, result in such passenger growth that the buses also have greater numbers of passengers despite the diversion of many of their previous users to LRT. The rail factor is thus very significant and obvious from the experiences in cities like San Diego, Calgary, St. Louis, Dallas, Denver, and Salt Lake City.

For a summary of the differences between these two modes, their major characteristics are listed and verbally described in Table 12.8. The superior mode with respect to each characteristic is listed in the last column.

The summarized evaluation and comparison in this table again shows that BRT and LRT are different in many aspects. Generally, BRT requires lower investment and shorter implementation time than LRT, but LRT provides higher service quality, attracts more passengers, and has stronger and more permanent positive impacts on its served area and often on an entire city. BRT has physical limitations: in cities with high density where exclusive busways or lanes cannot be provided, it cannot enter the urban core. LRT can do that either by using short tunnels (with higher investment) or by operating on various types of ROW B, including running through pedestrian areas.

The present strong promotion of BRT often includes some misconceptions about its relationship to LRT that can generate confusion, particularly among the lay public. Three such misconceptions are discussed here.

Misconception I: BRT Can Provide the Same Services as LRT at a Much Lower Cost. The preceding comparisons of the two modes clearly show that they do not produce the same services; LRT is superior in many features and provides a much stronger and more permanent role in the city than BRT can ever provide.

Table 12.8 Comparison of main BRT and LRT features

Characteristic \ Mode	Bus Rapid Transit (BRT)	Light Rail Transit (LRT)	Superior Mode
Vehicle performance and passenger comfort	Good	Excellent	LRT
Investment cost	High	Very high	BRT
Implementation time	Short	Medium	BRT
Operating cost	Lower for low passenger volume	Lower for high passenger volume	~
System image and passenger attraction	Good	Excellent	LRT
Air pollution and noise	Considerable	No direct pollution, very low noise level	LRT
Interaction with land development	Limited	Very good	LRT

Many technical papers about BRT, such as JPT (2002), discuss how BRT must approach the features of LRT. Thus, the more similar a BRT system is to LRT, the better its performance is. There are many differences between the two modes, and some of them can be reduced, but not eliminated. The measures that upgrade buses to acquire some features of LRT, such as higher quality of ROW design, larger stations, and very sophisticated vehicles, require additional investment, however. Upgrading of BRT then reduces its advantage over LRT with respect to lower investment costs.

Misconception II: Rail Modes Do Not Attract More Passengers than Buses Because This Attraction Depends on Service Parameters Such as Speed, Reliability, and Comfort, in Which Buses Can Match Rail. This statement is self-contradictory. If it is claimed that passenger attraction depends on service parameters, that actually proves that LRT has a substantial advantage over BRT because it has inherently higher service parameters than buses. In other words, the fact that passenger attraction depends on service parameters actually proves that LRT is superior to BRT with respect to passenger attraction. Actual experiences with passenger attraction by these two modes were described above.

Misconception III: LRT Is Superior to BRT in All Applications. This is another incorrect overgeneralization. The fact that LRT provides higher quality of service does not by any means make that mode better suited to all applications than BRT. There are many situations where BRT's advantages, such as lower investment cost and easier implementation, are much more important than the higher comfort and capacity of LRT. The ability of LRT to operate in tunnels is irrelevant in small to medium-sized cities that can provide ROW category B on their major arterials.

In conclusion, BRT and LRT are quite different modes and each one has a significant domain in urban transportation. Any publications that attempt to prove that one of these modes is always superior to the other, or that one of them has no place in urban transportation (such papers against rail transit are frequently produced and widely distributed in the United States), are products of special interest groups and have no professional validity.

12.2.4.2 Light Rail Transit Compared to AGT.

This comparison is rather difficult to make because AGT covers a very broad range of modes. If AGT represents mostly small- to medium-capacity rubber-tired AGT systems, the differences between the two modes are summarized in the box.

In some countries, particularly in France and Japan, which have several AGT producers, there are intensive debates about mode selection between LRT and AGT. LRT has been selected in most cities in Europe and North America, usually because of its considerably

LRT systems compared to rubber-tired AGT systems have the following advantages (+) and disadvantages (−):

+ LRT is not limited to ROW A only; it can utilize ROW categories B and C also.
+ LRT can fit into urban and pedestrian zones and enhance their environment.
+ LRT requires much lower (two to three times) investment cost per km of line than AGT.
+ Rail vehicles offer considerably better riding comfort than two-axle rubber-tired guided vehicles.
+ LRT has a good image and is very popular as a symbol of the city.
+ Rail systems are generic, produced competitively by many manufacturers; AGT modes are proprietary, produced by one supplier only.
− LRT cannot be operated automatically, unless it has only ROW A.
− LRT has lower speed and much lower frequency of service than AGT.
− LRT has somewhat lower safety than AGT.
− LRT schedules cannot be as quickly adjusted to unexpected changes as AGT schedules.

lower investment cost and greater physical ability to fit into a urban environment without either tunneling or using aerial structures that are not acceptable in high-density areas. In many Asian cities, particularly in Japan and Singapore, AGTs have prevailed when large funds were made available and cities are not so environmentally sensitive that aerial structures would be objectionable. Overall, LRT systems are used in many more cities and have much more extensive networks than AGT systems.

12.3 HIGH-PERFORMANCE TRANSIT MODES

The basic feature for provision of high-performance transit service is operation on ROW category A only. With exclusive use of protected ROW, guided modes represent the logical technology because guidance allows use of high-capacity electric trains with signal control. By far the most dominant modes in this category are rail rapid transit and regional rail systems. This section will, however, include several other high-performance modes, starting with LRRT operating exclusively on ROW A and moving to rubber-tired rapid transit and monorails and finally to the highest-performance mode, regional rapid transit.

12.3.1 Light Rail Rapid Transit (LRRT) Modes

This group of modes includes systems that have modified LRT rolling stock operating on alignments with ROW category A over their entire length. When these LRT trains are operated by drivers, the mode is known as light rail rapid transit (LRRT). The same mode but with fully automated operation is known as automated (or advanced) light rail transit (ALRT). The fact that it is automated indicates that this system has only ROW A, i.e., it actually represents an automated LRRT.

12.3.1.1 Light Rail Rapid Transit. Rolling stock of this mode has several differences from regular LRT stock. LRRT stations usually have high platforms and cars with several doors for fast passenger boarding and alighting. The trains have high operating speed and fail-safe signalization, the same as metro systems.

Thus, LRRT can be considered the highest type of LRT, or a smaller-scale "mini-metro."

There are not many LRRT systems because in most cases transit lines are built with ROW A when there are large passenger volumes, and in such cases long trains and other metro features are needed. Only in special cases is this highest type of ROW provided for lines that carry only moderate passenger volumes. Examples of this mode include the Norristown Line in the Philadelphia suburbs, line 8 in Göteborg, Sweden, and several recently built fully separated LRT lines, such as those in Los Angeles, Manila, the Philippines, and Frankfurt and Essen-Mülheim, Germany.

With the increasing demand for high-quality transit services, increasing use of the LRRT mode is likely in the future. Conversion of commuter railway lines to this mode, to achieve considerably lower operating costs and better level of service, is a logical development in many cities.

12.3.1.2 Automated Light Rail Transit. This mode is a fully automated version of LRRT, or it may be considered an AGT system with rail technology. It involves higher investment cost than LRRT because of automation and certain operational provisions that such operation necessitates (platform supervision, detection of objects fallen on track, emergency procedures, etc.), but its operating costs are lower, even for operation with shorter headways than LRRT, because they do not involve driver's wages. Without train crews, operating cost of, for example, four single-car trains is the same as that of one four-car train (at four times longer head-

Photo 12.13 *Light rail rapid transit (Green Line) in Los Angeles.*

RRT/metro compared to LRT has the following advantages (+) and disadvantages (−):

+ With ROW A only and full signal control, RRT has higher speed, reliability, and safety.
+ Due to long trains and rapid passenger exchange at stations, RRT has much higher line capacity.
+ With its high performance and strong image, RRT has stronger passenger attraction.
+ Distinct image of its stations and system permanence give RRT greater potential for positive impacts on urban development than any other transit mode.
− RRT requires substantially higher investment cost, causes more disruption due to construction, and requires longer implementation time than LRT.
− RRT requires more rigid alignment than LRT and cannot penetrate pedestrian areas.
− LRT is more conducive to construction in stages, utilizing different ROW categories, while RRT is limited to ROW A only.

ways). The other benefits of full automation, discussed in Section 12.1.2.4, also apply in this case.

The ALRT system, opened in Vancouver in 1986 and named Skytrain, is the best-known example of this mode. Operating up to four-car trains at headways as short as 2.5 minutes, the theoretical capacity of the Skytrain reaches 28,860 sps/h, considerably greater than the capacities of rubber-tired AGT systems. The Skytrain has propulsion by linear-induction motor, but that feature is independent of the basic system concept. The same type of ALRT with conventional electric motor propulsion is used on Detroit's Downtown People Mover, a collector-distributor loop in the CBD.

Another major ALRT system is the London Docklands system, a line planned and built as a part of a major urban development. It has been subsequently expanded into a network. The Docklands Line has an attendant on each train, but it can still be considered ALRT because driving is automatic and the attendant is on the train to assist passengers with information and fares. In 2003, another ALRT was opened in Copenhagen.

ALRT is a mode that will have increasing importance in the future because technical and operational problems of automatic driving have been solved and tested, and high quality of service (mainly short headways) is increasingly important for passenger attraction on all high-performance transit systems.

12.3.2 Rail Rapid Transit (Metro)

Rail rapid transit (RRT), popularly better known as metro, utilizes high-capacity electric trains with high acceleration and braking rates. It represents the highest-performance transit mode with the lowest operating cost per space-km. Its implementation requires very high investment and extensive construction, including disruption of areas along future lines. However, RRT systems have virtually unlimited life and exert a strong, permanent impact on mobility of population, urban form, and character.

A comparison of RRT with its closest widely used neighbor in the family of modes, standard LRT, is presented in the box.

This comparison shows that RRT has a significantly higher PCP than LRT. RRT actually represents the highest-performance transit mode that requires the highest investment and has the most permanent positive impacts on the city and region. The differences in these features between typical RRT systems and typical systems of other modes are considerable, but variations in designs of different modes have shown that LRT and regional rail systems can be designed to such high standards that they approach rapid transit in their characteristics. As already mentioned, the innovations in transit system designs in recent decades have created a nearly continuous family of transit modes. This is particularly the case among rail and other guided transit modes.

Growth of cities, increasing demand for travel, and the growing need for high-performance transit that is independent of frequently congested urban streets have resulted in construction of metros in a large number of cities during recent decades. In 1950, only 17 cities in the world had metros; in 2005, that number exceeds 100 cities.

The dominant reason for building many metro systems is provision of high transporting capacity in major urban corridors and arterials. In an increasing number of cities, however, metros are being built not only for greater capacity, but for provision of high-quality service. The reasons for these objectives are the growing expectations of high-quality transit services and the need for such services to attract urban travelers from their automobiles. Improved mobility, intensified economic activities, and enhanced urban livability are, of course, additional factors leading to investment and efforts to build metro systems.

Sometimes considered a fixed concept, RRT systems actually have a broad range of variations in their physical and operational features. Several features with their variations are described here.

With respect to ROW types, RRT systems usually have tunnel alignments in large central city areas, such as in New York, Paris, and Tokyo. The underground operation gives these systems a special image and often their names: "subway" in the United States; underground" in England, "U-Bahn" in Germany, and "T-Bana" in Sweden. However, RRT systems have very diverse rights-of-way: although always fully separated, they can be at-grade, on embankment, on aerial viaduct, in a cut, or in tunnel. Actually, because tunnels require by far the highest investment of all alignment types, many RRT systems, including those in Chicago, Hamburg, London, and New York, have larger portions of their lines above ground than in tunnels. This diversity of ROW types shows the ability of RRT to have alignments adjusted to physical conditions, urban environment, and financial resources.

By network extensiveness and station spacings (see Chapters 4 and 5), RRT systems vary from local metro lines in dense urban areas with close stations (Berlin, Madrid, Paris, Philadelphia) to regional lines, which, similar to regional rail systems, have few stations and much higher operating speeds (San Francisco, Washington).

Vehicles and train sizes also vary greatly. Lyon Metro has trains with only three 18-m-long cars, while San Francisco BART operates trains with up to ten 21-m-long cars. Thus, maximum TU lengths vary from 54 to 210 m. The shortest trains, operated during night hours on the PATCO Line in Philadelphia, consist of a single 21-m-long car.

Operating speeds of RRT go from about 30 km/h on short inner city lines with short station spacings to 60 km/h on regional metro lines such as San Francisco BART and Washington Metro, where on suburban sections with long station spacings operating speeds may reach even 80 km/h.

Line capacities have a similarly wide range, from about 16,000 sps/h in Lyon to over 70,000 sps/h in New York and Hong Kong.

Construction cost depends mostly on vertical alignment, as well as local conditions. In very general terms, the lowest investment is required for ground-level ROW, particularly if a transportation corridor already exists (railway line, freeway median). Other ground-level alignments may cost considerably more. Aerial structures are two to four times more expensive to build than ground-level alignment tracks, while the ratio of average costs of tunnels is again two to three times higher than the average cost of aerials. Most of the costs for underground alignment go to station construction.

It should be mentioned that these investment costs for metro line construction must be weighed carefully not only against operational aspects of the lines (tunnels have the advantage of total weather protection), but even more so against the desirability of alternative alignments. In many cases, the most direct alignments, serving areas of intensive activities, are the most expensive to build, but they bring the highest permanent benefits to the riders and the city in general. On the other hand, the lowest-cost alignments may be the least desirable ones for rail transit passengers.

Good examples of this alignment desirability are transit alignments in freeway medians. Since freeway

Photo 12.14 *Rail rapid transit being upgraded to full automation—Berlin.*

alignments, if correctly planned, go around and between areas of major activities (CBDs, suburban towns, etc.), by definition they do not follow corridors with heavy transit travel. RRT, to the contrary, should optimally follow major activity streets and boulevards and go into urban and suburban centers: it is a major element of their vitality and competitiveness with suburban sprawling, auto-based developments. The optimal alignments of the two therefore usually do not coincide. RRT can, however, effectively use medians on sections of radial freeways, then leave them to enter the high-density areas, usually in tunnels. Examples of such solutions are found in the San Francisco Bay Area (the Concord-Bay Point Line) and Chicago (the O'Hare Airport Line).

In conclusion, although RRT is a mode with extensive infrastructure, it provides many variations and different choices with respect to its ROW positions, types of stations, rolling stock, and performance to fit a variety of local conditions. The more than 100 rapid transit systems in the world illustrate a wide range of options that the designers of new rapid transit systems can use.

12.3.3 Rubber-Tired Rapid Transit (RTRT)

This is actually a metro system with a different support/guidance technology than rail. Its cars have trucks with four large rubber tires that carry the vehicle and four small tires that steer the truck, guided by lateral vertical surfaces. The trucks also have steel wheels that provide guidance through track switches and support the truck in the case of flat supporting rubber tire.

The track consists of two paths (which may be wood, concrete, or steel) on which supporting tires run, two vertical guiding surfaces, and standard track with steel rails. Thus, RTRT represents a technical subalternative of the rail rapid transit mode.

Introduced in Paris in 1959, this technology was designed to reduce noise in metro tunnels, increase acceleration/deceleration, and enable operation with shorter headways, increasing line capacity. As traditional rail rapid transit was modernized through welded rails, track construction preventing and absorbing noise, and new rolling stock designs, measurements have shown that most new rail systems are usually at least as quiet as RTRT systems. Acceleration and deceleration rates are also similar on rail and rubber-tired systems because their maximum values are limited by the comfort and safety of standing passengers. With considerably more complex guidance mechanism and guideways and higher energy consumption and heat production, RTRT technology has been considered but not selected by most new rapid transit system designers for the reasons summarized in the box.

The transit agency in Paris, RATP, converted several of its Metro lines to RTRT technology but retained conventional rail technology on other lines. French designers have, however, prevailed with the selections of RTRT technology on several new metro systems they designed and financed. At present, in addition to several metro lines in Paris, RTRT operates successfully on entire metro networks in Lyon and Marseilles, France; Montreal, Canada; Santiago, Chile; and, with a different design of guidance, in Sapporo, Japan. By far the largest metro system with rubber-tired technology is in Mexico City.

In the Mexico City application, however, some limitations of this technology are rather obvious. With small cars (15 m long, 2.50 m wide) due to rubber-tire limited load capability, nine-car trains provide a capacity of only about 35,000 prs/h, which is exceeded on several lines, causing serious congestion. The crowded conditions result in avoidance of the system by a considerable number of potential riders, particularly auto owners.

12.3.4 Monorails

Popularized by science-fiction writers many decades ago, monorails tend to enjoy the image of being the transit mode of the future. There are a number of monorail designs, starting from the classical and very popular Schwebebahn in Wuppertal, Germany (operating successfully since 1901) and including several other suspended and supported types. The most common

RTRT, compared with standard RRT technology, has the following advantages (+) and disadvantages (−):

+ Greater adhesion under normal (dry) conditions, allowing negotiation of steeper grades and use of trailers
+ Lower noise and vibration production in sharp curves
= The same acceleration rates
− Greater vulnerability to rain, snow, and ice conditions, limiting its applications to cities with a warm climate or to tunnels only (Montreal)
− Constraint on vehicle size due to the limited carrying capacity of rubber tires
− Higher energy consumption, mainly due to greater rolling resistance
− Higher heat production and fire hazard, requiring stronger ventilation and fire protection
− Higher investment and operating costs because of greater technical complexity and higher energy consumption

Photo 12.15 *Rubber-tired metro—Paris.*

monorails today are based on the Alweg design from the 1950s, with many improvements. Such monorails operate on long transit lines in Tokyo (from Hamamatsucho station in the city to Haneda Airport), the city of Kitakyushu on Kyushu Island, and several other Japanese cities. In Seattle, a short shuttle line has been in operation since the World's Fair in 1962 and is being replaced by a much longer line. A new line with the same technology has been opened in Las Vegas with the assistance of private capital, while Disneyland in California and Disney World in Florida have operated very popular smaller-scale monorails of the same technology for several decades. More than a dozen monorails with different technologies, supported and suspended, are operated in other cities, mostly in Japan.

A serious disadvantage of monorails is the nature of their guideways and switches. Unlike rail systems, monorail beams cannot cross each other, and their switches are slow and take a large area, so that their yards and other maneuvering areas take large space. Another disadvantage is that they have much larger profile than rail and other guided modes, so that they are not conducive to tunnel operations. Their aerial structures, consisting of columns and guiding beams, are, however, smaller and less imposing on the surroundings than aerial structures of rail and AGT modes.

These characteristics make monorails more conducive to single lines than to networks. That is confirmed by the fact that all existing monorails consist of single lines; none of them represents a network. They are used mostly on wide streets and boulevards and in open areas, where tunneling is not required. The presently designed new monorail line in Seattle will test the acceptability of aerial guideways in streets with high buildings and other urban settings.

In addition to their operational complexity and dependence on aerial alignments, monorails require investment much higher than LRT requires, in the range of costs of RRT systems. For example, the Kitakyushu monorail in Japan and the Newark Airport and Las Vegas monorails required investments in the range of \$85–125 M/km.

Thus, monorails represent a technological option for rapid transit lines, and in some cases they may represent an attractive solution. The factors favoring their applications include their exotic/novelty image, the need to follow horizontally and vertically curved alignments (the guideway does that efficiently, as demonstrated on the Tokyo-Haneda Line), and the need for minimum visual intrusion of aerial guideways along urban arterials. However, their technical characteristics present operational complexities. This is particularly the case in applications for networks of lines and alignments in built-up urban areas and tunnels, where monorails are distinctly inferior to conventional rail systems, LRT, or RRT.

12.3.5 Review of Guided Modes and Their Automation

The preceding sections have shown that rail and other guided transit modes have a wide variety of characteristics. Classifications of modes by their ROW categories, types of vehicles and TUs, operational characteristics, and degrees of automation show various overlaps in many of these elements. To provide an overview of the modes with ROW A with emphasis on their technical characteristics, particularly guidance, propulsion, and automation, typical systems of different categories of high-performance modes with added AGT modes are listed in Table 12.9.

The sequence of these modes generally follows their historic evolution. As the table shows, there has been an increase in automation, starting from AGT systems as shuttles and then on several regular transit lines. At the same time, conventional rail systems began to apply ATO controlled by the driver, followed later by full automation without driver (first on ALRT, then on RRT).

A note of caution about transit system automatic operation is called for here. Fully automated train operation, without drivers, should be expected to be increasingly used. However, its advantages are more important on medium-capacity modes than on large rapid transit systems such as in New York, Hong Kong, and Mexico City, where presence of a person (driver or attendant) is desirable for many reasons, and the cost of an employee in a train carrying hundreds or thousands of passengers is easy to justify. Thus, fully automated train operation is not necessarily a desirable target for all rapid transit systems.

Progress toward full automation has had its successes as well as problems. Many AGT systems could not operate in any other way: automation makes their frequent service feasible. However, the designers of many of these systems have failed to use two main advantages of full automation: provision of high service frequency without increased operating cost, and tailoring TU size to demand to reduce operating costs. Only the Skytrain in Vancouver and a few other AGT systems use these advantages of automation. On the other hand, all AGT systems in Japan and automated metro systems in Paris and Singapore operate fixed TU consists, forgoing the main operational benefits of operations without crews. The future progress of automation will depend on a more comprehensive systems approach that leads to better utilization of potential benefits from full guided transit automation.

12.3.6 Regional Transit Modes

As cities grow into metropolitan areas or urbanized regions, there is an increasing need to provide regional transit systems. These are characterized by long lines from the central urban area to suburbs and, in large metropolitan areas, among suburbs. Station spacings

BOMBARDIER
TRANSPORTATION

Photo 12.16 *Modern monorail system—Las Vegas (Courtesy of Bombardier Transportation).*

Table 12.9 Classification of RRT, ALRT, and AGT systems by design and operation with respect to automation

Category	*City*	*Support/ Guidance*	*Propulsion*	*Driving*	*Crews*	*Line Capacity*
Conventional RRT	New York Subway	Rail	EM	M	D+A	Very high
	Mexico City Metro	RT	EM	M	D+A	High
Upgraded conventional RRT	Hamburg U-Bahn	Rail	EM	M, ATO	D	High
	Moscow Metropolitan	Rail	EM	M, ATO	D, A	Very high
Contemporary RRT (since 1970)	San Francisco BART	Rail	EM	ATO	D	Very high
	Munich U-Bahn	Rail	EM	ATO	D	High
AGT (shuttles)	Atlanta (Hartsfield)	RT	EM	FA	N	Medium
	Singapore MRT feeder	RT	EM	FA	N	Low
AGT (urban)	Lille VAL	RT	EM	FA	N	Low
	Taipei VAL	RT	EM	FA	R	Medium
LRRT	Philadelphia (Norristown)	Rail	EM	M	D	Low
	Toronto (Scarborough)	Rail	LIM	ATO	D	Low
ALRT	Vancouver	Rail	LIM	FA	N	Medium
	London (Docklands)	Rail	EM	FA	A	Medium
Automated Metro/RRT	Lyon Metro Line D	RT	EM	FA	N	Medium
	Paris Metro Line 14	RT	EM	FA	R	High
	Singapore–NE Line	Rail	EM	FA	N	High

RT: Rubber tired
EM: Electric motor
LIM: Linear induction motor
M: Manual
ATO: Automatic train operation
FA: Fully automated
D: Driver
A: Attendant
R: Roving attendant
N: No crews

are long, so that travel speed on regional network is considerably greater than on urban transit. Serving mostly long trips among areas and subcenters, regional transit is usually superimposed on regular transit that serves local travel within central city and suburban areas.

Regional transit services are provided by bus, LRT, and commuter/regional rail modes. Regional express buses, operating in mixed traffic or on special lanes, cannot be defined as high-performance modes because they perform commuter services only and their reliability is dependent on traffic conditions. Yet they are briefly included here because of their important role as the only mode providing regional services in many cities.

12.3.6.1 Regional Express Buses. To provide regional service at speeds competitive with auto travel in U.S. metropolitan areas, buses usually must utilize major arterials and freeways. Most U.S. cities have more extensive freeway networks than their counterparts in other countries, and their express buses represent the most common, often the only, regional transit.

Typically, regional express buses run on streets in suburban areas, then get on a freeway for nonstop travel into center city, where they usually again get on

Photo 12.17 *First fully automated rapid transit—Lyon Line D.*

streets for distribution of passengers, transfers to other lines, etc. There are extensive networks of express bus lines utilizing HOV facilities in a number of cities, most notably Dallas, Denver, Houston, Los Angeles, Minneapolis, and Seattle. A large network of regional buses is operated also in the New Jersey suburbs of New York City, supplementing the NJT regional rail lines also serving that region. In New York City, these lines terminate in the Port Authority Terminal, the largest bus terminal in the world, with direct transfer to the subway lines. Similar networks terminate and have transfers to rail lines on the Manhattan side of the George Washington Bridge and in San Francisco in the Transbay Terminal.

Most regional bus lines in U.S. cities represent commuter transit rather than regular transit. They offer very limited, often no, services during off-peak periods. Among the exceptions to this are three leading bus systems in North America: Ottawa, Pittsburgh, and Portland. These systems are mentioned in Section 12.2.1.

12.3.6.2 Commuter Rail. In many cities around the world, railway lines going through suburbs and entering the city have been used for commuter trains, bringing mostly daily commuters into and out of the city. These lines typically terminate in stub stations on the fringes of central city. For example, London has about a dozen stations where commuter lines operated by British Rail and its successors terminate and connect with the Underground (RRT) lines. Commuter rail lines terminate in similar stub stations in Boston, Chicago, San Francisco, and Paris.

Commuter rail plays an extremely important role in bringing workers into large cities of both industrialized countries (Buenos Aires, Milan, Moscow, New York City, Osaka, Tokyo, etc.), as well as in developing

Photo 12.18 *Tokyo subway with linear induction motor.*

countries (Bombay, Cape Town, Johannesburg, São Paulo, and many others).

Commuter rail systems in many cities represent by far the most efficient mode for bringing and returning extremely peaked loads of daily commuters from suburbs. They offer high speed, high comfort, and reliability and provide high line capacities. Their disadvantages are that their headways are long and irregular, making their use in off-peak hours difficult or totally nonexistent. Moreover, in some cities the railway agency is not sufficiently interested in these services because its main focus is intercity passenger and freight trains. With the increasing demand for high-quality regional transit, the trend has been to upgrade commuter rail into regional rail lines or networks.

12.3.6.3 Regional Rail. These railway systems belong either to the railway or to the transit agency. They are specially designed as regional passenger rail networks, running through central city and providing better coverage than radial commuter lines, which serve only one or a few central stations. Moreover, the growth of polycentric urban areas with major activity centers in suburbs creates demand for high-performance transit networks that provide links among such suburban centers, rather than only radial lines to center city. These polycentric links are provided mostly by buses, but in several cities there is an increasing interest in upgrading such circumferential or tangential lines to rail systems. The highly successful link between Sacramento and San José in California is tangential to San Francisco, and similar lines are being considered in Chicago, Los Angeles, and Philadelphia.

Typical regional rail modes are the S-Bahn systems in Berlin and Hamburg, JR lines in Tokyo and other Japanese cities, and RER in Paris. With electric traction, lines through central cities with many stations, and operation with short headways, they are more sim-

Photo 12.19 *Push-pull diesel commuter rail service—Toronto.*

ilar to RRT than to commuter rail systems terminating in stub-end stations. Systems called "commuter" in Boston, Chicago, and San Francisco (Caltrain) are actually between these two modes. They are diesel-powered but provide hourly services throughout the day, although usually hourly only.

As described in Chapter 4, many cities (Paris, Brussels, Oslo, Philadelphia, Manchester) have upgraded their commuter networks into regional rail by connecting stub-end terminals across central cities, electrifying the lines, and integrating them with regular transit.

Similar to other neighboring modes, there is no sharp boundary between commuter and regional rail modes. Yet a typical commuter rail system, such as diesel-traction TriRail in Miami and Metrolink in Los Angeles, is very different from typical regional rail lines, such as Hamburg S-Bahn, Paris RER, and Sydney CityRail. The former, commuter lines, operate several inbound trains in the morning and outbound in the afternoon; the latter, regional rail lines, operate with 5–20-minute headways throughout the day. SEPTA's Regional Rail System in Philadelphia, NJ Transit Railroads, Metro-North, and the Long Island Rail Road fall in between these two modes.

Both commuter and regional rail modes are strong candidates for expansion or introduction in many cities, particularly in North America, due to their extensive suburban developments and congested freeways.

12.3.6.4 Regional Rapid Transit. Several cities or urbanized regions in the United States, such as the San

Photo 12.20 *Electric regional rail—S-Bahn in Germany.*

Francisco Bay Area, Washington, Philadelphia, and Atlanta, which decided in recent decades to build high-performance transit systems, designed rail systems with networks similar to regional rail: long lines into suburbs with long station spacings, high operating speed, and heavy reliance on bus feeders, park-and-ride, and kiss-and-ride. Unlike regional rail, however, they penetrate center city through tunnels and have a number of stations in the central area with extensive transfers to other transit modes, including LRT and buses. Another feature of these systems is that their operation is the same as that of RRT: enclosed stations with high platforms allow short headways and high capacity. These systems thus represent regional rapid transit, or an actual bridge between regional rail and RRT modes. In a way, these systems have metro-type in-city networks extending into regional rail suburban lines.

12.3.7 Trends in Regional Rail Transit Development

With rapid spatial growth of cities and urbanized areas, regional transit is growing in importance and has been the scene of intensive developments. Major developments in this group of transit modes are listed and briefly described here.

1. *Activation of abandoned railway lines or shared use of freight railway lines for transit services.* These actions require investments as well as organizational arrangements for shared use of

Photo 12.21 *High-speed regional rapid transit with high comfort level effectively competes with auto travel—San Francisco BART.*

tracks. In some cases, such as in Los Angeles, the transit agency has purchased hundreds of kilometers of lines from freight railroads.

2. *Upgrading of commuter rail into regional rail lines.* This involves introduction of regular headways, usually of 20–30 minutes, which greatly facilitate transfers and integration of lines of the same mode or among modes.
3. *Integration of regional rail with city transit.* This includes such activities as joint stations, coordination of schedules, joint fares, integrated information, and marketing.
4. *Building connecting lines between stub-end railway stations at the fringes of city centers.* With excellent results from these connections built in Munich, Oslo, Paris, Philadelphia, Zürich, and other cities, building such center city connections is a logical step in upgrading regional rail networks in Boston, New York, and many other world cities.
5. *Upgrading of diesel to electric traction.* This facilitates faster operations, ability to operate in tunnels, and use of shorter trains without locomotives.
6. *Physical integration of regional rail with RRT:* joint use of tracks by different operators, used in several Japanese cities.
7. *Tram-train concept: Integration of LRT with regional rail and even with intercity rail services.* Examples are Karlsruhe and Saarbrücken in Germany, Manchester in Great Britain, and the River Line in southern New Jersey.

Photo 12.22 *Light regional rail with diesel propulsion—River Line, southern New Jersey (courtesy of James Aslaksen).*

Photo 12.23 *(a) Dual system light rail/regional rail on center city alignment—Saarbrücken; (b) dual system light rail/regional rail on intercity railway line—Saarbrücken.*

8. *Development of self-powered diesel passenger cars* capable of running on railway as well as on transit lines. This rolling stock enables low-cost operation on regional railway lines with low to medium passenger demand, expanding the feasible domain of regional rail services at its lower end.

With the increasing need for high-performance regional transit services, further experience and innovations in this field should be expected in the future.

EXERCISES

12.1. What is the most significant single physical improvement of transit needed to increase its performance and competitiveness with auto, and why?

12.2. Compare rights-of-way B and C:

a. What is required to upgrade a transit line from operating on ROW C to ROW B in terms of physical changes, investments, and possible external impacts?

b. List and explain the potential benefits that can be obtained by such upgrading. Include changes in LOS, operating costs, impacts on line image and ridership, and possible change in modal split.

12.3. Define the differences between transit lines with ROW B and those with ROW A. Under what conditions are the advantages of ROW A particularly important? What are the limitations on providing ROW A?

12.4. Exclusive ROW for transit:

a. What are its economic, operating, and system characteristics?

b. How does it influence transit ridership?

c. How does it influence the role of transit in urban transportation?

d. Where should it be used?

12.5. Select two different modes, such as RB and LRT, BRT and AGT, or LRT and RRT, in a city you know well and descriptively define their PCPs. Which elements comprise their Performance in the selected case? Why should these two systems or modes be compared by their PCPs rather than by their costs only?

12.6. Compared to steered highway vehicles, what characteristics do guided modes have? Which conditions favor the introduction of guided systems?

12.7. List and discuss the advantages and disadvantages of electric as compared to ICE (diesel) traction. Then answer the following questions:

- Under what conditions is electrification not used?
- Which conditions or type of alignment require electric traction?
- Why do high-performance modes always use electric traction?

12.8. Buses are often described as a "flexible mode." What are the advantages and disadvantages of that modal characteristic?
How does upgrading an RB line into BRT affect its "flexibility"?

12.9. What is the rationale for converting two lanes on a street from general use to bus lanes? What are the advantages of such a conversion for bus users? Under what conditions are bus lanes justified, effective, and permanent?

12.10. Describe two different measures or innovations introduced in recent decades to upgrade bus services that have been successful, and two that have failed. Explain the reasons for the failures.

12.11. Why was the introduction of SSFC very significant for LRT systems?

Which other innovations led to the tenfold increase in crew productivity on LRT systems since the 1960s?

12.12. Compare LRT, ALRT, and standard RRT modes with respect to their technical, operational, and cost characteristics. You may want to use a comparison method similar to the one in Table 12.6.

12.13. Some persons believe that LRT is seriously impeded by its grade crossings or street running, so that it can never have satisfactory speed, reliability, safety, and other performance elements. Rail rapid transit (RRT) has no such impedance and is therefore a superior mode. Others claim that LRT has greater flexibility of alignment, and that when grade crossings and street running are properly regulated, this mode can nearly match RRT in performance. Therefore, LRT is better than RRT.

Briefly define the main characteristics of the two modes and compare them systematically to show which one of these two views (or a third view) is correct.

12.14. What are the advantages and disadvantages of fully automated transit systems (without drivers) as compared to conventional systems?

What is the basic condition with respect to ROW category required for full automation?

12.15. Is full automation more important for high-volume or for medium-volume transit lines?

Discuss and explain why.

12.16. You are planning a rail transit line (LRT or metro) in a heavily travelled corridor in a major city (Atlanta, Seattle, or Singapore). Your plan is being criticized by a group of economics professors who claim that your plan is wasteful: Buses operating on a busway and branching out to many streets would be superior. Buses would be cheaper and are flexible, while rail is rigid. They would deliver people closer to their destinations, require fewer transfers, and be more adjustable to changing urban development patterns.

Assuming the conditions typical for the above-mentioned cities, what would be your arguments countering this criticism? Be sure to include such items as system capacity, reliability, image, operating cost, attraction of passengers, etc. Specifically, discuss the characteristic "flexibility": how is it defined and is it always desirable?

12.17. A large city (over 5 million population) in a developing country is considering construction of a rail transit system. Assuming conditions typical for such cities (e.g., Cairo, Lagos, Bogotá, Karachi):

a. List the reasons for this project.

b. List the main obstacles that must be overcome.

12.18. Compare standard rail rapid transit (Lindenwold, BART, Munich) with rubber-tired rapid transit (Montreal, Mexico City) technology. Determine all items that are different between the two, and briefly discuss each item and then present your findings in a summary table:

	Advantage of		
Affected Group	*Factor*	*Rail Rapid Transit*	*Rubber-Tired Rapid Transit*
Passengers			
Operator			
Society-community			

12.19. Which characteristics distinguish a typical regional rail from a typical commuter rail system?

12.20. Which features of commuter and regional rail have resulted in an extensive development of these modes in a number of North American urbanized areas?

12.21. Describe the concept of LRT operating on regional rail and intercity rail tracks into the suburbs, the so-called tram-train system.

- Which major problems (technical and legal) must be resolved to make this operation feasible?
- What advantages are obtained from the tram-train concept as compared to LRT lines with transfers to regional rail?

12.22. What is the basic purpose in developing single-unit diesel powered regional rail vehicles similar to LRT vehicles for operation on lightly traveled rail lines?

12.23. (True-False) Which of the following statements comparing LRT with BRT operating on ROW B are correct? Mark each statement T for true or F for false.

LRT can offer better collection-distribution in suburbs	T	F
LRT offers lower riding comfort	T	F
LRT is less reliable	T	F
LRT requires lower investment cost	T	F
LRT is cheaper to operate at high volumes	T	F
The main item in operating cost differences between the two modes is the cost of energy	T	F
LRT has higher acceleration rates	T	F
LRT has higher capacity	T	F
BRT serves more stations along the trunk line	T	F

12.24. Marking "S" for superior and "I" for inferior, compare LRT (with feeder buses) with buses on busways (assuming optimal configurations for each mode), for service on a radial urban corridor with respect to the following items:

Level of service (comfort, reliability, frequency, etc.)	_____
Network extensiveness	_____
Downtown operation	_____
Labor productivity (number of operators/100 seats offered)	_____
Number of passengers who must transfer between modes	_____
Noise, exhaust, and other environmental impacts	_____
Operation in tunnels	_____

REFERENCES

Federal Transit Administration (FTA). 2000. *Transit Profiles.* 2000 Transit Database Report Year. Washington, DC: FTA.

Girnau, G., A. Müller-Hellmann, and F. Blennemann, eds. 2000. *Light Rail in Germany* (in German and English). Cologne: Verband Deutscher Verkehrsunternehmen.

Hondius, H. 1995. State-of-the-Art of bus, hybrid bus, trolleybus and intermediate systems. *Public Transport International* 4 and 6:24–31.

Rabinovitch, J., and J. Leitman. 1996. Urban Planning in Curitiba. *Scientific American* 274(3):26–33.

Regie Autonome des Transports Parisiens (RATP). 1977. Autobus en Site Propre. Paris: RATP.

Verband Öffentlicher Verkehrsbetriebe (VÖV) and VDA. 1979. Bus-Verkehrssystem. Düsseldorf: Alba Verlag,

Vincent, B., and B. Walsh. 2003. *The Electric Rail Dilemma: Clean Transportation from Dirty Electricity?* Washington DC: Breakthrough Technologies Institute.

Vuchic, V. R. 1971. Concept of Flexibility in Transportation Systems Analysis. *High Speed Ground Transportation Journal* 5(1):53–61.

Vuchic, V. R. 1981. *Urban Public Transportation Systems and Technology.* Englewood Cliffs, NJ: Prentice-Hall.

Vuchic, V. R. 2001. Comparison of LRT with bus semirapid transit system. In *Proceedings of UITP 5th LRT Conference.* Melbourne. Brussels: UITP.

Vuchic, V. R. 1985. O-Bahn–description and evaluation of a new concept. *TR Record 1011,* 8–15.

Vuchic, V. R., and S. Kikuchi. 1994. The Bus Transit System: Its Underutilized Potential. Report DOT-T-94-20 to Federal Transit Administration. Washington: FTA.

Wilbur Smith and Associates 1975. Bus Use of Highways. *NCHRP Report 155.* Washington: Highway Research Board.

FURTHER READING

International Association of Public Transport (UITP). 2002. *Proceedings of 6th International Light Rail Conference.* Nantes, France. Brussels: UITP.

Journal of Public Transportation (JPT) 2002. 5(2). Special issue on BRT.

Litman, T. 2004. *Comprehensive Evaluation of Rail Transit Benefits.* Victoria, BC: Victoria Transport Policy Institute.

Middleton, W. D. 2003. *Metropolitan Railways.* Bloomington: Indiana University Press.

Parsons Brinckerhoff. 2001. *The Effects of Rail Transit on Property Values: A Summary of Studies.* NEO Rail II Project.

Pushkarev, B., J. Zupan, and R. Cumella. 1982. *Urban Rail in America.* Bloomington: Indiana University Press.

Sproule, S. J., S. W. Lynch, and E. S. Neumann, eds. 1998. *Automated People Movers VI: Creative Access for Major Activity Centers.* Proceedings of the Sixth International Conference on Automated People Movers, Las Vegas. Reston, VA: American Society of Civil Engineers.

Transportation Research Board (TRB). 2003. *Proceedings of 9th National Light Rail Transit Conference.* Portland, OR. Washington DC: Transportation Research Board.

BIBLIOGRAPHY

Major publications about urban transit

Beimborn, E. A., A. Horowitz, J. Schuetz, and G. Zejun. 1993. *Measurement of Transit Benefits*. Report DOT-T-93-33. Washington, DC: Urban Mass Transportation Administration.

Black, A. 1995. *Urban Mass Transportation Planning*. New York: McGraw-Hill.

Canadian Urban Transit Association (CUTA). 1993. *The Canadian Transit Handbook*. Toronto: CUTA.

Cervero, R. 1998. *The Transit Metropolis*. Washington, DC: Island Press.

Delucchi, M. 1998. Summary of the Nonmonetary Externalities of Motor Vehicle Use. UCD-ITS-RR-96-3(9). University of California, Institute of Transportation Studies, Davis, CA, September.

Girnau, G., A. Müller-Hellmann, and F. Blennemann, eds. 1997. *Sustainable Mobility: People in Motion* (in German and English). Cologne: Verband Deutscher Verkehrsunternehmen.

———. 2000. *Light Rail in Germany* (in German and English). Cologne: Verband Deutscher Verkehrsunternehmen.

———. 2001. *Telematics in Public Transport in Germany* (in German and English). Cologne: Verband Deutscher Verkehrsunternehmen.

Gray, G. E., and L. A. Hoel, (eds.) 1992. *Public Transportation*, 2nd ed., Englewood Cliffs, NJ: Prentice-Hall.

Homburger, W. S. 1982. Urban transit. In *Transportation and Traffic Engineering Handbook*, 2nd ed., ed. W. S. Homburger, Chap. 7. Englewood Cliffs, NJ: ITE/Prentice-Hall.

Lehner, F. 1978. *Der maximale Wirkungsgrad des Personaleinsatzes* (Maximum efficiency of personnel assignments). Düsseldorf: Alba Buchverlag.

Mees, P. 1999. *A Very Public Solution: Transport in the Dispersed City*. Melbourne: Melbourne University Press.

Meyer, M. D., and E. J. Miller. 2001. *Urban Transportation Planning*, 2nd ed. New York: McGraw-Hill.

Molinero, A. R. M., and L. I. S. Arellano. 1997. *Transporte publico: planeacion, diseno, operacion y administracion* (Public transport: planning, design, operation, and administration). Universidad Autonoma del Estado de Mexico.

Newman, P., and J. Kenworthy. 1989. *Cities and Automobile Dependence: An International Sourcebook*. Aldershot, UK: Grover.

———. 1998. *Sustainability and Cities: Overcoming Automobile Dependence*. Washington, DC: Island Press.

Parkinson, T., and I. Fisher. 1996. *Rail Transit Capacity*. TCRP Report 13. Washington, DC: Transportation Research Board.

Pucher, J. 1995. Urban passenger transport in the United States and Europe: a comparative analysis of public policies. *Transport Reviews* 15(2):89–107 and 15(3):261–283.

Risch, C., and F. Lademann. 1957. *Der öffentliche Personennahverkehr* (Urban Public Transportation). Berlin: Springer-Verlag.

Rüger, S. 1974. *Betriebstechnologie städtischer öffentlicher Personenverkehr* (Operation of Public Transport). Berlin: Transpress VEB Verlag für Verkehrswesen.

Smerk, G. M. 1991. *The Federal Role in Urban Mass Transportation*. Bloomington: Indiana University Press.

Transportation Research Board (TRB). 2000. *Highway Capacity Manual*. Washington, DC: TRB.

Vuchic, V. R. 1981. *Urban Public Transportation Systems and Technology*. Englewood Cliffs, NJ: Prentice-Hall.

———. 1999. *Transportation for Livable Cities*. New Brunswick, NJ: CUPR Press, Rutgers University.

Websites

http://www.aashto.org (AASHTO).

http://www.apta.com/ (APTA).

http://www.gobrt.org/ (BRT advocacy).

http://www.busrapidtransit.net/ (BRT information).

http://www.fta.dot.gov/ (FTA).

http://www.fta.dot.gov/7639_7662_ENG_HTML.htm (FTA's BRT information site).

http://www.lightrail.com/ (LRT information).

http://www.lightrailnow.org/ (LRT advocacy).

http://www.lrta.org/ (LRT advocacy).

http://www.metropla.net/ (international rapid transit information).

http://mercurio.iet.unipi.it/ (European rail information).

http://www.monorails.org/ (monorail advocacy).

http://www.nationalcorridors.org/ (rail transit advocacy).

http://www.nycsubway.org/ (rapid transit information).

http://www.trb.org/ (TRB).

http://www.trolleycar.org/ (rail and general transit advocacy).

http://www.uitp.com (international transit information).

http://www.urbanrail.net/ (rail transit information).

APPENDIX I

SI AND ENGLISH UNITS AND CONVERSION FACTORS

The International System of Units (Système International d'Unités, SI) recognizes seven basic units from which all others are derived. They are:

Number	*Quantity*	*Name of Unit*	*Symbol*
1	Length	meter	m
2	Mass	kilogram	kg
3	Time	second	s
4	Electric current	ampere	A
5	Temperature	kelvin	K
6	Luminous intensity	candela	cd
7	Amount of substance	mole	mol

Multiples of units are designated by the following names:

Amount	*Prefix*	*Symbol*	*Amount*	*Prefix*	*Symbol*
10^1	deca	da	10^{-1}	deci	d
10^2	hecto	h	10^{-2}	centi	c
10^3	kilo	k	10^{-3}	milli	m
10^6	mega	M	10^{-6}	micro	μ
10^9	giga	G	10^{-9}	nano	n

This appendix presents the SI and English (Imperial) units, which are used in various areas of transportation engineering and their conversion factors. Some metric units and factors, which are not listed in the SI, but are commonly used and remain internationally recognized, are also included. Examples are the units of centimeter (10^{-2} m), hour (3.6×10^3 s) and kilowatt hour (3.6 MJ), and multiples of 10, 10^2, 10^{-1} and 10^{-2}.

For more extensive coverage of units, the following book is suggested: Cardarelli, François. *Scientific Unit Conversion.* New York: Springer Verlag, 2003.

Units and conversion factors within each and between the two systems of measures are presented here for each dimension in a separate box. The format followed for each dimension is shown in the first box:

DIMENSION

SI unit (symbol)		English unit (symbol)
	Conversion factors (for conversion in the direction of the arrow multiply by the number provided)	

LENGTH

SI unit (symbol)	English unit (symbol)
millimeter (mm)	inch (in, ")
centimeter (cm)	foot (ft, ')
meter (m)	yard (yd)
kilometer (km)	mile

10^3 mm = 10^2 cm = 1 m
10^3 m = 1 km

36 in = 3 ft = 1 yd
5280 ft = 1760 yd = 1 mile

SI unit	Conversion factor	English unit
mm	← 25.40	in
mm	0.0394 →	in
m	← 0.3048	ft
m	3.281 →	ft
m	← 0.9144	yd
m	1.094 →	yd
km	← 1.609	mile
km	0.6214 →	mile

AREA

square millimeter (mm^2)
square centimeter (cm^2)
square meter (m^2)
hectare (ha)
square kilometer (km^2)

$10^6\ mm^2 = 10^4\ cm^2 = 1\ m^2$
$10^6\ m^2 = 10^2\ ha = 1\ km^2$

cm^2	← 6.452	0.1550 →	in^2
m^2	← 0.0929	10.76 →	ft^2
m^2	← 0.8361	1.196 →	yd^2
ha	← 0.4047	2.471 →	acre
km^2	← 2.590	0.3861 →	$mile^2$

square inch (in^2)
square foot (ft^2)
square yard (yd^2)
acre
square mile ($mile^2$)

$1296\ in^2 = 9\ ft^2 = 1\ yd^2$
$3{,}097{,}600\ yd^2 = 640$ acres $= 1\ mile^2$

VOLUME

cubic millimeter (mm^3)
cubic centimeter (cm^3 or cc)
liter (l)
cubic meter (m^3)

$10^6\ mm^3 = 10^3\ cm^3 = 1\ l$
$10^6\ cm^3 = 10^3\ l = 1\ m^3$

cubic inch (in^3)
quart (qt)
gallon* (g)
cubic foot (ft^3)
cubic yard (yd^3)

$46{,}656\ in^3 = 27\ ft^3 = 1\ yd^3$
$4\ qt = 1\ gal = 231\ in^3$
$1\ qt = 57.75\ in^3$

cm^3	← 16.39	0.0610 →	in^3
l	← 3.785	0.2642 →	gal
m^3	← 0.0283	35.32 →	ft^3
m^3	← 0.7646	1.308 →	yd^3

*British Commonwealth countries also use the imperial gallon = 1.201 U.S. gallons.

MASS

gram (g)
kilogram (kg)
(metric) tone or tonne (t)

ounce (oz)
pound (lb) (pounds mass) (lbm)
(short) ton (tn^+ or s tn), long ton (t n)

10^3 g = 1 kg
10^3 kg = 1 t

g	← 28.35 ←	oz
	→ 0.0353 →	
kg	← 0.4536 ←	lb
	→ 2.205 →	
t	← 0.9072 ←	tn
	→ 1.102 →	
t	← 1.016 ←	l tn
	→ 0.9842 →	

16 oz = 1 lb
2000 lb = 1 tn
2240 lb = 1.12 tn = 1 ltn

DENSITY

grams/cubic centimeter (g/cm^3)
grams/liter (g/l)
kilograms/cubic meter (kg/m^3)

ounces/cubic inch (oz/in^3)
pounds/cubic inch (lb/in^3)
pounds/cubic foot (lb/ft^3)

10^3 kg/m^3 = 10^3 g/l = 1 g/cm^3

16 oz/in^3 = 1 lb/in^3 = 1728 lb/ft^3

g/cm^3	← 1.743 ←	oz/in^3
	→ 0.5737 →	
kg/m^3	← 16.02 ←	lb/ft^3
	→ 0.0624 →	
km/m^3	← 0.5933 ←	lb/yd^3
	→ 1.686 →	

SPEED

meters/second (m/s)
kilometers/hour (km/h)

feet/second (ft/s)
miles/hour (mph)

1 m/s = 3.6 km/h

m/s	← 0.3048 ←	ft/s
	→ 3.281 →	
km/h	← 1.609 ←	mph
	→ 0.6214 →	

88 ft/s = 60 mph (exact)
(1.467 ft/s = 1mph)

ACCELERATION

meters per second squared (m/s^2)
(km/h/s is not used)

$g = 9.807\ m/s^2$

feet per second squared (ft/s^2)
miles per hour per second (mphps)
$g = 32.18\ ft/sec^2$
1 mphps = $1.467\ ft/s^2$

From/To	Factor	To/From
m/s^2 ←	0.3048	← ft/s^2
m/s^2 →	3.281	→ ft/s^2
m/s^2 ←	0.4470	← mphps
m/s^2 →	2.237	→ mphps

JERK

meters per second cubed (m/s^3)

feet per second cubed (ft/s^3)

Factors are the same as the corresponding ones for acceleration.

FORCE, WEIGHT

dyne (dyn)
newton (N)
kilonewton (kN)
(metric) ton or tonne (not in SI)

ounce (oz)
pound force (lbf)
10^3 pounds (kip)
(short) ton force (tnf or s tnf)
32,000 oz = 2000 lbf = 1 tnf

10^8 dynes = 10^3 N = 1 kN
Force = mass × acceleration

(1 N = 1 kg-m/s^2)

From/To	Factor	To/From
N ←	4.448	← lbf
N →	0.2248	→ lbf
kN ←	8.896	← tnf
kN →	0.1124	→ tnf

$$\text{Force} = \frac{\text{Weight} \times \text{Acceleration}}{\text{Gravity Const. } (g_c)}$$

$$\left[g_c = 32.18 \frac{\text{lbm} - \text{ft}}{\text{lbf} - \text{s}^2} \right]$$

ENERGY

erg
joule (J)

British thermal unit (Btu)
foot-pound force (ft-lbf)

10^7 ergs = 1 J

1 Btu = 778.2 ft-lbf = 118.98 hp-h

J ← 1055 ← Btu
J → 9.482×10^{-4} → Btu

J ← 1.356 ← ft-lbf
J → 0.7376 → ft-lbf

Joint units:
watt-hours (Wh)
kilowatt-hours (kWh)
10^3 Wh = 1 kWh

J ← 3.6×10^6 ← kWh ← 0.3776×10^{-6} ← ft-lbf
J → 0.2778×10^{-6} → kWh → 2.665×10^6 → ft-lbf

POWER

ergs per second (erg/s)
horsepower (metric) (HP)

foot-pounds force per second (ft-lbf/s)
horsepower (British) (hp)

1 HP (metric) = 75 kg-m/s

1 hp (British) = 550 ft-lbf/s

Joint units:
watt (W) [1 W = 1 J /s = 10^7 ergs/s]
kilowatt (kW)

HP (metric) ← 1.360 ← kW ← 0.7457 ← hp (British)
HP (metric) → 0.7355 → kW → 1.341 → hp (British)

FUEL CONSUMPTION (C) AND EFFICIENCY (E)*

(C) liters per 100 kilometers (*l*/100 km)
(C) joules per kilometer (J/km)
(E) kilometers per liter (km/*l*)

(C) gallons per mile (gpm, gal/mile)
(C) British thermal units per mile (Btu/mile)
(E) miles per gallon (mpg, miles/gal)

km/*l* ← 0.4251 ← miles/gal*
km/*l* → 2.352 → miles/gal*

l/100 km ← 235.2 ← gal/mile*
l/100 km → 0.00425 → gal/mile*

J/km ← 0.6557×10^3 ← Btu/mile
J/km → 1.525×10^{-3} → Btu/mile

*The conversion factors between volume and energy depend on the type of fuel.

APPENDIX II
LIST OF ABBREVIATIONS

Abbreviation	Stands for	Introduced or Defined in Section
AASHTO	American Association of State Highway and Transportation Officials	11.2
ADA	Americans with Disabilities Act	6.5.1.2
AFC	Automatic fare collection	6.4.3.3
AI	Artificial intelligence	6.4.3.1
AGT	Automated guided transit	1.1.1
ALRT	Automated light rail transit	12.3.1
ALS	Area licence scheme (Singapore)	8.2.2.1
APC	Automatic passenger counting	1.4.7
APM	Automated people mover	12.3.2
ASHRAE	American Society of Heating, Refrigeration and Air-Conditioning Engineers	11.3.1
ATO	Automatic train operation	1.1.7
AVL	Automatic vehicle location	1.2.2
BART(D)	San Francisco Bay Area Rapid Transit (District)	1.1.5
BRT	Bus rapid transit	4.2.4
CAA	Clean Air Amendment	12.2.1.2
CAD	Computer-aided dispatching	1.4.7
CBD	Central business district	1.2.1
CCTV	Closed-circuit television	6.4.3.1
DBOM	Design-build-operate-maintain	8.4.2
DOT	Department of Transportation	9.5.2.1
E & H	Elderly and handicapped	6.5.1
EIS	Environmental impact statement	11.3.2
ERP	Electronic road pricing (Singapore)	10.4.2
FS	Far-side (bus stop)	2.2.2.4
FTA	Federal Transit Administration	6.5.1.2
GM	General manager	6.1.2
GO	General obligation (bonds)	8.2.3
GPS	Global positioning system	1.2.2
HHA	Hamburger Hochbahn AG	3.4
HOV	High-occupancy vehicle	3.2.2
ICE	Internal combustion engine	11.3.2
ISTEA	Intermodal Surface Transportation Efficiency Act	8.3.1
ITS	Intelligent transportation system	6.4.3
K+R	Kiss-and-ride	4.1.1
LOS	Level of service	1.3.1
LRRT	Light rail rapid transit	12.1.1
LRT	Light rail transit	1.1.1
LRV	Light rail vehicle	6.4.3.1

Abbreviation	Stands for	Introduced or Defined in Section
LTA	Land Transport Authority (Singapore)	10.4.2
MARR	Minimum allowable rate of return	11.1
MDBF	Mean distance between failures	6.3.5
MLS	Maximum load section	1.1.3
MPO	Metropolitan planning organization	10.1.3.1
MRT	Mass rapid transit	12.1.1
MTBF	Mean time between failures	6.4.4
NPV	Net present value	11.1
NS	Near-side (bus stop)	2.2.2.4
O-D	Origin-destination	4.1.1
OR	Operations research	3
P+R	Park-and-ride	4.1.1
PATCO	Port Authority Transit Corporation (Philadelphia-New Jersey)	4.2.4
PATH	Port Authority Trans-Hudson (New York-New Jersey)	8.2.3
PCP	Performance-cost package	12.1
PDA	Personal digital assistant	6.4.3.3
PHC	Peak hour coefficient	1.4.2
POP	Proof of payment	7.2.3
PRT	Personal rapid transit	12.2.3
RB	Regular bus	12.1.5
RGR	Regional rail	2.1.4
RRT	Rail rapid transit	12.3.2
RTRT	Ruber-tired rapid transit	12.1.1
RTS	ROW category, technology, and type of service	11.4
RUCUS	Run-cutting and scheduling	1.4.7
SCAG	Southern California Association of Governments	9.2.3
SMSA	Standard metropolitan statistical area	10.1.3.2
SOV	Single-occupancy vehicle	11.3.2
SSFC	Self-service fare collection	7.2.3
TEA-21	Transportation Equity Act for the 21st Century	8.3.2
TOD	Transit-oriented development	10.1.3.1
TIP	Transportation imporvement program	10.1.3.1
TRB	Transportation Research Board	9.6.1
TSM	Transportation system management	12.2
TTS	Timed transfer system	4.4.1
TU	Transit unit	1.1.2
UITP	International Association of Public Transport	6.7.2
UMTA	Urban Mass Transportation Administration (now FTA)	6.5.1.2
VKT	Vehicle-kilometers traveled	8.2.2.1
VMT	Vehicle-miles traveled	8.2.2.1

APPENDIX III

DEFINITIONS OF TRANSIT SYSTEMS TERMS

This Appendix contains definitions of selected important concepts and terms from the book. The list is not intended to be either comprehensive or exhaustive for the entire transit field. For more comprehensive terminology, the reader is referred to the following publications:

Dictionary of Public Transport (English, German, and French). Düsseldorf: Alba Buchverlag, 1981.

Gray, B. H., ed. 1989. *Urban Public Transportation Glossary.* Washington, DC: Transportation Research Board.

Parkinson, T., and I. Fisher. 2000. *North American Public Transport Glossary*. Vancouver: Transport Consuling Ltd.

ACCELERATED SERVICE—Transit line operation in which successive TUs travel without stopping through alternate sets of stations on a predetermined schedule.

- Express/Local Service—Transit line with long spacings between stations that has high operating speed and serves primarily long trips. Often operates on the same ROW as a local line with more frequent stopping, which serves short trips and acts as a collector/distributor for the express.
- Skip-Stop—Service where alternate TUs do not stop at two different sets of stops/stations.
- Zonal—Service with subsequent TUs serve different groups of stops/stations along a line. There are usually two or three zones.

AREA COVERAGE (TRANSIT)—Percent of the total urban area which is within 5-min (primary) and 10-min (secondary) walking distance of transit stations. The basic element of transit system availability.

AUTOMATED GUIDED TRANSIT (AGT)—Any guided rubber-tired or rail transit mode with fully automated operation (no crews on TUs). The term usually refers to guided modes with small or medium-sized vehicles. AGT is also known as automated people mover (APM) and people mover system (PMS).

AUTOMATIC TRAIN OPERATION (ATO)—Control system that performs all functions of train travel along a transit line.

BUS (SERVICE)—Transit services provided by buses, including:

- *Express:* express transit service operated by buses.
- *Regular* (RB) or *local:* common urban bus routes serving all stops.
- *Semirapid:* Regular or high-performance buses operating on routes that include substantial sections of ROW categories B or A and have long station spacings and distinctive designation (see also BUS RAPID TRANSIT).

BUS LANE—Traffic lane for dominant or exclusive use by buses. It may be:

- *Contraflow* (CBL): the same as regular bus lane, but with buses operating in the opposite direction from other traffic.
- *Exclusive* (EBL): bus lanes (usually two) for bus use only, physically separated (by curbs or barriers) from other traffic. See BUSWAY.
- *Regular* (RBL): lane on urban streets or freeways reserved for bus use only, separated from other lanes by pavement markings, signs, or rubber cones, but not by fixed physical barriers.

BUS RAPID TRANSIT (BRT)—Bus lines with regular or high-performance vehicles operating mostly on ROW categories B or A, with long stop spacings and distinct designation (popular name for semirapid bus).

BUSWAY—Roadway reserved for buses only (ROW category B or A).

CAPACITY (TRANSIT)—(1) Static: Total number of spaces or persons a vehicle can accommodate. (2) Dynamic: The maximum number of TUs, vehicles, spaces, or persons that can be transported on a transit line past a fixed point in one direction per unit of time (usually one hour). Major concepts of both static and dynamic capacities are:

- *Line:* the maximum number of spaces that transit vehicles on a line can transport past a fixed point in one direction per unit of time.
- *Offered:* line capacity in sps/h, or theoretical maximum ability of line to carry persons, as distinguished from utilized, practically achievable capacity in prs/h.
- *Scheduled line:* number of spaces transported on a line per direction per hour with a given line operating schedule.
- *Station:* maximum number of spaces transported in vehicles stopping at a station per direction per hour.
- *TU line:* maximum number of TUs that can pass a fixed point per direction per hour; also, maximum frequency of service.
- *Utilized:* maximum number of persons transit vehicles on a line can transport past a fixed point per direction per hour.
- *Vehicle line:* maximum number of vehicles that can pass a fixed point per direction per hour (as single vehicles or in TUs).
- *Way:* Maximum number of spaces that can be transported in vehicles past a point per direction per hour without stopping.

COMMUTER TRANSIT—Transit services operated during peak hours only, primarily serving work trips. See also REGIONAL RAIL.

CONSUMPTION RATE—Ratio of the quantity of a certain resource utilized in a process (e.g., of transporting) to the quantity of a specific output of the process, such as kWh/veh-km.

DEMAND (TRANSIT)—Volume of passengers utilizing given service with a given fare.

- *Latent:* potential but not actually performed travel: trips that could be attracted if service is improved, fares reduced, etc.
- *Elasticity:* the percent change in passenger volume divided by the percent change in the price (fare).

DEMAND ESTIMATION—Process for estimating urban travel.

- *Four-step model:* process consisting of steps: trip generation, trip distribution, modal split (see MODAL SPLIT) and trip assignment.
- *Trip assignment:* predicting the specific routing of trips on the basis of their origins, destinations, and travel conditions on the network of each mode.
- *Trip distribution:* process of connecting the origins and destinations obtained from trip-generation estimates.
- *Trip generation:* estimation of the number of trips originating from or destined to each zone of study area in a designated period.

DEMAND-RESPONSIVE TRANSIT—Paratransit services with routings and schedules partially or fully determined by travel desires of individual passengers. Examples: dial-a-ride, taxi.

DEREGULATION (TRANSIT)—Reduction or complete elimination of government regulations.

DIRECTNESS OF TRAVEL—Ratio of passenger-km traveled on the network to passenger-km along straight O-D lines.

DOWNTOWN PEOPLE MOVER (DPM)—AGT system used on short shuttle or circular lines in central city area (Detroit, Miami).

EFFICIENCY (TRANSPORTATION)—Ratio of output to input consumed in the transportation process, such as vehicle-km/employee.

EXPRESS SERVICE—*see* ACCELERATE SERVICE, Express/Local.

FARE—Passenger payment for a transit ride.

- *Flat:* constant fare regardless of the distance a passenger travels on a single vehicle or throughout the transit network.
- *Level:* absolute amount of fares; has direct impacts on ridership and revenues.
- *Sectional:* graduated fare increasing with the number of line sections traveled.
- *Structure:* fare amount in relation to the distance traveled; may be flat or graduated.
- *Zonal:* graduated fare increasing with the number of fare zones traveled through.

FAREBOX REVENUE—Revenue transit agency obtains from fares.

FARE COLLECTION

- *Automated:* fares collected and controlled by automatic machines and barriers.
- *Self-service:* fare payments where passenger must prepay the fare and walk on without control at entry. Spot-check control is used.

FARE-FREE ZONE—An area with fare-free transit services; common for center city, where it is intended to reduce automobile use in the area.

FREQUENCY OF SERVICE—The number of TUs passing a point in one direction during one hour. Inverse of headway.

GUIDEWAY—Travel way (rail track, guide beam, riding and guiding surfaces, and others) that physically guides vehicles specially designed to travel on it. Guideways always require ROW category A, except rail, which can be placed in any category (A, B, or C).

HEADWAY—The time interval between the moments two successive TUs pass a fixed point on a transit line in the same direction.

- *Clock:* headway divisible into 60, so that departure times are the same every hour.
- *Minimum:* shortest headway on a line determined by the physical characteristics of the system (technology, method of driving and control, degree of safety required) and station operation (rate of boarding/alighting, departure control, etc.)
- *Policy:* the longest headway scheduled for a line determined by the minimum level of service considered acceptable.
- *Pulse:* joint headway for all or most lines in a timed transfer system.
- *Station:* shortest headway that can be operated at stations (usually critical).
- *Way:* shortest headway achievable on open line without stops (seldom critical).

HIGH-OCCUPANCY VEHICLE (HOV)—Vehicle of any type (automobile, van, bus) that carries at least a certain prescribed number of passengers (usually four). Concept used for reserved HOV lanes.

LEVEL OF SERVICE (TRANSIT)—Set of travel components affecting passengers, such as speed, reliability, comfort, or others; influences the ability of transit to attract passengers.

LIGHT RAIL TRANSIT (LRT)—Electric rail vehicles, usually articulated in one- to three-car trains operating mostly on ROW category B, but also on A (e.g., tunnels) and C (in pedestrian zones). The wide range of its designs goes from tramway-type lines with priority treatments to small-sized rapid transit (see LRRT).

LIGHT RAIL RAPID TRANSIT (LRRT)—LRT system on ROW A only with correspondingly higher performance than LRT, but with smaller TUs and stations than metro-type systems.

LOAD FACTOR—Ratio of utilized to offered capacity in a TU or on a line.

MAINTENANCE (VEHICLE)

- *Predictive:* maintenance based on testing conditions of components.
- *Preventive:* programmed maintenance procedure based on service intervals.
- *Reactive:* maintenance of vehicles when parts which failed or are about to fail.

MAXIMUM LOAD SECTION (MLS)—Transit line interstation spacing with the maximum passenger volume that determines required line capacity. Often incorrectly called "maximum load point."

MEAN DISTANCE BETWEEN FAILURES (MDBF)—Average distance vehicles travel between two breakdowns in service.

MEAN TIME BETWEEN FAILURES (MTBF)—Average time vehicles operate between two breakdowns in service.

MODAL SPLIT—Percent distribution of trips among modes of transportation.

OPERATING RATIO—Ratio of revenue from transit fares to the costs of operating the service.

PASSENGER FLOW—Rate of passenger travel in persons per time.

PASSENGER VOLUME—Passenger flow in persons per hour.

PEAK HOUR COEFFICIENT—The ratio of the highest 15-min to hourly passenger volume.

PERFORMANCE-COST PACKAGE—Ratio of performance (capacity, speed, reliability, etc.) of a transit mode to its investment cost per km of line.

PERSONAL RAPID TRANSIT (PRT)—A theoretical transit mode that would consist of small-capacity (two to six seat) cabins traveling automatically over an elaborate system of guideways.

PREMETRO—LRT system designed with provisions for easy conversion into RRT (metro).

PRODUCTIVE CAPACITY—Product of line (or vehicle) capacity and its operating (or technical) speed. Dimension: sps-km/h^2. A good composite representative of transit system performance.

PRODUCTIVITY—The quantity of (transportation) output per unit of consumed resource. For example, veh-km/driver/hour, or prs-km/unit cost of operation.

RAIL RAPID TRANSIT (RRT); also METRO—Rapid transit utilizing rail technology.

RAIL TRANSIT—Family of transit modes with rail technology. Major ones, generally in ascending order of performance, are streetcars/tramways, light rail transit, rail rapid transit/metro, and regional rail.

RAPID TRANSIT (RT)—Generic class of electrically powered transit modes that operate exclusively on ROW category A and have high speed, capacity, reliability, and safety. Includes RRT, RTRT, LRRT, and most RGR systems.

Note: Since rapid transit is known as a high-performance system, various modes are often given that designation although they technically do not belong in that class of modes. Examples: in some cities, bus and LRT systems are popularly designated as "rapid transit," although they are actually semirapid transit because no bus line has ROW category A only. The correct designation of BRT and most LRT systems is semirapid transit.

REGIONAL RAIL (RGR)—Regional passenger service usually provided by railway agencies, which consists of electric or diesel-powered trains on grade-separated railroad lines (sometimes with protected grade crossings). Characterized by very high performance and service quality.

REGIONAL TRANSIT—Long rail or bus transit lines with few stations and high operating speeds that primarily serve long trips within metropolitan regions, as distinguished from city transit and short-haul transit.

REGULAR or ALL-DAY TRANSIT SERVICES—Transit lines operated during most hours, typically 16 to 18 hours per day.

REGULATION (TRANSIT)—Government rules that transit agency must follow. May include safety, right and obligation to operate, economic (fares), and service specifications.

RIDING HABIT—Per capita transit rides in a served area.

RIGHT-OF-WAY (TRANSIT), (ROW)—(1) Strictly defined, a legally and physically separated strip of land for exclusive use by transit vehicles; crossings may be allowed. (2) Broadly defined, any path or way on which transit vehicles travel. Based on the latter definition, transit rights-of-way are classified in three categories:

- *Category A:* fully controlled ROW without grade crossings or any legal access by other vehicles or persons; also called "private" or "exclusive" ROW. It can be a tunnel, aerial (elevated), or at grade level.
- *Category B:* ROW types that are longitudinally physically separated (by curbs, barriers, grade separation, etc.) from other traffic, but have grade crossings for vehicles and pedestrians, including regular street intersections.
- *Category C:* surface streets with mixed traffic. Transit may have preferential treatment, such as reserved, but not physically separated, lanes, or travel with other traffic.

ROLLING STOCK—Collective term for a fleet of rail (transit) vehicles.

RUBBER-TIRED RAPID TRANSIT (RTRT)—The same as rail rapid transit, except that the vehicles ride on and are guided by rubber tires on a specially designed guideway.

RUN-CUTTING—Determination of work assignments for individual drivers or crews during the day.

SCHEDULING (TRANSIT)—Determination of TUs and their travel times along a line during one day.

SEMIRAPID TRANSIT—Genetic class of modes utilizing mostly ROW category B, but C or A may also be used on some sections. By performance it is between street transit and rapid transit. Examples: BRT and LRT.

SERVICE QUALITY (SQ)—Qualitative elements of service affecting passengers, such as convenience of using the transit system, riding comfort, aesthetics, and cleanliness. Part of level of service.

SKIP-STOP SERVICE—*see* ACCELERATED SERVICE, Skip-Stop.

SPEED (TRANSIT)

- *Access:* average speed of a passenger travel to and from transit stop or station.
- *Legal:* maximum speed TUs can legally operate on a given section of line.
- *Line design:* maximum speed TUs can achieve on a given section of line with adequate comfort and safety when physical conditions govern.
- *Maximum technical:* highest speed a TU is physically capable of achieving on a straight horizontal way under normal weather conditions.
- *On-line travel:* operating speed on the line section that a passenger utilizes.
- *Operating/travel:* average speed of a TU traveling on a transit line.
- *Origin-destination:* average speed of passenger travel along his/her path from origin to destination.
- *Programmed:* speed TUs can operate meeting given standards of safety, economy, and vehicle performance.
- *Running:* average speed TUs achieve from leaving one station (stop) to arriving at the next one.
- *Station-to-station:* average speed of travel between moments a TU leaves two adjacent stations.

STATION (TRANSIT)—Off-street at-grade, under, or above street level facility for stopping of transit vehicles to pick up and drop off passengers. Usually has boarding/alighting platforms, waiting area, fare collection, information, and related facilities.

- *Terminal:* a transit station at which one or more lines terminate.
- *Transfer:* joint station for two or more lines at which passengers can transfer between lines.

STOP (TRANSIT)—On-street location where transit vehicles stop to pick up and discharge passengers.

STOPPING REGIME—Type of stopping of TUs along a line.

- *Demand:* TUs stop at any point passengers require.
- *Fixed:* all TUs stop at fixed stops/stations.
- *On-call:* TUs stop at fixed stops/stations only when passengers require.

STREETCAR, TRAMWAY, or TROLLEY—Street transit mode consisting of electrically powered rail vehicles operating in one- to three-car TUs, mostly on ROW category C or B.

STREET TRANSIT—Generic class of modes operating on streets with mixed traffic (ROW category C). Examples: regular bus, trolleybus, and streetcar/tramway.

SUBWAY—Rapid transit or metro line operating exclusively or mostly in tunnels. Sometimes used as a popular, although incorrect, name for any rapid transit (New York City, Tokyo).

TECHNOLOGY (TRANSIT)—Physical features of vehicles and ways, primarily vehicle support, guidance, propulsion, and control.

TIMED TRANSFER SYSTEM (TTS)—A network consisting of transit lines and one or several transit centers at which TUs from all intersecting lines arrive simultaneously, allowing easy passenger transfers in all directions.

TRAMWAY—See STREETCAR.

TRANSFER—Change between vehicles of the same or different modes (intra- and intermodal, respectively) in the course of passenger travel.

TRANSIT (URBAN)—Public transportation services for travel within a city or region with fixed routes and schedules, such as bus, trolleybus, and rail services. Also called mass transit, mass transportation, or urban public transportation.

TRANSIT LINE—Transit units operating on a prescribed alignment with a predetermined schedule. Also TRANSIT ROUTE, usually for buses.

- *Branch:* line section that separates from one or more others.
- *Feeder:* independent line that brings passengers to the main line.
- *Independent:* line not connected with other lines.
- *Integrated:* line that overlaps with other lines on some sections.
- *Trunk:* main section of a line, which then has several branches.
- *Types:* geometric forms of lines, such as radial, diametrical, circle, tangential, etc.

TRANSIT NETWORK—Set of transit lines with coordinated transfers. May be single-mode or multimodal (e.g., rail-bus).

TRANSIT MODE CLASSES—Groups of transit modes classified by their basic characteristics, such as ROW category: street transit C, semirapid transit B, and rapid transit A.

TRANSIT FEDERATION—An umbrella organization unifying different public or private agencies to provide integrated service to the public. Used in many European cities.

TRANSIT PRIORITY—Activation of traffic signals to reduce or eliminate waiting of transit vehicles passing through the intersection.

TRANSIT SERVICE AREA—The area a transit system is required to serve.

TRANSIT UNIT (TU)—One or more transit vehicles traveling together as a physical unit. Joint term for single vehicle and train of several vehicles.

TRAVEL TIME—Duration of individual time intervals in transit system or in passenger travel.

- *Access:* time required for approach from the trip origin to a transit stop or departure from a stop to the trip destination.
- *Cycle:* total round trip time on a line.
- *Deadhead:* travel when the TU is not in passenger service, usually between storage yard and transit line.
- *Operating:* scheduled time interval between departure of a TU from one terminal and its arrival at the other terminal on the line.
- *Origin-destination:* total time a passenger spends to make a trip from origin to destination.
- *Platform:* total time a TU is in operation, i.e., from leaving a depot to returning to it (includes deadheading).
- *Running:* time interval between TU's starting from one station (stop) and stopping at the next.
- *Station standing/dwell:* duration of TUs standing at a station for the purpose of boarding and alighting of passengers.
- *Station-to-station:* time interval between TU's departures from two adjacent stations.
- *Terminal:* the time a TU spends at a line terminal.
- *Transfer:* time used for transferring between different lines or modes.
- *Waiting:* time between passenger arrival at a stop and the time of TU departure.

TROLLEYBUS (TB)—Electrically propelled bus that obtains power via two trolley poles from two overhead wires along routes. It can travel a limited distance on battery power or auxiliary ICE.

TURNOUT, TRACK—Two-track section on a single-track rail line for meeting of TUs.

URBAN PUBLIC TRANSPORTATION—*see* TRANSIT (URBAN)

USER TAXES (IN TRANSPORTATION)—Taxes collected from travelers related to the transportation service they use.

UTILIZATION (in transportation)—Ratio of output to input of the same or similar unit, such as person-km/space-km or vehicles operated/vehicles owned.

UTILIZATION COEFFICIENT (IN TRANSIT SERVICE)—Ratio of passengers to offered spaces on a line. Also called LOAD FACTOR.

VEHICLE LIFE

- *Economic:* vehicle life that results in the lowest life-cycle cost.
- *Functional:* vehicle life until its design, technology, and operation are superseded by new technology or operational requirements.
- *Physical:* years during which vehicles can be operated with acceptable physical reliability and maintenance procedures.

ZONAL SERVICE—*see* ACCELERATED SERVICE, Zonal.

APPENDIX IV

ANSWERS TO SELECTED EXERCISE QUESTIONS

Answers presented here give numerical results to exercises, in some cases partial ones, allowing the reader to check that the computing procedure used is correct, yet requiring that all computations must be performed. Selected exercises requiring essay-type answers give the key concepts and, in some cases, augment the material in the body of this book.

CHAPTER 1

1.1. Average passenger access distance increases by 100 m and access time t_a increases by 75 s. Bus travel time per km t_k decreases from 220 to 170 s/km. Total passenger travel time for a distance of x (km) is ($t_a - x \cdot t_k$). Passenger travel time on a 4-km trip decreases by 125 s, on 8 km by 325 s. The distance on which travel time does not change is 1.5 km.

1.2. **a.** $\Delta s_a = 100$ m/prs

b. $\Delta t_a = 1.33$ min/prs

c. $\Delta t_t = 5.00$ min/prs

d. $\Delta t = 2.33$ min/prs

e. $L_a = 2.42$ km

f. $\Delta N = 5$ buses

1.3. $V_o = 18.82$ km/h

1.4. *Hints:* Compute T_o and T. Then draw a diagram showing the scale for time t from 0 to 1500 s, and distance s from 0 to 2500 m. Then plot the schedule on the diagram for the two vehicles with a headway (distance on the abscissa) of 4 min or 240 s.

1.5. **a.** $T = 120\ L/V_o + 2\ t_t = 152$ min $N = [T/h]^+ = 16$ TU
$T = N \cdot h = 160$ min $V_c = 120\ L/T = 10.5$ km/h

b. $N' = 14$ $T' = N' \cdot h = 140$ min
$V_c' = 120\ L/T' = 12$ km/h $V_o' = 120\ L/(T' - t_1 - t_2) = 13.33$ km/h

c. $C = 60\ C_v/h = 480$ prs/h $\alpha_{max} = P \cdot h/(60\ C_v) = 0.625$

1.6. **a.** $N = 20$ TUs; $\gamma = 0.27$

b. $\Delta t_t = 1.5$ min; $\gamma = 0.20$

c. $\Delta V_o = 2.46$ km/h

e. $\alpha = 0.50$

1.7. Partial results of scheduling are given here. Provide answers to questions *e* and *f*.

Schedule		α	f (TU/h)	h (min)	L (km)	T_o (min)	t_t (min)	T (min)	N_{TU}	$\bar{\alpha}$
I		0.65								0.42
II	Long-haul		10	6	16.48	48	12	60	10	
	Short-haul	0.65								0.56
	Joint									
III	Long-haul									
	Short-haul	0.87	6	10	7.91	23	17	40	4	0.7
	Joint		12	5	7.91					

1.8. $h_{av} = 63$ s; $\alpha = 0.64$

1.9. **a.** $h_C = 7.5$ min

b. $h_T = 84$ s

c. $N_f = 136$ cars

d. $\Delta N_f = 12$ cars

e. $V_o^B = 24.22$ km/h

1.10. **a.** $h^{\text{Blue}} = 12$ min

b. $N_f^{\text{Red}} = 11$ TUs

1.11. **a.** Route AA′: 3 min (I), 4 min (II)

b. Types of buses: AA′: II, BB′: II, CC′: I

c. Route CC′: $T = 111.1$ min; $N = 19$ buses; $T' = 114$ min; $V_c = 13.7$ km/h

d. $h_{QA'} = 1.22$ min

e. Route AA′: 18 buses; total buses for all three lines: 92 buses

1.12. $N_{DR} = 15$. Total number of trains for all lines: 67

1.13. **a.** $h_A = 3$ min, $h_D = 5$ min

b. $h_T = 47$ s

c. $T_B = 60$ min, $N_B = 15$. Total rolling stock for all four lines: 72

CHAPTER 2

2.2. Present: $C = 5773$ prs/h; $N_{TU} = 16$; $N_C = 32$
Required increased capacity: $C' = 7216$ prs/h

a. $C = 8659$ prs/h; $N_{TU} = 16$; $N_C = 48$. Increased investment and operating costs, headways unchanged

b. $C = 8659$ prs/h; $N_{TU} = 24$; $N_C = 48$. Greater operating cost increase, shorter headways

c. $C = 5773$ prs/h; $N_{TU} = 14$; $N_C = 28$. Present capacity, reduced cost

Measures **a** and **b** would meet the required capacity increase.

2.11. **a.** $h = 10$ min, $T_c = 50$ min, $N_{TU} = 5$, $t_t = 3.6$ min, $S_{to} = 1.92$ km

b. $h = 15$ min, $T_c = 60$ min, $N_{TU} = 4$, $t_t = 8.6$ min; $S_{to} = 2.88$ km

2.12. **a.** $t_s' = 19$ s

b. $V_o = 25.6$ km/h; $V_o' = 28.2$ km/h

c. $V_o' = 31.4$ km/h; $T' = 36$ min; $N' = 9$ TUs

d. $h' = 3$ min

2.14. **a.** $T_M = 60$ min

b. $T_N^i = 60$ min; $T_N^{ii} = 54$ min $t_M = 7.5$ min; $t_N^i = 13.2$ min; $t_N^{ii} = 7.2$ min

CHAPTER 3

3.8. **a.** **(a)** 9.4 km/h

(b) 11.4 km/h

3.11. For a given $P_{\max}$, increasing TU size results in decreasing operating cost but increasing passenger travel time due to longer headways.

3.17. The lowest cost obtained among the four values is \$2083/h

CHAPTER 4

4.3. **(a)** 375 m and 1125 m

(b) 187.50 m and 375 m

4.10. **(a)** 24

(b) 56

(c) 50

4.13.

Line	$2T_o$ *(min)*	$2T_o + t_t$ *(min)*	*Final* T (min)	$N_{TU} = T/20$	t_t^{inner} (min)	t_t^{outer} (min)
A		44	60	3	5	17
B						
C						

4.14.

	h_p = *15 min*							h_p = *20 min*						
Line	*A*	*B*	*C*	*D*	*E*	*F*	$\sum$	*A*	*B*	*C*	*D*	*E*	*F*	$\sum$
Desired *T* (min)	38							38						
Desired *h* (min)	20							20						
Offered *h* (min)	15							20						
Offered *T* (min)	45							40						
Δh (min)	5						15	0						15
ΔT (min)	7							2						
N	3						13	2						13
$N \cdot \Delta T$ (min)	21	8	12	16	2	24	84	4	12	12	36	24	18	106

4.20. **(a)** 45

(b) 53

(c) 19

4.21.

		Network I	*Network II*
Number of stations	Single	15	
	Terminal	3	
	Transfer	1	
	Total	19	
Number of interstation spacings		18	
Number of OD pairs	With transfer		
	Without transfer		
	Total	171	595

4.22.

Terminals	*Transfers*	*Total Stations*	*Interstation Spacings*	*Total Trips*	*Trips With Transfers*	*Trips Without Transfers*
6	3	40	37	780	484	296

CHAPTER 5

5.4.

V_o *(km/h)*	*5*	*15*	*40*
G_j (m)	212.2		
H_j (m)	272.8		
G_k (m)			236.1
H_k (m)			983.9

5.6. *Hint:* Use Eqs (5.7), (5.19), and (5.21).
In case *ii*: $T_\ell = 37$ s, $\gamma = 0.026$, $\rho = 0.083$, $S_1^* = S_2^* = 0.784$ km.
In case *v*: $S_1^* = 2.656$ km, $S_2^* = 2.739$ km.

5.7. Use the given tabulation to compute and list passenger volumes traveling along the line; then compute Stop and Go values for each station, and base your answers on them. Partial answers are given in the table within the question.

5.9. $V_o^{A-G} = 53.8$ km/h; $V_o^{A-L} = 51.7$ km/h; $V_o^{A-Q} = 49.8$ km/h

CHAPTER 6

6.3. Many cases of inadequate and inequitable wages, lack of working conditions standards, insecure jobs, inadequate benefits, no seniority benefits.

6.6. Prevention of strikes, which are highly damaging, particularly to the third party, transit users and society.

6.7. Wages have been reduced, work assignments are more efficient without labor rules. However, training and qualifications of workers have been relaxed, job security drastically reduced, attraction of employees is more difficult, and high turnover is a problem.

6.8. Transit network size, number of stations, and measures of services reflect the quantity and quality of transit available to population. Indicators relating these measures to area and population served allow comparative analysis with peer cities of similar as well as different size.

6.9. Efficiency is output of a process divided by input. The two may be expressed in different units (passengers-km per operating employee). Consumption is the inverse of efficiency. Utilization is the same as efficiency, but measured in the same or similar units (cars on the line divided by cars owned).

6.10. Ratio of revenue from fares to operating costs. Agencies with policies of maximizing use of transit tend to keep fares low and accept lower operating ratio than the agencies that set up fares to achieve the maximum revenue.

6.11. Offered work, computed as $W_s = N \times n \times C_v \times V_c$, is 26,054 sp-km/h for buses and 151,603 sp-km/h for LRT.
Utilized work, computed as $W_P = \alpha \cdot W_s$, is 10,195 prs-km/h for the bus line and 61,363 prs-km/h for the LRT line.

6.12. Using the same equations as in problem 6.11, offered work is 645,120 sps-km/h for the metro and 292,301 sps-km/h for the regional rail line. Utilized work is 361,267 sp-km/h for the metro and 111,074 sps-km/h for the regional rail line.

6.13. Productivity in prs-km/h on the Red, Blue, and Yellow Lines is, respectively, 203,861, 135,182 and 371,008. Average speed for the entire network, based on offered work in car-km/h, is 33.5 km/h.

6.16. Supervision and control of operations on lines; reaction to special situations and emergencies; fare collection; information for the public.

6.20. Increased supervision; selection of vandal-proof materials; immediate cleaning; high penalties for vandals; education for public property protection in schools.

6.21. Full accessibility should provide for access by the disabled, elderly, children in strollers and bicycles (subject to physical feasibility of such accommodations).

6.22. As a social measure, accessibility for special groups should be financed by special government funds.

CHAPTER 7

7.1. Maximum ridership; maximum revenue; convenience of collection and control; equity and other social goals.

7.2. Maximum ridership and maximum revenue, usually mutually conflicting.

7.5. SSFC greatly facilitates fare selection because of simplified collection and control.

7.17. Basic purpose: enforce payments and discourage fraud. Difficulties in enforcement of penalty payments. With flat fares controlled at entry, spot-check control prevents avoidance at entry or by boarding on other doors. No control is needed on fully enclosed systems, such as automatic tickets checked at station entrance and exit.

CHAPTER 8

8.1. To keep fares low or moderate and facilitate commuting. To shift some travel from one mode to another, with lower social costs (e.g., from auto to transit). To stimulate development of an urban or suburban area.

8.2. Amtrak is seriously underfunded and neglected compared to all other modes, such as highways and air transport. The requirement that Amtrak become self-supporting is not applied to any other transportation mode, nor to railways in any other country.

8.4. Public policy supporting transit results in greater financial assistance for transit, which allows maintenance of lower fares.

8.7. Each of these views considers only one role of transit and ignores others. The correct view is that both of these roles must be satisfied to some degree and reasonably balanced against other roles and requirements.

8.8. Transit services include many impacts (benefits and costs) to users that are not quantifiable. They also benefit nonusers and the entire city. Many benefits are long-range or permanent ones. None of these impacts and values can be easily expressed in monetary values, so that the free market cannot internalize or even consider them.

8.12. Retail sales, personal income, and property taxes. Each has different impacts, efficiency and equity aspects, and amounts of collected revenue.

8.16. Revenue bonds have higher risk, therefore higher interest rates (see also Section 9.2.3).

8.18. Local participation ensures that the potential transit system users and beneficiaries are committed and willing to contribute to the project implementation.

8.21. Generally, investment in the transit system should improve efficiency (decrease cost per unit of output) and/or improve quality of service. Unit costs may increase, however, if the quality of service is drastically improved. In such cases, the improved service may result in passenger willingness to pay higher fares, offsetting a portion of the investment.

CHAPTER 9

9.8. Ability to tax allows selling of GO bonds at lower interest rates than revenue bonds.

9.9. A transit federation is the most logical solution where a number of transit agencies cannot be merged. If regional transit is operated by a single agency, there is no need for it.

9.13. Passengers are protected from excessive or inequitable fares, frequent changes, and manipulations. Operators are protected from cutthroat competition.

9.16. Excessive, broadly applied, and rigid regulation may cause unnecessary costs, prevent efficient solutions, and reduce adjustment of design and operating practices to specific local conditions. Examples can be quoted from the regulatory requirements with respect to rail vehicle design, station accessibility standards, or various labor rules in transit operations.

9.17. The primary difference is the consideration of transit with respect to its public service role versus financial aspects of its operations.

9.18. Entry of many bus companies and individual operators is easier, competition forces economic efficiency, and usually no subsidies are given. Services are unstable and may be discontinued at any time; service quality and safety are often low; planning and system upgrading cannot be effectively made.

9.19. Regulation is needed to improve services and perform transportation planning. It is often opposed by operators, particularly large ones.

9.21. Gradual decrease of management's attention to user needs and innovations in services. In some cases, labor unions prevent changes and increases in economic efficiency.

9.24. Intensified analysis of passenger needs, treating them as customers who should be attracted by high-quality services.

9.27. Contractors' quality of service and attention to passenger needs must be monitored to prevent service deterioration to reduce costs.

9.29. Allowing joint operation of trains belonging to different companies—track sharing (successful). Forming a company that owns and maintains infrastructure and separate companies operating trains (applied to railway systems with mixed results: some failures, some successful so far). Dividing rail lines to different companies to achieve competition (mostly failures because rail transit is inherently optimal as an integrated network).

9.30. The purpose is to separate policy and strategic decisions from operations, to increase coordination among modes and implement intermodal policies. Results have been very positive.

CHAPTER 10

10.2. The 3C requirement was introduced to cope with the complex metropolitan areas' transportation planning requirements. It ensures that the long-range planning is comprehensive with respect to the region's area and jurisdictional units, functional requirements, and changes during the planning time horizon.

10.6. Extrapolation of trends can be used in planning to examine future conditions if no planning is implemented, and it can be done only when forecasts rely on steady trends. Designing a transportation system based entirely on trend extrapolation can be counterproductive. Planning must be oriented toward the goals adopted for the city, rather than framed by extrapolated trends.

10.7. Such planning may lead to very undesirable conditions in the city and region.

10.16. The trip-end model was used mostly because of its simplicity, but it was insensitive to changes in transit system service quality, thus largely defeating the planning effort.

10.18. All-or-nothing is much simpler, but it can be applied only if the network operates below capacity conditions. With capacities reached, its results are incorrect.

10.19. User equilibrium condition is achieved when equal travel times on all used alternate paths

10.24. Basic modes of rapid transit: metro, regional rail, LRRT, and some AGT systems. Semirapid transit modes: LRT, some AGT and BRT (see Chapter 12).

CHAPTER 11

11.1. Goals include interests of system users, facility operator, and the public. Economic aspects, dominant in the private sector, are important, but many other factors usually also play a significant role. Evaluation of benefits and costs on the users and society are more difficult to evaluate than monetary ones.

11.2. NPV represents project value during its life discounted to the present.

11.9. Flexibility implies the tendency to change, including physical elements, operations, pricing, etc. Flexibility may be positive or negative. Opposites of flexibility are permanence, durability, reliability.

11.12. In preparing your answer, study and utilize Figures 3.4 and 6.2.

11.16. Comparison of different systems on the basis of a single value requires a *ceteris paribus* assumption. Due to many qualitative aspects of transit services that passengers experience, impacts on nonusers and future benefits and costs, different transit systems, particularly different modes, can hardly ever satisfy this condition.

CHAPTER 12

12.2. **a.** Land availability, investment in construction of street and intersections, possible reduction of lanes for general traffic.

b. Use the comparison in Section 12.1.1.

12.5. PCP defines the ratio of system performance (capacity, speed, reliability, LOS, safety, and other characteristics) and costs. Comparing costs only cannot be done for alternative systems with different characteristics.

12.9. The rationale for bus (transit) lanes is based on higher productivity and public service of buses and the desirable change in modal split. Also, separation improves traffic flow.

12.11. Upgrading of ROW C into B and A, articulated vehicles, longer TUs and SSFC.

12.15. Medium-volume modes benefit more because automation enables shorter headways. Reduction of labor cost results in greater cost saving per passenger than with high-capacity trains.

12.21. Find information about tram-train systems in Karlsruhe and Saarbrücken, Germany, and Manchester, UK.

12.22. To make services on lightly traveled existing railway lines economically feasible.

INDEX